THE MILKY WAY GALAXY

INTERNATIONAL ASTRONOMICAL UNION
UNION ASTRONOMIQUE INTERNATIONALE

THE MILKY WAY GALAXY

PROCEEDINGS OF THE 106TH SYMPOSIUM OF THE INTERNATIONAL ASTRONOMICAL UNION HELD IN GRONINGEN, THE NETHERLANDS 30 MAY – 3 JUNE, 1983

EDITED BY

HUGO VAN WOERDEN

and

RONALD J. ALLEN
Kapteyn Institute, Groningen, The Netherlands

and

W. BUTLER BURTON
Sterrewacht Leiden, The Netherlands

D. REIDEL PUBLISHING COMPANY
A MEMBER OF THE KLUWER ACADEMIC PUBLISHERS GROUP
DORDRECHT / BOSTON / LANCASTER

Library of Congress Cataloging in Publication Data
Main entry under title:

The Milky Way Galaxy.

At head of title: International Astronomical Union, Union Astronomique Internationale.
Includes index.
1. Milky Way–Congresses. I. Woerden, Hugo van. II. Allen, Ronald J. III. Burton, W. B. (William Butler), 1940- . IV. International Astronomical Union.
QB857.7.M55 1985 523.1'13 84-26294
ISBN 90-277-1919-5
ISBN 90-277-1920-9 (pbk.)

Published on behalf of
the International Astronomical Union
by
D. Reidel Publishing Company, P. O. Box 17, 3300 AA Dordrecht, Holland

Sold and distributed in the U.S.A. and Canada
by Kluwer Academic Publishers,
190 Old Derby Street, Hingham, MA 02043, U.S.A.

In all other countries, sold and distributed
by Kluwer Academic Publishers Group,
P. O. Box 322, 3300 AH Dordrecht, Holland

Printed in The Netherlands

TABLE OF CONTENTS

(R = Review, I = Invited Paper)

PREFACE

In June 1983 the Astronomical Institute of the State University of Groningen, founded by Kapteyn about 100 years ago, celebrated its one-hundredth anniversary. At the suggestion of its Chairman, R.J. Allen, the Kapteyn Institute invited the International Astronomical Union to mark the centenary by holding a Symposium on "The Milky Way Galaxy". The purpose of the Symposium was to review recent progress in the study of our Galaxy, to define current problems, and to explore prospects for future development. The Symposium programme would emphasize the large-scale characteristics of our Galaxy, and highlight both the historical development of our understanding of the Milky Way Galaxy and the importance of studies of external galaxies to this understanding.

The Symposium was sponsored by four IAU Commissions: 33 (Structure and Dynamics of the Galactic System), 28 (Galaxies), 34 (Interstellar Matter) and 41 (History of Astronomy). The Scientific Organizing Committee, listed on page xviii, represented a broad range of nationalities and of expertise, including two historians of science. A meeting of the Committee, held during the IAU General Assembly at Patras, provided an excellent opportunity to discuss plan and format of the Symposium, topics and speakers; thereafter, the Committee was regularly consulted by letter and telephone.

IAU Symposium 106 was held at Groningen on 30 May - 3 June 1983, in the new building occupied by the Kapteyn Institute since January 1983. There were about 200 participants, coming from as many as 25 countries. Thus, the meeting was a full order of magnitude larger than IAU Symposium No. 1, "Coordination of Galactic Research", held at Vosbergen near Groningen in June 1953 with 27 participants. The Symposium coincided with the anniversary of the Bosscha Observatory at Lembang, Indonesia, and this fact was marked by an exchange of congratulation telegrams.

The historical component of the programme - which occupied 4 hours out of a total of 27 - featured four major scholarly reviews, given by professional historians of science; these lectures were highly appreciated and provoked considerable discussion. In the remainder of the programme, observational aspects were given rather more time than theoretical aspects, in view of the fact that an IAU Symposium on Dynamics of Galaxies had just been held, at Besançon in 1982. Throughout the programme, results of research on external galaxies were blended with discussions of our own Galaxy.

Because of the breadth of the subject and the large number of participants, emphasis was placed on Reviews and Invited Papers. These two categories of papers, together numbering 43, took about 70 percent of the Symposium time. Among the very many Contributed Papers offered, 25 were selected for (in some cases very brief) oral presentation. About 80 papers were presented as posters, and the arrangements for these - which were administered by Mr. P.J. Teuben - proved highly effective. Considerable time (about 20 percent) was set aside for Discussions, and these were often very lively.

On Monday evening, participants were guests at a reception by the President of Groningen University, Dr. J. Borgman. This reception was followed by a visit to an exhibition about the History of Astronomy at Groningen University, featuring the work of Kapteyn and Van Rhijn; Professor A. Blaauw gave an introductory lecture to this exhibition. Dr. M.A. Hoskin gave a Public Lecture on "The Milky Way from Antiquity to Modern Times" in the University Aula on Tuesday evening. The Wednesday afternoon was spent on an excursion by boat on the Frisian lakes, followed by a buffet dinner at "Lauswolt" in Beetsterzwaag and an after-dinner lecture on "Life in the Galaxy?" by Dr. G.S. Shostak. A Salon Concert was given on Thursday evening.

The local arrangements were very competently handled by the Local Organizing Committee, headed by R.J. Allen, and the Conference Secretariat; these committees are listed on page xviii. Much help was also given by a number of young astronomers from Groningen and Leiden. The Partners' Programme was arranged by Mrs. Ria van Woerden. Financial support by IAU, by the Royal Netherlands Academy of Sciences, by the State University of Groningen, and by several private foundations - again listed on page xviii - contributed to the success of the meeting through travel grants to key speakers and to young astronomers.

I wish to record my great gratitude to all members of the Scientific Organizing Committee for their help in the preparation of the meeting. In particular, I wish to thank Michael Hoskin and Elly Dekker for their advice on historical matters, and my colleagues Butler Burton and Tjeerd van Albada for detailed consultations and discussions throughout the period of preparation. Martien de Vries held a key position in administering our files of correspondence, participants and papers.

The Kapteyn Institute is grateful that so many astronomers came to share in its jubilee, and that both speakers and participants made the Symposium so informative and stimulating.

Hugo van Woerden

Chairman,
Scientific Organizing Committee

EDITORIAL NOTES

This Proceedings volume presents the substance of the lectures, papers and discussions held or given during IAU Symposium 106, "The Milky Way Galaxy". It contains the full text of 43 Reviews and Invited Papers, which together form the backbone of this book. Some of these papers have been condensed, others extended by the authors; thus, their length varies greatly. The book further contains extended Abstracts (usually two pages each) for 81 Contributed Papers; some of these had been presented orally, but the majority were displayed as posters during the Symposium.

The order of papers in this volume is not fully chronological, but follows a more logical pattern. The chronological order may be more or less reconstructed from the title pages of individual sections.

Most of the oral presentations at the Symposium were followed by discussions. These were partly recorded by participants on sheets handed out during the meeting. However, the bulk of the discussions was transcribed by the senior Editor from tape recordings. In making the transcript, he was assisted by the discussion log kept by young astronomers from Groningen and Leiden. Discussions thus reconstructed were edited and submitted to the speakers for approval or modification. Hence, the discussions as printed here represent an edited version - sometimes condensed, sometimes extended and clarified - of what was actually said.

A book containing 124 papers, each beginning on a right-hand page, tends to have a large number of blank pages. We have used most of this space for photographs taken during or between sessions, during the excursion or at the conference dinner. Thus, this volume commemorating the centenary of the Kapteyn Institute is illustrated with historical photographs showing the Symposium at work and the people who made the Symposium a success. The photographs are individually credited to the Centrale Fotodienst of Groningen University (CFD), to Loek Zuyderduin of Leiden Observatory (LZ), or to Seth Shostak (GSS). We thank the photographers for their excellent contribution to these Proceedings.

In our editorial work, we have been ably assisted by the secretaries at Groningen and Leiden: Joke Nunnink, Marijke van der Laan, Gineke Alberts, Ineke Rouwé and Hilda Mulder; Wanda van Grieken and Lena Cijntje, and also by Wim Brokaar, photographer at Leiden Observatory. The indexes were compiled by Remo Tilanus, Bart Wakker and Janet Sellwood. We thank Janet in particular for her large share in the final stages of the work.

We regret and apologize for the serious delays in our work, caused largely by illness.

We thank Dr. G. Lyngå of Lund Observatory for permission to reproduce the Lund Chart of the Milky Way on the dust jacket and on the Symposium poster.

Hugo van Woerden
W. Butler Burton
Ronald J. Allen

COMMITTEES AND SPONSORS

Sponsoring IAU Commissions

33 Structure and Dynamics of the Galactic System
28 Galaxies
34 Interstellar Matter
41 History of Astronomy

Commission representatives on the Scientific Organizing Committee are identified below.

Scientific Organizing Committee

A. Blaauw, Norg, Netherlands
W.B. Burton, Leiden, Netherlands (33)
C. Cesarsky, Saclay/Paris, France
E. Dekker, Boerhaave Museum, Leiden, Netherlands
J. Einasto, Tôravere, Estonia, USSR
M.W. Feast, Capetown, South Africa
K.C. Freeman, Canberra, Australia
M.A. Hoskin, Cambridge, UK (41)
D. Lynden-Bell, Cambridge, UK
M. Peimbert, Mexico D.F., Mexico (34)
V. Radhakrishnan, Bangalore, India (34)
T.S. van Albada, Groningen, Netherlands
B. Westerlund, Uppsala, Sweden (28)
R. Wielen, Berlin, F.R. Germany (33)
C.G. Wynn-Williams, Honolulu, Hawaii, USA
H. van Woerden, Groningen, Netherlands, Chairman

Local Organizing Committee

R.J. Allen, Chairman
W.J. Bosman-Noteboom
J.A. de Boer, Treasurer
M. de Vries

Symposium Secretariat

J.H.M. Nunnink
G.G.A. Rouwé
J. Renner
M. van der Laan

Supporting Organizations

International Astronomical Union
University of Groningen
Koninklijke Nederlandse Akademie van Wetenschappen
Leids Kerkhoven-Bosscha Fonds
Sterrenkundig Studiefonds "Kapteyn"
Groninger Universiteits Fonds

LIST OF PARTICIPANTS

J.S. Albinson, Stichting Radiostraling van Zon en Melkweg, Dwingeloo, NL
S.M. Alladin, Astronomy Department, Osmania University, Hyderabad, India
R.J. Allen, Kapteyn Instituut, Groningen, NL
E. Athanassoula, Observatoire de Besançon, Besançon, France
P.D. Atherton, Kapteyn Instituut, Groningen, NL
T.M. Bania, Astronomy Department, Boston University, Boston, MA, USA
F.N. Bash, University of Texas, Austin, TX, USA
B. Baud, Laboratorium voor Ruimteonderzoek, Groningen, NL
R. Beck, Max Planck Institut für Radioastronomie, Bonn, B.R. Deutschland
K. Begeman, Kapteyn Instituut, Groningen, NL
C.A. Beichman, Jet Propulsion Laboratory, Pasadena, CA, USA
E.M. Berkhuijsen, Max Planck Institut für Radioastronomie, Bonn, B.R. Deutschland
G. Bertin, Scuola Normale Superiore, Pisa, Italia
A. Blaauw, Kapteyn Instituut, Groningen, NL
L. Blitz, Astronomy Program, University of Maryland, College Park, MD, USA
J.B.G.M. Bloemen, Werkgroep Kosmische Straling and Sterrewacht, Leiden, NL
W.H.W.M. Boland, Stichting ASTRON, Den Haag, NL
T.R. Bontekoe, Kapteyn Instituut, Groningen, NL
A. Bosma, Sterrewacht, Leiden, NL
F. Boulanger, Kapteyn Instituut, Groningen, NL and Observatoire de Paris, Meudon, France
J. Brand, Sterrewacht, Leiden, NL
B.J. Brett, Mathematics Department, Plymouth Polytechnic, Plymouth, UK
K. Brink, Sterrenkundig Instituut Anton Pannekoek, Amsterdam, NL
E. Brinks, Sterrewacht, Leiden, NL
L. Bronfman, Astronomy Department, Universidad de Chile, Santiago, Chile
W.N. Brouw, Stichting Radiostraling van Zon en Melkweg, Dwingeloo, NL
B.F. Burke, Physics Department, Massachusetts Institute of Technology, Cambridge, MA, USA
W.B. Burton, Sterrewacht, Leiden, NL
R. Caimmi, Istituto Astronomico, Universita di Padova, Padova, Italia
C. Carignan, Kapteyn Instituut, Groningen, NL
R.G. Carlberg, Astronomy Department, University of Toronto, Toronto, Canada
S. Casertano, Scuola Normale Superiore, Pisa, Italia
C.J. Cesarsky, Service d'Astrophysique, Centre d'Etudes Nucléaires de Saclay, Gif-sur-Yvette, France
R. Chini, Max Planck Institut für Radioastronomie, Bonn, B.R. Deutschland
S.V.M. Clube, Royal Observatory, Edinburgh, Scotland, UK
R.S. Cohen, Goddard Institute for Space Studies, New York, NY, USA
F. Combes, Observatoire de Paris, Meudon, France
M. Crézé, Observatoire de Besançon, Besançon, France
J. Crovisier, Observatoire de Paris, Meudon, France
T.M. Dame, Goddard Institute for Space Studies, New York, NY, USA

R.D. Davies, Nuffield Radio Astronomy Laboratories, Jodrell Bank, UK
J.A. de Boer, Kapteyn Instituut, Groningen, NL
K.S. de Boer, Astronomisches Institut, Tübingen, B.R. Deutschland
A.G. de Bruyn, Stichting Radiostraling van Zon en Melkweg, Dwingeloo, NL
T. de Graauw, Laboratorium voor Ruimteonderzoek, Groningen, NL
T. de Jong, Sterrenkundig Instituut Anton Pannekoek, Amsterdam, NL
C.P. de Vries, Sterrewacht, Leiden, NL
M. de Vries, Kapteyn Instituut, Groningen, NL
P.T. de Zeeuw, Sterrewacht, Leiden, NL
J. Denoyelle, Koninklijke Sterrenwacht, Brussel, België
D. Despois, Observatoire Floirac, Bordeaux, France
E.R. Deul, Sterrewacht, Leiden, NL
J.M. Dickey, Astronomy Department, University of Minnesota, Minneapolis, MN, USA
B.G. Elmegreen, Astronomy Department, Columbia University, New York, NY, USA
D.M. Elmegreen, IBM Thomas J. Watson Research Centre, Yorktown Heights, NY, USA
D. Engels, Sternwarte der Universität, Bonn, B.R. Deutschland
S.M. Fall, Institute of Astronomy, Cambridge, UK
M.W. Feast, South African Astronomical Observatory, Capetown, South Africa
J.V. Feitzinger, Astronomisches Institut, Ruhr-Universität, Bochum, B.R. Deutschland
K.C. Freeman, Mount Stromlo and Siding Spring Observatories, Canberra, ACT, Australia
B. Fuchs, Institut für Astronomie und Astrophysik, Technische Universität, Berlin, B.R. Deutschland
M. Fujimoto, Department of Physics, University, Nagoya, Japan
R. Gathier, Kapteyn Instituut, Groningen, NL
T.N. Gautier, Jet Propulsion Laboratory, Pasadena, CA, USA
G.F. Gilmore, Royal Observatory, Edinburgh, Scotland, UK
O. Gingerich, Center for Astrophysics, Harvard University, Cambridge, MA, USA
W.M. Goss, Kapteyn Instituut, Groningen, NL
R. Güsten, Max Planck Institut für Radioastronomie, Bonn, B.R. Deutschland
H.J. Habing, Sterrewacht, Leiden, NL
R.H. Harten, Stichting Radiostraling van Zon en Melkweg, Dwingeloo, NL
F.D.A. Hartwick, Department of Physics, University of Victoria, Victoria, B.C., Canada
U.A. Haud, W. Struve Astrophysical Observatory, Tôravere, Eesti, SSSR
W. Hermsen, Werkgroep Kosmische Straling, Huygens Laboratorium, Leiden, NL
L.A. Higgs, Dominion Radio Astrophysical Observatory, Penticton, B.C., Canada
R.W. Hilditch, University Observatory, St. Andrews, Scotland, UK
P.T.P. Ho, Center for Astrophysics, Harvard University, Cambridge, MA, USA
P.W. Hodge, Astronomy Department, University of Washington, Seattle, WA, USA

M.A. Hoskin, Churchill College, Cambridge, UK
Hu Fu-Xing, Purple Mountain Observatory, Nanjing, China and Kapteyn Instituut, Groningen, NL
A.N.M. Hulsbosch, Sterrenkundig Instituut, Katholieke Universiteit, Nijmegen, NL
G.D. Illingworth, Kitt Peak National Observatory, Tucson, AZ, USA
F.P. Israel, European Space Technology Centre, ESA, Noordwijk, NL
W. Iwanowska, Institute of Astronomy, Copernicus University, Torun, Polska
M. Iye, Institute of Astronomy, Cambridge, UK
T.S. Jaakkola, University Observatory, Helsinki, Suomi
P.D. Jackson, Physics Department, York University, Downsview, ONT., Canada
C.J. Jog, Princeton University Observatory, Princeton, NJ, USA
L.E.B. Johansson, Onsala Space Observatory, Onsala, Sverige
T.C. Johns, Department of Applied Mathematics and Astronomy, University College, Cardiff, UK
N. Kaifu, Astronomical Observatory Osawa Mitaka, Tokyo, Japan
F.J. Kerr, University of Maryland, College Park, MD, USA
J. Kormendy, Dominion Astrophysical Observatory, Victoria, B.C., Canada
J.M.E. Kuijpers, Sterrenwacht, Utrecht, NL
M.L. Kutner, Physics Department, Rensselaer Polytechnic Institute, Troy, NY, USA
C.G. Lacey, Institute of Astronomy, Cambridge, UK
F. Lebrun, Centre d'Etudes Nucléaires de Saclay, Gif-sur-Yvette, France
D.T. Leisawitz, Goddard Institute for Space Studies, New York, and Astronomy Department, University of Texas, Austin, TX, USA
J.R.D. Lepine, Departamento de Astronomia, Universidade, São Paolo, Brasil
R.S. le Poole, Sterrewacht, Leiden, NL
C.C. Lin, Department of Mathematics, Massachusetts Institute of Technology, Cambridge, MA, USA
H.S. Liszt, National Radio Astronomy Observatory, Charlottesville, VA, USA
J. Lub, Sterrewacht, Leiden, NL
D. Lynden-Bell, Institute of Astronomy, Cambridge, UK
G. Lyngå, Observatoriet, Lund, Sverige
H.M. Maitzen, Universitätssternwarte, Wien, Österreich
R.N. Martin, Institut de Radio Astronomie Millimétrique, Grenoble, France
E.N. Maurice, European Southern Observatory, La Silla, Chile
A. May, Kapteyn Instituut, Groningen, NL
M. Mayor, Observatoire de Genève, Sauverny, Suisse
W.H. McCutcheon, Physics Department, University of British Columbia, Vancouver, B.C., Canada
A.D. McFadzean, University Observatory, St. Andrews, Scotland, UK
K.N. Mead, Physics Department, Rensselaer Polytechnic Institute, Troy, NY, USA
J. Milogradov-Turin, Institute of Astronomy, University, Beograd, Yugoslavija

I.F. Mirabel, Physics Department, University of Puerto Rico, Rio Piedras, PR,USA
Mo Jing-Er, Purple Mountain Observatory, Nanjing, China and Kapteyn Instituut, Groningen, NL
C.A. Murray, Royal Greenwich Observatory, Hailsham, UK
A. Natta, Astronomia Infrarossa, CNR, Firenze, Italia
A.H. Nelson, Department of Applied Mathematics and Astronomy, University College, Cardiff, UK
C.A. Norman, Sterrewacht, Leiden, NL
J.D. North, Filosofisch Instituut, Universiteit, Groningen, NL
H. Okuda, Institute for Space and Astronautical Science, Tokyo, Japan
J.H. Oort, Sterrewacht, Leiden, NL
T. Oosterloo, Kapteyn Instituut, Groningen, NL
J.P. Ostriker, Princeton University Observatory, Princeton, NJ, USA
E.R. Paul, Mathematical Sciences Department, Dickinson College, Carlisle, PA, USA
J.W. Pel, Kapteyn Sterrenwacht, Roden, NL
P. Pişmiş, Instituto de Astronomia, Universidad Nacional Autonoma, Mexico D.F., Mexico
S.R. Pottasch, Kapteyn Instituut, Groningen, NL
T.H. Pwa, Kapteyn Instituut, Groningen, NL
V. Radhakrishnan, Raman Research Institute, Bangalore T.N., India
E. Raimond, Stichting Radiostraling van Zon en Melkweg, Dwingeloo, NL
J.M. Rankin, Physics Department, University of Vermont, Burlington, VT, USA
M.J. Rees, Institute of Astronomy, Cambridge, UK
I.N. Reid, Royal Greenwich Observatory, Hailsham, UK
W. Renz, Institut für Theoretische Physik, Aachen, B.R. Deutschland
H.A.G. Robe, Institut d'Astrophysique, Université de Liège, Cointe-Ougrée, Belgique
W.W. Roberts, Department of Applied Mathematics, University of Virginia, Charlottesville, VA, USA
A. Robin, Observatoire de Besançon, Besançon, France
P.R. Roelfsema, Kapteyn Instituut, Groningen, NL
G. Rydbeck, Onsala Space Observatory, Onsala, Sverige
E.M. Sadler, European Southern Observatory, Garching/München, B.R. Deutschland
K. Särg, Observatoriet, Lund, Sverige
G.N. Salukvadze, Astrophysical Observatory, Abastumani, Georgia, SSSR
R. Sancisi, Kapteyn Instituut, Groningen, NL
D.B. Sanders, Five-College Radio Astronomy Observatory, University of Massachusetts, Amherst, MA, USA
R.H. Sanders, Kapteyn Instituut, Groningen, NL
A.I. Sargent, Owens Valley Radio Observatory, California Institute of Technology, Pasadena, CA, USA
R.T. Schilizzi, Stichting Radiostraling van Zon en Melkweg, Dwingeloo, NL
M. Schmidt, Astronomy Department, California Institute of Technology, Pasadena, CA, USA
U.J. Schwarz, Kapteyn Instituut, Groningen, NL
P. Schwering, Sterrewacht, Leiden, NL

N.Z. Scoville, Astronomy Department, University of Massachusetts, Amherst, MA, USA
P.E. Seiden, IBM Thomas J. Watson Research Center, Yorktown Heights, NY, USA
G.S. Shostak, Kapteyn Instituut, Groningen, NL
F.H. Shu, Astronomy Department, University of California, Berkeley, CA, USA
W.L.H. Shuter, Physics Department, University of British Columbia, Vancouver, B.C., Canada
R.W. Smith, History Project, Space Telescope Science Institute, Baltimore, MD, USA
Y. Sofue, Nobeyama Radio Observatory, Nagano, Japan
L.S. Sparke, Institute for Advanced Study, Princeton, NJ, USA
J. Spicker, Astronomisches Institut, Ruhr-Universität, Bochum, B.R. Deutschland
A.A. Stark, Bell Laboratories, Holmdel, NJ, USA
B.G. Strömgren, NORDITA, Kobenhavn, Danmark
R.G. Strom, Stichting Radiostraling van Zon en Melkweg, Dwingeloo, NL
A.W. Strong, Istituto di Fisica Cosmica, Milano, Italia
J.P. Swings, Institut d'Astrophysique, Université de Liège, Cointe-Ougrée, Belgique
P. te Lintel Hekkert, Sterrewacht, Leiden, NL
C. Terzides, Department of Astronomy, University, Thessaloniki, Hellas
P.J. Teuben, Kapteyn Instituut, Groningen, NL
K.O. Thielheim, Kernphysisches Institut, Universität, Kiel, B.R. Deutschland
J. Tinbergen, Kapteyn Sterrenwacht, Roden, and Sterrewacht, Leiden, NL
C.F. Trefzger, Astronomisches Institut der Universität, Basel, Schweiz
B.A. Twarog, Physics and Astronomy Department, University of Kansas, Lawrence, KS, USA
M. Urbanik, Observatorija Astronomiczne, Jagiellonski University, Krakow, Polska
E.A. Valentijn, European Southern Observatory, Garching/München, B.R. Deutschland
G.D. van Albada, Kapteyn Instituut, Groningen, NL
T.S. van Albada, Kapteyn Instituut, Groningen, NL
H.C. van de Hulst, Sterrewacht, Leiden, NL
E.P.J. van den Heuvel, Sterrenkundig Instituut Anton Pannekoek, Amsterdam, NL
J.M. van der Hulst, Radiosterrenwacht, Westerbork, NL
H. van der Laan, Sterrewacht, Leiden, NL
E.F. van Dishoeck, Sterrewacht, Leiden, NL
W. van Driel, Kapteyn Instituut, Groningen, NL
H. van Woerden, Kapteyn Instituut, Groningen, NL
F.M.M. Viallefond, Observatoire de Paris, Meudon, France
J.V. Villumsen, Institute for Advanced Study, Princeton, NJ, USA
W.H. Waller, "Science News" and University of Massachusetts, Amherst, MA, USA
R.A.M. Walterbos, Sterrewacht, Leiden, NL
R.H. Warmels, Kapteyn Instituut, Groningen, NL

J.M. Weisberg, Physics Department, Princeton University, Princeton, NJ, USA
P.R. Wesselius, Laboratorium voor Ruimteonderzoek, Groningen, NL
B.E. Westerlund, Observatoriet, Uppsala, Sverige
R. Wielebinski, Max Planck Institut für Radioastronomie, Bonn, B.R. Deutschland
R. Wielen, Institut für Astronomie und Astrophysik, Technische Universität, Berlin, B.R. Deutschland
H. Wolff, Kernphysisches Institut, Universität, Kiel, B.R. Deutschland
J.G.A. Wouterloot, European Southern Observatory, Garching/München, B.R. Deutschland
S. Wramdemark, Observatoriet, Lund, Sverige
J.S. Young, Five-College Radio Astronomical Observatory, University of Massachusetts, Amherst, MA, USA
C. Yuan, NASA Ames Research Center, Moffet Field, CA, USA and Physics Department, City College of New York, NY, USA
A.M. Zvereva, Crimean Astrophysical Observatory, Nauchnij, Krim, SSSR

JACOBUS CORNELIUS KAPTEYN (1851-1921)

After a portrait painted by Jan Veth in February 1918.

Kapteyn was born in Barneveld (Netherlands) on 19 January 1851, studied at Utrecht, and obtained a doctorate in Mathematics and Physics on 24 January 1875, on a thesis, "Onderzoek van Trillende Platte Vliezen". He was professor of Astronomy, Probability Theory, and Mechanics at Groningen University from 1878 to 1921.
The portrait in the upper right corner is one of Sir David Gill, Royal Astronomer at the Cape.

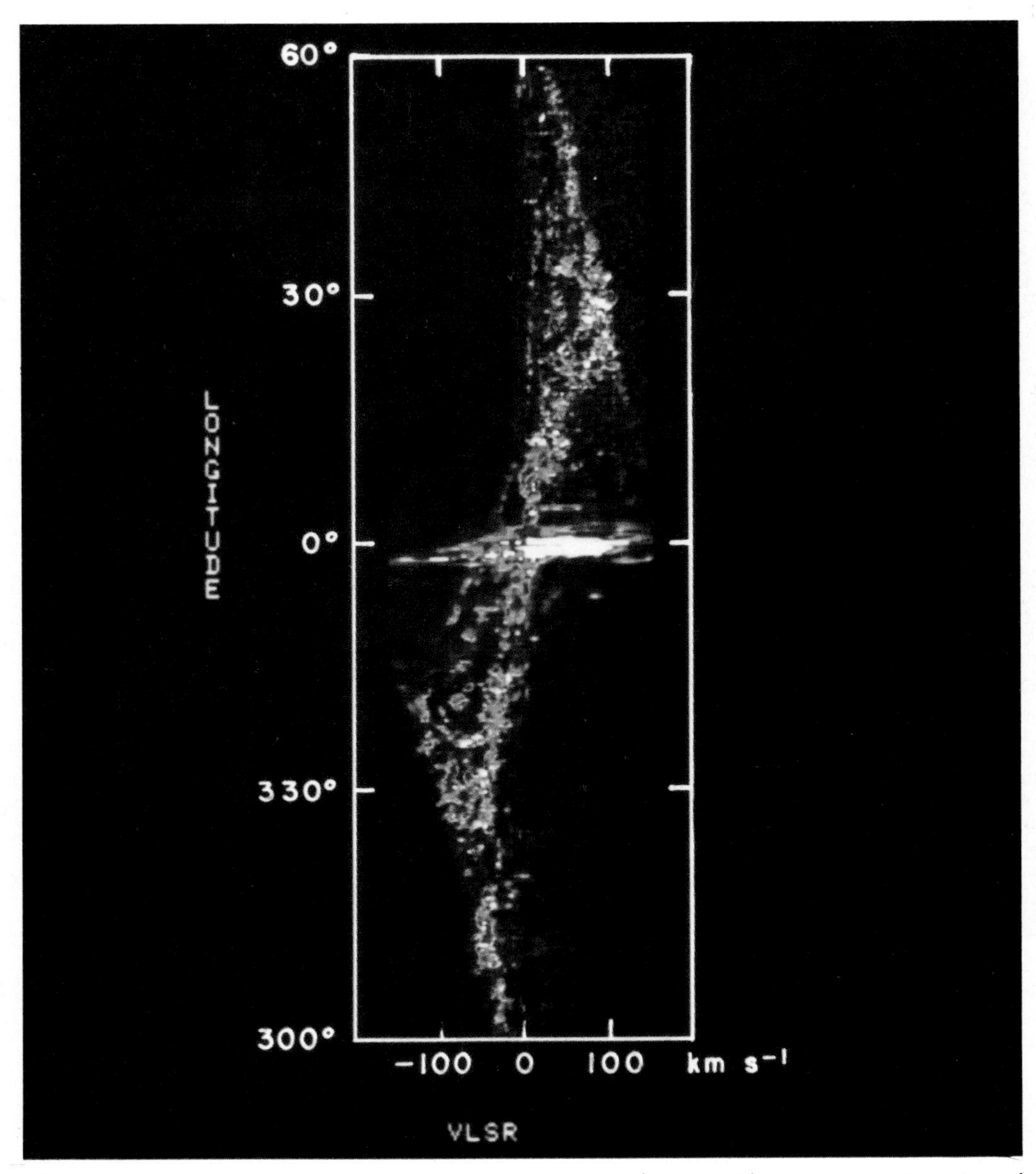

Figure 1 (Cohen, Thaddeus and Bronfman, Section II.3). False-color longitude-velocity plot of CO emission from the galactic equator. The data from $\ell = 12^{\circ}$ to 60° are from our fully sampled New York survey. The data from $\ell = 348^{\circ}$ to 12° are only sparsely sampled. From $\ell = 330^{\circ}$ to 348° spectra were taken every beamwidth. For the Northern Hemisphere the bandwidth is 166 km s^{-1} and the noise per 500-kHz channel is 0.45 K (RMS); for the Southern data the bandwidth is 333 km s^{-1} and the noise is 0.12 K. In both cases baselines are excellent: only straight lines have been removed.

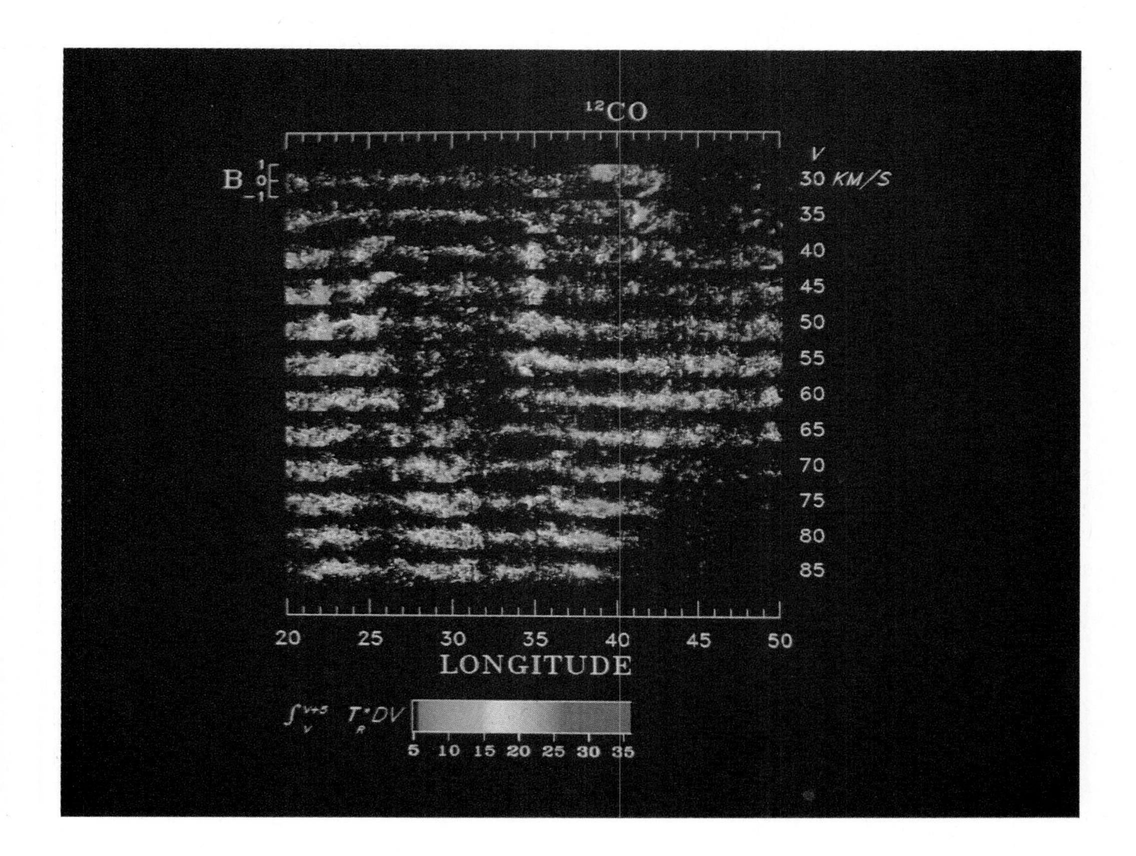

Figure 1 (Sanders, Clemens, Scoville and Solomon, Section II.6). Longitude-latitude strip maps of integrated CO emission, $\int T_R^*(CO)\,dv$, in 5 km s^{-1} bins. Each strip map reaches from -1° to +1° latitude. The integrated emission is coded in colour; the scale is in K km s^{-1}. Single giant molecular clouds (GMC) have line widths of 5-15 km s^{-1}, and hence usually appear in 2 or 3 adjacent strips. Each strip contains emission from both the near and far sides of the inner galactic disk. At low velocities these regions have distance ratios typically a factor 2-5, thus far-side clouds will be a factor 4-25 smaller in solid angle. Near the maximum velocity allowed by galactic rotation, blending of emission features increases due to the decreasing radial-velocity gradient.

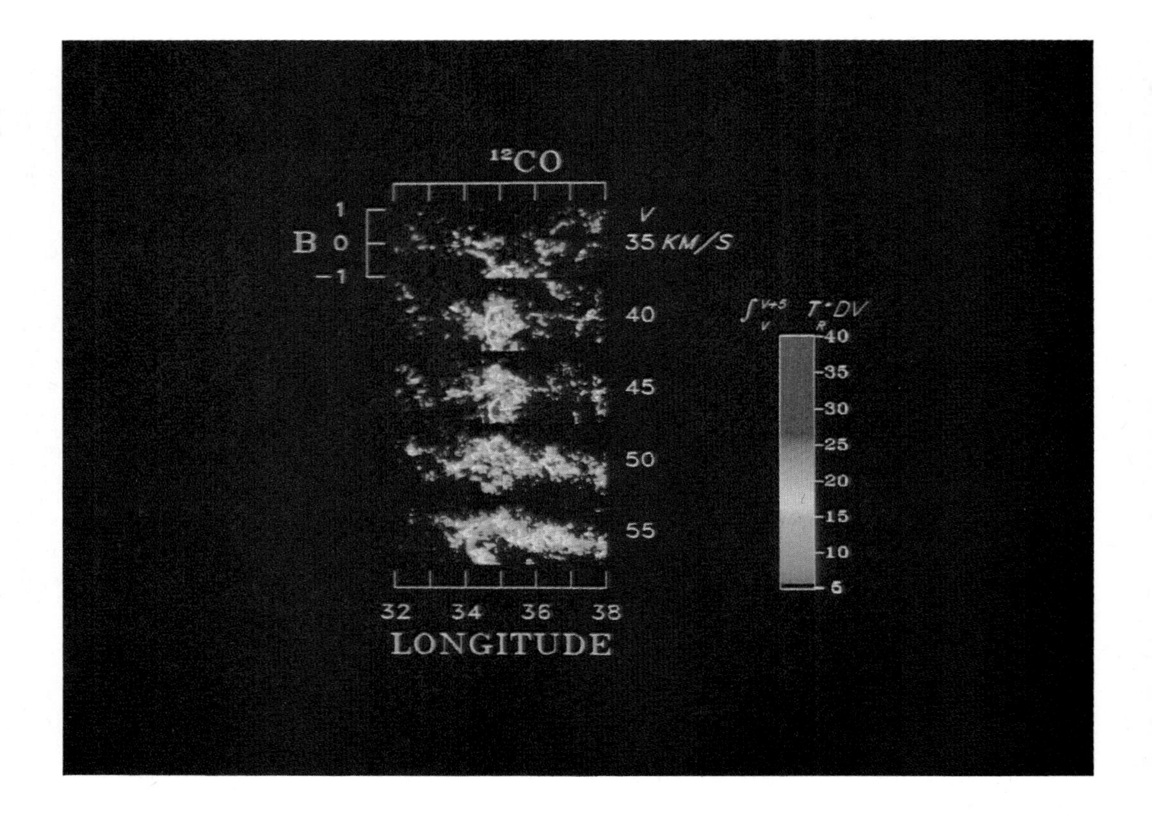

Figure 2 (Sanders, Clemens, Scoville and Solomon, Section II.6). CO emission from the cluster of GMC associated with the W44 region. This map represents a small section of Figure 1.

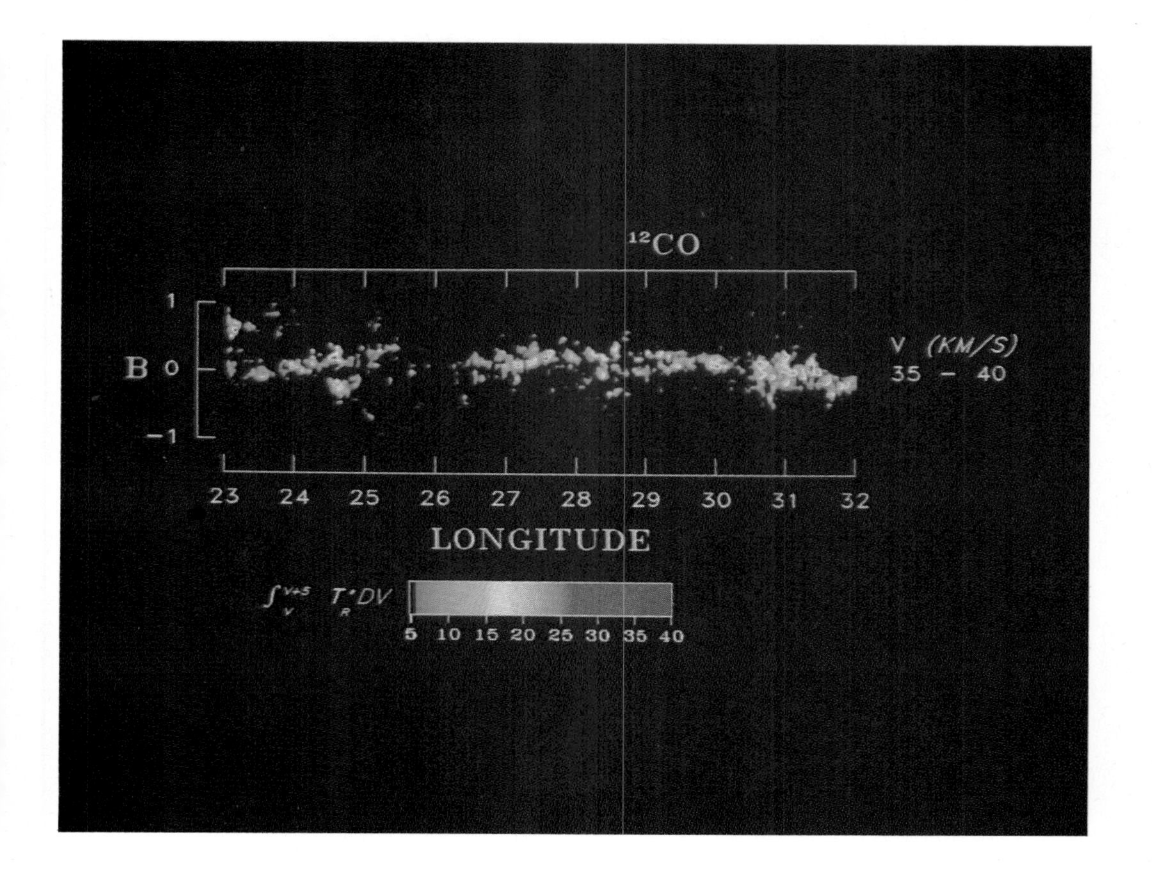

Figure 3 (Sanders, Clemens, Scoville, and Solomon, Section II.6). A chain of 'far-side' molecular clouds between longitudes 23° and 32°, located approximately 14 kpc from the Sun. The colour scale is in units (K km s^{-1}) of CO integrated intensity, $\int T_R^*(CO)\, dv$, in 5 km s^{-1} bins.

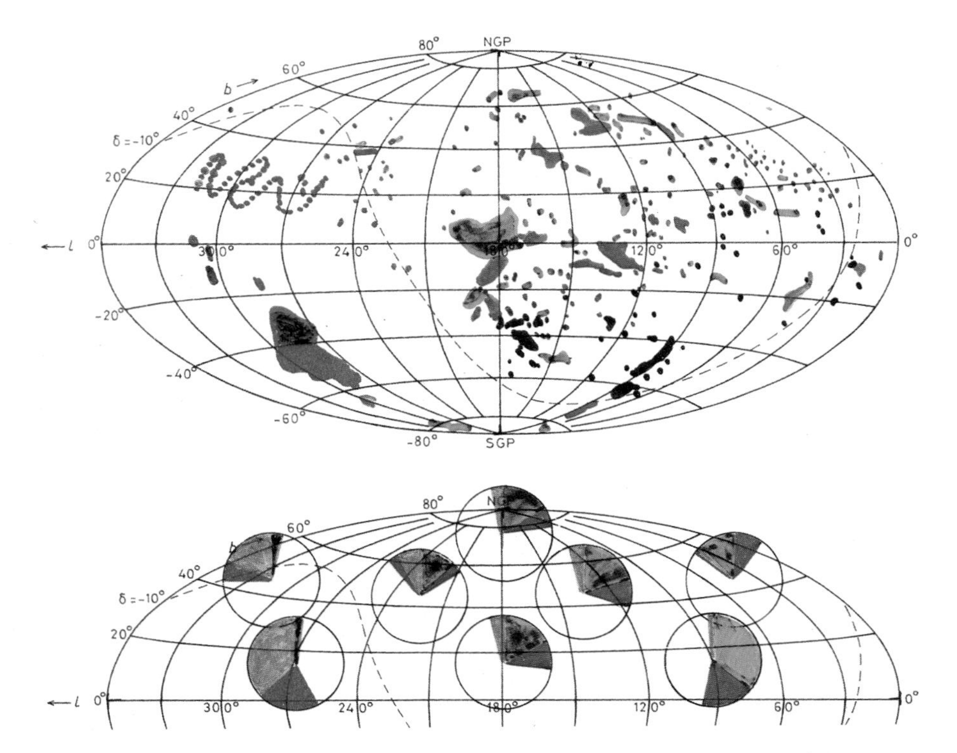

Figure 1. (Van Woerden, Schwarz and Hulsbosch, Section II.8)
Upper panel (a): Velocities of high-velocity clouds (HVCs) as summarized by Mirabel (1981a). Outlines of HVCs are only roughly sketched. Velocities (in km/s, relative to the local standard of rest, LSR) coded as follows: black, V < -220; blue, -220 < V < -130; green, -130 < V < -80; orange, +80 < V < +130; red, +130 < V < +220; brown, +220 < V. The brown patch at ℓ ∿ 280°, b ∿ -33° is the Large Magellanic Cloud.
Lower panel (b): HVC velocities predicted by galactic-fountain model (nr. E1) of Bregman (1980, Figure 5). Predicted frequency distribution of velocities in regions of ∿ 30° x 30° shown; colour code as for upper panel. Velocities |V| < 80 km/s, the majority, are left blank. Predicted distributions at b > 0 and b < 0 are equal.

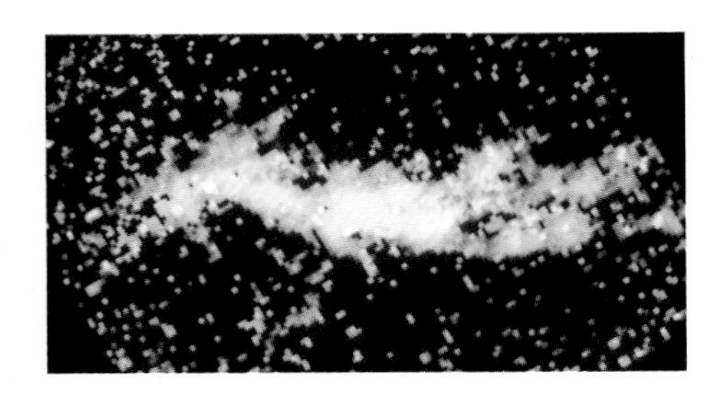

Figure 6. (Van Woerden, Schwarz and Hulsbosch, Section II.8) Small-scale structure and velocity field (colour-coded) in HVC 132+23-210, as measured by Schwarz at Westerbork. Resolution 50", field size dℓ x db shown 45' x 24'. Note long filament at roughly constant latitude, with several condensations. Velocities vary little in this filament and in the faint one crossing it.

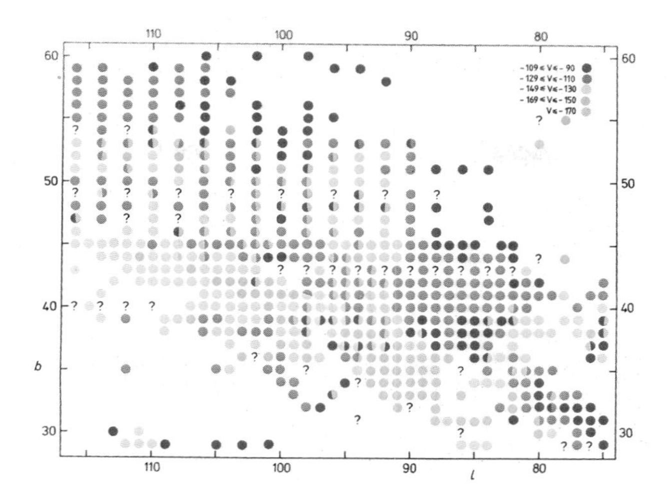

Figure 4. (Van Woerden, Schwarz and Hulsbosch, Section II.8) Velocity field in complex C, as measured by Hulsbosch at Dwingeloo on a 1° grid. Telescope beamwidth 0.6 degrees, velocity resolution 16 km/s. Velocities (in km/s, relative to LSR) coded as shown in top-right corner. Question marks denote points not yet measured. Above b = +45°, longitude spacing is 2°. Note presence of two components at many positions.

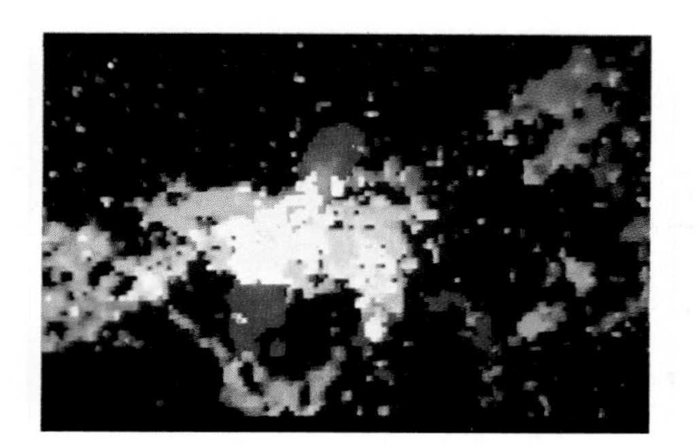

Left: Figure 7. (Van Woerden, Schwarz and Hulsbosch, Section II.8) Velocity field in HVC 139+28-190, cloud AI in Chain A, as measured by Schwarz and Oort (1981) at Westerbork. Field size shown 40 x 28 arcmin, angular resolution 50 arcsec, velocity resolution 2.1 km/s. Velocity scale runs from -209 km/s (red) to -173 km/s (blue). Note overlap on sky of 2 or 3 components in many places, shown by white or by mixed colours.
Right: Figure 8. (Van Woerden, Schwarz and Hulsbosch, Section II.8) Velocity field in HVC 153+39-178, cloud AIV in Chain A, as measured at Westerbork by Schwarz (in preparation). Angular resolution 50 arcsec, velocity resolution 1.7 km/s. Velocity scale runs from -192 km/s (blue) to -162 km/s (red); its length corresponds to 24 arcmin East-West; the vertical (declination) scale is a factor 1.9 compressed. Note transverse velocity gradients in several filaments. East at right in this figure.

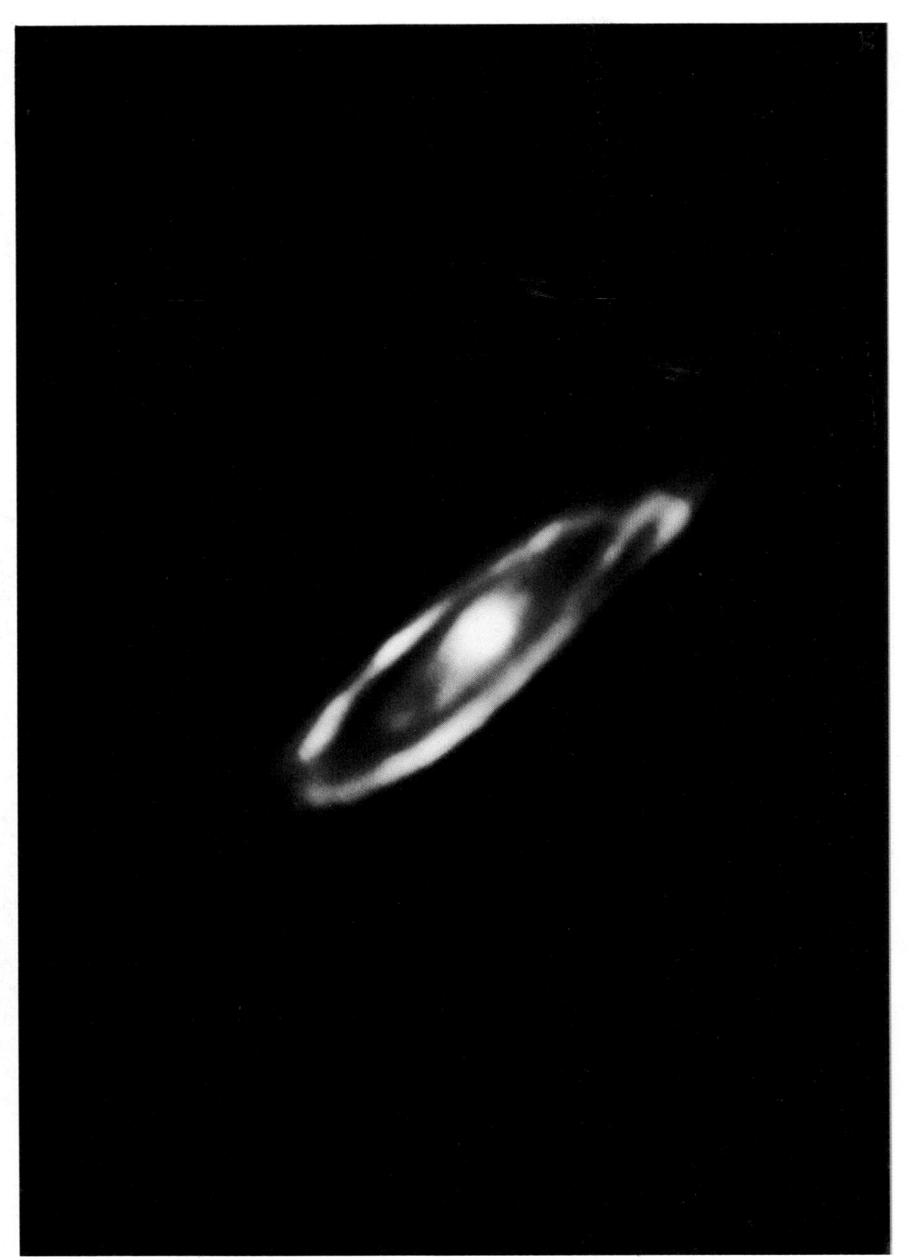

IRAS map of the Andromeda Nebula (Habing, Section II.9, Figure 1)

This colour picture combines the surface-brightness distributions at 12 micron (blue), 60 micron (green), and 100 micron (red). The first two maps have been convolved to the angular resolution of the 100-micron map. In the lower half of the picture the blue image is that of an unrelated, foreground star. It indicates the response to a point source. The white colour of the central region indicates that the emitting dust there is hotter than the dust in the ring.

PART I

HISTORY OF GALACTIC RESEARCH

Tuesday 31 May, 2045 - 2215

Public lecture by M.A. Hoskin
Chairman: H. van Woerden

Monday 30 May, 1450 - 1545

Lecture by E.R. Paul
Chairman: M.A. Hoskin

Thursday 2 June, 1130 - 1220

Lecture by R.W. Smith
Chairman: M.A. Hoskin

Tuesday 31 May, 1430 - 1505

Lecture by O. Gingerich
Chairman: B.E. Westerlund

At the exhibition about Kapteyn and Van Rhijn, in front of Kapteyn's portrait (page 1): Kapteyn's grandson, with A. Blaauw (left) and Dr. W.H. Koops, Director of University Museum. Blaauw was Professor of Astronomy at Groningen in 1957-1970.

CFD

THE MILKY WAY FROM ANTIQUITY TO MODERN TIMES

Michael Hoskin
University of Cambridge

ABSTRACT: The paper outlines the history of attempts to explain the Milky Way, from Antiquity to the early-twentieth century, with special reference to the eighteenth and nineteenth centuries. Also discussed is the relationship of the Galaxy to other star systems, and particularly the question of whether there are other galaxies in the visible universe.

1. INTRODUCTION

The other historical papers in this volume discuss the late-nineteenth-century and early-twentieth-century background to contemporary scientific debates concerning the Galaxy. In this paper our task is to outline the broader historical context. We shall concentrate on the eighteenth and nineteenth centuries, because whereas earlier attempts to make sense of the motions of the planets led to the creation of Newtonian dynamics, early discussions of the Milky Way proved largely sterile and are of concern to the historian rather than to the practising astronomer interested in the genesis of his science.

The late medieval picture of the world was dominated by the teaching of Aristotle, according to whom the spherical Earth is at the centre of a spherical heavens. On and near the Earth is constant change and decay, coming-to-be and passing-away. By contrast, the heavens are changeless except for the eternal cycling of the stars and planets. Comets, since they change, belong to the terrestrial world, and are discussed by Aristotle under 'meteorology' rather than 'astronomy': they are exhalations from the Earth that ascend to the sphere of fire. The Milky Way, partly because of the similarity of its appearance to an extended comet, is also part of meteorology rather than astronomy.

Aristotle's was by no means the only theory of the Milky Way proposed in Antiquity. Among the others was that of Democritus, preserved for us most completely in the words of Macrobius: "Democritus's explanation was that countless stars, all of them small, had been compressed into a mass by their narrow confines, so that the scanty spaces lying between them

H. van Woerden et al. (eds.), The Milky Way Galaxy, 11–24.

were concealed; being thus close-set, they scattered light in all directions and consequently gave the appearance of a continuous beam of light." By the sixteenth century there was considerable support for this view that the Milky Way was celestial rather than terrestrial, so much so that when Galileo published in 1610 an account of his first discoveries with the newly-invented telescope, his resolution of the Milky Way into "nothing but a congeries of innumerable stars grouped together in clusters" aroused little interest: the Milky Way was accepted as the optical effect of great numbers of small stars.

This being so, one might have supposed that in the century of Descartes and Newton, the century when the stars were recognised as distant suns and the Sun as merely our local star, and when the closed cosmos of Aristotle was replaced by the infinite space of the geometers, there would be attempts to discover the three-dimensional distribution of stars that would bring about the optical effect we see as the Milky Way. But this would be to underestimate the legacy of the many centuries in which the 'fixed' stars had been nothing more than a backcloth, a reference frame, for the challenging motions of the planets. Until towards the end of Newton's life, no single star had been known to alter its position in the sky relative to the other stars since records began in Antiquity, and the stars -- including those of the Milky Way -- continued to be of minimal interest. Newton's *Principia* almost totally ignores the stars, and he addressed himself to the question of whether the stars are finite or infinite in number only when challenged on the matter by a theologian. At no time did Newton give more than passing consideration to the Milky Way.

2. WRIGHT, KANT AND LAMBERT

In the middle decades of the eighteenth century, three speculative thinkers who lay outside the mainstream of astronomy turned their minds to the phenomenon of the Milky Way. The oldest of these, and the first to go into print on the subject, was Thomas Wright (1711-86) of Durham in the north of England. Wright came from a modest home and was largely self-taught. He earned a living giving popular lectures on science, and assisting aristocratic families with the care of their estates and the construction of new buildings. His interest in the Milky Way stemmed from his life-long desire to produce a unified cosmology, a vision of the universe that began close at home with the astronomer's account of the visible universe, and then extended this limited picture by the use of symmetry and of the principles of Wright's theology. A manuscript from 1734, apparently for a public lecture with elaborate illustration, describes a universe in which, at the centre, is heaven, the abode of God and of the blessed. Far away, in all directions, is the outer darkness, "the shades of Darkness & Dispare supposed to be The Desolate Regions of ye Damnd". In between these is "the Gulfe of Time or Region of Mortality", a spherical shell of space within which all the stars (including the Sun) move in orbit in different directions, each circling around heaven, which is at the centre of the spherical shell (see Figure 1). This led Wright

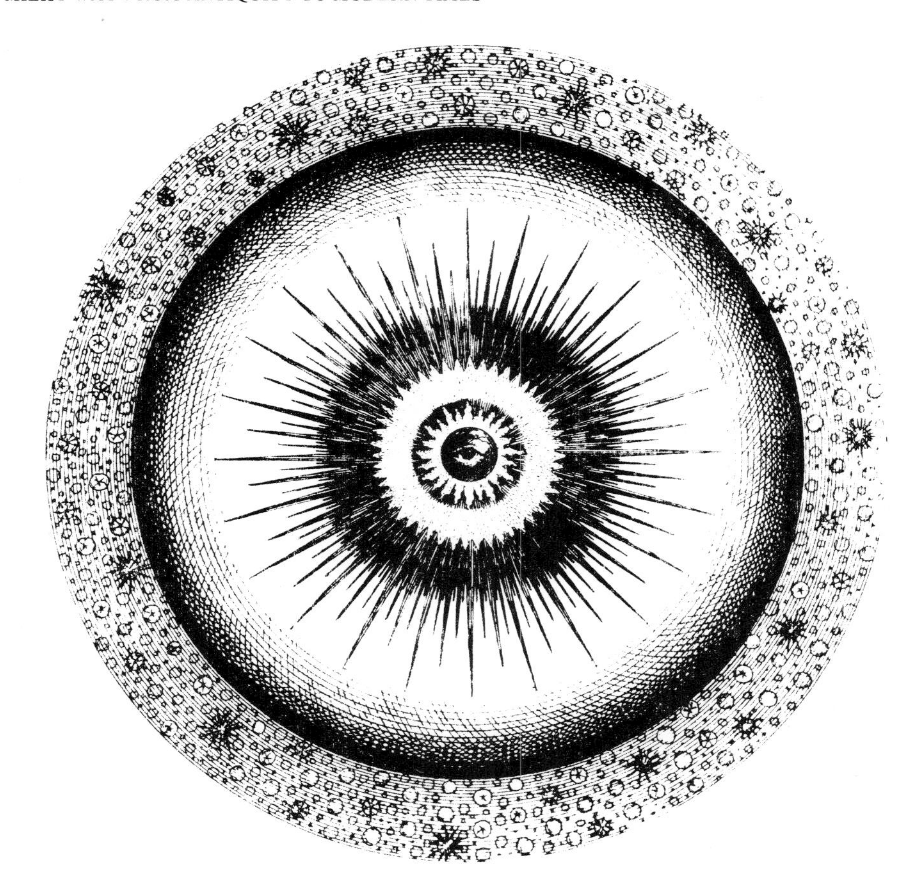

Fig. 1. Thomas Wright's basic conception of the universe. At the centre is heaven, denoted by the Eye of Providence. All the stars, including the Sun, occupy a shell of space surrounding heaven (and here represented in cross-section), and they orbit within this shell. (Plate XXV of An Original Theory.)

to postulate the motion of the Sun and of every star, and he was later gratified to discover in Philosophical Transactions for 1718 the paper by Halley giving news of the discovery of the first proper motions.

In order to bring home to his audience their personal involvement in Wright's cosmos, he allowed himself artistic license when picturing our immediate corner of the universe. The immense drawing he displayed to his audience showed a cross-section of the universe, in which the sphere of the stars was of course represented by a ring of stars surrounding heaven. But in portraying the Sun and the planets, he drew them as they are seen by us, rather than as seen by a distant observer. He took the same license in drawing the visible stars, first the nearest and brightest, then the fainter, until "at a certain distance from ye Sun equal to a vissual ray of ye smallest visible star is a faint circle of light terminating the utmost extent of ye visible creation, in a finite view from ye Earth...".

Wright believed that in this way he had succeeded in explaining the Milky Way, the "faint circle of light"; though in fact his drawing showed an *arbitrary* cross-section of the creation, whereas the plane of the Milky Way is unique. Sometime in the 1740s he realised this, and considered what modifications of his world-picture were required. His conclusions form the centre-piece of his handsome quarto volume, *An Original Theory or New Hypothesis of the Universe* (London, 1750), on which his fame chiefly rests. Although the explanation of the Milky Way is a source of great pride to Wright, the book is in fact a stage in Wright's life-long attempt to reconcile his theological world-picture with the observations of astronomers. Specifically, our Sun and the other stars of our system -- now one among many -- are in orbit around our (local) Divine Centre. These stars of our system may, as before, move within a spherical shell of space, but if so -- and this is the difference from 1734 -- the shell is very thin. When we look inwards or outwards, we quickly see past the individual stars that are our neighbours, and then we look into empty space; but when we look tangentially to the spherical shell, which has a radius so large that it curves almost imperceptibly, then we see so many distant stars that together they have a milky appearance. The plane of the Milky Way, in other words, is the tangent plane to the spherical shell occupied by the Sun and the other stars of our system, at the point occupied by the observer.

As an experienced teacher, Wright introduces his readers gently to this concept of a spherical shell of space with radius so great that the curvature is almost imperceptible to astronomical observers. He does so by discussing first the (hypothetical) situation where the radius of curvature is infinite -- where, that is, the stars would be located within two parallel planes. The inclusion of the related illustration in *An Original Theory* (see Figure 2) has misled many subsequent writers into believing that this was Wright's picture of the actual universe, and he has been credited with being the first to teach that the Galaxy has a disk-like structure. As we have seen, Wright's fundamental belief that the stars of our system orbit a Divine Centre made a disk-like Galaxy entirely in the natural order unthinkable for him.

Wright offered an alternative picture of the Galaxy that was to be misunderstood -- creatively -- by the philosopher Immanuel Kant (1724-1804). In this alternative picture, the space occupied by the Sun and the other stars of our system was not spherical but planar and shaped like a hollow disk. The stars would orbit about the Divine Centre within the plane of the disk, and the star-system would therefore look rather like the rings of Saturn, with Saturn itself replaced by the Divine Centre. (Indeed Wright speculated that Saturn's rings were "no other than an infinite Number of lesser Planets".) The *visible* stars would then occupy a (continuous) disk of space which was a small fragment to one side of the complete *and hollow* disk.

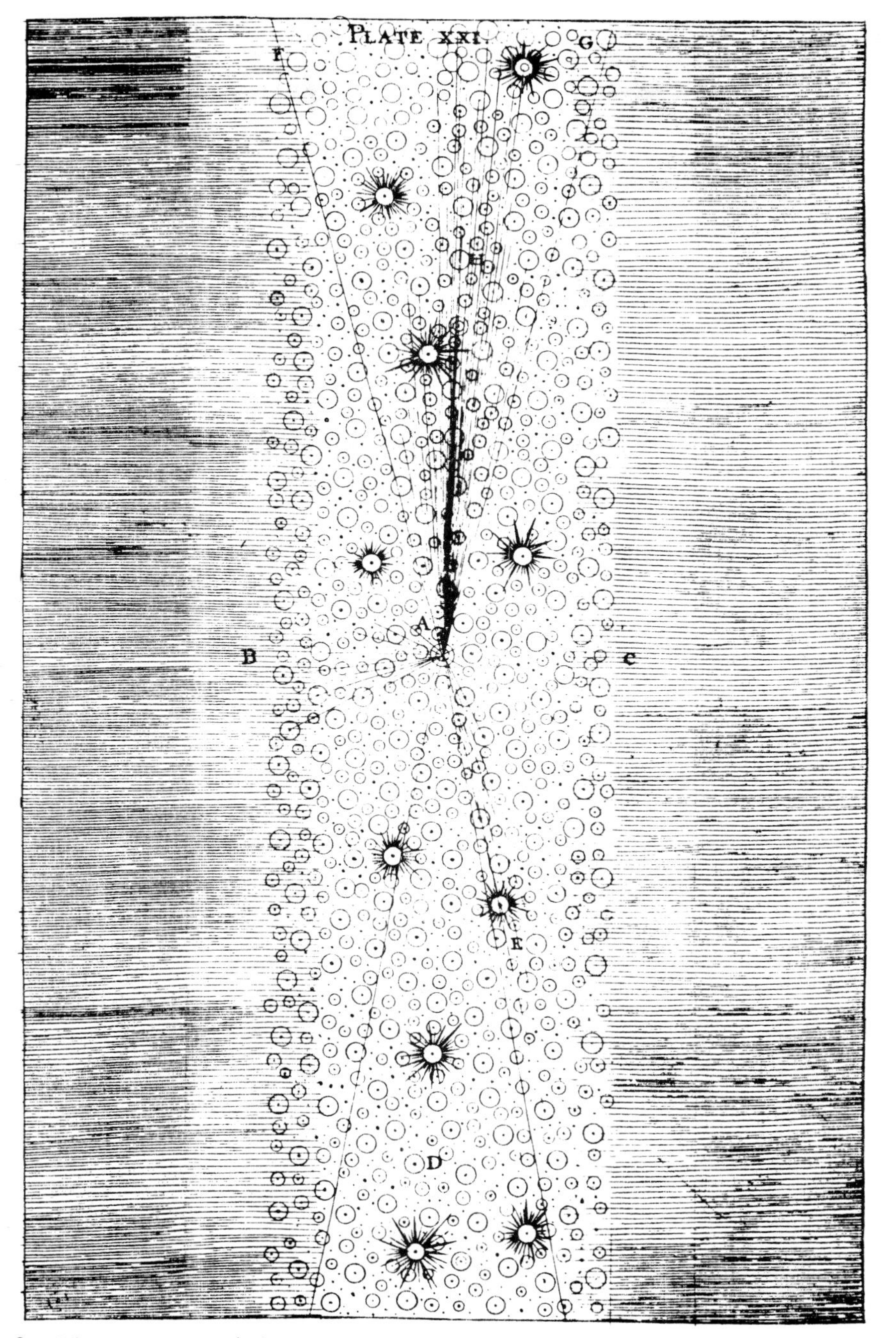

Fig. 2. Plate XXIII (misnumbered XXI) of Wright's An Original Theory, explaining the appearance of the sky as seen by an observer within a (hypothetical) star system bounded by parallel planes.

We can be sure that this second alternative picture was not that preferred by Wright, since the spherical symmetry is lost. But when a summary of Wright's book appeared (without illustrations) in a Hamburg periodical in 1751 and came to the eyes of Kant, it was not clear to Kant that every single star system was arranged about its own Divine Centre. He therefore saw no reason why, in the 'Saturn's rings' alternative, the stars should not extend from the outer edge on one side of the disk right across, without interruption, to the other. Such a system, viewed from a distance, would appear either circular or elliptical in outline; Wright's spherical shells, however, would always appear circular. As Kant believed that Maupertuis had observed nebulae that were elliptical in outline, and as Kant thought these nebulae were analogous star systems, he rejected Wright's spherical shells but gladly accepted a disk-model of the Galaxy (but without the hollow centre and therefore entirely in the natural order). This, the first genuine disk-model of the Galaxy, in which the Milky Way is seen as the 'ecliptic' plane of the stars, was included by Kant in his _Allgemeine Naturgeschichte und Theorie des Himmels_ (Königsberg, 1755). Kant envisaged -- as had Wright -- moons, planets, stars, star systems as forming steps in an hierarchy. But for Wright the hierarchy ended there, as it moved from the natural to the supernatural. Kant, being free of such limitations, allowed the hierarchy to extend upwards infinitely, so that just as the Milky Way is the visible appearance of the disk-shaped Galaxy or star system to which the Sun belongs, so the Galaxy is but one component of a larger system, and so on.

Simultaneously, but independently, the Alsatian physicist Johann Heinrich Lambert (1728-77) was also reflecting on the phenomenon of the Milky Way. In 1749, as he later told Kant, "I went into my room after the evening meal, and looked through the window at the stellar sky, and especially at the Milky Way. The insight, which I had then, to see it as an ecliptic of the fixed stars, I wrote down on a quarto page". Lambert was so struck by this analogy that he became convinced that the stars of our system, just like the planets, lie close to a given plane and are all in orbit about the centre of the system. He outlined his conception in his _Photometria_ (Augsburg, 1760), and elaborated it the following year in his _Cosmologische Briefe_ (Augsburg, 1761). Greatly influenced by Leibniz and therefore committed to a universe with all the stability and permanence of the Sun and the solar system, he too believed in a hierarchy of systems, but a hierarchy of finite extent. The Sun and the other stars that we see as remote from the plane of the Milky Way, together with the brighter (and nearer) stars in the plane of the Milky Way, form one of several clusters that together make up the Galaxy and orbit about its centre. Whereas Kant believed that at each stage of his hierarchy a luminous body lay at the centre of the system (Sirius being perhaps the body at the centre of the Galaxy), Lambert thought that the stars gave all the light necessary and the central bodies in the higher orders might well be dark: the variable light seen in Orion (actually the Orion Nebula) might be the central body of our cluster, variably illuminated by nearby stars as they orbited around it. From the details of the appearance of the

Milky Way he infers that our cluster lies "not only somewhat outside the plane of the Milky Way but also closer to its periphery than to its centre". Only when the Briefe were drafted did Lambert learn that some stars were known to have proper motions; until then he relied solely on his theory for proof that the stars are in orbit, reasoning (as Newton had failed to do) that it was because the stars are very distant that their proper motions had not yet been detected.

3. WILLIAM HERSCHEL

Wright, Kant and Lambert had been led to theorise about the Milky Way because of their unorthodox interests in astronomy, and it was partly for this very reason that their speculations had little impact. William Herschel (1738-1822), who was to make cosmology part of the science of astronomy, owned a copy of An Original Theory but probably obtained it late in his career. Kant's work he seems not to have known -- not surprisingly, as its publication was blighted by the bankruptcy of the publisher. Lambert's Briefe he encountered for the first time only in 1799, when he was asked for an opinion concerning a proposed English translation. Therefore, while we cannot exclude the possibility that some hint deriving from one or other of these works (and especially that of Wright, who was living near Durham when Herschel was an organist in the north of England) reached him and later resurfaced in his mind, it is likely that Herschel owed nothing to his speculative predecessors. Herschel himself saw no harm in speculation -- indeed he gave public notice of his own intention to speculate too much rather than too little -- but his speculations were based on heroic campaigns of observational astronomy, using monster telescopes that he had designed and built with his own hands and which were unavailable to any other astronomer. In his unorthodox commitment to discovering "the construction of the heavens" Herschel distanced himself from the professional astronomers of his day, as he did by embarking on a natural history of the heavens, collecting huge numbers of specimens of nebulae, double stars and so forth; and because he alone had access to the evidence, other astronomers did not see how to confirm or refute his novel theories. But whatever the ambiguities surrounding his impact on his contemporaries, he was given most generous use of the pages of Philosophical Transactions for the publication of his observations and theories, and this ensured that his work was available world-wide, both then and in the future. Herschel's attack on the problem of the Milky Way was therefore decisive in making the question a regular part of the science of astronomy.

The attack came in two major papers on the construction of the heavens, published in Philosophical Transactions in 1784 and 1785. In 1781, in the course of a systematic examination of all the brighter stars, Herschel had come across an object that he had recognised at once as an unknown member of the solar system. It proved to be a major planet, now known as Uranus, and the fame of the discovery enabled his allies in the English court to lobby successfully on his behalf for financial support. By 1782 the refugee musician from Hanover with an amateur enthusiasm for

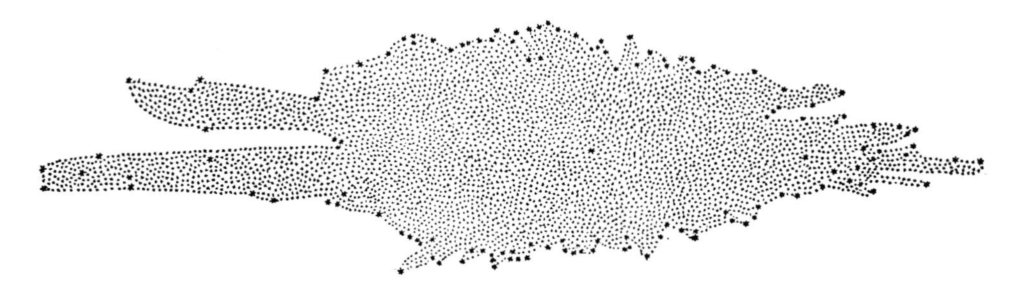

Fig. 3. William Herschel's cross-section of the Galaxy (from Phil. Trans. for 1785).

astronomy found himself installed near Windsor Castle, a professional astronomer whose duties were confined to showing the heavens on occasion to the royal family. The following year he completed his most successful reflector, with mirrors of 18 inches diameter and of 20ft focal length and -- most importantly -- with a stable and convenient mounting. His 1784 and 1785 papers presented the first major results from this instrument with its unrivalled capacity to reach distant and faint objects. Herschel almost takes it for granted that the Milky Way is the optical effect of our immersion in a layer of stars -- indeed, this virtually follows from his working principle that the apparent brightness of a star is an excellent index of its distance. By no means content with a merely qualitative result, however, Herschel asks himself how he can actually map the outline of the Galaxy. He concludes that this can be done with the help of two assumptions. First, that his telescope can indeed penetrate to the borders of the Galaxy in all directions -- for otherwise the task is hopeless. Second, that within the borders of the Galaxy the space is uniformly stocked with stars -- in other words, that the number of stars visible in his telescope at any one time is a reliable guide to the length of the axis of the cone whose vertex is the eye of the observer and whose base is the border of the Galaxy within the observer's field of view.

In what was the first major exercise in stellar statistics -- a technique which he virtually created for the purpose -- Herschel began to count stars. Time would not permit him to examine the whole of the sky that was accessible to him, so he chose a great circle on which to concentrate his efforts. In most directions he counted the stars in ten neighbouring fields of view and took the average. The resulting map of this cross-section of the Galaxy (see Figure 3) confirmed his visual impression that in the directions of the Milky Way there are indeed more stars than usual, and it incorporated quantitative evidence that our system extends further in the galactic plane.

Meanwhile Herschel was devoting his major effort to a systematic search for nebulae. To 'sweep' the whole of the sky visible from Windsor would take him years, and as he collected hundreds of specimens of nebulae so the problems of their classification, and of their physical nature, grew more acute. In his first modest exercises as an observer in the late 1770's, Herschel had become convinced that he had observed changes in the Orion Nebula and that this therefore could not be a distant star system. Some nebulae were therefore truly nebulous and not formed of the familiar stars, but others were undoubtedly star systems in disguise, appearing nebulous only because the telescope used to examine them was insufficiently powerful to 'resolve' them into stars. In his 1784 paper he took the view that 'true' nebulosity presented a smooth, milky appearance to the observer, whereas the 'resolvable' nebulosity of distant star systems appeared uneven and mottled. Soon after the paper had been sent for publication, he came across two nebulae in which both kinds of nebulosity were present, the one merging into the other. This convinced him that he had been on the wrong track. Ignoring the 'changes' he had himself observed in the Orion Nebula, he now took the view that the difference between 'milky' and 'resolvable' nebulosity was simply one of distance; both were star systems, and a star system that appeared 'resolvable' would appear 'milky' if removed to a greater distance. This being so, the Orion Nebula and other nebulae that appeared 'milky' must be very distant star systems; and if they nevertheless appeared extended across a wide area of sky, they must be of enormous extent and may well "outvie our Milky Way in grandeur" -- in other words, be galaxies larger than our own Galaxy.

In 1790 Herschel came across a 'nebulous star' (actually the planetary nebula NGC 1514, which has a prominent central star), and he was forced to admit that the star appeared to be condensing out of the surrounding nebula (by gravitational accretion). This implied that 'true' nebulosity existed after all. The Orion Nebula was now demoted to being a nearby (and changeable) cloud of nebulosity, and Herschel could no longer point to any nebula and declare it to be a galaxy to rank with our own -- for any such galaxy would be indistinguishable from a cloud of nebulosity. Worse still, his continued searches for nebulae had introduced him to many star clusters which were evidence of how non-uniform is the distribution of stars within our Galaxy. This undermined one of the two assumptions on which his map of the Galaxy was based, and he now accepted that a high star-count was a sign of clustering rather than of greater distance to the border of the Galaxy. His other assumption had been put into question by the recent completion of his monster reflector of 40ft focal length, which had brought into view many stars invisible in the 20ft. He had therefore been mistaken in assuming that the 20ft could reach the borders of the Galaxy in all directions, and there were no grounds for arguing that the 40ft could do so either.

The upshot of all this was that Herschel had to withdraw his map of a cross-section of the Galaxy (though this did not prevent it from being reproduced in textbooks long after his death), and he could not with confidence point to any nebula and declare it to be a galaxy independent of and comparable to our own vast system with its unknown extent. He might consider it unreasonable for anyone to argue that our Galaxy is unique in an infinite universe; but the confident theorizing of the late 1780s, with our Galaxy mapped in outline and compared to other galaxies, had had to be abandoned.

4. THE NINETEENTH CENTURY

The boldness of William Herschel's theorizing, and the sudden reversals of opinion that were forced upon him by new evidence, produced an inevitable reaction. The next generation of astronomers, his son John (1782-1871) among them, were much more cautious, and indeed the infant (Royal) Astronomical Society of London, of which William was nominally the first president in his extreme old age, was careful to distance itself from the recent spate of speculations. Part of the problem lay in the growing evidence that the stars differ greatly from one another in their physical characteristics. William had adopted as a working hypothesis the assumption that the stars are highly uniform, so that apparent brightness is a reliable guide to distance: since faint stars were more distant than bright ones, the very faint stars in the Milky Way were proof that the Galaxy extends very far in those directions. But now it appeared that some stars were intrinsically small and faint in comparison with others; and if so, then the Milky Way might be a true ring of small stars surrounding a central cluster containing the Sun and other large stars. Between 1834 and 1838 John Herschel took his father's refurbished 20ft reflector to the Cape of Good Hope, to extend his father's surveys to the southern skies that William had never seen. This gave John the opportunity for a leisurely examination of the Milky Way. He noted dark regions devoid of stars, and it seemed to him more reasonable to assume that these were gaps in a ring or other structure of limited extent, than that they were extended cylinders of empty space whose axes changed to be pointed directly towards the observer. He found many places where the stars were projected against a perfectly black background, which indicated that the system was of finite extent in those directions; and other places where bright stars were projected against a background of small ones. All this resulted in a view of the structure of the Galaxy that was more firmly grounded in dispassionate observation than the theories of William, but it was necessarily vague: "...our situation as spectators is separated on all sides by a considerable interval from the dense body of stars composing the Galaxy, which in this view of the subject would come to be considered as a flat ring of immense and irregular breadth and thickness, within which we are excentrically situated, nearer to the southern than to the northern part of its circuit" (_Outlines of Astronomy_ (London, 1849), art. 788).

In 1845 the third Earl of Rosse (1800-67) at Birr Castle in Ireland completed a monster reflector with mirrors 6ft in diameter. Within a month the new telescope had made what was to be its most significant discovery, that of the spiral structure of the nebula M51. In the years to come observers at Birr found spiral structure in several more nebulae (and claimed to find it in still more); and at the very end of the century, at Lick Observatory in California, long-exposure photographs showed that spiral nebulae exist in enormous numbers. Some astronomers suggested that our Galaxy has a spiral structure, and they had little difficulty in devising suitable spiral arms to reproduce the observed meanderings of the Milky Way. Especially influential were the drawings published by the Dutch astronomer Cornelis Easton (1864-1929) around the turn of the century.

Whether the spiral nebulae were galaxies was a matter of much controversy. Lord Rosse and his colleagues were convinced they had 'resolved' into stars the brighter nebulae they had examined, and some astronomers saw this as grounds for rejecting the existence of 'true' nebulosity in any form. But in 1864 William Huggins (1824-1910) used the infant science of spectroscopy to prove that some nebulae are indeed gaseous.

In the long debate over the status of the nebulae, two observational tests had been of central importance. The first was, whether nebulae had altered shape over the years, for rapid changes of shape would not be possible in galaxies of enormous extent. William Herschel, as we have seen, believed he had seen changes in the Orion Nebula; but later observers, notably his son John, were more sensitive to the danger of spurious changes ascribed to nebulae but in fact occasioned by changes in seeing conditions, in the power of the telescopes, in the skills of the artists in sketching the nebulae, and so on. The second was, whether particular nebulae had been 'resolved' into stars -- and here a danger not fully appreciated at the time was that condensations of light that were of star-like appearance might be taken for stars. One of the few to recognise this danger was Otto Struve (1819-1905), who with his colleagues at Pulkova in Russia believed (erroneously) that changes had been observed in the Orion Nebula that prevented it from being a huge star system. Writing in 1869 to Birr, where the nebula had supposedly been resolved into stars, he urged them to be more cautious and to say "there is a tendency of the nebulous matter to form itself in separate knots sometimes in this, sometimes in an other direction".

In addition to these ongoing observational questions, in the late nineteenth century two new observational facts encouraged the belief that our Galaxy is unique in the observable universe. The first was the clear recognition that those nebulae that were candidates for the status of galaxies (or "island universes") were mostly found well away from the Milky Way, which became known as the "zone of avoidance". Why, it was asked, should independent island universes arrange themselves in space so as to avoid the plane of the Galaxy? The second was the new star that flared up in the Andromeda Nebula in 1885 (S Andromedae). It was estimated that this one star had rapidly increased to become equal in brightness to one-tenth of the entire nebula -- easily explained if the nebula was a cloud that had encountered the star, but physically incomprehensible if the nebula was a vast galaxy of millions of stars.

5. THE EARLY TWENTIETH CENTURY

As our story reaches the twentieth century, we begin to trespass on the subject matter of the other historical papers in this volume. Nevertheless, an outline of the events leading to the final recognition of our Galaxy as one vast system among many may be of help.

The force of the arguments based on the zone of avoidance and the 1885 nova in the Andromeda Nebula was greatly weakened in the 1910s, when Heber D. Curtis (1872-1942) resumed the programme of nebular photography at Lick Observatory that had been cut short in 1900 by the untimely death of the then-director. Curtis found many examples of edge-on nebulae with dark bands of obscuring matter in their central planes, and he realised that similar obscuring matter in our own galactic plane would account for the zone of avoidance: spiral nebulae (in particular) are not seen near the galactic plane simply because they are hidden from us. Further, in 1917 Curtis found in past photographs of spiral nebulae additional examples of new stars, though his investigations were somewhat overtaken by events when G.W. Ritchey (1864-1945) of Mount Wilson announced that a nova was currently visible in the nebula NGC 6946. These novae were all much fainter than the 1885 star, which began to be recognised as wholly exceptional, and therefore an unsafe basis for theorising on the nature of nebulae.

Meanwhile, dramatic developments were taking place in the theory of the Galaxy. As related in the paper by R.W. Smith, Harlow Shapley (1885-1972) at Mount Wilson was using the powerful new technique of Cepheid variable stars to measure great distances, in particular the distances to the globular clusters, of which he was making a detailed study. These clusters, which other astronomers had already noted were concentrated to one half of the sky, he took to be grouped around the true centre of the Galaxy, whose position and distance he could now establish. On this dramatic new theory, the Galaxy was many times bigger than previously thought, and the Sun was far from the centre: earlier investigations, supposedly of the Galaxy, had in fact been studies of the stars in our immediate neighbourhood. Furthermore, on Shapley's view, since our Galaxy was so enormous, it was all the more unlikely that the spiral nebulae were independent island universes.

By no means all astronomers were convinced by Shapley's arguments, especially because earlier sizes for the Galaxy seemed perfectly satisfactory. Furthermore, believers in the island-universe theory of spiral nebulae (such as Curtis) had recently been much encouraged by the careful studies of radial velocities of spirals carried out at Lowell Observatory by V.M. Slipher (1875-1969), which had revealed speeds much larger than that of any known star. These studies, like the discovery of novae in spirals and the evidence in favour of obscuration in the galactic plane, fitted well with the theory that the spirals are independent star systems.

But unexpected opposing evidence now appeared. We have seen that in the debate over the status of the nebulae, the question of whether nebulae have changed shape was long recognised as fundamental, but that prudent observers had accepted that pencil sketches were not to be relied upon. However, this criticism did not apply to _photographs_ of nebulae. It chanced that at Mount Wilson one of Shapley's friends was the Dutch astronomer Adriaan van Maanen (1884-1946), who was noted for his meticulous measurement of photographic plates. In 1916 van Maanen used a stereo-comparator to (in effect) superimpose two plates of the spiral M101, and he concluded that the spiral had changed (by rotation) in the interval between the two photographs. Early in the 1920s van Maanen came to similar conclusions about several more spirals. It was common ground among astronomers that if van Maanen's results were reliable, then the spirals could not be island universes, for that would require their outlying parts to move with more than the speed of light; and it is hard to imagine an investigation that could stay closer to the basic evidence than these stereocomparator measurements. Shapley believed his friend; Curtis did not. The two men met in a famous encounter in Washington in April 1920, and agreed to differ.

In 1923, Edwin P. Hubble (1889-1953) began a photographic study of the Andromeda Nebula with the 100-inch telescope at Mount Wilson. He quickly found what at first he took to be a nova, but which proved to be a variable star. Plotting its light curve by examining plates going back to 1909, he found it was a Cepheid variable star. This proved that it was a true star, and not a star-like condensation. And because it was one of the stars that Shapley and others were using for measurement of great distances, Hubble could use Shapley's own theory to derive a distance for the nebula of around one million light years -- far outside our Galaxy even on Shapley's reckoning. Hubble waited until in February 1924 he had photographed the nebula on successive nights and confirmed that his plates showed the characteristic upward leap of the light of a Cepheid variable, and then he began to share his discovery with other astronomers with whom he was in correspondence. But he hesitated to publish his result in print because it implied that the Andromeda Nebula, and presumably other spirals, were independent island universes, in contradiction to the measurements of his colleague (but no friend!) van Maanen.

At the end of 1924 Hubble was persuaded to break silence, and as a result the existence of island universes (whether or not comparable with our Galaxy) was accepted by almost all astronomers. There remained however the problem of van Maanen's measures. At last, around 1930, Hubble determined to remove the anomaly once and for all. With the aid of colleagues he remeasured some of van Maanen's plates, and more besides, and he and his allies could find none of van Maanen's changes. This meant that either van Maanen was in error, or that Hubble was in error and in every case by exactly the amount needed to get a null result. Obviously the fault must lie with van Maanen, and Hubble prepared for publication long papers making this abundantly clear. His director, however, would not tolerate a public squabble between members of his staff, and imposed a compromise. Each man wrote a mild and brief paper for the 1935 volume of Astrophysical Journal, but it was clear that the last obstacle to the acceptance of island universes -- galaxies -- had been removed.

FURTHER READING

As this is a survey article, I have not given the references appropriate to a research paper, but the reader who wishes to pursue the subject further will find detailed discussion in my "Stellar Astronomy: Historical Studies" (Science History Publications, Halfpenny Furze, Mill Lane, Chalfont St Giles, Bucks, U.K., 1982). The most complete study of the history of theories of the Milky Way is Jaki, S.L., "The Milky Way: An Elusive Road for Science" (Neale Watson Academic Publications, New York, 1972), of which the earlier chapters are conveniently summarised by Jaki in J. Hist. Astron. (1971, 2, 161-7, and 1972, 3, 199-204). The best available account of the modern period is in Smith, Robert W., "The Expanding Universe: Astronomy's 'Great Debate' 1900-1931" (Cambridge University Press, 1982).

EDITOR'S NOTE

The extensive discussion following this Public Lecture was not recorded.

Michael Hoskin telling tales to Ria van Woerden at conference dinner. Further around table: Hugo van Woerden, Victor Clube, Gerry Gilmore; at right: Bernard Burke LZ

KAPTEYN AND STATISTICAL ASTRONOMY

E.R. Paul
Mathematical Sciences Department
Dickinson College, Carlisle, PA

1. INTRODUCTION

If Jacobus Cornelius Kapteyn (1851-1922) were here today, he would undoubtedly be among those asking the new questions on the cutting edge of contemporary astronomy. It is likely that he would even find a survey of his own contributions to the study of the Milky Way irrelevant. Nevertheless, Kapteyn's life-long interest in the Milky Way shaped the work of many astronomers, including, of course, his students, Willem de Sitter, H.A. Weersman, Pieter van Rhijn, and through van Rhijn: Jan Oort, Bart Bok and many others. But Kapteyn's influence extended far beyond his native Holland. After Kapteyn became a close colleague of George Ellery Hale and a Research Associate at the Mount Wilson facilities in 1908, Hale began to employ a number of Dutch astronomers, including van Rhijn, Adriaan van Maanen, and Kapteyn's Danish future son-in-law, Ejnar Hertzsprung. Moreover, Kapteyn's astronomical colleagues world-wide found his enthusiasm and penetrating insights infectious.

Equally significant were the important methods Kapteyn developed to investigate the complexities of the Milky Way. The techniques and models he obtained provided astronomy with tools and concepts needed by its practitioners to explore with increasing reliability the Milky Way system. Kapteyn's interests were motivated principally by his desire to solve the sidereal problem, viz., what is the arrangement of the stars in space? All of his efforts were directed to the realization of this one great, astronomical project. In this review paper, I will survey Kapteyn's contributions to the rise of statistical astronomy, principally his studies of systematic stellar motions and his analysis of the sidereal problem, with some attention devoted to his great star cataloguing efforts.

2. EARLY DEVELOPMENTS

Kapteyn's attack on the sidereal problem may be divided roughly into two periods, both of which represent a somewhat different conceptua-

H. van Woerden et al. (eds.), The Milky Way Galaxy, 25–42.

lization of the same problem. Until about 1904, Kapteyn was interested primarily in using stellar-motion data to understand the distribution of the stars in space. These efforts culminated in his dicovery of "star-streaming", first publicly announced in 1904. This discovery represented a watershed in his thinking. Not only did he confirm the existence of a preferential, dynamic stellar system, which Eddington considered at the time one of the five most important events in the history of astronomy during the past century, but afterwards he began to realize the inadequacy of using only stellar motion studies to assess the sidereal problem. Thus, after this discovery he emphasized the luminosity and density functions vis-a-vis the stellar velocity law as the key element for unraveling the arrangement of the stars. Let me not overemphasize this distinction, though, since his earlier motion work was essential to his later studies of the luminosity and density functions.

Kapteyn graduated in 1875 with a physics doctorate. Although jobs were difficult to find, he obtained a position as observer at the Leiden Observatory. Trained mostly in the rarefied atmosphere of mathematics and physics, Kapteyn immediately set about learning the practical necessities of his new profession. In his new post, his abilities were soon recognized, and early in 1878 he was elected to the newly instituted Chair of Astronomy and Theoretical Mechanics at the University of Groningen. He chose as his opening address "The Parallax of the Fixed Stars", a topic that showed he already regarded stellar distances as requisite knowledge for an understanding of the sidereal problem. Recognizing early the importance of a broader-based set of stellar data, Kapteyn hungrily searched the star catalogues. Even the great *Durchmusterung* catalogues of Argelander and Schoenfeld were limited, not only to magnitudes and position, but equally they lacked the important stellar motion data.

Thus, it is fitting that the International Astronomical Union, the founding of which Kapteyn ironically did not initially endorse, celebrate the 100th anniversary of the founding of Kapteyn's Astronomical Laboratory. For it is the concept of Kapteyn's Laboratory that perhaps best represents the kernel of Kapteyn's success. Kapteyn recognized early that he possessed neither the financial resources nor the proper geographical climate to undertake successful observational studies. Hence, if he were to explore the nature of the Milky Way system, it would be necessary to possess massive amounts of the right kind of data. Thus, as he set his mind on finding the solution to the sidereal problem, he recognized the necessity of good, reliable data. Coincidentally, it was during this period that Sir David Gill, then the leader in practical astronomy and the director of the observatory at the Cape of Good Hope, was attempting to fill the hiatus created by Schoenfeld's delimitation of the southern *Bonner Durchmusterung* to -22 degrees declination. Recognizing the importance of Gill's work, in 1886 Kapteyn offered without solicitation his aid to Gill for measuring the many photographic plates and cataloguing the numerous stars. It took nearly thirteen years of Kapteyn's constant attention before the *Cape Photographic Durchmusterung* was finally published between 1896 and 1900.

Thus began Kapteyn's widely publicized efforts at international co-operation in astronomical research. By the time the Cape project had gone to press, Kapteyn's laboratory had become institutionalized, and christened the "Astronomical Laboratory at Groningen". After his Plan of Selected Areas received international sanction, Kapteyn's astronomical laboratory continued to receive a flood of data for reduction and analysis. Thus, his American colleague, Frederick H. Seares (1922), later wrote that "Kapteyn presented the unique figure of an astronomer without a telescope. More accurately, all the telescopes of the world were his".

Equipped now with extensive experience with the data of his profession, during the 1890's Kapteyn published a series of papers on the nature of stellar motion, all with the idea of eventually solving the sidereal problem. Ever since William Herschel in the eighteenth century enunciated his project, the "Construction of the Heavens", an understanding of the arrangement of the stars in space had been a major problem. A central theoretical concern dealt with the kind of data that could be used accurately to measure stellar distances. Herschel had suggested stellar brightnesses, even though he was aware by 1817 of the contradiction implied in unequally luminous members of binary star systems. But many others throughout the nineteenth century expressed similar concerns, and sought for an alternative measure of stellar distances.

Beginning with Edmond Halley's discovery in 1718 of the motion of stars and Herschel's investigations of preferential stellar motions toward the solar apex, astronomers mostly during the latter half of the nineteenth century began to use stellar-motion data as a means to represent distances to the stars. Using Gauss' least-squares technique, astronomers, such as Friedrich Argelander and George Airy, began to distinguish random fluctuations in stellar motions. Eventually they concluded that random irregularities were due, not to any systematic errors of observations, but rather to the peculiar motions of the stars. Thus, they among others recognized the preferential nature of proper-motion data, and emphasized the need to reduce the data to peculiar motions. During the last half of the century there was wide-spread use of the assumption of random motions among the real motions of stars.

This assumption was coupled with the belief that proper-motion data, to be useful as a measure of stellar distances, must be correlated to a definite stellar yardstick. Following Bessel's discovery of stellar parallax in 1838, and after enough parallaxes had become available, many astronomers argued that a star's distance is inversely proportional to its proper motion. The determination of stellar distances using parallactic techniques would thus be greatly enhanced, since proper-motion data were available relatively abundantly. For each known parallax there were, of course, scores of measured proper motions. The publication in 1888 of the Auwers-Bradley Catalogue containing 3,200 reliable proper motions stimulated these efforts. Still, prior to Kapteyn's ground-breaking studies of stellar motions during the 1890's, no one had succeeded convincingly in relating a distance measure based on a few thousand proper motions to the demands of the large survey catalogues, such as the Bonner Durchmusterung, containing hundreds of thousands of stars. A statistically precise relationship between proper motions, parallaxes, and apparent magnitudes was complex, to say the least.

Although Kapteyn preferred the actual distances derived from parallaxes, he recognized, even with improved photographic techniques, that the scope of the sidereal problem demanded a much broader base than that allowed by the earth's orbit. Thus his emphasis on stellar motions was motivated by both practical and theoretical considerations: that by using proper motions understood correctly the base of parallaxes could be extended on the ever-increasing base line of the sun's motion through space, and that for an understanding of the structure of the stellar system knowledge of only the mean distances of groups of stars, rather than the absolute distances of individual stars, was necessary. To complicate matters further, Kapteyn and the Irish astronomer W.H.S. Monck discovered independently in 1892 that there was a direct relationship between proper motion and spectral type.

Within this context, Kapteyn achieved a major synthesis, both in method and conceptually, when he derived the "mean parallax relationship", a statistical law that formed the basis of nearly all his attempts to understand the arrangement of the stars in space. Examining proper motions in the Auwers-Bradley Catalogue, Kapteyn correlated known parallaxes with proper motions and magnitudes, and generalized this correlation among large numbers of stars to form a "mean" distance relationship. Published in 1901, the "mean parallax" relationship not only culminated years of close analysis of stellar motion data, but it also led directly to Kapteyn's luminosity function, which he first published a few months later. In terms of the achievements of classical statistical astronomy, the "mean parallax" formula has been over-shadowed only by the importance of the luminosity and density laws, and the "fundamental equation of stellar statistics".

3. STAR-STREAMING

Also within this context Kapteyn made his great "star-streaming" discovery. During the 1890's, Kapteyn's investigations were predicated on the supposition that stellar motions were the key element in understanding the distribution of the stars. These developments were codified in the so-called velocity law, a relationship, Kapteyn argued, that not only would provide an understanding of the stellar system, but that also would lead to the derivation of the density and luminosity laws. In turn the latter would yield a detailed understanding of the Milky Way system.

About 1895, Kapteyn developed a mathematico-statistical theory that related star-counts, the density function, and a Gaussian probability function of proper motions. In his theory he assumed the traditional view that stellar motions are randomly distributed. Co-authored with his mathematician brother, Willem, the complete discussion of his theory was published in 1900 as No. 5 in the famous series of Groningen Publications. In the introduction he succinctly stated their purpose:

> In what follows, an attempt will be made to deduce from the observations, what, for the sake of brevity, I will call the law of velocities, i.e., the law by which is defined the number of stars having a linear velocity equal to, double, triple, ..., half, a third, ... that of the solar system in space, or shorter: the law by which the frequency of a linear velocity is given as a function of its magnitude. The fundamental hypothesis on which this derivation rests is the following: ... The real motions of the stars are equally frequent in all directions.

The observational evidence supporting the theory was earmarked for No. 6 of the Groningen series. Though it represented the most up-to-date views of the velocity law, the theory turned out to be so wide off the mark that not even a comparison with the observational evidence could be made. The reason for the discrepancy between theory and observation was the invalidity of the fundamental hypothesis of random motions. Although most nineteenth-century astronomers considered this hypothesis as a priori valid, in 1895 Hermann Kobold showed that a random distribution did not represent the observed motions of nearby stars in the Auwers-Bradley Catalogue; soon Kapteyn was to provide an explanation of this startling fact in terms of preferential motions. The anomaly between theory and evidence represented a critical problem for Kapteyn's program, because such a basic discrepancy affected one of his stated aims: the derivation of the density and luminosity laws from the allegedly more fundamental velocity relationship.

When it became clear that the evidence needed to support the theory was not forthcoming, he deduced several hypotheses to explain the alleged discrepancy: (1) preferential stellar motions; (2) incorrect apex value; and (3) incorrect proper motion values. Although he showed theoretically that the last two explanations could account for the failure of his theory, he concluded that both the apex value and proper motions had been calculated correctly.

Of all the numbers of the Groningen Publications, a series forming, in the words of Eddington, "one of the most often consulted works in an astronomical library, No. 6--the one which has never been written--[is] perhaps the most famous of them all...." How could an unwritten document be so significant? Precisely because Kapteyn's failure to harmonize observation with theory reaffirmed the anomalous nature of stellar motions, and put him onto the track that culminated in his discovery of the two star-streams.

Despite the importance of theory in directing Kapteyn's research program, he also claimed to be an inductivist in his scientific methodology. "My studies", wrote Kapteyn to George Ellery Hale in 1915, "have made of me more and more of a statistician and for statistics we must have great masses of data". Kapteyn's procedure was to combine both deductive and inductive approaches. Commenting on the importance of an inductive approach in his letter to Hale, Kapteyn illustrated his point regarding the star-streaming discovery:

> Deduction sets in too soon and too much is still expected from it. To illustrate what I mean take the star-streams as an example. ... Schoenfeld was led, I think by analogy, to consider the question: May there not be a rotating motion of the Milky Way as a whole? He made the necessary computations, but found practically nothing. Other men tried a rotating motion of all the stars in orbit in the Milky Way, not necessarily all with the same period. Some, I believe, tried to adhere to a common direction of motion .... Now all this seems to me too much deductive. We began by making a wild guess, deduce its consequences and see whether it agrees with the observations. How long might we have guessed before we ... came to put the question: Are there two star streams? I blundered along for a long time in the same mistaken way, till one day I swore to go along as inductively as I could. I made drawings showing at a glance the observed data for each point of the sky. There showed very decided deviations from what was to be expected according to existing theory [i.e., random motions]. Considering these deviations as perturbations I tried to isolate these perturbations: I superimposed all the drawings belonging to Zones in which, according to existing theory, there ought to be equiformity and took averages. The result was a figure pretty well in conformity with existing theory. This drawing I then took to represent the undisturbed form and subtraction [of the solar motion] from the individual figures then gave the isolated perturbations. There showed at once a great regularity, which regularity was almost at once seen to consist in a convergence of the lines of symmetry to a single point of the sphere. From this to the recognition of two star streams. Thus the inductive process led in a very short time to a result which others, myself included, had tried in vain to bring out in a more deductive way, for ever so long.

Within a short time after publishing his velocity theory in 1900, Kapteyn rejected his theory and by 1902 had disovered star-streaming. Finding that the stars tended to move in two distinct and diametrically opposite directions, Kapteyn suggested that this phenomenon resulted from two once distinct but now intermingled populations of stars moving relative to one another.

Kapteyn first announced his new theory of stellar motions before the St. Louis World Exhibition in 1904, and again more importantly before the 1905 meeting of the British Association for the Advancement of Science. In both cases Kapteyn argued that without exception all the stars belong to one of the two streams. The over-riding consideration, in Kapteyn's opinion, was not a reevaluation of the reality of the phenomenon, but the necessity to <u>confirm</u> the theory, that is, that there exist two independent streams of stars passing through one another in opposite directions with different mean motions relative to the sun. In this regard, he suggested to his BAAS colleagues that radial-velocity observations might prove to be the most convenient data to test the theory:

> I suspect that the materials for a crucial test of the whole theory by means of these radial velocities are even now on hand in the ledgers of American astronomers--alas not yet in published form. It is this fact which long restrained me from publishing anything about these systematic motions, which, in the main, have been known to me for three years [since 1902].

He had in mind the Lick Observatory people and particularly W.W. Campbell, who had been doing radial-velocity work since about 1900 and therefore possessed the data needed to prove Kapteyn's hypothesis conclusively.

After Kapteyn's discovery, the number of theories and studies of preferential stellar motions increased quickly. Kapteyn, himself, participated relatively little in these developments, since, in the main, studies of the velocity law and star-streaming could not, in Kapteyn's view c. 1905, add directly to a detailed understanding of the arrangement of the stars in space. He remained keenly aware of the newer work, however, particularly the explanations of star-streaming by Schwarzschild and Eddington, returning to these ideas only in his last major paper in 1922.

4. DISTRIBUTION OF STARS

With the failure to derive the velocity law and the subsequent discovery of star-streaming, Kapteyn increasingly turned his attention to a derivation of the luminosity and density laws as conceptual tools to understand the form and structure of the stellar system. This approach was also based on his earlier work on the mean-parallax relationship. Thus, even though his efforts at detailing the nature of the Milky Way system were frustrated by his failed velocity-law studies, he continued to use both the basic data and results to understand the larger system.

In 1898, Kapteyn's contemporary Hugo von Seeliger derived the "fundamental equation of stellar statistics". From 1898 to 1920, Seeliger presented his results utilizing the "fundamental equation". In most of these studies, however, Seeliger had assumed an arbitrary probability function for the luminosity law. In addition to various solutions of the sidereal problem, Seeliger was interested in developing the mathematics that would allow for an analytic solution. On the other hand, while recognizing the importance of the form of this relationship in relating the star counts, the density law, and the luminosity law, Kapteyn realized that an accurate representation of the arrangement of the stars in space would require a precise understanding of the luminosity relationship.

After deriving the mean-parallax relationship, Kapteyn developed a numerical technique for deriving the luminosity function that would allow one to relate the magnitudes of stars to their motions, and hence their mean distances. Briefly his method entailed placing the catalogued stars in cells corresponding to their apparent magnitude and probable proper motion. The limiting characteristics of the cells corresponded to spherical concentric shells about the sun. Utilizing his "mean parallax" formula, which expressed a dependency between calculated parallaxes, on the one hand, and apparent magnitudes and proper motions, on the other, Kapteyn calculated the mean parallaxes corresponding to each cell. The results, of course, represented only mean parallaxes. In actuality the stars were distributed according to the laws of probability defined by some Gaussian function. The exact shape of the Gaussian curve was derived from a determination of the spread of 58 stars, with precisely known proper motions, apparent magnitudes, and measured parallaxes. Using this computed probability distribution, Kapteyn calculated the spread of parallaxes for the stars within each cell. This resulted in a two-dimensional table in which the catalogued stars were distributed by magnitude class and mean distances. The magnitudes were normalized by conversion to their absolute magnitude using the magnitude-distance relationship.

This technique was first published in 1901 in his classic paper "On the Luminosity of the Fixed Stars". Others, notably Gylden, Schiaparelli, and particularly Seeliger, had noted the importance of using a Gaussian function to represent the luminosities. But Kapteyn alone succeeded in actually deriving such a function. Thus Kapteyn introduced the term "luminosity-curve" into astronomical parlance as "the curve which for every absolute magnitude gives the number of stars per unit of volume". Since his luminosity table expressed distances from the sun to the stars of various magnitudes, it was a simple matter of dividing the numbers of stars within each shell by its volume to calculate the relative density. Thus Kapteyn's procedure also yielded the relative density distribution of stars in the local solar neighborhood. Since stars limited only to magnitude 9.5 were used, the density relationship was tentative at best.

During most of the two decades preceding Shapley's work and the emergence of the "new" astronomy of the 1920's, statistical astronomers generally believed that the density and luminosity laws would be sufficient to explain the arrangement of the stars in space. In addition to the fundamental equations of stellar statistics, Kapteyn's "mean-parallax" formula and his luminosity-curve provided the basic coneptual tools needed in this work. When Kapteyn's luminosity paper appeared in 1901, the possibilities inherent in a rigorous approach to statistical investigations were greatly changed. This was more the beginning, however, than the end. Kapteyn, himself, considered his 1901 paper as providing only a first approximation to the sidereal solution, a problem which, in his opinion, "must be solved by successive approximations".

As a "first attempt" to solve the sidereal problem, however, Kapteyn's research on the general luminosity function made several critical assumptions, that in following years defined key problem areas for statistical astronomy. His 1901 results assumed: (1) negligible light absorption; (2) a sun-centered stellar system; (3) a luminosity-curve uniform throughout the entire stellar system; (4) a luminosity-curve distributed according to Secchi's type I and II stellar spectra; (5) a density relationship independent of galactic longitude and latitude; and (6) true parallaxes of stars distributed about their mean in a Gaussian symmetric form. Let me briefly treat each of these assumptions.

The question of the transparency of space had been discussed by many nineteenth-century astronomers, including William Herschel, F.G.W. Struve, Olbers, and Kapteyn's contemporary Seeliger. Kapteyn recognized that the existence of an interstellar absorbing medium could seriously alter the form of both the luminosity and density functions and thus fundamentally change the parameters describing his stellar system. Although space does not permit detailed analysis of his work on the absorption problem, Kapteyn in 1904 found little evidence for absorption. By 1909, to rationalize away a sun-centered cosmology, he suggested a value of 0.3 magnitudes per kilo-parsec. By 1915, the work of Shapley showed that Kapteyn's value, however, "must be from ten to a hundred times too large ... and the absorption in our immediate region of the stellar system must be entirely negligible." Moreover, Walter S. Adams, also on the Mount Wilson staff, had results that seemed to show hydrogen absorption does not occur in space, but that the stars themselves are responsible for changes in stellar intensity.

Lacking a viable alternative, statistical astronomers had generally assumed that the solar system was centrally located in the universe. To be sure, the ad-hoc nature of this assumption made many feel uneasy; yet, as a workable hypothesis, it was the only really defensible position. As indicated, by 1915 studies tended to confirm the lack of an absorbing medium, which supported a maximum density-function value in the solar neighborhood. Kapteyn's star-streaming work also gave credence to this view, for, as he expressed to Hale in 1915, "... the stream velocity increases with decreasing distance from the sun. The result seems to me to be well established. One of the somewhat startling consequences is, that we have to admit that our solar system must be in or near the centre of the universe, or at least to some local centre."

Kapteyn's third assumption, the uniformity of the luminosity-curve, was theoretically independent of the question of interstellar absorption. It had been derived for the local solar neighborhood where, it was argued, absorption (even if present) was essentially negligible. Thus Kapteyn adopted the view that the luminosity-curve is the same for different distances from the sun. Hence, regardless of galactic position or distance from the sun, all regions of space exhibit the same distribution of luminosities within the same unit volume. Since the derivation of his luminosity function required empirical knowledge of parallaxes and proper motions, it could only be determined for the local region of space.

Kapteyn had noted the importance of spectral type in his 1901 studies of the stellar system. But with the revolutionizing developments in spectral classification early in the century, Kapteyn continually emphasized the importance of close spectral studies. Writing to Adams in 1912, Kapteyn noted: "In my mind, the most important problem in sidereal astronomy would be: the study of the arrangement of stars in space (including star streams) separately for stars of different spectral type." Thus, in writing to Hale a few years later in 1916, Kapteyn again noted: "... we can find the distributions in space of nearly all the Helium stars [, and] ... that there is a gradual transition in every direction from the Helium stars to the other types.... All this finished I will have to come to the A stars, which in the main I find to behave like the Helium stars. If I finish them too I think I may hope to solve the many riddles that remain for the rest."

Though the Herschels, Struve, and others had noted the dependency of the number of stars on galactic latitude, it was Seeliger who, in the 1880's, first rigorously demonstrated this fact. In later studies Kapteyn recognized Seeliger's work on this point, and noted that although the luminosity-curve is independent of galactic latitude, since it was not an absolute measure, but a distribution function, the density law is a function of latitude. A dependent relationship between density and galactic longitude was not rigorously confirmed until 1917, and thus was generally not taken into account in these early studies.

Finally, Kapteyn continued to assert throughout his investigations the reliability of the dispersion of the measured parallaxes about their mean. In 1920 he wrote in his classic paper on the so-called "Kapteyn Universe": "It has been shown in G.P. 11 [1901] that widely differing assumptions as to the dispersion law lead to results that differ but little .... Therefore, we have not deemed it necessary to derive this law anew, but have adopted the one found and tabulated in G.P. 8 [1901]." This was an important point since the dispersion determines the parameters of the mean-parallax formula.

Between 1904 and 1906, Kapteyn had proposed his _Plan of Selected Areas_ as a coordinated international effort to collect the kind of data needed to resolve these assumptions and verify his star-streaming hypothesis. Although his _Plan_ was not completed until 1920, Kapteyn continued to examine these problems closely in a whole series of important research papers.

5. KAPTEYN'S UNIVERSE

By 1914, the basic concerns of statistical astronomy had been further refined, and focused on the following problems: the relationship between spectral type and stellar distribution; the relation between mean-parallax and factors as proper motions and apparent magnitudes; the nature of the velocity law and star-streaming in general; the analytical form of the star-count function, including particularly its maximum magnitude value; the relation between stellar distribution and galactic latitude and longitude; the mathematical form of the luminosity function, and the absolute magnitude at which it obtains a maximum; and the analytic form of the density law.

After 1914, Kapteyn and others continued vigorously both to refine the empirical support needed to quantify these questions and to investigate the mathematico-statistical basis of the sidereal problem. Provisional answers to these questions came together conceptually in two classic papers Kapteyn published in 1920 and 1922, the year of Kapteyn's death. Briefly, the 1920 paper, co-authored with van Rhijn, described a transparent, ellipsoidal stellar system in which star density at low galactic latitudes diminishes in all directions with increasing distance from the sun. Star-density at 600 parsecs was about 60 percent of that near the sun; at 1,600 parsecs about 20 percent; at 4,000 parsecs only 5 percent; and at its perimeter, about 9,000 parsecs from the sun, star-density was less than 1 percent of that in the solar region. At high galactic latitudes, Kapteyn's results were closer to the actual state of things. Although his 1920 system was nearly sun-centered, he was not, despite increasingly stronger evidence from others' research, willing to relinquish this assertion easily.

Two years later, in 1922, Kapteyn developed a dynamical theory of the stellar system, in which he attempted to explain stellar distributions and motions in terms of gravitational forces. The force exerted by the Milky Way system was calculated at various distances perpendicular to its plane. From this calculation an estimate of the total mass density per volume was derived for the vicinity of the sun. Within the plane of the stellar system, he assumed a general rotation about the polar axis with the two star-streams accounting for the motion. Centrifugal forces plus random motions were balanced by the gravitational field.

In actuality, the 1920 model represented Kapteyn's lifetime achievements dealing with the sidereal problem; the 1922 theory was a provisional attempt to relate Kapteyn's (1920) model of the distribution of the stars with his earlier discovery of star-streaming. Taken together, Kapteyn's 1920/22 theory of the stellar system came to be known as the "Kapteyn Universe".

6. CONCLUSION

Obviously, Kapteyn was not solely responsible for the emergence of statistical astronomy prior to the "new" astronomy of the 1920's. But Kapteyn, and to a lesser degree Seeliger, continued as leaders in what promised to become an extremely fruitful research endeavor. With the beginning of their statistical studies in the 1890's until the early 1920's, Kapteyn defined, clarified, and devised many of the major research problems dominating statistical astronomy.

Not only problems of substance were explored, but new methods were developed all of which provided grist for the mills of many statistical astronomers. In these endeavors, the role of statistical theory increasingly came into prominence. In a sense these scientists considered themselves as some sort of latter-day Kepler. Just as they thought Kepler had derived empirically three planetary laws, so too the statistical astronomers believed they were seeking stellar laws, as true of the galactic system as Kepler's are about the planetary system. Furthermore, they were utterly convinced that an understanding of these relationships would yield universal laws of nature, not just statistical relationships. After the derivation of the luminosity function in 1920, Kapteyn expressed it this way: "It is difficult to avoid the conclusion that we have here to do with a law of nature, a law which plays a dominant part in the most diverse natural phenomena."

Not only was Kapteyn a supremely gifted scientist, but he managed, as nearly all leaders do, to stimulate international cooperation. We have already noted Kapteyn's involvement with Gill's _Cape Photographic Durchmusterung_. His _Plan of Selected Areas_ was perhaps the first truly _multi-national_ astronomical effort. His long association with numerous astronomers, and particularly with the Mount Wilson Observatory, only further high-lights Kapteyn's abilities to achieve great success with limited resources. Perhaps the clearest indication of Kapteyn's success was noted, shortly after Kapteyn received the Bruce Medal in 1913, when George Ellery Hale, in reference to Kapteyn's stellar studies, wrote: "You must not suppose for a moment that there was any mistake made in awarding you the Bruce Medal. In my opinion, no astronomical work of the past generation has been more significant or important than your own, and it is a compliment to the other men who have received the Medal to claim them with you."

ACKNOWLEDGEMENTS

I gratefully express my appreciation to H. van Woerden, R.J. Allen, and J. de Boer of the Kapteyn Astronomical Laboratorium for their hospitality and for inviting me to participate in IAU Symposium No. 106. Permission has been received to quote from the Harvard University Archives (Harlow Shapley Papers) and the Mount Wilson and Las Campanas Observatories (George Ellery Hale Papers). Research for this paper has been partially supported by the U.S. National Science Foundation under grant SES 82-06416.

REFERENCES

Eddington, A.S.: 1922, The Observatory 45, 265
Hale, H.E. to Kapteyn, J.C.: 18 Feb 1913, Hale Papers
Kapteyn, J.C.: 1905, Brit. Assoc. Report, section A, 264
Kapteyn, J.C.: 1922, Astrophys. J. 55, 302
Kapteyn, J.C. to Adams, W.S.: 11 Nov 1912, Hale Papers
Kapteyn, J.C. to Hale, G.E.: 23 Sept 1915, Hale Papers

Kapteyn, J.C. to Hale, G.E.: 26 Mar 1916, Hale Papers
Kapteyn, J.C. and van Rhijn, P.: 1920, Astrophys. J. 52, 29, 33
Kapteyn, J.C. and Kapteyn, W.: 1900, Groningen Publ. 5, 1
Seares, F.H.: 1922, Publ. Astron. Soc. Pacific 34, 233
Shapley, H. to Moulton, F.: 7 Jan 1916, Shapley Papers

DISCUSSION

The Chairman, M.A. Hoskin: : Dr. Clube has prepared a Discussion Contribution.

S.V.M. Clube: History has been less than fair to Kapteyn. Thus, it is well known that his analysis of proper motions led to the discovery of the two star streams and that he then embarked on a major programme to delineate the so-called Kapteyn Universe. His discovery was followed, however, by Schwarzschild's suggestion that the proper-motion data could be equally well represented by a velocity ellipsoid. This proposal was enthusiastically endorsed by Eddington (1914), but any mathematical convenience arising from the idea that a single population of stars was experiencing forces that perturbed them along a preferred axis (now recognized as the galactic centre-anticentre line), was not originally seen as denying the physical reality of Kapteyn's two streams - one of which (Stream I) moved relative to the other (Stream II) at around 35 km s^{-1} away from what is now recognized as the galactic centre direction. As it turned out, however, Shapley's discoveries around 1918 led to the sensational collapse of the Kapteyn Universe, and it now looks as though the ensuing loss of confidence in Kapteyn's programme of research led also to an (irrational) decline in interest in the two streams. Whatever the exact sequence of events, the velocity ellipsoid as represented by Stream I alone (generally an intrinsically brighter and younger population) soon graduated to become a primary observational base for Lindblad's and Oort's development of galactic rotation theory, whilst Stream II (generally a fainter and older population) was relegated to the historical dustbin. This is unfortunate, since modern surveys of nearby stars tend to demonstrate the validity of Kapteyn's division. Stream II for example has virtually no representatives among the local young population: see the space-motion studies of A stars by Eggen (1963) and Me dwarfs by Upgren (1976). These latter, comprising Stream I, have motions in the mean like that of the nearby gas and thus approximate closely to the currently adopted local standard of rest. The older and more widely dispersed M dwarfs on the other hand are representative of Stream II (cf. Clube 1978; also these proceedings), and such stars with well-mixed orbits evidently define a physically preferable but entirely different l.s.r. By continuing to overlook Stream II, we not only unbalance our understanding of galactic dynamics but deny Kapteyn's discovery its proper place in history.

REFERENCES

Clube, S.V.M.: 1978, Vistas Astron. 22, p.77
Eddington, A.S.: 1914, "Stellar Movements and the Structure of the Universe", London: MacMillan
Eggen, O.J.: 1963, Astron. J. 68, p. 697
Upgren, A.R.: 1976, Bull. Amer. Astron. Soc. 8, p. 542

The Chairman: Are there any comments to Dr. Clube's contribution?

H. van Woerden: Is it not true that the average motion of older, well-mixed stars is influenced by the radial density gradient in the Galaxy?

Clube: You are referring to the so-called Stromberg drift. The relative motion of Kapteyn's star-streams is orthogonal to that. All I am suggesting at the moment is that we have a clash of information given by the nearby young stars and the nearby old stars. And I would have thought it is more reasonable to be suspicious of the young stars rather than the old ones.

A. Blaauw: If there is a problem about these streams that both are nearby, should one not look at larger distances, where local effects are more smeared out and one has a better overall view? If I remember well, the star streaming or ellipsoidal distribution is shown just as well by the faint stars as by the bright ones, or even better.

Clube: I agree with your comment. I believe also that we ought to look at the behaviour of the more distant stars a great deal more carefully than has been done. Tomorrow I will describe some more recent observations that I think lead one to suspect that we may have misunderstood the more distant material as well as the nearby stars.

The Chairman: We shall now discuss Dr. Paul's paper.

H. C. van de Hulst: The word "sidereal problem", which you used many times, is not common in modern literature. Is it derived from the older literature, or is it a word you coined yourself?

Paul: The phrase "sidereal problem" appears about 1900. Kapteyn may have used it first in his 1904 survey paper published in Science. He there states that he considers the arrangement of the stars in space the central problem in astronomy, and calls it the sidereal problem. He does not indicate who coined the term, and I have been unable to find that out, but it is very prominent in the English-speaking literature.

M. Schmidt: As you indicated, Kapteyn adopted a gaussian distribution of the logarithm of the parallax around the mean value for given magnitude, proper motion, etc. Do you know whether this gaussian distribution was based on observations, or was it an assumption?

Paul: Kapteyn derives the distribution with a complex, numerical technique from the raw data and it drops out in a Gaussian form.

Schmidt: I wonder whether any of those present here that are senior to me knows the answer to this question?

Clube: I hesitate to claim to be more senior, but I would point out that this matter is discussed at length by Eddington in his 1914 book on "Stellar Movements and the Structure of the Universe". It is very interesting that in the end he clearly sits on the fence, and would not choose between the two-stream hypothesis - which would amount to some non-Gaussian distribution - and the Schwarzschild ellipsoidal distribution; that is, he regarded the issue as unresolved by the observations in 1914.

Schmidt: However, I think that the ellipsoidal distribution of velocities is not identical at all to the question that I am asking about the assumption of a gaussian distribution of log π around its mean value.

Clube: It is clear that you needed somebody more senior than you yourself.

Hoskin invites further discussion, after Clube has answered Schmidt.
CFD

R.H. Sanders: What in your opinion was the essential observation that overthrew the Kapteyn Universe?

Paul: I think you are all familiar with the debate about the globular clusters. One of the key elements was that Shapley had determined that the short-period Cepheids were giant stars. Kapteyn and Van Rhijn had assumed that they were dwarfs. A lot of the argument hinges on that particular point. If they are dwarfs, they are much closer - in fact, Kapteyn argues that they should be 8-10 times closer to the Sun rather than in the expanded system that Shapley argues for. There are other elements in the issue.

A. Blaauw: I always thought that the crucial point had been the influence of interstellar absorption. Was not the main point at that time, whether its influence was so large that one should discard the Kapteyn density distribution in the plane? I know that Kapteyn himself looked into this question very carefully; and in fact Van Rhijn's thesis (1915) investigated the possible effects of interstellar absorption by checking for reddening, and the answer was negative.

Paul: Certainly the absorption question was a major thing. Kapteyn vacillates somewhat on this. He has a 1904 paper responding to the American astronomer Comstock, in which he concludes that there is essentially no absorption. In 1909 he comes up with some absorption. In 1915 he is back down to no absorption essentially. The tradeoff is

A. Blaauw, with E.F. van Dishoeck and T.P. de Zeeuw CFD

that, if there is no absorption and the luminosity function is derived for the solar neighbourhood, the parameters defining the system are relatively small (about 9000 pc diameter), but the problem is that you have to accept the central location of the Sun. Alternatively, if one allows absorption - and indeed Kapteyn asks: if we assume a certain amount of absorption, what will the system look like? - then for large distances the density function increases enormously. So the tradeoff for absorption is that you get a very extended system, and Kapteyn finds that an unacceptable hypothesis.

R.G. Carlberg: Did Kapteyn have any opinion on the nebulae, the external galaxies?

Paul: People who assumed the small galactic system, in general asserted the Island Universe theory. Seeliger does argue that the spiral nebulae are island universes. I believe Kapteyn does as well. Robert, do you know?

R.W. Smith: He changes. At the start of the century he goes along with the old notion that there are no visible external galaxies. Later on I think he makes the spirals external galaxies.

B.F. Burke: Shapley's model of the Galaxy was well-known by the time Kapteyn's final papers on the "Kapteyn Universe" were published. What was Kapteyn's opinion of Shapley's model?

Paul, flanked by Clube and Hoskin, invites Smith to comment. CFD

Paul: The so-called Kapteyn Universe was published in 1920. The Universe presented in 1922 is substantially the same, with the Sun moved off about 650 pc from the centre, but its shape is almost identical. The 1920 Universe itself is not much different from earlier models that Kapteyn had presented in 1908 and 1914. So Kapteyn's model was already very well known when Shapley published his work. It appears that Kapteyn did not have much regard for Shapley's work, partly on the grounds of the luminosity of the Cepheids.

O. Gingerich: Most of us think of the Shapley model as it was given by H.S. Plaskett in his Halley lecture of 1935. However, at the time when Shapley presented his model, he linked together a large number of star clouds into a Galaxy that was quite different from our concept of today. So his model was not so very incompatible with Kapteyn's. He accepted Kapteyn's Universe as being the local star cloud that was swallowed up in his much grander system. Hence the problems of absorption and scale did not immediately enter the argument at that time. The two men were talking on two relatively different grounds, and in their meetings at Mount Wilson Kapteyn always seemed relatively uninterested in what Shapley was doing -- at least that is what Shapley told me several times.

Paul: In response to that: Shapley went to the IAU meeting in Rome in 1922, and on his return from Rome he stopped in Groningen or Leiden to speak with the Dutch face to face about this particular issue. In the private correspondence there is evidence of mutual respect, but there was a serious theoretical disagreement here at that time. Seeliger's model is in some respects very similar to that of Kapteyn; and one of Seeliger's students, Hans Kienle, becomes a very early advocate of Shapley's model. In 1924 Kienle edits a Festschrift for Seeliger, and in that book he publishes a paper by Shapley on the globular clusters. So we already see the next generation of people, including Van Rhijn, beginning to recognize that the Kapteyn- or the Seeliger Universe refers simply to the local solar neighbourhood.

J.H. Oort: About the attitude of Van Rhijn and Kapteyn to Shapley's work: if I remember correctly, it was mainly - in the beginning at least - a doubt about the distance scale, whether Shapley had got the distances or the absolute magnitudes of the globular clusters correctly. There was some reason to feel uncertain about that. The other point concerns the absorption. The reasons why Kapteyn and Van Rhijn decided to neglect the absorption was, in a way, a very sound one: they used Shapley's data in fairly high galactic latitudes to indicate that the absorption per kpc was negligibly small. This was correct in all latitudes above 10 or 15 degrees, and so one can say that the Kapteyn system was essentially correct for all directions that were not exactly in the plane of the Galaxy. But Kapteyn and Van Rhijn did not realize sufficiently at that time - and quite understandably so - that one had to go exactly in the galactic plane to find the real extent of the Galaxy, and there of course the absorption was all-important.

The Chairman: Thank you, Dr. Oort. The historians are especially happy to have you with us today. In spite of the many hands being raised, we must now close this discussion.

STUDIES OF THE MILKY WAY 1850-1930: SOME HIGHLIGHTS

Robert W. Smith
Space Telescope History Project

"The Copernicus of the sidereal system is not to be expected for many generations". So wrote R.A. Proctor[1] in his Essays in Astronomy in 1872. Indeed things did look bleak at this time for those who hoped for a good understanding of the size and structure of the Galaxy. Why was this so, and why was there to be such an astonishing transformation of this situation between 1918 and 1930? Certainly these twelve years saw the widespread acceptance of no less than six fundamentally new ways of viewing the Galactic System. These profound shifts, occurring in such a short time, form, I would suggest, one of the most exciting chapters in the entire history of astronomy. And in this paper I shall attempt to describe and analyze what these changes were, what led up to them, as well as to examine the events surrounding them.

EARLY IDEAS

But to put the developments of the 1910's and 1920's into context, let's first travel back to the years around 1850 and then work our way forward. How, then, was the Galactic System viewed in 1850? An astronomer who wanted an authoritative account of the latest thinking of the Galaxy would quite likely have turned to John Herschel's volume Outlines of Astronomy[2]. Here our astronomer would have found that Herschel had emphasized the complexity and irregularity of the Milky Way, the consequence of his having spent many hours observing in both the northern and southern hemispheres. Herschel further identified four great clouds of stars that he argued were distant extensions of the Milky Way. One of them was the Orion Nebula. The conviction that the Orion Nebula was a huge star cloud stemmed chiefly from the observations made with Leviathan of Parsonstown, Lord Rosse's 72-inch telescope in Ireland. Rosse, it seemed, had resolved the Orion Nebula into stars, and given its apparent size, we can understand Herschel's belief that the Nebula was a giant star cloud. This seems very odd now, but we need to remember that in the 1850's it was widely accepted that all nebulae were star systems.

H. van Woerden et al. (eds.), The Milky Way Galaxy, 43–58.

The Orion Nebula and other similar great star clouds were moreover employed by Herschel as models for the Galaxy itself, since, as he wrote, "could we view [the Galaxy] as a whole, from a distance such as that which separates us from these objects, [it] would very probably present itself under an aspect quite as complicated and irregular." But Herschel's thinking on the matter was not settled, for at other times he inclined toward a ring model. In this the Sun was in a relatively empty region of space, separated from a denser ring of stars. However, the important point here is that Herschel's conception of the Galaxy was very dependent on his beliefs about other stellar systems. Here indeed is a theme that runs through the history of galactic studies between 1850 and 1930, and the theme is that the way astronomers viewed the Galaxy depended intimately on their opinions and beliefs about other galaxies.

Now there were a number of different views to John Herschel's, and there was no consensus in the early 1850's on the nature and size of the Galaxy. Without good distances progress in galactic studies seemed a remote goal, and this realisation helps explain why during the nineteenth century astronomers were, in general, little concerned with the distribution of the stars. For most of them, for most of the time, it was sufficient that the stars' positions be catalogued, in order that the motions of Solar-System objects be followed more exactly. Those who researched the structure of the Galaxy or the nature of the nebulae, were thereby placing themselves somewhat out of the mainstream of astronomy. For example, in a text-book published in 1852, the Reverend Robert Main, First Assistant at the Royal Observatory, Greenwich, devoted only 14 of the book's 155 pages to the stars and nebulae and he referred to the Milky Way only in passing. The nineteenth century thus ended with astronomers knowing for sure little more than William Herschel's starcounts had shown in the 1780s: that is, that the plane of the Milky Way contains more stars than are to be found in other parts of the sky.

There had, nevertheless, been some developments towards the end of the century that, with the benefit of hindsight, we can see sowed the seeds of a spectacular blossoming of galactic astronomy.

First, there was renewed interest in the idea that the Galaxy is a spiral. This had first been proposed in the middle of the nineteenth century, but owed most to the Dutch amateur astronomer Cornelis Easton. In 1900, for example, he wrote on 'A new theory of the Milky Way'.[3] Here he contended that the latest observations of nebulae had demonstrated the spiral to be a much commoner form than had previously been supposed. This was due in part to the habit of astronomers in the late nineteenth and early twentieth century to see 'spirality' in all sorts of objects. Might the spiral structure, Easton asked, be the plan on which the Galaxy was designed? He fleshed out his hypothesis with a sketch of the stellar system as a spiral, though he warned that it was not intended to give even an approximate representation of the actual Galaxy.

Easton's hypothesis gained in popularity during the first two decades of the century. The major reason for this was that many astronomers were ready to admit the existence of visible external galaxies, or island universes as they were sometimes called. The candidate island universes were the spiral nebulae, a remarkable turn around from the position near the turn of the century, when they had generally been believed to be merely proto-solar systems. For example, when one of Isaac Roberts' photographs of the Andromeda Nebula had been shown at a meeting of the British Royal Astronomical Society in London in the late 1880s, it had caused a sensation. Many years afterwards one witness vividly remembered the reaction of the Society's Fellows: "One heard ejaculations of Saturn, the nebular hypothesis made visible, and so on". However, by the 1910s, the spiral nebulae, including the Andromeda Nebula, were widely claimed as external galaxies, and since they seemed to possess a spiral structure, many astronomers inferred that the Galaxy itself was a spiral.

The second major development around the turn of the century to affect galactic astronomy involved a handful of astronomers who were giving rise to a new kind of astronomy that some hoped would eventually enable them to discover the true arrangement of the stars. This was statistical astronomy. Dr. Paul elsewhere in this volume describes the evolution of statistical astronomy; so I shall say little about it here except to note that one important product of the endeavours of the statistical astronomers was to help bring studies of the Galaxy to a more central place in astronomy, to a place where its problems would become the concern of an increasing number of astronomers.

Now despite their hopes of achieving a better picture of the structure and size of the Galaxy, Kapteyn and other statistical astronomers laboured under a major handicap. As Kapteyn himself admitted in 1909,[4]

> Undoubtedly one of the greatest difficulties, if not the greatest of all, in the way of obtaining an understanding of the real distribution of the stars in space lies in our uncertainty about the amount of loss suffered by the light of the stars on its way to the observer.

Astronomers were particularly uneasy because the observed change in density seemed to place the Sun in a nearly central, and apparently privileged, position. It had been known since the time of William Herschel that dark regions exist in the Milky Way. During the nineteenth century it had been generally accepted that they were genuine holes or rifts, not dark clouds of obscuring matter. By the mid-1910s, many astronomers had moved toward the view that, while localised obscuring clouds did exist, the general absorption of star light is significant. This shift was due mainly to the examination by Harlow Shapley, then at Mount Wilson Observatory, of the colours of stars in globular clusters. In 1915 Shapley had estimated the Hercules globular cluster to be about 100 000 light-years away,[5] a very much larger distance that the contemporary estimates of the diameter of the Galaxy. But Shapley found no indication that the cluster stars had been reddened, and despite the fact that other astronomers believed that they had detected such an effect for distant stars, Shapley claimed that the extinction and reddening of star-light could effectively be ignored in researches on the Galaxy, and Shapley's result swayed many people.

Shapley's studies certainly freed Kapteyn of any reservations he had over assigning a nearly central position to the Sun. As Kapteyn himself told Shapley:[6] "Unless there be still a systematic error in your color indices, it seems that we may really, at least provisionally, neglect the consideration of this absorption in the study of the structure of the Milky Way system. It is almost too good to believe...I congratulate you on this achievement with all my heart."

Let's pause briefly to consider how the Galaxy was viewed in 1917, an apparently arbitrary date, but which I shall show is not so. A common view of the Galaxy was well summarized by Eddington.[7] In his widely-read Stellar Movements and the Structure of the Universe he had argued that, to give a general idea of the scale of the main part of the stellar system, "it may be stated that in directions towards the galactic poles the density continues practically uniform up to a distance of about 100 parsecs; after that at 300 parsecs it is only a fraction (perhaps a fifth) of the density near the Sun. The extension in the galactic plane is at least three times greater. These figures are subject to large uncertainties." This was one view, but just about all astronomers in 1917 regarded the Galaxy as a flattened or lens-shaped structure, as certainly no larger than 30 000 light-years in diameter, and perhaps considerably smaller. Almost all accepted that the Sun was close to the centre, while some suspected that the Galaxy's stars might be distributed in a spiral pattern.

SHAPLEY'S MODEL

Imagine, then, the astonishment that greeted Shapley's[8] announcement in 1918 that the Galaxy has a diameter of about 300 000 light-years, a staggering increase on the then current sizes. Morever, Shapley placed the Sun in an eccentric position tens of thousands of light-years from the galactic centre, a centre he argued was defined by the globular clusters.

Shapley's model was the product of a prodigious amount of intensive work, and its origins can be traced back at least to 1916.

Before 1916 Shapley had assumed, along with everybody else, that the Sun was roughly central, and that the stellar system had a radius of the order of a few thousand light-years, but in November 1916 Shapley found faint blue stars in some galactic star clusters. He calculated that if they were ordinary stars of types B and A, then the accepted dimensions of the Galaxy would need revising. Shapley was also well aware that the globular clusters were crowded into one section of the sky, a fact so well known at the time that Eddington had even called the Sagittarius region the "home of the globular clusters". Early in 1917 Shapley reported on his research to Kapteyn.[9] He described how "the work on clusters goes on monotonously - monotonous as far as labor is concerned, but the results are a continual pleasure. Give me time enough and I shall get something out of the problem yet." Certainly Shapley was making progress for, as his investigations of the colours and magnitudes of stars in globular clusters had advanced, he had been able to secure the distances of an increasing number of clusters. In consequence, by late 1917 all the main elements for Shapley's model were present: (1) his concern for the highly asymmetrical distribution of the globular clusters across the sky; (2) values for the distances of the clusters that he believed were reasonably accurate; (3) a suspicion that the Galaxy was much larger than his contemporaries conceded, and (4) a conviction that the existing galactic models were inadequate. Sometime late in 1917 these seemingly disparate elements became meshed together in Shapley's mind and he invented a startling galactic model: the Big Galaxy. By January 1918, Shapley could write to a correspondent that "with startling suddenness and definiteness" the globular clusters had elucidated the "whole sidereal structure".[10] He now had values for all the globular clusters, and he had found that the equatorial plane of the system of globular clusters was identical with the galactic plane, and so he was now proposing that the stellar system and the huge system of globular clusters had the same centre and were co-extensive, the globular clusters actually outlining the Galaxy.

It is worth emphasizing, because present-day text-books ignore this, that the model embodied an imaginative vision, for Shapley was also proposing an evolutionary theory for the Galaxy. He hypothesized that the Galactic System "may have originated in the combination of two clusters and has grown, as it appears to be growing now, by the accretion of other stellar systems-adding the smaller units such as the globular clusters with ease, and the larger ones such as the Magellanic Clouds with some difficulty, if at all. It appears to be an example on a grand cosmic scale of survival of the fittest, that is, survival of the most massive and most stable."

How was Shapley's model received? There is no doubt that it soon met with some strong support. For example, by late 1918 Eddington was telling Shapley:[11] "I think it is not too much to say that this marks an epoch in the history of astronomy, when the

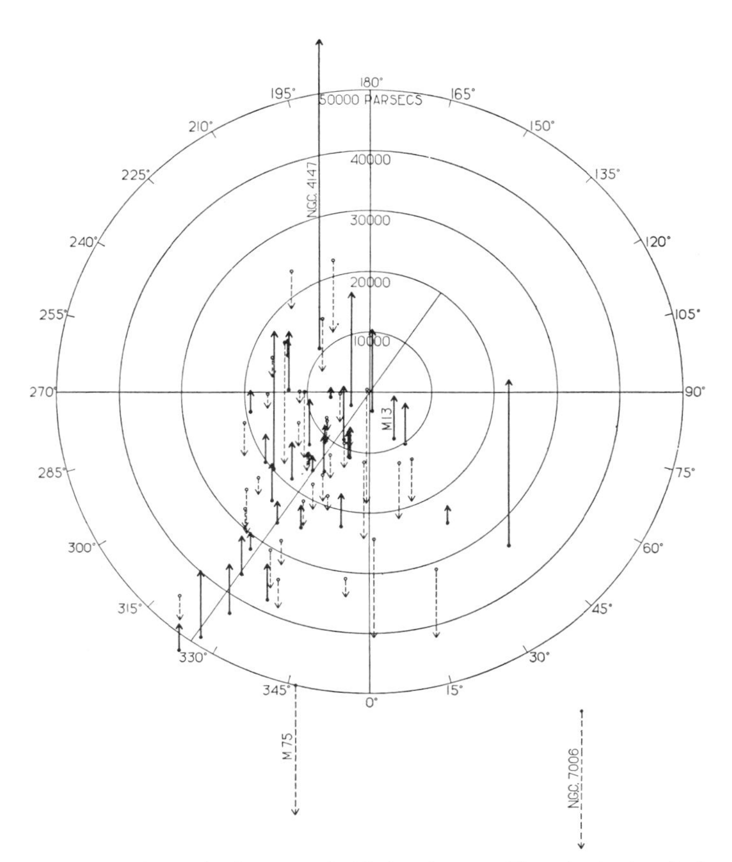

FIG. 2.—Distribution in space of globular clusters. The galactic plane is the plane of the diagram; distances above and below are shown to scale by full-line and broken-line vectors, respectively. Galactic longitudes are indicated in the margin and the scale of distances along the vertical radius. The sun is at the origin of co-ordinates. The diagram illustrates the remarkable distribution in longitude, with a maximum frequency at 325°, and by the absence of very small or zero vectors shows that globular clusters are not found within 1000 parsecs of the plane of the Milky Way. Cf. Fig. 1 of the twelfth paper.

Shapley's 1918 plot[12] of the distribution of globular clusters. Reprinted by permission of The Astrophysical Journal, published by the University of Chicago Press; copyright 1984 The American Astronomical Society.

boundary of our knowledge of the Universe is rolled back to hundred times its former limit". But Shapley's model of the Galaxy was radical. It was also very ambitious, for he was attempting to solve many of the problems of galactic astronomy in one very broad attack. Not surprisingly, the daring set of proposals that his model embodied met with some hostility, and even his strongest supporters were critical of certain aspects of it. Walter Baade was later to describe the reception of Shapley's model:[13] "I have always admired the way in which Shapley finished the whole problem in a very short time, ending up with a picture of the Galaxy that just about smashed up all the old school's ideas about galactic dimensions. It was a very exciting time, for these distances seemed to be fantastically large, and the old boys did not take them sitting down."

Why was this so? First, Shapley's model was unexpected. There was no sense of crisis, no feeling that galactic astronomers were widely off the mark, and that something drastic needed to be done to put galactic astronomy onto a sound basis. Hence most astronomers had no inclination to demolish the contemporary notions of the size and structure of the Galaxy. Thus Shapley's model did not halt the publication of research founded on the more traditional approaches, and in the early 1920s Kapteyn wrote two papers, in part with P.J. Van Rhijn, that capped his life's work in galactic astronomy.[14] In the second Kapteyn wrote what he described as a "First attempt at a theory of the arrangement and motion of the sidereal system". Now he calculated that the limits of the Galactic System were found at roughly 1700 pc at right angles to the plane and 8500 pc along the plane. Kapteyn thereby advanced dimensions that, though small compared to Shapley's, were far larger than the estimates of the mid-1910s, and so the Kapteyn Universe, as this model became known, was itself a notable departure from the previously prevailing hypothesis.

The second reason for the sometimes hostile reception of Shapley's model was what astronomers saw as weaknesses and flaws in Shapley's distance scale. Shapley obtained his distances with the aid of three interlocking methods. Some of the larger, and apparently closer, globular clusters contained stars that Shapley identified as Cepheids. Shapley argued that he could thereby secure the distance from the period-luminosity relationship. To reach those clusters that contained no visible Cepheids, Shapley examined the thirty brightest stars in a cluster or a cluster's apparent diameter.

As so often, an argument in galactic astronomy was revolving around the accuracy of the available distance indicators, for it was to the initial Cepheid calibration that astronomers objected most strongly. If Shapley had erred at this point, then the distances he claimed for the globular clusters, upon which his model rested, would be undermined. The reality of the period-luminosity relation was far from generally accepted and it would be several years before it would be so. Hence Shapley began his distance determinations with what some saw as a very dubious step indeed, particularly as his calibration relied on the statistical analysis of only eleven stars.

One of the astronomers who disputed Shapley's model was H.D. Curtis of Lick Observatory. And of course Shapley and Curtis met in the so-called 'Great Debate' on 'The scale of the Universe'. Until recently it was assumed that the papers published after the Debate were a verbatim record of what transpired that night in 1920 at the National Academy of Sciences in Washington D.C. In fact, it seems likely that the previous accounts bear little relation to actual events. First, the debate nearly did not take place, as a number of other topics were discussed as possible subjects. At one time vivisection was considered, and the Prince of Monaco was shortlisted as a speaker on oceanography. Secondly, it seems almost certain that Shapley was apprehensive that his encounter with Curtis might hinder his chances of becoming Director of the Harvard College Observatory, a post then vacant. Shapley reasoned that if Curtis crushed his arguments he was unlikely to be offered the post, and Shapley's actions, before and during the debate, are much more intelligible when we take this belief into account. Anyway, the meeting took place on 26 April 1920. To minimise the possibility that Curtis, an experienced public speaker, would demolish his case, Shapley spoke at a non-technical level. Although members of the public were admitted to the meeting, it is hard to see what other explanation can be offered for the facts that Shapley only reached the definition of a light-year after seven pages of his script of nineteen pages, and that he devoted the last three pages to an intensifier he had developed to photograph very faint stars. The intensifier had little bearing on the theoretical argument, but probably Shapley reasoned that it would impress those members of the audience, like Mr. Agassiz of the Harvard College Observatory Visiting Committee, concerned with the future activities of the Harvard College Observatory. Curtis, who had expected a more technical presentation, was left throwing his verbal blows at a non-existent opponent. It was only in their correspondence after the debate and with their papers in a _Bulletin of the National Research Council_ that Shapley and Curtis finally got to grips with each other's arguments.[15] I should stress that there was no animosity between Curtis and Shapley–far from it. Before their encounter they conducted a cordial correspondence about the meeting's procedure and their post-debate letters were also friendly, this despite Curtis's suggestion that during the meeting they should go at each other "hammer and tongs", and that he would be wielding his shillelah! Rather, Curtis's quarrel was with Shapley's distance indicators. In particular, he argued that the existence of the period-luminosity relation was very uncertain, and at this time many astronomers agreed with him. Furthermore, Curtis inclined towards a mixture of the spiral and ring theories of the Milky Way.

Indeed there was a strong move to the spiral theory in the 1910s and early 1920s. As mentioned earlier, this was a consequence of the revival of the island-universe theory, a revival due to a series of new observations of spiral nebulae made principally at the observatories in the West of the United States. The spiral nebulae that had for so long been seen as members of the Galaxy, were thus often viewed by the early 1920s as likely island universes. It was then extremely tempting to contend that, if observed from a great distance, our own stellar system might itself be seen as a spiral.

Yet, as the rival views of Shapley and Curtis help to show, there was no consensus in early 1920 on the size, nature, or form of the Galaxy. Nor was there any close agreement on the best ways to tackle these problems, no agreement on whether the key lay with the techniques of statistical astronomy, those employed by Shapley, or some fusion of the two.

Within a few years however, there were to be several developments that would drastically affect this situation. First, in late 1923, amidst the continuing confusion about the true nature and distance of spiral nebulae, Edwin Hubble detected a Cepheid in the Andromeda Nebula.[16] Hubble's momentous discovery set him on a course that would soon bring the long-standing debate on the existence of external galaxies to a swift end. It thus became accepted that many galaxies do have spiral shapes, and so this lent credence to the theory of our own Galaxy as a spiral. Moreover, Hubble based his investigations of the nearby galaxies on the period-luminosity relationship. By doing so, he helped it to gain acceptance. And if the period-luminosity relationship was not the spurious product of meagre data, then Shapley's distances to the globular clusters had to be taken seriously.

ROTATION

Hubble's findings were soon followed by others that supported the main structural features of Shapley's model. During 1927 and 1928 there was a rapid acceptance by many, probably the majority of, astronomers that the Galaxy rotates differentially. That the Galactic System rotates had long been suspected, and its flattened form seemed to be a natural consequence of rotation. The spectrographic measurements of the rotation of spiral nebulae in the 1910s and 1920s had further assured astronomers that the Galaxy, which many accepted was a spiral, itself rotated. But suspicion is a very long way from proof. Where did the proof come from? In 1924, Bertil Lindblad had been driven to consider a rotation of the Galaxy through his attempts to interpret star-streaming. By 1925 he had decided that the motions of the constituents of the Galaxy were explicable on the hypothesis that the Galaxy is divided into a series of sub-systems, each of which has rotational symmetry about a common rotational axis.[17] Each sub-system has the same equatorial extent, but possesses a different speed of rotation and hence a

M101, as photographed with the 60-inch reflector at Mount Wilson in March 1910 (Courtesy of Mount Wilson and Las Campanas Observatories, Carnegie Institution of Washington).

different degree of flattening. He reasoned, that while high-velocity stars do not belong to the same dynamical system as those of low velocity, they must be related to the rest of the Galaxy, since their motions are symmetrical with respect to the galactic plane. Lindblad then explained the motions of the so-called high-velocity stars by arguing that the Sun and other low-velocity stars in fact have high speeds of rotation about a remote centre, and that the so-called high-velocity stars have much smaller speeds of rotation (and so form a more nearly spherical system), thereby falling behind as the Sun overtakes them. The high-velocity stars will thus appear to move asymmetrically. So, once again, a crucial development arose from looking at a set of well-known observations in a new way, this time seeing the high-velocity stars as actually slow-moving stars. Further, Lindblad (as Oort was soon to do) calculated the dynamical centre of the Galaxy to be very close in galactic longitude to the centre of the globular-cluster system as determined by Shapley.

Jan Oort was deeply influenced by Lindblad's researches, and in 1927 he announced that, through his attempt to verify directly Lindblad's theory of galactic dynamics, he had secured firm evidence of a differential rotation of the Galaxy.[18] Oort had found that the proper and the radial motions of the nearby stars exhibited the small but systematic effects to be expected of differential rotation. The genesis of Oort's analysis has recently been recalled by Bart Bok:[19] "...Jan Oort was presenting (on Monday afternoon at four) a series of seminars for Doorn, Kuiper, Oosterhoff and Bok on Lindblad's theories of galactic rotation. As I remember it - others may have different recollections!-, Jan told the four of us one Monday that he had got bogged down in Lindblad's complex mathematics and that there would be no lecture the next Monday afternoon. And, as I remember it, there were no lectures for two Monday afternoons to follow. And then there came the first Monday after the crisis, a lecture in which Jan Oort basically developed the simple formulae for the double sine-wave effect of galactic rotation in radial velocities and the corresponding formulae for the effects in proper motions. The four of us realised that we were listening to an amazing new step in the understanding and interpretation of stellar motions...." Moreover, the validity of Oort's inferences about galactic rotation was corroborated by other astronomers, particularly by the Canadian J.S. Plaskett who analyzed the radial motions of hundreds of O- and B- type stars. The detection of galactic rotation had been "in the air" and Oort had presented what was generally seen as its observational proof. In addition, Oort and Lindblad had seemingly shown that the Galaxy rotates about a point that lay in almost the same direction as Shapley's proposed direction to the galactic centre. Nevertheless, there was one point where the researches of Oort and Lindblad did not mesh with Shapley's model: the distance of the Sun from the centre of the Galaxy. Oort had reckoned that the distance to the centre about which the stars rotated was roughly 6000 pc, about one-third of the size of Shapley's estimate. But despite this discrepancy, the discovery of differential galactic rotation swept away the opposition to the eccentric position of the Sun within the Galaxy. Shapley and Oort's estimates were, furthermore, soon to be brought into a close agreement by the demonstration of the existence of a general interstellar absorption.

ABSORPTION

In the late 1910s Shapley had found no evidence of significant absorption in his examinations of the globular clusters, and he had proceeded to argue that, except for isolated dark clouds, space is effectively transparent. This view continued to be very influential until the implications of R.J. Trumpler's study of the open clusters within the Galaxy had been fully grasped.

As his chief working hypothesis Trumpler had taken the open clusters of similar constitution to have, on average, the same dimensions; by comparing the observed angular diameter of a cluster with the average linear diameter of the sub-class to which it belonged, he derived a value for the distance to the cluster. Trumpler had also examined the magnitudes and spectral types of the stars within the clusters. Then by constructing for a cluster the Hertzsprung-Russell diagram and comparing the observed diagram with a standard diagram, Trumpler secured another value for the cluster's distance. He thereby found that the two distance indicators gave systematically different answers: the more distant the cluster, the more the two distance values differed. Trumpler argued (and astronomers soon agreed) the the reason for this deviation was interstellar absorption. Although, with the benefit of hindsight, a number of earlier investigations can be seen to have pointed towards the existence of a general interstellar absorption, it was Trumpler's analysis of the open clusters - probably because it was more extensive and complete than earlier researches - that convinced astronomers of the presence of obscuring matter throughout the Galaxy.

In his calculation of the size of the Galactic System Shapley had not allowed for this dimming effect. As a result he had overestimated the distance of the Sun from the centre of the Galaxy. Yet the confirmation of galactic rotation and interstellar absorption was instrumental in bringing into wide acceptance the two central structural features of Shapley's Big Galaxy - the eccentric position of the Sun and the role of the globular clusters in outlining the Galaxy. But there was still no agreement on whether or not it possessed a spiral structure, or on the Galaxy's size. The resolution of the spiral problem would have to await the 1950s, as Gingerich shows elsewhere in this volume. The size problem was particularly puzzling because diameter estimates of about 100 000 light-years were now common. This meant that it was very much larger than any other galaxy. For example, Hubble argued that the mean diameter of galaxies ranged from 360 parsecs for E0's to 2500 parsecs for Sc's. This puzzle led a few astronomers to propose that the Galaxy is in truth an assemblage of galaxies. But others adopted a more sceptical attitude. One of these was Eddington. In his words,[20]

> The lesson of humility has so often been brought home to us in astronomy that we almost automatically adopt the view that our own Galaxy is not specially distinguished - not more important in the scheme of nature than the millions of other island galaxies. But astronomical observation scarcely seems to bear this out. According to the present measurements the spiral nebulae, though bearing a general resemblance to our Milky Way system, are distinctly smaller...Frankly, I do not believe it; it would be too much of a coincidence. I think that this relation of the Milky Way to the other galaxies is a subject on which more light will be thrown by further observational research, and that ultimately we shall find that there are many galaxies of a size equal to and surpassing our own.

CONCLUSIONS

This outstanding anomaly should not blind us to the fact that, towards the end of the nineteenth and in the first three decades of the twentieth century, there were great strides taken in galactic astronomy. First, the Galaxy is much larger than had been realised. Second, the Sun is eccentrically placed. Third, the globular clusters surround the Galaxy. Fourth, there is a general interstellar medium. Fifth, the Galaxy rotates differentially. And sixth, the Galaxy is not alone in space, but there are vast numbers of other galaxies. These developments arose largely from new ways in which astronomers sought to discover the size and structure of the Galaxy. And it is these tools and techniques, both observational and theoretical, that were, I think, the most important fruits of galactic astronomy between 1850 and 1930.

NOTE

For a fuller treatment of the topics dealt with here, see Robert W. Smith's The Expanding Universe: Astronomy's 'Great Debate' 1900-1931 (Cambridge, England, and New York, 1982), and the references cited therein. But see also Oort and the Universe: A sketch of Oort's Research and Person (Dordrecht, 1980), edited by H. van Woerden, W.N. Brouw, and H.C. van de Hulst.

REFERENCES

1) Proctor, R.A.: 1872, 'Essays in Astronomy', Longmans, Green and Co., London, p. 240
2) Herschel, J.: 1849, 'Outlines of Astronomy', Lea and Blanchard, Philadelphia, First Edition. See particularly sections 786-792
3) Easton, C.: 1900, Astrophys. J. 12, pp. 136-158
4) Kapteyn, J.C.: 1909, Astrophys. J. 29, pp. 46-54, see p. 46
5) Shapley, H.: 1915, Contr. Mount Wilson Solar Obs., No. 116
6) Kapteyn, J.C. to Shapley, H., 23 September 1915, Widener Library, Harvard University
7) Eddington, A.S.: 1914, 'Stellar Movements and the Structure of the Universe', Macmillan, London, p. 31
8) See, for example, Shapley, H.: 1918, Publ. Astron. Soc. Pacific, 30, pp. 42-54
9) Shapley, H. to Kapteyn, J.C., 6 February 1917, Widener Library, Harvard University
10) Shapley, H. to Eddington, A.S., 8 January 1918, Widener Library, Harvard University
11) Eddington, A.S. to Shapley, H., 24 October 1918, Widener Library, Harvard University
12) Shapley, H.: 1918, Astrophys. J. 48, pp. 154-181, see p. 169
13) Baade, W.: 1963, 'Evolution of Stars and Galaxies', Harvard University Press, Cambridge, Mass., p. 9
14) Kapteyn, J.C. and Van Rhijn, P.J.: 1920, Astrophys. J., 52, pp. 23-38, and Kapteyn, J.C.: 1922, Astrophys. J. 55, pp. 302-328
15) Shapley, H.: 1921, Bull. Nat. Research Council 2 (part 2), pp. 171-193, and Curtis, H.D.: 1921, Bull. Nat. Research Council, 2 (part 3), pp. 194-217
16) See Berendzen, R. and Hoskin, M.A.: 1971, Astron. Soc. Pacific Leaflet No. 504
17) See Lindblad, B.: 1925, Astrophys. J. 62, pp. 191-197, and Lindblad, B.: 1925, Meddelanden Astron. Obs. Uppsala, Series C, 1, No. 3
18) Oort, J.H.: 1927, Bull. Astron. Inst. Netherlands, 3, pp. 275-282
19) Bok, B.J.: 1980, in Van Woerden, H., Brouw, W.N., and Van de Hulst, H.C. (eds), 'Oort and the Universe: A sketch of Oort's Research and Person', Reidel, Dordrecht, p. 56
20) Eddington, A.S.: 1933, 'The Expanding Universe', Cambridge University Press, Cambridge, p. 4

DISCUSSION

J.H. Oort: In connection with your remarks on the correspondence between Kapteyn and Shapley concerning the smallness of the absorption in the Galaxy, shown by the absence of change in colour of clusters with increasing distance, I should like to draw attention to other evidence in Shapley's system of clusters which in my opinion (even in those early years) gave convincing proof of the existence of strong absorption in the galactic plane, viz. the striking deficiency of clusters at galactic latitudes less than about 3°. Shapley himself thought that the absence of clusters close to the galactic plane was due to their being disrupted by encounters with the abundant stars in that region. But it is (and was at that time, at least to me) clear that such perturbation would be entirely insufficient to disrupt the clusters during the short time of their passage through the galactic layer.

H. van Woerden: Was this argument, that the dip in the latitude distribution of globular clusters must be due to insterstellar absorption, made in the literature?

Oort: No.

Van Woerden: About the sizes of galaxies compared to the Milky Way, I remember a lecture in 1945/46 at Leiden, where Oort pointed out that the Andromeda Nebula and especially the Galactic System were among the very biggest galaxies known. (And when I asked whether that was not peculiar, Oort blushed.) Only the changes in the extragalactic distance scale, in 1952 (Baade) and thereafter, have changed this; by 1945 we still had a Hubble constant of 500 km s^{-1} Mpc^{-1}.

J.V. Villumsen: When you go to the southern hemisphere and look up at the sky, you can see the galactic bulge clearly. How did Kapteyn explain that?

Smith: Indeed, if you look at Sagittarius, it seems obvious that you have the centre there. Kapteyn had, in his 1922 model, the Sun slightly displaced from the centre, but the direction of the centre was still towards Cygnus. I do not know precisely why. For Kapteyn, the most important evidence was the mathematical evidence rather than looking up at a brightness distribution in the sky.

F.J. Kerr: Do you know who first introduced the word "parsec", and when?

Smith: There was a great deal of debate in the 1910s and early 1920s. Kapteyn played a role in this, but I cannot remember the details. The reason for jumping between "parsec" and "lightyear" in my talk was that I was following the original writings. The matter of parsec vs. lightyear certainly had not been settled by 1920.

M. Schmidt: Is it correct that Lindblad proposed galactic rotation in order to understand the ellipsoidal distribution of peculiar velocities?

Smith: Indeed. He was trying to understand star-streaming, and this led him on to consider galactic rotation. Kapteyn had explained star-streaming by having the two streams revolve around the Galaxy in different directions.

J.H. Oort: The explanation of the ellipsoidal distribution of velocities was one of the most important developments besides galactic rotation.

A. Blaauw: I think that the Strömberg relation between asymmetric flow and velocity dispersion was an essential thing in the development of Lindblad's theory.

Smith: That is true. Similarly to what you said in your article in "Oort and the Universe", I was striding with large boots from mountain top to mountain top; given time, I should have mentioned Strömberg. He certainly pointed out the asymmetry in high-velocity stars, although during the 1920s he tied these in with a rather strange idea about velocity limitations. This was when relativity was still very new, and he thought this asymmetry might be some sort of relativistic effect, that stars would have particular limitations on their velocities; and in fact Strömberg even considered a very crude kind of velocity-distance relationship for galaxies around the same time.

M.A. Hoskin (Chairman): Before relinquishing this chair, let me express the appreciation of the historians to the Organizing Committee for the welcome we have been given, and for the opportunity to take part in this splendid week.

Robert Smith and Gösta Lyngå playing checkers during Wednesday boat trip. In background: Katrin Särg, Ria and Hugo van Woerden. LZ

THE DISCOVERY OF THE SPIRAL ARMS OF THE MILKY WAY

Owen Gingerich
Harvard-Smithsonian Center for Astrophysics

Abstract Attempts in the 1930s and 1940s to determine the spiral structure of the Milky Way by star counting methods, essentially the continuation of the work of the Kapteyn Astronomical Laboratory, failed to reach this goal. A new foundation for the search was laid by Walter Baade in his studies of stellar populations. With the recognition that highly luminous objects, especially H II regions, would outline the spiral structure, W.W. Morgan and his young associates Sharpless and Osterbrock carried out the observational program that first delineated, in 1951, the nearby arms of the Milky Way. The full paper was never published, so the historical details have remained somewhat vague, primarily because the 21-cm discoveries so quickly overtook the optical researches.

In 1937, in his The Distribution of Stars in Space, Bart J. Bok concluded:

> Working models for the galactic system have, at various past stages of development, been of value for the co-ordination of existing knowledge and the effective planning of future research. Shapley's model, in which the local system played the role of an important subsystem, has proved eminently satisfactory for the past twenty years. The time has now come to go one step farther and consider a more detailed working model. Several astronomers have stressed the probable similarity in structural features between the galactic system and some of the larger spiral nebulae. Seares has suggested that our stellar system may well have a structure similar to that of Messier 33, the well-known spiral nebula in Triangulum. Our sun would then be located in one of the spiral knots, at a distance equal to two-thirds of the radius of the nebula from the center.
>
> It is surprising to note how well such a model agrees with the general impression obtained when the Milky Way is viewed from the tropics. The best view may be had at sidereal time 15-16 hours, when the Carina region is setting, Sagittarius is well up in the sky,

H. van Woerden et al. (eds.), The Milky Way Galaxy, 59–70.

> and the cross of Cygnus is rising above the horizon. No one who had the privilege of thus seeing the Milky Way in all its grandeur would ever deny that the Sagittarius cloud marks the central region of our galactic system. Individual stars do not stand out particularly against the brilliant continuous background of the Milky Way in Sagittarius; but the impression is quite different for the Cygnus and Carina clouds, in which a multitude of individual stars is seen projected against a faintly luminous background. The observer in the tropics should not find it difficult to accept as a working model for our Milky Way system one with a distant center in Sagittarius and in which a spiral arm passes from Carina through the sun toward Cygnus.

Bok had originally been inspired to study astronomy by the popular writings of Cornelis Easton in _Hemel en Dampkring_. Easton, a journalist and skilled amateur observer, had made his mark on astronomy by his careful drawings of the Milky Way and his speculations that spiral structure could be discerned visually. At the instigation of Kapteyn, Easton received an honorary doctorate from Groningen in 1903. In 1913 his Milky Way studies culminated with an article in the _Astrophysical Journal_.[1] Although he depicted the sun just off the center of a spiral whose nucleus lay in the direction of Cygnus, he was careful to state that "I am well aware that the great problem of the Milky Way can never be solved in this way, and that we may aim only at a plausible interpretation of known facts, and at a working hypothesis... The figure in the center of our plate does not pretend to give even an approximate representation of the galactic system, but only to indicate in a general way how the stellar accumulations might be arranged so as to produce the phenomenon of the Milky Way--on the supposition of a spiral galaxy."

Within a few years Harlow Shapley proposed an entirely different layout for the Milky Way, and Easton abandoned his earlier ideas of visible spiral structure. As a consequence of Easton's writings, Bok too became an enthusiastic Shapley supporter. At Leiden Bok studied under Ehrenfest and Oort, at the time when Oort worked out his famous equations for the rotation of the Milky Way, which turned out to be one of the most convincing arguments for Shapley's model. In 1929 Bok took a fellowship at Groningen under Kapteyn's successor, P.J. van Rhijn; then at Harvard he mined the observatory plate collection for his dissertation research on Eta Carinae, but he actually received his doctorate in Groningen for this work. Since both Oort and van Rhijn had been students of Kapteyn, Bok can be considered a third-generation astronomer of this distinguished Dutch school. More than anyone else he developed and applied the numerical methods originated by Kapteyn toward the problem of Milky Way structure, and in particular toward the delineation of its spiral features.

The history of astronomy shows repeatedly that well-defined problems are often solved by totally unexpected lines of research, and this proved to be the case for the spiral structure of our galactic system. Over a decade of devoted starcounting and analysis, particularly under Bok's direction at Harvard, failed to disclose the expected stellar density concentrations that could be identified with the spiral arms. The solution to this puzzle lay elsewhere, with the observational analysis of the Andromeda Nebula and other nearby galaxies. This work was carried out by Walter Baade at Mount Wilson, and it took particular advantage of the dark skies produced by the wartime blackout of Los Angeles and Hollywood.

Like Bok, Baade had been inspired by the work of Harlow Shapley, in his case by the physical nature of pulsating stars. In 1931 Baade received an offer to join the staff at Pasadena, an opportunity he had accepted immediately. He eventually applied for American citizenship, but lost the papers, and with a characteristic disdain for bureaucracy never reopened the matter. Thus, when the United States entered World War II, he was classified as an enemy alien, unfit for war work, and consequently he had free rein with the 100-inch reflector during those dark years. Baade pushed the Hooker telescope to its very limits in order to resolve the inner portions of M31 and its satellite galaxies M32 and NGC 205. The results were published in 1944 in a famous article in ApJ 100, in which he distinguished between two stellar populations and introduced the terms Type I and Type II. He concluded by stating that[2]:

> Although the evidence is still very fragmentary, there can be no doubt that, in dealing with galaxies, we have to distinguish two types of stellar populations, one which is represented by the ordinary H-R diagram (type I) the other by the H-R diagram of the globular clusters (type II). Characteristic of the first type are highly luminous O- and B-type stars and open clusters; of the second, globular clusters and short-period cepheids.... Both types coexist, although differentiated by their spatial arrangement, in the intermediate spirals like the Andromeda nebula and our own galaxy.

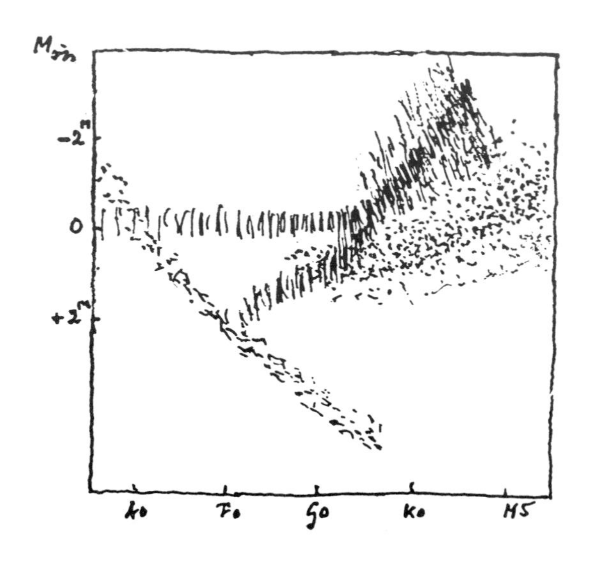

A sketch of the H-R diagram of the two stellar populations sent by Walter Baade to Cecilia Payne-Gaposchkin in 1947. Giant branches of type I are shown by the dots, of type II by the vertical hatching.

Baade also added that the same two types of stars had been recognized, from their differing motions, by Oort in 1926. Because of their common interests in galactic structure, a lively correspondence ensued after the publication of Baade's classic paper. In 1946 Baade wrote to Oort[3]:

> You mention in one of your remarks that the classical cepheids would be objects par excellence from which to determine the spiral structure. I think it is not certain yet that the longer period cepheids are especially concentrated in the spiral arms (they occur in the same regions in which the arms occur). But the B-stars of high luminosity are strongly concentrated in the spiral arms as my UV-exposures of the outer parts of M31 show most convincingly. I am therefore wondering, after reading Blaauw's fine paper about the Scorpius-Centaurus cluster whether this extraordinary aggregation of B-stars is not in reality a short section of a spiral arm, the more so because in its orientation and motion it would fit perfectly into the expected picture (the arms trailing).

Early in 1947 Oort replied, saying[4]:

> I quite agree that a study of the early B-type stars would be one of the most important steps for finding the spiral structure of the Galactic System. I have been discussing this subject with Van Rhijn for some time, and when Van Albada left Holland in order to pass a year at Cleveland we suggested to him that he should try to start a program with the Schmidt camera for finding faint B-type stars in the Milky Way (Van Albada, by the way, is a very intelligent and original young astronomer, a former student of Pannekoek). This is a large programme, however, and I don't think the Warner and Swasey people are sufficiently interested yet to start it on a sufficiently big scale. How about future possibilities with the large Schmidt cameras on Mt. Palomar?[16]

But, in fact, J.J. Nassau at the Warner and Swasey Observatory did eventually start just such a large-scale program, of which more presently. Although Baade himself did not get involved with such a program at Palomar, he began to appreciate that this must be the direction for solving the problem of spiral structure, rather than by less discriminating methods of star counting. On this latter topic he wrote rather scornfully to Leo Goldberg in 1949. Baade had been asked to give the opening address at the dedication of the Curtis Schmidt telescope at Michigan, and to advise on a related symposium. "The idea to celebrate the dedication of your new Schmidt with a symposium on galactic structure seems to me most appropriate," he replied, adding[5]:

> I shall be glad indeed to pitch in to the best of my ability. I have only some doubts whether I would be the proper man for the proposed opening lecture. People expect on such occasions to be edified and uplifted by tales of heroic achievements and I fear I could not accomodate them in this respect re galactic structure if I went much beyond old William Herschel. But the main thing is the symposium

> and I hope you can arrange the program in such a way that it revolves around the really fundamental questions. No papers about the stellar distribution in a field in Cepheus, etc. What we all would like to know is: What is the large scale structure of our galaxy and which roads appear promising at present.... I realize that with the inadequate equipment of their observatories many astronomers were simply forced to restrict their research to our immediate solar neighborhood with the natural result that the fin de siècle ideas regarding what constituted the problem of galactic structure survived longer than they otherwise would have had. But with large Schmidts coming up now everywhere it is time to reassess the situation and your symposium would offer a splendid opportunity to do just that.

Indeed, the symposium was convened in June 1950, with a nucleus of the most eminent researchers present. Baade had particularly expressed his hope to Goldberg that Oort would accept an invitation, and Oort's absence was perhaps the most disappointing feature. Baade gave both the opening lecture and the first symposium paper itself, on "Galaxies--Present Day Problems." He addressed a wide variety of issues before he

Chief participants at the Michigan Symposium in June, 1950. Front row (l. to r.): F.D. Miller, K.G. Henize, H. Shapley, W. Baade, J.J. Nassau, and L. Goldberg. Back row: G. Abetti, B. Lindblad, W.W. Morgan, A.N. Vyssotsky, N.U. Mayall, R. Minkowski, J. Stebbins, and S.W. McCuskey.

came to his final point, "Our Galaxy as a Spiral Nebula."[6] Baade wrote, "We have, I think, convincing evidence now that our galaxy is an Sb spiral, because it has a nucleus similar to that of the Andromeda nebula and not to that of M33." He described first his attempt with Sergei Gaposchkin to probe for the variable stars embedded in the galactic bulge. He reiterated his view that our galaxy had a nuclear lens made up of population II stars. Then he turned to the spiral structure itself as another problem ready for attack. "The procedure in this case is obvious. Since the supergiants of the population I are restricted to the spiral arms, we have to study their spatial arrangement in the solar neighborhood. The most promising stars for a first test are undoubtedly the O and early B stars, on account of their high frequency in spiral arms. But we will need for each star accurate data on the following, in order to determine its position relative to the sun: apparent magnitude, absolute magnitude, and color excess. Since apparent magnitudes and color excesses of most of the O and early B stars brighter than 7.5 visual and north of declination -30° are already known, their individual absolute magnitudes are the only remaining desideratum. W.W. Morgan's spectroscopic luminosity criteria for O and B stars should fill this gap and it is, I think, no secret that Morgan and Nassau are now engaged in a large program of determining the absolute magnitudes of O and B stars by this method."

It was, of course, no secret that William W. Morgan of Yerkes Observatory and Jason J. Nassau at Warner and Swasey were at work on this problem, for they reported their results at the same symposium.[7] With respect to the spiral arms, however, their paper took a very conservative stance. In this case it is fascinating to examine the report of the Nassau-Morgan paper given in the _Sky and Telescope_ article on the symposium. There we read[8]:

> The search yielded over 900 OB stars, but for the majority of them the distances are undetermined. However, for 49 relatively nearby OB stars and for three groups shown on the diagram below, Dr. Morgan has collected the required data. Combining the results with already existing knowledge of many facts about the galaxy and other galaxies, these astronomers suggested that the sun is located near the outer border of a spiral arm. The arm extends roughly from the constellation Carina to Cygnus. The fact that many faint and hence distant OB stars are found toward Cygnus indicates that we are observing the stars in the extension of this arm beyond the clustering in that constellation, that is, beyond 3,000 light years.
>
> The part of the spiral arm near our sun contains a large cloud, or groups of small clouds, of interstellar dust and gas which obscures the distant stars and divides the Milky Way into two branches, easily visible to the unaided eye. This obscuring cloud or rift is in the shape of a slightly bent cigar and is over 3,500 light years long. At one end of it is the southern Coalsack and at the other the brilliant group of OB stars of the Northern Cross... Dr. Nassau cautioned, however, that the evidence is insufficient to preclude

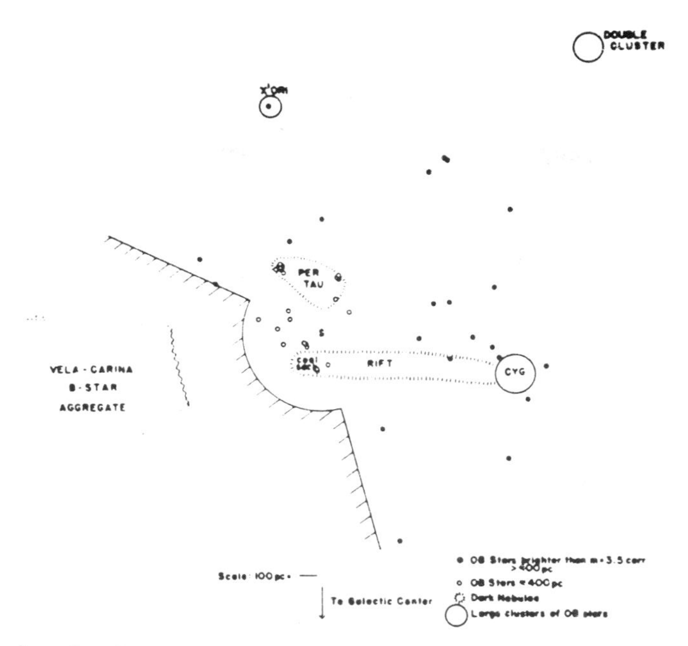

The plot by Nassau and Morgan of 49 OB stars and 3 OB groups for which distances were determined, from the Michigan Symposium volume (p. 50). The position of the sun is shown by S and the cross-hatching designates the limit of the survey.

the hypothesis that a great disorganization exists in the galaxy and that the star groupings do not trace definite spiral arms.

An entirely different reception to these studies came approximately 18 months later at the 1951 Christmas meeting of the American Astronomical Society in Cleveland, where Morgan presented new results based partly on the OB stars but largely on an investigation of H II regions. By examining the distribution of emission nebulae, Morgan, together with Stewart Sharpless and Donald Osterbrock, was able to delineate segments of two spiral arms, one that passes through the sun and the other at a distance of over 6,000 light years in the direction away from the galactic center, that is, about twice as far as the limits of the earlier Nassau-Morgan work. Concerning the Morgan-Sharpless-Osterbrock paper, Otto Struve wrote[9]:

> Astronomers are usually of a quiet and introspective disposition. They are not given to displays of emotion. Moreover, they tend to be cautious--more often than not they take plenty of time to weigh the evidence of any new and startling development before they accept it. But in Cleveland, Morgan's paper on galactic structure was greeted by an ovation such as I have never before witnessed. Clearly, he had in the course of a 15-minute paper presented so convincing an array of arguments that the audience for once threw

caution to the wind and gave Morgan the recognition which he so fully deserved.

From a historical perspective we need to look more closely at the research pattern that made the difference between the Michigan Symposium and the Cleveland AAS meeting. Although, as Baade had stated, the procedure was obvious, carrying it out was not. Finding the 900 OB stars was only the first stage. Determining luminosity criteria and finding the really faint specimens was more difficult. Morgan had long appreciated that for studies of galactic structure, accurate spectral types with luminosity classifications would be required to get absolute magnitudes, and to that end he had produced in 1943 with P.C. Keenan and E. Kellman the MKK *Atlas of Stellar Spectra*. Furthermore, high luminosity objects would be crucial, and for these accurate color indices would also be essential in order to correct the photometric distances for absorption. Thus, Morgan had correctly analyzed the approach for finding the large-scale galactic structure, and had established the basis for a successful program. What remained was to find the truly distant high-luminosity objects.

In an oral history interview made a few years ago at Yerkes Observatory by David DeVorkin of the American Institute of Physics,[10] Morgan remarked that he had *two* papers at the Michigan Symposium, a joint one given by Nassau on the arrangement of the B stars in space, "which at the time had not gone far enough to show anything but a beautiful Gould belt... But my own paper in it was a description of what was called natural groups in stellar spectra." In that second paper Morgan had coined the expression "OB stars" and had described these as a natural group with little spread in luminosity. "It made it possible, by just a glance, a few seconds at each spectrum ... to tell if a star was located in this area [of the H-R diagram].... Now this was the crucial conceptual development. This was then applied to a program which Dr. Nassau and his associates and helpers worked on. I used to go to Cleveland for a week or so every few months, for a number of years. Nassau and I did all of the classifying.... We had a belt I believe 10 degrees wide, as far south as we could get around the sky, and this [furnished] the basic catalogue that was used here [i.e., at Yerkes] for taking slit spectrograms of as many of those stars as possible. Anyway, in the fall of 1951 I was walking between the observatory and home, which is only 100 yards away. I was looking up in the northern sky, just looking up in the region of the Double Cluster, and it suddenly occurred to me that the Double Cluster in Perseus and then a number of stars in Cassiopeia and even Cepheus, that along there I was getting distance moduli of between 11 and 12. Well, 11.5 is two kiloparsecs, and so I couldn't wait to get over here and really plot them up. It looked like a concentration.... but the hardest thing is to know what's going on if you're in the middle of something. So when I plotted out the Perseus arm, I then plotted out the other stars, and it turned out through the sun there was this narrow lane parallel to the other one. So that's the way it happened. It was a burst of realization. It was not a question of a reasoned process of steps."

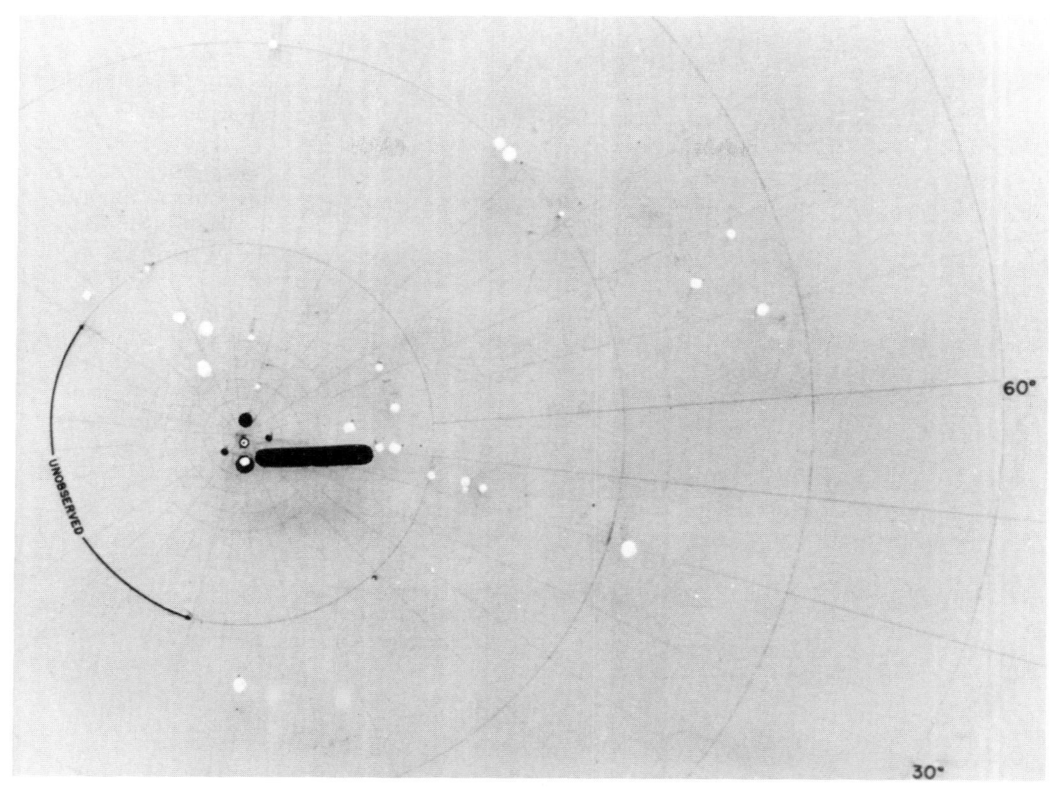

The model of the three spiral arms, as shown by Morgan on a lantern slide at the Cleveland AAS meeting in December, 1951.

Simultaneously with that work, another line of evidence was leading to the same conclusion. At the Michigan Symposium Baade had shown a very suggestive illustration, of the H II regions in M31, and at some point Baade sent the original plate to Morgan. Around this time Baade must have formulated the analogy, which I heard him use a few years later, that the spiral arms in M31 are much like the candles and frosting on a birthday cake--all show and little substance. In other words, there was not much of a stellar density difference in the spiral arms, and hence little to be found by star counting methods. One had instead to look for bright and showy spiral indicators. The existence of the H II regiions convinced Morgan that they should be the pointers for the faint-distant OB stars and hence the key for tracing the spiral structure.

About a year ago one of Morgan's young collaborators, Donald Osterbrock, provided me with details concerning the Yerkes search for the galactic H II regions.[11] As assistants for Morgan, Osterbrock and his fellow graduate student Stewart Sharpless set up the wide-angle Henyey-Greenstein camera for their survey. This complex optical system had been designed during World War II as a wide-angle projector for training aerial gunners, but it could also be used in reverse as a camera to image 140° of the sky onto a circle 2 cm in diameter on a photographic plate. At this scale distant extended H II regions appeared stellar, so they were detected by a microscopic comparison of pairs of red and blue plates. Altogether they took about 50 to 100 plates in 1950-51, and they found 20 or 30 H II regions, including NGC 2244, the well-known nebula in Monoceros, which had not been previously recognized as a giant H II region. Concerning finding the spiral arms Osterbrock told me, "Morgan wanted to do it, he wanted to do it himself, and I guess part of his fear was that somebody else would tumble to the idea before he got it done in what he considered the right way. And I must say that Morgan involved Sharpless and me in every stage of it. Yet our major contribution was really in taking the plates. Most of the H II regions that were found he found, although we had looked very hard for them too. I've always felt that he gave us an awful lot of credit for two young graduate students whose contribution was quite minor. Many other investigators, I think,

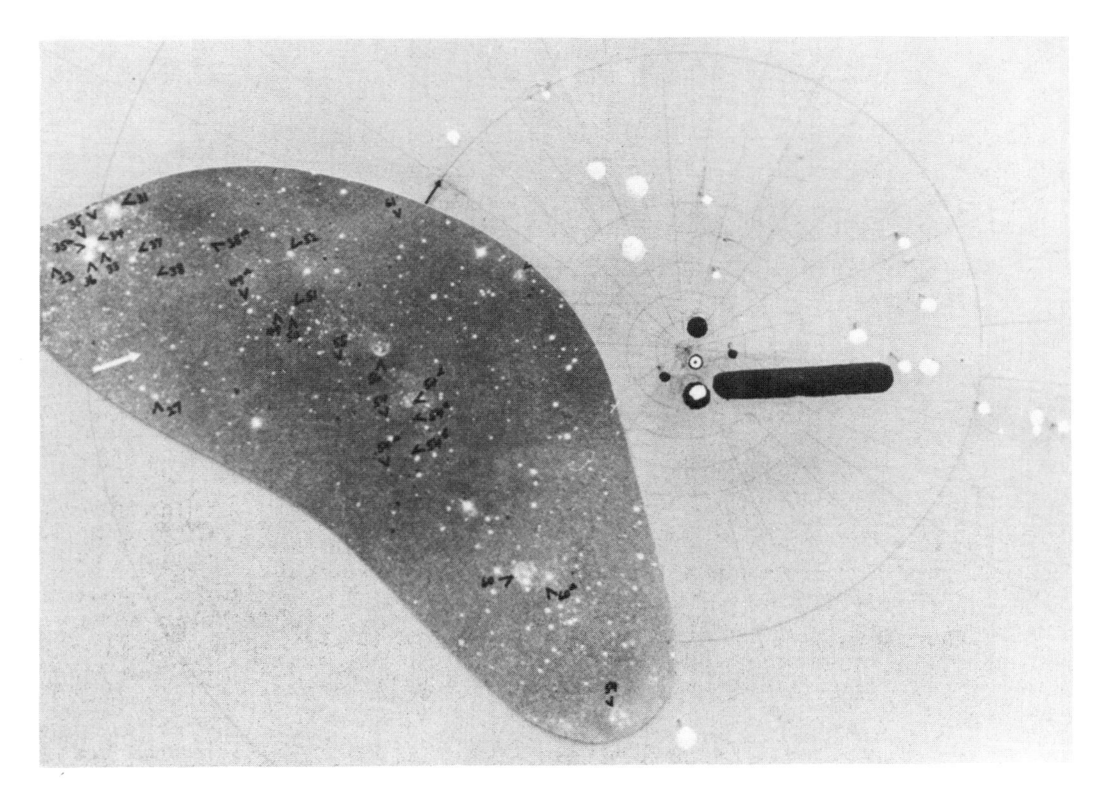

The cover for the April 1952 Sky and Telescope cropped off the Perseus arm, but included a detail of the H II regions from Baade's plate of M31.

would have written the paper themselves. It was an idea that he had had for many, many years; he was asked to go to a meeting to deliver a paper on it, and yet he felt it was right to include us as authors."

Thus it was that Morgan took to Cleveland a jointly authored paper that proved to be the sensation of the December, 1951, AAS meeting. This time Jan Oort came to the meeting, and in fact chaired the session. In the oral history interview Morgan described the situation[12]: "Oort had introduced me, and when he sat down to listen, he sat down in my seat. It was one of those steeply sloping classrooms at Case with the seats all the way up high. Well, when I got through, the first thing was that I had no place to sit down. The second thing was people started to applaud by clapping their hands, but then they started stamping their feet. It was quite an experience."

Remarkably enough, the full paper describing the first discovery of the Milky Way's spiral arms was never published and it is necessary to go to the April, 1952, Sky and Telescope to find the best account of it.[13] In a poignant letter to me, W.W. Morgan has written,[14] "The reason for this was that I had a collapse in the spring of 1952 and spent the summer in Billings Hospital in Chicago in a helpless condition. When I returned to Yerkes in October, I had my partially written paper waiting for me, begun in the early part of the year; I was unable to work on it and complete it; instead, I wrote the UBV paper with Harold Johnson. The rapid growth of radio astronomy resulted in my never finishing and publishing the original paper."

In this way a problem that had stood at least three decades, and which had been one of the critical goals of the kind of studies in galactic structure pioneered at the Kapteyn Astronomical Laboratory, finally found its solution by a quite different avenue from the numerical star-counting procedures. Yet the analysis that grew out of the earlier Dutch studies soon found another and even more powerful application in the interpretation of the line profiles of the 21-cm radiation from neutral hydrogen, radiation that had been discovered in the same year, 1951, as the optical discovery of the spiral structure from the ionized hydrogen. Almost immediately Oort and Bok and their students began a vigorous investigation of the Milky Way structure using radio wavelengths, and by 1952 the spiral structure had been confirmed and extended using the radio methods.[15]

Nevertheless, I think there is some larger justice in the circumstance that the optical studies, on which so many decades of effort had been spent, narrowly won the race with the new and powerful radio astronomy to establish the fact that our galaxy really did have spiral arms, as had long been conjectured.

NOTES AND REFERENCES

1. C.W. Easton, Astrophys. J. 37, pp. 105-118 and Plate III, 1913; A.

Blaauw, "C.W. Easton," in Dictionary of Scientific Biography 4, pp. 272-273, (New York, 1971); Oral history interview with B.J. Bok, American Institute of Physics (AIP) -- I wish to thank Prof. Bok and the Niels Bohr Library, AIP, for permission to use this interview.

2. W. Baade, Astrophys. J. 100, pp. 137-150, 1944.

3. W. Baade to J. Oort, 23 September 1946. I wish to thank Prof. Oort and the Niels Bohr Library, AIP, for permission to examine and use the microfilm of this correspondence.

4. J. Oort to W. Baade, 11 January 1947.

5. W. Baade to L. Goldberg, 18 May 1949. I wish to thank Prof. Goldberg for the use of this material.

6. W. Baade, Publ. Univ. Michigan Obs. 10, pp. 7-17, 1951.

7. J.J. Nassau and W.W. Morgan, Publ. Univ. Michigan Obs. 10, pp. 43-50, 1951.

8. Sky Telesc. 9, p. 244, 1950.

9. O. Struve, Leaflet Astron. Soc. Pacific No. 285, January, 1953.

10. Oral history interview with W.W. Morgan, American Institute of Physics. I wish to thank Prof. Morgan and the Niels Bohr Library, AIP, for permission to quote from this interview. The quoted passage has been somewhat condensed.

11. Oral history interview with D.E. Osterbrock, on deposit at American Institute of Physics. I wish to thank Prof. Osterbrock for permission to quote from this interview and for his comments on the draft of this article. Photographs of the H II regions made with the Henyey-Greenstein camera are found in S. Sharpless and D. Osterbrock, Astrophys. J. 115, pp. 89-93, 1952.

12. Interview with Morgan [note 10].

13. Sky Telesc. 11, p. 138, 1952; reprinted in K. Lang and O. Gingerich (eds.), Source Book in Astronomy and Astrophysics, 1900-1975, pp. 638-642, (Cambridge, Mass., 1979); the abstract of the Morgan, Sharpless, Osterbrock paper appeared in Astron. J. 57, p. 3, 1952.

14. W.W. Morgan to O. Gingerich, 25 March 1982.

15. See K. Lang and O. Gingerich, op. cit. [note 13], pp. 643-651.

16. The Van Albada mentioned here is G.B. van Albada, later Director of the Lembang Observatory and of the Astronomical Institute at Amsterdam. (Editor.)

PART II

COMPOSITION, STRUCTURE AND KINEMATICS

Musician Betty Schwarz and dynamicist Stefano Casertano discuss musical composition during Wednesday outing. LZ

SECTION II.1

GALACTIC CONSTANTS, ROTATION AND MASS DISTRIBUTION

Tuesday 31 May, 0900 - 1020

Chairman: R. Wielen

Three generations. At the Kapteyn-Van Rhijn exhibition, Maarten Schmidt has a conversation with Mrs. Reina van Rhijn, widow of Pieter J. van Rhijn, professor of astronomy at Groningen in 1921-1956. At right: a portrait of Jan H. Oort, painted in 1938 by Dirk Nijland.

Schmidt was a student of Van Rhijn at Groningen and of Oort at Leiden, where he obtained his doctorate in 1956. Oort was a student of Kapteyn and obtained his doctorate from Van Rhijn in 1926. CFD

MODELS OF THE MASS DISTRIBUTION OF THE GALAXY

Maarten Schmidt
California Institute of Technology
Pasadena, California, U.S.A.

1. INTRODUCTION

Mass models of the Galaxy play an important role in studies of the structure of the Galaxy. The various populations or components combine to yield a gravitational field that produces the observed rotation curve. For the spheroid and disk this requirement can be used to set limits on some of their properties. The properties of the dark corona are entirely defined this way.

Early models of the mass distribution were primarily based on the rotation curve interior to the Sun (see Schmidt 1965). Since that time, observations of late-type galaxies have shown that their rotation curves are flat or rising beyond a few kiloparsecs from the center (Krumm and Salpeter 1979; Rubin et al. 1980, 1982). This suggests that the overall density of matter decreases approximately as R^{-2}.

Due to our location and dust absorption in the galactic plane, we cannot reliably derive the distribution of stars in the disk of our Galaxy. Here, too, external galaxies have supplied important evidence: the disk component of their luminosity distribution exhibits an exponential profile (Freeman 1970).

We review briefly in the next section published mass models that have approximate exponential disks and flat rotation curves. Then, we report on some test models to investigate what range of model parameters is permissible for a given rotation curve. Finally, we comment on some of the properties of the dark corona which is postulated to explain the flat rotation curve.

2. MASS MODELS

Recent mass models of the Galaxy have been published by Clutton-Brock et al. (1977), Sinha (1978), Einasto (1979), Miyamoto et al. (1980), Caldwell and Ostriker (1981), Rohlfs and Kreitschmann (1981),

H. van Woerden et al. (eds.), The Milky Way Galaxy, 75–84.

and Bahcall et al. (1983). A useful comparison between some of these models is given by Caldwell and Ostriker (1981, see Table 4).

In general, the models include three components: a) Spheroid--corresponding to the halo population (population II) such as subdwarfs and globular clusters. Densities usually follow a Hubble or de Vaucouleurs law. Shape approximately spherical. b) Disk--usually a flat disk with an exponential density law. The disk has sometimes a central hole. c) Dark Corona--a usually spherical component added to yield an approximately flat rotation curve.

The most interesting properties of the model components are the core radius and local density of the dark corona, for which there is no independent evidence, and the local density of the spheroid for which only an approximate lower limit is known, as discussed below. The total range of these properties in the above mentioned models is:

Spheroid: local density = (2-35) x $10^{-4} M_\odot/pc^3$
Dark Corona: local density = 0.001-0.011 $M_\odot/pc^3$
core radius = 2-15 kpc.

It is difficult to trace in detail the origin of the very large ranges shown. One would suspect that a substantial part of the variation may be caused by different density laws and parameters of the components, different adopted rotation curves, etc. In order to investigate this aspect, we explore in the next section models made up of components with a given density law and based on a given rotation curve.

3. EXPLORATION OF TEST MODELS

On the basis of the evidence from external galaxies, we assume that the rotation curve of the outer parts of the Galaxy is essentially flat. Following a review by Knapp (1980) of available rotation curves,

we fit at	0.5	2.5	8.5	∞	kpc
to circular velocities of	240	195	220	220	km/sec.

We use R_o = 8.5 kpc and impose an asymptotic velocity of 220 km/sec at large distances.

For bulge and spheroid we use the following density law

$$\rho \sim R^{-1.5}(b^n+R^n)^{-1}$$

with n = 1.5 or 2. This approximately represents the $R^{-1.8}$ density law observed for IR/OH sources in the central bulge (Isaacman and Oort 1981) and at larger distances fits the R^{-3} or $R^{-3.5}$ density law observed for RR Lyrae stars and globular clusters (see Oort and Plaut 1975; Harris 1976).

The disk is represented by an inhomogeneous spheroid in which surfaces of constant density are oblate spheroids of constant excentricity. The density is given by a four-term polynomial such that the

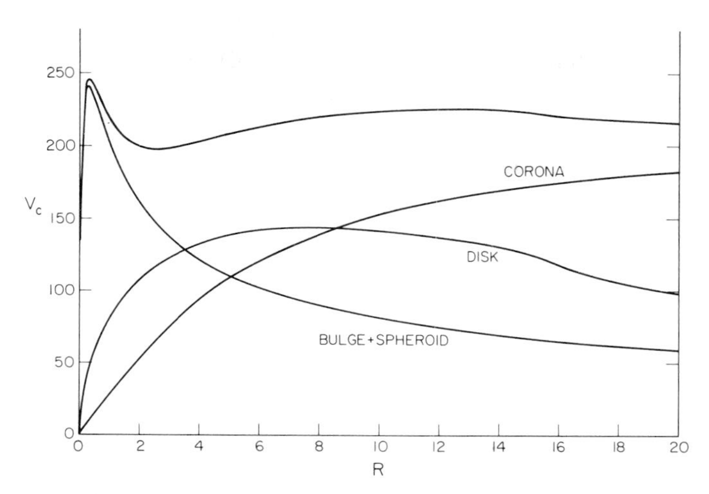

Figure 1. Rotation curve for a three-component mass model of the Galaxy. The contribution of each of the three components is shown separately.

surface density is approximately exp(-R/H). The polynomial is cut off at zero density, at R = 4.5 H, in approximate agreement with that observed by van der Kruit and Searle (1982) in edge-on galaxies. For the dark corona we assume a spherical shape and density law

$$\rho \sim (a^2+R^2)^{-1}$$

which yield an asymptotically flat rotation curve at large distance.

We show in Figure 1 the rotation curve corresponding to a model given by:

bulge + spheroid	b = 0.3 kpc	$\rho_o = 0.0002\ M_\odot/pc^3$
disk	H = 3.5 kpc	$\sigma_o = 50\ M_\odot/pc^2$
dark corona	a = 4.6 kpc	$\rho_o = 0.010\ M_\odot/pc^3$

where we used for the spheroid density law n = 2. The quantities ρ_o, σ_o are the local values (at R_o = 8.5 kpc) of volume and surface densities.

The rotation curve for R = 0 - 2 kpc is primarily determined by the bulge+spheroid component. For n = 2, corresponding to a $R^{-3.5}$ law in the spheroid, the local density in the model is $2 \times 10^{-4}\ M_\odot/pc^3$. If we take n = 1.5, or a R^{-3} density law in the spheroid, then the local model density is $6 \times 10^{-4}\ M_\odot/pc^3$. These densities may be compared to that based on the luminosity function of high-velocity stars which yields $1.7 \times 10^{-4}\ M_\odot/pc^3$ (Schmidt 1975), and that derived from star counts, namely $(0.4\text{-}0.6) \times 10^{-4}\ M_\odot/pc^3$ (Bahcall _et al._ 1983).

The star count derivation involves a considerable extrapolation of the mass function, since it is based on relatively bright stars (M_V = 4-8) which carry a small fraction of the total mass. The density determined from the high-velocity stars is based on the assumption that the median tangential velocity of spheroid stars is 250 km/sec. This value of the median tangential velocity may be somewhat high (Richstone and Graham 1981). If, instead, we adopt a median tangential velocity of 200 km/sec, then the corresponding mass density increases to 2.7 x 10^{-4} $M_\odot/pc^3$ (Schmidt 1975).

It appears that the local model density in the spheroid of 2 x 10^{-4} $M_\odot/pc^3$ for n = 2 is consistent with the observational evidence. For n = 1.5, the local model density is 6 x 10^{-4} $M_\odot/pc^3$ and a substantial part of the mass would have to be in stars below the hydrogen-burning mass limit, or in other dark objects.

Caldwell and Ostriker (1981) used a Hubble law in which the density falls approximately as R^{-3}, and derived a local model density of 11 x 10^{-4} $M_\odot/pc^3$. The difference with our R^{-3} model is probably caused by a combination of higher central velocity peak, a larger adopted value of R_o, a central hole in the disk component, and a somewhat different density law. I conclude that for a given density law, the local density in different models may differ by a factor of two, and that for a $R^{-3.5}$ spheroid density law, the local density of the spheroid is in approximate agreement with that derived from high-velocity stars.

In the range 2-8 kpc, the disk is the main contributor to the rotation curve. In the model shown in Figure 1 we used H = 3.5 kpc following de Vaucouleurs and Pence (1978) and a local surface density of 50 $M_\odot/pc^2$. Based on current estimates of the density of stars and gas, I would estimate at the present time a mass surface density of 39 $M_\odot/pc^2$. In order to compare this observed density to that based on the K_z determination by Oort (1960), I use (for this purpose only) for a spherical component of the Galaxy a pseudo surface density that equals 1000 pc times the local volume mass density. The surface density corresponding to Oort's K_z determination is 80 $M_\odot/pc^2$. The dark corona has a pseudo surface density of about 10 $M_\odot/pc^2$, and the spheroid contributes a negligible surface density, so we would <u>expect</u> for the disk about 70 $M_\odot/pc^2$. The difference with our estimated <u>observed</u> surface density of 39 $M_\odot/pc^2$ illustrates the local hidden-mass problem.

The model of Figure 1 employed a disk of 50 $M_\odot/pc^2$ and H = 3.5 kpc. The adopted points of the rotational velocity curve can still be fitted within about 5 km/sec for a surface density as large as 65 $M_\odot/pc^2$, if the core radius of the dark corona is increased to a = 6.5 kpc. Similarly, the exponential scale length H can be increased to 6 kpc, in which case a = 2.8 kpc. An H value as low as 3 kpc is only possible if the local disk surface density drops to 40 $M_\odot/pc^2$, in which case a = 4.4 kpc.

The dark corona is gravitationally the dominant component outside the solar radius. The local density and core radius are essentially set dynamically, through the models, as illustrated above. In the various models with different disk parameters discussed above, we find local densities in the range (0.008-0.011) $M_\odot/pc^3$ and a = 2.8-6.5 kpc.

We can summarize the results of this exercise in modelling as follows. Even with a given rotation curve and given density laws for the three components, there are a variety of acceptable solutions. The scale lengths of disk and dark corona have an allowable range of about a factor of 2, the local disk density a factor of 1.5, and the local spheroid density a factor of 3. The even larger range of properties shown by published models of the Galaxy (discussed in the beginning of this paper) is no doubt a consequence of the additional effect of different model components, different rotation curves, etc. We conclude that the complexities of mass modelling are such that properties derived from such models should be viewed with caution.

4. THE DARK CORONA

Since the dark corona is not seen but only felt through its gravitational effect, its properties are less well defined than those of disk and spheroid. We briefly discuss three questions and one comment:

1) Is a dark corona needed?
2) What is the evidence for an R^{-2} density law?
3) Is the dark corona spherical?
4) A comment about the balance between dark corona, disk and spheroid.

4.1. Is a dark corona needed?

Tests with the models discussed in the preceding section show that an essentially flat rotation curve out to 30 kpc can be obtained with disk parameters H = 8 kpc and σ_o = 200 $M_\odot/pc^2$. From local conditions in the solar neighborhood, we know that this surface density is far too large. With a realistic mass model, such as those discussed in the preceding section, the rotation velocity near the sun corresponding to disk and spheroid is no larger than 170 km/sec. Clearly, we need another mass component to boost the rotation velocity locally to its actual value, which probably lies between 200 and 250 km/sec.

4.2. What is the evidence for an R^{-2} law?

The notion of an R^{-2} density law is based on the flatness of the observed rotation curves. However, as Figure 1 shows, spheroid and disk account for a substantial fraction of the rotation curve inside the Sun. Outside, we balance the decreasing contributions of spheroid and disk with an increasing contribution of the dark component--but this gives little support for the need of an asymptotic R^{-2} law.

The situation might be different in those external galaxies where the flat rotation curves are observed out to large distances, where the effect of spheroid and disk might be thought to be small. This is, however, not the case as illustrated by Bahcall et al. (1982) who show that a dark corona with a local logarithmic density gradient of about -2.7 can yield an essentially flat rotation curve in the distance range 40 to 60 kpc. We conclude that on the basis of the observations there is little direct evidence for a R^{-2} or $(a^2+R^2)^{-1}$ density law.

4.3. Is the dark corona spherical?

We have assumed that the dark corona is spherical, in which case the local model density is 0.010 $M_{\odot}/pc^3$. If the corona is spheroidal in shape, then the density is larger by a factor equal to the inverse axial ratio. For an axial ratio of around 1/4, the local mass density of the dark corona would be of the same order as that of the hidden mass in the solar neighborhood needed to interpret K_z. This opens the possibility that the galactic dark mass and the local hidden mass could be the same material, unless there exist theoretical arguments why the corona should be spherical.

4.4. The balance between dark corona, disk, and spheroid.

Figure 1 shows that the spheroid dominates within the first few kiloparsecs. Beyond that, the disk is the main contributor to the rotational velocity interior to the Sun, and the dark corona dominates at larger distances. Each of these three mass components contributes about the same rotational velocity over their respective ranges of dominance. As a consequence, the rotation curve is essentially flat over a large galactocentric distance range.

Assuming that this situation holds for most external galaxies, the fact that they have essentially flat rotation curves suggests that there is a balance between spheroid, disk, and dark corona. If such a balance did not exist, there should be cases where the rotation velocity changes considerably with galactocentric distance, but few if any such cases are seen in the rotation curves obtained by Rubin et al. (1980, 1982). Further study of this apparent conspiracy between the different mass components of galaxies is warranted.

REFERENCES

Bahcall, J.N., Schmidt, M., and Soneira, R.M.: 1982, Astrophys. J. (Letters), 258, L23
Bahcall, J.N., Schmidt, M., and Soneira, R.M.: 1983, Astrophys. J. 265, 730
Caldwell, J.A.R., and Ostriker, J.P.: 1981, Astrophys. J. 251, 61
Clutton-Brock, M., Innanen, K.A., and Papp, K.A.: 1977, Astrophys. Space Sci. 47, 299
de Vaucouleurs, G., and Pence, W.D.: 1978, Astrophys. J. 83, 1163
Einasto, J.: 1979, in IAU Symposium 84, The Large Scale Characteristics of the Galaxy, ed. W.B. Burton (Dordrecht: Reidel), p. 451

Freeman, K.C.: 1970, Astrophys. J. 160, 811
Harris, W.E.: 1976, Astron. J. 81, 1095
Isaacman, R., and Oort, M.J.A.: 1981, Astron. Astrophys. 102, 347
Knapp, G.R.: 1980, private communication
Krumm, N., and Salpeter, E.E.: 1979, Astron. J. 84, 1138
Miyamoto, M., Satoh, C., and Ohashi, M.: 1980, Astron. Astrophys. 90, 215
Oort, J.H.: 1960, Bull. Astron. Instit. Netherlands 15, 45
Oort, J.H., and Plaut, L.: 1975, Astron. Astrophys. 41, 71
Richstone, D.O., and Graham, F.G.: 1981, Astrophys. J. 248, 516
Rohlfs, K., and Kreitschmann, J.: 1981, Astrophys. Space Sci. 79, 289
Rubin, V.C., Ford, W.K., and Thonnard, N.: 1980, Astrophys. J. 238, 471
Rubin, V.C., Ford, W.K., Thonnard, N., and Burstein, D.: 1982, Astrophys. J. 261, 439
Schmidt, M.: 1965, in Galactic Structure, ed. A. Blaauw and M. Schmidt (Chicago: University of Chicago Press), p. 513
Schmidt, M.: 1975, Astrophys. J. 202, 22
Sinha, R.P.: 1978, Astron. Astrophys. 69, 227
van der Kruit, P.C., and Searle, L.: 1982, Astron. Astrophys. 110, 61

DISCUSSION

D. Lynden-Bell: What evidence have you that the spheroid is round?

Schmidt: You mean: that it is a sphere? I believe the work on RR Lyrae stars by Oort and Plaut indicated that the distribution was very round, in fact almost prolate (which, however, was not proposed). Work at Lick by Wirtanen and Kinman in other directions seemed to indicate axial ratios of perhaps 0.7. I think in the present discussion it would not make much difference - it is a rather relaxed model, and if one flattens the spheroid a bit, the density would just go up in proportion.

J.H. Oort: But we know the globular-cluster distribution also.

Schmidt: And what would you say about that? Round?

Oort: Yes, except perhaps for the innermost part.

Lynden-Bell: What about the light of external galaxies?

Schmidt: There the axial ratio is typically around 0.7.

J.P. Ostriker: I have two comments of possible interest.
1) I would guess that in your model the total quasi-spherical mass interior to the Sun is of order 1/3 - 1/2 of the total, as in other published models. It is useful that this ratio is so invariant, since it insures (barely) the gross stability of the Galaxy.
2) I would propose that you could reduce the corona significantly, if you used only the constraints on the rotation curve interior to the Sun and were willing to take up the slack with the other components, leaving the M/L ratio of the spheroid free.

Ostriker and Schmidt during the boat trip. Foreground: Denoyelle; background T.S. van Albada and (partly hidden) Illingworth, Fujimoto and Norman. LZ

Schmidt: I doubt it, Jerry. Even with an R^{-3} law, the central velocities in the spheroid would come out much too high compared to observation.

Ostriker: I leave it for you as an exercise while you play with these parameters. I bet you can get rid of the corona and make a model which....

Schmidt: You do it, Jerry.

Oort: What is the evidence for a minimum in the rotation curve at 2 kpc?

Schmidt: The review by Knapp (1980) indicates such a minimum; I have not myself looked critically at the evidence. If one does not stick to this minimum, then surface densities and scale lengths beyond the ranges indicated by me are possible.

Oort: Why do you think there is no good evidence in the Galaxy for a flat rotation curve outside the solar circle? My impression was that the evidence is fairly convincing.

Schmidt: My point is that present knowledge of the rotation curve out-

Mrs. Oort, Oort, Schmidt, De Zeeuw and Lacey at the conference dinner.
LZ

side the Sun, and of the interior components of the Galaxy, is not precise enough to base an R^{-2} law for the dark corona on it.

B.F. Burke: Evidence for dark coronae may best be found in external systems. J.M. Mahoney, J.M. van der Hulst and I are engaged in a study of simple interacting pairs of galaxies, using both the morphological data and HI radial-velocity fields. The presently observed tidal distortions are, in principle, a fossil record of the past orbital history, and if the interaction has been close enough, a large massive halo should modify the tidal interaction sufficiently to leave a dynamical record. Simple systems, in which only one encounter has occurred, would be preferred. The well-known "antennae", NGC 4038/39, are such a system, with simple, well-ordered structures, and we have completed VLA observations of this object. Model studies, now in progress, should evaluate the promise of the method. Perhaps those who study the tidal interactions of the Magellanic Clouds with the Milky Way can find evidence for our own corona if their work becomes sufficiently quantitative.

W.B. Burton: The problem of the minimum in the rotation curve at 2 kpc or somewhat less centres on the nature of the rotation curve at smaller galactocentric distances. There seem to me to be good reasons to think that the rotation curve does not change as abruptly as in the earlier interpretations of the observations. In the earlier observations absorption effects gave the appearance that there was no HI at negative velocities in the first quadrant, and that led to a steeply rising curve with very high rotation speeds. Molecular data which are not influenced by absorption effects indicate a much slower variation of velocity across the galactic centre, hence a less steep rotation curve in the interior part of the Galaxy. It remains difficult to derive an accurate rotation curve in the inner parts, partly because of lack of knowledge about the exact form of the potential. Present evidence should not be interpreted as requiring a steeply rising rotation curve.

Schmidt: As I said before, this will mean that the models are even less restrictive in (so that one will be able to use reasonable values for) scale length, disk mass and density.

Lynden-Bell: So you do not believe in that 250 km/s peak at R = 0.5 kpc?

Burton: Indeed, that peak is entirely open to question.

Schmidt: I do not necessarily disagree with you.

A MODEL OF OUR GALAXY

U. Haud, M. Jôeveer, and J. Einasto
Tartu Astrophysical Observatory

1. INTRODUCTION

The construction of models is the most effective tool for a synthesis of various observational data and for a quantitative study of physical and dynamical structure and evolution of stellar systems. Classical models of spiral galaxies were based on rotational velocities, which were identified with circular velocities. They were designed to represent the galactic attraction force in the radial direction.

Significant advances are currently taking place in a wide variety of observational approaches which will greatly clarify our picture of the Galaxy's large-scale structure. In this report we present a new model of the Galaxy. It has been constructed using the most recent data available.

2. THE METHOD OF MASS MODELLING

By a model of the Galaxy we mean a set of functions and parameters which quantitatively describe the principal properties of the Galaxy and its populations. By a population we mean the family of stars or other objects having similar physical properties (age, chemical composition, etc.) and similar parameters of spatial distribution and kinematics (Einasto 1974). The structure of the galactic populations may be described by their gravitational potential, ϕ, and its radial and vertical derivatives K_R, K_z, the spatial density, ρ, the projected density, P, velocity dispersions, σ_R, σ_θ, σ_z, and the centroid velocity, V.

On the basis of the existing data we may assume that the Galaxy is well-relaxed, that its populations are physically homogeneous, and that equidensity surfaces of the galactic populations are similar concentric ellipsoids or they can be represented in the form of sums of such ellipsoids. Under these assumptions simple relations hold between all the descriptive functions (Einasto 1974).

H. van Woerden et al. (eds.), The Milky Way Galaxy, 85–94.

The most convenient way of determining a model is to use a certain analytic expression for the density of the galactic populations. Our experience has shown that the best representation can be obtained by the use of an exponential function (Einasto 1974):

$$\rho(a) = \rho(0) / \exp (a/ka_o)^{1/N} \tag{1}$$

where $a = (R^2 + z^2/\varepsilon^2)^{\frac{1}{2}}$ is the major semiaxis of the equidensity ellipsoid, ε is the axial ratio of the ellipsoid, a_o is the harmonic mean radius of the population, $\rho(0) = hM/(4\pi\varepsilon a_o{}^3)$ is the central density, M is the mass of the population, N is a structural parameter of the model, and h and K are dimensionless normalizing constants. The density distribution in the massive corona can be represented by a modified isothermal model (Haud and Einasto, 1983).

To build a model with a hole in the centre of the disk, the spatial density of disk and flat-population objects can be expressed as a sum of two spheroidal mass distributions:

$$\rho(a) = \rho_+(a) + \rho_-(a) \tag{2}$$

Both of them can be approximated with an exponential law (1), but the second component $\rho_-(a)$ of the disk has a negative mass (Einasto et al., 1980). If one adopts a disk model with a zero density at the axis R = 0 and a non-negative spatial density $\rho(a) \geq 0$, one will have the following relations between the parameters of the both components: $\varepsilon_- = \kappa\varepsilon_+$, $a_{o-} = a_{o+}/\kappa$, $M_- = M_+/\kappa^2$, where $\kappa > 1$ is a parameter which determines the amount of the hole in the centre of the disk. The structural parameter N should be identical.

3. GALACTIC MODEL

Recently a new computer program was completed at Toravere. It enables automatic construction of models of galaxies on the basis of almost all observational data available on the object under consideration. By means of this program models of M31, M32, M81, M87, M100, M104 and our own Galaxy have been constructed.

The model of the Galaxy consists of six populations. They are the nucleus, the bulge, the halo, the disk, the flat population and the massive corona. Parameters of these populations, found by fitting observational data with the model by means of the method of least squares, are given in Table 1. They will be discussed in detail in the following sections.

a) The nucleus. Its structure is determined on the basis of the infrared data (Becklin and Neugebauer 1968, 1975) at effective wavelengths of 2.2μ on the central part of our Galaxy. The estimate of the mass of the nucleus is derived from the observations of the [Ne II] fine-structure line at 12.8μ (see Oort 1977).

Table 1. Parameters of galactic populations

Population	ε	κ	a_o (kpc)	M ($10^{10}M_\odot$)	N	h	K
Nucleus	0.87	0.0	0.001	0.001	1.4	9.1206	0.21613
Bulge	0.4	0.0	0.206	0.646	4.37	7.333×10^3	3.667×10^{-5}
Halo	0.64	0.0	1.096	0.457	7.072	4.155×10^6	1.935×10^{-9}
Disk	0.1	4.3	4.926	8.36	1.202	6.032	0.33278
Flat	0.02	1.67	5.372	0.7	0.546	1.699	1.0626
Corona	1	0.0	60	200	0.5	8.3206	0.25817

b) The bulge. Here we have two kinds of observational data on the structure of this population. In the inner regions there exist 2.2μ infrared data (Becklin and Neugebauer 1968, 1975) on the distribution of the surface density of the bulge. In the outer regions some information may be obtained from the first maximum of the rotation curve. The mass of the population can be derived from the observed mean velocity dispersion in the bulge. The comparison of the model surface-density distribution with the observed one is given in Figure 1. The observed (see Mould 1982) and computed velocity dispersions are consistent within 5%.

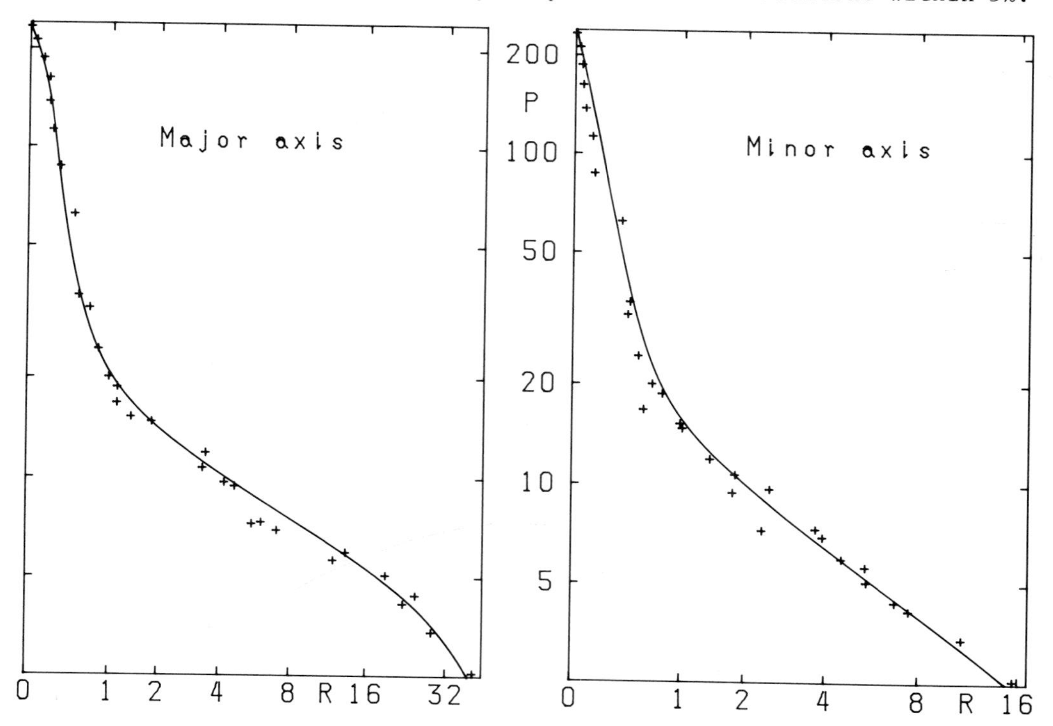

Figure 1. 2.2μ brightness distribution in the central region of our Galaxy. Crosses - observations, solid line - model.

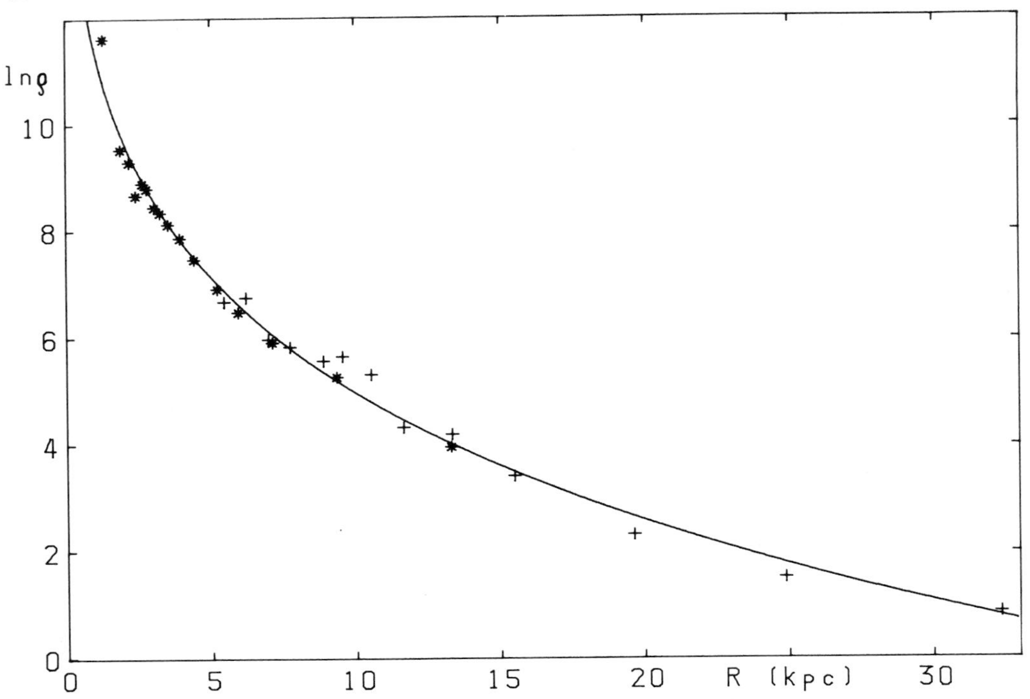

Figure 2. Density distribution in the halo. Asterisks - RR Lyrae stars, crosses - globular clusters, solid line - model.

c) The halo. The structure of the halo is determined from the spatial distribution of globular clusters (Harris 1976) and RR Lyrae stars (Plaut 1966, 1968a, b, 1970, 1971, 1973a, b; Oort and Plaut 1975; Kinman et al. 1965, 1966; Meinunger 1977). The mass of the halo is estimated on the basis of 1) the total number of globular clusters in the halo (228 in our model), 2) the mean mass of globular clusters, about 2×10^5 $M_\odot$ (Mihalas and Binney, 1981, p. 122) and 3) the fraction of the halo mass in globular clusters, about 1% (Woltjer 1975). The comparison of our model with observations is given in Figure 2.

d) The disk. The mass and structure of the disk are determined from the rotation-velocity curve. Our adopted rotation curve is corrected for the effects of radial motions of the gas in our Galaxy (Haud 1979, 1983). As follows from these corrections, our rotation curve may be relatively inaccurate in the regions $R < 3$ kpc and $R > 10$ kpc. Therefore, the value of κ cannot be determined very precisely. Moreover, as the disk represents galactic populations over a wide range of axial ratios between the flat ($\varepsilon \approx 0.02$) and the intermediate ($\varepsilon \approx 0.4$) population objects, its axial ratio, $\varepsilon = 0.1$, is only a compromise. The comparison of the model rotation curve with the observed one is given in Figure 3.

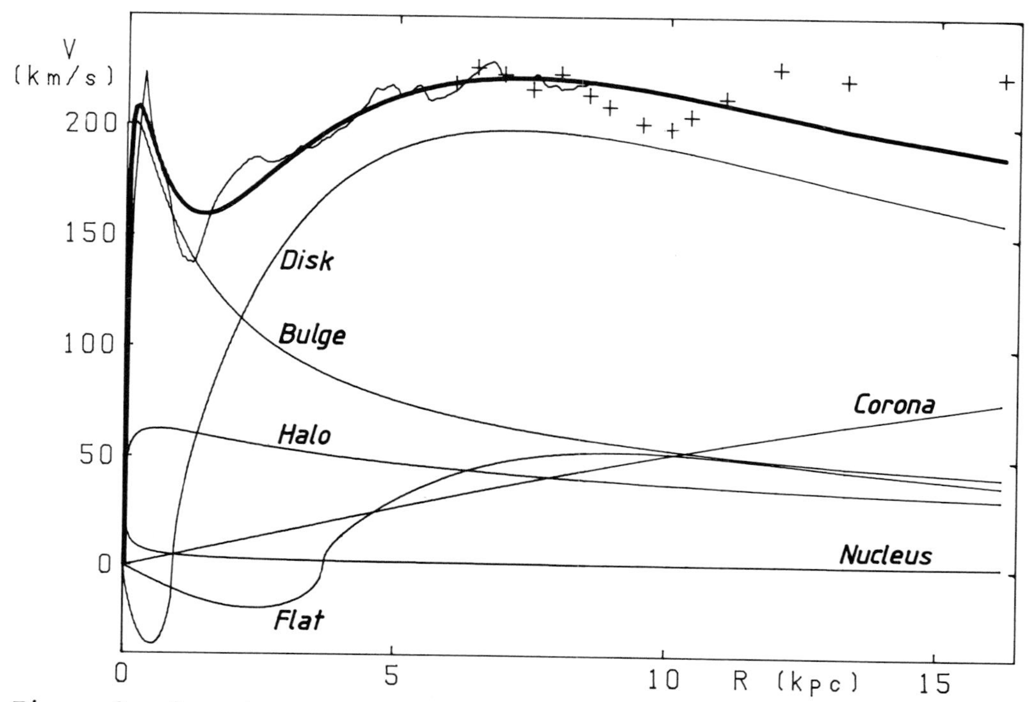

Figure 3. Rotation curve of our Galaxy. Wavy curve - HI observations, crosses - HII observations, thick line - model.

e) The flat population. This population represents the interstellar gas and young stars. Its structural parameters are estimated from the distribution $n(H) = 2\ n(H_2) + n(HI)$ (Gordon and Burton 1976). The axial ratio of the equidensity ellipsoids is found from the z-distribution of the gas (Burton and Gordon 1976, Celnik et al. 1979). The mass of this population was determined on the basis of density estimates of the gas and young stars.

f) The corona. Visible elements of the corona (galactic companions) form a flat disk. The form of the invisible corona is at present unknown; in the model we adopt a spherical corona (Haud and Einasto 1983). The determination of its parameters is described in our earlier papers (Einasto et al. 1976; Einasto and Lynden-Bell 1982).

4. THE SYSTEM OF GALACTIC CONSTANTS

An independent check of the reality of the model can be obtained by comparing the observed local galactic constants with those of the model. Table 2 summarizes the mean values of recent independent determinations of these constants. Here R_o is the solar distance from the galactic centre; V_o the local circular velocity; $W = 1/2\ dU/dx$ where U is the maximum relative radial velocity of rotation in the inner parts

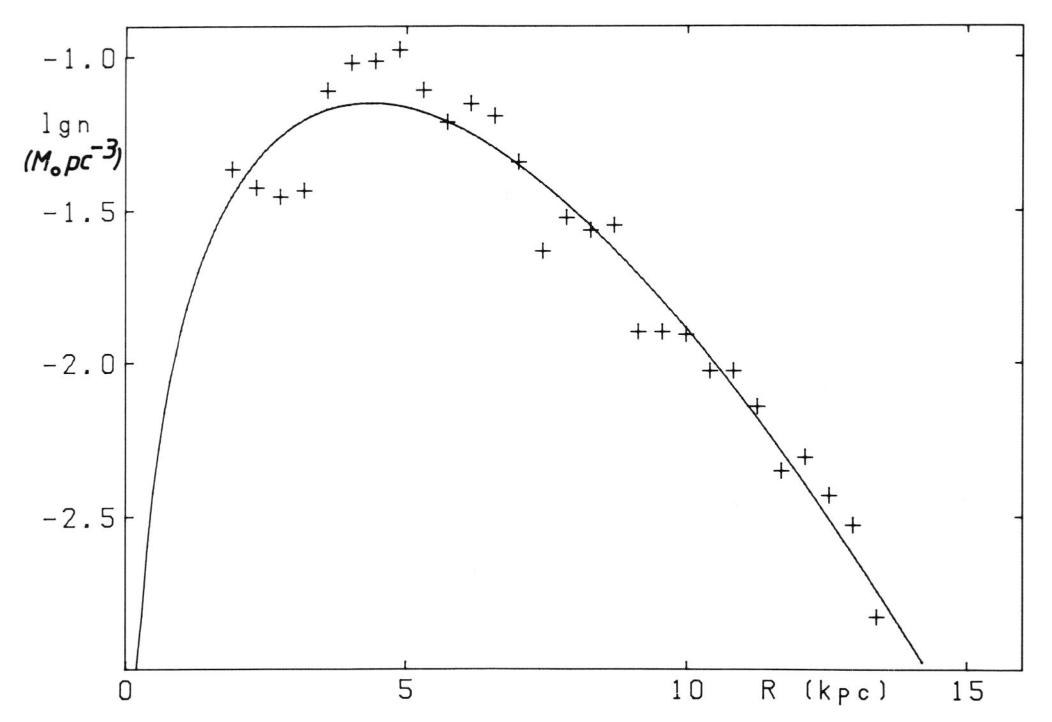

Figure 4. Density distribution in the flat population. Crosses - observations, solid line - model.

of the Galaxy and $x = R/R_o$; A is Oort's constant; Ω the angular velocity; ρ the mean mass density.

The observed values of the constants are quite similar to our previous compilation (Einasto 1979), probably most constants are known with 5-10% accuracy. Now the only ill-defined local constant is evidently the mass density. Here, preferring the studies with relatively low values of ρ, we kept in mind the discussion by Joeveer and Einasto (1976). However, the accuracy of density determinations is low and the values of $\rho \approx 0.15\ M_\odot\ pc^{-3}$ reached by Oort (1965) and others cannot be definitely excluded.

The observed values of the galactic constants are subject to random and undetected systematic errors. That is why they do not exactly satisfy the equations connecting individual galactic constants with each other. To reduce the role of errors considered, we have found by the method of least squares a smoothed and mutually concordant system of galactic constants, where the equations valid in stationary stellar systems are exactly fulfilled (for details see Einasto and Kutuzov 1964). This system of galactic constants is presented in the last but one column of Table 2, next to the corresponding constants from the model calculations.

Table 2. Galactic Constants

Constant	Unit	Observed Value	References	Smoothed Value	Model Value
R_o	kpc	8.7±0.7	1-4	8.4±0.3	8.5
V_o	km s^{-1}	220±10	5,6	218±5	220
W	km s^{-1}	120±12	7	130±4	122
A	km s^{-1} kpc^{-1}	15.6±1	8,9	15.4±0.4	14.4
Ω	km s^{-1} kpc^{-1}	26.2±2	10-12	25.8±0.7	25.9
K_z		0.282±0.02	13	0.289±0.008	0.307
ρ	$M_\odot$ pc^{-3}	0.1±0.03	14,15		0.10

1. Oort & Plaut, 1975. 2. Harris, 1976. 3. Quiroga, 1980. 4. Glass & Feast, 1982. 5. Einasto et al., 1979. 6. Gunn et al., 1979. 7. Haud, in press. 8. Balona & Feast, 1974. 9. Crampton & Georgelin, 1975. 10. Asteriadis, 1977. 11. Fricke, 1977. 12. Dieckvoss, 1978. 13. Einasto, 1972. 14. Woolley & Steward, 1967. 15. Joeveer, 1974.

The comparison of these systems of constants reveals a satisfactory agreement. It should be noted that only in the case of V_o special efforts were taken to equalize the observed value with the model one. The difference of the observed Oort constant A (and also of K_z) from the corresponding model constant is caused by the local minimum in the observed rotation curve, which is not represented by the smooth rotation curve of the model (compare the run of rotation curves at distances $7 < R < 10$ kpc, Fig. 3).

5. COMPARISON WITH OTHER MODELS

We confine our brief comparison to two recent studies by Rohlfs and Kreitschmann (1981, hereafter RK) and Caldwell and Ostriker (1981, hereafter CO), in which three-component galactic models were constructed. All three models have an essential, similar feature. To explain the kinematics of galactic objects at large distances from the centre a new population, the massive dark corona, is introduced into the model. Besides the corona RK distinguished inner-bulge and disk populations, CO spheroidal and disk populations. The mathematical techniques used in the construction of the models are different, but the basic observational data are quite similar. For the most critical parameters, R_o and V_o, almost identical values were adopted in this study and by RK (8.5 and 8.5 kpc; 220and 225 km/s, respectively). CO arrived at somewhat larger values: 9.1 kpc and 243 km/s. The rotation curves in the outer parts of the Galaxy remain at a rather constant level in all models and do not show a Keplerian decrease. The RK model has the largest decrease.

The comparison of the parameters of our model and the RK model reveals satisfactory agreement, for example the total masses of the corresponding populations are at least of the same order. The mass of the bulge component ($0.56 \times 10^{10} M_{\odot}$) in the RK model is comparable with the sum of the bulge and halo components in our model ($1.1 \times 10^{10} M_{\odot}$). The disk mass in the RK model and the sum of the disk and flat-component masses in our model are 7.63×10^{10} and $9.06 \times 10^{10} M_{\odot}$, respectively.

More disturbing is the comparison of our model and that of RK with the CO model. Here moderate agreement obtains only for the disk population mass ($6.6 \times 10^{10} M_{\odot}$), whereas for the spherical population CO deduced a much larger mass value ($6.4 \times 10^{10} M_{\odot}$) in comparison with our and RK models. The density of matter at the solar radius in the spheroidal (not coronal) component of the CO model is $1.1 \times 10^{-3} M_{\odot}\ pc^{-3}$, far in excess of that found by Schmidt (1975) and other authors from analysis of observed high-velocity stars.

Owing to rather different model-constructing techniques it is hard to indicate the individual data and deduction steps which lead to the contradictory results. Nevertheless, the mentioned contradiction once more stresses that our knowledge about the role of population II objects in the Galaxy is quite poor yet.

REFERENCES

Asteriadis, G.A.: 1977, Astron. Astrophys. 56, 25

Balona, L.A., and Feast, M.W.: 1974, Monthly Notices Roy. Astron. Soc. 167, 621

Becklin, E.E., Neugebauer, G., 1968: Astrophys. J. 151, 145

Becklin, E.E., Neugebauer, G.: 1975, Astrophys. J. (Letters) 200, L71

Burton, W.B., Gordon, M.A.: 1976, Astrophys. J. (Letters) 207, L189

Caldwell, J.A.R., Ostriker, J.P.: 1981, Astrophys. J. 251, 61

Celnik, W., Rohlfs, K., Braunsfurth, E.: 1979, Astron. Astrophys. 76, 24

Crampton, D., and Georgelin, Y.P.: 1975, Astron. Astrophys. 40, 317

Dieckvoss, C.: 1978, Astron. Astrophys. 62, 445

Einasto, J.: 1972, Thesis, Tartu

Einasto, J.: 1974, Proc. First European Astr. Meeting 2, 291

Einasto, J.: 1979, IAU Symp. No. 84, 451

Einasto, J., Haud, U., and Joeveer, M.: 1979, IAU Symp. No. 84, 231

Einasto, J., Haud, U., Joeveer, M., Kaasik, A.: 1976, Monthly Notices Roy. Astron. Soc. 177, 357

Einasto, J., Kutuzov, S.A.: 1964, Tartu Astron. Obs. Teated, 10, 1

Einasto, J., Lynden-Bell, D.: 1982, Monthly Notices Roy. Astron. Soc. 199, 67

Einasto, J., Tenjes, P., Barabanov, A.V., Zasov, A.V.: 1980, Astrophys. Space Sci. 67, 31

Fricke, W.: 1977, Heidelberg Ver. 28, 1

Glass, I.S., and Feast, M.W.: 1982, Monthly Notices Roy. Astron. Soc. 198, 199

Gordon, M.A., Burton, W.B.: 1976, Astrophys. J. 208, 346

Gunn, J.E., Knapp, G.R., and Tremaine, S.D.: 1979, Astron. J. 84, 1181

Harris, W.E.: 1976, Astron. J. 81, 1095

Haud, U.: 1979, Pisma v Astron. Zhurn. 5, 124

Haud, U.: 1984, Astrophys. Space Sci. 104, 337
Haud, U., and Einasto, J.: 1983, IAU Symp. No. 100, 81
Joeveer, M.: 1974, Tartu Teated 46, 35
Joeveer, M., Einasto, J.: 1976, Tartu Astron. Obs. Teated 54, 77
Kinman, T.D., Wirtanen, C.A., Janes, K.A.: 1965, Astrophys. J. Suppl. Ser. 11, 223
Kinman, T.D., Wirtanen, C.A., Janes, K.A.: 1966, Astrophys. J. Suppl. Ser. 13, 379
Meinunger, I.: 1977, Astron. Nachr. 298, 171
Mihalas, D., Binney, J.: 1981, Galactic astronomy, Structure and kinematics. 2d Edition. San Francisco, W.H. Freeman and Company. 13+597 pp
Mould, J.R.: 1982, Annual Rev. Astron. Astrophys. 20, 91
Oort, J.H.: 1965, in Stars and Stellar Systems. Vol. 5: "Galactic Structure", A. Blaauw, M. Schmidt (Editors). Chicago, London, The University of Chicago Press, p. 455
Oort, J.H.: 1977, Annual Rev. Astron. Astrophys. 15, 295
Oort, J.H., Plaut, L.: 1975, Astron. Astrophys. 41, 71
Plaut, L.: 1966, Bull. Astron. Inst. Netherlands, Suppl.Ser. 1, 105
Plaut, L.: 1968a, Bull. Astron. Inst. Netherlands, Suppl. Ser. 2, 293
Plaut, L.: 1968b, Bull. Astron. Inst. Netherlands, Suppl. Ser. 3, 1
Plaut, L.: 1970, Astron. Astrophys. 8, 341
Plaut, L.: 1971, Astron. Astrophys. Suppl. Ser. 4, 75
Plaut, L.: 1973a, Astron. Astrophys. 26, 317
Plaut, L.: 1973b, Astron. Astrophys. Suppl. Ser. 12, 351
Quiroga, R.J.: 1980, Astron. Astrophys. 92, 186
Rohlfs, K., Kreitschmann, J.: 1981, Astrophys. Space Sci. 79, 289
Schmidt, M.: 1975, Astrophys. J. 202, 22
Woltjer, L.: 1975, Astron. Astrophys. 42, 109
Woolley, R., and Steward, G.: 1967, Monthly Notices Roy. Astron. Soc. 136, 329

DISCUSSION

J.P. Ostriker: What is the difference in the definition of the bulge and halo components in your model?

Haud: There is a difference in metallicity.

Ostriker: Isn't there a smooth variation from the one to the other seen in other galaxies?

Haud: There may be a smooth variation, but we define populations as physically homogeneous components and then, if there is some variation, we must represent this variation as the sum of two different populations.

S.M. Alladin: You have put the mass of the corona at 2×10^{12} $M_\odot$. Within which radius is this? And what is the truncated radius?

Haud: It is the total mass of this component, within the truncated radius of 390 kpc.

R. Wielen: What radius would contain half of the total mass? Is that 60 kiloparsec or more?

Haud: This radius is 96 kpc.

T.M.Bania: How did you correct the rotation curve for radial motions in the Galaxy?

Haud: This is explained in two papers: Haud (1979, Soviet Astron. Letters 5, 68) and Haud (1984, Astron. Astrophys., in press).

(Left to right) Salukvadze, Zvereva and Haud at the reception by the President of Groningen University CFD

MASS MODELS OF THREE SOUTHERN LATE-TYPE DWARF SPIRALS

C. Carignan
Mount Stromlo and Siding Spring Observatories
and
Kapteyn Astronomical Institute, Groningen

Although rotation-curve studies of spiral galaxies have unambiguously established the presence of dark matter, and theoretical studies have shown that its location is likely to be in a separate spheroidal halo component (Binney, 1978; Tubbs and Sanders, 1979; Monet, Richstone and Schechter, 1981), very little is known about its spatial distribution and its nature. Recently, Faber and Lin (Faber and Lin, 1983; Lin and Faber, 1983) have shown that, if one can get a rough idea of fundamental parameters like the halo scale length and the halo-to-disk ratio, it is also possible to put strong constraints on the nature of non-luminous matter.

One way to determine its spatial distribution, is to try to probe the gravitational potential as far out as possible through rotation velocities, and then, by using mass models, to subtract the contribution of the luminous matter to the potential. Assuming a constant M/L for the luminous disk, this can be done since the light distribution can then be transformed directly into a mass distribution. However, high-sensitivity HI observations are necessary, since optical velocities rarely extend past the disk-dominated region (e.g. see Kalnajs, 1983 for NGC 7217 and NGC 4378).

Late-type dwarf spirals are ideal candidates for determining the basic halo parameters for these galaxies. The contribution by dark matter to the total mass in the region that can be surveyed by 21-cm line emission is an order of magnitude larger than that of luminous matter. Moreover, since these systems have almost no bulge, one can, in principle, trace the distribution of dark matter over almost the entire galaxy; when a bulge component is present, it will tend to make V(r) flat in the inner regions, and it is difficult to determine reliably the contribution of the disk and of the bulge to the potential field, since each component has different M/L and mass distribution.

One-component models (Kalnajs, 1983) using Carignan's photometry (1983) and two-component disk-halo models (exponential disk and iso-

H. van Woerden et al. (eds.), The Milky Way Galaxy, 95–96.

thermal halo) are presented for the three Sculptor Group Sd galaxies NGC 7793, NGC 247 and NGC 300.

Only in the case of NGC 7793 does a one-component disk model with M/L = 2.5 (mass = 5.5×10^9 $M_\odot$) fit the observed rotation curve reasonably well. However, only optical velocities (Davoust and de Vaucouleurs, 1980) are available, and no conclusion can be reached before we get HI data to greater radii.

On the other hand, the two-component disk-halo models fit the three observed rotation curves fairly well with the following parameters: M/L_B = 2.0, 5.0, 2.5 for the luminous disk, core radius r_c = 2.5, 3.0, 4.0 kpc and one-dimensional velocity dispersion σ_h = 75, 80 and 60 km/sec for the dark halo of NGC 7793, NGC 247 and NGC 300 respectively. The mass of the luminous disk is almost identical for the three systems, at about 5.0×10^9 $M_\odot$. In these calculations, we have assumed distances of 3.13, 2.10 and 1.85 Mpc for NGC 7793, NGC 247 and NGC 300.

These models yield interesting quantities such as halo-to-disk mass ratios which at the Holmberg radii are 1.8, 3.3 and 0.8, and halo-to-disk scale-length ratios of 2.5, 1.3 and 2.0 in the same order.

More details on these models will be published elsewhere (Carignan and Freeman, 1983) and other pure-disk galaxies are being observed in HI at Westerbork.

REFERENCES:

Binney, J. 1978, M.N.R.A.S. 183, 779
Carignan, C. 1983, PhD thesis, Australian National University
Carignan, C. and Freeman, K.C. 1983, in preparation
Davoust, E. and de Vaucouleurs, G. 1980, Ap. J. 242, 30
Faber, S.M. and Lin, D.N.C. 1983, Ap. J. Lett. 266, L17
Kalnajs, A. 1983, in Internal Kinematics and Dynamics of Galaxies, ed. E. Athanassoula, I.A.U. Symposium no. 100, p. 87
Lin, D.N.C. and Faber, S.M. 1983, Ap. J. Lett. 266, L21
Monet, D.G., Richstone, D.O. and Schechter, P.L. 1981, Ap. J. 245, 454.
Tubbs, A.D. and Sanders, R.H. 1979, Ap. J. 230, 736.

THE GALACTIC CONSTANTS--AN INTERIM REPORT

F. J. Kerr
University of Maryland, College Park, Md., USA

At the IAU General Assembly in Patras in August 1982, Commission 33 set up a Working Group on the Galactic Constants. The Working Group is charged with developing a critical review of the values of the main galactic constants, for publication before the General Assembly in 1985. It has not been specifically charged to come up with a proposal for a revised set of values, although it can do so if it wishes.

The members of the Working Group are W. B. Burton, J. Einasto, M. W. Feast, F. J. Kerr (Chairman), D. Lynden-Bell (Vice Chairman), M. Mayor, M. Schmidt, R. Wielen. In addition, a number of consultants have been invited to join in the activity; these are L. Blitz, S. V. M. Clube, K. C. Freeman, G. R. Knapp, C. A. Murray, J. P. Ostriker, B. J. Robinson, V. C. Rubin, and A. R. Sandage.

A meeting of the Working Group was held in Groningen on May 29, the day before Symposium 106 began. This was attended by 7 members of the Group, 4 consultants, and 3 visitors. The purposes of the meeting were to review the present status of the various galactic constants, and to plan the development of the Working Group's report.

After a consideration of recent work on R_o, the Sun-center distance, it was clear that its value is probably lower than the presently accepted 10 kpc, but there is not yet a consensus of what it should be. Similarly, the circular velocity at the Sun, V_c, is probably lower than 250 km s^{-1}, but a value of 250 is still considered tenable by some. The Oort constants A and B are approaching agreed values, and two recent determinations approximate to $A = -B$, as is required for a flat rotation curve. Any system to be recommended should also follow the relation $V_c = (A - B) R_o$.

We also considered recent solar-motion solutions, and some local kinematical problems. The motions of young and older stars are clearly different, indicating that they are not in well-mixed orbits, a fact that affects any large-scale interpretation. Another major difficulty, especially in the study of galactic rotation, is that the Galaxy is

H. van Woerden et al. (eds.), The Milky Way Galaxy, 97–98.

asymmetrical in the outer parts, but it is not clear whether the asymmetry is structural or kinematical.

Some time was spent on discussing the purpose of any defined system of constants, and who will use the system. The current system was set up as a useful standardizing tool, for the presentation of data and for dynamical calculations. Any system can be expected to be most useful in the area of "applied kinematics," for facilitating comparisons, rather than for fundamental studies.

After hearing summaries of the present scientific evidence, and discussing the purposes of a standardized system, the May 29 meeting considered three possible options: (i) we know enough now to recommend a new system; (ii) we know enough now to recommend an alternative system for people to use in parallel with the present system, (iii) we cannot make a recommendation yet--we must collect more information, and reconsider the question in 1985. The meeting preferred option (iii).

The Working Group will now concentrate on preparing a report (of about 10 pages), in time for circulation to Commission 33 members, and perhaps publication, before the next General Assembly. The report will be developed by F. J. Kerr and D. Lynden-Bell, with the aid of contributions from other members. The next meeting of the Working Group will be in New Delhi in November 1985, probably just before the General Assembly.

One of the main reasons for setting up the Working Group was to encourage new work on any aspect of the galactic-constants problem. We therefore issue an appeal for information on recent work that we do not yet know about, on plans for future work in this field, and on new results as they come along.

DISCUSSION

G. Lyngå: Sometimes one needs the distance from the Sun to the plane of symmetry of the disk. While you are collecting information, would you mind including this item and making a recommendation?

Kerr: Sofar we had not thought of that, but indeed that is a good quantity to add on. Thank you. Incidentally, an observational system of galactic coordinates has to go through the Sun, by definition. It is convenient in fact that the Sun is close enough to the mean plane, or it was 25 years ago, that this z distance could be neglected.

J.P. Ostriker: On making A, B, R_o and V_c consistent, I think we want to preserve the virtues of arithmetic to better than the least-squares sense.

Kerr: That sounds like a wise remark.

(Above) Kerr pondering, and (below) Lynden-Bell chatting with Oort and Van der Laan, during the excursion.
LZ

Chini giving his paper CFD

GALACTIC ROTATION OUTSIDE THE SOLAR CIRCLE

R. Chini
Max-Planck-Institut für Radioastronomie
Bonn, West Germany

New radial velocities and optical distances for partly very distant HII regions give evidence for a rising rotation curve in the outer parts of the Galaxy.

The differential rotation in the outer parts of the Galaxy is a difficult problem of great consequence which has created two competing opinions: The model of a flat rotation curve is represented by Schmidt (1965), whereas recent results by Blitz et al. (1982) propose an increasing rotation curve. The data shown in the present paper contribute to this subject. The results, however, are preliminary in many respects and will be improved by additional observations.

The measurements of the radial velocities were carried out at the 100-m telescope of the MPIfR, Bonn, by observing the H112α recombination line. Radio-continuum maps or cross-scans have been superimposed on optical pictures (POSS) of the HII regions to locate possible candidates for the exciting stars. All stars lying within reasonable limits of the radio contours were investigated by UBVRI photometry at the 2.2-m telescope of the DSAZ, Spain, and at the 1.5-m telescope of the Steward Observatory, Arizona. Those stars which turned out to be of early type by the UBV photometry were studied spectroscopically at the 2.3-m telescope of the Steward Observatory, Kitt Peak. In this way 34 O and early B stars could be identified within 17 HII regions. The BVRI colours served as a check whether the normal interstellar reddening law can be applied to compute the visual extinction (see Chini and Krügel, 1983). It turned out that the normal value of R=3.1 could be used for all regions. The conversion of MK spectral types into absolute visual magnitudes was done by the calibration of Schmidt-Kaler (1981).

Table 1 gives the observational results. Along with the source name and galactic coordinates the distance d from the sun and the radial velocity v are given. The errors in d are estimated to be better than 20%; the accuracy of v is ±1 km s^{-1}. For some regions, where data on their exciting stars exist, optical observations have been taken from literature. For reasons of homogeneity their distances have been recomputed by using the luminosity calibration cited above.

H. van Woerden et al. (eds.), The Milky Way Galaxy, 101–104.

Table 1: Observations

Source	l	b	d[kpc]	v[km s^{-1}]
S100	70°.3	+1°.6	9.0[1]	-23.8
S121	90.2	+1.7	8.5	-55.3
S124	94.6	-1.5	4.0	-37.0
S127	96.3	+2.6	13.7	-96.8
S128	97.6	+3.2	6.9	-75.4
S142	107.1	-0.9	2.5[2]	-39.1
S148	108.3	-1.1	5.0[1]	-54.9
S152	108.7	-0.7	3.5[1]	-50.1
S155	110.2	+2.6	0.8	-7.8
S157	111.3	-0.7	4.7	-47.3
S159	111.6	+0.4	5.4	-67.2
S162	112.2	+0.2	3.8	-44.2
S168	115.8	-1.6	3.8[3]	-47.2
S184	123.1	-6.2	2.5[2]	-28.0
S201	138.5	+1.6	5.5	-34.3
S206	150.6	-0.9	3.3[4]	-25.2
S207	151.2	+2.1	9.8	-38.0
S208	151.3	+2.0	8.9	-30.9
S209	151.6	-0.2	11.3	-47.7
S211	154.6	+2.5	7.3	-35.1
S212	155.4	+2.5	4.8	-42.1
S217	159.2	+3.3	8.3	-21.2
S219	159.3	+2.6	5.1[4]	-30.2
S228	169.2	-0.9	4.2	-11.2
S285	213.9	-0.6	8.5[4]	+45.2
S288	218.7	+1.8	7.5	+56.8

Notes to Table 1: optical data from

1 = Crampton et al. (1978)
2 = Georgelin & Georgelin (1976)
3 = Georgelin et al. (1973)
4 = Moffat et al. (1979)

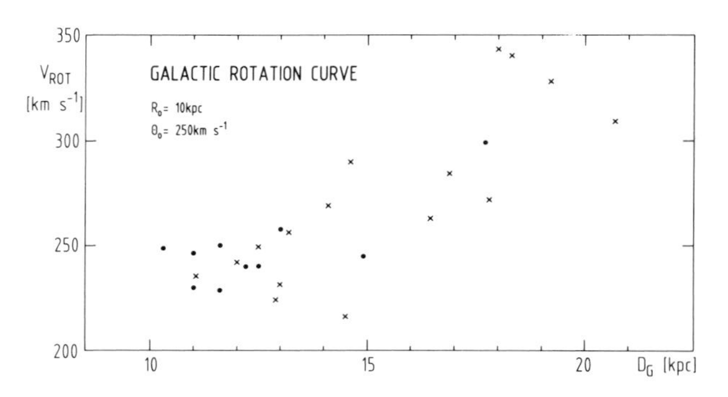

x = *present work*, • = *optical data from lit.*

Fig. 1: Rotation curve from galactic HII regions.

Figure 1 shows the rotation velocities of the HII regions as a function of the distance from the galactic center. If the observed velocities are due to galactic rotation, then Figure 1 clearly demonstrates that the rotation curve increases beyond 15 kpc from the center. There is no evidence that systematic errors in stellar distances could produce this effect. Non-circular motions, however, and the fact of incomplete galactic cover might be important. Likewise, a change of the rotation constants could produce a flat rotation curve.

ACKNOWLEDGEMENT. It is a pleasure to thank the director of the Steward Observatory, P. Strittmatter, for the generous allocation of observing time.

REFERENCES

Blitz, L., Fich, M., Stark, A.A.: 1982, Astrophys. J. Suppl. 49, 183
Chini, R., Krügel, E.: 1983, Astron. Astrophys. 117, 289
Crampton, D., Georgelin, Y.M., Georgelin, Y.P.: 1978, Astron. Astrophys. 66, 1
Georgelin, Y.M., Georgelin, Y.P.: 1976, Astron. Astrophys. 49, 57
Georgelin, Y.M., Georgelin, Y.P., Roux, S.: 1973, Astron. Astrophys. 25, 337
Moffat, A.F.J., Fitzgerald, M.P., Jackson, P.D.: 1979, Astron. Astrophys. Suppl. 38, 197
Schmidt, M.: 1965, in "Stars and Stellar Systems", Vol. 5, U. Chicago Press (Chicago), p. 513
Schmidt-Kaler, Th.: 1981, Landolt-Börnstein, Group VI/2d

DISCUSSION

P. Pismis: Have you checked whether there is a difference between the rotation curves obtained from the northern and southern regions of the anticentre? In the northern region there is a marked discrepancy between the rotation speeds in the Perseus Arm and the general rotation curve. Have you noted any such discrepancy?

Chini: From our material of observed HII regions we cannot find any difference between northern and southern objects. There are no HII regions in our sample which belong to the Perseus Arm.

L. Blitz: Have you reobserved any of the exciting stars previously observed by, for example, Moffat et al. as an overall check of your distance scale?

Chini: We have reobserved several HII regions. The results concerning the spectral types of the exciting stars agree with those of Moffat et al. Small differences occur in the distances obtained, but these are due to the different calibrations of spectral type vs. absolute magnitude used by Moffat et al. and by myself.

K.S. de Boer: How would your plot look if you had taken a galactocentric distance R_o = 8.5 kpc for the Sun, and a local rotation speed of Θ_o = 220 km/s?

Chini: As shown by Blitz, such a change of the rotation constants would steepen the rotation curve even further.

M.L. Kutner: The scatter in the results is larger than the error budget you outlined.

Chini: The large scatter in the diagram reflects, in my opinion, our poor knowledge of the absolute magnitudes of the earliest spectral types.

F. Bash: Have you assumed circular orbits for your HII regions?

Chini: Yes.

Bash: If, in fact, HII regions do not move on circular orbits, then the rotation curve you get will depend on the distribution of your sample of HII regions in longitude and in distance from the Sun.

B.G. Elmegreen: Would you have noticed a decrease in the ratio of total-to-selective extinction at larger distances from the galactic centre, as might be expected if the grain size is smaller in proportion to the lower metal abundance? You mentioned that the reddening appeared to be normal.

Chini: If you plot the observed colours of the stars, e.g. V-I vs. B-V, and the extinction law is normal, all stars earlier than about MO (without intrinsic excesses) lie on a straight line. Whenever a star does not follow that line, it might be due to a deviation from the normal extinction law. From our UBVRI observations we could check several colour-excess ratios; we did not find any deviations from the normal extinction law.

P. Pişmiş in discussion with Chini.
Foreground: T.P. de Zeeuw and E.F. van Dishoeck.
Beside Mrs. Pismis: S. Wramdemark, C.C. Lin and K. Begeman.
Behind: J.M. Dickey and H.S. Liszt.
In background: Ch. Terzides and H. van der Laan.

CFD

YOUNG GALACTIC CLUSTERS AND THE ROTATION CURVE OF OUR GALAXY

J.Hron and H.M.Maitzen
Institut für Astronomie der Universität Wien

The most recent determination of Oort's galactic rotation constant 'A' from open clusters was made by Taff and Littleton (1972). They obtained A=15 km/s/kpc, but unfortunately they omitted a detailed description of their cluster sample. For the present work we compiled a catalogue of O-B3 clusters for which radial velocities and distances are available. The individual cluster references given by Janes and Adler (1982) have been used to find best distances for the clusters. Radial velocities were taken from the list of Dr. Wramdemark (private communication) and from Hron et al. (1984). The catalogue contains 105 clusters distributed over the four galactic quadrants as 16:27:35:27.

Assuming circular motions in our Galaxy, the relation between the difference of the angular velocities $w(R)$ and $w(R_o)$ at the galactocentric radii R and R_o (=at the Sun) and the observed radial velocity corrected to the Local Standard of Rest V_r(LSR) is:

$$w(R)-w(R_o) = V_r(\mathrm{LSR})/(R_o \sin l \cos b).$$

'A' was computed from a weighted least-squares fit:

$$w(R)-w(R_o) = \mathrm{const} - 2(A/R_o)(R-R_o) - 2(\alpha/R_o)(R-R_o)^2$$

The weighting factor $\sin^2 l$ assigns low weight to clusters at longitudes where small deviations from circular motion or observational errors in V_r have large influence on $w(R)-w(R_o)$.

The results for 'A' depend sensibly on the interval of $R-R_o$ chosen for the fit:

The whole sample yields $A = 16.9 \pm 1.3$ and a statistically significant curvature term $\alpha = -2.1 \pm 0.8$. However, if we consider only clusters with $R-R_o > -1.5$ kpc and exclude those with lowest quality of V_r and distance, α becomes statistically insignificant and 'A' drops down to about 13 km/s/kpc.

H. van Woerden et al. (eds.), The Milky Way Galaxy, 105–106.

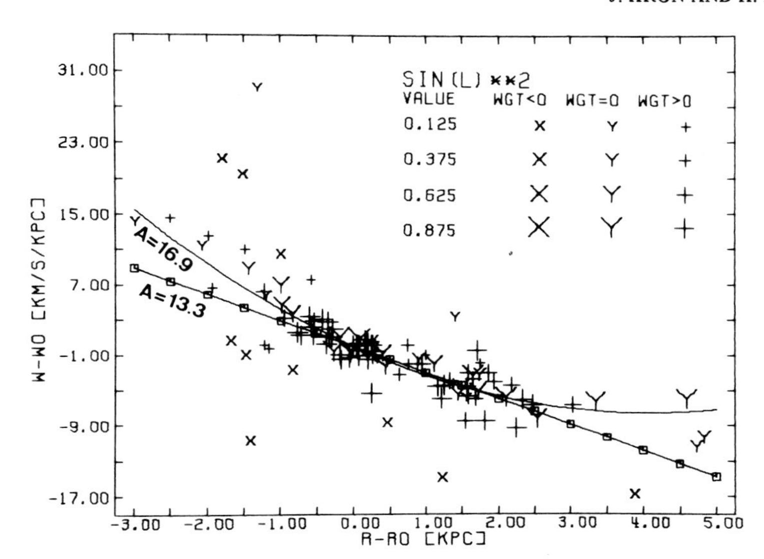

Fig. 1. The galactic rotation curve $w(R)-w(R_o)$ from young open clusters (R_o = 9 kpc and Standard Solar Motion adopted). Symbol size is proportional to $\sin^2\ell$ as indicated. The symbols denote 3 classes of clusters:
'Y' - clusters of lowest quality in V_r and distance.
'X' - clusters deviating more than 3 s.d. in the preliminary fit and excluded from the final fit.
'+' - clusters used in the final fit.
The two lines represent the cases described in the text.

Fig. 1 illustrates this situation: points left of $R-R_o = -1$ kpc exhibit a marked increase of scatter, which is largely due to the low sin l values associated. Nevertheless, due to the relatively high values of $w(R)-w(R_o)$ in this region, the curvature term must obviously become significant.

Whereas $A = 13.3 \pm 1.5$ derived for $-1.5 < R-R_o < 2.5$ kpc can be regarded as good local value of 'A', a more global definition of this galactic rotation constant (taking into account the curvature of $w(R)-w(R_o)$) should be based on an enlarged sample of clusters with larger $|R-R_o|$ and better quality than currently available for them.

REFERENCES:

Hron,H.,Maitzen,H.M.,Moffat,A.F.J.,Schmidt-Kaler,Th.: 1984 (in prep.)
Janes,K.,Adler,D.: 1982, Astrophys.J.Suppl.Ser. 49,425
Taff,L.G.,Littleton,J.E.: 1972,Astron.Astrophys. 21,443

HOW WELL DO WE KNOW THE ROTATION CURVE OF OUR GALAXY?

Paris Pişmiş
Instituto de Astronomia
Universidad Nacional Autonoma de Mexico

The existence of variations from a smooth curve, in the form of waves, in the rotation curves of galaxies was pointed out earlier, and an interpretation was proposed based on the argument that the waves were the manifestation of the coexistence of different populations in a galaxy (see for example Pişmiş 1965, 1974). Observations in the past few years have shown that "undulations" in the rotation curve of spiral galaxies are rather common phenomena; maxima and minima occur roughly at arm and interarm regions, respectively. The velocity fields of the majority of the 23 galaxies compiled by Bosma (1978) exhibit well-defined waves. In particular the velocity field in the 21-cm HI line of M81 by Visser shows clearly the correlation of the waves with the spiral structure.

It is reasonable to expect that our Galaxy will also have undulations in its rotation law. In fact radial velocities of HII regions (optical and CO velocities) are consistent with this expectation; moreover they show that the Sun is located close to a minimum.

Now, (i) we accept the above to be true and (ii) make the plausible assumption that the spiral structure can be represented by a pair of symmetrically located logarithmic spirals, with equations $R = a \exp b\lambda$ and $R = a \exp b(\lambda+\pi)$, respectively. Here R is the galactocentric distance and λ the galactocentric polar angle. From (i) and (ii) it follows immediately that the rotation curve of the Galaxy (and of other galaxies) is not unique, as it will be a function of λ. As λ varies the waves will gradually be displaced. The rotation curve may be smoothed out only when $\Delta\lambda$ reaches the value of 180° (see Fig. 11 in Pişmiş, 1981). In external galaxies an average rotation curve where the waves are smoothed out can be obtained easily, as one can observe the object at all central angles. But in our Galaxy this is not possible; the eccentric position of the Sun allows observation of the terminal velocities within $\Delta\lambda = 120^{\circ}$ at best.

H. van Woerden et al. (eds.), The Milky Way Galaxy, 107–108.

It is true that an overall rotation law can be estimated by having recourse to distances of galactic objects,but it will conceal the λ-dependence of the rotation curve. We suggest further that the cause of the well-known "north-south asymmetry" of the rotation curve should be sought for in the light of the arguments brought forth above.(For more details we refer to a review paper by Pişmiş, 1981).

REFERENCES

Bosma, A. 1978, Ph.D. Dissertation, University of Groningen
Pişmiş, P. 1965, Bol. Obs. Tonantzintla y Tacubaya 4, 8.
Pişmiş, P. 1974, Proceed. First European Astr. Meeting, Ed. B. Barbanis and J.D. Hadjidemetriou, p. 133.
Pişmiş, P. 1981, Rev. Mex. Astr. y Astrof. 6, 65.

P. Pişmiş and H.C. van de Hulst during lunch-break. In background: D. Leisawitz and K. Mead discuss a poster. CFD

GALACTIC ROTATION AND VELOCITY FIELDS

W.L.H. Shuter and A. Gill
University of British Columbia

Using unpublished 21-cm data from Jackson and Kerr of the entire Galaxy at $b = 0°$, we have determined the terminal velocities, v_m, to be (Gill & Shuter, 1983)

$$v_m = -173 - 72 \sin(\ell) + 88 \sin^2(\ell) \quad \text{km/s} \qquad 270° < \ell < 330°$$
$$= 162 - 31 \sin(\ell) - 116 \sin^2(\ell) \qquad 30° < \ell < 90°.$$

To obtain a rotation curve from the above expression for v_m, we first express the southern-hemisphere equation in equivalent northern-hemisphere terms (ie. change the signs of the first and third terms), and then average. Next we remove a constant offset of 13 km/s (as in Gunn, Knapp, & Tremaine, 1979). Finally, since at the tangent points

$$v_m = v_{rot} - v_\odot \sin(\ell), \qquad \text{and } R/R_\odot = \sin(\ell),$$

we find

$$v_{rot} = 154 + (v_\odot - 52)\, R/R_\odot - 102\, (R/R_\odot)^2 \quad \text{km/s} \qquad 0.5 < R/R_\odot < 1,$$

where v_{rot} is the circular velocity of rotation.

We can next substitute this equation into the Bottlinger expression for the line-of-sight velocity, v_1,

$$v_1 = (v_{rot}(R)\, R_\odot/R - v_\odot) \sin(\ell)$$

to get the VELOCITY FIELD

$$v_1 = (-52 + 154\, R_\odot/R - 102\, R/R_\odot) \sin(\ell) \quad \text{km/s} \qquad 0.5 < R/R_\odot < 1.$$

This is a very convenient expression to work with in analyzing 21-cm and CO data, as one need not know the values for $v_\odot$ and $R_\odot$ in order to apply the equation.

Two illustrations of the use of this expression for v_1 follow.

H. van Woerden et al. (eds.), The Milky Way Galaxy, 109–110.

Figure 1(a) shows the velocity field of the previous expression in a circle of radius 0.48 $R/R_\odot$ about the Sun. Figure 1(b) shows the velocity field of 990 B stars (Ovenden, Pryce, & Shuter, 1983) with the same velocity contours for a circle of 4.8 kpc. The reasonable match suggests $R_\odot \sim 10$ kpc.

In Figure 2, we have plotted the concentrations of gaseous material in the galactic disk. In this diagram, the galactic centre is at the centre and the Sun is at the top of the circle of radius $R_\odot$. In the outer Galaxy ($R > R_\odot$), we have used the previously mentioned 21-cm data. In the inner Galaxy ($R < R_\odot$), we have used CO data from the Columbia Survey in the northern hemisphere and the CSIRO survey in the southern hemisphere. No data has been plotted within ± 10° of the centre-anticentre line, for $R < 0.4\ R_\odot$, and in the inner Galaxy for $60° < \ell < 90°$ and $270° < \ell < 300°$. In the inner Galaxy, we have plotted the material at both kinematic distances. Finally, we have corrected for an outward motion of the LSR of 5 km/s.

REFERENCES

Gill, A., & Shuter, W.L.H.: 1983, MNRAS, in press.
Gunn, J.E., Knapp, G.R., & Tremaine, S.D.: 1979, Astron. J., 84, 1181
Ovenden, M.W., Pryce, M.H.L., & Shuter, W.L.H.: 1983 in "Kinematics, Dynamics and Structure of the Milky Way", ed. Shuter, W.L.H., (Dordrecht, Holland : Reidel), p. 67-72.

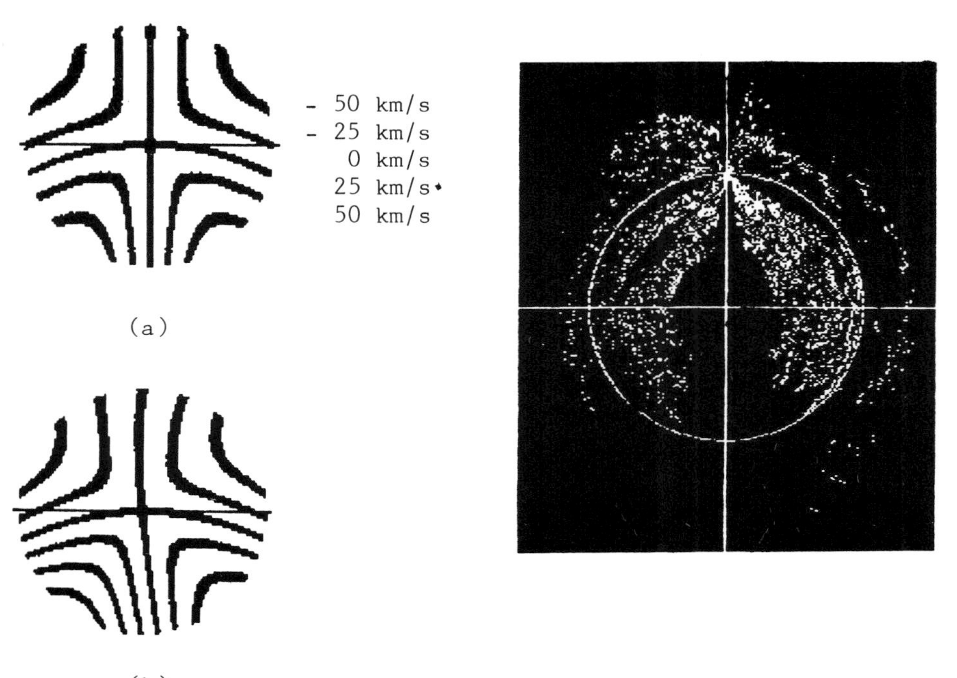

(a)

(b)

Figure 1

Figure 2

SECTION II.2

THE STELLAR COMPONENT

Tuesday 31 May, 1020 - 1320

Chairman: M.W. Feast

M.W. Feast pouring wine at conference dinner. With him, counter-clockwise around the table: G.P. Illingworth, R.H. Sanders, J.V. Feitzinger, N. Reid, C. Carignan, E. Sadler, Mme. Carignan LZ

THE OLD POPULATION

K.C. Freeman
Mount Stromlo and Siding Spring Observatories
Research School of Physical Sciences
The Australian National University

ABSTRACT

We review the kinematical and chemical properties of the old population of our Galaxy. Comparison is made with the properties of the old population in other disk galaxies.

INTRODUCTION

Disk galaxies like the Milky Way have two visible components, the disk and the spheroidal component. Both contain old stellar populations, which are the subject of this review. There is probably also an invisible component, which is probably old. For the Milky Way, this invisible component is discussed in the papers by Schmidt and Lynden-Bell. In some ways, this component is easier to study in other galaxies: see the paper by Carignan, for example.

THE SPHEROIDAL COMPONENT

In other galaxies, the spheroidal component (or bulge) appears as a single component. Its surface brightness distribution follows roughly the $R^{\frac{1}{4}}$-law, and it usually shows a radial color gradient, which is interpreted as an outward decrease in the mean metallicity. For the Galaxy, the nuclear bulge and the outer metal-weak halo are sometimes regarded as separate components. I suggest that it is more useful to consider the spheroidal component as a single dynamical system, with a chemical gradient. The inner more bound parts are relatively metal-rich, while the more energetic objects of the outer parts are more metal-weak.

Table 1 lists some of the gross properties of the Milky Way, for comparison with other galaxies. The estimates in this Table come from a compilation by de Vaucouleurs (1982: dV) from many sources. From these estimates, the Galaxy appears normal for its type. The absolute magnitudes

H. van Woerden et al. (eds.), The Milky Way Galaxy, 113–122.

TABLE 1

THE MILKY WAY

R_o	8.5 ± 0.5 kpc	sun-center distance
$V(R_o)$	220 ± 15 km s^{-1}	rotational velocity
$\sigma(0)$	130 ± 7 km s^{-1}	central bulge dispersion
$M_T(B)$	-20.2 ± 0.15	total integrated magnitude
$(B-V)_T$	0.53 ± 0.04	and color
$M_I(B)$	-18.2 ± 0.3	bulge integrated magnitude
$(B-V)_I$	0.65 ± 0.05	and color
$R_e(I)$	2.7 ± 0.3 kpc	bulge effective radius

of the disk and bulge are close to the average for Sbc systems (Simien and de Vaucouleurs, 1983), and the rotational velocity V_o of the disk and the central velocity dispersion $\sigma(0)$ of the bulge are close to those expected from the absolute magnitudes of these two components (dV; Kormendy and Illingworth 1982). We see that the spheroidal component is fairly small: it provides only about 15 percent of the Galaxy's integrated blue luminosity, and its effective (half-light) radius is only 2.7 kpc.

We now look at the kinematics of the spheroidal component in more detail. Table 2 gives the line-of-sight velocity dispersion for objects in the spheroidal component, at different distances from the galactic center. Sources for most of this data can be found in dV. Two of the measurements are recent. The value for the Sgr star clouds comes from direct measurement of the velocity dispersion from the integrated light of patches in these star clouds (de Vaucouleurs, Freeman and Wainscoat, to be published). The dispersion for giants at R = 25 kpc comes from Ratnatunga's study of giants of the outer halo (also to be published). We see from Table 2 that the spheroidal component is nearly isothermal out to about 60 kpc from the galactic center ($>$ 20 effective radii).

From the work of Kormendy and Illingworth (1982), we would expect the inner parts of the spheroidal component to rotate; the maximum expected mean rotational velocity is about $0.7\sigma(0)$ = 90 km s^{-1}. This is roughly consistent with the kinematics of the planetary nebulae near the galactic center (Kalnajs and Webster, to be published). We also have an estimate of the rotational velocity of the spheroidal component near the sun, from observations of stars in the solar neighborhood. The sun is about 3 effective radii from the center, so we would expect the rotation to be small near the sun. Also, we would expect the spheroidal

Table 2

VELOCITY DISPERSION OF SPHEROIDAL COMPONENT
(one-component σ)

OBJECT	R(kpc)	σ(km s^{-1})
Planetary nebulae	< 1	130
OH/IR sources	near center	135
Long-period variables	" "	112
Late M stars	" "	113
Sgr star clouds	" "	120
RR Lyrae stars	2	125
" " "	8	140
Metal-weak giants, subdwarfs[1]	8	120
Globular clusters		120
Distant giants	25	124
Distant Palomar clusters, dwarf spheroidals[2]	60	125

[1]From Norris, unpublished

[2]From Hartwick and Sargent (1978)

component stars near the sun to be relatively metal-weak. This is nicely illustrated by Yoshii and Saio (1979). Their figure 1b shows the V-velocity of stars (i.e. the component in the direction of galactic rotation) against the ultraviolet excess $\delta(0.6)$. For the metal rich stars (low values of δ), fairly rapid rotation and low velocity dispersion are seen; these are the stars of the old disk. At $\delta(0.6) = 0.15$, a fairly abrupt transition occurs, to a much hotter population with low mean rotation and lower mean metal abundance. This population is the spheroidal component. The value of the ultraviolet excess at which the transition from old disk to spheroidal component occurs near the sun corresponds to an abundance [Fe/H] = -0.6. Although many of the stars in the Yoshii-Saio sample were discovered kinematically, this does not appear to introduce a serious bias on the estimates of the mean rotational velocity for the spheroidal component stars in the solar neighborhood. A recent study by Norris (unpublished) of the kinematics of a large sample of spectroscopically selected metal-weak stars shows that their mean rotational velocity is very similar to that for the kinematically selected subdwarfs and also for the globular-cluster system (Frenk and White, 1981). All of these objects show a low mean rotational velocity of about 50 km s^{-1} (referred to a nonrotating frame).

The globular clusters and the metal-weak stars are both part of the spheroidal component. It is interesting to compare their chemical and kinematical properties. We have already seen that their mean rotational velocities are similar. There is an apparent difference in the distributions of [Fe/H] of the clusters and the stars. Field stars have been found with [Fe/H] less than -3, while the most metal-weak clusters have [Fe/H] ≃ -2.4. Is this difference significant ? Hartwick (1982) argues that it is probably statistical; both distributions are consistent with his chemical-evolution model, which has a zero initial metal abundance. (There are some well-known CN-abundance anomalies in globular-cluster stars, which are not so evident in the corresponding metal-weak field stars: see Kraft (1982). These anomalies are not yet fully understood. They may have to do with local processes in globular clusters, and are probably not relevant to this discussion).

Table 3 compares the velocity dispersion components for the nearby halo subdwarfs, halo giants, RR Lyrae stars and the globular-cluster system (from Hartwick, 1982). For the stars, the velocity dispersion is clearly anisotropic. On the other hand, Frenk and White's (1981) solution for the globular-cluster system produces an isotropic velocity dispersion. How seriously should we take this kinematical difference

TABLE 3

VELOCITY DISPERSIONS FOR GALACTIC HALO OBJECTS
(km s^{-1})

OBJECT	σ_R	σ_ϕ	σ_θ
Subdwarfs	178 ± 22	111 ± 39	106 ± 32
Giants	140 ± 16	108 ± 23	55 ± 31
RR Lyrae stars	145 ± 19	124 ± 22	71 ± 26
Globular clusters	118	118	118

(See Hartwick 1983 for references)

between the clusters and the halo field stars ? Fairly seriously, I believe: a quite independent analysis by Seitzer and Freeman (1982) showed that the distribution of orbital eccentricities for the clusters is consistent with an isotropic velocity distribution. They used the clusters' tidal radii to estimate the perigalactic distances R_{min}. The distribution of R_{min}/R for the clusters (where R is the present galactocentric distance) is a sensitive estimator of the distribution of orbital eccentricities. It is not yet clear that this kinematical difference between field stars and globular clusters was set up at the time of their formation. It is possible that the clusters originally had an anisotropic velocity dispersion also, and that the clusters in the more radial orbits have been destroyed subsequently by the galactic tidal field, as suggested to me by Ostriker and others.

The question of a radial abundance gradient in the spheroidal component is not yet settled. Among the globular clusters and RR Lyrae stars, the more metal-rich objects are found near the galactic center. Beyond about 8 kpc from the center, however, there is not much evidence for a radial gradient in their mean abundance. However there is a large spread in their abundance distributions in these outer regions; see Sandage (1981) for a summary. Some indirect evidence for a chemical gradient comes from the subdwarfs in the solar neighborhood. Eggen (1979) shows how the apogalactic distance for these subdwarfs increases with decreasing metal abundance. The more metal-rich subdwarfs mostly have small apogalactic distances, close to 10 kpc. Large apogalactic distances are seen only for the metal-weakest stars. This is consistent with a chemical gradient in the outer spheroidal component, but does not necessarily imply one. For example, it could be that the metal-weak stars with large apogalactic distances also have large orbital eccentricities (so that some are seen in the solar neighborhood), and that there are relatively metal-rich stars in low-eccentricity orbits far from the galactic center; these stars would of course not be seen near the sun.

The most direct way to settle the question of the abundance gradient in the outer spheroidal component is to study the chemical properties of halo stars that are now in these outer regions of the galaxy. Ratnatunga is now completing a program to discover halo giants spectroscopically, at distances of up to 40 kpc from the sun, to measure directly their chemical abundances and motions. This program provides a sample of distant halo objects, independent of the globular clusters and RR Lyrae stars. Preliminary results suggest that there is a significant gradient in the [Ca/H] values for these stars.

Finally we consider again the globular cluster system. We have seen so far that the system rotates slowly, has an apparently isotropic velocity distribution, and includes some relatively metal-rich clusters in the inner 8 kpc. The frequency distribution of abundances for the cluster system is clearly bimodal (see Freeman and Norris 1981, Figure 3). There is some evidence (see Zinn 1980 for example) that the clusters of the metal-richer mode belong to a disklike subsystem in the inner parts of the galaxy. Comparison with the cluster system of M31 supports the reality of this disklike system of metal-rich globular clusters. In M31, the metal-rich clusters lie in a rapidly rotating disk, within 10 kpc of the center. The metal-weaker clusters are in a slowly rotating system, as in the Milky Way (see Freeman 1983 for a review).

For $R > 8$ kpc in the Galaxy, we see only clusters of the metal-weaker mode: this mode includes clusters with $-1.2 > [Fe/H] > -2.3$. In this region of the Galaxy, there is a striking dependence of the orbital properties of the clusters on their abundance. Seitzer and Freeman (1982) showed (using the cluster tidal radii) that the more metal-rich of these clusters are all in highly elongated orbits; the more metal-poor clusters have orbits of all eccentricities. A similar effect is seen in M31, among the clusters of the slowly rotating subsystem mentioned above. From the frequency distribution of the cluster radial velocities, it appears that the metal-richer clusters of this subsystem are again in more elongated orbits than the metal weaker clusters.

In summary, the gross properties of the galactic spheroidal component are fairly much like those for similar spirals. The spheroidal component appears to be closely isothermal out to at least 60 kpc from the galactic center. Its abundance gradient is not yet clearly established. There are some interesting kinematical differences between the field stars and the globular clusters, and also between the metal-richer and metal-weaker clusters: these should be helpful for understanding the formation and evolution of the globular cluster system.

THE OLD DISK

The old disk is a major part of the old population of the galaxy. We will discuss some aspects of its structure here. First we should look at the structure of the disks of other galaxies. van der Kruit and Searle (vdKS: see van der Kruit and Searle, 1982, for references) have made surface photometry of several edge-on galaxies, and found a semi-empirical law to describe the luminosity volume-density distribution in their disks. This law has the form

$$L(R,z) = L_o \exp(-R/h)\ \mathrm{sech}^2(z/z_o) \qquad R < R_{max}$$
$$= 0 \qquad R > R_{max}$$

where h and z_o are radial and z-lengthscales. For $z/z_o >> 1$, $\mathrm{sech}^2(z/z_o) \sim \exp(-2z/z_o)$, so the usual exponential z-scaleheight is about $z_o/2$. This disk is exponential radially. In the z-direction, it has the structure of a locally isothermal sheet, if we assume that the mass density $\rho(R,z)$ also follows this semi-empirical law. The local z-velocity dispersion σ_z is then given by

$$\sigma_z^2 = z_o^2\ 2\pi G\rho(R,0).$$

This luminosity distribution $L(R,z)$ is an excellent fit to the photometry of bulgeless disk galaxies. It is particularly interesting that the z-lengthscale z_o is independent of radius: this must be explained by theories for the heating of the stellar disk. Even from galaxy to galaxy, z_o does not change much. Table 4 gives values for some of the photometric parameters, for several galaxies studied by vdKS. V is the maximum observed rotational velocity, which varies from 95 to 255 km s^{-1} for this sample. Despite the range in V (and hence in total luminosity), z_o is relatively uniform at about 700 pc, which corresponds asymptotically to an exponential scaleheight of about 350 pc. In Table 4, $L_{o,J}$ is the value of L_o in the blue J-band, and the column headed "G" gives

TABLE 4

PHOTOMETRIC PARAMETERS FOR DISK GALAXIES

(from vdKS)

NGC	4244	5907	5023	4565	891	G
Type	Scd	Sc	Sc	Sb	Sb	Sbc
$L_{o,J}$ (10^{-2} $L_\odot$ pc^{-3})	3.2	3.3	4.1	4.5	2.4	4.0
h (kpc)	2.6	5.7	2.0	5.5	4.9	5.0
z_o (kpc)	0.6	0.8	0.5	0.8	1.0	0.7
h/z_o	4.5	6.9	4.3	7.0	5.0	7.1
R_{max}/h	5.3	3.4	3.9	4.5	4.3	4.4
V (km s^{-1})	115	210	95	255	225	220

estimates for our Galaxy: see vdKS for details.

For edge-on disk galaxies with even small bulges, the L(R,z) law above is no longer an excellent fit at all values of z. Near the galactic plane, the law fits well. However at larger z-heights, there is a clear excess of light above the $\text{sech}^2(z/z_o)$ law, even at large values of R, far from the central bulge. This excess of light appears similar to the thick disks identified by Burstein (1979) for edge-on SO galaxies. From the work of vdKS, this thick disk is seen in edge-on spirals only when a spheroidal component is also seen, so it is probably associated in some way with the spheroidal component. The thick disk could be the spheroidal component itself, responding to the flat potential of the disk (Jarvis and Freeman, preprint) or it could be an intermediate population that formed as the spheroid formed. There is some evidence from vdKS that the thick disk is slightly bluer (or more metal-weak) than the regular old disk. It seems important to understand what this thick disk is dynamically, and how it fits in to the galactic formation and disk heating pictures. The main point of this discussion here, however, is to warn us what to expect for the vertical structure of the Galaxy near the sun.

We turn now to the star-count data of Gilmore and Reid (1983) at the SGP. They measured I magnitudes and V-I colors for 12500 stars to I = 18 and, assuming that all these stars are mainsequence stars, they constructed the vertical number-density profile N(z). One interpretation of the structure seen in this N(z) profile is that it is the sum of two exponentials: one has a scaleheight of about 300 pc and dominates up to about z = 1 kpc, and the other has a scaleheight of about 1450 pc and contributes about 2 percent to the local density. Gilmore and Reid suggest that these two components may represent the regular old disk and the thick disk, respectively. The scaleheights would then be comparable to those seen in other galaxies. This particular analysis and interpretation remains contentious, however. Bahcall et al.(1983) argue that such a thick disk is inconsistent with other star-count data. It seems important now to resolve this question; if there is indeed a thick disk locally, then we have an excellent opportunity to study its chemical and kinematical properties directly. This would be a great help in understanding the nature of the thick disks seen in other disk galaxies.

Let us proceed, however, and assume that there are really two disklike components, with scaleheights of 300 and 1450 pc near the sun. The ratio of their z-velocity dispersions should then be about 2.2. If the velocity dispersion for the regular (300 pc) old disk is 20 km s^{-1}, then we would expect the 1450 pc component to have a velocity dispersion of about 45 km s^{-1}. Recently, Hartkopf and Yoss (1982) have published a chemical and kinematic survey of G and K stars at the galactic poles. While this survey does not claim to be complete, it makes some interesting points. The survey includes stars up to about 5 kpc from the galactic plane. Within a few hundred parsecs of the plane, all their stars have [Fe/H] > -0.5, and they identify this abundance range as normal (old disk ?). These normal-abundance stars are found not only near the plane, but also up to at least 5 kpc from the plane. They then look at the run of velocity dispersion with z for the normal-abundance ([Fe/H] > -0.5) stars and also for the metal-weaker stars ([Fe/H] < -0.5). The normal-abundance sample is almost isothermal up to about 3 kpc from the plane, with a velocity dispersion of 20 km s^{-1}. This is close to the usual old-disk value. The isothermal behaviour of this old disk is very interesting: recall the locally isothermal model of van der Kruit and Searle for the old disks of other spiral galaxies. The metal-weaker sample of Hartkopf and Yoss is also isothermal up to about 5 kpc, with a velocity dispersion of about 45 km s^{-1}. (Beyond 5 kpc the dispersion for this sample increases, probably due to a preponderance of population II giants in the sample at this distance.)

The close agreement of these two velocity dispersions (20 and 45 km s^{-1}) for the metal-rich and metal-weaker samples with the prediction at the beginning of the previous paragraph is interesting. It tempts us to identify the two isothermal subpopulations of Hartkopf and Yoss with the thin disk - thick disk structures of Gilmore and Reid and vdKS. The ratio of the velocity dispersions is in accord with the Gilmore and Reid scaleheights, and the chemical abundances correspond at least qualitatively to the vdKS color differences between thin and thick disks.

Finally we discuss briefly the presence of normal abundance stars far from the galactic plane. Hartkopf and Yoss found normal G and K giants up to 5 kpc from the plane. However there are also younger stars of normal abundance at similar heights. Rodgers (1971) studied 62 A stars at the SGP, with z-distances between about 1 and 4 kpc. About half of these have [Ca/H] > -0.5. Detailed spectrophotometry showed that they are not field horizontal-branch stars, but rather main sequence or slightly evolved stars, with ages of about 10^9 years. Hartkopf and Yoss propose a common origin for these high-z normal-abundance A and GK stars, perhaps resulting from an encounter with a third Magellanic system, as suggested by Rodgers et al. (1981). We recall, however, that the normal-abundance sample of GK stars has an isothermal velocity dispersion of about 20 km s^{-1} up to at least 3 kpc from the plane. At large z, this dispersion is of course determined by the normal-abundance high-z stars which we have just discussed. If their origin is different, as these authors suggest, then it would be worth understanding why their velocity dispersion is so similar to the dispersion of the other normal abundance GK stars near the galactic plane.

REFERENCES

Bahcall, J.N., Schmidt, M., Soneira, R.M. 1983. Ap.J., 265, pp 730-47.
Burstein, D. 1979. Ap.J., 234, pp 829-36.
de Vaucouleurs, G. 1982. Preprint.
Freeman, K.C. 1983. Internal Kinematics and Dynamics of Galaxies (IAU Symposium 100), ed E. Athanassoula (Reidel, Dordrecht) pp 359-64.
Freeman, K.C., Norris, J.E. 1981. Annual Reviews Astron. Astrophys., 19, pp 319-56.
Frenk, C.S., White, S.D.M. 1980. M.N.R.A.S., 193, pp 295-311.
Gilmore, G., Reid, N. 1983. M.N.R.A.S., 202, pp 1025-47.
Hartkopf, W.I., Yoss, K.M. 1982. A.J., 87, pp 1679-1709.
Hartwick, F.D.A. 1982. Preprint.
Hartwick, F.D.A., Sargent, W.L.W. 1978. Ap.J., 221, pp 512-20.
Kormendy, J., Illingworth, G.D. 1982. Ap.J., 256, pp 460-80.
Kraft, R. 1982. Preprint.
Rodgers, A.W. 1971. Ap.J., 165, pp 581-97.
Rodgers, A.W., Harding, P., Sadler, E. 1981. Ap.J., 244, pp 912-8.
Sandage, A. 1981. A.J., 86, pp 1643-57.
Seitzer, P., Freeman, K.C. 1982. To be published.
Simien, F., de Vaucouleurs, G. 1983. Internal Kinematics and Dynamics of Galaxies (IAU Symposium 100), ed E. Athanassoula (Reidel, Dordrecht) pp 375-6.
van der Kruit, P., Searle, L. 1982. Astron.Astrophys., 110, pp 61-78.
Yoshii, Y., Saio, H. 1979. P.A.S. Japan, 31, pp 339-68.
Zinn, R. 1980. Ap.J., 241, pp 602-17.

DISCUSSION

J.P. Ostriker: The present isotropy of the globular-cluster velocity distribution (as compared to say the RR-Lyrae distribution) may indeed well be due to the destruction of clusters on eccentric orbits which pass near the centre of the Galaxy. In papers with Spitzer and Tremaine several years ago I found that massive clusters which come near to the centre would be dragged in by dynamical friction, and the low-mass (low-density) clusters destroyed by tidal shocks.

A.A. Stark: If a thick disk exists in our Galaxy, would it be as bright as the thick disks seen in other galaxies? Or are you implying that there could be thick disks in all galaxies but they are just too faint to be seen?

Freeman: A thick disk makes no major contribution to the local density, only about 2%; but to the column density it is more like 10%.

Stark: Then if we looked from M31, could we see our thick disk?

Freeman: Yes, we would see our thick disk as we see it in other galaxies.

M.W. Feast: In your Table 2, the OH-IR stars have a velocity dispersion similar to that of the spheroidal component. However, I believe they are much more concentrated to the plane.

Freeman: An abstract by Habing and others (unpublished) claims that the OH-IR sources near the Galactic Centre are part of the disk population. But one may say the same for the planetaries. I just don't know. Maybe the disk has a velocity dispersion of about 100 km/s near the centre?

(Left to right) Clube, Van der Laan, Freeman and Hartwick during boat-trip. LZ

NEAR-INFRARED STUDIES OF THE MILKY WAY

H. Okuda
Institute for Space and Astronautical Science
Komaba Tokyo 153, Japan

The inner region of the Galaxy has been explored by means of near-infrared observations; the distribution and population of the stars are studied from the near-infrared brightness mapping and star counts in the Milky Way, while the magnetic-field configuration is probed by the near-infrared polarimetry.

1. DISTRIBUTION AND POPULATION OF THE STARS

A series of balloon observations of the 2.4μm brightness distribution in the galactic plane have thrown light on the stellar distribution in the inner Galaxy (e.g. Okuda 1981). An interesting result was the fact that there exists an active arm located at about 5 kpc from the Galactic Center. From the extremely strong concentration of the emission into the galactic plane, it has been suggested that the contributing stars are a young population, rich in M-supergiants (Maihara *et al.* 1978, Hayakawa *et al.* 1977, 1981). It was however hard to identify them to specific type of stars only from the observations of integrated radiation in the wide field of view, and in a single band.

As a follow-up observation, we have made near-infrared source counts by using a multi-color photometer with I, H, K, and L bands that should be more informative for the identification. The source counts were carried out in 17 strips across the galactic plane between $\ell=350^{\circ}$ and $\ell=45^{\circ}$. Preliminary results were reported elsewhere (Kawara *et al.* 1982).

The results of the analysis are as follows:
1) General behaviour of the number density distribution of the sources is similar to the surface-brightness distribution observed by the balloons.

2) A conspicuous enhancement in the number density is found in the direction of the Galactic Center. This must be associated with the nuclear bulge, and if so, the absolute K-magnitude is estimated to be brighter than -8 mag, applying the distance modulus of 15 mag to the observed K-magnitude, 6-7. This is compatible with very luminous M-giants. They

H. van Woerden et al. (eds.), The Milky Way Galaxy, 123–126.

are more tightly concentrated towards the Galactic Center than OH/IR sources or planetary nebulae.

3) At the midst of the Galactic Center, within $\pm$ 0.2^{o}, there is a spiky concentration of the sources, mostly with large reddening. Their absolute K-magnitude should be brighter than -10 if corrected for the distance modulus and the reddening. This indicates a clustering of M-supergiants in the galactic nucleus.

4) As was found in the 2.4μm mapping, a remarkable concentration of the sources is seen in the region ℓ=26° - 28°. The sources concentrate more strongly to the galactic plane (FWHM 2°) than the 2.4μm surface brightness. A considerable fraction of the sources are highly reddened. After correction for reddening, the absolute K-magnitude of the sources is brighter than -9, compatible with very luminous M-giants or M-supergiants.

In order to make these conclusions more quantitative, we have tried to build a comprehensive model to explain the observed results of the source counts as well as the 2.4μm mapping (Kawara and Okuda, in preparation). After several trials, we find that the following assumptions are necessary for a reasonable fit to the observations.

1) In the nuclear bulge, there exists an additional concentration of luminous M-giants ($M_K \leq -9$); their relative frequency is 100 times that in the solar neighbourhood.

2) An overpopulation of luminous M-giants ($M_K \leq -9$) by a factor of 20 is also present in the 5-kpc arm.

3) The scale height of the z-distribution of the stars decreases towards the Galactic Center, where it is almost halved relative to that in the 5-kpc arm.

In Figure 1, the observed results for the latitude dependence of the surface number density are compared with the model calculations without (dotted line) and with (solid line) the extra components assumed above. The enhancement in the bulge and the clustering in the region ℓ=26°-28° cannot be explained without the additional bright sources. From these analyses, it is concluded that the stars with high infrared luminosity are overabundant in the bulge as well as in the 5-kpc arm. Their luminosity indicates that they correspond to M-supergiants or to the upper end of the asymptotic branch of M-giants.

2. INTERSTELLAR POLARIZATION

Interstellar polarization has been used for diagnoses of the magnetic field in the Galaxy. However, the observations so far made have been mostly in the visible range and limited to the solar neighbourhood, due to strong interstellar extinction.

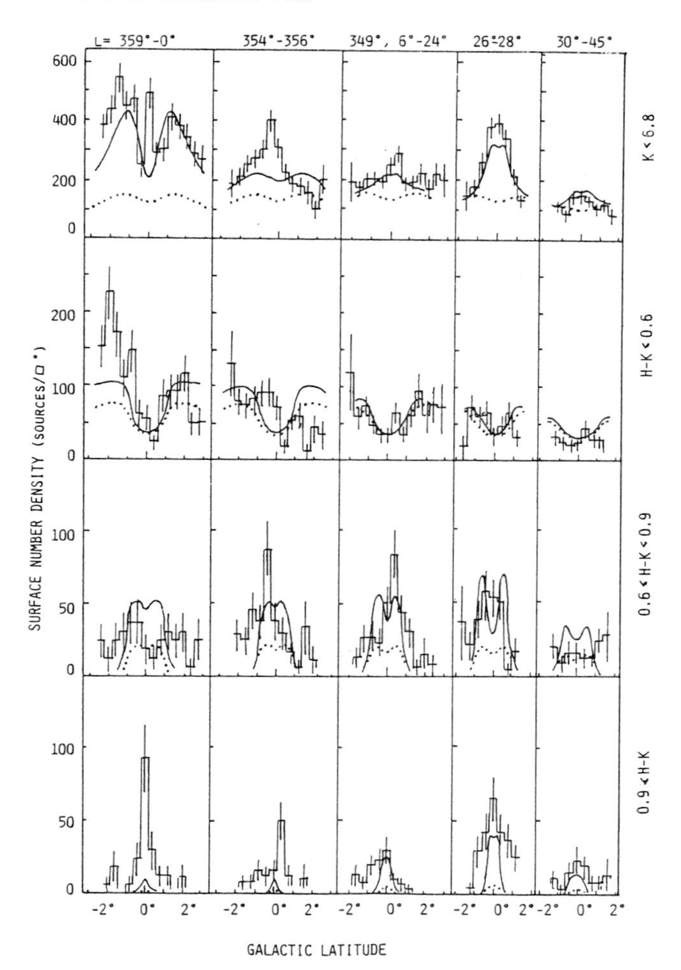

Figure 1. Surface density of infrared sources as a function of galactic latitude in various longitude intervals (top scale). Uppermost panels: total numbers of bright infrared sources. Other panels: distributions for sources in various infrared-colour ranges. Dotted and solid lines represent model calculations (see text).

By expanding the observations into the near infrared, we have succeeded to look into deep space as far as the central region of the Galaxy. Polarization was measured in the K-band for the sources in the galactic plane at $\ell=0^{\circ}$, 20°, and 30°, together with photometry in the J, H, K, and L-bands.

The observed polarizations are preferentially aligned parallel to the galactic plane and well correlated with the interstellar extinction, represented by the H-K index. The dependence of the polarization on H-K varies with galactic longitude as shown in Figure 2.

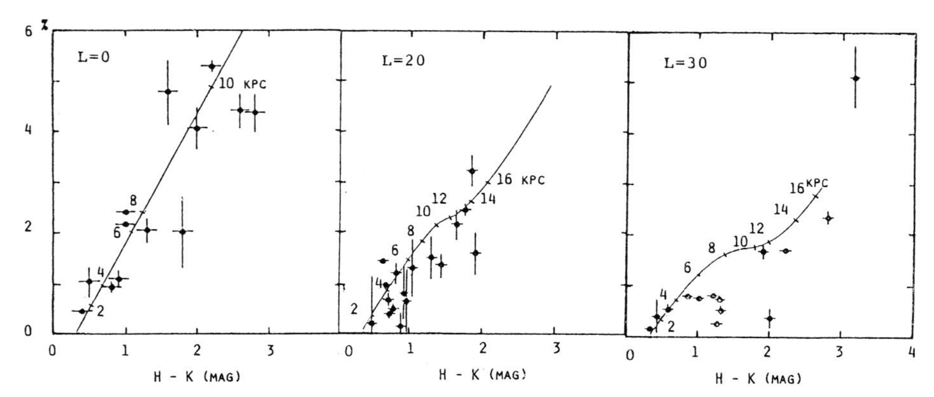

Figure 2. H - K dependence of the observed polarization in 3 longitude regions. The expected dependences are shown by curves, with distance scales estimated from the HI and CO data.

The variation can be well explained if we assume that the magnetic field runs concentric around the Galactic Center, and that the polarizing efficiency depends on the relative angle of the magnetic line of force and the line-of-sight. By using the gas density distribution in the inner Galaxy, estimated from the HI and CO observations, we correlate the H-K index with the distance, and it is found that our observations reach distances far beyond 10 kpc. This is the first evidence for the presence of a uniform magnetic field in the inner Galaxy, similar to that found in the solar neighbourhood (Kobayashi *et al*., in preparation).

Since our Galaxy is the nearest and largest edge-on galaxy, it should be an irreplaceable sample for detailed studies of galactic structure, particularly in the z-direction. As is shown above, near-infrared observations can become a powerful tool for such investigation.

REFERENCES

Hayakawa, S., Ito, K., Matsumoto, T., and Uyama, K.: 1977, Astron. Astrophys. 58, 325

Hayakawa, S., Matsumoto, T., Murakami, H., Uyama, K., Thomas, J.A., and Yamagami, T.: 1981, Astron. Astrophys. 100, 106

Kawara, K., Kozasa, T., Sato, S., Kobayashi, Y., and Okuda, H.: 1982, Pub. Astron. Soc. Japan 34, 389

Maihara, T., Oda, N., Sugiyama, T., and Okuda, H.: 1978, Publ. Astron. Soc. Japan 30, 1

Okuda, H.: 1981, in "Infrared Astronomy", IAU Symp. No. 96, review p. 247.

INFRARED SCANNING OF THE GALACTIC BULGE

R.M. Catchpole, P.A. Whitelock, I.S. Glass,
South African Astronomical Observatory,
P.O. Box 9, Observatory 7935, Cape, South Africa.

ABSTRACT

We find values of the interstellar absorption from J, H and K scans of a 6.8 x 200 arcmin strip of sky extending from the Sgr I field to the Galactic Centre. The visual absorption, excluding dark clouds, increases from 3 to 30 magnitudes in this region.

Three InSb detectors, mounted behind J, H and K (1.2, 1.6 and 2.2μ) filters in a single dewar, view the sky through 6 x 12 arcsec apertures at the Cassegrain focus of the 1.9m reflector at SAAO Sutherland. The DC output of the detectors is measured every 40 ms as the sky drifts past at the sidereal rate. Between scans the telescope is moved 8 arcsec in

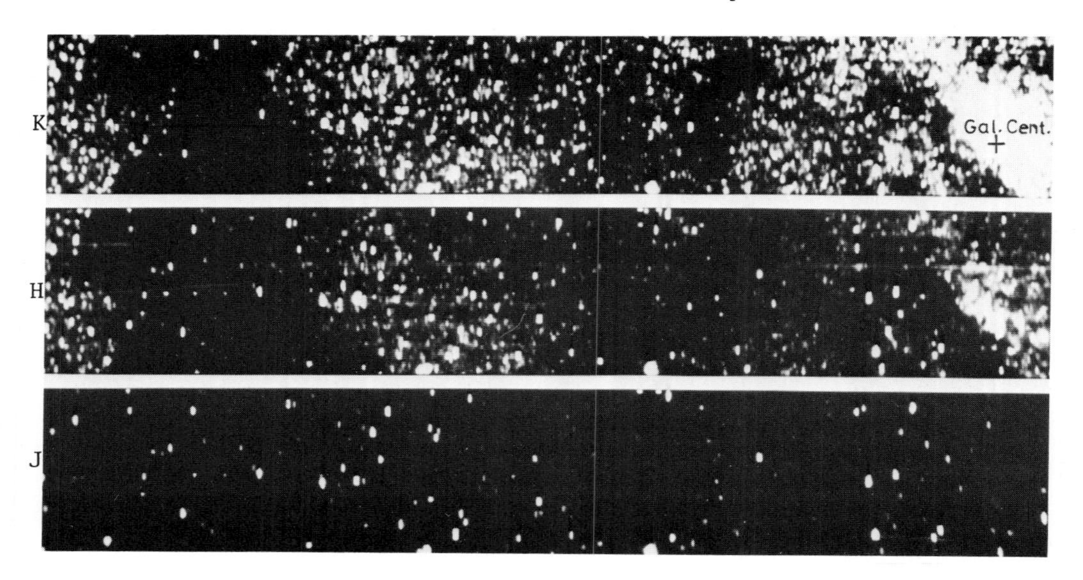

Fig. 1. A 6.8 x 44 arcmin area of sky at K, H and J. N is up, E to left.

H. van Woerden et al. (eds.), The Milky Way Galaxy, 127–128.

declination. JHK magnitudes are obtained by first removing background variations from all the scans to reduce them to a common zero and then integrating the flux along a short strip centered on each star. Stars are automatically found and centered on the K scans. The scans are also combined to produce monochromatic pictures of the sky, as shown in Fig.1.

Individual K against (H-K) diagrams, which show a well-defined giant branch, are constructed for consecutive 11 arcmin sections of the scan. The observed giant branch is then compared with the de-reddened 47 Tuc giant branch from Frogel et al. (1981) placed at the distance of the Galactic Centre (9.2 kpc, Glass and Feast (1982)). This allows us to deduce mean values of A_V for each field using the ratios $A_V : A_H : A_K = 1.0 : 0.141 : 0.088$. We empirically extend the 47 Tuc reference giant branch to K = 6.0 by using our photometry of Lloyd Evans' (1976) Sgr I long-period variables, de-reddened by $A_V = 2.01$ (Glass and Feast (1982)). Fig. 2 shows the resulting model values of A_V, for various de-reddened magnitude intervals, as a function of galactic latitude. We estimate that there is an uncertainty of $\pm$ 0.3 in A_V due to the as yet poorly defined colour transformation between our photometry and that of Frogel et al. The solid line in Fig. 2 is van Herk's (1965) galactic absorption law:

$$A_V = 0.14 \text{ Cosec } b \ (1 - \exp(-10 \ r \text{ Sin } b)) \text{ with } r = 9.2 \text{ kpc.}$$

The scans show patches of very high obscuration, several of which can be seen in the K scan in Fig. 1 and which have $A_V > 40$. These dark clouds only appear on our scans out to about 1° from the Galactic Centre, which corresponds to the angle subtended by the galactic disk (150 pc) at that distance. This may imply that the obscuring clouds occur in the disk mainly at the distance of the Galactic Centre.

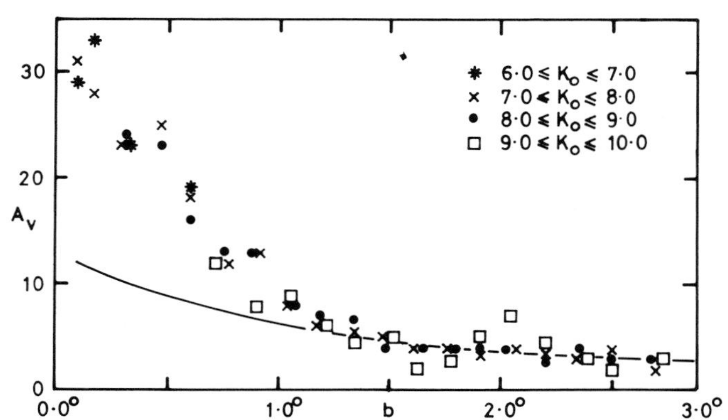

Fig. 2. A_V as a function of galactic latitude.

REFERENCES

Frogel, J.A., Persson, S.E., Cohen, J.G.: 1981, Astrophys. J. 246, pp. 842-865.

Glass, I.S., Feast, M.W.: 1982, Month. Not. R. astr. Soc. 198, pp. 199-214

van Herk, G.: 1965, Bull. astr. Inst. Neth. 18, pp. 71-105.

Lloyd Evans, T.H.H.: 1976, Month. Not. R. astr. Soc. 174, pp. 169-184.

INFRARED STUDIES OF THE STELLAR POPULATION IN BAADE'S WINDOW

R.A. Ruelas-Mayorga and A.R. Hyland
Mount Stromlo and Siding Spring Observatories
The Australian National University

T.J. Jones
University of Minnesota

An IR-scan of the Baade's Window ($l \simeq 0\overset{\circ}{.}0$, $b \simeq -4\overset{\circ}{.}0$) area (BW) has been obtained (Ruelas-Mayorga et al., 1983). The Cumulative Counts Function (CCF) at 2.2 μm (No. of sources per sq. degree down to a given K magnitude) down to $K \simeq +13.5$ was formed by combining 1.9-m telescope scans with Anglo Australian Telescope (AAT) scans.

With the aid of a theoretical exponential-disk model (Jones et al., 1981) and observations at $l = 20^{o}$, $b = -5^{o}$ and $l = 10^{o}$, $b = -5^{o}$, we decomposed the observed CCF into disk-CCF and bulge-CCF components. The bulge-CCF is steeper than the disk-CCF in the range $+5.0 < K < +11.0$, showing a relative depletion of high-mass stars with respect to the disk. The contribution of the bulge component towards BW is significant only at $K \simeq +9.5$ or fainter; the bright end of the CCF is dominated by the disk.

The bulge CCF has been compared with those of several globular clusters (47 Tuc, M3, M13, M92) and with that of the open cluster M67. The bright end of the globulars' CCFs have similar slopes to that of the bulge, suggesting that the stellar population of the bulge may be similar in age and metallicity to the globular clusters.

Figure 1 shows the observed CCF (solid dots are from JHK photometry of individual stars, triangles are from photometry derived from the 1.9-m scan, open squares are from photometry from the AAT scan). The theoretical CCF for an exponential disk towards BW (solid line) and the derived CCF for the bulge (open dots) are also illustrated.

Photometric studies of a bright-K subsample (135) of the 578 sources found in BW down to $K \simeq +11.0$ were made. Several sources with mild IR-excesses were found and later were confirmed spectroscopically as Mira variables. The majority of the sources lie along the reddening line at $E(J-K)=0.27$ from the solar-neighbourhood intrinsic giant sequence. The reddening agrees very well with the value $E(B-V)=0.45$ (van den Bergh 1971) obtained by optical techniques.

H. van Woerden et al. (eds.), The Milky Way Galaxy, 129–130.

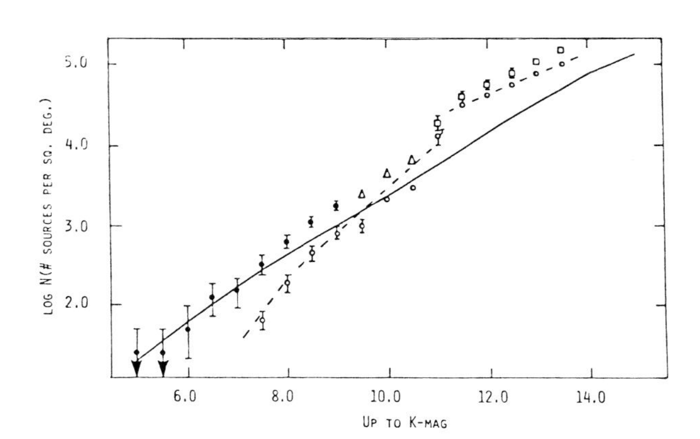

Figure 1. Cumulative Counts Function at 2.2 μm for Baade's Window. Full JHK photometry (solid dots), 1.9-m scan (triangles), AAT scan (squares). Theoretical CCF for and exponential disk towards BW (solid line). The bulge CCF (open circles).

Spectroscopic studies of 67 stars in our photometric sample were made with the Circular Variable Filter (CVF) facility at the AAT in the wavelength range $1.96<\lambda< 2.4$ μm. In this range the absorption bands of H_2O (shortward of 2.1 μm) and CO (longward of 2.3 μm) are commonly seen for stars of late spectral types.

In the interpretation of our photometric and spectroscopic observations it was convenient to divide the BW stars into 3 groups according to the strengths of their CO bands. It is shown that those stars with normal and strong CO bands may be consistently interpreted as disk stars. We suggest that the CO-weak stars may be true bulge members. The relative number of CO-weak objects to the total number of stars is consistent with the bulge and disk CCF data discussed earlier.

On the K vs. J-K diagram the sources in our photometric subsample lie above the giant-branch tips of 47 Tuc and M92. If the giant branches (GB) of these clusters were extrapolated to higher K brightnesses, a sizeable fraction of our sample would lie between them. This also suggests that its metallicity lies in the range between that of M92 and that of 47 Tuc. For those sources with redder J-K colours than the 47 Tuc GB and with magnitudes brighter than K = +8.5, an even higher value of metallicity is required. However, on the basis of the CCF results discussed above, these sources appear to be disk members, hence their high metal content should not be surprising.

REFERENCES

Jones, T.J., Ashley, M., Hyland, A.R. and Ruelas-Mayorga, R.A.: 1981, M.N.R.A.S. 197, p. 413
Ruelas-Mayorga, R.A., Hyland, A.R. and Jones, T.J.: 1983, Ap.J. (submitted
van den Bergh, S.: 1971, Astron. J. 76, p. 1082

ON THE NATURE OF OH/IR STARS

Dieter Engels
Sternwarte der Universität Bonn
Bundesrepublik Deutschland

ABSTRACT

OH/IR stars are the infrared counterparts of galactic OH maser sources which show a characteristic double-peaked emission-line profile. Their strong radio emission can be detected at large distances, making them excellent tracers of distribution and kinematics of evolved stars in the Milky Way. The OH maser profile is typical for line emission from an expanding circumstellar shell. The circumstellar shells of OH/IR stars absorb the optical emission of the central star nearly completely and reemit the energy in the infrared. Having luminosities $\sim 10^5$ $L_\odot$ and energy distributions peaking around 10μm, they may make a major contribution to the interstellar radiation field beyond 5μm. With mass loss rates of 10^{-5} to 10^{-4} $M_\odot$/yr they lose several solar masses in a few hundred thousand years. OH/IR stars are therefore important objects for recycling stellar matter into the interstellar medium.

Progress has been made in understanding the nature of OH/IR stars. They are Mira-like large-amplitude variables with periods up to 5 years long. It is proposed that they are stars of intermediate mass (2-10 $M_\odot$) on the asymptotic giant branch (AGB). They have not only larger masses than Mira variables proper, but also longer periods of pulsation and larger mass loss rates. As a result optically thick circumstellar dust shells are formed, which prevent the detection of these more massive Mira-like variables at optical wavelengths. Radial pulsation (Mira variability) is thus thought to occur for all intermediate-mass stars in the course of their evolution on the AGB. In view of their high mass-loss rates, these stars may be key objects in the study of the formation of planetary nebulae.

H. van Woerden et al. (eds.), The Milky Way Galaxy, 131.

Gösta Lyngå playing checkers at the conference dinner. To his left: Elaine Sadler and Claude Carignan; to his right: a guest, Mme. Carignan and Ken Freeman LZ

THE STELLAR DISK COMPONENT: DISTRIBUTION, MOTIONS, AGE AND STELLAR COMPOSITION

Gösta Lyngå
Lund Observatory
Lund, Sweden

ABSTRACT

The distribution of the stellar disk component is studied with particular emphasis on the properties of open clusters. A scale height of 80 pc is found for clusters younger than 10^9 years while the scale height is 200 pc for older clusters. A radial metallicity gradient in the disk is confirmed but there seem to be significant abundance variations apart from this. There are radial gradients in age as well as in linear diameter. Kinematically, there are a number of regions in the Milky Way with systematic radial velocity residuals.

1. INTRODUCTION

Because of the distances of the objects and even more because of the extinction of light by interstellar dust, optical studies are usually confined to the nearest one percent of the galactic disk. If we concentrate on luminous stars and are prepared to accept some selection effects we can reach about ten percent of the disk. It is the knowledge about these rather nearby stars that I will review to-day, and it is one of the shaky assumptions of this field that we may draw conclusions regarding the whole of our Galaxy from the nearby stars. I will start by discussing the z distribution of some components, then we shall look at the distribution in the plane of the Galaxy. Various aspects of the grand design will be discussed and then I will review the meagre observational data that exist concerning the distribution of objects with various abundances. A lot of knowledge has lately been gained about the kinematics of stars of various types. I shall only discuss some points here quite briefly - other aspects will be presented by other speakers at this symposium.

There are three reasons why I will discuss open clusters at some length. The main one is that clusters are the part of the stellar component which is most accurately known; distances, radial velocities and abundances are easier to determine accurately for clusters than for

H. van Woerden et al. (eds.), The Milky Way Galaxy, 133–142.

single stars. The second reason is that much new information has become available lately through the efforts of many people. The third and perhaps the most subjective reason is that I have just finished a collection of quite a lot of data for open clusters enabling me to back up some of my comments about the disk structure. That catalogue is presented as a poster at this meeting.

2. THE Z DISTRIBUTION OF CLUSTERS AND OF DUST

Although professor Strömgren will shortly discuss the z distribution of field stars, I would like to give a short report on what cluster data can give. I will limit the discussion to the closest 1 kpc for two reasons: Firstly, the selection effects increase with distance and secondly, the undulations of the galactic disk make the plane approximation less valid. Closer than 1 kpc there are 78 young open clusters (log (age) < 8.0) with a mean z distance of -15 pc. This is interpreted as due to the sun's position being 15 pc above the plane of symmetry of the Galaxy, quite in accordance with other determinations. The scale height of the distribution is 60 pc. For 86 older clusters we get a scale height of 80 pc. The difference between these values is barely significant as can be seen from Figure 1.There is no clear trend in z distance with age except for the very oldest clusters (log (age) ≧ 9.0) which have a much wider distribution - the scale height is 200 pc for these 38 clusters.

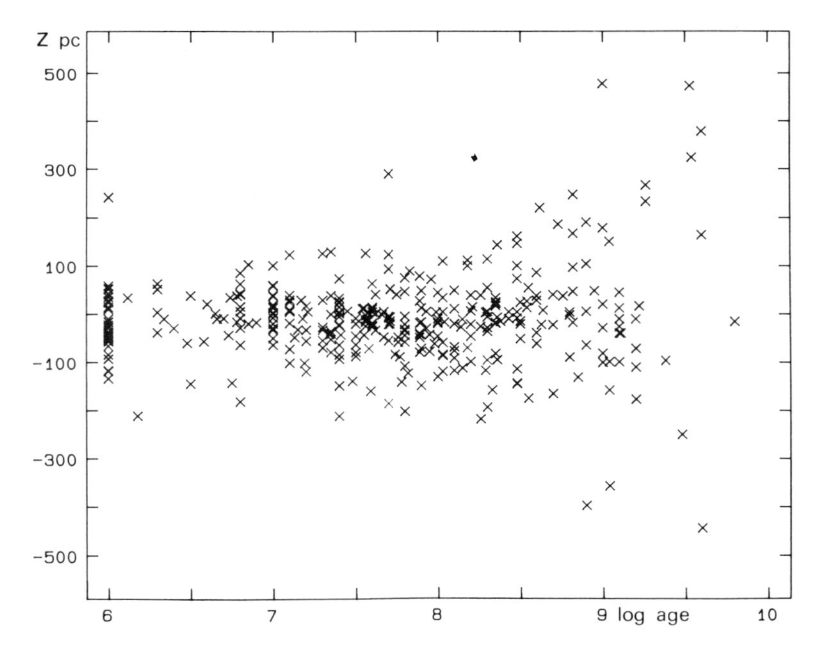

Figure 1. Distance from the galactic plane as a function of log (age).

The thickness of the layer of interstellar extinction can be estimated from the reddening of clusters at varying z distances (Lyngå, 1979, 1982) and this method applied to the new material gives a scale height of 130 pc with an average extinction in the plane of 0.75 magnitudes per kpc. This is in good accordance with earlier known values.

3. THE DISTRIBUTION OF CLUSTERS AND OTHER OBJECTS IN THE PLANE

Several people have discussed the lack of old open clusters in the inner parts of the galactic disk (Van den Bergh et al., 1980; Lyngå, 1980; Janes and Adler, 1982). The effect seems to be real; in any case it has remained in spite of the addition of new data. Present knowledge about ages and positions gives the plot shown in Figure 2. This may be interpreted as an effect of shorter relaxation time in the denser inner disk than in the outer disk. Perhaps the massive molecular clouds have tidal effects and thus shorten cluster lives.

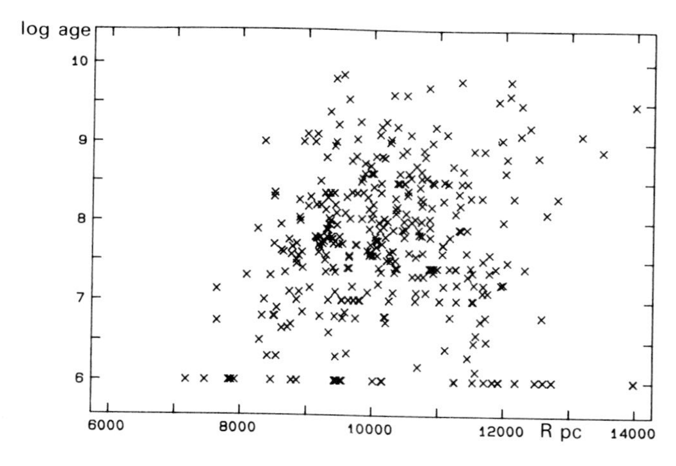

Figure 2. Log (age) as a function of galactocentric radius.

In an attempt to relate the small scale age distribution to a possible grand design, Balazs and myself (1984) try to relate the distances of clusters from recognized spiral arms to the ages of the clusters; naturally the solution is dependent on the grand-design parameters: arm inclination, pattern speed and distance scale. The most surprising of our first results is the arm inclination 7° which is quite different from other results of spiral-arm tracing as given by Humphreys (1979). The existence of a solution does not necessarily mean that the model obtained is the only possible. It does not even follow that a spiral pattern is necessary. What it says is merely that there is an age stratification, which would result as long as the stellar velocity is different from the velocity of the pattern.

The simplified model of spiral structure does not seem sufficient to fully explain the distribution of young objects. Figure 3 shows a graph of the positions of young clusters on the galactic plane. It seems that the large clumps of objects are quite a characteristic property of the distribution. Let me point out that I have limited the graph to a distance of 4 kpc. It is quite possible to find similar features at larger distances but then the selection effects caused by interstellar extinction may be severe. We can compare such features, of the order of 1 kpc, with the groupings of interstellar clouds (Lucke, 1978) or the cepheid complexes (Efremov, 1978). The understanding of such complexes has lately been made possible through the works of Elmegreen and Elmegreen (1983) and others.

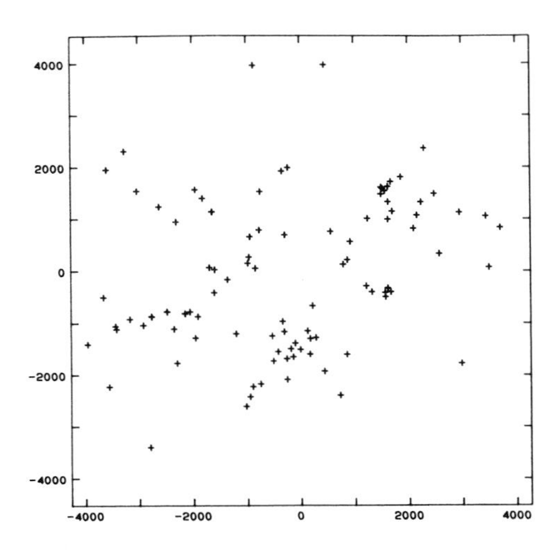

Figure 3. Plot on the galactic plane of positions of young clusters. The Sun is at the origin.

In an earlier investigation (Lyngå, 1982) I found that linear diameters of open clusters are as an average smaller in the inner part of the Galaxy than in the outer part. The trend is slight and the more homogeneous set of diameter data in my updated catalogue does not show it very well. If, however, only the very youngest clusters are considered there is a trend (see Figure 4), which may be interpreted in terms of variation of the Jeans radius through the Galaxy (Burki, 1980).

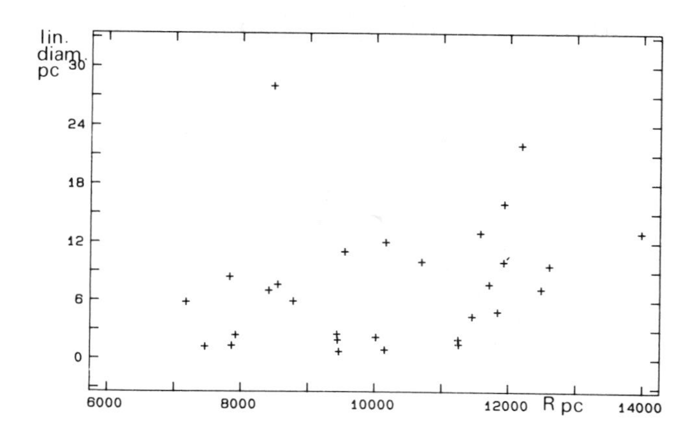

Figure 4. Radial gradient of linear diameter for young clusters.

4. THE METAL ABUNDANCE GRADIENT

That an abundance gradient with galactocentric radius is present for the gas component of the disk has been shown in the review by Peimbert (1979). It would be expected that the same effect is present for young stars; whether it is also present for older stars is a question that bears on the radial mixing properties of the disk. Let us review the evidence:

- Mayor (1976) drew conclusions from uvby studies of nearby stars, the birthplaces of which he found by kinematic evidence. Relating metallicity to galactocentric radius of the birthplace, he obtained a gradient of -0.05 kpc^{-1} from 1000 stars; the young stars gave a steeper gradient.

- Luck (1982) discussed abundances of supergiants on the basis of model-atmosphere calculations; his data give a slope of d[Fe/H]/dR = -0.13. Christian and Smith (1983) add a couple of distant, anticentre F giants with low abundance consistent with a gradient of -0.06 dex.

- Harris (1981) has studied a great number of cepheids, determining a metallicity index from Washington photometry. A gradient of -0.07 dex is found.

- Janes (1979) has indicated that stars in open clusters show a gradient in metallicity of -0.05 kpc^{-1}.

- Panagia and Tosi (1981) have studied young clusters particularly and found a metal gradient of -0.095 kpc^{-1}.

On the question of abundances of clusters I believe that my own catalogue of open-cluster data (Lyngå, 1983) has the most comprehensive information. Figures 5 and 6 are based on data collected there and show plots of abundance against galactocentric radius. We find a gradient of $-0.13\ kpc^{-1}$ whether all clusters or only the young ones are used; while there is clear evidence of a gradient, I believe there is more to it than that. Typical imprecision values are, for nearby clusters, about 0.1 dex, and the scatter in these diagrams is much larger.

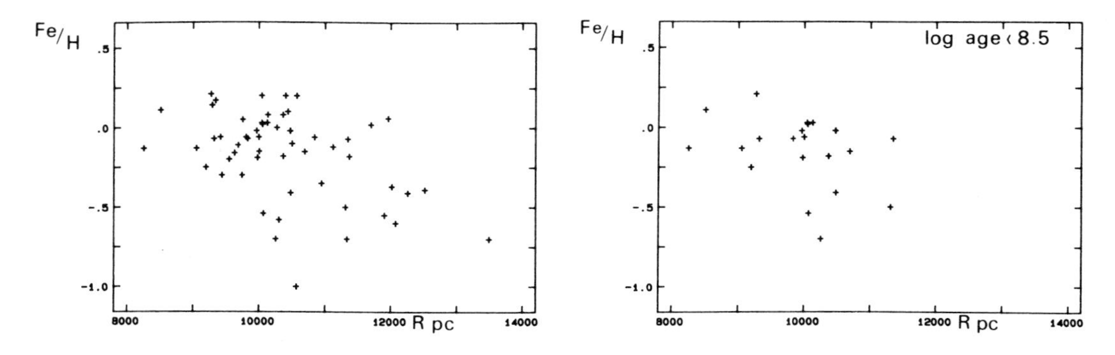

Figure 5 (all clusters) and Figure 6 (young clusters) show that the radial abundance gradient is similar for the two age intervals considered.

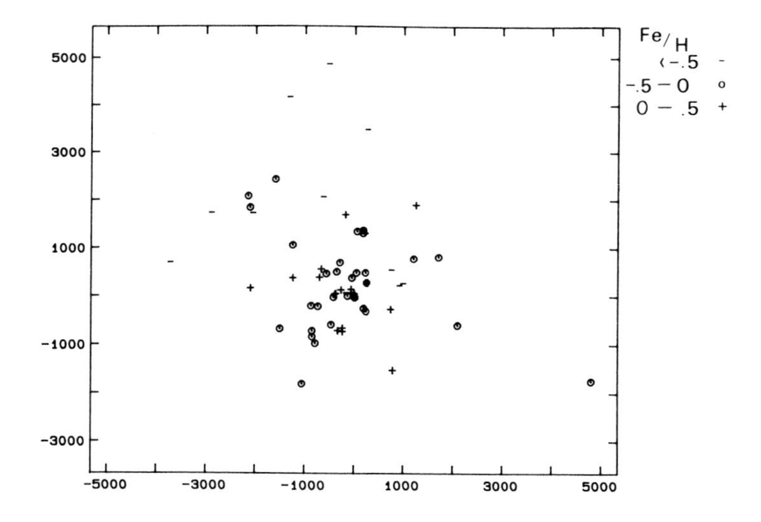

Figure 7. Plot on the galactic plane of [Fe/H].

Figure 7, which is a plot of the cluster data on to the galactic plane, shows a more complex picture. To me it appears that there are rather large regions of similar metallicity and other regions of other metallicities, not in a one-to-one relation with galactocentric radius.

Such a situation could easily arise if the star formation rate varies throughout the disk. Again we have indications of homogeneity inside certain regions. Wramdemark and myself (1983) have just studied a few clusters in the Gould Belt region with uvby photometry. Clusters that are kinematically and positionally in Gould's Belt also show the same metal abundances; ages for these are around 4×10^7 years, i.e. comparable to the expansion age of Gould's Belt according to Olano (1982).

5. HELIUM DISTRIBUTION

Akin to this picture but actually concerning primordial abundances is the distribution of helium. Nissen (1980) has reported significant differences between the He abundances in relatively young clusters. From λ 4026 photometry he found that Sco-Cen has a value of $Y = 0.28$ while Cep OB III and h + χ Per have $Y = 0.19 \pm 0.01$, extremely helium deficient. Scatter between cluster members was small. Not only is a primordial abundance difference necessary to explain these differences; the finding also requires that mixing in the galactic disk is quite limited. The question whether a helium-abundance gradient is present in the gaseous component does not seem to have been resolved. Peimbert (1979) has reviewed the possibilities; whatever gradient is indicated is, however, much smaller than is needed to explain the differences between the clusters.

6. KINEMATICS OF THE OPEN CLUSTER SYSTEM

Recent determinations of Oort's constants and of the rotation curve of the Galaxy from various data have been summarized by Knapp (1983). Not wishing to repeat this here, I shall only discuss some local deviations from the large-scale pattern and add the results obtained for the radial velocities for 114 clusters which have been calculated by Wramdemark (1983) and are included in my cluster catalogue (Lyngå, 1983).

Systematic deviations of velocities from circular motions have been demonstrated by several authors among whom are Humphreys (1976), Ardeberg and Maurice (1981) and Ardeberg et al. (1985). Interpretations have included effects of massive spiral arms. It is also well established (Lindblad, 1983) that the kinematics of young stars closer than 400 pc are strongly affected by the expansion of the Gould belt complex.

Let us now study systematic effects on the basis of the cluster catalogue. In Figure 8 we have corrected the radial velocities to the local standard of rest defined by B stars and also corrected for differential galactic rotation if a flat rotation curve is assumed (other rotation curves give qualitatively the same picture). We see that there are residuals showing systematic trends. Most notable is the positive excess near us, consistent with the expansion in the Gould Belt system and the predominantly negative residuals at l ~ 285°. These clusters have a systematic velocity excess of around -10 km/s which may be compared with the residual velocity for stars and interstellar gas amounting to +6 km/s at l ~ 295° (Ardeberg and Maurice, 1981).

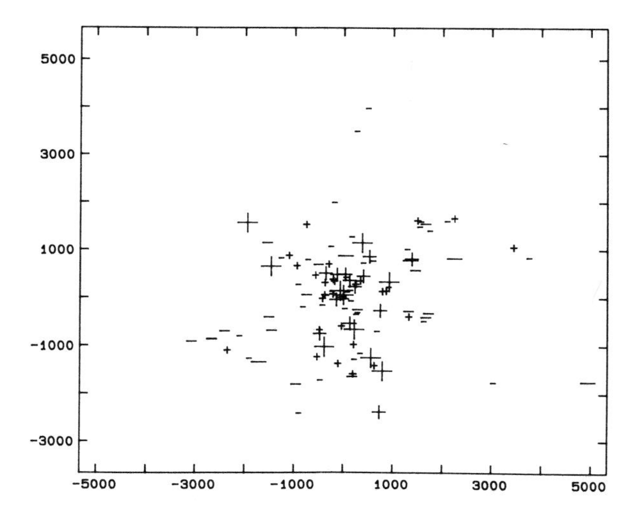

Figure 8. Residuals in radial velocity on the galactic plane.

Symbols are			
	>	20 km/s	+
	10 to	20 km/s	+
	0 to	10 km/s	+
	-10 to	0 km/s	-
	-20 to	-10 km/s	–
	<	-20 km/s	—

7. PROBLEMS AND PROSPECTS

What we seem to know well enough is the thickness of the galactic disk as defined from objects < 10^9 years old. There has not been any significant variation in thickness during that time. However, older clusters have a much wider distribution, one which is also shared by the G and later-type stars.

The distribution of abundances can be approximated with a galactocentric gradient but this does not tell the full story. Some regions exist, inside which the metallicity does not vary much but for which it deviates from the metallicity predicted by a gradient. Also the widely differing helium abundances between clusters that Nissen (1980) has pointed out are raising problems. At the time being there may not be enough observations to get a firm hold on any of these questions. However, there are enough clusters to study, so I have hopes for fruitful collection of data in these fields during the next few years.

Whether we should consider the grand design spiral structure as the most important characteristic of our Galaxy, or a clumpy distribution where homogeneous regions of about 1 kpc are typical, is at the moment an open question. Some people favour the one picture, some favour the other. I would rather say: both!

The large-scale kinematics of the disk contains problems which I have only touched upon. However, it is clear from the radial velocities of clusters that systematic deviations from circular motions exist.

REFERENCES

Ardeberg, A., Lindgren, H., Maurice, E.: 1985, this volume
Ardeberg, A., Maurice, E.: 1981, Astron. Astrophys. 98, 9
Balazs, B.A., Lyngå, G.: 1984, in preparation
Burki, G.: 1980, in 'Star Clusters', ed. J.E. Hesser, IAU Symp. 85, p. 169
Christian, C.A., Smith, H.A.: 1983, Publ. Astron. Soc. Pacific 95, 169
Efremov, Yu.M.: 1978, Pis'ma Astron. Zh. 4, 125
Elmegreen, B.G., Elmegreen, D.M.: 1983, preprint
Harris, H.C.: 1981, Astron. J. 86, 707
Humphreys, R.M.: 1976, Astrophys. J. 206, 114
Humphreys, R.M.: 1979, in 'The Large-Scale Characteristics of the Galaxy', ed. W.B. Burton, IAU Symp. 84, p. 93
Janes, K.A.: 1979, Astrophys. J. Suppl. 39, 135
Janes, K.A., Adler, D.: 1982, Astrophys. J. Suppl. 49, 425
Knapp, G.R.: 1983, in 'Kinematics, Dynamics and Structure of the Milky Way', ed. W.L.H. Shuter, Vancouver Workshop, p. 233
Lindblad, P.O.: 1983, in 'Kinematics, Dynamics and Structure of the Milky Way', ed. W.L.H. Shuter, Vancouver Workshop, p. 55
Luck, R.E.: 1982, Astrophys. J. 256, 177
Lucke, P.B.: 1978, Astron. Astrophys. 64, 367
Lyngå, G.: 1979, in 'The Large-Scale Characteristics of the Galaxy' ed. W.B. Burton, IAU Symp. 84, p. 87
Lyngå, G.: 1980, in 'Star Clusters', ed. J.E. Hesser, IAU Symp. 85, p. 13
Lyngå, G.: 1982, Astron. Astrophys. 109, 213
Lyngå, G.: 1983, Computer-based catalogue of open cluster data, 3rd ed., CDS, Strasbourg; see also description in this volume
Mayor, M.: 1976, Astron. Astrophys. 48, 301
Nissen, P.E.: 1980, in 'Star Clusters', ed. J.E. Hesser, IAU Symp. 85, p. 51
Olano, C.A.: 1982, Astron. Astrophys 112, 195
Panagia, N., Tosi, M.: 1981, Astron. Astrophys. 96, 306
Peimbert, M.: 1979, in 'The Large-Scale Characteristics of the Galaxy', ed. W.B. Burton, IAU Symp. 84, p. 307
Van den Bergh, S., McClure, R.D.: 1980, Astron. Astrophys. 88, 360
Wramdemark, S.: 1983, Private communication
Wramdemark, S., Lyngå, G.: 1983, Astron. Astrophys., in press

DISCUSSION

J.P. Ostriker: You could calculate from your data a mortality rate for clusters as a function of age in the same way as actuaries do, if you assume an essentially constant birthrate. What fraction of clusters that exist at age 10^7 years still exist at age 10^8 years and at age 10^9 years?

Lyngå: In 10^8 years the number of clusters falls by roughly a factor e. After 10^9 years there should not be any left. Clusters older than 10^9 years are probably truly different.

J.V. Feitzinger: Do you find a correlation between the cluster-diameter and the reddening values? Absorption effects might cause different cluster diameters in different directions.

Lyngå: I find no such correlation.

M.W. Feast: Can I ask a related question? What effect could interstellar absorption have on the clumping you see?

Lyngå: Well, it could have an effect, but not so much out to 2 kpc. Further out, of course, it certainly will. But if a cluster is heavily absorbed, that fact is known. Another selection effect is that people tend to study those clusters that they think are more interesting. I think my catalogue is quite complete out to 1 kpc, and about 30 or 40% complete out to 2 kpc. Beyond that distance I don't really want to do this sort of thing.

R. Güsten: I am somewhat puzzled by the small He abundance you mentioned for some of these open clusters. Could you explain how one derives the He abundance from one line only?

Lyngå: I must refer you to the paper by Nissen (1980).

M.L. Kutner: With regard to the clumpiness in the abundance variations, I note that interstellar isotope-abundance ratios, determined from interstellar molecules, show significant source-to-source variations which can probably not be explained by radiative-transfer effects, chemical fractionation or an overall galactic gradient.

P. Pismis: 1) Have you detected any relationship between the ages and masses of clusters, or between the age and a parameter that may specify the mass? 2) Is there a relationship between the age and the morphology of clusters, for example the smoothness or symmetry in the distribution of stars in a cluster?

Lyngå: 1) Yes, there is quite a clear correlation. If I consider only the bound ones, then the younger clusters have higher masses.
2) I have not looked for a relationship between age and morphology.

COMPUTER-BASED CATALOGUE OF OPEN-CLUSTER DATA

Gösta Lyngå
Lund Observatory
Lund, Sweden

The third edition of the computer-based catalogue of open-cluster data has now been produced and is disseminated through the Centre de Données Stellaires, 11 rue de l'Université, F-67000 Strasbourg, France. It is also available through World Data Center A, NASA, Greenbelt, MD 20771, USA.

The following exhibits are available at the poster session:

I. An ASCII print-out of the cluster catalogue, corresponding to file 2 of the magnetic-tape version (blue folder)

II. ASCII print-outs of the following files (grey folder):

Introduction and description (file 1)
References (file 3)
Alias lists (file 4)
Clusters in order of longitude (file 5)

The aim of this catalogue is to give salient data for all known open star clusters in our Galaxy. As far as possible only published data values have been quoted; for some of the parameters these values have been selected from references, which are listed.

In particular, for the 1983 edition I have added new information based on inspection of the Palomar, ESO and SERC surveys. For each cluster that could be identified on the available charts I have estimated the diameter, made a new classification in Trumpler's system and given an estimate of the number of member stars. The inspection has also been the basis for statements in the alias list that certain clusters are dubious. Those clusters that are dubious both according to my inspection and according to the CSCA catalogue, have been removed from the catalogue but they still appear in the alias listing.

The definitions of the data values are in most cases unambiguous. However, the following points should be clarified:

H. van Woerden et al. (eds.), The Milky Way Galaxy, 143–144.

Cluster diameter. In the literature some quite convincing cases have been made for the existence of large coronas around open clusters. This may be a general phenomenon, although for most clusters only the diameter of the cluster nucleus is known. For reasons of homogeneity, this is used even when the existence of a cluster corona is likely. For most clusters two diameters are given. My own estimates from survey prints should form a reasonably homogeneous sample. In addition, I have quoted the most reliable, often the only, value available in the literature.

Interstellar extinction. Even when interstellar extinction is known to vary across a cluster I have chosen to quote the average value of A(V). The references given provide more details. If the reference gives only colour excess, I have derived the extinction using the relations A(V) = 3.1 E(B-V) and A(V) = 4.28 E(b-y) for the UBV and uvby systems. Extinction and (B-V) turn-off values are given according to Janes and Adler (1982). When that compilation includes multiple references, Dr K. Janes has made the selection.

Integrated properties. The total V magnitude and the integrated B-V colour have been calculated by Mr. B. Skiff, Flagstaff, Arizona, using available photometric data. The number of observed stars used is sometimes different from my own estimate of membership from survey charts. Reasons for such difference can be inclusion of non-members or limited resolution on the survey charts.

Cluster age. When a cluster shows star formation during an extended period, I have given the beginning of that period, i.e. the highest age. In my selection of age data I have preferred results based on isochrone evaluations. Nevertheless, most of the catalogue values for ages are from colour-magnitude morphology.

Metallicity. As far as possible, metallicity values refer to [Fe/H] data such as can be derived from spectral analysis or from uvby photometry.

Radial velocities. The data for cluster radial velocities have been collected by Dr. S. Wramdemark of Lund Observatory. They represent weighted mean values for member-star velocities as published up to 1980. The weighting procedure takes into account the number of stars in the cluster for which radial velocity had been measured, the accuracy (Wilson class) for each stellar radial velocity and the calculated imprecision in the mean radial velocity for the cluster. Five weight classes are identified by w = 1, 2, 3, 4 and 5 in order of increasing precision.

REFERENCE

Janes, K.A., Adler, A.: 1982, Astrophys. J. Suppl. 49, 425

THE KINEMATICS OF NEARBY STARS AND LARGE-SCALE RADIAL MOTION IN THE GALAXY

S.V.M. Clube
Royal Observatory, Edinburgh, Scotland

ABSTRACT: The concept of star streams originally due to Kapteyn is revived. It is suggested that the existence of Stream II may have been overlooked, yet it is of crucial importance in specifying the true local standard of rest.

Kapteyn's star streams (SI, SII) in the solar neighbourhood have been generally overlooked for > 50 yr: see my remarks following E.R. Paul (these proceedings), and Eddington (1914), Eggen (1963), Upgren (1976), and Clube (1978). The present indications are that SI is a young population representative of the nearby spiral arm, with some contamination from the slightly older stellar groups pervading the solar neighbourhood; its mean velocity w.r.t. the Sun is $(u_I, v_I, w_I) \simeq (-23, -19, -8)$ km s^{-1}. SII on the other hand is a mostly late spectral-type population made up of stars with well-mixed orbits corresponding to the older galactic disc: $(u_{II}, v_{II}, w_{II}) \simeq (+16, -16, -7)$ km s^{-1}. Although the evidence for a Stromberg drift is well known, the dominant sense of the streaming here, SI relative to S II, is in a roughly orthogonal direction; indeed, if we take the not unreasonable view that SII defines the preferred l.s.r. (zero radial motion), then SI has a local non-circular motion directed outwards, $(u_{II}- u_I)$, of 39 km s^{-1}. There is some uncertainty in this figure of course, depending on the precision with which the stream motions have been determined (e.g. see Eddington, 1914), but probably no worse than ±5 km s^{-1}. On this hypothesis, it is only because SI is more dominant among all spectral types (50-100%) within the immediate solar neighbourhood (< 300 pc) that the currently adopted l.s.r. with $(u_s, v_s, w_s) \simeq (-10, -16, -7)$ km s^{-1} inclines towards SI rather than SII.

Since the nearby young stars (≃SI) produce a deviated velocity ellipsoid, it is expected that merged stellar populations beyond 300 pc will produce a u-value that tends towards that of SII and a less deviated "ellipsoid". Both these effects are borne out in a recent study (Clube et al. 1983) of the kinematics of ~ 1000 O,B-type stars within 2000 pc. Similarly, beyond 300 pc, we might expect to see otherwise unexplained changes in mean $\bar{u}$ along the anticentre and centre directions, reflecting the greater presence or otherwise of SI. Two recent studies

H. van Woerden et al. (eds.), The Milky Way Galaxy, 145–147.

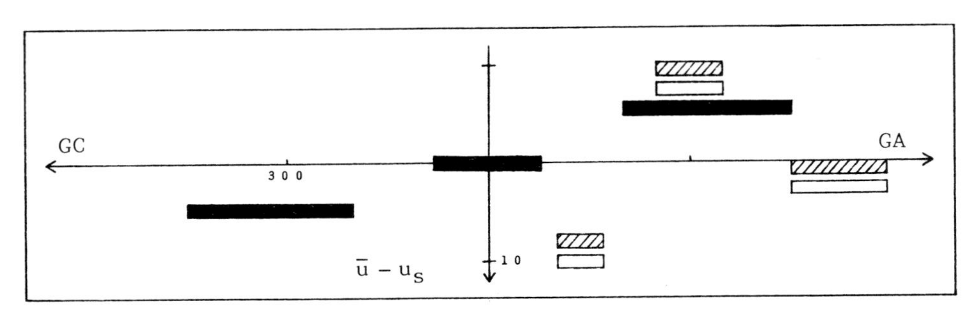

Figure 1. Trend in $\bar{u}$ through solar neighbourhood.

of ninth-magnitude K stars (Woolley et al. 1977, Griffin et al. 1983), mostly giants, seem to bear out these predictions. Thus, in the figure are plotted values of $\bar{u}$ inferred from Woolley et al. (1977) at various distances for groups of stars along anticentre lines of sight towards $b = \pm 20°$. The strong correlation above and below the plane suggests a real effect and there is also a striking displacement of $\bar{u}$ towards SI at ~ 350 pc where the nearest O,B associations in the Orion spur congregate (Blaauw 1985). No further associations are encountered within 1500 pc in these directions. Also in the figure are plotted values of $\bar{u}$ for two groups of K-stars of similar apparent magnitude spanning the regions ($l = 0° \pm 60°$, $20° < |b| < 40°$) and ($l = 180° \pm 60°$, $20° < |b| < 40°$). Again there is a significant difference with the u-value of the galactic-centre sample tending towards that of SII. In both cases also, $|v| > |v_S|$, the anticentre sample reflecting differential galactic rotation and the centre sample apparently reflecting the Stromberg drift of an older population.

If these observations are correctly interpreted and SII truly corresponds to the preferred l.s.r., a redetermination of the mean motion of the halo is important. It has been suggested previously for example that the globular-cluster system is locally infalling (Kinman 1959), but this effect may well have been underestimated due to the influence of expanding motions near the centre of the Galaxy (Clube and Watson 1979). Of similar importance are direct observations of the Galactic nucleus (see Clube et al. 1983): thus the significance of Kapteyn's SII appears enhanced by the fact that the cores of the H_2CO, Ne II and 511 MeV electron-positron annihilation lines are centred on Sgr A West, and by the fact they are all receding from the conventional l.s.r. at 40-50 km s^{-1}. The molecular ring of the Galaxy also now appears to be bounded on its inner face (Bania 1985) by the 3 kpc expanding arm (-55 km s^{-1}) and the +135 arm, and symmetry here would similarly suggest the centre is receding at 40 km s^{-1}.

The existence of Kapteyn's Stream II is evidently a question of fundamental importance to our understanding of solar-motion variations, and should it now be re-established, there may be far-reaching implications for our understanding of the kinematics of the Galaxy as a whole.

REFERENCES

Bania, T.M.: 1985, these Proceedings, Section II.3
Blaauw, A.: 1985, these Proceedings, Section II.6
Clube, S.V.M.: 1978, M.N.R.A.S. 184, 553
Clube, S.V.M., and Watson, F.G.: 1979, M.N.R.A.S. 187, 863
Clube, S.V.M., Watson, F.G., and Zhao, J.L.: 1983, Astron. Astrophys. in press
Eddington, A.S.: 1914, Stellar Movements and the Structure of the Universe, London: MacMillan
Eggen, O.J.: 1963, Astron. J. 68, 697
Griffin, R.F., Zhao, J.L., and Clube, S.V.M.: 1983, in preparation
Kinman, T.D.: 1959, M.N.R.A.S. 119, 559
Upgren, A.R.: 1976, Bull. Amer. astr. Soc. 8, 4, 542
Woolley, R.v.d.R. et al.: 1977, M.N.R.A.S. 179, 81

A. Blaauw (centre, background) introduces R.J. Allen, Chairman of Department of Astronomy, and J. Borgman, President of Groningen University, to descendants of J.C. Kapteyn CFD

(Left to right) Murray, Reid, Okuda and Brink discussing posters CFD

THE GALACTIC RADIAL GRADIENT OF VELOCITY DISPERSION

Michel Mayor
Geneva Observatory

Edouard Oblak
Besançon Observatory

The kinematical study of the stars in the solar neighbourhood allows, via the equation of the so-called asymmetrical drift, to deduce the sum of the gradients of the density and the velocity dispersions, $\frac{\partial \ln \rho}{\partial \varpi} + \frac{\partial \ln \sigma_u^2}{\partial \varpi}$. In order to deduce the density gradients in the solar neighbourhood, the second term is generally supposed to be zero. This kind of hypothesis, certainly wrong, comes from the old "ellipsoidal theory". A velocity dispersion independent of $\bar{\varpi}$ is not compatible with the Toomre's local stability. On the contrary, if we suppose

$$Q = \frac{\sigma_u(\bar{\varpi})}{\sigma_u(\bar{\varpi})_{min.}} \simeq \text{cte, we estimate } \frac{\partial \ln \sigma_u^2}{\partial \bar{\varpi}} \simeq -0.2, \text{ a non-}$$

negligible value compared with $\partial \ln \rho / \partial \varpi$ (Mayor, 1974). Using Vandervoort's (1975) hydrodynamical approach, Erickson (1975) obtains a similar value for the local velocity-dispersion gradient.

In the following we briefly describe a new method for deducing the local value of $\partial \ln \sigma_u^2 / \partial \varpi$ using also the local distribution of residual velocities.

The galactic potential allows us to transform the observed residual-velocity distribution $f_\odot(U, V, W)$ to a local distribution of the orbital eccentricity e, the mean orbital radius $\bar{\varpi}$ and the velocity perpendicular to the galactic plane W (at z = o), $N_\odot(e, \bar{\varpi}, W)$.

The local distribution $N_\odot(e, \bar{\varpi}, W)$ results from the epicyclic centre distribution $\Sigma(\bar{\varpi})$, from the eccentricity and W distributions at different places in the galactic disk $g_1(e, \bar{\varpi})$, $g_2(W, \bar{\varpi})$ and from the probability to observe the star in the solar neighbourhood $p(\varpi_\odot, e, \bar{\varpi}, W)$.

H. van Woerden et al. (eds.), The Milky Way Galaxy, 149–150.

$$N_{\odot}(e,\bar{\varpi},W) = cte \cdot \Sigma(\bar{\varpi})\ g_1(e,\bar{\varpi})\ g_2(W,\bar{\varpi})\ p(\bar{\varpi}_{\odot},e,\bar{\varpi},W)$$

Locally $g_1(e,\bar{\varpi}_{\odot}) \propto e \cdot \exp\left(-e^2/e_o^{\,2}(\bar{\varpi}_{\odot})\right)$

and $g_2(W,\bar{\varpi}_{\odot}) \propto \exp\left(\dfrac{-W^2}{\sigma_W^{\,2}(\bar{\varpi}_{\odot})}\right)$

On one hand we suppose these kinds of distribution valid not too far from $\varpi_{\odot}$ (for example ± 3 kpc) but with varying $e_o(\bar{\varpi})$ and $\sigma_W(\bar{\varpi})$

$$e_o(\bar{\varpi}) = e_{o_{\Theta}} + (\bar{\varpi}-\bar{\varpi}_{\odot})\ \alpha_1 + \ldots\ldots.$$

$$\sigma_W(\bar{\varpi}) = \sigma_{W_{\Theta}} + (\bar{\varpi}-\bar{\varpi}_{\odot})\ \alpha_2 + \ldots\ldots.$$

On the other hand $\Sigma(\bar{\varpi})$ can be expressed as

$$\Sigma(\bar{\varpi}) \propto \bar{\varpi}^{-n}$$

or $$\Sigma(\bar{\varpi}) \propto \bar{\varpi}\ \exp\cdot(-^{\bar{\varpi}}/\beta).$$

The three free parameters (α_1, α_2 and β) are adjusted to fit at best the locally observed distribution $N_{\odot}(e,\bar{\varpi},W)$.

The α_1 and α_2 parameters can be related to the velocity gradients

$$\frac{\partial \ln \sigma^2_u}{\partial \varpi} \text{ and } \frac{\partial \ln \sigma^2_W}{\partial \varpi}.$$

In order to avoid possible local perturbations, only regions with e and W not too small can be used in the fitting procedure.

A detailed description for the application of this method to different stellar samples of the solar neighbourhood will be published in Astronomy and Astrophysics.

REFERENCES

Erickson, R.R.: 1975, Astrophys. J. 195, 343
Mayor, M.: 1974, Astron. & Astrophys. 32, 321
Vandervoort, P.O.: 1975, Astrophys. J. 195, 333

STELLAR CHEMICAL-ABUNDANCE GRADIENT IN THE DIRECTION OF THE SOUTH GALACTIC POLE - PRELIMINARY RESULTS

Ch. F. Trefzger
Astronomical Institute, Basel, Switzerland
J.W. Pel and A. Blaauw
Kapteyn Astronomical Institute, Groningen, The Netherlands

ABSTRACT

Using the Walraven VBLUW photometric system, we have studied the metal content of 89 F and G stars in the Galactic South Pole field SA141. Our sample is based on the Basel survey of RGU photometry in Selected Areas, and it contains all stars in SA141 with $V_J < 14^m.5$ and $(G-R) < 1^m.15$ (spectral types earlier than about G7). The observations were made with the VBLUW photometer and the 90-cm Dutch Telescope at ESO, La Silla.

For unreddened intermediate-type stars the VBLUW photometry enables us to separate the effects of temperature, gravity, and metallicity (cf. Lub and Pel, 1977). Since reddening is negligible in SA141, we can therefore determine these three parameters for each program star once the photometric indices are calibrated in terms of T_{eff}, log g, and [Fe/H]. The latter calibration was made in a semi-empirical way, using VBLUW observations of stars with spectroscopic analyses in combination with theoretical colors based on the model spectra by Kurucz (1979). We used the Hyades main-sequence as a zeropoint, adopting [Fe/H] = +0.15 for this cluster.

The results in the (V-B)-(B-L) diagram are shown in Fig.1. This diagram is very sensitive to metallicity, but almost gravity-independent. Fig.1 indicates that most program stars have metallicities in the range $-1 \leqq [Fe/H] \leqq 0$. The distribution of the program stars in the gravity-sensitive (V-B)-(L-U) diagram is very narrow, log g = 4.2 ± 0.3, which means that these stars are probably all dwarfs, with only very few possible subgiants.

The absolute-magnitude calibration was derived by using the data of Cayrel de Strobel et al. (1980) and of Cayrel de Strobel and Bentolila (1983). From their [Fe/H]-catalogue we took all stars with known distances, and within the parameter range of our program stars, to construct an empirical M_V - T_{eff} relation. This relation was used to derive distances for the stars in SA141.

These distances are plotted against [Fe/H] in Fig.2. The diagram clearly shows the correlation between distance and metallicity, suggesting for this particular sample a gradient of -0.6 in [Fe/H] over the first 500 pc. Similar values were found in the Basel RGU program (cf. Trefzger, 1981) and by Blaauw and Garmany (1975).

H. van Woerden et al. (eds.), The Milky Way Galaxy, 151–152.

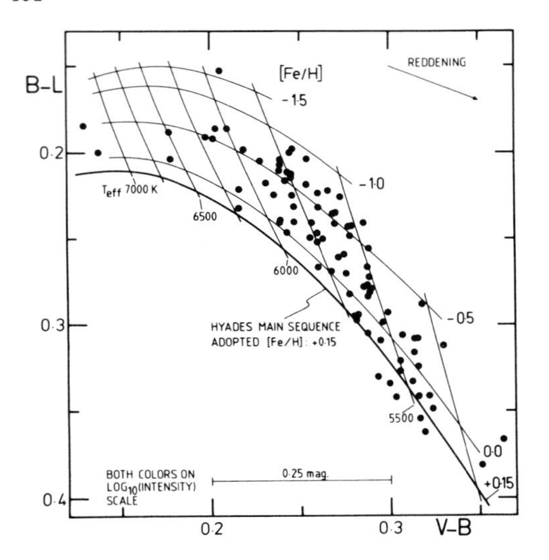

Fig.1. The program stars in the (V-B)-(B-L) diagram. The temperature-metallicity calibration is indicated. Gravity effects are very small in this diagram.

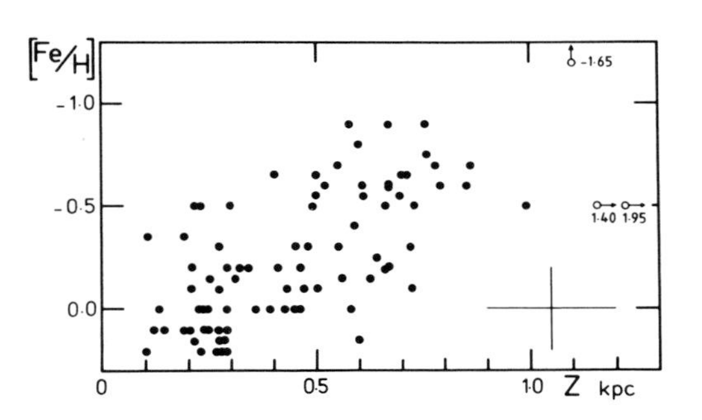

Fig.2. [Fe/H] versus distance from the galactic plane for the F and G dwarfs observed in SA141. The sample is complete up to 500 parsec. The cross corresponds to typical uncertainties of σ[Fe/H] = = ± 0.2 and σM_V = ± $0^m.5$ at z=700 pc.

REFERENCES

Blaauw, A., Garmany, C.D.: 1975, in "Proceedings of the Third European Astronomical Meeting", ed. E.K. Kharadze, Tbilisi, p.3.

Cayrel de Strobel, G., Bentolila, C., Hauck, B., Curchod, A.: 1980, Astron. Astrophys. Suppl. 41, 405.

Cayrel de Strobel, G., Bentolila, C.: 1983, Astron. Astrophys. 119, 1.

Kurucz, R.L.: 1979, Astrophys. J. Suppl. 40, 1.

Lub, J., Pel, J.W.: 1977, Astron. Astrophys. 54, 137.

Trefzger, Ch.F.: 1981, Astron. Astrophys. 95, 184.

STAR COUNTS, LOCAL DENSITY AND K_z FORCE

Bengt Strömgren
NORDITA, Copenhagen, Denmark

ABSTRACT

The approach by Bahcall and Soneira to the determination of galactic parameters through the use of star counts is referred to, and tests of the Bahcall-Soneira Galaxy model based on additional observational data are discussed.

The determination of the local mass density by Hill, Hilditch and Barnes through studies of A and F stars in the region of the North Galactic Pole is briefly discussed, as is a recent investigation of the problem by Bahcall.

In the determination of the galactic force K_z and the local mass density from the density distribution $\nu(z)$ and the distribution $f(W)$ of velocities at right angles to the galactic plane for a group of tracer stars, it is important to secure homogeneity of the tracer group. This has led Hill, Hilditch and Barnes in a continuation of their investigation to use photoelectric uvby photometry to segregate homogeneous groups of F stars. A similar approach is followed by Danish astronomers, whose work is briefly described.

STAR COUNTS, THE BAHCALL-SONEIRA GALAXY MODEL, TESTS OF THE MODEL

During the last few years star counts based on determinations of magnitudes in selected areas reaching to magnitudes 22^m-23^m have become an important tool in galactic research (cf. Kron (1978), Tyson and Jarvis (1979), Peterson et al. (1979) and Kron (1980)). Automated methods for evaluation of deep photographic plates have been developed and used to determine magnitudes for large numbers of faint stars. Fields in high and intermediate galactic latitudes, where interstellar absorption presents no great problems, have been studied, and here methods for distinguishing faint images of galactic stars and galaxies have been developed so that star counts to magnitude $B = 22\overset{m}{.}5$ are not appreciably affected by misclassification of galaxy images. This is a significant extension beyond the 18^m limit of the classical star counts.

H. van Woerden et al. (eds.), The Milky Way Galaxy, 153–160.

Guided by results concerning light distribution in external galaxies and by a large amount of available information regarding density distribution and luminosity function in our own Galaxy, Bahcall and Soneira (1980) have proposed a relatively simple two-component Galaxy model consisting of an axisymmetric, flattened disc and a spheroid, in their standard model assumed to have the three axes equal. For the disc and the spheroid, fairly simple density-distribution laws and luminosity functions are assumed, the choices being guided by the observational data just referred to. The number of free parameters in the density-distribution laws and luminosity functions is fairly small. They are adjusted so that agreement between predictions of the model and the results of the star counts is achieved as far as possible.

It is clearly important to test the predictions of the Galaxy model with data other than star counts. Kron (1978) has determined the colour-index distribution for stars in the selected area SA57 ($b=86^{\circ}$) for stars in the V-magnitude range $19^{m}.75$ to $22^{m}.0$, and found a bimodal distribution with well-separated maxima near $B-V = 0^{m}.5$ and $B-V = 1^{m}.5$. The distribution predicted from the Bahcall-Soneira Galaxy model agrees very well with the observed in this case. It is clear from the model prediction that the maximum at $B-V = 0^{m}.5$ is contributed mainly by spheroid stars, while the maximum at $B-V = 1^{m}.5$ comes from red main-sequence disc stars. Recent, as yet unpublished, results by Kron on star counts and colour-index distribution in selected areas at lower galactic latitudes do not agree as well with the predictions (Kron, personal communication).

In a paper on the significance of deep star counts for models of the Galaxy Gilmore (1981) has considered proposed Galaxy models, including the Bahcall-Soneira model, and the comparison of their predictions with observation. In particular, Gilmore discusses the question of the luminosity function for the spheroid component. In the Bahcall-Soneira model the spheroid luminosity function is put equal to the assumed disc luminosity function times a factor which is taken to be 1/800. The choice of this factor is based on a determination by Schmidt (1975) of the local density of spheroid stars from a study of local high-velocity stars with spheroid-star (i.e. halo-star) kinematics. Since the disc stars are mainly Population I stars, while the spheroid stars are extreme Population II stars, there are arguments for a different choice of luminosity function. The question of luminosity function of spheroid stars is further discussed by Bahcall, Schmidt and Soneira (1983).

In an extensive investigation Gilmore and Reid (1983) have determined the density distribution in the direction of the South Galactic Pole for a number of groups of main-sequence stars, divided according to absolute magnitude. The basic material consists of magnitudes V and colour indices V-I for a complete sample of 12500 stars brighter than $I = 18^{m}$ in an area of 18.24 square degrees towards the South Galactic Pole. The chosen procedure for determination of individual distances depends on the assumption that the stars are on the ZAMS, so that a previously established calibration of absolute magnitude M_V versus V-I can be used. A relatively small dependence of M_V upon metal content is taken summarily into account.

As a result a two-dimensional array can be constructed that gives the number of stars in bins defined by limits in log z and M_V, the log z range covered extending from 2.05 to 3.85, while the range of M_V is from $3^m.5$ to $12^m.5$.

Among the results obtained is a determination of the luminosity function at a distance z from the galactic plane of about z = 1000 pc. Here the absolute-magnitude range from 3.5 to 8 is well covered, and it is shown that the luminosity function in this range does not differ significantly from that valid near the galactic plane. Since the relative proportion of Population II is undoubtedly higher at z = 1000 pc than in the plane, this suggests that there is no great difference between the luminosity functions for Population I and Population II in the absolute-magnitude range in question.

Using the data for the groups with M_V in the ranges 4-5 and 5-6, the density distribution is determined out to z = 4000 pc and 3000 pc respectively. The authors conclude that there is, in addition to the standard disc with a scale height of 300 pc, a "thick disc" with a scale height of about 1350 pc, the latter locally comprising about 2 per cent of the stars. This result differs markedly from that of other investigations, and in particular it disagrees with the assumptions of the Bahcall-Soneira Galaxy model. Further studies are desirable. The authors have planned continued investigations of the faint stars that are indicators of the thick disc, through determination of further indices to strengthen the absolute-magnitude determinations (Gilmore, personal communication).

With regard to the assumption that the ZAMS absolute magnitude, as indicated by V-I, can be adopted for the stars in the important M_V magnitude range 4-5, some further information is already available. We note that the "thick-disc" stars must presumably be of Population II. Now E.H. Olsen (1983), from a photoelectric uvby survey of all A5-G0 stars in the Henry Draper catalogue which are brighter than $V=8^m.3$, has established an unbiased list of Population II stars, selected according to metal content as determined from the uvby photometry (cf. the last section of this contribution). The stars are practically all within 100 pc, and the list is complete to about 40 pc. For stars in the colour-index range (b-y) $0^m.29$-$0^m.39$, which corresponds to the ZAMS M_V range 4-5, E.H. Olsen (personal communication) has determined individual M_V- values. In addition to ZAMS stars the sample of about 400 stars contains a quite substantial number of evolved stars including subgiants, and the range of M_V for given colour index is well over one magnitude. Taking this effect into account for the Gilmore-Reid stars would influence the computed run of density with distance from the galactic plane.

Referring again to work discussed in the last section, the importance of forthcoming material of space velocities for the stars in E.H. Olsen's list of Population II stars should be mentioned. The sample is large enough for a fair determination of $\sigma(W)$, the r.m.s. velocity at right angles to the galactic plane, as a function of metal content for the stars of intermediate Population II, and an analysis will show whether indeed a fraction of all stars (in the absolute magnitude range considered) as high as 2 per cent has $\sigma(W)$-values large enough for the explanation of the "thick disc" feature.

In discussing tests of galactic models, in particular with regard to the spheroid component, we shall turn to two investigations which are based on identification of red giants of the spheroid population with the help of observational methods particularly developed for the purpose.

H.E. Bond (1980) has carried out a survey of extremely metal-deficient red giants using search on objective-prism plates followed by photoelectric uvby photometry. One objective-prism survey (Bond 1970) covers about 4000 square degrees of the northern hemisphere, with limiting B magnitude 10^m-$10^m.5$. A deeper survey, to B~$11^m.5$, covers 2300 square degrees near the North Galactic Pole and 2200 square degrees near the South Galactic Pole. These surveys are believed to be essentially complete for red giants with [Fe/H] $\leq$ -1.5 to -1.8. The selected sample of red giants thus has a metal content similar to that of member stars of the globular cluster M92. Next, photoelectric uvby photometry was carried out and used to determine individual values of [Fe/H], using the m_1-index, and further to isolate red giants, with (b-y) > $0^m.42$ and more luminous than $M_V \sim 2^m$ according to the c_1-index. Finally radial velocities were obtained for about half of the selected stars, and it was found that although selected without kinematic bias essentially all of the stars have halo-type motions. It is clear that the observational material just described is of considerable importance in tests of any galactic model with regard to its predictions concerning red giants in the spheroid population within distances 1000-2000 pc from the Sun.

A study by K.U. Ratnatunga (1982) aims at investigation of extremely metal-deficient K giants out to much larger distances and is therefore concerned with much fainter stars, to about V = 17^m. In a preliminary survey an area of 2.8 square degrees in SA141 close to the South Galactic Pole was investigated. Stars in the magnitude range 13^m<V<16^m and colour index range $0^m.9$<(B-V)<$1^m.4$ were selected. The subsequent procedure for finding metal-deficient giant K stars used low-resolution spectrophotometric observations for the determination of the strength of the MgH and Mg b feature around 5100 Å, which according to Clark and McClure is a good discriminant when the aim is to separate metal-deficient K giants from disc dwarfs. The preliminary survey of 2.8 square degrees showed that out of 126 stars in the V-magnitude range 13^m to 16^m and the B-V colour-index range $0^m.9$-$1^m.4$ a total of 23 were metal-deficient halo giants, and that of this sample 6 stars had distances larger than 10 kpc, while only 1 star was at a distance >20 kpc.

On the basis of these findings preparations have been made for an extensive survey to limiting magnitude V=$17^m.5$ in 5 fields at high galactic latitude, each covering 36 square degrees. It is expected that the required luminosity discrimination can be carried out using objective-prism Schmidt telescope plates with a dispersion of 900 Å per mm at Hγ.

It is clear that surveys of this type could contribute data of great value in the study of the properties of the spheroid component.

LOCAL DENSITY AND K_z FORCE

Oort (1965) has reviewed the problems in connection with the determination of the galactic K_z force and the local density from observational data for selected tracer groups of stars on the density distribution function $\nu(z)$ and the distribution function $f(W)$ of the space velocity component at right angles to the galactic plane, as determined for stars near the plane. It is essential that the tracer groups should be homogeneous so that the characteristics of the member stars do not vary with distance from the plane. King (1983) has emphasized this point and concluded that main-sequence stars of spectral classes F and G are much more suitable as tracer groups for the determination of the K_z force than either A stars or giant K stars.

Hill, Hilditch and Barnes (1979) have carried out a determination of the local density ρ_o, based on radial-velocity and photometric data for A and F stars. The results for F stars are of particular interest. A theoretical model due to Camm (1950, 1952) was used in deriving ρ_o. The density distribution $\nu(z)$ for the F stars for $z>100$ pc resulted from a revision of the analysis by Upgren (1962, 1963) of the Slettebak-Stock (1959) catalogue of spectral classes for stars in a North Galactic Pole region. Available data were used for a determination of the velocity dispersion $\sigma(W)$ near the galactic plane. Here the material concerning the intermediate-Population-II component among the late-type F stars was weak at the time, and it is of interest to note that quite adequate information on this point will soon be available (cf. the last section of this paper). Hill, Hilditch and Barnes (1979) emphasize the advantages that would be gained if the determination of $\nu(z)$ for F stars were based on uvbyβ photometry rather than on fairly crude spectral classification, and mention that they have embarked on an observational program of this type for stars within 15° of the North Galactic Pole. Their conclusion in the 1979 paper is that $\rho_o=0.14\ M_\odot\ pc^{-3}$, while they derive a value of the local density due to known matter in the solar neighbourhood equal to $0.108\ M_\odot\ pc^{-3}$. Thus their derived value for the "missing mass" is lower than that found in previous investigations, being reduced to $0.03\ M_\odot\ pc^{-3}$.

Bahcall (1984) has undertaken a broad-based discussion of the determination of the total amount of matter near the Sun. The combined Poisson-Boltzmann equation for the gravitational potential as a function of z is solved for Galaxy models composed of a number of isothermal components, the properties of the stellar components being assumed according to information on disc luminosity function and velocity dispersion $\sigma(W)$, cf. particularly Wielen (1974). The calculated potential is used to fit the distribution function $\nu(z)$ of F stars reported by Hill, Hilditch and Barnes (1979), by varying ρ_o as a parameter. As in previous determinations, the result depends on the assumptions made concerning the velocity dispersion $\sigma(W)$ for the "missing mass". For the preferred hypothesis Bahcall derives $\rho_o = 0.185\ M_\odot\ pc^{-3}$. He concludes that the "missing mass" amounts to about one-half of the total mass, and further that this "missing mass" must be largely concentrated to the galactic disc. Alternative models are discussed, but the results obtained do not change the main conclusions.

Further discussions in which improved determinations of $\nu(z)$ and $\sigma(W)$ for main-sequence F and G stars (cf. above, and also the following section) are utilized will clearly be of importance.

THREE uvbyβ SURVEYS AND THEIR ROLE IN CONNECTION WITH THE K_z PROBLEM

E.H. Olsen (1983) has carried out an all-sky uvby survey of nearly all Henry Draper stars of types A5 to G0 and brighter than visual magnitude $8^m_.3$. For about 2000 of the most metal-weak stars found, additional uvby and Hβ photometry has been obtained. The resulting catalogue presents the results of 27096 uvby and 7273 β measurements of 14816 stars.

A list of stars of intermediate Population II, as defined through the metal-content criterion $0^m_.045 < \delta m_1(b-y) < 0^m_.080$, was established on the basis of the catalogue. It contains about 600 stars forming a complete, magnitude-limited, kinematically unbiased sample of F stars belonging to intermediate Population II. An examination of the sample has fully confirmed a previous conclusion based on a much smaller sample, namely, that there is a sharp limit in the (b-y)-distribution corresponding to a turn-up at $(b-y)=0^m_.29$. This shows that the members of the group of stars defined through the metal-content criterion are practically all old stars - older than about 10×10^9 years. The cut-off in the program-star selection at Harvard spectral class G0 corresponds to $(b-y)\sim0^m_.40$, and the ZAMS visual absolute-magnitude range of the sample is therefore $4^m_.0-5^m_.0$.

J. Andersen and Mayor have undertaken to determine radial velocities for the stars on the intermediate Population II list. The observations, made with the Coravel radial-velocity instrument, have been completed, and a catalogue can be expected to be ready fairly soon (J. Andersen, personal communication). For most of the stars three Coravel determinations will be available, so that spectroscopic binaries can be segregated. Combined with proper motions and photometric distances derived from the uvbyβ photometry, the radial velocity will yield space velocities for the great majority of the 600 stars of the list mentioned, and consequently the correlations between age, chemical composition and kinematics can then be analyzed on the basis of information from an adequate sample.

With regard to the kinematics of F stars of Population I, in particular determination of $\sigma(W)$, reference is made to an investigation by Dennis (1966) which yielded $\sigma(W)$ as a function of stellar age, as derived from uvby observations of stars brighter than $V=6^m_.5$. A similar investigation of a somewhat larger sample, also to $V=6^m_.5$, by E.H. Olsen has given results close to those derived by Dennis (E.H. Olsen, personal communication). J. Andersen and B. Nordström have started observations on a radial-velocity program comprising a much larger sample of Population-I F stars drawn from E.H. Olsen's catalogue of 14816 stars. Ultimately this program should yield excellent information on the kinematics, in particular on $\sigma(W)$, for very homogeneous groups of stars selected on the basis of uvbyβ photometry, and thus contribute to K_z-force determinations.

J. Knude (personal communication) is obtaining photoelectric uvbyβ for A5-G0 stars within 20° of the North Galactic Pole. A list of 5500 A5-F0 program stars, complete to B about $11^m.5$, has been put at disposal by T. Oja. The spectral classes determined by Oja are from plates taken with a Schmidt telescope at the Uppsala-Kvistaberg Observatory for the purposes of a wider-scope spectrophotometric program for North Galactic Cap stars. J. Knude has carried out one-half of the photometric observations and expects that the program will be completed in 1984. Of the various applications of the resulting uvbyβ photometry the following is of importance in the context of the present article, namely, the derivation of very satisfactory distribution functions $\nu(z)$ for selected homogeneous groups of Population-I F stars out to distances of 500 pc.

T.B. Andersen (personal communication) has carried out photoelectric uvbyβ photometry in an area of 40 square degrees in the South Galactic Cap. The program consists of all stars brighter than $V=15^m.2$ and with colour index $(U-V)<0^m.60$. The selection of the program stars was made using Palomar Schmidt-telescope plates taken by B. Strömgren, these plates showing an ultraviolet-yellow image pair for each star. The evaluation of the plates was carried out by T.B. Andersen, who had the opportunity to use COSMOS at the Royal Observatory Edinburgh, with a program developed at ROE that was particularly suited to the purpose of the selection of the program stars. A pilot project by Crawford et al. (1979) showed that the selection of program stars to 15^m according to (U-V) leads to lists that contain well over 60 per cent Population II stars, a conclusion that was confirmed by T.B. Andersen's photometric results.

T.B. Andersen's project will be continued through further observations in the North Galactic Cap. It is of interest, however, to compare the results already obtained, and which pertain to stars out to distances between 1000 pc and 2000 pc, with the results found by E.H. Olsen for stars within 100 pc. The following table shows the distribution of the metal index δm_1 (b-y) for two samples, both with (b-y)-limits $0^m.29$-$0^m.39$ and δc_1 $(b-y)<0^m.10$, the E.H. Olsen sample (3949 stars) being complete to $V=8^m.2$, the T.B. Andersen sample limited to the V-magnitude range $14^m.0$-$15^m.0$ (114 stars). The material was kindly put at disposal by T.B. Andersen and E.H. Olsen prior to publication.

δm_1(b-y) unit $0^m.001$	[Fe/H]	Percentage of stars, EHO	Percentage of stars, TBA	Ratio TBA/EHO
-15 to 23	0.0	44 %	6.1 %	0.14
24 to 44	-0.2	45	26	0.58
45 to 80	-0.6	10.6	59	5.6
>80	<-0.9	0.23	8.8	38

In further work a more refined comparison procedure will be in order. However, the present "coarse analysis" clearly shows that the star density drop from the plane to volumes located somewhat beyond 1000 pc is a steeply varying function of the metal content. This fact should be of importance in future discussions of the K_z force problem, particularly when the aim is to determine K_z to larger distances z, using selected groups of Population-II stars as tracer samples.

REFERENCES

Bahcall, J.N., 1984, to appear in Ap.J. 276
Bahcall, J.N., and Soneira, R.M., 1980, Ap.J. Suppl. 44, 73
Bahcall, J.N., Schmidt, M., and Soneira, R.M., 1983, Ap.J. 265, 730
Bond, H.E., 1970, Ap.J. Suppl. 22, 117
Bond, H.E., 1980, Ap.J. Suppl. 44, 517
Camm, G.L., 1950, MNRAS 110, 305
Camm, G.L., 1952, MNRAS 112, 155
Crawford, D.L., Mavridis, L.N., and Strömgren, B., 1979, Abh. der Hamburger Sternwarte 10, 82
Dennis, T., 1966, Ap.J. 146, 581
Gilmore, G., 1981, MNRAS, 195, 183
Gilmore, G., and Reid, N., 1983, MNRAS, 202, 1025
Hill, G., Hilditch, R.W., and Barnes, J.V., 1979, MNRAS, 186, 813
King, I., 1983. Proceedings of the Vancouver Conference on the Milky Way (Reidel 1983, ed. W.H. Shuter)
Kron, R.G., 1978, Ph.D. thesis, University of California, Berkeley
Kron, R.G., 1980, in Two Dimensional Photometry (ESO Workshop) edited by P. Crane and N. Kjär, p. 349
Olsen, E.H., 1983, Astron. and Astrophys. Suppl. 54, 55
Oort, J.H., 1965, in Galactic Structure, ed. A. Blaauw and M. Schmidt (University of Chicago Press), p. 455
Peterson, B.A., Ellis, R.S., Kibblewhite, E.J., Bridgeland, M.T., Hooley, T., and Horne, D., 1979, Ap.J. (Letters) 233, L 109
Ratnatunga, K.U., 1982, Proc. ASA 4 (4), 422
Schmidt, M., 1975, Ap.J. 202, 22
Slettebak, A., and Stock, J., 1959, Hamburger Sternwarte 5, No. 5
Tyson, J.A., and Jarvis, J.F., 1979, Ap.J. (Letters), 230, L153
Upgren, A.R., 1962, A.J. 67, 37
Upgren, A.R., 1963, A.J. 68, 194
Wielen, R., 1974, in Highlights of Astronomy, Vol. 3, ed. G. Contopoulos (Dordrecht, Reidel), p. 395

THE STELLAR DISTRIBUTION IN THE GALACTIC SPHEROID

Gerard Gilmore
Royal Observatory, Edinburgh, and Visiting Associate,
Mount Wilson and Las Campanas Observatories

Our Galaxy is the only galaxy in which the 3-dimensional distribution of visible mass, chemical abundances and the stellar velocity field are all directly measurable. A project to determine these properties is currently underway, utilising direct photographic plates from the UK Schmidt telescope and the Las Campanas du Pont reflector, and the COSMOS and APM automated measuring machines. These provide reliable number-magnitude-colour distributions for complete samples of stars to V = 19 in 20 square degrees, and V = 22 in 1 square degree, in each of eight directions. These data may then be directly interpreted to determine the density profile, shape and luminosity function of the Galactic spheroid.

The chosen fields lie near $(l,b) = (0^{\circ},-90^{\circ})$; $(0^{\circ},-45^{\circ})$; $(90^{\circ},-45^{\circ})$; $(180^{\circ},-45^{\circ})$; $(270^{\circ},-45^{\circ})$; $(0^{\circ}-30^{\circ})$; $(0^{\circ},-60^{\circ})$; $(40^{\circ},-50^{\circ})$ and photometric results for the first and last of these fields are now available.

The complete sample of stars with photometrically determined absolute magnitudes $M_V > 14$ has been identified from the visual photometry, and every such star has been observed in the infrared (JHK). This provides both independent confirmation of the absolute magnitude, and a luminosity classification from the J-H/H-K diagram. These data have also been used to derive a revised effective-temperature scale for M dwarfs, which confirms that all known very late M dwarfs lie on the hydrogen-burning main-sequence. No "brown dwarfs" have yet been discovered. This same sample provides the first determination of the stellar main-sequence luminosity function which is complete to the minimum mass for hydrogen burning near $M_V = 19$. Integration of this function determines the total mass density in main-sequence stars in the solar neighbourhood to be $0.04\ M_{\odot}\ pc^{-3}$, and the total mass to light ratio, in solar visual units, to be ~ 1.2.

The available number-magnitude-colour data from the first two fields for $12 < V < 18$ are well fit by a three-component model with a disk, a "thick disk" and a spheroid, but are in poor agreement with the model published by Bahcall and Soneira (1981). The old disk follows an ex-

H. van Woerden et al. (eds.), The Milky Way Galaxy, 161–162.

ponential perpendicular to the plane with scale height ∿ 100 pc for M_V < 4 and ∿ 300 pc for M_V > 4. The luminosity function follows that of Wielen, with a dip for 6 < M_V < 9, a broad maximum near M_V = 12, and a slow decrease to M_V = 19. The "thick disk" is modelled with the luminosity function and colour-magnitude diagram of 47 Tuc, and the spheroid with the luminosity function and colour-magnitude relation for M92. The density law of the "thick disk" for $1 \leq z \leq 5$ kpc is equally well described by an exponential in z distance with a scale height near 1.5 kpc, and by a flattened $r^{1/4}$ spheroid with axial ratio near 1:4. In both cases approximately 2% of all stars in the solar neighbourhood belong to this population.

These results are in excellent agreement with recent studies of several edge-on spiral galaxies, thought to be similar to the Galaxy, which show pronounced "thick disk" structure perpendicular to their disks (e.g. van der Kruit and Searle, 1981), with scale heights and mass fractions relative to their thin disks very similar to those derived here.

Complete subsamples of stars from several fields are currently also being observed to determine the metallicity distribution and the velocity structure as a function of position in the spheroid, and to redetermine the local K_z force law and mass density.

Further details are available in Gilmore and Reid (1983) and references therein, and several forthcoming papers.

REFERENCES

Bahcall, J., and Soneira, R.: 1981, Ap. J. Suppl. 47, 357.
Gilmore, G., and Reid, N.: 1983, M.N.R.A.S. 202, 1025.
van der Kruit, P., and Searle, L.: 1981, Astron. Astrophys. 95, 105.

DISCUSSION

B. Strömgren: It will be extremely interesting to follow this work up through four-colour photometry and radial-velocity work, to determine the composition and kinematical properties of the stars that form the "thick disk". The suggestion that it represents 2% of the local population, and that its velocity dispersion is about 60 km/s, does not check with Intermediate Population II (IP2) as a whole, but it might represent just the lower-metal-content fringe of IP2 and the higher-metal-content fringe (the 47-Tucanae stars) of the halo-population II as generally defined. I think that further work will soon clear up this point.

STUDIES OF O-F5 STARS AT THE GALACTIC POLES

R.W. Hilditch and A.D. McFadzean
University Observatory, St Andrews, Scotland
Graham Hill
Dominion Astrophysical Observatory, Victoria
J.V. Barnes
Kitt Peak National Observatory, Tucson

We report progress on a spectroscopic and photometric programme devoted to the study of the dynamics of O-F5 stars within 15° of the North and South Galactic Poles. The aims of the programme are to test dynamical and chemical evolution models of the Galaxy by establishing velocity dispersions as a function of z-distance for stars of different population groups. We are also able to investigate the interstellar reddening at the poles and the kinematic properties of apparently normal early-type stars found more than 1 kpc from the galactic plane.

An initial survey of ∿300 stars at the NGP was completed in 1976 and a discussion of the observed and dynamical local mass densities was published by Hill et al. (1979 and see refs. therein). This survey was extended to include all O-F5 stars within 15° of the NGP in the AGK3 catalogue and the objective-prism survey of Slettebak and Stock (1959). At the SGP, we initiated a survey of all HD stars in that spectral-type range.

Completed uvbyβ photometry of a further ∿700 stars at the NGP is now published (Hill & Barnes 1982a,b) together with a catalogue of cross-reference star names and positions (Hill 1982) and a catalogue of derived intrinsic colours, distances and classifications (Hill et al. 1982a). A reddening map together with an intrinsic-colour calibration for B9-A3 stars is given by Hilditch et al. (1983). This map refines the original 1976 data and confirms the zero point of the HI/GC method of Burstein & Heiles (1982) for determining total interstellar extinction through the Galaxy at intermediate and high latitudes. At the SGP, uvbyβ photometry of 572 O-F8 stars has been completed by McFadzean et al. (1983). Interstellar reddening has been shown to be negligible and a number of horizontal-branch stars, subdwarfs etc have been identified.

Combining the uvbyβ photometry on these ∿1600 stars at the galactic poles has allowed us to study the relative proportions of intermediate- and halo-population II stars (as defined by Stromgren 1966) as a function of z-distance (Table I). These data show essentially constant relative proportions of ipII and pII stars out to ∿400 pc, after which they begin

H. van Woerden et al. (eds.), The Milky Way Galaxy, 163–164.

to increase. This result reflects the scale height of the population I disk stars.

Spectroscopic observations (at 30 and 80 A mm^{-1}) are now completed (∿3 spectra per star) for the ∿1000 NGP stars and for ∿300 SGP stars. The determination of radial velocities from these spectra is being carried out using an interactive graphics package REDUCE (Hill et al. 1982b) and we expect these velocities to be completed by mid-1984.

Table 1
Relative proportions of population groups

	A stars		F stars	
distance (pc)	ipII/pI	pII/pI	ipII/pI	pII/pI
0- 100	0.04	0.00	0.12	0.02
101- 200	0.03	0.01	0.08	0.01
201- 300	0.03	0.00	0.07	0.02
301- 400	0.00	0.00	0.10	0.02
401- 500	0.04	0.00	0.25	0.04
501-1000	0.16	0.01	0.44	0.06

REFERENCES

Burstein, D., and Heiles, C.: 1982, Astron. J. 87, 1165.
Hilditch, R.W., Hill, G., and Barnes, J.V.: 1983, Mon. Not. Roy. Astron. Soc. 204, in press.
Hill, G.: 1982, Pub. Dom Astrophys. Obs. 16, 87.
Hill, G., and Barnes, J.V.: 1982a, Publ. Dom. Astrophys. Obs. 16, 71.
Hill, G., and Barnes, J.V.: 1982b, Publ. Dom. Astrophys. Obs. 16, 81
Hill, G., Hilditch, R.W., and Barnes, J.V.: 1979, Mon. Not. Roy. Astron. Soc. 186, 813.
Hill, G., Barnes, J.V., and Hilditch, R.W.: 1982a, Publ. Dom. Astrophys. Obs. 16, 111
Hill, G., Fisher, W.A., and Poeckert, R.: 1982b, Publ. Dom. Astrophys. Obs. 16, 43
Mc Fadzean, A.D., Hilditch, R.W., and Hill, G.: 1983, Mon. Not. Roy. Astron. Soc. 204, in press
Slettebak, A., and Stock, J.: 1959, Hamburger Sternw. V. No. 5
Stromgren, B.: 1966, Ann. Rev. Astr. Astrophys. 4, 433

GALAXY POPULATION STRUCTURE FROM PROPER MOTIONS

N. Reid
University of Sussex

C.A. Murray
Royal Greenwich Observatory

We are currently engaged in astrometry of Kapteyn Selected Area photographic plates covering the period 1908 to the present day using the GALAXY measuring engine at RGO. Our intention is to use the apparent motions to investigate the stellar velocity structure within a kiloparsec of the Sun. We have completed a preliminary analysis of SA 68 (ℓ = 111°, b = 46°), deriving positions and proper motions using the central overlap method on measures of plates spanning the period 1909 to 1980. These include the original (1909) and second epoch (1925) plates taken with the Radcliffe refractor at Oxford, and used in the compilation of the Radcliffe Catalogue (Knox-Shaw and Scott-Barrett, 1934), as well as more recent plates taken using the 26-inch refractor at Herstmonceux. For stars present on all thirteen exposures (including double exposures on the same plate) the internal residuals in positions in each co-ordinate are 0"09 per plate, giving annual motions accurate to ~1 milli-arcsecond over the 71 year baseline. Since our plates have a limiting magnitude of only B ~15, we have used 17 stars in common with Chiu (1980) to transform our relative motions to the absolute frame. Our motions are in very good agreement with those by Chiu, with a scatter of less than 3 milliarcseconds in each co-ordinate and linearity of scale between the two datasets.

From density-law considerations, 95 percent of the stars with colours in the range $0.4 < (B-V) < 0.8$ are expected to be disk dwarfs. Hence we have used photometric parallaxes to determine distance moduli for these stars and convert the observed tangential motions to linear velocities. Since the average distance above the Plane is ~350 parsecs including a reasonable metallicity gradient changes distances by less than 5 percent. We derive mean motions of

$$\bar{V}_\alpha = -29 \text{ km/sec} \qquad \bar{V}_\delta = -26 \text{ km/sec}$$

with dispersions of

$$\sigma_1 = 50.0 \text{ km/sec} \qquad \sigma_2 = 40.9 \text{ km/sec}$$

along the principal axes of the projected velocity ellipsoid.

H. van Woerden et al. (eds.), The Milky Way Galaxy, 165–166.

This compares with

$$\bar{V}_\alpha = 17.5 \text{ km/sec} \qquad \bar{V}_\delta = -9.5 \text{ km/sec}$$

$$\sigma_1 = 35 \text{ km/sec} \qquad \sigma_2 = 25 \text{ km/sec}$$

expected from basic solar motion and the velocity ellipsoid of GK dwarfs in the Plane. While the larger dispersions are expected for stars at higher z, the change in the solar motion is not. Since this sample includes 80 stars, the mean motions are particularly well determined - although systematic errors in the conversion to absolute motions cannot be excluded. However, even zero point errors as large as 0.005 arcsec only change the centroid of motion by 12 km/sec. This result will be further investigated using data from other fields.

The most significant result from this preliminary analysis of SA68 is shown in the reduced proper-motion diagram, given by plotting

$$H = M + 5 + 5 \log \mu = M + 5 \log V_t - 3.378 \text{ versus } (B-V).$$

This represents a composite H-R diagram, where the zero point in H for a given population is set by the mean transverse velocity, V_t. From our data it is evident that there is a well-populated giant branch extending from (H = 10; (B-V) = 0.8) to (6;1.2). These stars have apparent magnitudes of V ~12-14, and hence are at distances of 2.5 - 7 kpc. The conventional interpretation of these stars as metal-poor (M92, M13) halo giants requires that they have unreasonably high mean tangential motions (650 and 540 km/sec respectively). Alternatively, these stars could be represented as disk-metallicity giants with V_t ~100 km/sec, or as 47 Tucanae-type giants with $V_t \sim$ 180 km/sec. Ratnatunga and Freeman have reported the discovery of substantial numbers of the latter stars in their objective-prism surveys (see Freeman, this conference). This implies the presence of significant numbers of relatively metal-rich stars in the halo population of our Galaxy, and we tentatively identify these stars with the extended stellar component recently discussed by Gilmore and Reid (1983). Spectroscopic observations of these stars are in progress to confirm their luminosity class and metallicity and to determine radial velocities, while we are currently extending our proper-motion survey to cover other selected areas.

A more detailed discussion of these data will be submitted to Monthly Notices of the RAS.

REFERENCES

Chiu, L-T. G.: 1980, Astrophys. J. Suppl. 44, 31
Gilmore, G., and Reid, N.: 1983, Mon. Not. r. Astr. Soc. 202, 1025
Know-Shaw, H. and Scott-Barrett, H.G.: 1934, The Radcliffe Catalogue of proper motions in the selected areas 1 to 115. Oxford Univ. Press

SECTION II.3

THE GASEOUS COMPONENT: LARGE-SCALE DISTRIBUTION

Monday 30 May, 1100 - 1245 and 1400 - 1445

Chairmen: V. Radhakrishnan and W.M. Goss

Radhakrishnan (top) and Goss (bottom), absorption observers and southern sea sailors, share the chair in session on Gas in the Galaxy. CFD

LARGE-SCALE DISTRIBUTION AND MOTIONS OF GAS

Unfortunately, the text of this review paper is not available. For recent reviews, we refer the reader to the following papers:

REFERENCES

Burton, W.B. 1976, Ann. Rev. Astron. Astrophys. **14**, 275

Henderson, A.P., Jackson, P.D., Kerr, F.J. 1982, Astrophys. J. **263**, 116

Kerr, F.J. 1983, in "Kinematics, Dynamics and Structure of the Milky Way" ed. W.L.H. Shuter, Dordrecht, Reidel, p. 91

Knapp, G.R. 1983, in "Kinematics, Dynamics and Structure of the Milky Way", ed. W.L.H. Shuter, Dordrecht, Reidel, p. 233

Kulkarni, S.R., Blitz, L., Heiles, C.E. 1982, Astrophys. J. (Letters) **259**, L63

Lockman, F.J. 1979, in "Large-Scale Characteristics of the Galaxy", ed. W.B. Burton, IAU Symp. 84, p. 73

Sanders, D.B., Solomon, P.M., Scoville, N.Z. 1984, Astrophys. J. **276**, 182

H. van Woerden et al. (eds.), The Milky Way Galaxy, 169.

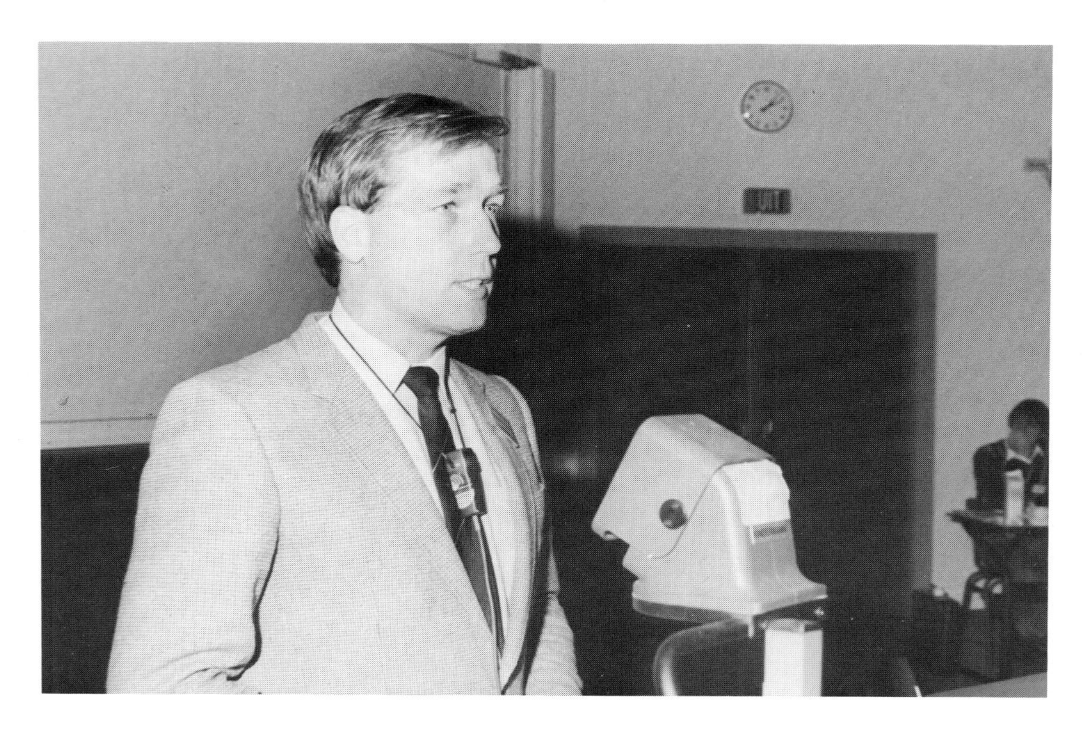

Hermsen (top) and Gautier (bottom) discuss the large-scale distribution of gas on the basis of observations of gamma-rays (COS-B, page 213) and dust emission (IRAS, page 219) CFD

SURVEY OF GALACTIC HI EMISSION AT $|b| \leq 20^o$

W.B. Burton and P. te Lintel Hekkert
Sterrewacht, Leiden

We have used the NRAO 140-ft telescope to survey galactic HI near the galactic equator. The observations cover the part of the sky north of $\delta = -40^o$ in the latitude range $b = -20^o$ to at least $+20^o$; in some regions the survey extends to $b = +33^o$. The longitude coverage in the equator ranges from $\ell = 340^o$ to 270°. The data represent an improvement over existing large-scale HI surveys in terms of velocity coverage (± 250 km s^{-1}), velocity resolution (1 km s^{-1}), sensitivity ($3\sigma = 0.2$ K), and extent of sky coverage, but not in density ($\Delta\ell \times \Delta b = 1^o \times 1^o$) of coverage.

The data are amenable to a wide variety of studies pertaining to galactic structure and the interstellar medium. In Leiden we are directing our attention to (i) the parameters specifying the galactic warp and flare, (ii) the integrated HI properties, especially those which reveal optical-depth characteristics, (iii) the identification of the phenomenon of HI supershells as proposed by Heiles, and (iv) the extent of asymmetries in the galactic velocity field.

The survey is being published as a series of ℓ,v maps at integral values of b from $b = -20^o$ to $+20^o$; b,v maps at integral values of ℓ from $\ell = 0^o$ to 359°; and ℓ,b maps representing intensities integrated over 2.5 km s^{-1} velocity intervals centered every 2.5 km s^{-1} from v = -150 km s^{-1} to +150 km s^{-1}. An example of an ℓ,v map and one of an ℓ,b map are given on the following page. The profiles will be made available on request in FITS or NRAO T-POWER format.

Figure 1. Arrangement on the plane of the sky of the HI emission integrated over the range $+30.0 < v < +32.5$ km s^{-1}. The contours represent levels 0.8, 1.6, 3.0, 5, 8, 14, 20, 30, 40, 50, 60, 80, 100, 120, 150, ... K km s^{-1}; the lowest grey levels are at 0.2, 0.3, and 0.4 K km s^{-1}.

Figure 2. HI intensities in ℓ,v coordinates at $b = +11^o$. The contours represent antenna temperatures at levels 0.2, 0.4, 0.8, 1.5, 2.5, 4.0, 7, 10, 15, 20, 25, 35, ... K.

H. van Woerden et al. (eds.), The Milky Way Galaxy, 171–172.

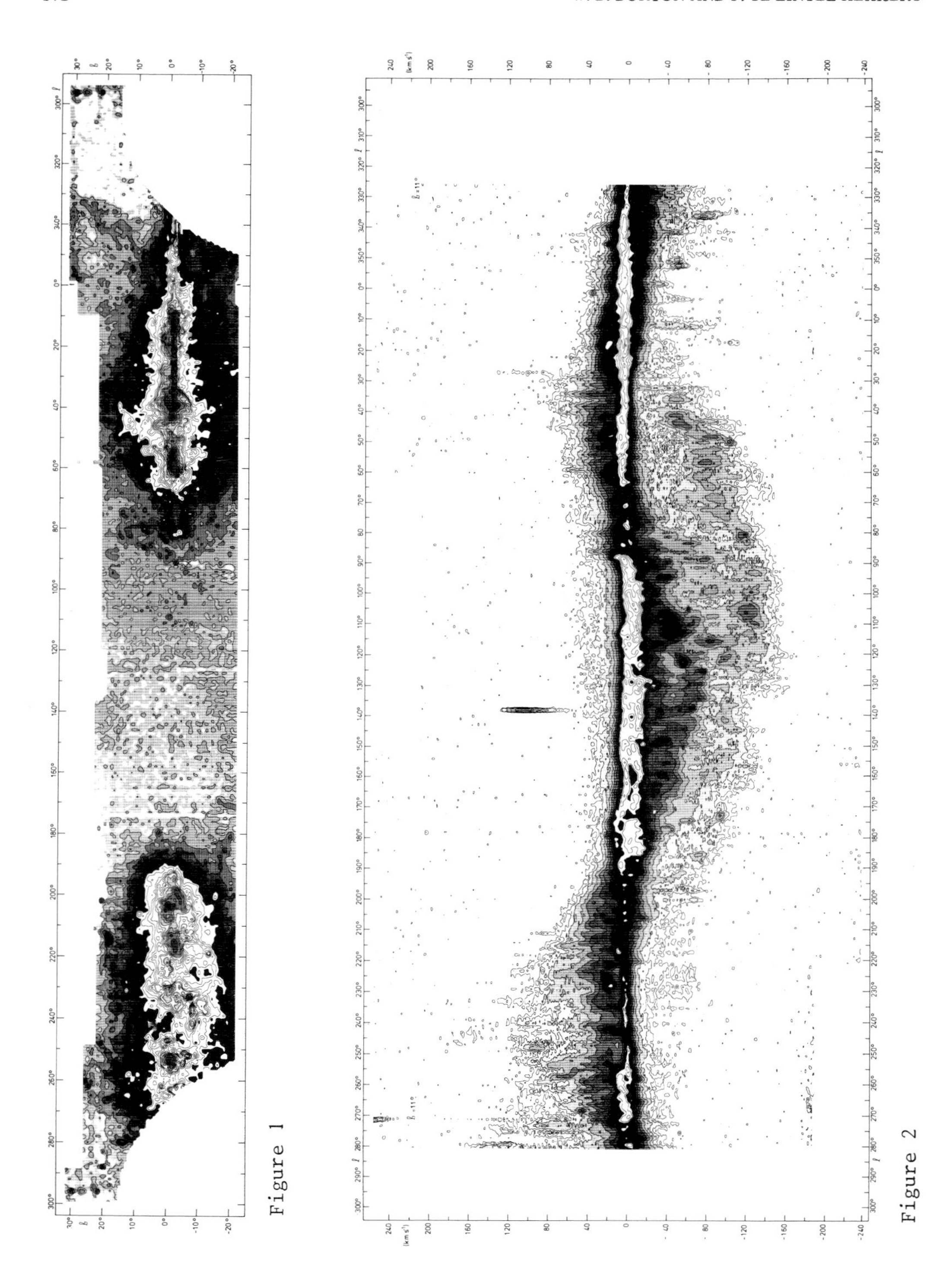

Figure 1

Figure 2

THE VERTICAL DISTRIBUTION OF GALACTIC HI: THE ARECIBO-GREEN BANK SURVEY

T.M. Bania
Department of Astronomy, Boston University, Boston, MA

F.J. Lockman
National Radio Astronomy Observatory, Charlottesville, VA

We have measured the vertical distribution of HI in the first quadrant of galactic longitude between tangent-point radii, $R_t = R_o \sin \ell$, of 2.5 kpc and 9.75 kpc at 250 pc intervals of tangent-point distance, using the 305m Arecibo telescope and the 91m and 43m NRAO telescopes. The survey is complete to at least $z = \pm 500$ pc and it was sampled with half-beamwidth spacings; the angular resolutions at 21cm wavelength are 4, 11 and 20 arcmin for these instruments.

The HI distribution at the 4' resolution shows many discrete emission minima caused by absorption of hot background HI emission by cooler foreground gas. The special geometry required for self-absorption provides a resolution of the kinematic-distance ambiguity for cool HI clouds at positive LSR velocities in our survey. We have compiled a catalog of ≃200 self-absorbing HI clouds with such kinematic distances. These objects provide a new population of dense, cool clouds for further study. (The sample is not biased towards discrete sources of continuum radiation, either extra-galactic objects or pulsars, for example.) Accurate masses and sizes can in principle be derived for these clouds with known distances.

We have used this survey to study the vertical structure of both the HI distribution and galactic rotation. There is 21cm emission from corotating HI in the inner Galaxy to $|z| \geqslant 1500$ pc from the plane. At least 10% of the HI emission at the subcentral point comes from gas at $|z| \geqslant 500$ pc. The HI terminal velocity is often not symmetric about the plane, but varies linearly over hundreds of parsecs in the manner of a vertical shear. Typically, dV_t/dz is -10 km s^{-1} kpc^{-1} between 5 and 8 kpc from the galactic center. This implies that the HI below the plane is rotating faster than the HI above the plane.

H. van Woerden et al. (eds.), The Milky Way Galaxy, 173.

Butler Burton relaxes on the lakes. GSS

NEW LIGHT ON THE CORRUGATION PHENOMENON IN OUR GALAXY

J.V. Feitzinger and J. Spicker
Astronomical Institute, Ruhr-University Bochum

This investigation presents a total picture of the well-known corrugation-phenomenon for the (heliocentric) longitude range $10° \leqslant l \leqslant 240°$ as derived from HI-studies. For each spiral arm of the spiral pattern of Simonson (1976), we derived the centroid of the HI distribution from the 21-cm line surveys of Weaver and Williams (1974), Sinha (1979), and Westerhout and Wendlandt (1982). The three-component mass model of Rohlfs and Kreitschmann (1981) was used to derive a radial-velocity field, which was supplemented by a radial expansion field and by density-wave kinematics. This combined field served to calculated kinematic distances. The warp was taken into account according to Henderson et al. (1982) and Kulkarni et al. (1982).

Fig. 1 shows the corrugated spiral arms as viewed from the NGP. Plus sign denotes above, minus sign below, circle in the plane. Fig. 2 gives the detailed configuration for the Perseus Arm (-I).

The corrugation scale length λ is defined as a wavelength, i.e., from maximum to maximum. The corrugation scale amplitude Δ is defined like a wave amplitude, i.e., between extreme values. Table 1 summarizes optical and radio observations in azimuthal direction, Table 2 the same in radial direction. In azimuthal direction, Schmidt-Kaler and House (1976) predict values of λ = 1.2 - 1.4 kpc and Δ = 200 pc for arm -I. In radial direction, Nelson (1976, 1980) predicts hydrodynamical oscillations with λ = 1 - 2 kpc and Δ = 140 pc. In azimuthal direction, three corrugation scales are clearly visible in almost every arm: 1) $1 \leqslant \lambda \leqslant 2$ kpc, 2) $4 \leqslant \lambda \leqslant 7$ kpc, 3) $\lambda \geqslant 10$ kpc. These scale lengths do not depend on the distances R_0 from the centre and have nearly the same value for every arm. The scale amplitude increases with increasing R_0, perhaps exponentially. In radial direction (across the galactic plane), there may be two different scale lengths ($\lambda \lesssim 3$ kpc and $\lambda \gtrsim 4$ kpc), one of which is not completely covered by our observations. Difficulties arise, because λ and Δ are of the same order in radial and in azimuthal direction, so that some sort of interference occurs. In conclusion, a corrugation effect is clearly visible in spiral arms as well as in the plane itself.

H. van Woerden et al. (eds.), The Milky Way Galaxy, 175–178.

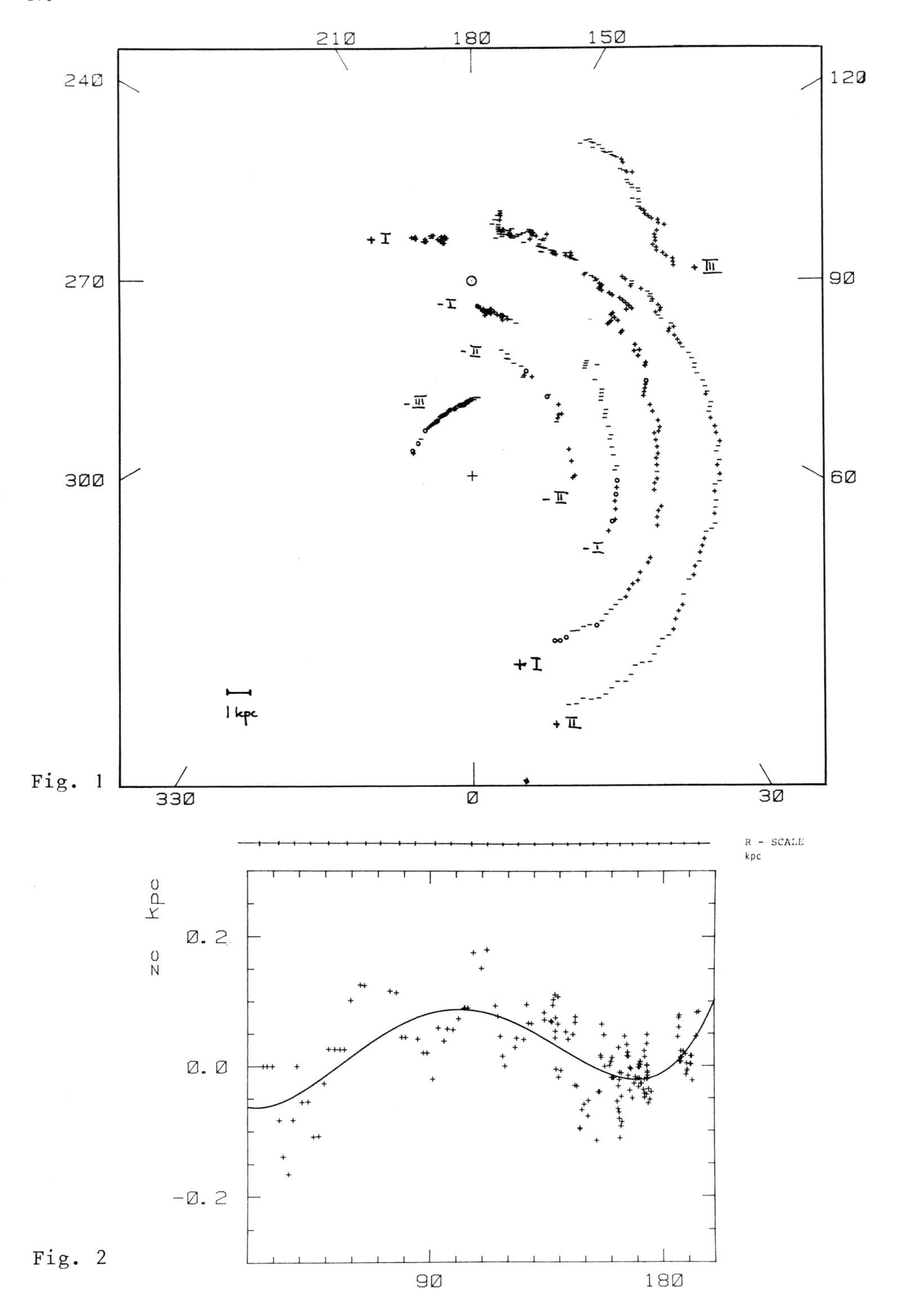

210
180
150
240
120
270
90
300
60
330
0
30
+I
+III
-I
-II
-III
-II
-I
+I
+II
1 kpc
R - SCALE
kpc
z0 kpc
0.2
0.0
-0.2
90
180

Fig. 1

Fig. 2

TABLE 1
Corrugations in azimuthal direction

Reference	Spiral Arm	Object	L-Range	λ kpc	Δ pc	Scale Assignment
This paper	-III	HI	350- 10	1.0	60	1
This paper		HI		4.2	90	2
This paper	-II	HI	30- 60	15.6	110	3
Schmidt-Kaler, Schlosser 1973	-I	Clusters, OB-assoc.		1.2	70	1
This paper		HI	30- 60	1.8	40	1
This paper		HI		6.0	125	2
This paper		HI		13.6	160	3
Dixon 1967	0	B0-B2 Stars		1.7	300	1
Lyngå 1970		OB Stars		0.8		1
Quiroga, Schlosser 1977		HI	60- 90	1.6	200	1
				1.3	300	1
Kolesnik, Vedenisheva 1979, 1981	+I	O-B2 Stars	80-270	2.0	150	1
Hidayat et al. 1982		WR-Stars		2.3	300	1
This paper		HI	10-250	1.7	140	1
Kolesnik, Vedenisheva 1979, 1981		O-B2 Stars	70-185	4.5	200	2
This paper		HI	10-250	6.0	180	2
This paper		HI		23.5	350	3
This paper	+II	HI	10-100	6.6	260	2
This paper	+III	HI	90-160	2.0	200	1
This paper		HI		4.3	350	2

REFERENCES

Dixon, M.E., 1967, Monthly Notices Roy. Astr. Soc. 137, 337
Henderson, A.P., Jackson, P.D., Kerr, F.J., 1982, Astrophys.J. 263, 116
Hidayat, B., Supelli, K., van der Hucht, K.A., 1982, IAU Symp. 99, p.27
Janes, K., Adler, D., 1982, Astrophys. J. Suppl. 49, 425
Kolesnik, L.N., Vedenisheva, I.P., 1979, Astron. Astrophys. 76, 124
Kolesnik, L.N., Vedenisheva, I.P., 1981, Astron. i Astrofys. 43, 67
Kulkarni, S.R., Blitz, L., Heiles, C., 1982, Astrophys. J. 259, L63
Lockman, F.J., 1977, Astron. J. 82, 408
Lyngå, G., 1970, Astron. Astrophys. 8, 41
Milne, D.K., 1979, Austr. J. Phys. 32, 83
Nelson, A.H., 1976, Monthly Notices Roy. Astr. Soc. 174, 661
Nelson, A.H., 1980, Monthly Notices Roy. Astr. Soc. 191, 221

TABLE 2
Corrugations in radial direction

Reference		Object	R-Range kpc	λ kpc	Δ pc
Varsavsky, Quiroga	1970	HI	4 - 10	2.8	80
				2.8	90
				2.2	85
Quiroga	1974	HI	4 - 9	2.5	150
				2.8	160
				2.6	170
				1.5	100
Quiroga	1977	HI	4 - 8	2.3	160
				2.6	190
Lockman	1977	HII	3 - 9	4.5	100
				3.8	110
				2.2	80
Quiroga	1978	CO	4 - 8	2.3	240
Milne	1979	SNR	3 - 13	6.0	280
Janes, Adler	1982	Open clusters	6 - 10	1.3	150
Sanders	1983	CO	3 - 8	4.2	40
This paper		HI	4 - 12	4.0	120
				4.2	180
			4 - 11	3.8	90
				4.3	90
			6 - 11	4.3	130

Quiroga, R.J., 1974, Astrophys. Space Sci. 27, 323
Quiroga, R.J., 1977, Astrophys. Space Sci. 50, 281
Quiroga, R.J., 1974, Astrophys. Space Sci. 53, 295
Quiroga, R.J., Schlosser, W., 1977, Astron. Astrophys. 57, 455
Rohlfs, K., Kreitschmann, J., 1981, Astrophys. Space Sci. 79, 289
Sanders, D.B., 1983, in: Shuter, W.L.H. (ed.), Kinematics, Dynamics and Structure of the Milky Way, Dordrecht, Reidel, p. 115
Schmidt-Kaler, Th., Schlosser, W., 1973, Astron. Astrophys. 25, 191
Schmidt-Kaler, Th., House, F., 1976, Astron. Nachr. 297, 77, 83
Simonson, S.C., 1976, Astron. Astrophys. 46, 261
Sinha, R.P., 1979, Astron. Astrophys. Suppl. 37, 403
Varsavsky, C.M., Quiroga, R.J., 1970, IAU Symp. 38, p. 147
Weaver, H.F., Williams, D.R.W., 1974, Astron. Astrophys. Suppl. 17, 1
Westerhout, G., Wendlandt, H.U., 1982, Astron. Astrophys. Suppl. 49,143

HI AT THE OUTER EDGE OF THE GALAXY AND ITS IMPLICATIONS FOR GALACTIC ROTATION

Peter D. Jackson
Department of Physics, York University
Downsview, Ontario Canada

In their analysis of global properties of the Galaxy based on the HI distribution outside of the solar circle, Knapp, Tremaine, and Gunn (1978 - hereafter referred to as KTG) have examined the extreme HI velocities for the outer edge of the Galaxy. They fit these values to a model based on an exponential drop-off of HI surface density with galactocentric radius, R, and a flat rotation curve. Adopting a value of 4 kpc for the scale length, L, of this drop-off, KTG fit the extreme velocities with various values of Θ_o (the circular rotation speed of the Local Standard of Rest), to obtain a best fit of Θ_o = 220 km/s. While KTG obtained sensitive observations with the NRAO 43-m radio telescope for the galactic-longitude range l = 80° to 225°, they had to rely on the HI survey by Kerr, Harten, and Ball (1976 - hereafter referred to as KHB), obtained with the Australian 64-m radio telescope at Parkes, for the balance of the Southern Milky Way.

A fully-sampled HI survey, using the Parkes 18-m telescope and extending to galactic latitudes ±10°, is now ready for publication (Kerr et al. 1983). The values for the extreme HI velocities for the outer Galaxy have been determined as a function of l and were found to be systematically higher than one would determine using KHB. The reason for this discrepancy is apparently that KHB used a filter-bank receiver with limited velocity range and, consequently, the upper velocity of their observing window was at a velocity where HI emission is still present. The 18-m survey, while also using a filter-bank receiver, used a wider bandpass which has definitely included all of the HI.

The extreme velocity at which HI appears at a brightness temperature >1 K has been determined for all galactic longitudes using both the 18-m survey and the Weaver and Williams (1973) HI survey. Latitudes up to $|b|$ = 10° were searched. Results are plotted in Figure 1 along with the predictions of KTG for their model with Θ_o = 220 km/s and 250 km/s. As reported elsewhere (Jackson and Kerr 1981), the southern Milky Way data fit with Θ_o = 250 km/s while the northern Milky Way data fit with Θ_o = 220 km/s. Figure 1 may be compared directly with Figure 12 of KTG.

H. van Woerden et al. (eds.), The Milky Way Galaxy, 179–180.

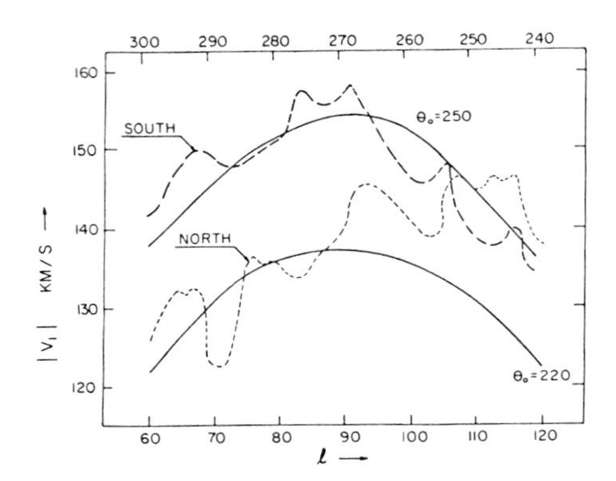

Figure 1. The largest velocity at which the HI brightness temperature exceeds 1 K plotted against l for the northern and southern Milky Way.

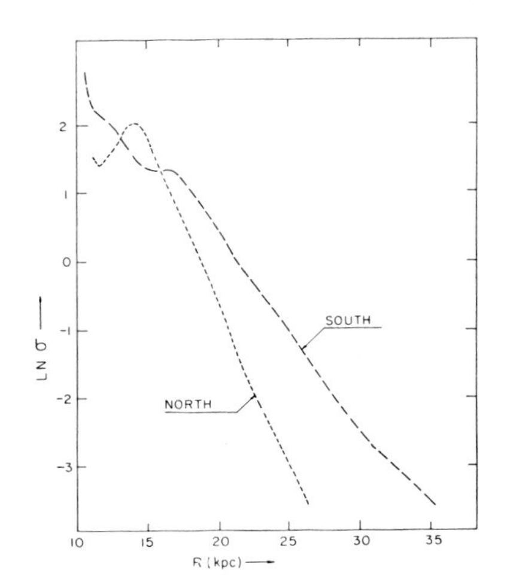

Figure 2. The neutral-hydrogen surface density in solar masses per square parsec plotted against galactocentric radius in kilo-parsecs.

More recently, an analysis (Henderson, Jackson, and Kerr 1982) of the 18-m and Weaver and Williams surveys has yielded a determination of the surface density, σ, of the HI as a function of R assuming a flat rotation curve, and is plotted in Figure 2 on a logarithmic scale. Straight-line fits yield L = 2.2 for the northern and L = 3.3 kpc for the southern Milky Way. Clearly, a steeper drop-off in the southern HI surface density than that used by KTG, assuming other model parameters are unchanged, means that they have underestimated Θ_o. In fact, KTG find that Θ_o = 250 km/s can fit their data if L = 3.2 kpc. New models should now be run with the steeper density gradient and, perhaps, also with a rising rotation curve in view of the results of Blitz, Fich and Stark (1980) and Chini (1983).

REFERENCES

Blitz, L., Fich, M., Stark, A.A.: 1980, in "Interstellar Molecules", IAU Symp. 87, ed. B.H. Andrew (Reidel, Dordrecht), p. 213

Chini, R.: 1983, IAU Symp. 106, "The Milky Way Galaxy"

Henderson, A.P., Jackson, P.D., Kerr, F.J.: 1982, Astrophys. J. 263, 116

Jackson, P.D. and Kerr, F.J.: 1981, Bull. Amer. Astron. Soc. 13, 538

Kerr, F.J., Bowers, P.F., Kerr, M., Jackson, P.D.: 1983 (in preparation)

Kerr, F.J., Harten, R.H., Ball, D.L.: 1976, Astron. Astrophys. Suppl. Ser. 25, 391

Knapp, G.R., Tremaine, S.D., and Gunn, J.E.: 1978, Astron. J. 83, 1585

Weaver, H. and Williams, D.R.W.: 1973, Astron. Astrophys. Suppl. Ser. 8, 1

THE ELECTRON DENSITY IN THE PLANE OF THE GALAXY

J.M. Weisberg
Department of Physics, Princeton University

J.M. Rankin
Department of Physics, University of Vermont

V. Boriakoff
NAIC, Cornell University

Pulsars provide probably the best probes of electron density in the plane of the Galaxy. The dispersion measure, the path integral of electron density along the line of sight from the pulsar to Earth, $\int n_e ds$, is directly measurable from multi-frequency pulse-timing observations. The distance to a pulsar, d, can be estimated from its HI absorption and emission spectra. The mean electron density along the line of sight is then just $\langle n_e \rangle = \int n_e ds/d$.

We have recently used the Arecibo 305-meter telescope to measure the HI absorption and emission spectra in the direction of nine pulsars. These measurements bring to 42 the number of pulsars with distance estimates and hence estimates of mean electron densities along the lines of sight. However, since most pulsars are weak radio sources at 21-cm wavelength, a significant fraction of the reported pulsar distance measurements are corrupted by noise or overzealous interpretation. We have reanalyzed all published pulsar distance measurements, rejecting those deemed marginal and estimating kinematic distances in a uniform fashion for the remainder of the sample (32 pulsars).

If we restrict our attention to the 16 lines of sight with both upper <u>and</u> lower distance limits (and hence lower <u>and</u> upper electron density limits), then we find that the electron density distribution peaks sharply near .03 cm^{-3}. The density enhancements previously noted in the direction of the Gum Nebula and the inner Galaxy (Ables and Manchester 1976, Weisberg et al. 1980) are less evident in our selected sample, although they are still marginally visible.

REFERENCES

Ables, J.G., and Manchester, R.N.: 1976, Astron. Astrophys. 50, pp.177-184
Weisberg, J.M., Rankin, J., and Boriakoff, V.: 1980, Astron. Astrophys. 88, pp. 84-93

H. van Woerden et al. (eds.), The Milky Way Galaxy, 181.

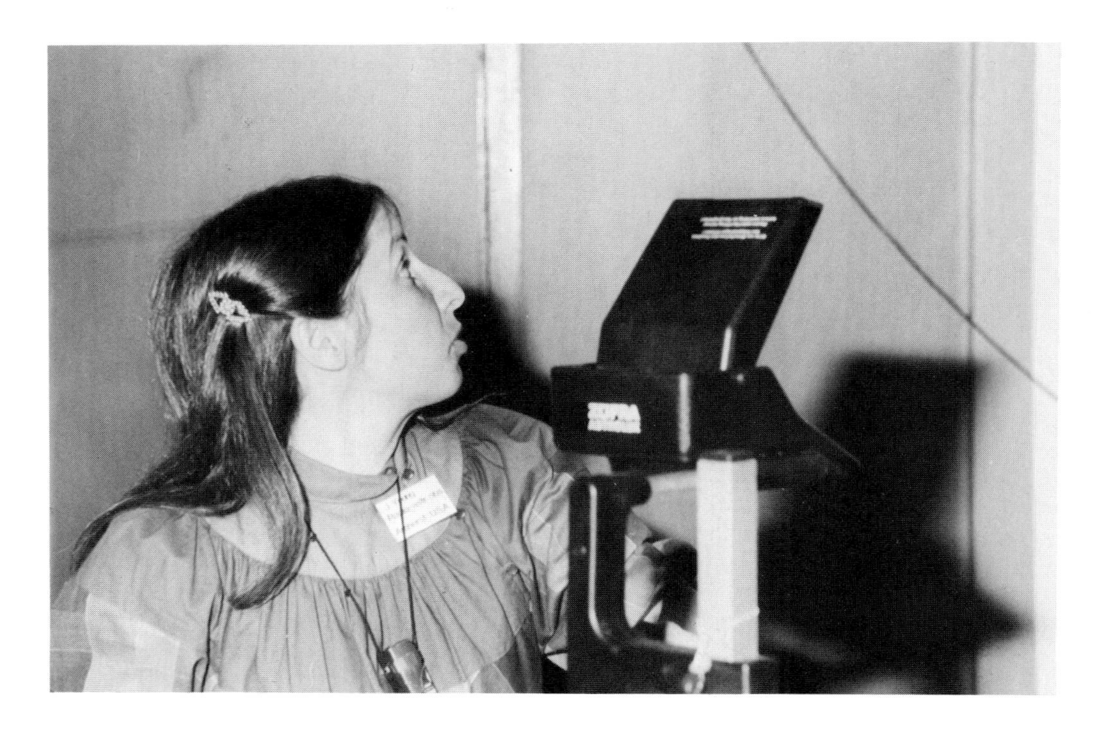

Judy Young presents distributions of molecules in galaxies CFD

MOLECULAR CLOUDS IN EXTERNAL GALAXIES

Judith S. Young
Five College Radio Astronomy Observatory, U. of Massachusetts

I. INTRODUCTION

Observations of the large-scale distribution of molecular clouds in external galaxies offer a unique opportunity for investigating galactic evolution. New generations of stars form in these dense regions, and the most massive of these stars recycle their processed interiors into the interstellar medium. Early observations of the CO distribution in the Milky Way (Scoville and Solomon 1975; Burton and Gordon 1976) indicated that there is intense emission at the center of our Galaxy, very little gas between 1 and 4 kpc radius, and a "molecular ring" feature between 4 and 8 kpc. Observations of molecular clouds in external galaxies of a variety of Hubble types and luminosities will enable us to more clearly understand the origin of this distribution. Although no other galaxies are observed to contain CO distributions precisely like that in the Milky Way, the differences which are present provide important clues to the structure and evolution of galaxies.

I have been conducting a large observational program in collaboration with Nick Scoville investigating the molecular contents of galaxies using the 14-m telescope of the Five College Radio Astronomy Observatory (HPBW = 50"). The aims of this program are to determine (1) the radial distributions of molecular gas in particular galaxies, (2) the relative CO content in galaxies as a function of Hubble type and luminosity, (3) the relative confinement of molecular clouds to spiral arms, and (4) the CO contents of active galaxy nuclei. To date we have observed 77 galaxies--including spirals, ellipticals and irregulars--detected 40 and mapped 23.

II. CO DISTRIBUTIONS IN SPIRAL GALAXIES

Of all spiral galaxies, the Sc's have been found to be the most abundant in molecular clouds. We have detected 21 out of 27 Sc galaxies observed out to the distance of the Virgo Cluster. Observations of several luminous, relatively nearby Sc galaxies--IC 342, NGC 6946, and

H. van Woerden et al. (eds.), The Milky Way Galaxy, 183–191.

M51--indicate that the CO peaks at the centers and decreases with radius, following the exponential luminosity profile of the disk out to ~ 10 kpc (Young and Scoville 1982a; Scoville and Young 1983). In each of these high-luminosity galaxies, the abundance of H_2 at the center is found to be much greater than that of HI, while in the outer parts the H_2/HI ratio falls close to or below unity. The molecular masses out to radii of ~ 10 kpc in these galaxies were found to be comparable to the HI masses of the entire disk (out to > 25 kpc radius).

While the HI distributions in late-type galaxies are all very similar (i.e. relatively flat profiles with surface densities of ~ 10^{21} atoms cm^{-2} across the disk; cf. Rogstad and Shostak 1972), this was not found to be the case for the molecular distributions. This is dramatically shown in Figure 1, illustrating the H_2 and HI distributions in the high-luminosity galaxies IC 342 and NGC 6946 as well as the low-luminosity Sc galaxies M33 and NGC 2403. Although the HI distributions in these galaxies are all very similar, the surface densities of H_2 vary tremendously.

All observations of Sc galaxies made by us or reported in the literature indicate that the strongest emission in these late-type galaxies originates at their centers. The systematics for Sb galaxies, however, are not as well defined. We have observed 21 galaxies of this

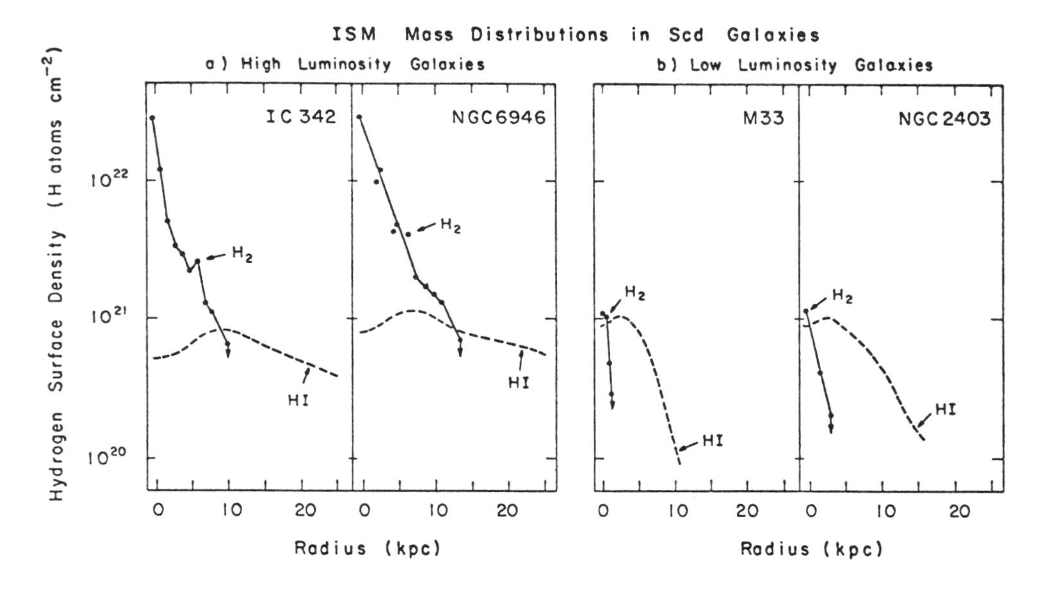

FIGURE 1. Comparison of H_2 and HI distributions in 4 Scd galaxies, two with high luminosities (NGC 6946 and IC 342) and two with low luminosities (M33 and MGC 2403). Although these galaxies all have similar HI distributions across their disks, the amounts of H_2 present vary by 2 orders of magnitude, such that only the high-luminosity galaxies have plentiful supplies of molecular clouds with which to form stars.

type and detected 11. Of the Sb galaxies whose radial distributions we have mapped, 2 were observed to have CO profiles which peak at their centers--NGC 3627 and NGC 3628 (Young, Tacconi and Scoville 1983) -- while the centers of NGC 7331 and NGC 2841 were found to contain less CO than at 5 kpc radius (Young and Scoville 1982b). Since the galaxies with central CO holes are also ones with significant nuclear bulges, we suggest that the CO hole in the early galaxies may be due to the depletion of gas in order to form stars in the nuclear bulge.

In order to test the hypothesis that central CO holes are present only in galaxies with significant nuclear bulges, we have observed a sample of 16 Sa and Sab galaxies and detected 6 of type Sab for a detection rate of 38%, relative to 78% for the Sc's. We did not detect CO in the center of NGC 4594, the Sombrero Galaxy. In two Sa galaxies which we mapped--NGC 3623 and NGC 7814--no CO was detected at any point in the disk. On the other hand, we detected and mapped the CO emission in the Sab galaxies NGC 4736 (M94) and MGC 4826 (M64). Curiously enough, in both galaxies the CO distributions show central peaks. Clearly, more observations of Sa galaxies with large bulges are needed to determine if central CO holes are present in Sa galaxies.

III. CO CONTENTS AS A FUNCTION OF HUBBLE TYPE AND LUMINOSITY

The correlation of the CO distributions and optical luminosity profiles within particular galaxies, and of the molecular content with galaxy luminosity led us to investigate a large sample of galaxies in order to elucidate the interplay between the molecular-gas contents of galaxies and their luminosities. We have observed the CO emission in a sample of Sc galaxies covering a wide range in luminosity both in the Virgo Cluster and in the field. Since it is most meaningful to compare regions of the same size in a variety of galaxies, we have mapped the CO distributions over the central 5 kpc in each galaxy and compared the CO luminosity with the optical luminosity in the same region. For the Sc galaxies we found that over 2 orders of magnitude in luminosity the optical luminosity of a galaxy increases in direct proportion to the CO content (Young and Scoville 1982c; Brady and Young 1982). If the CO emission is assumed to trace the abundance of H_2, and if the blue luminosity is taken to be mostly from Population I stars (thereby indicating the amount of star formation over the last $\sim 2\times10^9$ years), the correlation of CO intensity with B luminosity suggests that the amount of star formation is proportional to the amount of gas present. Thus, we infer that the star-formation rate per nucleon is approximately constant in Sc galaxies.

We have expanded this investigation of the interdependence of CO content and luminosity to include galaxies with earlier Hubble types. For a sample of 26 galaxies from the Virgo cluster--8 Sa's, 7 Sb's, 9 Sc's, and 2 ellipticals (M86 and M87)--we detected the centers in 13 at greater than the 4σ level (4 Sa's, 2 Sb's, and 7 Sc's). Within this sample we have compared the CO and optical luminosities in the central

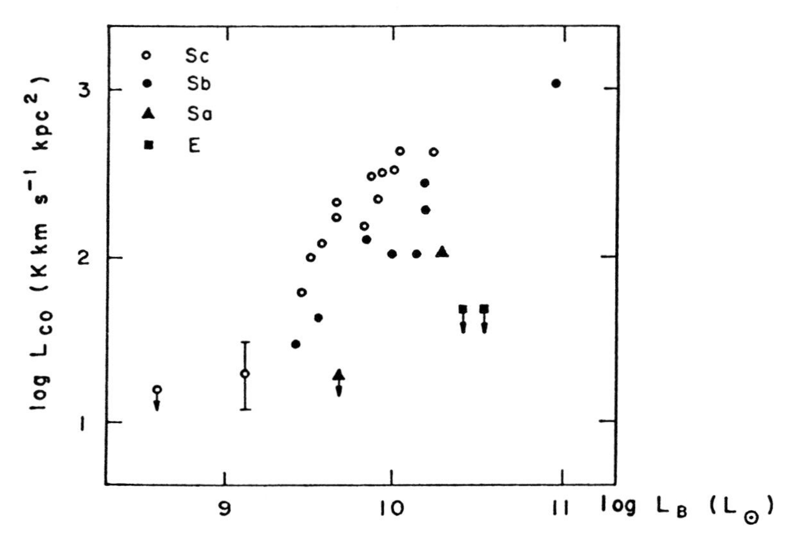

Figure 2. Comparison of CO and B luminosities of the central 5 kpc for galaxies in the Virgo Cluster and in the field. The Sb and Sc galaxies show a linear correlation over almost 2 orders of magnitude. For a given B luminosity the Sc galaxies have the most CO; for a given amount of CO the early-type galaxies are the brightest.

5 kpc of the galaxies for which B luminosities were available in the literature. Figure 2 shows this comparison for the Sa, Sb, Sc, and E galaxies in the Virgo Cluster and in the field. This figure dramatically illustrates several points. First, within each Hubble type the galaxies with high CO luminosities have the highest B luminosities. Second, at a given CO luminosity the E and Sa galaxies are more luminous than the Sb's and Sc's. And third, for a given optical luminosity the Sc galaxies have more CO than the Sa's and Sb's. The differences occurring from one type to another may simply be due to the presence of nuclear bulges -- the earlier types have higher luminosities at the centers due to the presence of the bulges (i.e., the blue luminosity is contaminated by Population II stars), and the gas contents may additionally be low as a result of gas consumption to form stars in the bulge. In order to determine the differences among the CO contents of the _disk_ components of the various Hubble types, it will be necessary to observe CO in the Virgo-Cluster galaxies outside of the nuclear-bulge regions, out to radii of $\sim$ 10 kpc.

IV. IRREGULAR GALAXIES

Although the molecular contents of spiral galaxies are becoming fairly well understood, the link between CO luminosity and large-scale star-formation properties in irregular galaxies remains less clear. In particular, irregular galaxies often display high star-formation rates and other young structural features such as OB associations and HII regions commonly found in spirals, but most irregulars have proven

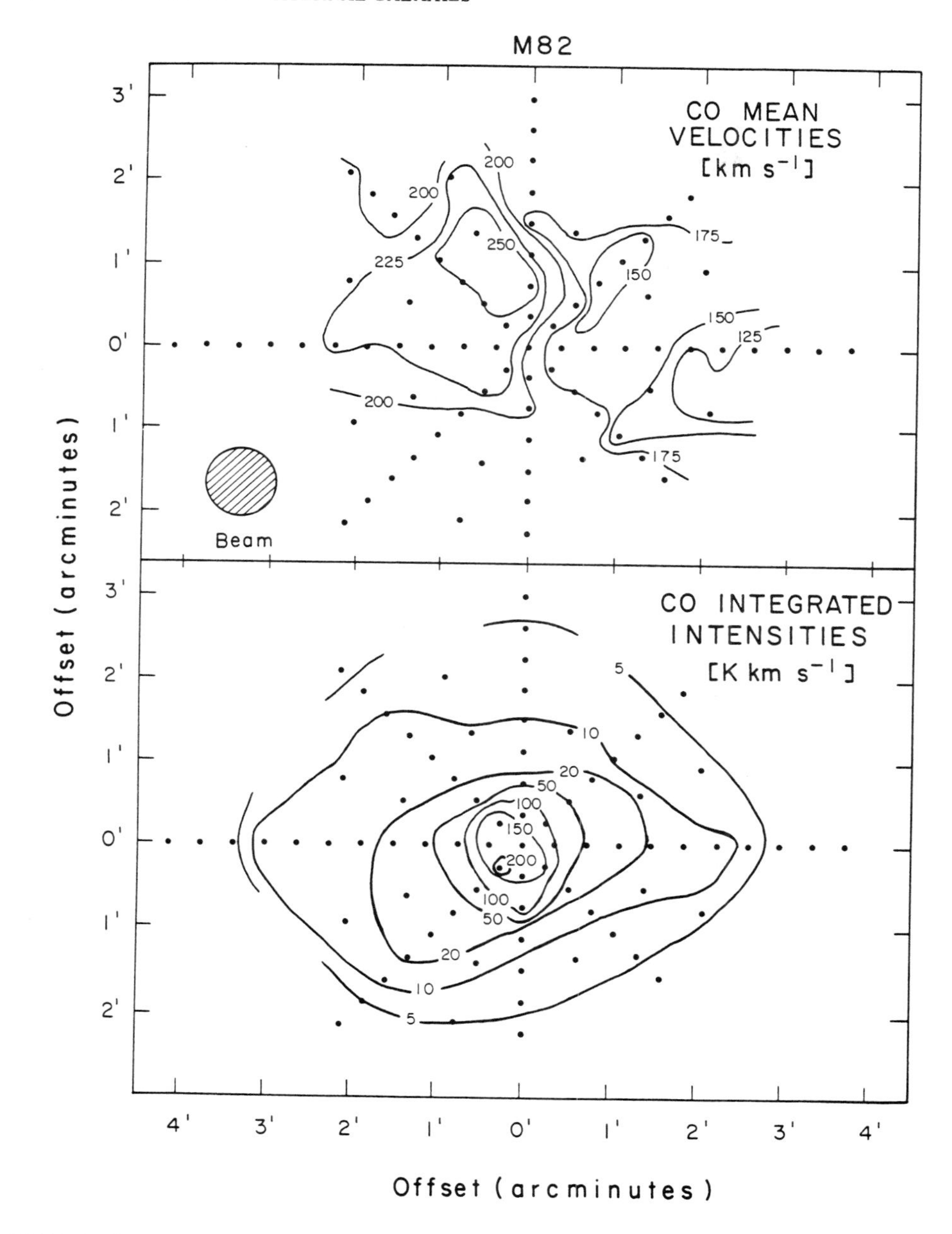

FIGURE 3. Contour maps indicating (a) the mean CO velocities and (b) the integrated intensities in M82. The velocity contours indicate that the kinematic major axis in the NE is rotated 45° to the optical major axis, an effect which can be caused by radial motions of ~ 50 km s^{-1} in the disk. The integrated intensities decrease smoothly away from the center in all directions, showing evidence for molecular clouds in the vicinity of the optical filaments located above and below the disk.

surprisingly deficient in CO emission (Elmegreen, Elmegreen, and Morris 1980). In order to more clearly define the relationship between CO luminosity and star-formation processes in irregulars, we have undertaken a program to observe CO in several irregular galaxies.

The only galaxy classified as an irregular which has been found to be abundant in molecular clouds is M82 (Rickard et al. 1977), and we have recently completed an 81-point map of the radial distribution out to 3' at 22" spacing. Figure 3 shows the mean CO velocities and integrated intensities as a function of position across M82. This irregular is similar to the high-luminosity Sc galaxies in that the H_2 abundance greatly exceeds that in HI out to at least 4' radius on the major and 3' on the minor axis. The CO emission in M82 also exhibits several peculiarities. First, emission is present not only along the major axis, but also off the disk along the minor axis in the vicinity of the tangled system of optical filaments, an aspect which was first observed by Stark (1982). Second, the mean velocities show distinct evidence for warps and irregularities, such that the overall velocity field in M82 is clearly not dominated by rotation. Since the highest velocities lie on an axis which is rotated 45° to the major axis, there must be radial motions present with a magnitude of at least 50 km s^{-1} at a radius of 1 kpc. These CO observations support the view that M82 is presently accreting material; this accretion may be responsible for the present high rate of star formation, radial motions and gas along the minor axis.

REFERENCES

Brady, E., and Young, J.S.: 1982, B.A.A.S., 14, 661
Burton, W.B., and Gordon, M.A.: 1976, Ap. J. (Letters), 207, L189
Elmegreen, B.G., Elmegreen, D.M., and Morris, M.: 1980, Ap. J., 240, 455
Rickard, L.J., Palmer, P., Morris, M., Turner, B.E., and Zuckerman, B.: 1977, Ap. J., 213, 673
Rogstad, D.H., and Shostak, G.S.: 1972, Ap. J., 176, 315
Scoville, N.Z., and Solomon, P.M.: 1975, Ap. J. (Letters), 199, L105
Scoville, N.Z., and Young, J.S.: 1983, Ap. J., 265, 148
Stark, A.A.: 1982, in Proceedings of Workshop on Extragalactic Molecules, ed. Blitz and Kutner, p. 77
Young, J.S., and Scoville, N.Z.: 1982a, Ap. J., 258, 467
Young, J.S., and Scoville, N.Z.: 1982b, Ap. J. (Letters), 260, L41
Young, J.S., and Scoville, N.Z.: 1982c, Ap. J. (Letters), 260, L11
Young, J.S., Tacconi, L., and Scoville, N.Z.: 1983, Ap. J., 269, 136

DISCUSSION

W.M. Goss: What do you know about the distribution of CO in M101?

Young: M101 has been mapped by Solomon et al. (1983). They have the radial distribution out to about 10 kpc. It shows the same type of fall-off that we see in the other external galaxies shown here.

J.V. Villumsen: What is the azimuthal variation in CO emission in your sample of external galaxies?

Young: The azimuthal variation was indicated by the vertical bars in my viewgraphs for IC 342 and NGC 6946. The azimuthal variation at one radius is greater than the uncertainties in the measurements. The variations are not entirely random: there are no very low values close to the centre, nor very high values in the outskirts.

M.L. Kutner: Could the interpretation of the correlation between CO luminosity and blue light be the opposite to yours? The sources of blue light may heat the clouds, and by measuring CO brightness one would then measure, in part, the average cloud temperature.

Young: The blue light comes not only from the extremely young stars, which are the ones that are heating the clouds; the light comes from stars that are up to 2 x 10^9 years old, which are well away from the molecular clouds. In addition most of the gas that we see is probably not in the process of forming stars, but rather it is in the outer envelopes of molecular clouds, and at 10 K only. Some gas will be hot, the gas which is associated with the stars presently forming, but that is a very small fraction of the mass of gas that we see.

Kutner: My poster paper shows in our Galaxy a radial gradient of the average temperature of molecular clouds. The conversion from CO luminosity to mass goes as $T^{1.3}$; hence small temperature variations can have large spurious effects on the mass derived.

Young: That's right: if the temperature varies, one will not infer the correct mass for the molecular clouds from the CO observations. On the other hand: we are discussing here the inner 10 kpc of galaxies which may extend much farther (say 25 kpc) in HI. While the temperature may vary with radius, it is not clear that it would vary from one galaxy to another.

H. van Woerden: Your central holes in Sb galaxies - can they be due to saturation (selfabsorption) of the CO emission?

Young: The two Sb's shown (NGC 7331 and 2841) indeed have relatively high inclinations. Solomon et al. find a central hole in the edge-on Sb NGC 891, and they think this may in fact be due to selfabsorption. However, because the rotation curve is so steep, the velocity profile measured at the centre is very wide; hence I think selfabsorption is not much of a problem in those Sb's.

A.I. Sargent: At the Owens Valley Radio Observatory we have observed several of the galaxies you have just described at 230 GHz in the J = 2→1 CO-line. Our resolution is of order 30 arcseconds and is therefore somewhat better than yours. From a preliminary inspection of the data, we agree with the trends you have noted in IC 342. However, in the case of M51 it is not completely certain that the gas falls off as you describe; there are indications that the line intensities may first fall off away from the nucleus, but rise again near the spiral arm. We fail to detect NGC 2403 although, on the basis of your results, it should have been detectable. In addition, our results suggest that the gas is not completely optically thick in several active galaxies, not only in M82.

R.J. Allen: There is a correlation of CO emission with radio continuum emission, as already pointed out by Morris and Rickard (1982). It is the best correlation with radio continuum that I have seen. As an example, NGC 6946 and M101, which have similar morphological types, have nonthermal radio-disk brightnesses differing by almost an order of magnitude - and they now turn out to have a similar difference in CO brightness.

Young: The radio brightnesses in NGC 6946, M51 and M101 indeed fall off with the same scale length as CO.

T.M. Bania: So, is the Milky Way the only galaxy with a molecular annulus?

Young: No: the Sb galaxies NGC 7331 and 2841 have molecular rings. Further, in NGC 253, a very highly inclined Sc galaxy, we have mapped CO out to 5 kpc radius. It shows something like a ring, a factor of 2 less intense than that in our Galaxy, but definitely a maximum in the radial distribution.

Infrared observations by Neugebauer et al. show a bar in the centre of NGC 253. The point has been often made that our Galaxy may have a bar. Maybe these ring distributions are related to the presence of a central bar.

L. Blitz: It is important to remember, when interpreting CO data, that you are using a tracer which has an abundance of $\sim 10^{-4}$ of the H_2 you wish to determine. Therefore small changes in CO abundance (due to metallicity gradients) and/or excitation (due to temperature gradients) will cause large changes in the derived H_2 surface density. What you are observing is a convolution of the number of molecular clouds in the beam with the average properties of the clouds in the beam. Therefore, when you observe CO radial gradients, you may be measuring gradients in the mean properties of the cloud ensemble, rather than gradients in the H_2 distribution.

Consider, for example, the plot (Figure D1) of CO emission versus radial distance for M101 published by Solomon et al. (1983). This galaxy shows the same decline of integrated CO emission with blue light that you have shown for other galaxies. The dots are [O/H] abundances from Smith (1975). The slope of the [O/H] gradient is about one half that of the CO gradient. If the [C/H] gradient is similar to that of [O/H], the CO abundance might be proportional to the square of the O abundance. The gradient of CO emission then could be due entirely to the metallicity gradient, and the H_2 surface density might be constant as a function of radius. Alternatively, possible temperature gradients (such as those observed in Maffei 2 by Rickard and Harvey) combined with [O/H] gradients could produce the observed CO distribution, even if the H_2 surface density is constant with radius. In the light of known and possible temperature and metallicity gradients, we can not now determine the radial H_2 distribution in galaxies with any confidence from CO data, and any conclusions based on an inferred H_2 distribution must be viewed with considerable scepticism.

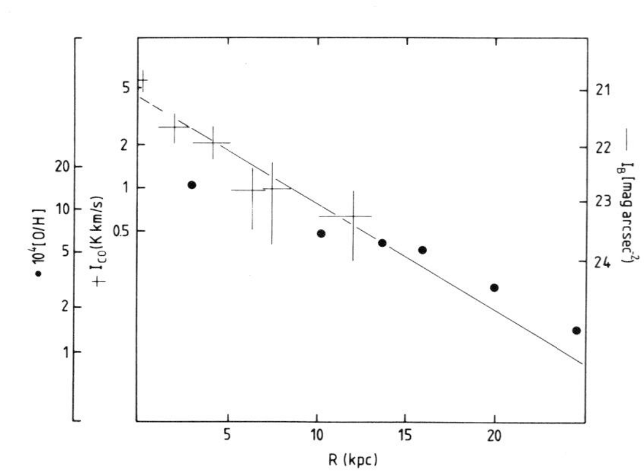

Figure D1. Radial distributions of CO emission (crosses: profile integrals in K km s^{-1}) and blue light (solid line: surface brightness in mag $arcsec^{-2}$), from P.M. Solomon et al. (1983). The dots represent [O/H] abundance ratios from H.E. Smith (1975).

Young: I shall leave it for Solomon to discuss the CO distribution in M101. As to the effects of abundance variations, let me make the following point. The metallicity at the centre is the same in M101, M51 and M83; yet M101 has a factor 5 lower CO brightness at its centre. Metallicity may have an effect, but we do not know what effect, and it can not have caused the difference between M101 and the other two galaxies. In addition, the metallicity in M33 is constant over the range in radius where we observe CO to fall off. In M51, the metallicity drops by a factor 5 only, the CO emisson by a factor 100. In general, the distributions of CO and O/H are not proportional like in M101. We need good metallicity gradients and a good understanding how C varies with O.

REFERENCES TO DISCUSSION:

Smith, H.E.: 1975, Astrophys. J. 199, 591
Solomon, P.M., Barrett, J., Sanders, D.B., de Zafra, R.: 1983, Astrophys. J. (Letters) 266, L103
Morris, M. and Rickard, L. J: 1982, Ann. Rev. Astron. Astrophys. 20, 517

Top: Anneila Sargent (right) discusses her CO observations of galaxies with Antonella Natta. CFD

Bottom: Leo Blitz in his discussion with Judy Young CFD

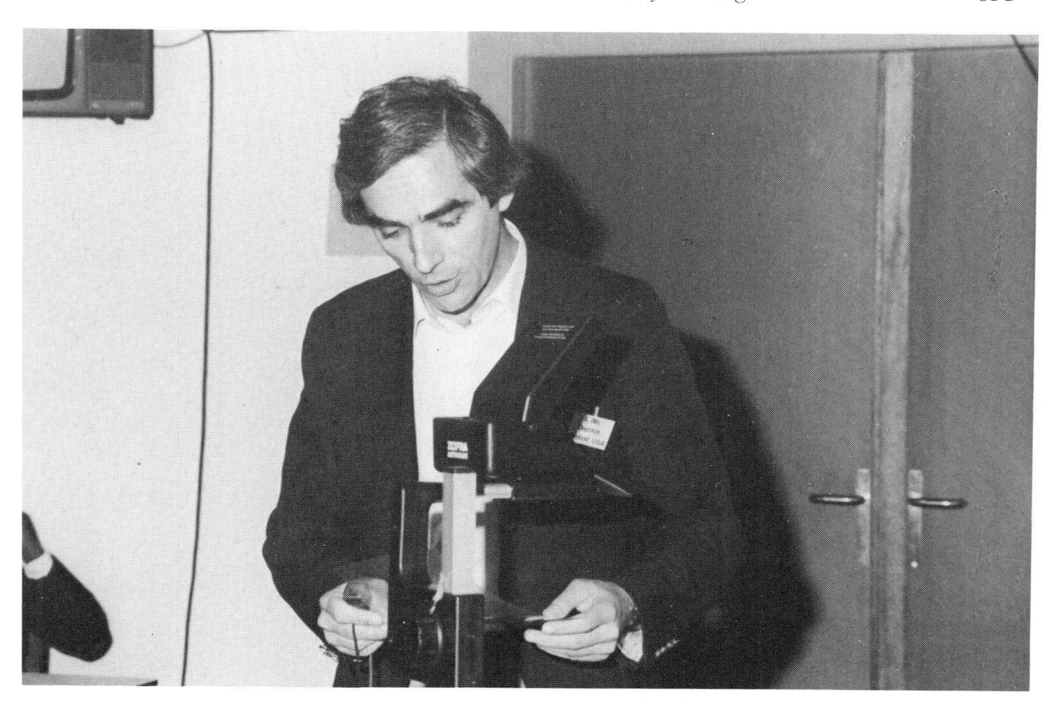

COMMENTS ON THE DISTRIBUTION OF MOLECULES IN SPIRAL GALAXIES

L. J Rickard
Howard University

Patrick Palmer
University of Chicago

1. AXISYMMETRIC STRUCTURE

Young and Scoville (1982) have argued that the CO distribution, and by inference the H_2 distribution, follows the shape of the blue luminosity of the disk, and is thus exponential. However, full maps of bright-CO galaxies (Rickard and Palmer 1981) show considerable structure, with real peaks and depressions on scales as small as the telescope beam. This means that the noise in the determination of the underlying structure is dominated not by the instrumental contribution but by the intrinsic noise of the structure itself. A correct analysis of the axisymmetric structure requires the use of statistical tests comparing the observations with different hypothetical distributions. One finds that equally good fits can be obtained with exponential distributions, r^{-1} distributions, or even flat disks with central nuclei. However, profiles of the full map data averaged over azimuth, with "error" bars determined by the structural variation with azimuth, show a clear deviation from the optical luminosity profile in the outer disk of NGC 6946. Furthermore, if the profiles for NGC 6946 and IC 342 are fit with exponentials, the scale lengths are rather larger than previously suggested (8.9 and 6.6 kpc, respectively). By sampling only a few radii, one can miss much of the emission of the outer parts of the galaxies, and also underestimate the intrinsic noise of the structure.

These questions involve the CO distribution alone. The step to the H_2 distribution requires additional assumptions about conversion factors. The empirical conversion factor is still uncertain to at least a factor of two; e.g. Frerking et al. (1982) find rather less H_2 per unit CO emission than do Young and Scoville. That factor can have drastic consequences on where the H_2 and HI distributions are found to be comparable, and what the actual shape of the total proton distribution will be. Furthermore, the conversion is probably not constant over the entire disk. Far-IR color temperatures (Rickard and Harvey 1983) and ^{13}CO data (Rickard and Blitz, in preparation) indicate changing characteristic cloud temperatures from the center outward, requiring a similarly varying conversion factor from ^{12}CO integrated intensity to H_2 surface density.

H. van Woerden et al. (eds.), The Milky Way Galaxy, 193–194.

2. IS THERE SPIRAL STRUCTURE IN THE CO DISTRIBUTIONS OF GALAXIES?

Spiral structure is clearly present in the CO distribution of M31. Linke (1982) reports a high arm-interarm contrast in the SW part of M31. Admittedly, the structure in this region is rather complex; e.g. Unwin (1980), in describing the HI data, prefers to call it a set of spiral-arm segments. But the fact of the contrast is indisputable. On the other hand, M31 is a rather different CO galaxy from most studied. It has no detectable central molecular source, and it has a very low ratio of H_2 to HI mass surface density (about 0.1). Spiral contrast may vary from galaxy to galaxy depending, among other things, on the H_2 to HI ratio.

A galaxy with well-defined, high-compression spiral structure, like M51, that also has a high H_2 to HI ratio (larger than 1 for M51, NGC 6946, IC 342, etc.) would be better for study. Yet full CO maps have so far shown no obvious spiral structure in such galaxies. There is, though, a resolution problem, even for the 45" FCRAO telescope. A convolution of the Tully (1974) Hα spiral pattern with a 1' beam produces a rather spiral-less shape for M51. In order to extract a pattern from the CO data, one must turn to statistical analysis.

Using an axisymmetric model of central source plus flat disk, one finds a best-fit arm-interarm contrast for M51 of about a factor of 4. However, the noise in the map - which is dominated by the intrinsic noise in the structure of the galaxy itself - results in an inability to exclude at the 95% confidence level either the case of a very large contrast or the case of no contrast at all.

IC 342, on the other hand, being at only half the distance, allows better resolution of the spiral pattern. Recent CO observations of the NW quadrant show one region of size about 2' x 3' that is enhanced above the surrounding disk. It coincides in position with a segment of the spiral pattern that is well defined in near-IR photographs (Elmegreen 1982). This feature has recently been confirmed by reobservation with the newly resurfaced NRAO 12-m telescope. The arm-interarm contrast (between the arm segment and the regions on either side at the same azimuth) is about a factor of 3.

REFERENCES

Elmegreen, D.M. 1982, Astrophys. J. Suppl. **47**, 229
Frerking, M.A., Langer, W.D. and Wilson, R.W. 1982, Ap. J. **262**, 590
Linke, R.A. 1982, in "Extragalactic Molecules", eds. L. Blitz and M.L. Kutner, Green Bank: NRAO, p. 87
Rickard, L. J and Harvey, P.M. 1983, Ap. J. (Letters) **268**, L7
Rickard, L. J and Palmer, P. 1981, Astron. Astrophys. **102**, L13
Tully, R.B. 1974, Astrophys. J. Suppl. **27**, 449
Unwin, S.C. 1980, Monthly Notices Roy. Astron. Soc. **190**, 551
Young, J.S. and Scoville, N.Z. 1982, Ap. J. **258**, 467

DISTRIBUTION AND MOTIONS OF CO IN M51

G. Rydbeck, Å. Hjalmarson, and O. Rydbeck
Onsala Space Observatory
S-439 00 Onsala, Sweden

We present results (partly preliminary) of an extensive map (73 positions) of CO (J=1-0) emission in M51 (Fig. 1). The spectra were obtained with the Onsala 20-m antenna (beam size 33"), equipped with a cooled mixer and a 512x1 MHz multichannel receiver. The data are not yet fully analyzed but our preliminary results are as follows:

1. A small central minimum in the CO emission is apparent (Fig. 2). The average radial CO distribution shows a maximum at $\sim$ 15" (corresponding to 0.7 kpc, for an assumed distance of 9.6 Mpc).
2. The above minimum results from what seems to be an oval ridge structure in the emission intensity around the center, with an extent of about 30" by 40" (Fig. 3). This could be the markings of an elongated hole or the beginning of the spiral arms. (The ridges are fairly well correlated with the innermost parts of the spiral arms as delineated by continuum observations by Segalowitz, 1976.)
3. It seems that a central oval velocity pattern is needed to explain a dip at 450 km s^{-1} in the central spectrum. (The dip is present in all and independent parts of our data.)
4. The outer parts of the galaxy have only been covered in strips (Fig. 1). Clearly there is structure in these strips, i.e. not only a radial decrease. The "on-arm" spectrum (88", 0") has for example a greater integrated and peak intensity than the "interarm" spectrum (77", 0"). Such arm-interarm contrast is not always clear, though. Observations covering a larger section of an arm-interarm region, where contrast has been seen, are being planned for the next observing season.
5. We also note that the observed apparent velocity differences between CO and ionized gas, reported by Rydbeck <u>et al</u>. (1983), persist. A fully convincing argument on this point (taking into account the finite beam etc.) requires some further model work, however.

REFERENCES

Rydbeck, G., Pilbratt, G., Hjalmarson, Å., Olofsson, H., Rydbeck, O.E.H.: 1983, in "Internal Kinematics and Dynamics of Galaxies", IAU Symp. 100, ed. E. Athanassoula, p. 53.

Segalovitz, A.: 1976, Astron. Astrophys. 52, 167.

H. van Woerden et al. (eds.), The Milky Way Galaxy, 195–196.

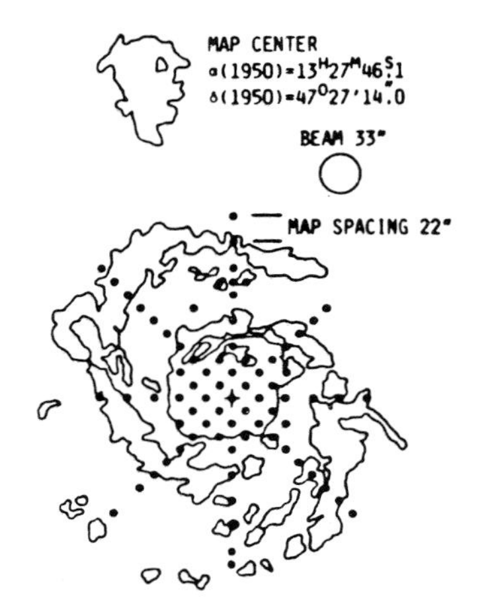

Figure 1. Observed positions in M51.

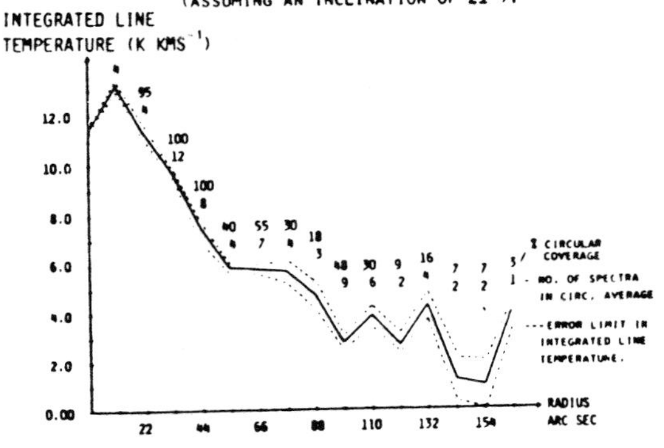

Figure 2. Radial CO distribution in M51.

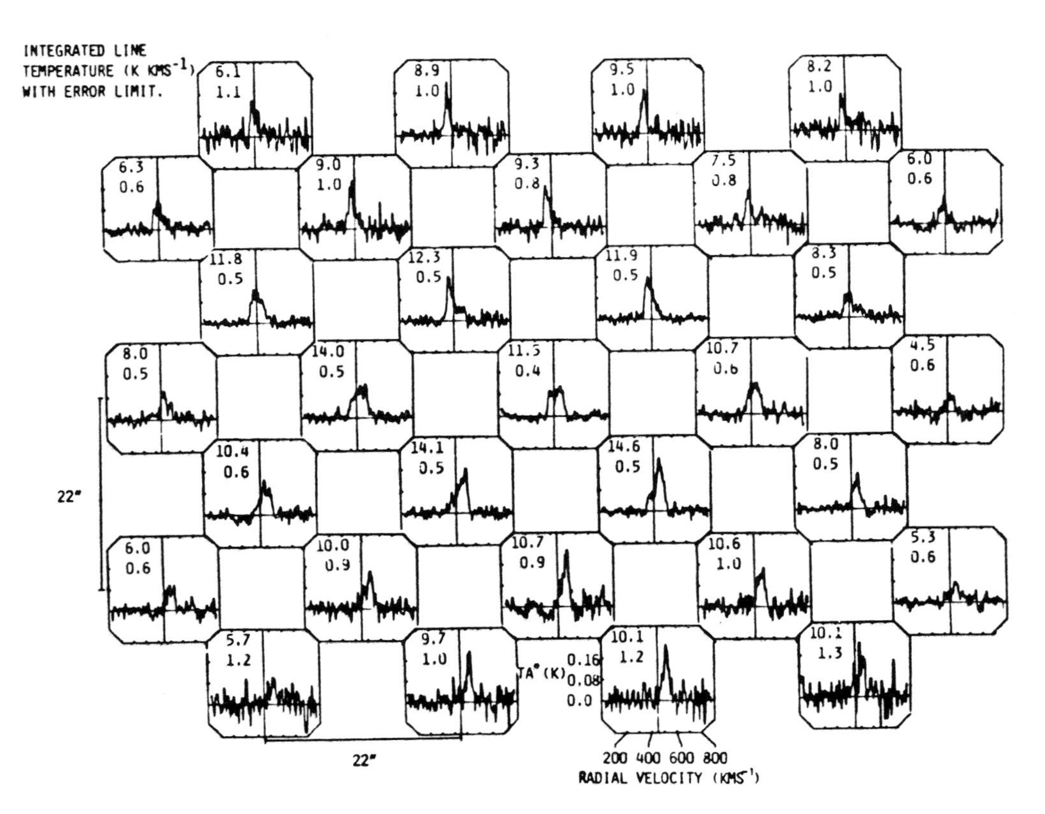

Figure 3. Spectra of J=1→0 transition of CO taken towards the central region of M51 (cf. Figure 1).

CO (2-1) OBSERVATIONS OF MAFFEI 2

Anneila I. Sargent, E.C. Sutton, C.R. Masson, T.G. Phillips and K.-Y. Lo
Owens Valley Radio Observatory, California Institute of Technology, Pasadena, CA U.S.A.

Maffei 2 is a highly obscured galaxy, probably of type Sbc, at a distance of 5 Mpc (Allen and Raimond 1972; Spinrad et al. 1973). Since it lies close to the Galactic plane, there is considerable confusion in infrared and 21-cm HI observations due to Galactic emission, but investigations of its structure can be carried out at millimeter wavelengths where the Galaxy contribution is confined to a limited velocity range. The high resolution (30") of our CO J=2-1 observations permits both a detailed examination of Maffei 2 and a study of the nature of the gas in its nucleus, through comparison with the CO J=1-0 observations.

The CO (2-1) observations were made with one of the 10.4 meter telescopes at the Owens Valley Radio Observatory in January, 1983, using an SIS receiver (Sutton 1983) and a 512-channel acousto-optical spectrometer (cf. Masson 1982). The frequency width of each channel was 1.03 MHz, 1.34 km s^{-1} at 230 GHz. The receiver temperature, T_r was 550 K SSB. Spectra, taken at spacings of 0.3' on a grid oriented along the major axis of the galaxy (PA = 30°), are shown in Figure 1. The central position, marked by

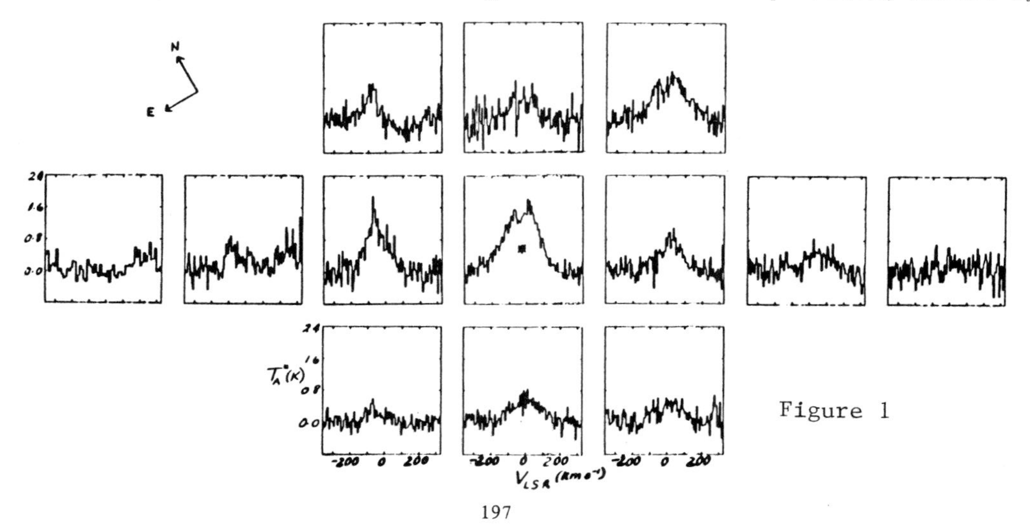

Figure 1

H. van Woerden et al. (eds.), The Milky Way Galaxy, 197–198.

an asterisk, is α (1950) = $02^h38^m08^s.5$, δ (1950) = $+59°23'24''$. At -7 km s^{-1} and -56 km s^{-1} there is some evidence of Galactic emission in the reference position (see the dip near zero velocity in Figure 2). Assuming the source to be just resolved, a beam coupling efficiency 0.52 was adopted in our determination of the corrected antenna temperature, $T_A^*(2-1)$. While variations in peak velocity are consistent with the rotation curve, CO(2-1) emission is much more confined than CO(1-0) and 21-cm HI radiation (Rickard et al. 1977; Bottinelli et al. 1972).

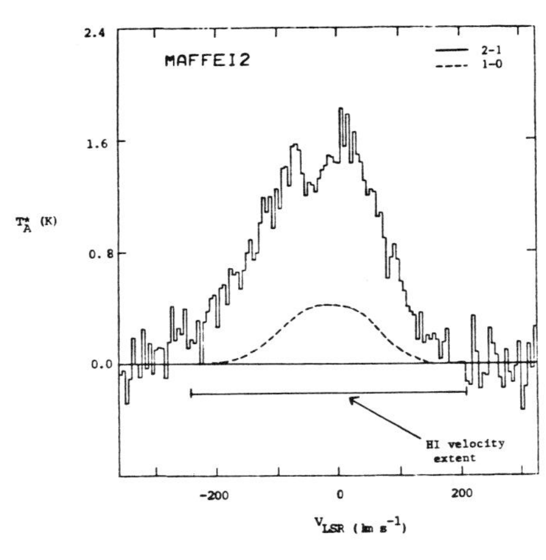

Figure 2

In Figure 2, our central CO(2-1) profile, the CO(1-0) profile from Rickard et al. (1977) and the velocity extent of the 21-cm HI emission from Bottinelli et al. (1972) are reproduced. Published values of $T_A^*(1-0)$ were corrected assuming a beam efficiency of 0.65 (Ulich and Haas 1976) and a source size 1' at CO(1-0). Peak $T_A^*(2-1)$ = 1.6K, significantly higher than peak $T_A^*(1-0)$ = 0.4K. If the nuclear source is ultra-compact, much of this discrepancy could be the effect of beam dilution, but it seems more likely that gas in the nucleus of Maffei 2 is at least partially optically thin, as in the active galaxy, M82 (Sutton et al. 1983)

Nuclear activity is often evident in galaxies with nearby companions. Maffei 1, a giant elliptical, is only 40' from Maffei 2 on the plane of the sky but, assuming a mass-to-light ratio of 30, is at a distance of 1 Mpc (Spinrad et al. 1971). Adopting instead M/L = 6 (Sargent et al. 1978) puts Maffei 1, like Maffei 2, at about 5 Mpc. Thus Maffei 1 and Maffei 2 may be companion galaxies and interaction between them may have influenced star formation in Maffei 2.

REFERENCES

Allen, R.J., and Raimond, E.: 1972, Astron. Astrophys. 19, 317

Bottinelli, L., et al.: 1972, Astron. Astrophys. 12, 264

Masson, C.R.: 1982, Astron. Astrophys. 114, 270

Rickard, L.J., Turner, B.E., and Palmer, P.: 1977, Astrophys. J. (Letters) 218, L51

Sargent, W.L.W., Young, P.J., Boksenberg, A., Shortridge, K., Lynds, C.R., and Hartwick, F.D.A.: 1978, Astrophys. J. 221, 731

Spinrad, H. et al.: 1971, Astrophys. J. (Letters), 163, L25

Spinrad, H. et al.: 1973, Astrophys. J. 180, 351

Sutton, E.C.: 1983, IEEE Trans. on MTT, in press

Sutton, E.C., Masson, C.R., and Phillips, T.G.: 1983, Astrophys. J., submitted

Ulich, B.M., and Haas, R.W.: 1976, Astrophys. J. (Supp.), 30, 247

CO SURVEY OF THE SOUTHERN MILKY WAY

R.S. Cohen
Columbia University
P. Thaddeus
Columbia University and Goddard Institute for Space Studies
L. Bronfman
Columbia University and Universidad de Chile

When the Columbia Southern Millimeter-Wave Telescope on Cerro Tololo, Chile, began operation in January 1983, one of the main goals was to make the first extensive out-of-plane survey of 2.6-mm CO emission in the Southern Milky Way. Because this telescope, a 1.2-meter Cassegrain, is a close copy of the Columbia telescope in New York -- with nearly identical antenna, feed, and calibration -- it will be very easy to join our northern and southern surveys to make the first homogeneous radio survey of the entire Galaxy. The Chile telescope does have one important improvement over the New York system: a liquid-nitrogen-cooled receiver with a single-sideband noise temperature of 385 K. Details of the telescope are given in Cohen (1983).

The basic galactic survey, a mirror image of the survey done with the New York telescope, will be completed in early 1984. It will run at least from $\ell = 300^\circ$ to 348° and from $b = -1^\circ$ to $+1^\circ$. Points will be spaced at most every two beamwidths over the entire range, and every beamwidth from $b = -0.5^\circ$ to $+0.5^\circ$, the latitudes containing the most intense CO emission. Because Cerro Tololo is such a good site for the galactic survey both in terms of weather and location (the galactic center transits 2° from the zenith), and because our liquid-nitrogen receiver is so fast, the final coverage will probably be somewhat better: we will extend the survey to higher latitudes and will increase the sampling. In addition, by late 1983 we will connect the northern and southern surveys with the first well-sampled survey of CO in the region of the galactic center.

To trace out the inner arms of the Galaxy as Dame (1983) did in the Northern Hemisphere -- by identifying the large molecular clouds and locating them in the Galaxy with respect to the classic spiral arms -- requires the full latitude coverage that is not yet available. Some interesting features are however already clearly apparent in our preliminary results. Figure 1 (these Proceedings, page 2) is a longitude-velocity map of the entire galactic equator from $\ell = 300^\circ$ through the center to $\ell = 60^\circ$. The most prominent feature is the well-known peak at the galactic center. The classic "4-kpc expanding arm" is visible as the line

H. van Woerden et al. (eds.), The Milky Way Galaxy, 199–202.

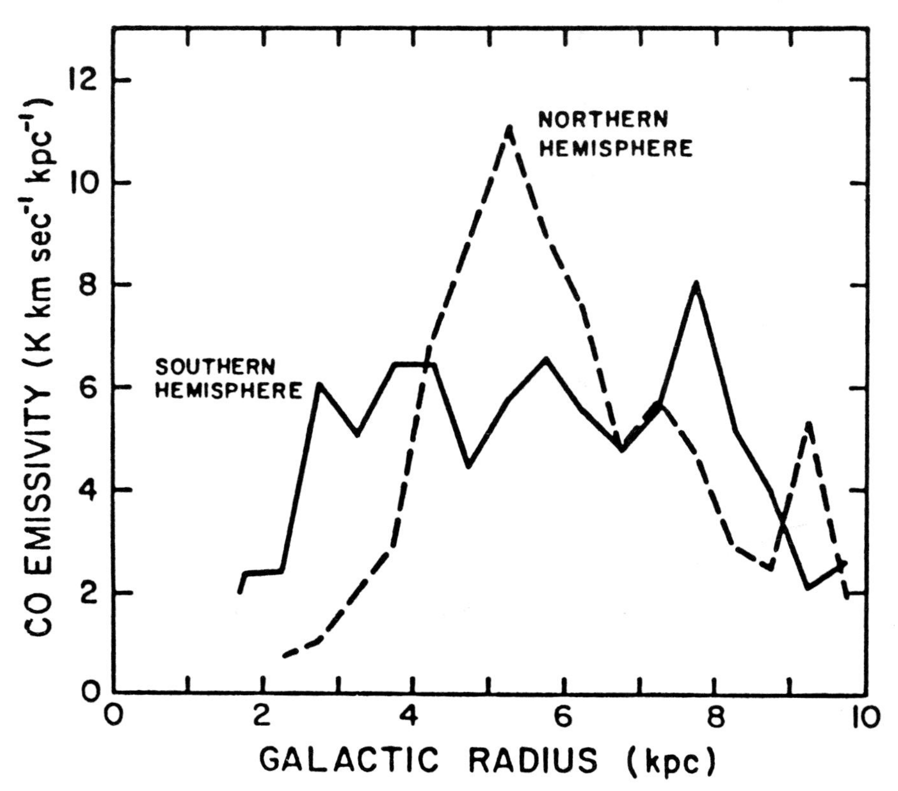

Figure 2. Emissivity of CO as a function of distance from the galactic center. The Northern Hemisphere data are from our New York survey. The sampling of the Southern data is still somewhat irregular, and the Southern Hemisphere curve should be considered preliminary.

of emission crossing $\ell = 0^o$ near $v = -50$ km s^{-1}. Further from the center, both north and south, the CO is organized into giant molecular clouds which in turn are clearly organized into larger structures on the scale of galactic spiral arms.

From the data available in May, 1983 we made a preliminary analysis of the distribution of CO as a function of galactocentric radius using a simple axisymmetric model of the Galaxy (Cohen et al., 1980). The results (Figure 2) are in general agreement with early Australian results (Robinson et al., 1983), and show that the galactic distribution of molecular clouds is plainly not axisymmetric. The well-known "molecular ring" -- the band of intense CO emission between 5 and 7 kpc from the galactic center -- has broadened into a wide band that begins, like the HI, abruptly at 4 kpc and then trails off by the solar radius. Because CO is a good tracer of spiral structure (Cohen et al., 1980), this is just what we would expect: the spiral model of Georgelin and Georgelin

(1976) based on HII regions shows that in the Northern Hemisphere the spiral arms are close together and lie in the molecular ring, while in the Southern Hemisphere they spread apart in galactic radius.

REFERENCES

Cohen, R.S., Cong, H., Dame, T.M., and Thaddeus, P.: 1980, Ap. J. (Letters) 239, p. L53

Cohen, R.S.: 1983, in Surveys of the Southern Galaxy, ed. W.B. Burton and F.P. Israel (Dordrecht, Reidel)

Dame, T.M.: 1983, Ph. D. Thesis, Columbia University

Georgelin, Y.M., and Georgelin, Y.P.: 1976, Astron. Ap., 49, p. 57

Robinson, B.J., McCutcheon, W.H., Manchester, R.N., and Whiteoak, J.B.: 1983, in Surveys of the Southern Galaxy, ed. W.B. Burton and F.P. Israel (Dordrecht, Reidel)

DISCUSSION

(There was no Discussion after Cohen's paper. The following comments were made later in the Symposium, in Section II6, after the papers by T.M. Dame and D.B. Sanders, but fit best here. - Editor)

A. Blaauw: A comment to the diagrams shown by Cohen and by Sanders. The surveys made by both are restricted to latitudes 0°±1°. This means that the material must be very incomplete at low velocities, where one deals with small distances. I never see in these beautiful diagrams any warning of this incompleteness. For instance, Orion sits about 100 pc below the plane; at a distance of 1000 pc, it would still be at 6° latitude and hence missing in the surveys by Cohen and by Sanders. One should either indicate this incompleteness due to latitude limits, or suppress the low velocities from the diagrams; otherwise the diagrams are misleading.

T.M. Dame: Indeed, the local emission may go up to very high latitudes. At velocities above 20 km/s we miss very little emission.

R.S. Cohen: Dame has done a wide-angle survey going 5° or 10° out of the plane to map nearby objects. But the local region is irrelevant to the problem of spiral structure.

F.J. Kerr: Still, Blaauw has given us an important warning.

(The spiral-structure implications of these surveys are discussed in Section II5 of these Proceedings. - Editor)

H. van Woerden to Sanders: What is your estimate for the total amount of molecular hydrogen in the Galaxy?

D.B. Sanders: About 3 x 10^9 solar masses.

M.L. Kutner: What is your estimate for the H_2 mass inside the solar circle?

Sanders: About 85% of the total, hence about 2.5 x 10^9 $M_\odot$.

Kutner: That seems high compared to Hermsen's estimate based on the γ-ray observations.

Sanders: The γ-ray group derive a conversion factor from CO to H_2 of about 3 × 10^{20} molecules cm^{-2} K^{-1} $(km/s)^{-1}$; our number is 3.6 × 10^{20}, hence there is good agreement. And high-resolution observations of individual clouds, giving virial masses and column densities from ^{13}CO, do not show any evidence for a lower conversion factor.

Cohen presenting his CO survey CFD

DISTRIBUTION OF 'CO IN THE SOUTHERN MILKY WAY AND LARGE-SCALE STRUCTURE IN THE GALAXY

W.H. McCutcheon
Department of Physics, University of British Columbia
Vancouver, B.C.

B.J. Robinson, R.N. Manchester, and J.B. Whiteoak
Division of Radiophysics, CSIRO
Sydney, Australia

The southern galactic-plane region, in the ranges $294° \leqslant l \leqslant 358°$, $-0°.075 \leqslant b \leqslant 0°.075$, has been surveyed in the J = 1-0 line of ^{12}CO with a sampling interval of 3' arc. Observations were made with the 4-metre telescope at the CSIRO Division of Radiophysics in 1980 and 1981. Details of equipment and observing procedure are given in Robinson et al. (1982, 1983); see also McCutcheon et al. (1983).

The variation of radial velocity with longitude shows well-defined terminal velocities whose locus matches fairly well (with deviations not exceeding about 10 km s^{-1}) rotation curves determined from HI and CO observations along the northern galactic plane. Over certain ranges of longitude, absence of CO emission near the tangential velocity is more apparent in the southern observations and strongly suggests that there are arm-like structures in the CO distribution. In particular, the HI 3.5-kpc Expanding Arm has a clearly defined CO counterpart. Combining our l-v diagram with that of Sanders (1982) and using a standard model of the Galaxy, we identify much of the CO emission with sections of four spiral-like arms, namely: Local, Sagittarius-Carina, Scutum and Perseus. There is good agreement with much of the distribution of the 1720-MHz OH sources (Turner 1983). Some of the CO gas and OH sources lie between the major arm-like features, indicating perhaps the presence of spurs and bifurcations. Using models with velocity deviations of ~ 10 km s^{-1} from circular velocities changes parameters of the spirals but does not destroy the overall pattern. Thus, while certain arm-like patterns appear to stand out well, we cannot specify a unique picture since there are deviations from circular velocities and also some uncertainties in the distances of emitting regions.

The radial distribution of CO displays two pronounced peaks: a sharp peak near R = 3.5 kpc and a broader peak near R = 7 kpc. This contrasts with the northern CO distribution, which shows only a broad peak, centred near R = 6 kpc.

H. van Woerden et al. (eds.), The Milky Way Galaxy, 203–204.

REFERENCES

McCutcheon, W.H., Robinson, B.J., Whiteoak, J.B., Manchester, R.N.: 1983, in "Kinematics, Dynamics and Structure of the Milky Way", ed. W.L.H. Shuter, Dordrecht: Reidel, p. 165
Robinson, B.J., McCutcheon, W.H., and Whiteoak, J.B.: 1982, Int. J. Infrared and Millimeter Waves 3, p. 63
Robinson, B.J., McCutcheon, W.H., Manchester, R.N., and Whiteoak, J.B.: 1983, in "Surveys of the Southern Galaxy", eds. W.B. Burton and F.P. Israel, Dordrecht: Reidel, p. 1
Sanders, D.B.: 1982, Ph. D. Thesis, State University of New York at Stony Brook
Turner, B.E.: 1983, in "Kinematics, Dynamics, and Structure of the Milky Way", ed. W.L.H. Shuter, Dordrecht: Reidel, p. 171

Asian millimeter astronomy: Kaifu (left) and Radhakrishnan CFD

A CO(2-1) SURVEY OF THE SOUTHERN MILKY WAY

H. van de Stadt
Sterrewacht Sonnenborgh, Utrecht

F.P. Israel, Th. de Graauw
Astronomy Division, Space Science Department, Estec

C.P. de Vries, J. Brand, H.J. Habing
Sterrewacht, Huygens Laboratorium, Leiden

J. Wouterloot
ESO-Garching, Muenchen

We have used the Estec/Utrecht heterodyne submillimetre receiver together with the 1.4 m ESO CAT at La Silla (Chile) to survey the southern Galaxy (l = 270 - 355°) in the CO(2-1) transition at 230 GHz (1.3 mm). The beam used had a HPBW size of 5.5 arcmin, overall system efficiency was 0.35, and the system temperature was 1400 - 1750 K (DSB). Our filterbanks had velocity resolutions of 1.3 and 0.325 km s^{-1} and velocity ranges of 333 and 83 km s^{-1} respectively.

The survey consists of three parts: 1. a survey of the galactic plane (b = 0°) in the range l = 270 - 355°; 2. a survey of 88 dark clouds with and without associated nebulosity; 3. a survey of 47 bright HII-region complexes. Some additional observations were made with the same receiver and the use of the ESO 3.6 m telescope (beamsize 2 arcmin HPBW). The following is a summary of the results obtained; they are published in more detail elsewhere (Israel et al. 1983; De Vries et al. 1983; Brand et al. 1983).

We obtained contour maps of CO(2-1) emission in the galactic plane, with an effective velocity resolution of 5.2 km s^{-1} and convolved to spatial resolutions of one and two degrees (see Israel et al. 1983). The CO distribution is very clumpy, and shows numerous holes. There is a lack of emission shortwards of l = 300°, due to the tilt of the Galaxy which places most material here below b = 0°; the CO complex associated with the Carina Nebula is just visible. In general, the CO(2-1) distribution is very similar to that observed in the CO(1-0) transition (c.f. McCutcheon et al., this volume). In addition, we find that in the south the 'molecular ring' is broader than in the north and also shows a double-peaked structure; that the cloud-cloud velocity

H. van Woerden et al. (eds.), The Milky Way Galaxy, 205–206.

dispersion is 4.5±0.5 km s^{-1} as it is in the northern hemisphere, and that the CO(2-1) terminal velocities closely follow the Sinha (1978) rotation curve.

We detected over 50 per cent of all dark clouds observed; these statistics are strongly influenced by beam-dilution effects (i.e. clouds not detected were usually significantly smaller than the beam). Dark clouds associated with nebulosity have measured velocity widths between 1 and 5 km s^{-1} with a mean around 2.8 km s^{-1}. This indicates that the majority of these clouds shows the influence of interaction between dark-cloud material and associated stellar energy sources. Particularly clear examples of such interaction are found in the dark clouds associated with Herbig-Haro objects HH 46/47, HH 52/54 and GGD 27/28. In contrast, dark clouds not associated with nebulosity, with very few exceptions show velocity widths between 1 and 2 km s^{-1}. Consequently, the majority of these clouds does not contain a hidden stellar energy source. A good example is given by the Coalsack globules (Tapia, 1973).

About a quarter of all dark clouds observed with sufficient signal-to-noise shows enhanced wing emission (and/or asymmetrical profiles), red wing emission being more common than blue by a factor of three. The number of clouds with enhanced wing emission is a lower limit because of the limited sensitivity of our observations. Thus mass outflow seems to be a very common phenomenon in relatively cool dark clouds, as it is on a more energetic scale in hot dark clouds associated with HII regions (c.f. Bally and Lada, 1983).

Out of 47 HII regions observed (mainly RCW sources) we detected 28. The results are generally consistent with those obtained earlier in the CO(1-0) and CO(2-1) transitions by other workers. Three objects were observed in more detail (RCW 36, G327.3-0.6/RCW 97 and W 48). Of these, RCW 97 turned out to be the most interesting. It shows a compact CO cloud core in which two active star-formation sites are embedded. The two sites are consistent with sequential star-formation models; the youngest of the two shows characteristics of strong bipolar mass outflow.

REFERENCES

Bally, J., Lada, C.J.: 1983, Astrophys. J. 265, 824.
Brand, J. et al.: 1983, Astron. Astrophys., submitted.
De Vries, C.P. et al.: 1983, Astron. Astroph., in press.
Israel, F.P. et al.: 1983, Astron. Astrophys., submitted.
Sinha, R.P.: 1978, Astron. Astrophys. 69, 227.
Tapia, S.: 1973, in IAU Symposium 51 "Interstellar Dust and Related Topics", p.43.

THE CARBON MONOXIDE DISTRIBUTION IN THE INNER GALAXY

T.M. Bania
Department of Astronomy, Boston University, Boston, MA

The latitude distribution of the emission from the ^{12}CO J=1→0 rotational transition has been surveyed for the region 350° ⩽ l ⩽ 25° at b = 0', ±10' and ±20'. Most of the ^{12}CO emission in the inner Galaxy, the region extending from the galactic center to 4 kpc radius, is produced by three large and massive objects: the nuclear disk/bar, the 3-kpc arm and the "+135 km s^{-1} feature". These structures all have observed HI counterparts and each shows extremely large deviations (50-180 km s^{-1}) from circular motion. Observations of ^{13}CO in selected directions show that the two structures outside the nuclear disk each span at least 2 kpc in length and that together they imply ⩾ 10^{55} ergs in kinetic energy of expansion away from the galactic nucleus.

The properties of the clouds within these inner-Galaxy features are quite similar to those derived for the Giant Molecular Cloud population residing beyond 4 kpc from the galactic center. The inner-Galaxy clouds are typically 100-150 pc in diameter with H_2 masses of ≃ 10^6 $M_\odot$ (inferred from $^{13}CO/^{12}CO$ data toward selected clouds). The average properties for the 3-kpc and +135 km s^{-1} arm clouds are summarized below. The most striking object is the "Clump 1" cloud at (l,b,v)=(355°, 0.4°,+100 km s^{-1}). It lies at the edge of an HII region, G354.67,+0.25, making it one of the only clouds in the inner Galaxy outside the nuclear disk which shows evidence for <u>recent</u> star formation.

AVERAGE PROPERTIES OF CO CLOUDS IN INNER-GALAXY EXPANDING FEATURES

	3-kpc Arm	+135 km/s Arm	Clump 1
Number of Clouds	10	9	1
Mass (H_2) ($M_\odot$)	$1.0 \pm 1.1 \times 10^6$	$9.6 \pm 8.2 \times 10^5$	2.6×10^6
Diameter (pc)	120	130	85
$N(HI)/2N(H_2)$	0.5	0.1	0.1
Arm Filling Factor	0.7	0.6	--
Arm Proton Mass ($M_\odot$)	$2\text{-}6 \times 10^7$	$2\text{-}6 \times 10^7$	2.9×10^6
E_{exp} (ergs)	$0.6\text{-}1.7 \times 10^{54}$	$0.3\text{-}1.1 \times 10^{55}$	5.2×10^{53}

H. van Woerden et al. (eds.), The Milky Way Galaxy, 207–208.

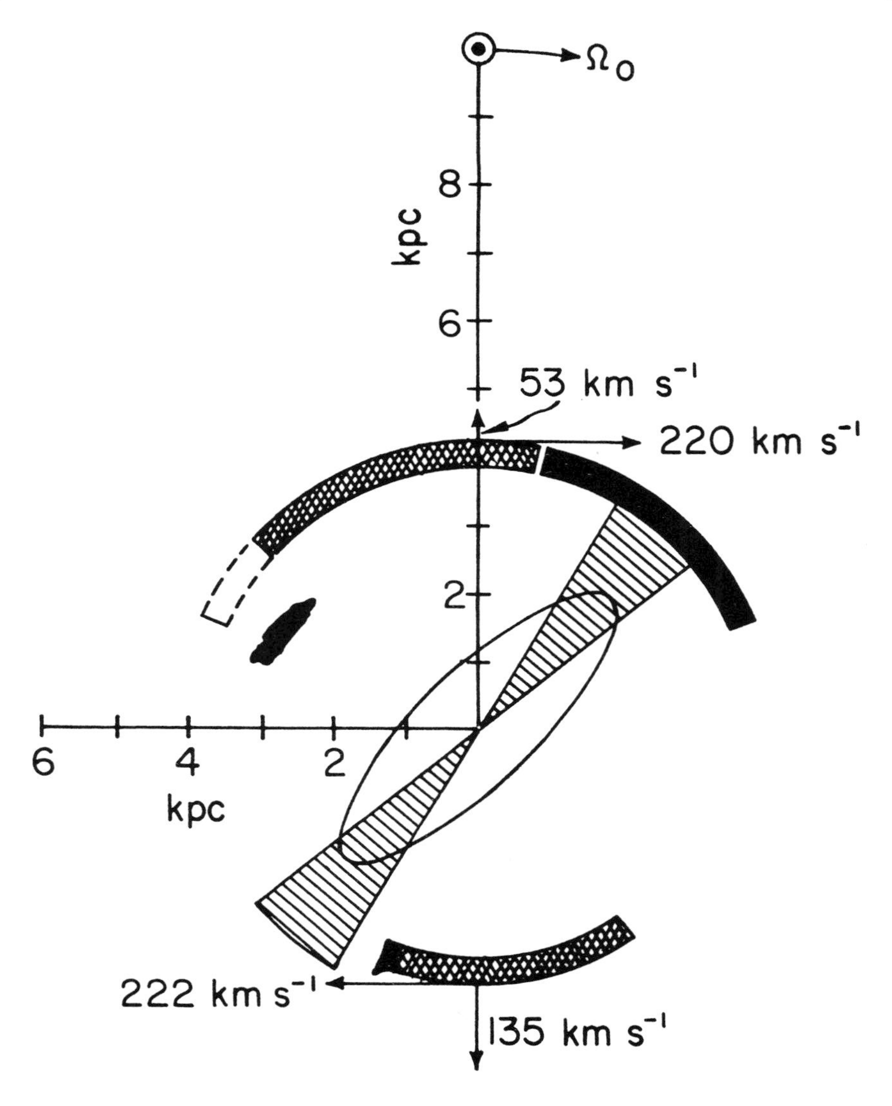

Figure 1: Schematic map of the distribution of the largest and most massive molecular objects in the galactic-center region based on ^{12}CO data (Bania: 1980, Ap.J. 242, 95). The positions of the ^{12}CO structures result from analyses which assume that the (l,v)-loci of the molecular clouds are produced by simple kinematic rings that have fixed radii and which also rotate and expand. The shaded ring segments are conservative estimates for the extent of the molecular gas in the two expanding arms. Outside the nucleus itself, there is evidence for HII gas in the inner Galaxy only for the fully shaded regions (this includes the remarkable Clump 1 cloud). Also shown is the ≃ 1.5 kpc radius nuclear disk/bar whose major-axis position angle coincides with the position angles (stripes) required by resonant-orbit models for the 3-kpc arm.

OUTER-GALAXY MOLECULAR CLOUDS

Marc L. Kutner and Kathryn N. Mead
Physics Department, Rensselaer Polytechnic Institute

Since our original report of CO emission from outside the solar circle in the first quadrant (Kutner and Mead, 1981) we have extended the observations in two ways: (1) We have improved latitude and longitude coverage. Preliminary results on the latitude distribution were reported by Kutner (1983). (2) We have extended our cloud mapping, giving us at least partial CO maps of 55 clouds, along with ^{13}CO, $C^{18}O$, CO (2-1), and 2-mm H_2CO observations of some clouds.

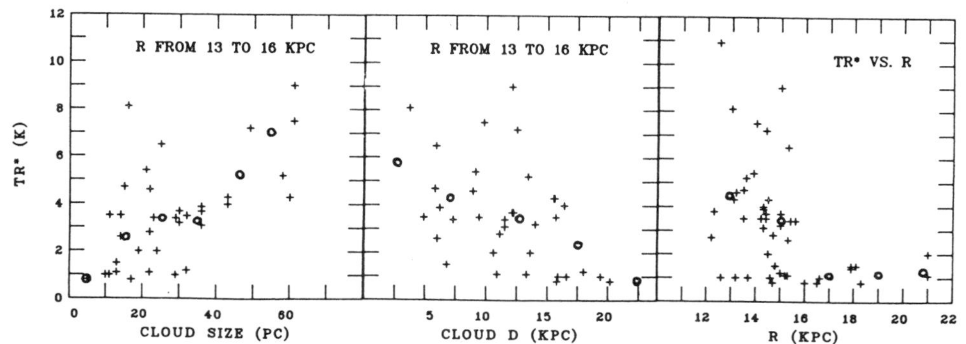

Fig. 1 presents statistics on clouds with sufficient mapping data. (Crosses represent data points and circles indicate binned averages.) (A) For clouds at the same distance, R, from the galactic center there is a correlation between peak temperature T_R^* and cloud size. (B) For clouds with similar R, we see a falloff in T_R^* with distance, d, from the Sun, suggesting that beam dilution becomes important in distant clouds. (C) There is a falloff in T_R^* with R, only weak clouds being seen beyond R ~ 16 kpc.

Using the CO and isotopic data, as well as CO (2-1) data taken at the Texas Millimeter-Wave Observatory (Dickman et al., in preparation), along with microturbulent radiative transfer models (Kutner and Leung, in preparation), we conclude that densities and column densities of GMCs in the outer Galaxy are essentially the same as in the inner Galaxy. However, GMCs in the outer Galaxy are cooler (6 K vs 13 K), probably due to a lower cosmic-ray flux.

H. van Woerden et al. (eds.), The Milky Way Galaxy, 209–210.

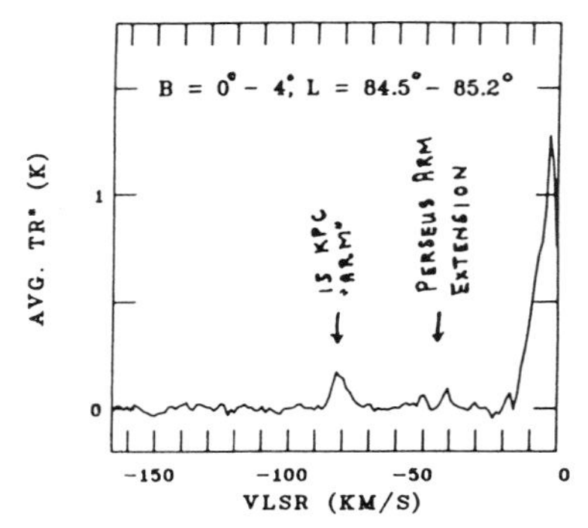

We have used a region of 0.8° in l by 4.0° in b, sampled at 0.1° intervals (for profiles see Kutner 1983) to get an idea of the radial distribution in the outer Galaxy. Fig. 2 shows the average of all 328 spectra. At -40 and -80 km/s it contains two features pointed out by Kutner and Mead (1981): an extension of the Perseus Arm into the first quadrant, and a feature about 15 kpc from the galactic center. The latter is also prominent in HI-data averaged over b (Blitz et al. 1981). The integrated intensity from -25 to -100 km/s in this average spectrum is 1.8 K km/s, corresponding to $N(H_2) \sim 7 \times 10^{20}$ cm^{-2}. If this slice is typical of the outer Galaxy, then the total mass of H_2 in the outer Galaxy would be 5×10^8 $M_\odot$.

The results of Kutner and Mead have been questioned by Solomon et al. (1983). We attribute the failure of Solomon et al. to reproduce our results to problems in calibration and sensitivity. In fact most of the disagreement is over the weak, small clouds that contribute little to the total mass. One must be careful in comparing integrated intensity along any given line of sight. First, the outer Galaxy subtends 6 times the solid angle of the inner Galaxy. Second, each galactic radius, R, contributes twice to each line of sight through the inner Galaxy. Finally, one must consider the variations with R in the conversion from CO intensity to mass (Kutner and Leung, in preparation). Despite all of the disagreement, Solomon et al. (using the same conversion factor from CO to H_2 as we did) get an estimate for the outer-Galaxy mass which differs by only a factor of two from ours. (This factor of two arises from their use of an assumed scale height vs our measured scale height). This supports the idea that most of the mass is in GMCs.

This work is supported, in part, by NSF grant AST81-20900 and by an anonymous gift to support student research in astrophysics at RPI.

REFERENCES

Blitz, K., Kulkarni, S., Heiles, C. 1981, BAAS, 13, 539
Kutner, M.L., Mead. K.M. 1981, Ap. J. (Lett.), 267, L29
Kutner, M.L. 1983, in "Southern Galactic Surveys", eds. W.B. Burton and F.P. Israel, Dordrecht: Reidel, 143
Solomon, P.M., Stark, A.A., Sanders, D.B. 1983, Ap. J. (Lett.) 267, L29

CH IN THE GALAXY

L.E.B. Johansson
Onsala Space Observatory
Sweden

The distribution of CH in the Galaxy has been investigated via its main-line transition in the $^{2}\Pi_{1/2}$, J=1/2 ground-state Λ-doublet at 3335 MHz. The galactic plane was observed with a spacing between adjacent points of 2°.5 in the longitude ranges 10°-60° and 310°-350°. The northern data (Johansson, 1979) were obtained with the 25.6-m telescope of the Onsala Space Observatory and the southern data with the Parkes 64-m antenna; the corresponding beamwidths are 15' and 6'.6, respectively.

The longitude-velocity plot of the CH emission (Fig.1) shows the same basic characteristics as found for CO: (i) the kinematics of the CH gas is almost completely governed by the general rotation of the Galaxy, and (ii) the lack of CH emission close to the terminal velocities within 15°-20° from the galactic centre. Fig.1 also indicates extended gaps in the CH emission that could be interpreted as interarm regions. The presence of large-scale structures is further emphasized in Fig.2,

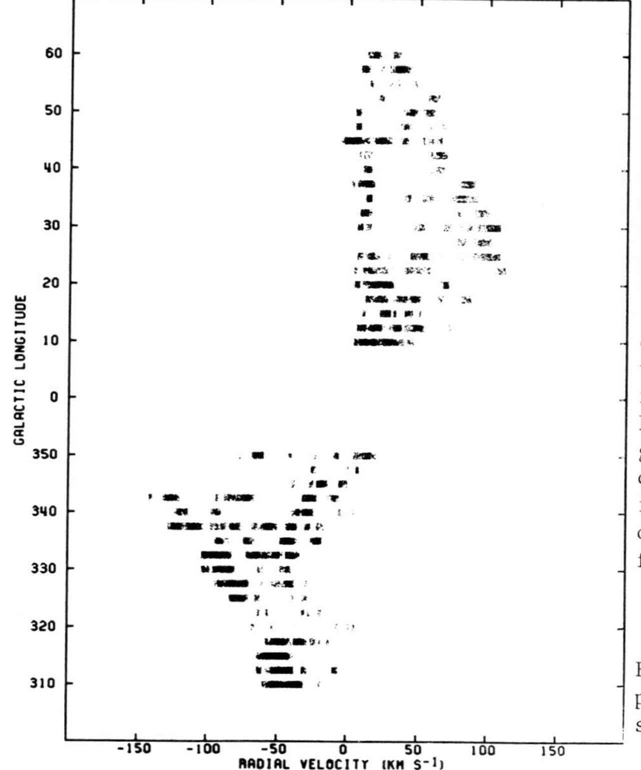

Figure 1. Longitude-velocity plot of CH emission in grey-scale representation.

H. van Woerden et al. (eds.), The Milky Way Galaxy, 211–212.

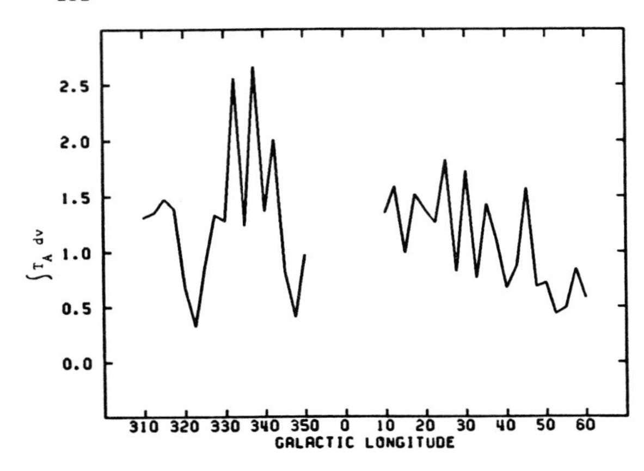

Figure 2. Integrated antenna temperature as a function of galactic longitude.

which is a plot of integrated antenna temperature (proportional to the column density) against longitude. The large variations, notably in the southern data, may originate from arm-interarm regions seen tangentially.

The radial distribution of CH in the Galaxy derived separately from the northern and southern data is shown in Fig.3. These distributions

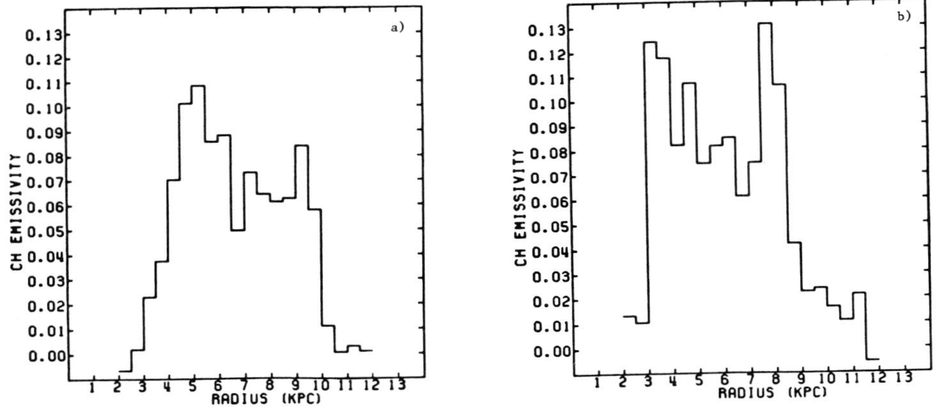

Figure 3. Radial distributions of CH emissivity as derived from the northern (a) and southern (b) data sets.

resemble those of CO: the pronounced maximum at about 5 kpc in the northern distribution and the two peaks at 3.5 and 8 kpc in the southern one. In deriving the radial distributions, the Burton and Gordon (1978) rotation curve was used.

REFERENCES

Burton, W.B., Gordon, M.A.: 1978, Astron.Astrophys. 63, pp. 7-27
Johansson, L.E.B.: 1979, Research Report No. 136, Res. Lab. of Electronics and Onsala Space Observatory

HIGH-ENERGY GAMMA RAYS AND THE LARGE-SCALE DISTRIBUTION OF GAS AND COSMIC RAYS

W. Hermsen
Cosmic Ray Working Group, Huygens Laboratorium, Leiden, The Netherlands

J.B.G.M. Bloemen
Sterrewacht and Cosmic Ray Working Group, Huygens Laboratorium, Leiden, The Netherlands

On behalf of the Caravane Collaboration

ABSTRACT

The COS-B gamma-ray survey is compared with ^{12}CO and HI surveys in a region containing the Orion complex and in the outer Galaxy. The observed gamma-ray intensities in the Orion region (100 MeV<E<5 GeV) can be ascribed to the interaction of uniformly distributed cosmic rays with the interstellar gas. Calibration of the ratio between H_2 column-density and integrated CO line intensity resulted in the value: $(3.0\pm0.7)\times10^{20}$ molecules $cm^{-2}K^{-1}km^{-1}s$. In the outer Galaxy HI column-density maps in three galacto-centric distance ranges are used in combination with COS-B gamma-ray data to determine the radial distribution of the gamma-ray emissivity. A steep negative gradient of the emissivity for the 70 MeV-150 MeV range and an approximately constant (within $\sim$20%) emissivity for the 300 MeV-5 GeV range is found. The result is interpreted as a strong decrease in the cosmic-ray electron density and a near constancy of the nuclear component.

1. INTRODUCTION

The diffuse component of galactic gamma radiation in the energy band discussed in this paper (E>70 MeV) has long been interpreted to be mainly the result of the interaction of cosmic-ray (CR) electrons (via bremsstrahlung, the dominant process at low energies, $E \lesssim 150$ MeV) and CR nuclei (via π°-decay, dominant at higher energies, $E \gtrsim 300$ MeV) with the interstellar gas. The produced gamma-ray intensity, I_γ, can be formulated as follows:

$$I_\gamma = (q_\gamma/4\pi)\,[\,N(HI) + 2N(H_2)]\,, \qquad (1)$$

in which q_γ is the gamma-ray production rate per H atom and N is the

H. van Woerden et al. (eds.), The Milky Way Galaxy, 213–218.

column density of atomic hydrogen (HI) and molecular hydrogen (H_2) respectively. Therefore, gamma radiation is a tracer of the product of the CR density and the interstellar gas density. Gamma-ray measurements can provide a diagnostic of the CR density in regions where the interstellar gas is well traced at other wavelengths, as well as a diagnostic of the total gas content in cases where the CR density can be assumed to be equal to the value for the solar neighbourhood (local value, $\lesssim 1$ kpc). The latter has been verified at intermediate latitudes using galaxy counts as a total-gas tracer for the local interstellar medium [see e.g. Lebrun et al. (1982), Strong et al. (1982) and Lebrun and Paul (1983)]. A good correlation was found between gamma-ray intensities and total-gas column densities

Sofar, the interstellar gas has been mapped on a large scale only for the atomic hydrogen component, using its characteristic 21-cm line. Large-scale surveys of molecular hydrogen are not possible and other molecules (especially CO) have to be used to trace this component. However, considerable ambiguity remains in the conversion of observed CO emission to H_2 column densities [see e.g. Lequeux (1981)]. Gamma-ray astronomy can provide new insight into this calibration problem.

In the actual comparison between measured gamma-ray intensities and atomic- and molecular-hydrogen column densities equation (1) becomes:

$$I_\gamma = (q_\gamma/4\pi)\,[N(HI) + 2XW_{CO}] + I_b, \tag{2}$$

where W_{CO} is the integrated temperature of the ^{12}CO 2.6-mm line (which is best mapped at the moment) and $X=N(H_2)/W_{CO}$ is the $N(H_2)$-to-W_{CO} conversion factor. I_b is the mainly instrumental background level. N(HI) and W_{CO} should be convolved with the energy-dependent point-spread function of the COS-B instrument (Hermsen, 1980) before making comparisons with the gamma-ray intensities. So, if the convolved distributions of N(HI) and W_{CO} show significant (and mutually different) structure, a multiple-linear regression method can be used to determine q_γ, X and I_b. In this paper a maximum-likelihood method, similar to that used by Lebrun et al. (1982), was applied on 1°x1° bins. We will concentrate on recent results obtained by the Caravane Collaboration for the COS-B gamma-ray satellite from the comparison of gamma-ray data with measurements at radio and millimetric wavelengths. The gamma-ray data used are those described by Mayer-Hasselwander et al. (1982), supplemented by later observations. HI column densities are determined from the Berkeley 21-cm line survey (Heiles and Habing, 1974; Weaver and Williams, 1973). The comparisons using ^{12}CO 2.6-mm line surveys are performed in collaboration with the Goddard Institute for Space Studies and Columbia University, who made the CO data available. For a detailed description and maps of the large-scale distribution of galactic gamma-ray emission see e.g. Mayer-Hasselwander et al. (1982) and Hermsen and Bloemen (1983), the latter together with large gamma-ray maps folded in the same volume.

2. THE ORION CLOUD COMPLEX

A region of the sky, away from the intense ridge of gamma radiation along the galactic plane, which is ideal to study eq. (2), is the region containing the nearby Orion cloud complex. Previous studies of one COS-B

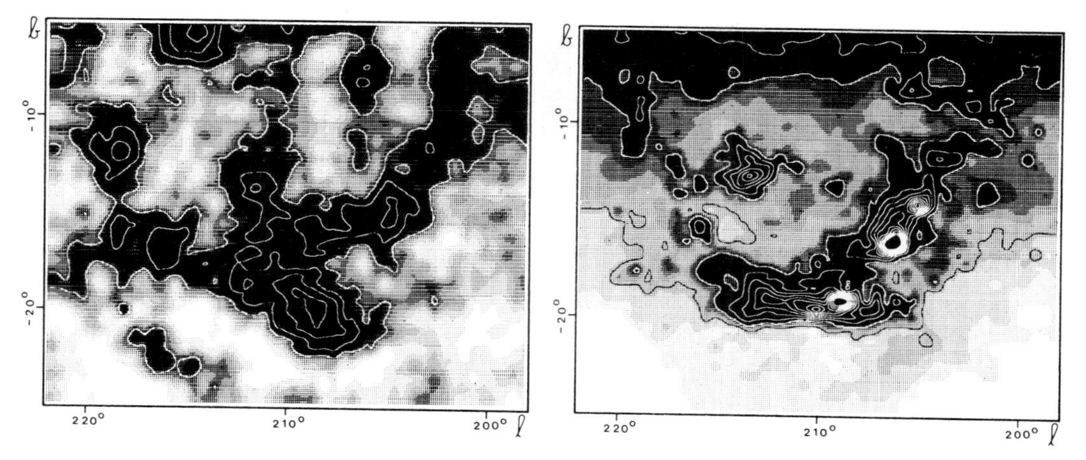

Figure 1: a) Observed gamma-ray intensities (100 MeV<E<5 GeV) in the Orion region. The isotropic background level ($\sim 5.5 \times 10^{-5}$ph cm^{-2}s^{-1}sr^{-1}) is not subtracted. Contour values: (12, 16, 20, ...)$\times 10^{-5}$ph cm^{-2}s^{-1}sr^{-1}. b) Total gas column densities determined from N(HI) and integrated intensities W_{CO} of ^{12}CO-line emission at 2.6 mm, using the N(H_2)-to-W_{CO} ratio found in the present analysis. Contour values: (2, 4, 6, ...) 10^{21}H at cm^{-2}.

observation of the Orion complex (Caraveo et al., 1980, 1981) did reveal the cloud structure in gamma rays and did yield a total mass of the complex in agreement with independent radio-astronomical evaluations. Now all data are available to perform a three-parameter fit following eq. (2). The gamma-ray map of the region of the sky analysed is shown in Figure 1a for energies >100 MeV. For this analysis the high-energy range (300 MeV-5 GeV) is selected, since only for these energies the three parameters can be determined independently, exploiting the better angular resolution at high gamma-ray energies (HWHM $\sim 1°$). The obtained value for the background level is consistent with other analyses of the same COS-B data base (e.g. Strong et al., 1982) and q_γ (300 MeV-5 GeV)/4π=(0.52±0.13)$\times 10^{-26}$ph H at^{-1}s^{-1}sr^{-1}, in good agreement with the average local values determined by Strong et al. (1982) [(0.59±0.15)$\times 10^{-26}$ph H at^{-1}s^{-1}sr^{-1}] using galaxy-count data at intermediate latitudes, and by Bloemen et al. (1984a and section 3) [(0.50±0.04) $\times 10^{-26}$ph H at^{-1}s^{-1}sr^{-1}] from the radial distribution of the gamma-ray emissivity in the outer Galaxy. For the average N(H_2)-to-W_{CO} ratio is found X=(2.6±1.2)$\times 10^{20}$molecules cm^{-2}K^{-1}km^{-1}s. Figure 1b shows the total mass distribution (HI plus H_2) using this conversion factor. It is verified that after convolution of the distribution in Figure 1b no significant differences exist between the structures in the convolved map and in Figure 1a.

For the energy range 100 MeV-5 GeV a three parameter fit is not possible, because after convolution with the broader COS-B point-spread function the N(HI) distribution resembles, over the largest part of the map, too closely the flat isotropic background level. Since in Caraveo et al. (1980, 1981) for E>70 MeV and above for E>300 MeV values for the CR density consistent with the local value are found, it seems reasonable to use the average local emissivity q_γ(100 MeV-5 GeV)/4π=1.7$\times 10^{-26}$ph H at^{-1}

$s^{-1}sr^{-1}$ [derived from Bloemen et al. (1984a) and Strong et al. (1982)] as an input value for the analysis. Then a value for the conversion factor is obtained, consistent with the result for the 300 MeV-5 GeV range, namely $N(H_2)/W_{CO}=(3.0\pm0.7)\times10^{20}$ molecules $cm^{-2}K^{-1}km^{-1}s$, in which the error of 0.7 includes systematic uncertainties. Full details of the analysis are given by Bloemen et al. (1984b).

3. RADIAL DISTRIBUTION OF GAMMA RAYS AND COSMIC RAYS IN THE OUTER GALAXY

In the outer Galaxy no extensive millimetre-wave surveys have been completed yet. However, for this region Bloemen, Blitz and Hermsen (1984) have shown that the gamma-ray intensity is proportional to the HI column density alone to within the uncertainty of the analysis (the H_2 mass at $R>R_\odot$ they found to be $<3\times10^8 M_\odot$). Therefore the W_{CO} term can be deleted from eq. (2) and q_γ and I_b can be determined fitting the gamma-ray and N(HI distribution. In addition, the kinematics of HI can be used to construct column-density maps in various galacto-centric distance ranges in the outer Galaxy. These maps can be used in combination with COS-B gamma-ray data to determine gamma-ray emissivities in these distance ranges. Bloemen et al. (1984a, details of the analysis are given in that paper) used the rotation curve of the outer Galaxy given by Blitz, Fich and Stark (1980) as modified by Kulkarni, Blitz and Heiles (1982) to determine distances beyond the

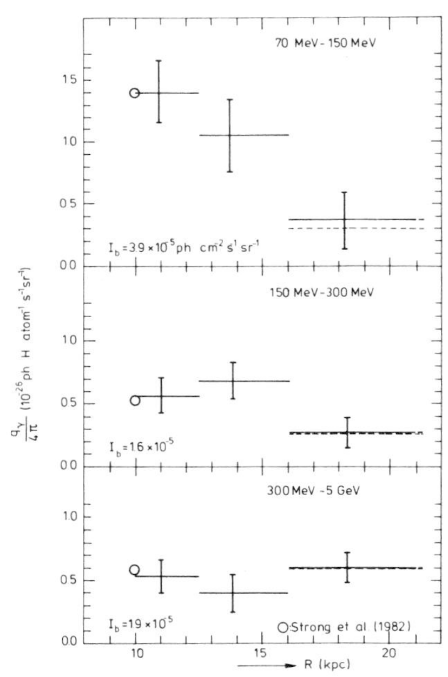

Figure 2. Radial distribution of q_γ in the outer Galaxy for three energy ranges. The resulting isotropic background levels I_b are given in the figures. The errorbars indicate formal 1σ errors. The dashed lines for R>16 kpc show the values of q_γ after correction for a π°-decay input spectrum.

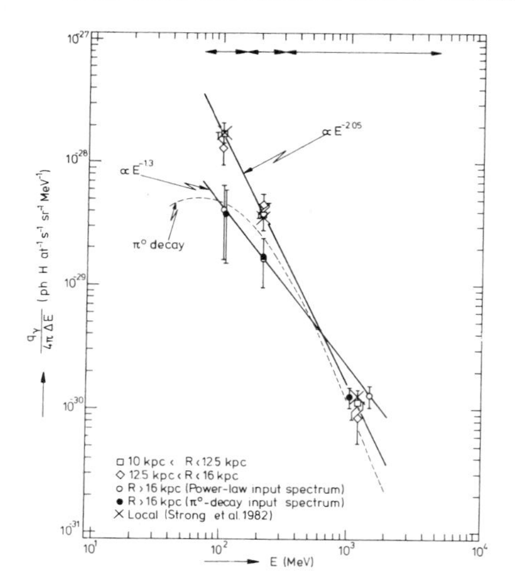

Figure 3. The gamma-ray emissivity spectrum for three distance ranges in the outer Galaxy. Formal (1σ) error bars are indicated. The used energy ranges are given at the top of the figure. Solid lines: power-law fits for R<12.5 kpc and R>16 kpc. Dashed curve: fit to a π°-decay spectrum (after Stephens and Badwahr, 1981).

solar circle. The analysis was performed in the longitude range covered by the Berkeley HI surveys for $95° \leq l \leq 245°$ and $|b| < 10°$. Gamma-ray intensity maps were derived in three energy ranges: 70 MeV-150 MeV, 150 MeV-300 MeV and 300 MeV-5 GeV. HI column-density maps have been constructed for the gas in three distance intervals: R<12.5 kpc (1), 12.5 kpc<R<16 kpc (2) and R>16 kpc (3).

The distribution of N(HI) in the three distance intervals is quite different. On a large scale this is due to the warp of the hydrogen layer, which is more pronounced for increasing galacto-centric distances [see e.g. Henderson, Jackson and Kerr (1982), Kulkarni, Blitz and Heiles (1982)]. On smaller scales the clumpiness of atomic hydrogen produces distinct differences in the distribution of N(HI) in the distance ranges selected. These differences in the projected distributions allowed the following two-dimensional correlation analysis. It was investigated which combination of gamma-ray emissivities in the three distance ranges best describes the observed gamma-ray distribution, using a relation of the form:

$$I_\gamma = (1/4\pi)\,[q_{\gamma,1}\,N(HI)_1 + q_{\gamma,2}\,N(HI)_2 + q_{\gamma,3}\,N(HI)_3] + I_b$$

where $N(HI)_1$, $N(HI)_2$, $N(HI)_3$ are the convolved HI column densities in the three distance ranges. A maximum-likelihood method, similar to that used in Section 2, was applied on 1°x1° bins to determine the three emissivities and I_b for each energy range. The resulting fit values show that the emissivity decreases with increasing R for the 70 MeV-150 MeV range, while the emissivity for high energies (300 MeV-5 GeV) remains approximately constant (within 20%) out to large distances (Figure 2). For the 70 MeV-

150 MeV range the likelihood was found to reduce by a factor of about 10 when a constant emissivity is assumed.

Figure 3 presents the resulting gamma-ray emissivity spectrum for the three distance ranges. There is a clear hardening of the spectrum for increasing galacto-centric distances outside the solar circle. The best power-law fits are indicated in the figure. For R>16 kpc, the spectrum is equally well fitted by a π°-decay spectrum and is thus the first measurement of a diffuse gamma-ray spectrum significantly different from the local gamma-ray spectrum and consistent with the π°-decay spectrum.

These results can be interpreted as 1) a decrease of the density of cosmic-ray electrons with energies up to several hundreds of MeV, such that at large ($\sim$18 kpc) galacto-centric distances the electron density is approximately zero, and 2) a near-constancy of the density of cosmic-ray nuclei with energies of a few GeV, out to large distances [see Bloemen et al. (1984a) for a detailed discussion].

The variation of the electron component is consistent with low-frequency radio-continuum observations [e.g. at 30 MHz Webber et al. (1980) and at 408 MHz Phillips et al. (1981)]. The results confirm a galactic origin of electrons with energies up to several hundreds of MeV. For cosmic-ray nuclei with energies of a few GeV either confinement in a large halo or an extra-galactic origin is suggested by the data.

REFERENCES

Blitz, L., Fich, M., and Stark, A.A.: 1980, in Interstellar Molecules, ed. B. Andrew, Reidel, Dordrecht, p.213
Bloemen, J.B.G.M., Blitz, L., and Hermsen, W.: 1984, Ap.J., in press
Bloemen, J.B.G.M., et al.: 1984a, Astron. Astrophys., submitted
Bloemen, J.B.G.M., et al.: 1984b, Astron. Astrophys., in prep
Caraveo, P.A., et al.: 1980, Astron. Astrophys., 91, L3
Caraveo, P.A., et al.: 1981, Proc. 17th Int. Cosmic Ray Conf., Paris, 1, p.139
Heiles, C., and Habing, H.J.: 1974, Astron. Astrophys. Suppl. 14, p.1
Henderson, A.P., Jackson, P.D., and Kerr, F.J.: 1982, Ap. J. 263, p.116
Hermsen, W.: 1980, Ph. D. Thesis, University of Leiden, The Netherlands
Hermsen, W. and Bloemen, J.B.G.M.: 1983, in Southern Galactic Surveys, eds. F.P. Israel and W.B. Burton, Reidel, Dordrecht, p.65
Kulkarni, S.R., Blitz, L., and Heiles, C.: 1982, Ap. J. Letters 259, L63
Lebrun et al.: 1982, Astron. Astrophys. 107, p.390
Lebrun, F., and Paul, J.A.: 1983, Ap. J. 266, p.276
Lequeux, J.: 1981, Comments on Astrophysics, 9, p.117
Mayer-Hasselwander, H.A., et al.: 1982, Astron. Astrophys. 105, p.164
Phillipps, S., et al.: 1981, Astron. Astrophys. 98, p.286
Stephens, S.A., and Badwahr, G.D.: 1981, Astrophys. and Space Science 76, p.213
Strong, A.W., et al.: 1982, Astron. Astrophys. 115, p.404
Weaver, H., and Williams, D.R.W.: 1973, Astron. Astrophys. Suppl. 8, p.1
Webber, W.R., Simpson, G.A., and Cane, H.V.: 1980, Ap. J. 236, p.448

DISCUSSION printed after Cesarsky's Review Paper in Section II4.

LARGE-SCALE MAPPING OF THE GALAXY BY IRAS

T.N. Gautier, III
National Research Council Resident Research Associate
Jet Propulsion Laboratory
M.G. Hauser
Laboratory for Extraterrestrial Physics
NASA/Goddard Space Flight Center

ABSTRACT

The Infrared Astronomical Satellite (IRAS), launched 1983 January 25, has been conducting a high-sensitivity, high-resolution all-sky photometric survey at wavelengths of 12, 25, 60, and 100 μm in the infrared. One of the data products from the survey will be a map of the entire Milky Way within latitude limits of 10 degrees at a resolution of 4 arcminutes. Since the IRAS detector system is DC-coupled and has demonstrated excellent stability, this map will contain reliable information on all spatial scales larger than the map resolution. The extremely high sensitivity of the IRAS instrument for the detection of interstellar material in the survey mode is illustrated here in terms of visual extinction and dust and gas column densities.

I. INTRODUCTION

The Infrared Astronomical Satellite (IRAS) has been designed to carry out an all-sky photometric survey in four wavelength bands in the spectral region from 8 to 120 μm. The satellite, developed and operated jointly by the United States (NASA), the Netherlands (NIVR), and England (SERC), contains a 60-cm aperture, liquid-helium cooled telescope and infrared instrumentation. IRAS was successfully launched into a 900-km altitude, near-polar orbit on 1983 January 25. Table 1 summarizes the characteristics of the instrument. In the survey mode, IRAS scans the sky in long tracks extending roughly from ecliptic pole to ecliptic pole at fixed solar offset angle. The survey is being conducted so that each point on the sky is scanned by the entire array a minimum of four times in two pairs of observations. The two observations within a pair are separated by a few hours and the pairs are separated by about one week.

One of the planned data products of the survey is a map along the galactic plane extending to a latitude of 10 degrees on both sides of the plane. To produce this map, the survey data will be binned into 2-arcmin square pixels to match the instrumental cross-scan sampling

H. van Woerden et al. (eds.), The Milky Way Galaxy, 219–222.

interval. The signals from the IRAS detectors are dc-coupled, and the system has shown excellent dc stability in flight. Hence, the galactic-plane map is expected to be photometrically meaningful on all spatial scales larger than the pixel size. In the remainder of this article we shall address the expected sensitivity in this map for interstellar material, and describe a very small sample of early data. A more extensive description of the IRAS instrumentation and survey method has recently been published by the IRAS Science Working Group (1983). Current plans call for release of the reduced IRAS survey data in the fall of 1984.

II. SENSITIVITY FOR DETECTION OF INTERSTELLAR MATTER

The noise - equivalent flux densities and intensities of typical IRAS detectors in flight are given in Table 1 for a frequency corresponding to spatial scales of a few tens of degrees.

TABLE 1. IRAS Passbands and Noise for a Typical Detector

Band	λ (μm)	$d\lambda/\lambda$	NEFD(Jy $Hz^{-1/2}$)	NEI_ν(Jy $sr^{-1} Hz^{-1/2}$)
1	12	0.5	0.075	2.6×10^5
2	25	0.4	0.075	2.6×10^5
3	60	0.5	0.16	2.7×10^5
4	100	0.3	0.80	6.3×10^5

This noise performance has been translated into some predictions of what should be visible in the survey data in terms of astrophysically significant quantities. To express the results in these terms, we have assumed that the following are typical properties for the interstellar dust and gas:

$$\tau_\nu = K_\nu M_D,\ K_\nu = 4.6(\nu/10^{12})^2\ cm^2 g^{-1},\ M_H/M_D = 10^2$$

$$N_H/A_V = 2 \times 10^{21}\ cm^{-2} mag^{-1},\ N_H = 3.2 \times 10^{20} \int T_a(^{12}CO)\ dv\ cm^{-2}$$

where M_H = gas mass column density, N_H = gas number column density, M_D = dust mass column density, A_V = visual extinction, K_ν = dust mass absorption coefficient, ν = optical frequency, v = velocity and τ_ν = optical depth. We consider detection of thermal emission from optically thin dust clouds of temperature T. The detectable dust optical depth is obtained directly from the system-noise-equivalent flux for any assumed dust temperature, and the result expressed in terms of column densities or extinction using the above assumptions. Figure 1 shows the 5σ detection levels for the 60- and 100-μm IRAS bands as a function of dust temperature for data smoothed to a 1/2-degree square beam (average of 16 detectors integrated for 8 seconds). For dust warmer than about 15K, IRAS sensiti-

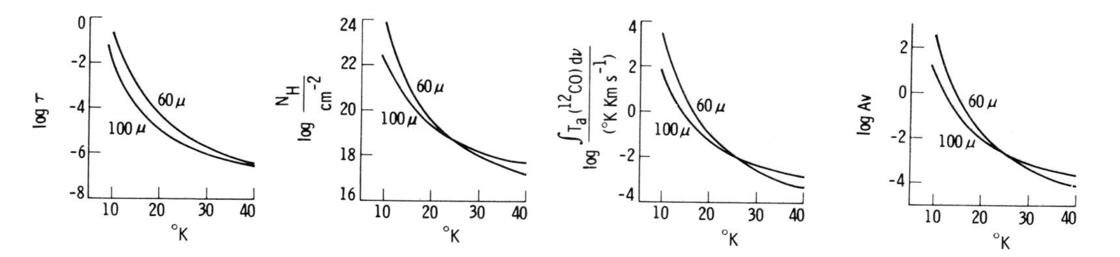

Figure 1. IRAS 5σ detection limits for various astrophysical parameters. Data integrated to 0.5-deg square beam.

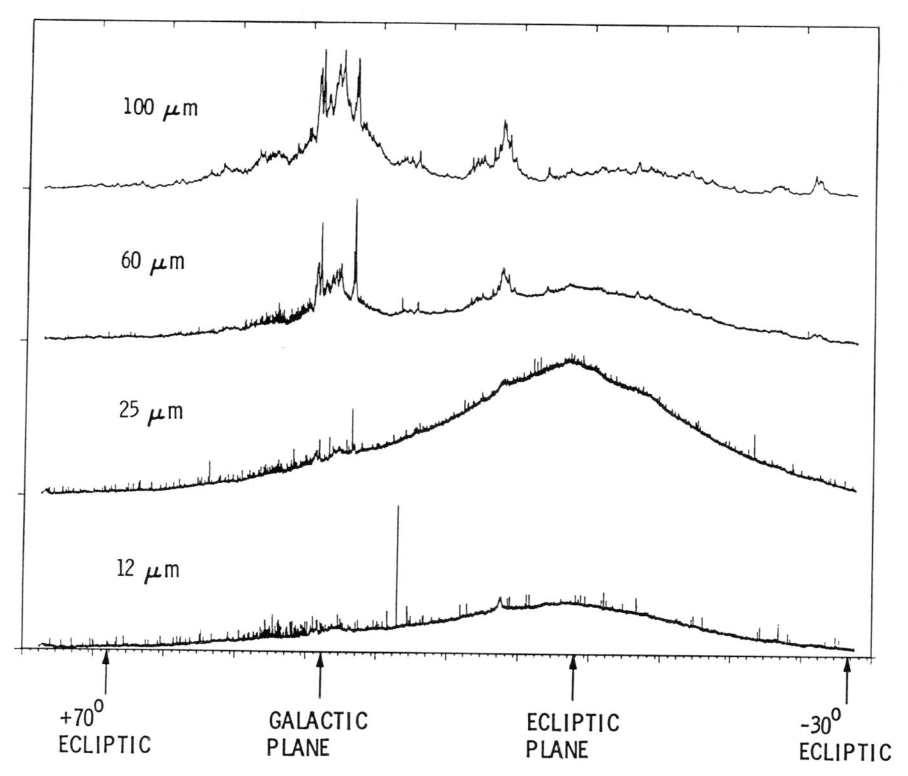

Figure 2. Time history for one detector from each band during one scan. Vertical axes are offset. The abscissa is labelled in ecliptic latitude.

vity to interstellar material is clearly at least comparable to that attained with other techniques.

III. GALACTIC-PLANE EMISSION

A sample of IRAS survey data is shown in Figure 2. These data are from a scan at about 90° from the Sun which crosses the Milky Way at an angle of 48° near galactic longitude 140°. Except for the section with many spikes near the Galactic plane, obtained when the satellite passed through the northern horn of the Van Allen belt, there is no noise visible in the figure. All features are due to the sky and reproduce to a high degree of accuracy from scan to scan. A notable feature in Figure 2 is the dominant contribution of the interplanetary dust emission at wavelengths shorter than 100 μm.

We shall eventually construct a detailed model of the "zodiacal emission" to permit discrimination between solar-system and galactic emission. Though these results are very fragmentary, it is already clear that the infrared maps of the Galaxy to be obtained by IRAS will be a major new resource for study of the structure and constituents of the Milky Way.

REFERENCE

IRAS Science Working Group 1983, Nature 303, 287.

DISCUSSION

R. Wielen: Could you describe the IRAS results on the Andromeda Nebula?

Gautier: Results on M31 will be discussed by Habing later during the Symposium (Section II.9).

H. Okuda: Do the detection limits achieved on IRAS in actual operation agree with the ultimate sensitivity levels obtained in the laboratory?

Gautier: We do indeed obtain noise levels in 2 arcmin windows consistent with those expected.

SECTION II.4

THE HIGH-ENERGY COMPONENT

Monday 30 May, 1615 - 1755

Chairman: H.C. van de Hulst

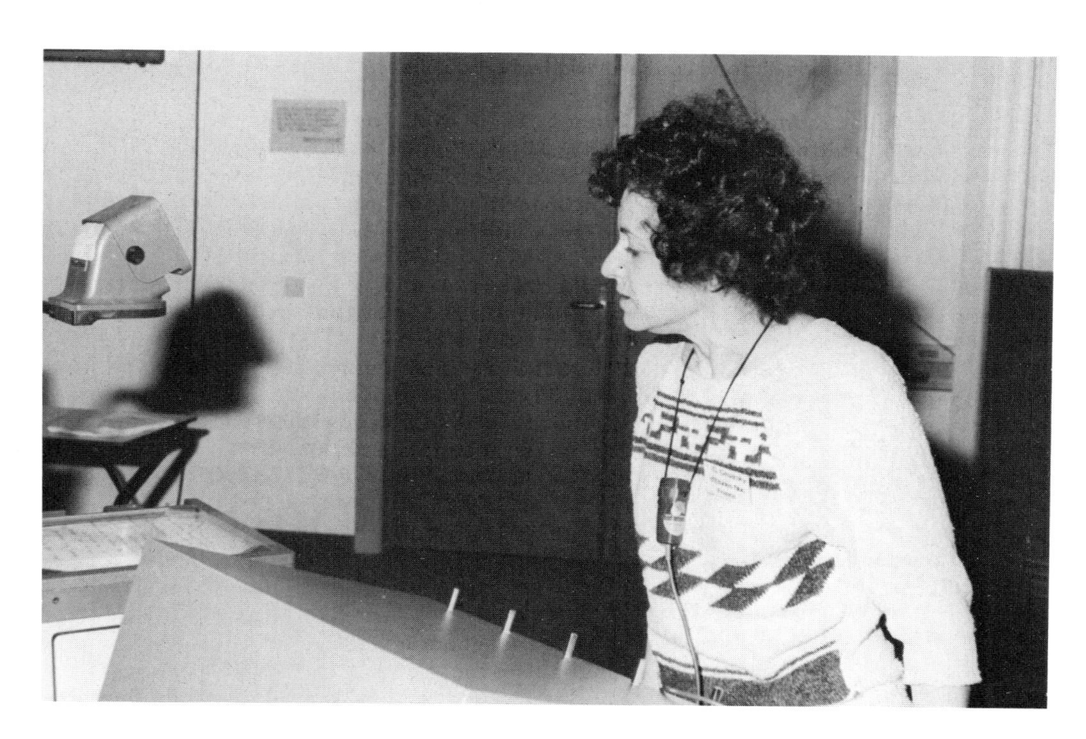

C.J. Cesarsky (top) and R. Beck (bottom) presenting their review papers
CFD

HIGH-ENERGY GALACTIC PHENOMENA AND THE INTERSTELLAR MEDIUM

Catherine J. Cesarsky
Service d'Astrophysique
Centre d'Etudes Nucléaires de Saclay, France

GAMMA RAYS AND THE GALACTIC DISTRIBUTION OF COSMIC RAYS

Gamma rays of energy in the range 30 MeV-several GeV, observed by the satellites SAS-2 and COS-B, are emitted in the interstellar medium as a result of interactions with gas of cosmic-ray nuclei in the GeV range (π° decay γ rays) and cosmic-ray electrons of energy > 30 MeV (bremsstrahlung γ rays). W. Hermsen has presented at this conference the γ ray maps of the Galaxy in three "colours" constructed by the COS-B collaboration; the information in such maps is supplemented by radio-continuum studies (see lecture by R. Beck), and is a useful tool for studying the distribution of gas, cosmic rays (c.r.) and magnetic fields in the Galaxy. The variables in this problem are many:large-scale ($\sim$ 1 kpc) and small-scale ($\sim$10 pc) distributions of c.r. nuclei, of c.r. electrons, of atomic and molecular hydrogen, of magnetic fields, fraction of the observed radiation due to localized sources, etc. Of these, only the distribution - or at least the column densities - of atomic hydrogen are determined in a reliable way. Estimates of the amount of molecular hydrogen can be derived from CO observations or from galaxy counts. The radio and gamma-ray data are not sufficient to disentangle all the other variables in a unique fashion, unless a number of assumptions are made (e.g. Paul et al. 1976). Still, the COS-B team has been able to show that :
a) there is a correlation between the gamma-ray emission from local regions, as observed at intermediate latitudes, and the total column density of dust, as measured by galaxy counts. The simplest interpretation is that the density of c.r. nuclei and electrons is uniform within 500 pc of the sun, and that dust and gas are well mixed. Then, γ rays can be used as excellent tracers of local gas complexes (Lebrun et al. 1982, Strong et al. 1982).
b) In the same way, the simplest interpretation of the γ-ray emission at energy > 300 MeV from the inner Galaxy, is that c.r. nuclei and electrons are distributed uniformly as well : there is no need for an enhanced density of c.r. in the 3-6 kpc ring ; on the contrary, even assuming a uniform density of c.r., the γ-ray data are in conflict with the highest estimates of molecular hydrogen in the radio-astronomy literature (Mayer-Hasselwander et al. 1982).

H. van Woerden et al. (eds.), The Milky Way Galaxy, 225–233.

c) In the outer Galaxy, the gradient of c.r. which had become apparent in the early SAS-2 data can now, with COS-B data, be studied in three energy ranges. A gradient in the c.r. distribution is only required to explain the low-energy radiation, which is dominated by bremsstrahlung from relativistic electrons (Bloemen et al., in preparation).

Thus, as the COS-B data have been gaining in statistical accuracy through the 7.5-year lifetime of the mission, they have not confirmed (but not really ruled out either) one of the main early findings of gamma ray astronomy : the existence of gradients in the distribution of galactic c.r. nuclei. Gamma-ray observations do not stand anymore in the way of the tenants of the hypothesis of the universality of c.r. nuclei.

GAMMA-RAY SOURCES : A NEW GALACTIC POPULATION ?

COS-B has also observed - and the COS-B team is still studying - a number of localized and extended γ-ray sources. The Orion molecular cloud complex is the best studied of the extended sources. The correspondence between the γ-ray map and the CO map is excellent ; the emission is as expected for a uniform c.r. density, equal to that present in the nearby clouds at intermediate latitudes (Caraveo et al. 1981).

SAS-2 had discovered three localized γ-ray sources in the sky : the Crab and Vela pulsars, and a mysterious source named "Geminga" (Fichtel et al. 1975). (For present day γ-ray telescopes, a faint source whose angular extent is up to 2° may appear as a point source.) The COS-B workers searched systematically for sources with a Vela-like profile, standing above the background (Swanenburg et al. 1981 ; see also Bignami and Hermsen 1983). Of the new 22 sources isolated, only one is clearly identified to a point-like source : the quasar 3C273. Another source in the COS-B catalogue is in the direction of the ρ Oph dark cloud. It appeared to emit $\sim$ 5 times more γ rays than expected, given its estimated mass. This source has stirred much controversy, and stimulated a great deal of theoretical and observational work - some of it very fruitful -. It has been proposed that the ρ Oph cloud encompasses or is close to a source of c.r. ; these c.r. could be shock-accelerated (see next section) by winds from OB stars (Cassé and Paul, 1980) or by a nearby supernova (Morfill et al. 1981). Alternatively, ρ Oph could contain a compact γ-ray source. In an attempt to locate the hidden source, Montmerle et al. (1983) pointed the Einstein satellite in the direction of the core of ρ Oph ; they discovered an X-ray Christmas tree, thus showing that the atmospheres of T Tauri stars are the site of numerous energetic flares - a fascinating result, even though it may be irrelevant for the interpretation of the high-energy γ-ray data. In the mean time, more recent observations of the ρOph region by COS-B have led to a revised estimate of the γ-ray luminosity, which now exceeds the expected value by a factor of only $\sim$ 2. Also, this γ-ray source is about to lose its "localized source" status : it has now been resolved by COS-B, and the correlation between the CO map and the γ-ray map is quite good (Hermsen and Bloemen, 1983).

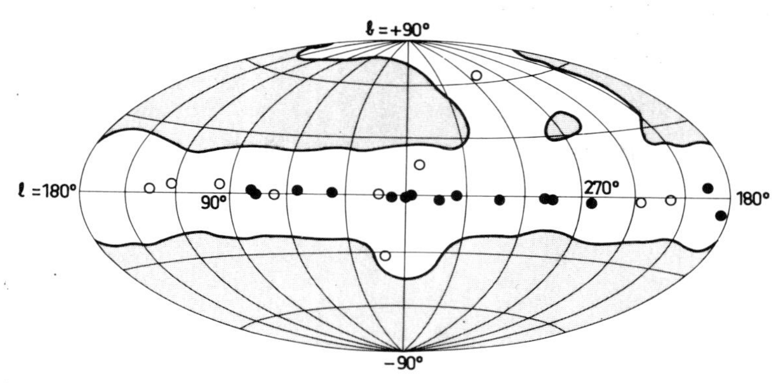

Figure 1 - Celestial distribution of the COS-B sources (Swanenburg et al. 1981). The closed circles denote sources with measured fluxes > 1.3×10^{-6} photons (>100 MeV) cm^{-2} s^{-1}; open circles denote sources below this threshold. Sources have been searched only in the unshaded area.

The remaining 21 γ-ray sources all lie at very low galactic latitudes (fig. 1). The fact that the latitude distribution is so narrow implies that they are concentrated along the galactic plane, and at distances from the sun much larger than their scale height ; this in turn suggests that they are associated to young galactic objects. Assuming a scale height H as small as that of the flattest (and youngest) galactic populations, H ∿ 40 pc, we find that the sources must be at least 2 kpc away. At the same time, the absence of a strong concentration of sources towards the Galactic Centre ($|\ell|<30°$) suggests that the observed sources are not too far either — probably not beyond 7 kpc from the sun. Their intrinsic luminosity, therefore, is in the range $0.4 - 5 \times 10^{35}$ ergs s^{-1}. A few of the COS-B sources may simply be a dense cloud or cloud complex, at intermediate distances (∿1 kpc), traversed by a normal flux of c.r.

A class of objects which are young and tend to lie at low latitudes are the supernova remnants. These are either directly visible in optical light, or are detected through their radio-emission or X-ray emission. Their radio-spectrum is a power law, characteristic of synchrotron emission by a power-law spectrum of relativistic electrons. The energy density of relativistic particles in these objects is much higher than in the general interstellar medium, so that they are interesting candidates for gamma-ray sources. Given their linear sizes of ∿ a few pc, their angular size is generally small enough that they are point sources for the present gamma-ray telescopes. A few positional coïncidences between some of the first gamma-ray sources discovered and supernova remnants had been pointed out, early on ; at present, with the gamma-ray source error box covering over 10 % of the central regions for b<1°, while 125 supernova remnants are known, it seems evident that an undiscerning search for positional coïncidences cannot bring fruitful results.

Strong γ-ray emission may be expected if a supernova remnant bites on a dense molecular cloud ; this is most likely to happen in regions of star formation. A subclass of supernova remnants has been defined (Montmerle, 1979), singling out the objects in whose direction there is also observational evidence for star formation, such as the presence of associations of young stars (OB associations), or of hot gas ionized by them. Thirty two such objects, called SNOBs, have been listed, and 7 to 9 COS-B sources coïncide with them. Several other SNOBs are situated in more extended regions of intense gamma-ray emission. The γ-ray emission from SNOBs can be accounted for quantitatively by bremsstrahlung interactions of the low-energy tail of the electron spectrum revealed by the synchrotron emission, and cloud matter of density $\sim 10^2$–10^3 cm^{-3} (Montmerle and Cesarsky, 1980).

Apart from the two known young pulsars and the conventional sources obeying to the equation gas + c.r. = γ rays, is there a more exotic population of γ-ray sources ? Despite many efforts spent attempting identifications, we do not know yet. The best studied source is also the most tantalizing one : Geminga, the second brightest γ-ray source in the sky. The COS-B error box of Geminga, which is only a half square degree, has been explored at various wavelengths. The first searches yielded purely negative results : almost no gas, no bright X-ray source, or radio source, nothing ! But with deep searches, two possible candidates have appeared :
a) Moffat et al. (1983) have discovered several faint radio sources in the Geminga error box. One of them is optically identified to a quasar of redshift 1.2. The ratio L_γ/L_{radio} would then be $\sim 5.10^5$. For 3C273, the "typical" and only γ-ray quasar, this ratio is only ~ 100.
b) Bignami et al. (1983) found four X-ray sources in the Geminga error box. The brightest source, which has an a priori probability of a few % to be in the COS-B error box, seems a particularly good candidate. Because of the lack of absorption in the soft X-ray spectrum, it is probably nearby - not much beyond 100 pc. This peculiar X-ray source has an optical counterpart : a very faint blue object, such that $L_x/L_{opt} \sim 250$ (Caraveo et al. 1983). There is no radio source associated with it, and the X-ray emission does not appear to be variable on scales going from 2.6 msec to hours. If the identification is correct, Geminga has $L_\gamma/L_x \simeq 10^3$. There is at present no obvious explanation for this object.

Will the search for Geminga lead to the discovery of a new class of high-energy sources, and help to explain a sizeable fraction of the COS-B sources ? Difficult to predict ! The only established characterictic of the γ-ray sources is that they lie at low latitudes ; if Geminga is a quasar far away or a very nearby X-ray source, it is at low latitude just by accident !

COSMIC-RAY ACCELERATION BY INTERSTELLAR SHOCKS

Let us now return to the cosmic rays, which we had abandoned in our search for exotic γ-ray sources. Two ideas on the origin of c.r. have been in favour for much longer than I have been in the business : that cosmic rays are accelerated by the Fermi mechanism, and that their energy source is supernovae. Fermi (1949) introduced the idea that c.r. acquire their considerable energies by colliding a large number of times with magnetized clouds, gaining only a small amount of energy at each encounter.

One of the main reasons why this process has enjoyed such an endurable popularity among astrophysicists is that, under very simple conditions, it predicts that the energy spectrum of the colliding particles should be a power law, and power-law spectra are extremely frequent in non-thermal sources of radiation all over the universe. But the great drawback is that it cannot explain why the power law exponents of c.r. in the galaxy and in radio sources fall almost invariably in the range 2-3. A possible explanation of the spectral index and a link between cosmic-ray acceleration and supernovae has been established recently with the study of particle acceleration by diffusive shocks (Bobalsky, 1977, 1978).

Let us consider a strong shock, propagating at a velocity V in the direction of the magnetic field lines. We assume that $V \gg V_A$, where V_A is the Alfvèn velocity ($V_A^2 = B^2/4\pi\rho^*$, where ρ^* is the density of ionized particles). In the shock frame, the gas is flowing in at a velocity $u_1 = V$. At the shock, the gas is compressed by a factor r, so that the velocity downstream, relative to the shock, is $u_2 = V/r$. The presence of scattering centres of cosmic rays is postulated, so that cosmic rays diffuse on both sides of the shock. The scattering centres act as cosmic-ray traps, ensuring that the particles will be reflected back and forth across the shock a large number of times. Every passage through the shock is equivalent to running head-on into a "magnetic wall" of velocity $V=u_1-u_2=V(1-1/r)$. Thus, particles increase their energy by a small amount every time they cross the shock, as in a Fermi (first-order) mechanism. But here the average number of passages of the particles through the shock, before escaping downstream, is, like the energy gain per passage, completely determined by the shock characteristics. In the time-independent limit, this mechanism generates a power-law spectrum whose spectral index depends only on the compression ratio r of the shock, and not at all on the shock velocity, on the diffusion coefficient (assumed "small enough") or on the dimensions of the scattering region (assumed "large enough"). For strong adiabatic shocks, r = 4 and the differential spectrum of relativistic particles predicted is proportional to $1/(\text{energy})^2$.

The study of shock acceleration of cosmic rays is now an active area of research [See reviews by Axford (1981), Drury (1983)]. A detailed application of the simple mechanism I have just described to the acceleration of galactic cosmic rays by supernova shocks has been presented by Blandford and Ostriker (1980) [See also Moraal and Axford, 1983, Bogdan and Völk, 1983]. The mechanism has also been applied to terminal shocks from stellar winds, but not without controversy [see review by Cesarsky and Montmerle, 1983]. Let us now mention some of the problems encountered by the linear, time-independent theory just described :

a) If cosmic rays extract so much energy from the shock, their pressure can become the dominant one. For instance, this will inevitably occur if cosmic rays are getting accelerated by a strong shock, to a spectrum E^{-2} for a sufficiently long time. Even if the shock is not so strong ($r<4$), the cosmic ray pressure can become dominant if the rate of injection of particles in the system is sufficiently rapid. The expectation is that, eventually, the cosmic rays broaden the shock, making it a less efficient particle accelerator ; if the shock becomes wider than the particle mean free path λ, particles only get a small amount of adiabatic acceleration as they traverse the compressed regions, but a power-law tail does not develop. While some progress has been made (Eichler 1979, Ellison 1981), the full problem of non-linear shocks, as well as the distinct, but coupled problem of particle injection into the acceleration mechanism, still poses many intriguing puzzles.

b) This problem has always been treated in the framework of the quasi-linear theory, which assumes that the turbulent energy in the hydromagnetic waves acting as particle scatterers is much less than the energy density of the magnetic field. However, the anisotropies induced by supernova shocks in the pre-existing population of galactic cosmic rays are sufficient to render these waves extremely unstable; the wave amplitudes predicted by the quasi-linear theory are too high to be fully consistent with the theory. In addition, the waves play a role in the hydrodynamics and the thermodynamics of the system (McKenzie and Völk, 1982).

c) Finally, this acceleration process is slow ; consequently, when applied to realistic shocks, which have a finite lifetime, the theory predicts a high-energy cut-off. Under optimum conditions, the maximum energy that a proton interaction with a supernova shock can attain is 10^5 GeV ; under more realistic conditions, this upper limit can be as low as 2000 GeV (Lagage and Cesarsky 1983). In contrast, the observed spectrum of cosmic-ray protons is a power law up to 10^6 GeV (see review by Webber, 1983).

SUPERNOVAE AND THE INTERSTELLAR MEDIUM

In addition to their possible effect on the cosmic ray component, supernova (s.n.) shocks have a profound effect in shaping up the interstellar medium. Cox and Smith (1974) first pointed out that, given the high rate of s.n. explosions in the Galaxy, a part of the gas heated by a blast wave does not have time to cool down before it is hit again by a shock. Thus, at any time, a sizable fraction of the interstellar medium should be filled by hot ($T \gtrsim 5.10^5$ °K) and tenuous ($n \lesssim 10^{-2}$ cm^{-3}) gas. Global models of the interstellar medium have been proposed (McKee and Ostriker 1977, Cox 1981) ; but uncertainties on the distribution of cloud sizes, on the possibility of thermal and mechanical exchanges between clouds and the hot medium surrounding them, on the filling factor of a neutral, warm intercloud medium, and on several other variables make it impossible to devise a definitive model as yet.

The presence of hot gas in the solar environment has been confirmed by observations :

a) The Copernicus satellite detected absorption lines of O VI in the direction of several stars, indicating the presence of gas with $T \lesssim 5.10^5$ °K (Jenkins 1978a,b).

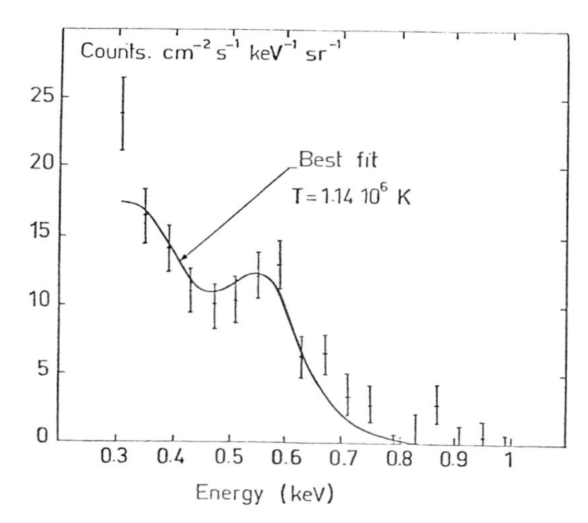

Figure 2 - Spectrum of the local hot bubble, obtained in the north galactic hemisphere with a solid-state detector Si(Li) with a large field of view (Rocchia et al. 1983). The feature around 550 eV is produced by the OVIII line emission. The large excess at low energies is attributed to a blend of CV and CVI lines. The solid curve is the expected spectrum of a thin plasma with T = $1.14\times10^{6}\,^{\circ}$K, in ionization equilibrium and with normal abundances.

b) Detailed maps of the soft X-ray sky in seven energy bands have been constructed over the last 7 years by the Wisconsin group (McCammon et al. 1983 and ref. therein). The interpretation of these data is far from straightforward : it is not easy to disentangle the relative positions of the hot gas, the cold clouds, the old s.n. remnants, the galactic halo. Still, a relatively isotropic, unabsorbed background of soft X rays seems to be present, which probably indicates that the solar system is embedded in a bubble of hot gas of radius 60 to 80 parsecs (Hayakawa et al. 1979), which is relatively empty of neutral gas. The thermal origin of this local X-ray emission has been confirmed by low-resolution spectra showing lines of stripped ions of iron, silicon, sulfur (Inoue et al. 1979) and carbon and oxygen (Rocchia et al. 1983, figure 2). If the elemental abundances are normal, the relative intensities of the lines are best understood if the gas is not in ionization equilibrium, which is what can be expected if a s.n. exploded nearby less than a few 10^5 years ago (Hayakawa et al. 1979, Cox and Anderson 1982, Arnaud et al. 1983).

It is a pleasure to thank Monique Arnaud, Pierre-Olivier Lagage, François Lebrun and Robert Rothenflug for useful discussions.

REFERENCES

Arnaud, M., Rothenflug, R., and Rocchia, R., 1983, Physica Scripta (in press).

Axford, W.I. 1981, Proc. 17th Int. Cosmic Ray Conf., Paris, 12, p. 155.
Bignami, G. and Hermsen, W. 1983, Ann. Rev. Astron. Astrophys., in press.
Bignami, G.F., Caraveo, P.A., Lamb, R.C. 1983, Astrophys. J; (Lett.), in press.
Blandford, R.D. and Ostriker, J.P. 1980, Astrophys. J. 237, p. [illegible]3.
Bobalsky 1977, 1978 is short for the following 4 papers:
    Blandford, R.D. and Ostriker, J.P. 1978, Astrophys. J. 221, p.L29.
    Bell, A.R., 1978 M.N.R.A.S. 1982, p. 147.
    Axford, W.I., Leer, E. and Skadron, G. 1977, Proc. 15th Int. Cosmic Ray Conf. Plovdiv 11, p. 132.
    Krymsky, G.F. 1977, Dokl. Akad. Nauk. SSSR, 234, p. 1306.
Bogdan, T.J., Völk, H. 1983, Astron. Astrophys. 122, p. 129.
Caraveo, P.A. et al. 1983, preprint.
Caraveo, P.A. et al. 1981, Proc. 17th Int. Cosmic Ray Conf., Paris, 1, p. 139.
Cassé, M. and Paul, J.A. 1980, Astrophys. J. 237, p. 236.
Cesarsky, C.J. and Montmerle, T. 1983, Space Sci. Rev. 36, p. 173
Cox, D.P. 1981, Astrophys. J. 245, p. 534.
Cox, D.P. and Anderson, P.R. 1982, Astrophys. J. 253, p. 268.
Cox, D.P. and Smith, B.W. 1974, Astrophys. J. 189, p. L105.
Drury, L.O.'C. 1983, Rep. Progr. Phys., in press.
Eichler, D. 1979, Astrophys. J. 229, p. 419.
Ellison, D. 1981, Ph.D. Thesis, Catholic University.
Fermi, E. 1949, Phys. Rev. 75, p. 1169.
Fichtel, C.E. et al. 1975, Astrophys. J. 198, p. 163.
Hayakawa, S. et al. 1979, in COSPAR X-Ray Astronomy, ed. W.A. Baity and L.E. Peterson, Pergamon Press, p. 319.
Hermsen, W. and Bloemen, J.B.G.M. 1983, Leiden Workshop on Southern Galactic Surveys, Reidel, p. 65
Inoue, H. et al. 1979, Astrophys. J. 227, L85.
Jenkins, E.B. 1978a, Astrophys. J. 219, p. 845.
Jenkins, E.B. 1978b, Astrophys. J. 220, p. 107.
Lagage, P.O. and Cesarsky, C.J. 1983, Astron. Astrophys., in press.
Lebrun, F. et al. 1982, Astron. Astrophys. 107, p. 390.
Lebrun, F. et al. 1983, Astrophys. J., in press.
Mayer-Hasselwander, H.A. et al. 1982, Astron. Astrophys. 105, p. 164.
McCammon, D. et al. 1983, Astrophys. J. 269, p. 107.
McKee, C.F. and Ostriker, J.P. 1977, Ap.J. 218, p. 148.
McKenzie, J.F. and Völk, H.J. 1982, Astron. Astrophys. 116, p. 191.
Moffat, A.F.J. et al. 1983, Astrophys. J. (Lett.), in press.
Montmerle, T. 1979, Astrophys. J. 231, p. 95.
Montmerle, T. and Cesarsky, C.J. 1980, Non-Solar Gamma Rays, Pergamon Press, p. 61.
Montmerle, T. et al. 1983, Astrophys. J. 269, p. 182.
Moraal, H. and Axford, W.I. 1983, Astron. Astrophys., in press.
Morfill, G.E. et al. 1981, Astrophys. J. 246, p. 810.
Paul, J., Cassé, M. and Cesarsky, C.J. 1976, Astrophys. J. 207, p. 62.
Rocchia, R. et al. 1983, submitted to Astron. Astrophys.
Strong, A.W. et al. 1982, Astron. Astrophys. 115, p. 404.
Swanenburg, B.N. et al., 1981, Astrophys. J. 243, p. L69.
Webber, W., 1983, Composition and Origin of Cosmic Rays, Reidel, p. 25.

DISCUSSION

J.P. Ostriker: The Wisconsin Group argue that the soft-X-ray evidence implies that we live in a hot bubble. If the three-phase model of the interstellar medium is correct, then every observer will think he lives in a hot bubble: soft X-rays come from nearby, because of absorption by interstellar clouds, while hard X-rays can penetrate the clouds and thus come from larger distances. However the [OVI] lines observed by Copernicus cover a much wider range of space, and they are thought to come from the surroundings of clouds evaporating into a hot medium.

Another comment concerns the cosmic rays. The new fact in Hermsen's paper is the evidence for a variation of the electron/proton ratio with galactocentric distance. That reminds one of the following long-standing puzzle. The supernova remnants, though ideal candidates for supplying the galactic cosmic rays, cannot be the source locally because they contain predominantly electrons, while the local cosmic rays consist predominantly of protons. Perhaps the γ-ray observations can give us a handle on this puzzle.

Cesarsky: Yes - except that the variation in the e/p ratio indicated by the γ-ray studies is only a factor of 5, while the discrepancy between supernovae and cosmic rays is a factor of 100. So we are still a factor of 10 short. Also, the tracing of cosmic rays by γ-rays favours regions of high density, where protons may be trapped and generate secondary electrons. If dense clouds are more frequent at smaller galactocentric distances, this might already explain part of the factor 5.

J.B.G.M. Bloemen: From the COS-B data, the radial gradient of the CR-electron density and the near-constancy of the CR-proton density (out to about 20 kpc) has sofar only been determined in the outer Galaxy. The analysis for the inner Galaxy remains to be done, but there are good indications that the results will be the same.

New COS-B observations of the ρ Oph complex show that the γ-ray source is extended along the two dust lanes.

J.V. Feitzinger to Hermsen: COS B has a very low angular resolution. This means that you run into problems of beam dilution at distances greater than about 5 kpc from the Sun. Can you comment on this problem?

W. Hermsen: At energies above 300 MeV, the resolution (FWHM) is ~1.5°. However, the angular response distribution is very sharp, also for lower gamma-ray energies for which the resolution is worse. The galactic large-scale structure is very different, in two dimensions, at small and at large distances from the Sun. Since the HI scale height is increasing with increasing distance in the outer Galaxy, the angular scales of structures at distances e.g. greater than 5 kpc from the Sun are still sufficiently large to be recognised in the gamma-ray distribution.

H.C. van de Hulst (Chairman): I am glad to hear that the COS-B quanta, which after all cost about a hundred dollars a piece, are paying off.

Wim Hermsen (left) and Hans Bloemen during boat-trip. Background: Jan Lub. Foreground: Harvey Liszt (left) and Hugo van Woerden discuss problems of mapping Our Galaxy's spiral structure. LZ

Below: Strong and Mayor discuss gamma-ray halo during poster session. CFD

THE HIGH-LATITUDE DISTRIBUTION OF GALACTIC GAMMA RAYS AND POSSIBLE EVIDENCE FOR A GAMMA-RAY HALO

A.W. Strong
Istituto di Fisica Cosmica del CNR, Milano, Italy
(on behalf of the Caravane Collaboration)

During its 6.7-year lifetime the COS-B experiment included about 15 months of observations towards latitudes $|b| > 20°$ and covered almost all latitudes from the South to North galactic poles. Studies comparing the local gamma-ray emission with the distribution of gas (Lebrun et al. 1982, Strong et al. 1982) have so far been limited to $10° < |b| < 20°$, where the correlation is found to be fairly good and the structured emission can therefore be attributed mainly to cosmic-ray interactions with gas. The extension of this type of analysis to higher latitudes is now possible using the COS-B database.

The latitude distribution of gamma rays was compared with that expected from gas using galaxy counts as the total gas tracer. It was found that this component is inadequate to account for the whole of the observed latitude variation and that an additional component having a wide latitude distribution is required. Figure 1 shows the latitude variation of the residual 70-5000 MeV intensity after subtraction of the estimated gas contribution, averaged over all available longitudes. One interpretation of the latitude dependence of the residual attributes it to a thick disk or 'halo' surrounding the Galaxy, with for example a radius of 15 kpc and scale height of a few kpc, although the parameters and form are not strongly constrained by the data. In a typical model the intensity from the 'halo' towards the galactic poles is $\sim 10^{-5}$ cm^{-2} sr^{-1} s^{-1}, and at latitudes $|b| < 30°$ it has a substantial longitude variation while its intensity is comparable to that of the gas emission.

An identical analysis has been performed on the SAS-2 data (taken from Fichtel et al., 1978) and this leads to essentially the same conclusions.

The origin of the 'halo' emission is unknown, but a plausible mechanism is inverse Compton scattering (ICS) by cosmic-ray electrons on starlight (Worrall and Strong 1977). In this case it may be related to the large scale-height diffuse X-ray emission seen in the UHURU (Protheroe et al. 1980), Ariel V (Warwick et al. 1980) and HEAO-1 (Iwan et al. 1982) data. If these X-rays are at least in part due to ICS on the 2.7K back-

H. van Woerden et al. (eds.), The Milky Way Galaxy, 235–236.

ground then gamma rays of $\sim$ 100 MeV will be produced by ICS on starlight. Taking an X-ray intensity at $|b| = 90°$ of 2×10^{-9} erg cm^{-2} sr^{-1} (2-20 keV) (Iwan et al. 1982) and a starlight energy density of 1 eV cm^{-3}, then the expected gamma-ray intensity is roughly equal to that required for the halo component.

Independent of the physical process involved the present geometrical interpretation attributes a part of the 'isotropic' background deduced from the SAS-2 data (Thompson and Fichtel 1982) to the 'halo'; however, in view of the inadequacy of the simple model, it is possible that all of the high-latitude component is in fact galactic.

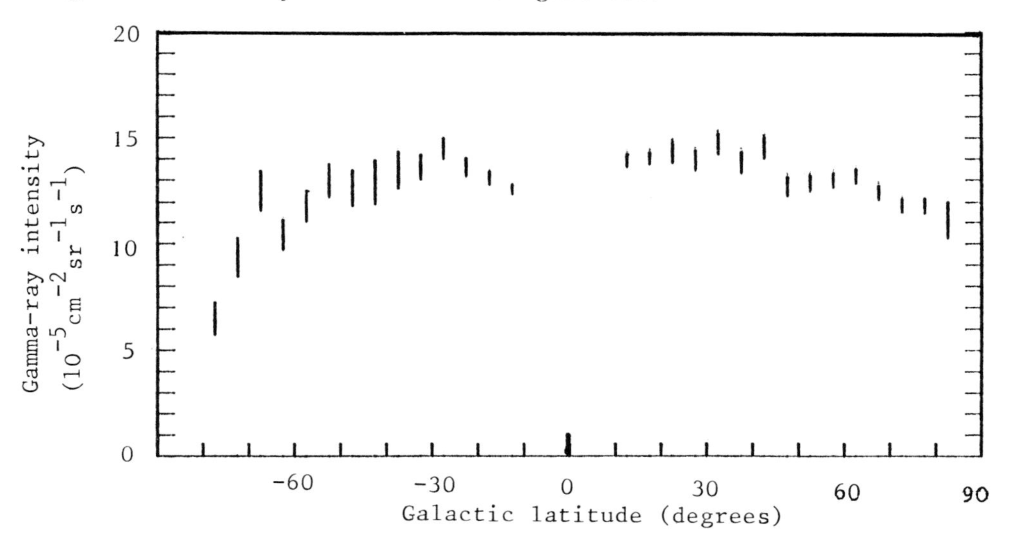

Figure 1. Latitude distribution of gamma-ray intensity (70-5000 MeV) after subtraction of the estimated contribution from cosmic-ray/gas interactions and averaged over all longitudes for which COS-B data are available. Instrumental background has not been subtracted.

REFERENCES

Fichtel, C.E. et al.: 1978, NASA Technical Memorandum No. 79650
Iwan, D. et al.: 1982, Astrophys. J. 260, 111
Lebrun, F. et al.: 1982, Astron. Astrophys. 107, 390
Protheroe, R.J. et al.: 1980, M.N.R.A.S. 192, 445
Strong, A.W. et al.: 1982, Astron. Astrophys. 115, 404
Thompson, D.A., and Fichtel, C.E.: 1982, Astrophys. J. 109, 352
Warwick, R.S. et al.: 1980, M.N.R.A.S. 190, 243
Worrall, D.M., and Strong, A.W.: 1977, Astron. Astrophys. 57, 229

ON THE ULTRAVIOLET BACKGROUND RADIATION OF THE GALAXY

A. Zvereva, A. Severny, and L. Granitsky
Crimean Astrophysical Observatory, Acad.Sci. USSR
G. Courtès, P. Cruvellier, and C.T. Hua
Laboratoire d'Astronomie Spatiale du C.N.R.S. France

Observations of the far-UV spectrum (1300-1800 Å) of the sky background at different galactic latitudes are presented. The measurements were made in deep space, at distances up to 2×10^5 km from the Earth with the photoelectric spectrometer ("Galaktika") on board the "Prognoz-6" satellite. The highly elongated orbit of the satellite permitted corrections for scattered L_α-light. The contribution of stars was eliminated using fluxes measured in the TD-I and OAO-2 experiments.

Figure 1 shows a map of the sky in celestial coordinates, indicating the location of our targets along the ecliptic (marked by serial numbers) with 36 square degree field-of-view for each. Some other interesting data are exhibited on this map: 1) The regions observed in other UV-background experiments, e.g. "Apollo-17"; "Aries-8" - triangles and D2B observations - shaded areas. 2) The mean position of two spurs or remnants of Supernovae (Shklovsky and Sheffer, 1971) is labelled by a dashed line. As displayed on the map observed regions coincide pretty well with the position of spurs.

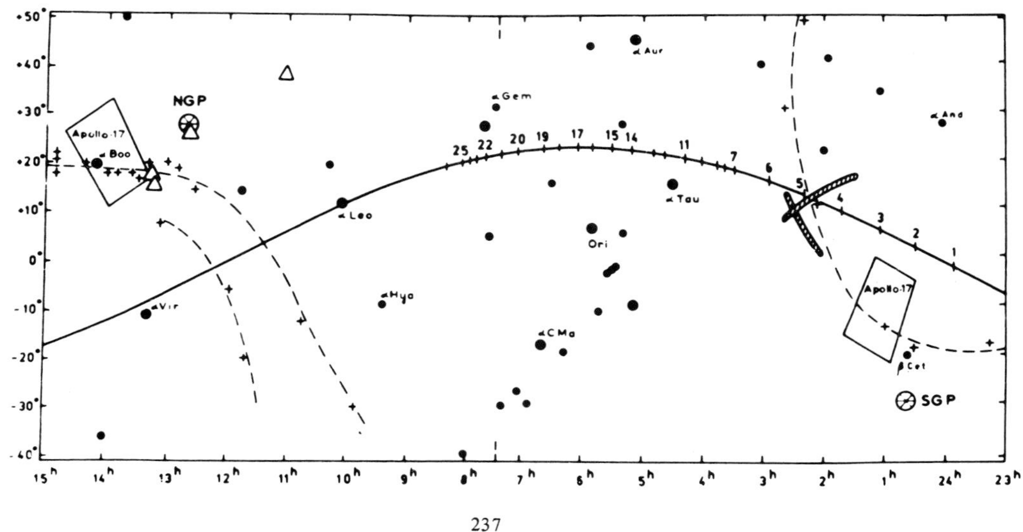

H. van Woerden et al. (eds.), The Milky Way Galaxy, 237–238.

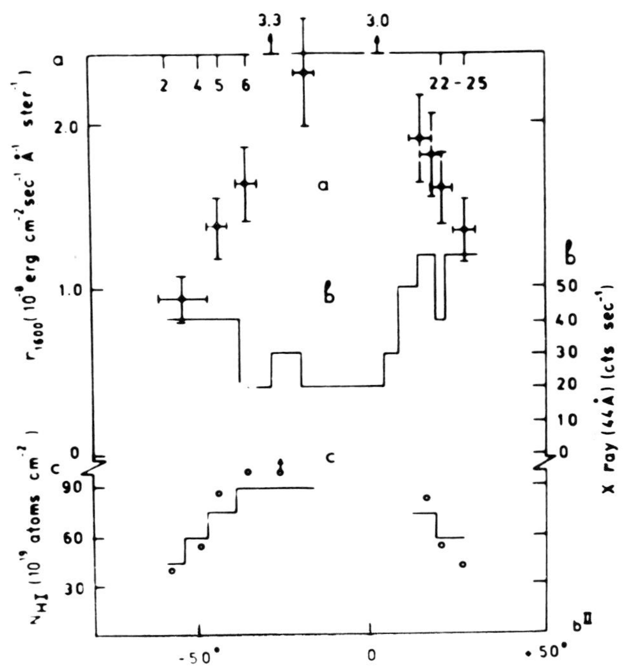

The UV-background radiation r_λ (at λ 1600) shows a good correlation with galactic latitude. For $|b^{II}| < 30°$ the dependence of r_λ on latitude is consistent with that predicted by DGL models for scattering (see Zvereva et al., 1982).

We note in Figure 2 the correlation between r_λ (a), enhanced soft X-ray brightness (b), and the deficient neutral hydrogen density (c). The background UV radiation in high - latitude areas might be interpreted as the emission from hot gas (Severny and Zvereva, 1983).

REFERENCES

Severny, A.B., and Zvereva, A.M.: 1983, Astrophys. J. (Letters) 23, 71
Shklovsky, I.S., and Sheffer, E.K.: 1971, Nature 231, 173
Zvereva, A.M., Severny, A.B., Granitsky, L.V., Hua, C.T., Cruvellier, P., and Courtès, G.: 1982, Astron. Astrophys. 116, 312

RADIO CONTINUUM EMISSION OF THE MILKY WAY AND NEARBY GALAXIES

Rainer Beck and Wolfgang Reich
Max-Planck-Institut für Radioastronomie, Bonn, FRG

ABSTRACT: The radio continuum emission of the Milky Way and nearby galaxies can be decomposed into a central region, a clumpy "thin disk", concentrated in the spiral arms, and a smooth "thick disk" (or flattened "halo"). The emissivity ratio of the two disks seems to be related to the magnetic field properties: Galaxies with strong radio spiral arms reveal a highly ordered field following the arm direction, while galaxies with diffuse disks contain a less ordered, smoothly distributed field. The degree of uniformity of the field seems to correlate with the total optical luminosity. The average magnetic field in the Milky Way is weak and turbulent compared to most of the nearby galaxies observed so far.

1. THE MILKY WAY

Our unfavourable position inside the Milky Way requires all-sky surveys in order to derive the general shape and structure of our Galaxy. Following the tradition of the "Bonner Durchmusterung" by F.W. Argelander more than 120 years ago, surveying the radio sky is being continued by radio astronomers at Bonn. An all-sky survey was completed recently at 408 MHz (λ73 cm) (Haslam et al., 1982) as well as the northern part of a companion 1420-MHz (λ21 cm) survey (Reich, 1982). Further studies at 2700 MHz (λ11 cm) have been started which concentrate on the galactic-plane region ($|b| \leq 20°$).

Radio continuum emission is a probe of the high-energy component of the interstellar medium. The 408-MHz survey reveals various components: the central region, the disk, spiral arm complexes, loops, spurs, supernova remnants, and HII regions. The separation of these components requires spectral-index information, e.g. from the 408-MHz and 1420-MHz surveys which match in angular resolution and sensitivity. The global spectral index of brightness temperature is close to 2.7. Loop structures, however, show a steeper spectral index. A local feature in the Perseus-Cassiopeia region ($115° \leq \ell \leq 165°$) exhibits a high-frequency spectral break which correlates with the bend at $\sim$10 GeV in the local cosmic-

H. van Woerden et al. (eds.), The Milky Way Galaxy, 239–244.

ray electron spectrum (Kallas et al., 1983). If spectral breaks frequently occur in our Galaxy, they would influence the global radio spectrum and affect the separation of thermal and nonthermal emission.

Surveys of the galactic plane need higher resolution in order to separate small-scale from large-scale structures (e.g. Altenhoff et al., 1978; Haynes et al., 1978; Kallas and Reich, 1980; Sofue et al., 1985). The Effelsberg 100-m telescope is involved in further sensitive surveys. Several old supernova remnants have been discovered so far (e.g. Reich et al., 1979; Reich and Braunsfurth, 1981). Improved SNR statistics are needed to compute their direct contribution to the total radio emission of the Milky Way.

The 408-MHz survey has been analyzed to obtain the large-scale structure of our Galaxy by Phillipps et al. (1981a,b) and by Beuermann et al. (1984) (see also Kanbach, 1983). The face-on view is in agreement with a 2-arm or a 4-arm spiral model with a pitch angle of 12 - 13°. The magnetic field strength is ∿ 5 μG on average, with about equal contributions of uniform and random fields. The radial scale length is ∿ 2 kpc for the thermal and ∿ 5 kpc for the nonthermal component (see Berkhuijsen and Klein, 1985). The edge-on view is differently modelled by the two groups: Phillipps et al. (1981b) use a thin disk plus a flattened, box-shaped "halo", while Beuermann et al. (1984) prefer a thin disk with a full width between half-intensity points increasing with distance from the Galactic Centre (250 - 690 pc) plus a thick disk with increasing full width (2 - 6 kpc) (Fig. 1). A similar two-disk model had already been suspected by Yates (1968) and discussed in detail by Ilovaisky and Lequeux (1972). The thin disk contains contributions from

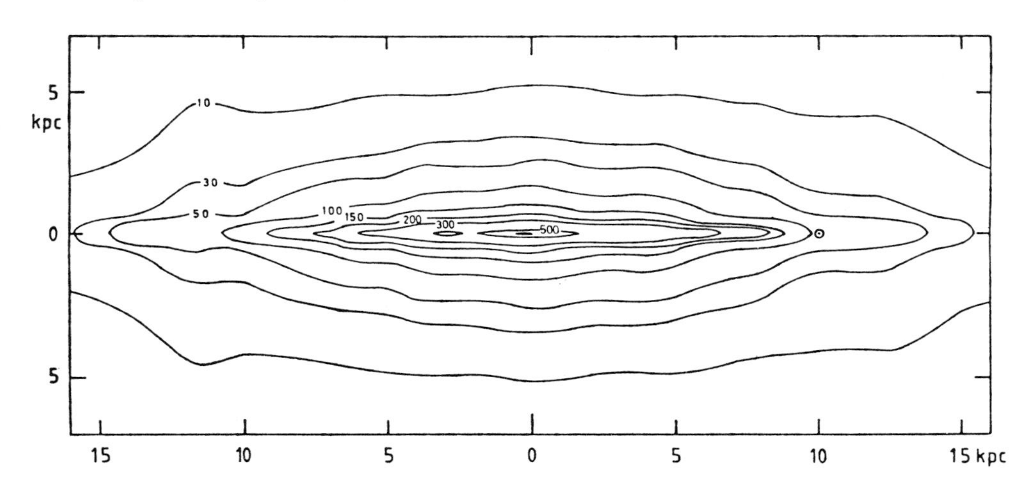

Fig. 1: Edge-on view of the Galaxy at 408 MHz as seen from the direction $\ell = 90°$. The picture includes the emission from the thin and thick disk. The contour intervals are given in degrees K; the 30 K level corresponds to about 1 K at 1415 MHz (from Beuermann et al., 1984)

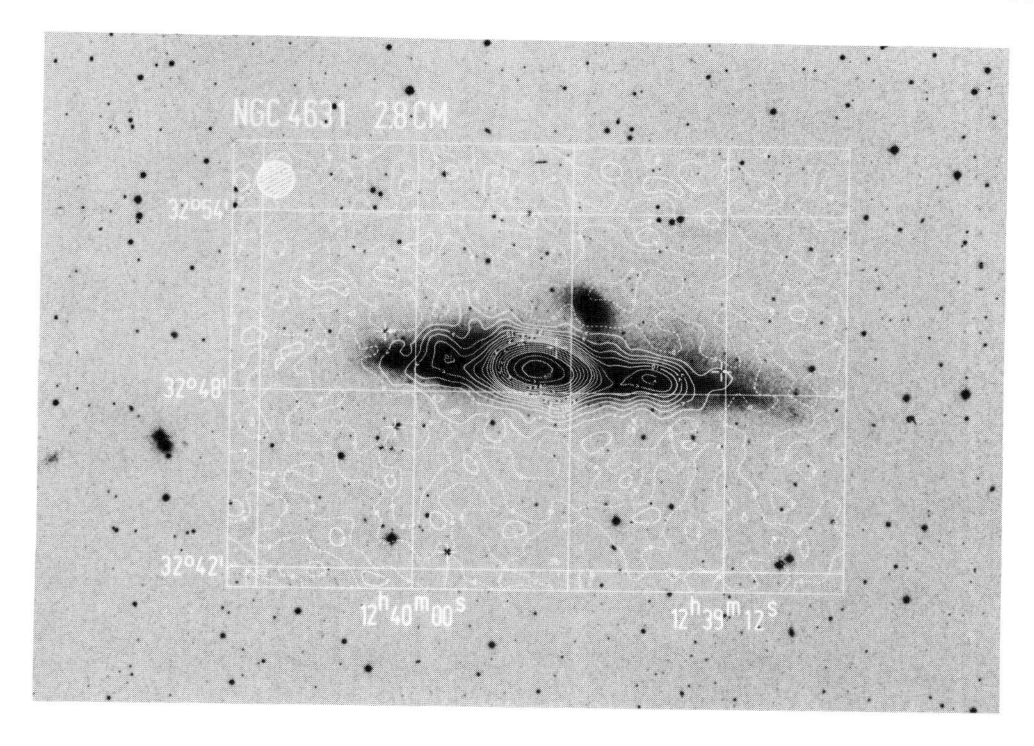

Fig. 2: Effelsberg map of NGC 4631 at 10.7 GHz (λ2.8 cm), superimposed onto an optical PSS plate. Contour levels are: 0 (dashed), 2, 4, ..., 10, 13, ..., 40, 50, 60 mJy/beam area (1 mJy/b.a. ≙ 1.9 mK). The angular resolution is 71 arcsec (hatched circle) (from Klein et al., 1984).

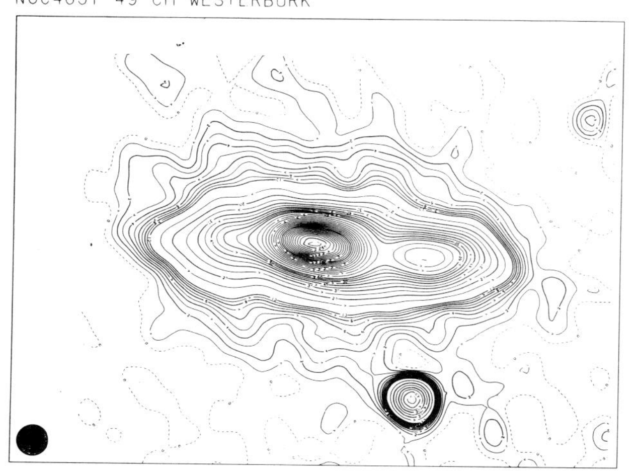

Fig. 3: New Westerbork map of NGC 4631 at 610 MHz (λ49 cm) by W. Werner. Contour levels are in mK. The map has been smoothed to the same angular resolution as in Fig. 2.

HII regions, supernova remnants and interstellar clouds, while the thick disk emits diffuse nonthermal radiation. The thick disk coincides with the volume occupied by the hot phase of the interstellar medium and by cosmic rays which are accelerated in supernova shock fronts (e.g. Axford, 1982; Bogdan and Völk, 1983).

2. NEARBY GALAXIES

The discrepancies between the present models will not be solved within the near future. Nearby galaxies are better suited to study the global radio properties of spiral galaxies. In M31 (NGC 224), thermal and nonthermal radio emission are concentrated in a ring-like structure with ~ 1 kpc full width (Beck and Gräve, 1982). No extended disk emission has been observed. Any halo emission must be weak and is difficult to separate from confusing Galactic emission (Gräve et al., 1981). The only known edge-on galaxy with a prominent thick radio-disk is NGC 4631 (Ekers and Sancisi, 1977). Recent observations with the Effelsberg and Westerbork radio telescopes (Figs. 2 and 3) confirm a spectral-index steepening with increasing distance from the plane and will be used to test galactic-wind models (Werner et al., in prep.). All other thick disks observed so far reveal an emissivity of only a few percent of the disk emissivity and a full width of only ~ 1 kpc (Hummel et al., 1984). The significantly stronger thick-disk emission of NGC 4631 may be connected with the finding that its magnetic field in the thin disk is more ordered than in other edge-on galaxies, e.g. NGC 253 (Klein et al., 1983).

In face-on galaxies two components can be distinguished in radio continuum: a clumpy component (thermal and nonthermal), concentrated in the spiral arms, and a smooth nonthermal component. They can be identified as the thin and thick disks observed edge-on. The ratio of emissivities of these components determines whether radio arms dominate (e.g. in M31 and M51) or strong diffuse disks (e.g. in NGC 6946; Klein et al., 1982). The different ratios between thin- and thick-disk emissivities have to be considered as the result of the interstellar magnetic field interacting with the processes of cosmic-ray acceleration and propagation.

The energy sources for the acceleration of cosmic-ray electrons must belong preferentially to the spiral-arm population (e.g. type II supernovae), because correlations were found between the radio-to-optical-flux ratio and the optical colour (Klein, 1982) and between the total radio and Hα fluxes (Kennicutt, 1983). The different radial scale lengths of the thermal and nonthermal radio emission (Sancisi and Van der Kruit, 1981; Berkhuijsen and Klein, 1985) require diffusion lengths for cosmic-ray electrons of a few kiloparsecs. If cosmic rays are scattered by magnetic-field irregularities, they cannot stream faster than with the Alfvén speed. Our understanding of this interaction demands better observational data about interstellar magnetic fields.

3. MAGNETIC FIELDS

The linearly polarized emission of a few nearby galaxies has been mapped with the Effelsberg telescope (Beck, 1982; Klein et al., 1982, 1983). Magnetic field lines follow spiral arms, but their overall structure is probably closed, in agreement with the theory of turbulent interstellar dynamos (Beck, 1983). The average degree of alignment of the field correlates with total optical luminosity (Fig. 4), indicating a connection between star formation and field ordering, e.g. by density-wave shock fronts. The average magnetic field strength, on the other hand, does *not* correlate with total luminosity. Strongest fields occur in galaxies with massive spiral arms. Hence, introduction of magnetic fields into the theory of stochastic star formation (Seiden, 1985) is promising.

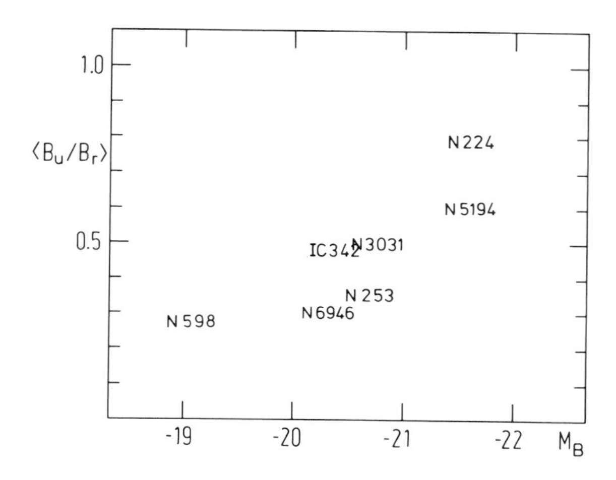

Fig. 4: The correlation between the average ratio B_u/B_r of the uniform and random magnetic field strengths and the total optical luminosity M_B. B_u/B_r refers to similar linear resolutions of ∿ 3 kpc.

The Milky Way fits into these relations: Its moderate luminosity ($M_B \cong -20$) indicates a low degree of alignment of the magnetic field, in agreement with the observations; its intermediate luminosity class ($L \cong II$) is in accordance with the low magnetic field strength observed. The field structure, however, is still under debate. A bisymmetric model was proposed by Simard-Normandin and Kronberg (1980) and Sofue and Fujimoto (1985) on the basis of rotation-measure data, while a closed configuration is favoured by Inoue and Tabara (1981) and Vallée (1983).

Radio observations of nearby galaxies have proved to be fertile to understand the physics of the interstellar medium in the Milky Way. In particular, the role of magnetic fields for the formation of disks and spiral structures has to be considered seriously.

REFERENCES

Altenhoff, W.J., Downes, D., Pauls, T., Schraml, J.: 1978, Astron. Astrophys. Suppl. 35, 23
Axford, W.I.: 1982, Proc. Int. School on Plasma Astrophys., Varenna (ESA SP-161), p. 425
Beck, R.: 1982, Astron. Astrophys. 106, 121
Beck, R.: 1983, Proc. IAU Symp. No. 100, p. 159
Beck, R., Gräve, R.: 1982, Astron. Astrophys. 105, 192
Beuermann, K., Kanbach, G., Berkhuijsen, E.M.: 1984, preprint
Berkhuijsen, E.M., Klein, U.: 1985, this volume
Bogdan, T.J., Völk, H.J.: 1983, Astron. Astrophys. 122, 129
Ekers, R.D., Sancisi, R.: 1977, Astron. Astrophys. 54, 973
Gräve, R., Emerson, D.T., Wielebinski, R.: 1981, Astron. Astrophys. 98, 260
Haslam, C.G.T., Salter, C.J., Stoffel, H., Wilson, W.E.: 1982, Astron. Astrophys. Suppl. 47, 1
Haynes, R.F., Caswell, J.L., Simons, L.W.J.: 1978, Australian J. Suppl. No. 45
Hummel, E., Sancisi, R., Ekers, R.D.: 1984, Astron. Astrophys. (in press)
Ilovaisky, S.A., Lequeux, J.: 1972, Astron. Astrophys. 20, 347
Inoue, M., Tabara, H.: 1981, Publ. Astron. Soc. Japan 33, 603
Kallas, E., Reich, W.: 1980, Astron. Astrophys. Suppl. 42, 227
Kallas, E., Reich, W., Haslam, C.G.T.: 1983, Astron. Astrophys. 128, 268
Kanbach, G.: 1983, Space Sci. Rev. 36, 273
Kennicutt, R.: 1983, Astron. Astrophys. 120, 219
Klein, U.: 1982, Astron. Astrophys. 116, 175
Klein, U., Beck, R., Buczilowski, U.R., Wielebinski, R.: 1982, Astron. Astrophys. 108, 176
Klein, U., Urbanik, M., Beck, R., Wielebinski, R.: 1983, Astron. Astrophys. 127, 177
Klein, U., Wielebinski, R., Beck, R.: 1984, Astron. Astrophys. (in press)
Phillipps, S., Kearsey, S., Osborne, J.L., Haslam, C.G.T., Stoffel, H.: 1981a, Astron. Astrophys. 98, 286
Phillipps, S., Kearsey, S., Osborne, J.L., Haslam, C.G.T., Stoffel, H.: 1981b, Astron. Astrophys. 103, 405
Reich, W.: 1982, Astron. Astrophys. Suppl. 48, 219
Reich, W., Braunsfurth, E.: 1981, Astron. Astrophys. 99, 17
Reich, W., Kallas, E., Steube, R.: 1979, Astron. Astrophys. 78, L13
Sancisi, R., Kruit, P.C., van der: 1981, Proc. IAU Symp. No. 94, p. 209
Seiden, P.E.: 1985, this volume
Simard-Normandin, M., Kronberg, P.P.: 1980, Astrophys. J. 242, 74
Sofue, Y., Fujimoto, M.: 1985, this volume
Sofue, Y., Hirabayashi, H., Akabane, H., Inoue, M., Handa, T., Nakai, N.: 1985, this volume
Vallée, J.P.: 1983, Astrophys. Letters 23, 85
Yates, K.W.: 1968, Australian J. Phys. 21, 167

LOOP I (THE NORTH POLAR SPUR) REGION - A QUASI RADIO HALO

Jelena Milogradov-Turin
Institute of Astronomy, Faculty of Sciences
University of Beograd, Studentski Trg 16, p.f. 550, 11000 Beograd
Yugoslavia

The distribution of total spectral indices between 38 and 408 MHz with a resolution of 7°.7 for $\delta > -25^{\circ}$ has the following main properties:

1. Relatively small variations of spectral indices over the sky.
2. High indices in the central region at high galactic latitudes.
3. Moderately low spectral indices in the anticentre region.
4. Lower indices in the low-brightness regions (cold holes).
5. The lowest indices in regions containing large amounts of HII.

These properties may be explained by distribution of the moderate-spectral-index emission from the galactic disk, high-spectral-index extragalactic emission, moderate-spectral-index emission from the galactic spurs, high-spectral-index emission from regions on the outer side of the galactic spurs, and absorption in HII regions. The disk and the spurs could be related, at least partly. The absence of a high-spectral-index radio halo is indicated by the low-spectral-index cold hole inside Loop I, centered at $\ell \sim 0^{\circ}$, $b \sim 0^{\circ}$. This hole in the Ser-Vir region, where the halo is expected to be bright, has almost as low an index as other cold holes (e.g. the hole centered at $\ell \sim 195^{\circ}$, $b \sim 45^{\circ}$).

The spectral-index distribution is similar to the distribution of the quantity T_g'. T_g' was defined generalizing the idea published by Webster (1978). T_g' is constructed formally in the same way as Webster's T_h', eliminating the radiation of a chosen spectral index β_c and the extragalactic radiation. T_g' is proportional to the brightness of the residual radiation dominant in the region concerned.

The computation of T_g' based on temperatures at 38 and 408 MHz shows that after eliminating the emission of any spectral index between 2.4 and 2.5, the T_g' distribution looks similar to the spectral-index distribution. If there were a radio halo, T_g' in the Ser-Lib region would be greater than T_g' at higher galactic longitudes. The values of T_g' calculated from the observations are much lower than those expected for the spheroidal radio halo in the Ser-Lib region (e.g. Figure 1). The similarity in shape between the contours of the spheroidal halo and the North Polar Spur (NPS) produces a good fit of the observed and theoretical distribution in the

H. van Woerden et al. (eds.), The Milky Way Galaxy, 245–246.

region lying more than 10° outside the main ridge of the NPS. This similarity arises from the coincidence that the centre of Loop I lies in the Sco-Cen association, near the direction to the galactic centre which is expected to be the centre of the halo.

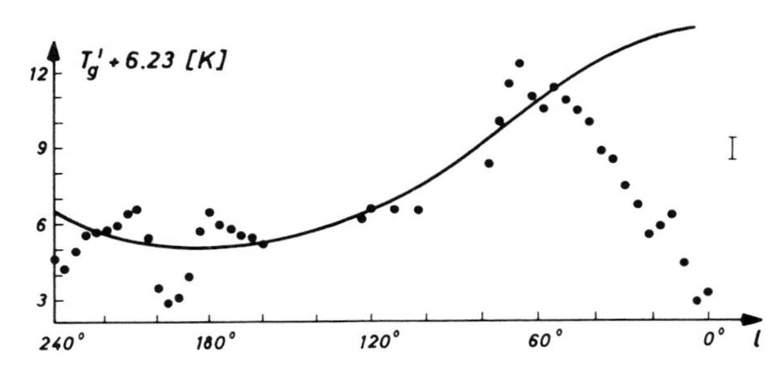

Figure 1. The T_g' distribution for b=38°. The dots are values of $T_g'+6.23$ computed from data at 38 and 408 MHz for $\beta_c=2.5$. The curve corresponds to the best-fit spheriod (19, 18.75 kpc) matched to the dots for $\ell>40^o$.

The high-spectral-index region outside of the NPS comes from a weak component having steep spectrum and related to the NPS. The existence of such a component explains naturally the splitting of T-T diagrams of the NPS. Furthermore, similar components should exist in other spurs as well. It was found that all spurs exhibit a splitting of their T-T diagrams which follows a "curvature rule": the points on the convex side of the spur lie on the upper part of a standard T-T diagram. The model, consisting of the four emission components described in connection with the total spectral indices, can interpret all observed characteristics of T-T diagrams. This "4v" (four vectors) model demands only an adequate distribution of the galactic components. The high-spectral-index feature south of the galactic plane in the region of Loop I is likely to be related to this loop also.

The spectrum of the NPS based on data in the range between 10 and 820 MHz is found to be flatter at lower frequencies and steeper at higher frequencies than the spectrum of the general galactic background. The flux-density spectrum of the NPS was derived to be:

$$\log S = a + 0.13 \log f \; [\text{MHz}] - 0.145 \log^2 f \; [\text{MHz}] .$$

The difference in indices between the low and high-frequency ends is 0.5. There is no maximum in the spectrum in this frequency range.

The author acknowledges Prof. F.G. Smith, Dr. C.G.T. Haslam, Dr. C.J. Salter, and Dr. A.S. Webster for help at various stages of this work. These results are part of the author's Ph.D. Thesis (1982).

REFERENCE

Webster, A.S.: 1978, Monthly Notices Roy. Astron. Soc. 185, pp. 507-519.

THE MAGNETIC-FIELD STRUCTURE AND DYNAMICS OF NGC 253

M. Urbanik
Astronomical Observatory
Jagiellonian University, Krakow, Poland

U. Klein, R. Beck and R. Wielebinski
Max-Planck-Institut für Radioastronomie
Bonn, West-Germany

ABSTRACT

Total-power and polarization observations of NGC 253 at 10.7 GHz have been performed with the 100-m MPIfR radio telescope. The observed arm/interarm polarization contrasts are discussed in the context of possible field configurations in spiral arms.

1. THE OBSERVATIONAL DATA

The nearly edge-on SABc galaxy NGC 253 was studied at 10.7 GHz with the 100-m MPIfR radio telescope. High-sensitivity total-power and polarization maps with a resolution of 1.2 arcmin have been obtained. Three components of the radio-continuum emission have been observed: the completely unpolarized, barely resolved central source, the high-brightness plateau associated with the brightest parts of the spiral structure, and the weak outer disk. Although the high inclination of $78\overset{\circ}{.}5$ made some details of the plateau difficult to resolve with our beam, the existence of a bright, weakly polarized radio bar seems unquestionable. Extensions along the inner spiral arms producing most of the polarized flux are also clearly visible.

The highest polarization, reaching 50 percent after correction for the thermal flux, has been observed close to tracers of a spiral shock in the NW arm. The integrated nonthermal emission from the disk of NGC 253 is however only 14 percent polarized. An even lower degree of polarization is expected in the interarm region and in the outer disk.

2. IMPLICATIONS FOR THE FIELD STRUCTURE

Depolarization effects at 10.7 GHz were found to be small. It is thus likely that the observed polarization contrast between the spiral

H. van Woerden et al. (eds.), The Milky Way Galaxy, 247–248.

arms and the remaining part of the disk represents true changes in the field structure. Due to the high inclination of NGC 253 that contrast would imply differences in the relative strength of the vertical component of the random magnetic field. The ratio of the vertical component to the total field, $\langle B_z \rangle / \langle B_t \rangle$, in a fully isotropic three-dimensional field is about 0.57. Computer simulations of the radiation from random fields showed that a value $\langle B_z \rangle / \langle B_t \rangle = 0.5$ is needed to account for a polarization as low as 14 percent. This ratio is likely to be even higher in the interarm region. The B_z component seems to be significantly suppressed in spiral arms where $\langle B_z \rangle / \langle B_t \rangle$ drops to about 0.2.

The formation of a rather flat arm field from the nearly isotropic interarm field is difficult to explain within the framework of pure hydrodynamic shocks allowing no significant compression along the z-axis (Tubbs, 1980; Soukup and Yuan, 1981). Models involving the development of pure Parker instabilities (Parker, 1966; Mouschovias et al., 1974) poorly explain the dramatic polarization increase in the spiral arms, unless we assume a significant change of properties of the instability in compressed regions. The upward migration of cosmic-ray electrons as needed for the formation of a "thick disk", perhaps efficient in weakly polarized interarm regions, must be in some way inhibited in spiral shocks. Such a condition is necessary to avoid too intense emission from highly inclined parts of the field structure, which would give too high an observed B_z in the spiral arms. A significant change in the nature of the instability itself in shocked regions cannot be excluded either (Elmegreen, 1982). Detailed models of radio emission from various types of curved Parker-type structures are now in preparation.

REFERENCES

Elmegreen, B.G., 1982, Astrophys. J. 253, p. 655.
Mouschovias, T.C., Shu, F.H., Woodward, P.R., 1974, Astron. Astrophys. 33, p. 73.
Parker, E.N., 1966, Astrophys. J. 145, p. 811.
Soukup, J.E., Yuan, C., 1981, Astrophys. J. 246, p. 376.
Tubbs, A.D., 1980, Astrophys. J. 239, p. 882.

DISCUSSION

G.D. van Albada: What is the reason to include B_z? You can probably depolarize the radiation by just having a random field in the plane?

Urbanik: Not for that inclination angle of $78\overset{\circ}{.}5$! Because for a completely random field in the plane you would still have 65% polarization everywhere. And depolarization by Faraday effect is completely insignificant: the maximum depolarization we could get for very extreme assumptions is of order 0.7.

MODELLING THE GALACTIC CONTRIBUTION TO THE FARADAY ROTATION OF RADIATION FROM EXTRA-GALACTIC SOURCES

B.J. Brett
Department of Mathematics, Statistics & Computing, Plymouth Polytechnic, Drake Circus, Plymouth PL4 8AA, Devon, UK

Faraday rotation occurs in the Galaxy according to the formula:

$$RM = k \int_L n_e H_{||} d\ell$$

where: RM is the rotation measure
n_e density of electrons
$H_{||}$ component of the magnetic field parallel to line of sight
L distance travelled through interstellar medium
k constant depending on units.

We have taken a collection of rotation measures of 552 extra-galactic sources, compiled by Simard-Normandin, Kronberg and Button (Preprint, 1980) and modelled them over the sphere using spherical harmonics. We hope in this way to model the dependence on galactic co-ordinates which will be due to the structure of our Galaxy (position of free electrons and direction and strength of the magnetic field).

The spherical harmonic functions are the solutions of Laplace's equation expressed in spherical polar co-ordinates and are the fitting functions for data on the surface of a sphere. We used a least-squares fitting procedure, starting with first-order harmonics, and extending to second, third, fourth and fifth. (The mean of the data, as a constant 'function', gives a zero-order model.)

We found that the first, second and fourth-order models were statistically most significant, and illustrate the first and second-order models here. The first-order model (Fig. 1) is a significant improvement over the mean model at 0.001 level, using the F-test. The second-order model (Fig. 2) is a significant improvement over the first-order model at 0.001 level using the F-test.

These are preliminary results, more detailed investigations are still in progress. The first-order harmonics support a simple model of a linear field in the solar neighbourhood, parallel to the plane of the

H. van Woerden et al. (eds.), The Milky Way Galaxy, 249–250.

Galaxy, and having a direction from $\ell^{II} = 112^{o}$ to $\ell^{II} = 292^{o}$. Higher-order models indicate large-scale deviations from this model.

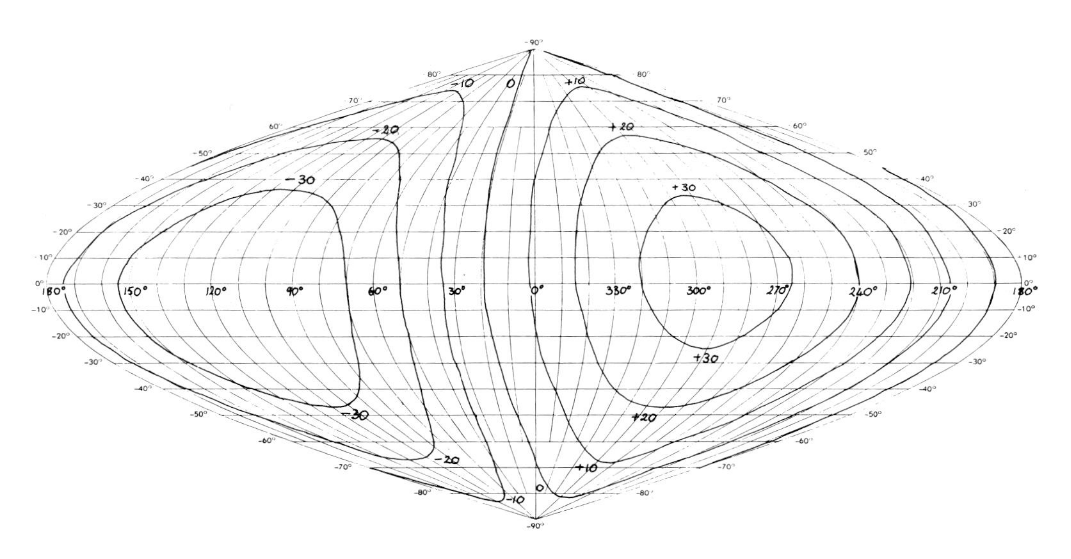

Figure 1: Rotation Measures, First-Order Harmonic Model

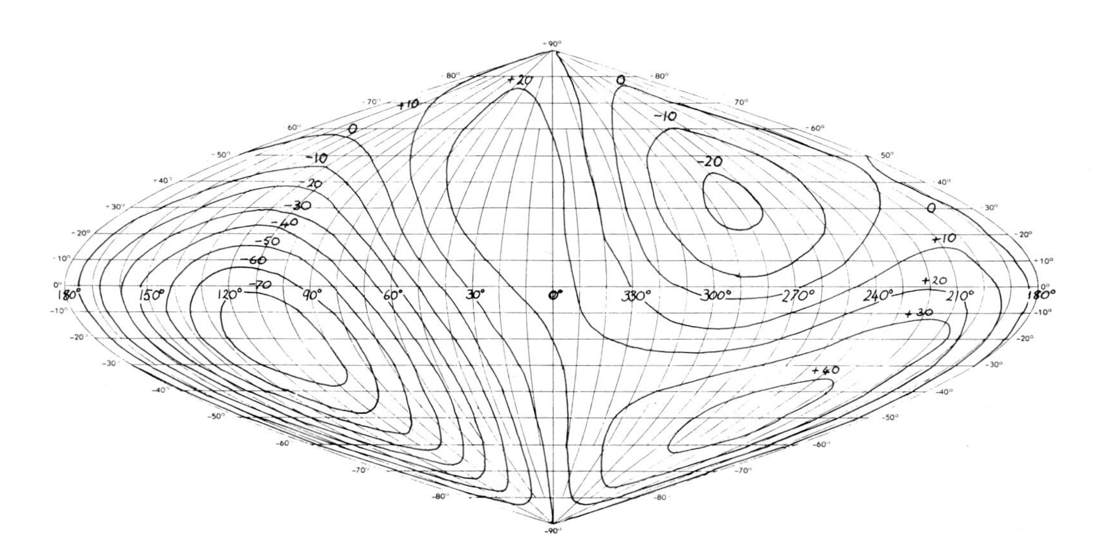

Figure 2: Rotation Measures, Second-Order Harmonic Model

A BISYMMETRIC SPIRAL MAGNETIC FIELD IN THE MILKY WAY

Y. Sofue
Nobeyama Radio Observatory, Tokyo Astronomical Observatory, and
A. von Humboldt Fellow at the MPI für Radioastronomie Bonn

M. Fujimoto
Department of Physics, Nagoya University

The distribution of Faraday rotation measure (RM) of extragalactic radio sources shows that a large-scale magnetic field in the Galaxy is oriented along the spiral arms. The field lines change direction from one arm to the next in the inter-arm region.

Figure 1 shows the distribution of RM for extragalactic radio sources using data by Tabara and Inoue (1981). Fig. 2a shows a smoothed RM distribution obtained by convolving the RM in Fig. 1 by a Gaussian beam of HPBW = 20°. Fig. 3 shows the RM variation along the galactic plane. The wavy RM variation suggests a reversal of the magnetic-field direction from one arm to the next as shown in Fig. 4. A model calculation of the RM distribution on the sky (Fig. 2b) based on a bisymmetric field configuration as shown in Fig. 5 reproduces well the characteristic features in Fig. 2a and the wavy variation along the galactic plane (Fig. 3) insofar as the RM at $|b| \lesssim 30^{\circ}$ is concerned. A full description is given in Sofue and Fujimoto (1983).

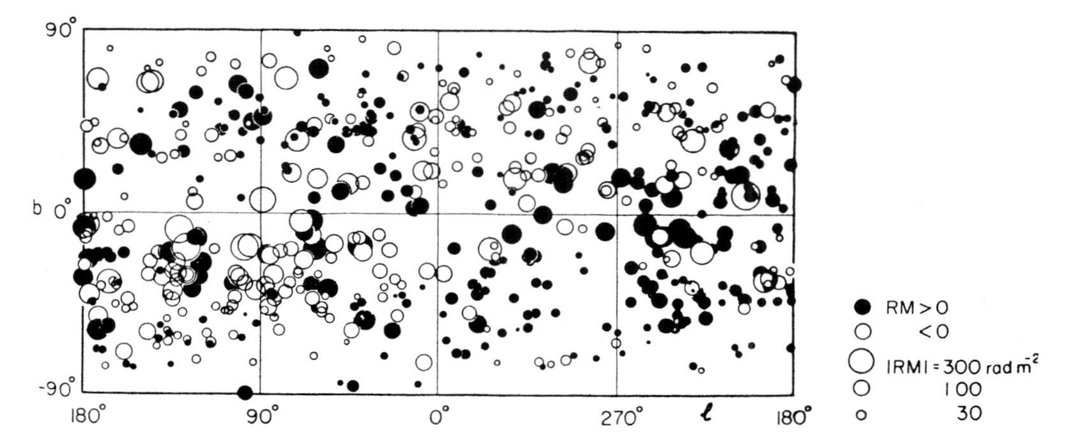

Figure 1. RM distribution on the sky. Positive RM (filled circle) shows field line toward the observer.

H. van Woerden et al. (eds.), The Milky Way Galaxy, 251–252.

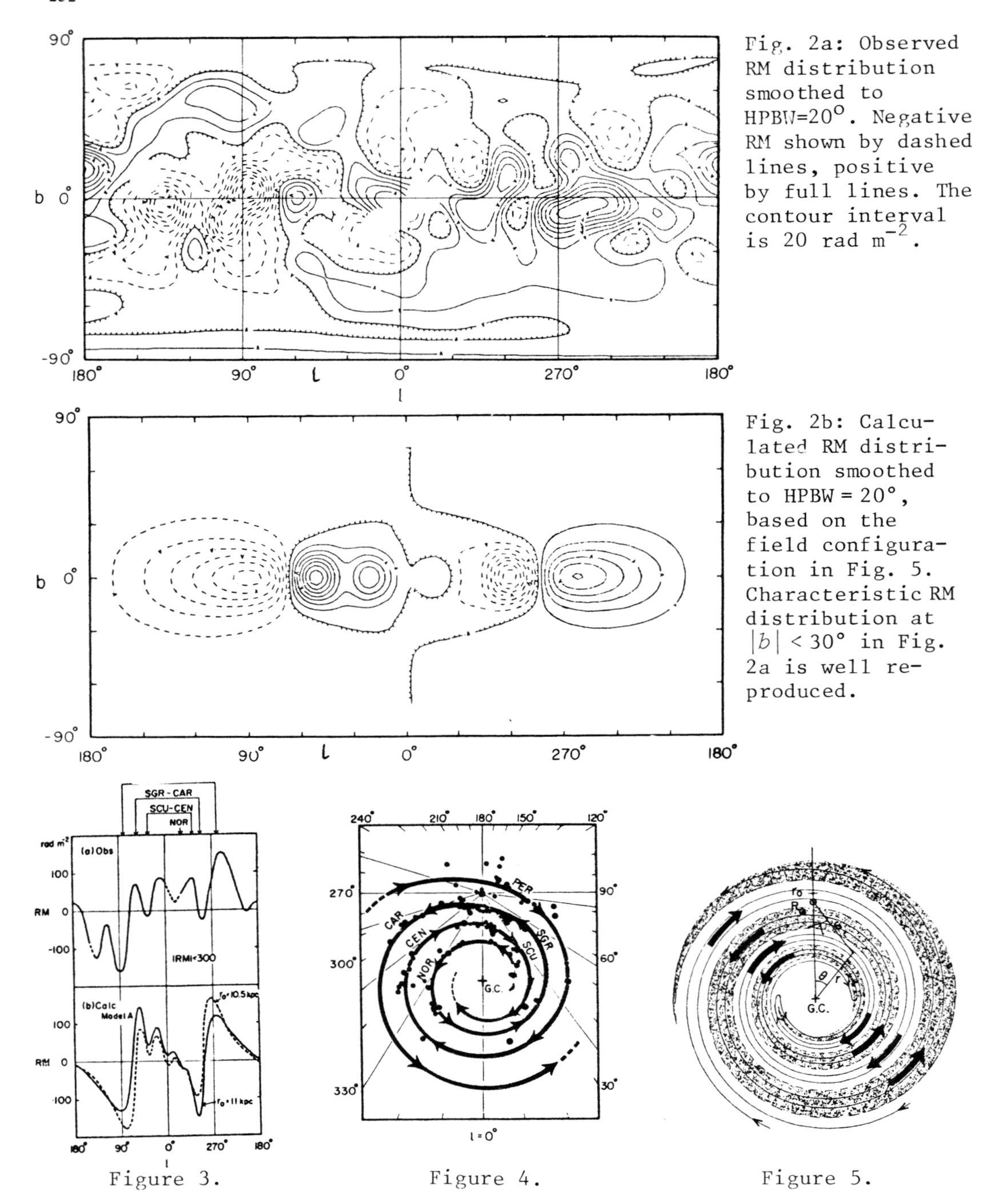

Fig. 2a: Observed RM distribution smoothed to HPBW=20°. Negative RM shown by dashed lines, positive by full lines. The contour interval is 20 rad m^{-2}.

Fig. 2b: Calculated RM distribution smoothed to HPBW = 20°, based on the field configuration in Fig. 5. Characteristic RM distribution at $|b| < 30°$ in Fig. 2a is well reproduced.

Figure 3. Figure 4. Figure 5.

Sofue, Y., and Fujimoto, M.: 1983, Astrophys. J. 265, 722
Tabara, H., and Inoue, M.: 1981, Publ. Astron. Soc. Japan 33, 603

SECTION II.5

SPIRAL STRUCTURE

Tuesday 31 May, 1505 - 1725

Chairmen: B.E. Westerlund and F.J. Kerr

F.J. Kerr (centre) catching up with I.F. Mirabel's research over pre-dinner drinks at Lauswolt. Left: C.A. Norman and B.F. Burke, right: L.A. Higgs

LZ

SPIRAL STRUCTURE OF THE MILKY WAY AND EXTERNAL GALAXIES

Debra Meloy Elmegreen
IBM Thomas J. Watson Research Center
P.O. Box 218, Yorktown Heights, NY 10598 USA

ABSTRACT

The spiral structure of the Milky Way galaxy has always been somewhat elusive because of our internal vantage point. This review will present methods and data for determining the overall pattern, and will summarize various models that have been proposed. Observations of spirals in external galaxies will also be discussed, because they can provide insight into the spiral structure of the Milky Way.

1. OBSERVATIONS OF SPIRAL STRUCTURE IN OUR GALAXY

1.1. Spectrophotometric distances

There are four common methods for determining the distributions of stars and gas. The first method, plotting known stellar distances, is the most direct way to determine the structure of our Galaxy. The only disadvantages of this method are that distances are usually accurate to only $\sim$10%, which leads to large uncertainties for distant stars, and that there is a relatively small number of objects to which this method may be applied. Morgan, Sharpless and Osterbrock (1952) first found that OB associations outlined spiral arms in the Milky Way. Open clusters with earliest spectral types less than B3 also prove to be good spiral tracers, as demonstrated by Becker and Fenkart (1970, 1971), Fenkart (1979), and Vogt and Moffat (1975). Figure 1 shows the local spiral structure obtained from these open clusters. Individual O stars are not perfect tracers, however. Lynds (1980a) found that only the hottest O stars trace the spiral arms; the later types occur in both the arm and interarm regions. A map of exciting stars in HII regions was made by Georgelin and Georgelin (1976). This will be discussed in §1.2. Long-period Cepheids (P>15 days) also tend to occur in spiral arms (Kraft and Schmidt 1963, Tammann 1970, Humphreys 1976, Efremov et al. 1981, and Berman and Mishurov 1981). Southern surveys reveal that A stars and early M stars show some concentration to spiral arms (Westerlund 1963, McCuskey and Houk 1971, McCuskey 1974); a small part of the Carina Arm is mapped out by A stars

H. van Woerden et al. (eds.), The Milky Way Galaxy, 255–272.

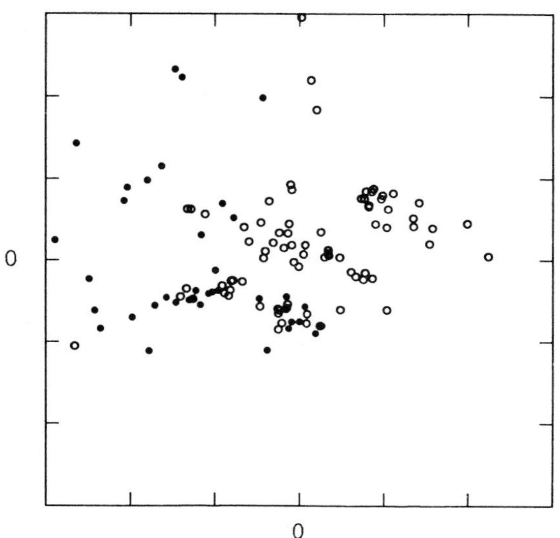

Figure 1. The local spiral structure is shown in the distribution of young open clusters plotted as open circles (Becker and Fenkart 1971) and closed circles (Vogt and Moffat 1975). The Sun is at (0,0). Tick marks are 2 kpc apart.

where no OB stars are present (Bok 1963). Other stars that appear to be loosely concentrated near spiral arms include Wolf-Rayet stars (Smith 1968), carbon N-stars (McCuskey 1970), and Be-stars (Kilkenny et al. 1975, Dolidze 1980).

1.2. Kinematic distances

Distances may also be determined from spectral-line velocities and the galactic rotation curve. This technique is usually applied to gaseous regions of HII, HI, and CO emission. Beyond the solar circle such distances are unambiguous, but inaccurate because of uncertainties in the rotation curve. Inside the solar circle, each velocity corresponds to two possible distances, so some additional information must be used to get the distance.

The pattern of optical HII regions in our Galaxy was observed by Georgelin and Georgelin (1976) and Georgelin, Georgelin and Sivan (1979); 20% of the distances were kinematic and 80% were spectrophotometric. Lockman (1979) and Downes et al. (1980) obtained kinematic distances to a large number of radio HII regions in the northern hemisphere. The near-far distance ambiguities were resolved by observing OH or H_2CO absorption features towards the HII regions. When an absorption line appears with a velocity much greater than that of the HII region, the HII region must be at the far kinematic distance. When no such high-velocity absorption is detected, then the HII-region distance is still ambiguous, but the near distance is usually assigned. The average error in this method is typically 10-20%, because (a) the systemic rotation curve is not perfectly well-known, (b) the emission-line velocity may have some random dispersion around the local systemic velocity, and (c) the emitting gas may be streaming along or between the spiral arms at unknown velocities. The distribution of HII regions obtained from these studies is shown in Figure 2.

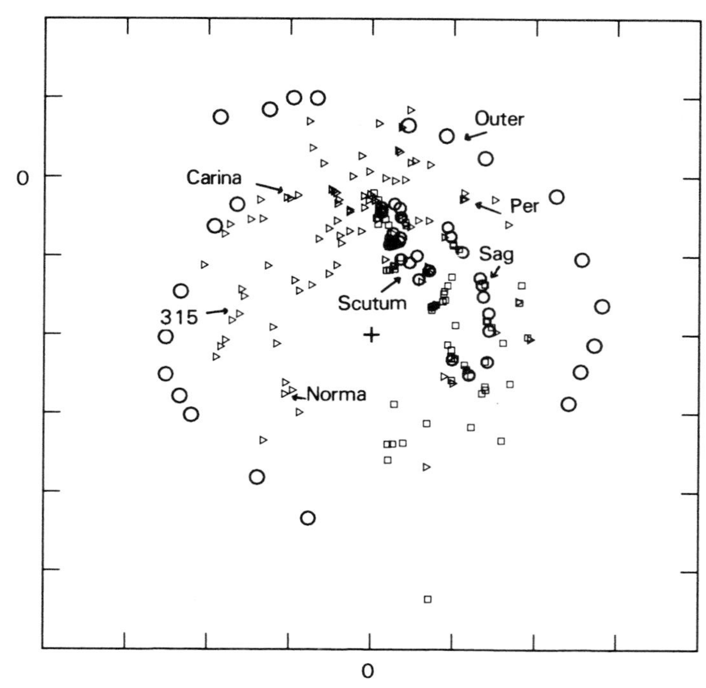

Figure 2. The global spiral structure is shown by the distribution of HI emission peaks (from Henderson, Jackson and Kerr 1982) plotted as large open circles, the CO clouds (Dame et al. 1983) plotted as small open circles, the optical HII regions (Georgelin and Georgelin 1976) plotted as open triangles, and the radio HII regions (Downes et al. 1980) plotted as open squares. The galactic center is indicated by the cross, and the Sun is at (0,0). Tick marks are 5 kpc apart.

Extensive HI surveys have been made of our Galaxy (Kerr 1968, and Burton 1974 and references therein, Weaver 1970, Weaver and Williams 1974, Kerr 1970, 1979). The distribution of HI in the outer regions of the Galaxy was determined recently by Henderson, Jackson and Kerr (1982). They obtained distances using a modification to the Schmidt rotation curve that accounts for N-S. The regions of maximum intensity, estimated from their contours, are shown in Figure 2. The structure in the inner region is more difficult to determine because of the distance ambiguity, and it also appears to be more complicated than in the outer region. The inner part of the Galaxy contains many spurs and branches, in addition to the prominent arms traced by the HII regions. The local distribution of HI is even less clear, because the random velocity dispersion in the local gas makes the kinematic distances very uncertain when the systemic radial velocity is small.

Carbon-monoxide emission from the northern hemisphere shows evidence for spiral-arm concentrations in the inner Galaxy. The recent study by Dame et al. (1983) indicates that the largest molecular clouds trace out the Sagittarius Arm and part of the Scutum Arm very clearly. Dame et al. resolved the distance ambiguity by using known distances to associated HII regions, H_2CO absorption features on the line-of-sight to background HII regions, and, for redundancy whenever possible, the cloud latitude extents for radii derived from a radius-linewidth relationship. The distribution of these large CO clouds is shown in Figure 2. Confinement of the general distribution of CO emission to spiral arms is not so obvious. Bash et al. (1977) and Cohen et al. (1980) proposed that most of the CO is in spiral arms, and, to the extent that most of the CO mass is in the largest clouds (Solomon and Sanders 1980; Dame 1983), their interpretation appears to be correct. Carbon-monoxide observations of the southern Galaxy also appear to show spiral arms (Robinson *et al.* 1982; Cohen 1985; Israel *et al.* 1985). However, the presence of emission between arms (Stark 1979), and the continuity of the terminal-velocity ridge on the longitude-velocity diagram of the first-quadrant emission, led Solomon *et al.* (1979), and Burton and Gordon (1978) to interpret the CO data as showing a more homogeneous distribution of intensity.

In the outer Milky Way, Kutner and Mead (1981, 1982, 1985) reported CO out to 27 kpc; they mapped an arm at a distance of 15 kpc from the galactic center, and an extension of the Perseus spiral arm in the first quadrant. Recent CO studies by Solomon, Stark and Sanders (1983) were unable to confirm many of these detections, however. Detections out to 18 kpc were reported by Blitz, Fich and Stark (1982), who noted that CO beyond 13 kpc was rare.

1.3. Tangent points from terminal velocities and continuum emission

Tangent directions to spiral-arm pieces may be found by noting the positions on terminal-velocity versus longitude curves where the slope suddenly changes. The difficulty with this method is that such inflections may sometimes result from a lack of emission at the tangent point, rather than from streaming motion in a spiral arm. Spiral-arm tangents for observations of HI (Henderson 1977) and CO (Robinson *et al.* 1982) have been determined using this method (see Table 1).

Spiral-arm tangent points may also be estimated from maxima in the radio-continuum intensity mapped along the galactic plane. The assumption is that the peak intensity corresponds to the edge of a spiral arm. This method has been applied by Mills (1963; see Table 1), Wielebinski *et al.* (1968), and Okuda (1985), who observed nonthermal and thermal radiation.

TABLE 1: Longitudes of Tangential Directions

Source	Ref.	Longitudes (degrees)					
HI	(1)	28-34	44-51	75-80	285-292	310-318	328-333
CO	(2)	31	52			309	327 337 341
HII	(3)	35	49		285	310	329
continuum	(4)	0 13 27.5	50	80	262.5 282.5	310	327 337 344

References: (1) Henderson (1977); (2) Robinson *et al.* (1982);
(3) Georgelin and Georgelin (1976); (4) Mills (1963)

2. GLOBAL FITS TO THE OBSERVED SPIRAL STRUCTURE

2.1. Global models

The proposed global fits to the observations of our Galaxy include either two or four main spiral arms, with pitch angles of 5° to 27°. In order to discuss the details of different models, it is convenient to transform the map of Figure 2 into (log R, θ) coordinates (Figure 3). The names of particular regions of the Galaxy have been indicated on each figure for comparison. Logarithmic spiral arms will appear straight on this plot, with a slope that depends on the pitch angle. The (log R,θ) plot emphasizes that the Milky-Way spiral arms are represented only by short segments; there is no obvious and unambiguous way to join the pieces. Possible branching structures are evident. The following paragraphs will summarize various attempts to fit the data.

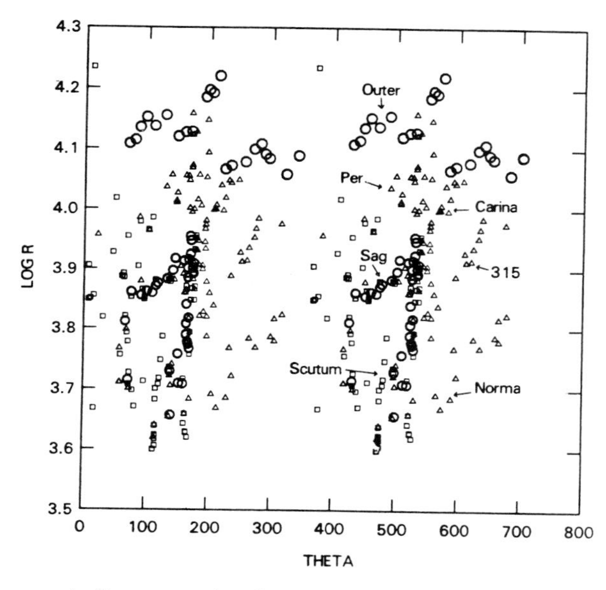

Figure 3. The global spiral pattern of Figure 2 is shown here in polar coordinates, with the same symbol convention as before. Logarithmic spiral arms would appear straight on this plot. The Sun is at log R = 4.0, theta = 180°.

A 2-arm spiral galaxy was the model proposed by many observers. Weaver (1970, 1974) suggested that a 2-arm global spiral with a 12°5 pitch angle could represent the major HI arms. Figure 4a shows straight lines representing his spiral fit superposed on the data points of Figure 3. He noted that the local and inner regions are complicated by many spurs and fragments. Simonson (1976) determined a global spiral pattern for HI data, with a pitch angle of 6-8° for 2 inner arms, which branch to multiple outer arms with pitch angles of 16°; the Orion arm appears as a spur with a pitch angle of about 20°. The Carina and Outer arms appear only as discrete arm segments in between the 2 major spiral arms, Sagittarius and Perseus. The Norma arm is suggested as an extension of the Sagittarius arm. Figure 4b shows his fit. Lockman (1979) obtained a 2-arm logarithmic spiral based on northern HI data; it does not fit the southern area as well. He noted that streaming motions would not affect his conclusions, since the HII regions which the model fits occur near the spiral-arm potential minima where streaming motions would be small. Models by Bash (1981), which include velocity dispersions and non-circular motions due to density waves, also show that a 2-arm global pattern might be consistent with the observations.

Several observers have interpreted the spiral structure as having four main spiral arms. Georgelin and Georgelin (1976) delineated 4 arm segments on the basis of their optical HII regions, as shown in Figure 4c. Henderson (1977) proposed a 4-arm global model with a 13° pitch angle and a 7 km s^{-1} streaming motion; this fit is shown in Figure 4d. He noted that inside the solar circle, the optical data near $\ell = 290^{\circ}$ are not very well fit by this model. Henderson, Jackson and Kerr (1982) and Kulkarni, Blitz and Heiles (1982) also suggested models with 4 major arms in the outer regions. Kulkarni *et al.* interpreted the arms as being coherent spiral features 20-25 kpc long, with pitch angles of $22\text{-}27^{\circ}$ based on the Blitz *et al.* (1982) revised rotation curve. Here, the Outer Arm (which they call Cygnus) and the Perseus Arm appear as major spiral arms. They could not interpret from their model whether the Orion feature is a spur or a long arm; its pitch angle was estimated to be 29°. They extended the outer 4-arm spiral in towards the 4-kpc region in this model, but noted that the extrapolated spiral does not fit the CO data in the inner regions everywhere (see Figure 4e). The preliminary analysis of CO data by Robinson *et al.* (1982) suggested 4 main arms with pitch angles of $11\text{-}13^{\circ}$. Three arms were indicated by the data, and the fourth, which occurs in an area difficult to map, was adopted to make the model symmetric (see Figure 4f).

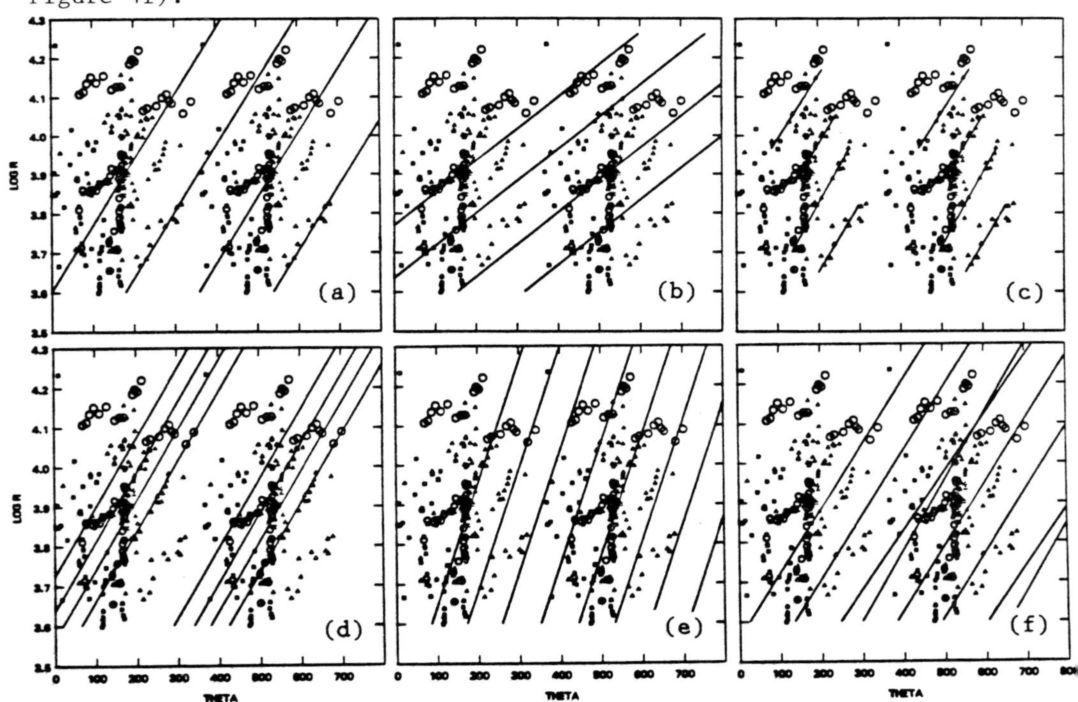

Figure 4. Fits to the spiral pattern are shown superposed on the data points of Figure 3. Details are discussed in the text. (a) Weaver (1970) 2-arm, $i=12^{\circ}.5$; (b) Lockman (1979) 2-arm, $i=6^{\circ}.2$ for inner regions; (c) Georgelin and Georgelin (1976) 4-arm, "eye-ball" fit; (d) Henderson (1977) 4-arm, $i=13^{\circ}$; (e) Kulkarni *et al.* (1982) 4-arm, $i=22\text{-}27^{\circ}$; and (f) Robinson *et al.* (1982) 4-arm, $i=11\text{-}13^{\circ}$.

For the inner parts of the Galaxy, Quiroga (1977) fit HI and optical data to four arms, with pitch angles of 13-17°. Mills (1963) stated that 2, 3, or 4 arms are possible within the resolution of the continuum-emission data. In the extreme outer Galaxy, Sills (1982) has analyzed high-velocity gas in the first and second quadrants, which indicated a spiral arm with a pitch angle of 16-24° at a distance greater than 20 kpc.

2.2. Interpretation

The models just discussed show the confusion which prevails in considering spiral structure. All of the models are reasonably consistent with the observations. The difficulty is that the determination of a global pattern is not unique. So far, it is only possible to define discrete arm-segments; data in localized regions may be connected in unambiguous ways. The overall structure of the Galaxy must be extrapolated from a relatively small number of data points. It is clear that more data are needed in order to be able to distinguish between two, four, or even three-arm models, between large and small pitch angles, or between models with and without global symmetry.

The HI in the outer Galaxy, which is not subject to near-far distance ambiguities, provides a good clue to the overall structure of the Galaxy; it appears to have a roughly bisymmetric pattern. It is not clear whether this pattern extends to two or four arms in the inner region; the inner region does not show any obvious symmetry. It is possible that the Milky Way has only spiral-like pieces in the inner regions and a symmetric grand-design spiral in the outer parts; this is typical of galaxies like NGC 2903. In the next section, observations of external galaxies will be examined for comparison.

3. OBSERVATIONS OF SPIRALS IN EXTERNAL GALAXIES

3.1. Properties of external spirals applied to the Milky Way

Observations of spirals in nearly face-on external galaxies are unhindered by a poor vantage point; a knowledge of their properties might lead to a better assessment of the spiral structure of the Milky Way.

The Hubble type of a galaxy might be expected to give an indication of the pitch angle of its arms. De Vaucouleurs and Pence (1978) examined the overall photometric properties of the Milky Way, and classified it is an SAB(rs)bc II galaxy; such galaxies typically have pitch angles of $\sim$12-15°. Kennicutt (1981) found, however, that measured pitch angles in other galaxies correlate only weakly with bulge-to-disk ratios, and that the pitch angles have a large dispersion. Thus, the Hubble type of the Milky Way does not define our global pitch angle unambiguously.

Kennicutt (1982) and Block (1982) found that arm widths correlate with galaxy luminosity for Sbc and Sc spirals. Based on these results, we might determine an expected arm width for our Galaxy, which could aid in distinguishing between an arm and an interarm region. De Vaucouleurs and

Pence (1978) estimate a brightness $M_T^o(B) = -20.2 \pm 0.15$ for the Milky Way, which implies an arm width of some 820 ± 100 pc based on Block's data. This width is consistent with the width of the local arms in the Milky Way, as estimated from the distribution of young clusters.

HII regions have traditionally been used to trace spiral structure in other galaxies. Lynds (1970) made an optical survey of HII regions in external spirals, and Hodge (1969) determined the positions of HII regions for over twenty spirals based on Hα photographs. Hodge and Kennicutt (1983) have an atlas of HII regions in 125 galaxies, and Hodge (1982) summarizes the available data for HII regions in 223 galaxies. These HII maps show that, while HII regions give the impression of defining spiral arms, a plot of their distributions in galaxies reveals much interarm activity; indeed, spiral arms are often difficult to trace based on the positions of HII regions alone. As an example, Athanassoula (1978) determined, on the basis of HII regions, that the best-fit model to the highly inclined M31 galaxy was a 1-arm leading spiral. Rumstay and Kaufman (1983) examined the distributions of HII regions in M33 and M83 with excitation parameters greater and less than 115 pc cm^{-2}, and found that only the giant HII regions are concentrated in spiral arms. This may be true in our Galaxy as well (Fich and Blitz 1982).

Kennicutt and Hodge (1976), Boeshaar and Hodge (1977), and Anderson, Hodge and Kennicutt (1983) have examined correlation functions for HII regions distributed across spiral arms in NGC 628, NGC 3631, and NGC 4321, respectively, but have found no conclusive evidence for clumping. Elmegreen and Elmegreen (1983a) studied galaxies that happen to have regularly spaced giant HII regions and HI clumps, and found that the separation is about 20% of the galaxy size; this would correspond to 2 kpc in the Milky Way. Such giant HII regions and HI conglomerates have, in fact, been observed in our Galaxy. Twenty-nine large HI clouds ($M \sim 10^7 M_\odot$) in 4 arms were observed by McGee and Milton (1964), and 15 giant CO clouds ($M \sim 10^6 M_\odot$) in the Sagittarius Arm were mapped by Dame _et al._ (1983). The spacings between the giant clouds in our Galaxy are similar to what is observed in other galaxies.

Considère and Athanassoula (1982), Krakow, Huntley and Seiden (1982), and Iye _et al._ (1982) produced Fourier transforms of galaxy images in order to study spiral shapes and to determine the modes of the spiral waves that may be present. Kennicutt (1981) found that logarithmic as well as hyperbolic spirals may be present in a galaxy.

Dust lanes have long been recognized as good spiral tracers. Lynds (1970) mapped spiral features from dust lanes in 23 galaxies. It is therefore reasonable to examine whether dust might be useful as a tracer in our Galaxy. Holmberg (1950) and Lynds (1970) estimated the gas density of a typical dust lane to be ~ 10 cm^{-3}. Elmegreen (1980) determined that visual extinctions in external dust lanes are 1-2 mag; the typical extinction of the ambient medium perpendicular to the galactic plane is 0.5 to 1 mag. Thus, dust lanes represent only a relatively small compression of the local interstellar medium. Dust lanes are probably difficult to find in the Milky Way; we would not expect to see a dust lane stand out in any obvious way in a 21-cm survey, unless its velocity is peculiar because of streaming motions.

We can speculate about the external appearance of the dust lanes in our Galaxy. Lucke (1978) and Krautter (1980) determined that the cloud distribution within 2-3 kpc of the solar neighbourhood is patchy, and Lyngå (1979) suggested that the Milky Way does not have well-developed dust lanes. Further evidence for this deduction comes from external galaxies. Lynds (1980b) developed a Dust Classification (1 through 5) for Sbc and Sc spirals, based on the continuity of the dust lanes. The dust class was correlated with the number of bright HII regions. In our Galaxy, there are at least 25 optical HII regions with excitations in excess of 100 pc cm^{-2}, corresponding to a Lynds dust class between 2 and 3. Galaxies in this class have dark clouds or pieces of dust lanes associated with major spiral arms, but dust-lane continuity over no more than 30° to 90°.

Spurs are very common in external galaxies. While they do not contribute to the global pattern of a galaxy, they may be prominent regional features. Elmegreen (1981) found that spurs tend to jut outward from the edges of spiral arms, with pitch angles of $63^{o} \pm 12^{o}$. Weaver (1970) suggested that our Sun is in a spur jutting outward from the Sagittarius Arm with a pitch angle of 20-25°. However, optical data showing a gap between Sagittarius and the Local Arm led Humphreys (1976, 1979) to suggest that the local feature is probably a spur jutting inward from the Perseus Arm. Blaauw (1985) discusses details of the Local Spur.

Gaseous components of other galaxies might be expected to trace spiral structure, as they do in the Milky Way. Aperture-synthesis maps made at 21 cm have revealed detailed spiral features with resolutions of 20-30". These results will be discussed by Allen during this conference. Whether CO shows a spiral pattern is not as certain, since for external galaxies the CO beamwidth is generally too large. The effective resolution may be improved by using information about the known rotation of a galaxy, coupled with the observed CO velocities (Elmegreen and Elmegreen 1982a, Scoville and Young 1983). CO emission has been found to be associated with bright HII regions and possibly spiral arms in MGC 4321 (Elmegreen and Elmegreen 1982a), IC 342 (Rickard 1983), and M51 (Rydbeck 1985). However, CO appears to be stronger in the dusty interarm regions of NGC 5248 than in the bright spiral arms, and it is very strong in the dust lanes in NGC 1068 (Elmegreen and Elmegreen 1982a). CO has been mapped in a prominent southern dust lane in M31 (Boulanger, Stark and Combes 1981; Stark 1985), and it appears to be a better tracer of the spiral structure than HI; but this galaxy is too highly inclined to determine spiral structure unambiguously.

3.2. Flocculent and grand-design spirals

The global arm structure of spirals has prompted much attention. Sandage (1961) and Kormendy (1977) noted that many early-type spiral galaxies, like NGC 2841 or NGC 5055 (Figure 5a), have a patchy arm appearance. Woltjer (1965) suggested that such galaxies be referred to as "spiral-like" to distinguish them from grand-design spirals (e.g., NGC 4321, Figure 5b) with long, continuous arms. Elmegreen and Elmegreen (1982b) developed an Arm Class system (1-12) in order to rank the coherence and symmetry of spiral arms; the term "flocculent" was adopted to describe the spiral-like galaxies.

It is now possible to study spirals quantitatively with photographic surface photometry; Schweizer (1976) was the first to apply such methods to several galaxies. Talbot, Jensen and Dufour (1979, 1981) made a detailed study of M83. The derived azimuthal profiles did not reveal the transverse color gradients initially expected from spiral density-wave theory. Such effects are probably obscured anyway by epicyclic motions of the young stars (Bash 1979, Yuan and Grosbøl 1981). Efremov and Ivanov (1982) detected an age gradient in M31 based on long-period Cepheids; Humphreys and Sandage (1980) found only vague hints of an age gradient in some regions of M33, based on the distributions of young galactic clusters.

Strom, Jensen and Strom (1976) examined spiral structure in the red passband, and suggested that density waves should be sought in the near-infrared. Schweizer (1976), Jensen (1977), and Elmegreen (1981) made photographs of galaxies in the near-infrared passband; the underlying spiral is prominent in grand-design galaxies, but still is not present in galaxies that look flocculent in the blue.

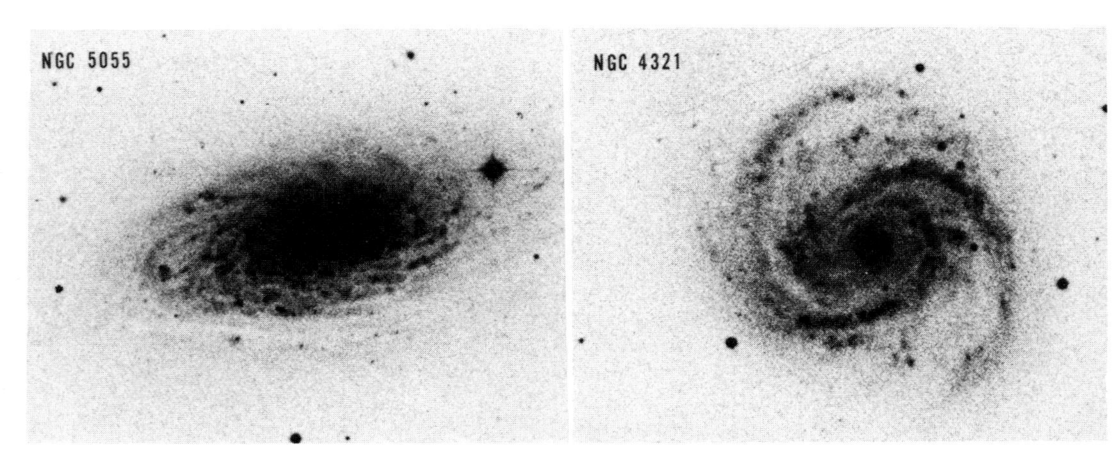

Figure 5. (a) Blue photograph of NGC 5055, a prototype of flocculent galaxies; (b) Blue photograph of NGC 4321, a prototype of grand-design galaxies.

Elmegreen and Elmegreen (1983b) acquired photographic surface photometry on 34 non-barred grand-design and flocculent spirals. For the grand-design galaxies, the arm-interarm contrast was found to be very large in the blue as well as in the near-infrared, often exceeding 1 magnitude. In flocculent galaxies, the arm-interarm contrast was large in the blue but small in the near-infrared. A plot of the arm-interarm contrast ratio for the blue divided by the contrast ratio for the near-infrared, versus the arm-interarm contrast in the near-infrared, reveals a pattern of arm features that correlates with arm class (see Figure 6). The large arm-interarm contrasts seen in the near-infrared for grand-design galaxies can be explained only if there is a significant stellar density enhancement in the arms. The enhancements are often larger than 40%, Elmegreen and Elmegreen (1983b). Flocculent spirals, on the other hand, may get their blue spiral structure from sheared star-forming regions. It therefore appears to be worthwhile to determine the distribution of older disk stars in the Milky Way in order to search for further clues of spiral structure.

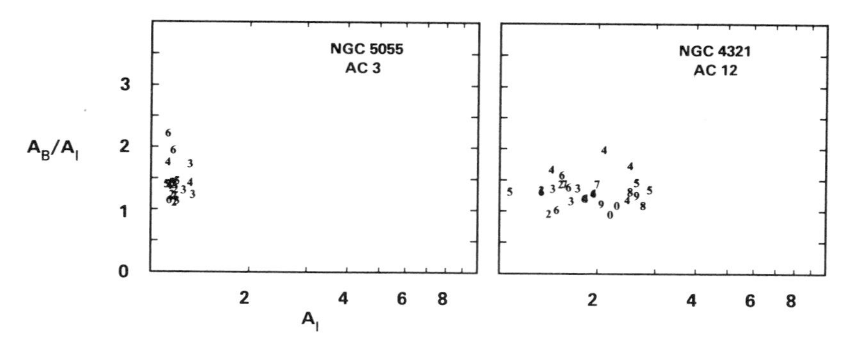

Figure 6. "Color-amplitude" plot for (a) NGC 5055 and (b) NGC 4321; the arm-interarm contrast in the blue divided by the arm-interarm contrast in the infrared is plotted versus the arm-interarm contrast in the near-infrared. Numbers correspond to tenths of R_{25}, the radius where the surface brightness is 25 mag arcsec^{-2}. Features to the right in the diagrams require density enhancements (Elmegreen and Elmegreen 1983c).

4. ORIGIN OF THE SPIRAL STRUCTURE IN THE MILKY WAY

A large number of spiral galaxies has been examined in order to understand how internal and external factors might contribute to spiral structure. These results may provide an indirect clue to the structure of the Milky Way galaxy.

Kormendy and Norman (1979) studied 54 spiral galaxies with known rotation curves, and found that galaxies with grand-design patterns had companions or bars. Madore (1980) found a weak correlation between the luminosity class of a galaxy and the presence of companions; spirals in small groups tended to be slightly more luminous than were isolated spirals. Thus, the presence of bars or companions appears to affect the way a spiral looks.

Elmegreen and Elmegreen (1982b, 1983c) determined spiral-arm classifications for over 300 spiral galaxies in the field and in groups. They found that, among non-barred galaxies, isolated spirals tend to be flocculent (68% ±10%), and spirals with companions tend to be grand-design (67% ±6%). Among barred or oval galaxies (types SB and SAB), 71% ±7% of isolated spirals, 72% ±4% of group-galaxy spirals, and 93% ±5% of binary-galaxy spirals are grand-design. Thus, symmetric spirals dominate disk galaxies except in isolated, non-barred, non-oval systems.

In a galaxy group, the percentage of non-barred spiral galaxies having grand-design structure correlates with the group-crossing rate. These results can be applied to the Local Group to estimate whether or not the Milky Way is likely to have a grand-design (symmetric) spiral. Four local galaxies are large enough to have been included in the surveys used in the previous studies (galaxies bigger than the SMC); M31 and M33 have a grand-design spiral structure. The Local-Group density is high enough that the Milky Way galaxy (and its companions, M31 and M33) has a 65% chance of being a grand-design spiral. The classification of the Milky Way as an SAB further strengthens the likelihood that our Galaxy has some kind of grand-design spiral over part of its disk. The Milky Way may have an arm class 7 or 9, which are the categories for NGC 2903 (Figure 7a), or M101 and NGC 1232 (Figure 7b), respectively.

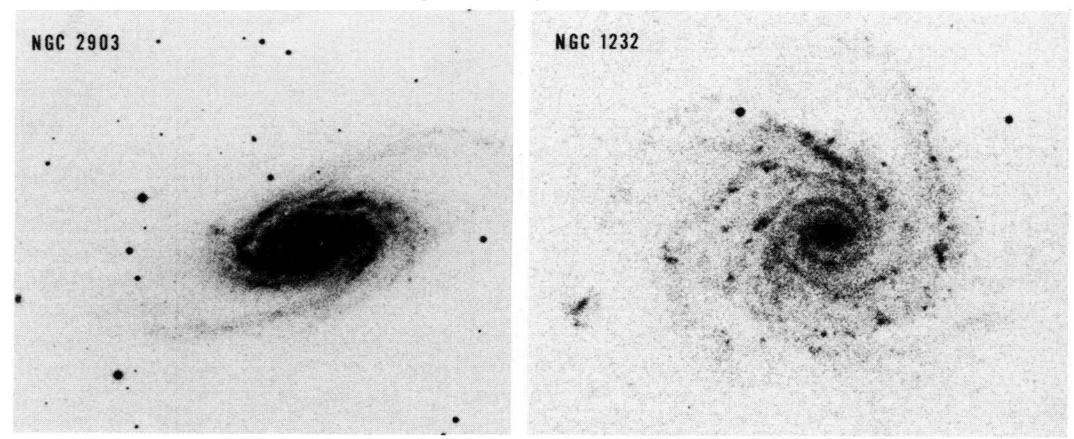

Figure 7. (a) Blue photograph of NGC 2903, which has a flocculent inner region and a grand-design outer region. (b) Blue photograph of NGC 1232, which has multiple arms and branches.

Toomre (1970) examined binary interactions and the possibility that the Milky Way structure was influenced by Local-Group galaxies. Fujimoto and Sofue (1976) also investigated the dynamical effects of the Large and Small Magellanic Clouds on the Milky Way. These calculations showed that at least the outer spiral structure of the Milky Way may be affected by its nearest neighbours.

The author is grateful to Dr. Bruce Elmegreen for many informative discussions and suggestions, and a critical reading of this manuscript.

REFERENCES

Anderson, S., Hodge, P., and Kennicutt, R.C.: 1983, Ap. J. 265, 132
Athanassoula, E.: 1978, Astron. Astrophys. 67, 73
Bash, F.: 1979, Ap. J. 233, 524
Bash, F.: 1981, Ap. J. 250, 551
Bash, F., Green, E., and Peters, W.: 1977, Ap. J. 217, 464
Becker, W., and Fenkart, R.: 1970, IAU Symp. No. 38, "The Spiral Structure of Our Galaxy", eds. W. Becker and G. Contopoulos (Dordrecht, Reidel), p. 205
Becker, W., and Fenkart, R.: 1971, Astron. Astrophys. Suppl. 4, 241
Berman, V.G., and Mishurov, Yu.N.: 1981, Soviet Astron. Lett. 7, 328
Blaauw, A.: 1985, IAU Symp. No. 106, "The Milky Way Galaxy", this volume
Blitz, L., Fich, M., and Stark, A.A.: 1982, Ap. J. Suppl. 49, 183
Block, D.L.: 1982, Astron. Astrophys. 109, 376
Boeshaar, G., and Hodge, P.W.: 1977, Ap. J. 213, 361
Bok, B.: 1963, IAU-URSI Symp. No. 20, "The Galaxy and the Magellanic Clouds", eds. F.J. Kerr and A.W. Rodgers (Canberra, Austr. Academy of Science), p. 147
Boulanger, F., Stark, A.A., and Combes, F.: 1981, Astron. Astrophys. 93, L1.
Burton, W.B.: 1974,"Galactic and Extra-Galactic Radio Astronomy" (New York, Springer-Verlag), p. 82
Burton, W.B., and Gordon, M.A.: 1978, Astron. Astrophys. 63, 7
Cohen, R.S.: 1985, IAU Symp. No. 106, "The Milky Way Galaxies", this volume
Cohen, R.S., Cong, H., Dame, T.M., and Thaddeus, P.: 1980, Ap. J. (lett.), 239, L53
Considère, S., and Athanassoula, E.: 1982, Astron. Astrophys. 111, 28
Dame, T.: 1984, Ph. D. thesis, Columbia University
Dame, T., Elmegreen, B.G., Cohen, R., and Thaddeus, P.: 1983, preprint
de Vaucouleurs, G., and Pence, W.D.: 1978, A. J. 83, 1163
Dolidze, M.V.: 1980, Soviet Astron. Lett. 6, 51 and 394
Downes, D., Wilson, T.L., Bieging, J., and Wink, J.: 1980, Astron. Astrophys. Suppl. 40, 379
Efremov, Yu.N., and Ivanov, G.R.: 1982, Ap. Sp. Sci. 86, 117
Efremov, Yu.N., Ivanov, G.R., and Nikolov, N.S.: 1981, Ap. Sp. Sci. 75, 407
Elmegreen, B.G., and Elmegreen, D.M.: 1983a, Monthly Not. Roy. Astron. Soc. 203, 31
Elmegreen, B.G., and Elmegreen, D.M.: 1983c, Ap. J. 267, 31
Elmegreen, D.M.: 1980, Ap. J. Suppl. 43, 37
Elmegreen, D.M.: 1981, Ap. J. Suppl. 47, 229
Elmegreen,D.M., and Elmegreen, B.G.: 1982a, A. J. 87, 626
Elmegreen, D .M., and Elmegreen, B.G.: 1982b, Monthly Not. Roy. Astron. Soc. 201, 1021
Elmegreen,D.M., and Elmegreen, B.G.: 1983b, Ap. J. Suppl., in press
Fenkart, R.: 1979, IAU Symp. No. 84, "The Large-Scale Characteristics of the Galaxy", ed. W.B. Burton (Dordrecht, Reidel), p. 101
Fich, M., Blitz, L.: 1983, "Kinematics, Dynamics, and Structure of the Milky Way", ed. W.L.H. Shuter (Dordrecht, Reidel), p. 151
Fujimoto, M., and Sofue, Y.: 1976, Astron. Astrophys. 47, 263
Georgelin, Y.M., and Georgelin, Y.P.: 1976, Astron. Astrophys. 49, 57

Georgelin, Y.M., Georgelin, Y.P., and Sivan, F.P.: 1979, IAU Symp. No. 84, "The Large-Scale Characteristics of the Galaxy", ed. W.B. Burton (Dordrecht, Reidel), p. 65

Henderson, A.P.: 1977, Astron. Astrophys. 58, 189

Henderson, A.P., Jackson, P.D., and Kerr, F.J.: 1982, Ap. J. 263, 116

Hodge, P.W.: 1969, Ap. J. 155, 417

Hodge, P.W.: 1982, A. J. 87, 1341

Hodge, P.W., and Kennicutt, R.C.: 1983, A. J. 88, 296

Holmberg, E.: 1950, Medd. Lund Astr. Obs., Ser. II, No. 128

Humphreys, R.: 1976, P.A.S.P. 88, 647

Humphreys, R.: 1979, IAU Symp. No. 84, "The Large-Scale Characteristics of the Galaxy", ed. W.B. Burton (Dordrecht, Reidel), p. 93

Humphreys, R., and Sandage, A.: 1980, Ap. J. Suppl. 44, 319

Israel, F., Brand, J., and de Vries, C.: 1985, IAU Symp. No. 106, "The Milky Way Galaxy", this volume

Iye, M., Okamura, S., Hamabe, M., and Watanabe, M.: 1982, Ap. J. 256, 103

Jensen, E.B.: 1977, Ph. D. thesis, University of Arizona

Jensen, E.B., Talbot, R.J., and Dufour, R.J.: 1981, Ap. J. 243, 716

Kennicutt, R.C.: 1981, A. J. 86, 1847

Kennicutt, R.C.: 1982, A. J. 87, 255

Kennicutt, R.C., and Hodge, P.W.: 1976, Ap. J. 207, 36

Kerr, F.J.: 1968, "Nebulae and Interstellar Matter", eds. B.M. Middlehurst and L.H. Aller (Chicago, Univ. of Chicago Press), p. 575

Kerr, F.J.: 1970, IAU Symp. No. 38, "The Spiral Structure of Our Galaxy", eds. W. Becker and G. Contopoulos (Dordrecht, Reidel), p. 95

Kerr, F.J.: 1979, IAU Symp. No. 84, "The Large-Scale Characteristics of the Galaxy", ed. W.B. Burton (Dordrecht, Reidel), p. 61

Kilkenny, D., Hill, P.W., and Schmidt-Kaler, Th.: 1975, Monthly Not. Roy. Astron. Soc. 171, 353

Kormendy, J.: 1977, "The Evolution of Galaxies and Stellar Populations", eds. B.M. Tinsley and R.B. Larson (New Haven, Yale University Printing Service), p. 131

Kormendy, J., and Norman, C.A.: 1979, Ap. J. 233, 539

Kraft, R.P., and Schmidt, M.: 1963, IAU-URSI Symp. No. 20, "The Galaxy and the Magellanic Clouds", eds. F.J. Kerr and A.W. Rodgers (Canberra, Austr. Academy of Science), p. 102

Krakow, W., Huntley, J.M., and Seiden, P.E.: 1982, A. J. 87, 203

Krautter, J.: 1980, Astron. Astrophys. 89, 74

Kulkarni, S., Blitz, L., and Heiles, C.: 1982, Ap. J. (Lett.), 259, L63

Kutner, M., and Mead, K.: 1981, Ap. J. (Lett.), 249, L15

Kutner, M., and Mead, K.: 1985, IAU Symp. No. 106, "The Milky Way Galaxy", this volume

Lockman, F.J.: 1979, Ap. J. 232, 761

Lucke, P.B.: 1978, Astron. Astrophys. 64, 367

Lynds, B.T.: 1970, IAU Symp. No. 38, "The Spiral Structure of Our Galaxy", eds. W. Becker and G. Contopoulos (Dordrecht, Reidel), p. 26

Lynds, B.T.: 1980a, Ap. J. 238, 17

Lynds, B.T.: 1980b, A. J. 85, 1046

Lyngå, G.: 1979, IAU Symp. No. 84, "The Large-Scale Characteristics of the Galaxy", ed. W.B. Burton (Dordrecht, Reidel), p. 87

Madore, B.F.: 1980, A. J. 85, 507

McCuskey, S.W.: 1970, IAU Symp. No. 38, "The Spiral Structure of Our Galaxy", eds. W. Becker and G. Contopoulos (Dordrecht, Reidel), p.189
McCuskey, S.W.: 1974, A. J. 79, 107
McCuskey, S.W., and Houk, N.: 1971, A. J. 76, 1117
McGee, R.X., and Milton, J.A.: 1964, Austr. J. Phys. 17, 128
Mead, K.N., and Kutner, M.L.: 1982, B.A.A.S. 14, 617
Mills, B.Y.: 1963, IAU-URSI Symp. No. 20, "The Galaxy and the Magellanic Clouds", eds. F.J. Kerr and A.W. Rodgers (Canberra, Austr. Academy of Science), p. 102
Morgan, W.W., Sharpless, S., and Osterbrock, D.E.: 1952, A. J. 57, 3
Okuda, H.: 1985, IAU Symp. No. 106, "The Milky Way Galaxy", this volume
Quiroga, R.J.: 1977, Ap. Sp. Sci. 50, 281
Rickard, L.J.: 1983, B.A.A.S. 14, 948
Roberts, W.W., Roberts, M.S., and Shu, F.: 1976, Ap. J. 196, 381
Robinson, B.J., McCutcheon, W.H., Manchester, R.N., and Whiteoak, J.B.: 1983, "Surveys of the Southern Galaxy", eds. W.B. Burton and F.P. Israel (Dordrecht, Reidel), p. 1
Rumstay, K.S., and Kaufman, M.: 1983, Ap. J., in press
Rydbeck, G.: 1985, IAU Symp. No. 106, "The Milky Way Galaxy", this volume
Sandage, A.: 1961, "The Hubble Atlas of Galaxies" (Washington, The Carnegie Institution of Washington)
Schweizer, F.: 1976, Ap. J. Suppl. 31, 313
Scoville, N., and Young, J.: 1983, Ap. J. 265, 148
Sills, R.: 1982, Ph. D. thesis, Univ. of California, Berkeley
Simonson, S.C.: 1976, Astron. Astrophys. 46, 261
Smith, L.F.: 1968, Monthly Not. Roy. Astron. Soc. 141, 317
Solomon, P.M., and Sanders, D.B.: 1980, "Giant Molecular Clouds in the Galaxy", eds. P.M. Solomon and M.G. Edmunds (New York, Pergamon), p. 41
Solomon, P.M., Stark, A.A., and Sanders, D.B.: 1983, Ap. J. (Lett.) 267, L29
Solomon, P.M., Sanders, D.B., and Scoville, N.Z.: 1979, IAU Symp. No. 84, "The Large-Scale Characteristics of the Galaxy", ed. W.B. Burton (Dordrecht, Reidel), p. 35
Stark, A.A.: 1979, Ph. D. thesis, Princeton University
Stark, A.A.: 1985, IAU Symp. No. 106, "The Milky Way Galaxy", this volume
Strom, S., Jensen, E.B., and Strom, K.: 1976, Ap. J. (Lett.) 206, L11
Talbot, R.J., Jensen, E.B., and Dufour, R.J.: 1979, Ap. J. 229, 91
Tammann, G.A.: 1970, IAU Symp. No. 38, "The Spiral Structure of our Galaxy", eds. W. Becker and G. Contopoulos (Dordrecht, Reidel) p.236
Toomre, A.: 1970, IAU Symp. No. 38, "The Spiral Structure of our Galaxy", eds. W. Becker and G. Contopoulos (Dordrecht, Reidel), p. 334
van den Bergh, S.: 1960, Ap. J. 131, 215
Vogt, N., and Moffat, A.F.J.: 1975, Astron. Astrophys. 39, 477
Weaver, H.: 1970, IAU Symp. No. 38, "The Spiral Structure of our Galaxy", eds. W. Becker and G. Contopoulos (Dordrecht, Reidel), p. 126
Weaver, H.: 1974, IAU Symp. No. 60, "Galactic Radio Astronomy", eds. F.J. Kerr and S.C. Simonson (Dordrecht, Reidel), p. 573
Weaver, H., and Williams, D.R.W.: 1974, Astron. Astrophys. 17, 1
Westerlund, B.E.: 1963, IAU-URSI Symp. No. 20, "The Galaxy and the Magellanic Clouds", eds. F.J. Kerr and A.W. Rodgers (Canberra, Austr. Academy of Science), p. 160

Wielebinski, R., Smith, D.H., and Garzon-Cardenas, X.: 1968, Austr. J. Phys. 21, 185
Woltjer, L.: 1965, Stars and Stellar Systems, Vol. 5 "Galactic Structure", eds. A. Blaauw and M. Schmidt (Chicago, University of Chicago Press), p. 531
Yuan, C., and Grosbøl, P.: 1981, Ap. J. 243, 432

DISCUSSION

L. Blitz: Since much of the mapping of our Galaxy is incomplete, it is very difficult to know whether we are connecting spurs or pieces of real long spiral arms.

D.M. Elmegreen: That's right.

M.L. Kutner: The spiral structure in our Galaxy may be more easily discerned in the outer parts. Hence: "If you want to work out the spiral structure of our Galaxy, start at the outside and work your way in".

D.M. Elmegreen: I think that would be dangerous. Often the spiral structure in the outer part of a galaxy does not trace continuously into the inner part. In a galaxy like NGC 2903, for example, it would lead to erroneous results.

P. Pismis: It is reasonable to expect that the existence or not of spiral structure in galaxies depends in some way on the initial global parameters with which a galaxy gets started. One such parameter is the total mass. The total mass of a galaxy is related to the Hubble type, in the sense that it decreases on the average as one goes from the Sa's to the Sc's and on to the irregulars. A. Meisels (1983, Astron. Astrophys. 118, p. 21) has recently reached a similar conclusion, without being aware that the relation I mentioned had already been discussed some five years earlier (P. Pismis and L. Maupomé 1978, Rev. Mexic. Astron. Astrofis. 2, p. 319). Thus there is independent confirmation that statistically the average masses are related to the morphology of spirals, although the scatter is large. It is worth emphasizing that spiral structure tends to be well developed in a fixed mass range and that below 10^{10} solar masses spirals are not well developed; they are rather of the Magellanic type at best. In addition it is found that the average mass of barred spirals falls below that of "normal" spirals for the same Hubble type.

D.M. Elmegreen: I agree that Hubble type may be loosely correlated with galaxy mass. But we find (Elmegreen and Elmegreen, 1982b) that the spiral-arm class, which is a measure of the symmetry and continuity of the spiral arms, is independent of Hubble type and therefore apparently independent of galaxy mass (except for the very-low-mass systems which are Magellanic types). Galaxies of Hubble types Sa through Sd may have all ranges of arm class, from flocculent to grand-design. For example, NGC 5055 is an SAbc which is very flocculent, and yet M51 which is also an SAbc has a grand design. There has also been some theoretical conjecture that the presence of a grand-design spiral structure may depend on the ratio of disk mass to halo mass. While this may be true in a statistical sense, it cannot be true for every galaxy. I call attention to Carignan's paper in Section II.1; he finds evidence that NGC 7793, which is a classic flocculent galaxy, apparently has no halo, whereas we might have expected it to have a large one. So I think it is not just the mass dependence that determines spiral structure.

R.J. Allen: I would like to ask a question about the change in morphology with position in a galaxy. We see in the inner regions of M101 the flocculent kind of spiral arms, and in the outer regions the heavy spiral arms with lots of HII complexes. Would you believe that the character of spiral structure is correlated with the surface density of dust?

D.M. Elmegreen: Some galaxies have inner flocculent structure and outer grand-design (e.g. NGC 2903), some have inner grand-design and outer flocculent structure (e.g. NGC 6946), some have grand-design everywhere (e.g. NGC 5194), and some are flocculent everywhere (e.g. NGC 5055). I have not yet noticed any clear correlation between the type of spiral structure and the column density of the dust.

T.M. Bania: To my eye the dominant patterns on your log R vs. Θ plot of spiral tracers for the Milky Way are directly vertical. Would you care to comment?

D.M. Elmegreen: The feature to which you are referring is the one near $\theta = 180°$. Nobody really knows what it is. It might be a spur. Other equally good arms have a more normal pitch angle of some 12°. Spurs with high pitch angles (about 60°) are also seen in external galaxies.

K.S. de Boer: I wish to point out that you build a second model on top of a first model by making fits in your Θ-log R diagrams. All radial distances from 21-cm HI data and for distant HII regions were derived by assuming a rotation curve for our Galaxy. Then you try to fit a second model: a logarithmic spiral!

D.M. Elmegreen: Of course all models of spiral structure require distance determinations, and, as I mentioned, the kinematic distance determination may contain systematic errors. I am not using the θ - log R plots to build a second model; I am merely using them as a convenient way of illustrating the distribution of spiral-arm tracers.

F.H. Shu: I wonder how many independent parameters it takes to characterize a spiral galaxy. Two are obvious: Hubble type and Van den Bergh luminosity class. A third may also be important, namely the fractional gas content. Do you know if there is a systematic difference in the gas content of flocculents and grand-designs?

D.M. Elmegreen: Yes, I think that is an important point. We have looked at that a little bit, but not in detail.

R.G. Carlberg: Would you care to say anything about the lifetime of any particular pattern?

D.M. Elmegreen: The correlation between group-crossing time and the fraction of galaxies with grand-design structure (Elmegreen and Elmegreen, 1983c) allows for the possibility that grand-design spiral patterns are transient. A galaxy in a flocculent state could go into a grand-design state for some time after an external perturbation. By the way, regarding Dr. Pismis' comment, this possible transience is another reason why we think that grand-design spirals should not correlate with galaxy mass.

Debra M. (right) and Bruce G. Elmegreen CFD

WHICH KIND OF SPIRAL STRUCTURE CAN FIT THE OBSERVED GRADIENT OF VERTEX DEVIATION?

E. Oblak and M. Crézé
Observatoire de Besançon

We analyse the distribution of peculiar velocities of stars within 200 pc of the Sun in terms of space variations of the velocity ellipsoid.

1. THE SAMPLES

A sample of 757 stars, most brighter than $m_V = 6.5$, have been selected on the basis of existing data: for all of them we got uvby beta photometry, proper motions and radial velocities. From the photometric data we derive distances and age estimates. Ages are based on evolutionary tracks of Hejlesen et al. (1972).

2. THE ESTIMATION METHOD

The sample has been divided according to ages as indicated in Table 1. Kinematic data in each subsample has been processed in a maximum-likelihood solution including the classical parameters of the velocity ellipsoid plus gradients of those parameters in the galactic plane.

3. THE RESULTS

Table 1 shows the decrease of the vertex deviation with σ_π and age.

TABLE 1

Sample	Size	log age yr	σ_π kms^{-1}	Vertex Dev.	Significant Gradient?
1	130	8.2 -8.4	16.2	10 deg	NO
2	225	8.4 -8.7	16.3	30	YES
3	113	8.7 -8.9	17.1	28	NO
4	56	8.9 -9.0	17.3	14	NO
5	103	9.0 -9.18	22.5	8	NO
6	130	9.18-9.5	24.7	0	NO

H. van Woerden et al. (eds.), The Milky Way Galaxy, 273–274.

Due to the very large error bars on derived gradients, no significant ones can be obtained but for sample 2. In this sample, the vertex deviation is shown to vary deeply within the small portion of space involved.

The estimated gradients can be used to draw lines of constant vertex deviation in the solar neighborhood. Whatever may be the dynamical cause of the quoted deviation, isodeviation lines ought to be interpreted as isoperturbation lines.

4. WHICH KIND OF PERTURBATION?

Spiral density waves are non-axisymmetric perturbations and have been considered as an acceptable cause of the vertex deviation. The observed isodeviation line turns out to follow more or less the direction $l = 30^\circ$, $l = 210^\circ (\pm 20^\circ)$. This is in acceptable agreement with the direction of the Orion Arm, but conflicts definitely with interpretations of the local kinematics in terms of perturbation by a tightly wound spiral wave.

One cannot reconcile local kinematic observations with the large-scale spiral structure unless admitting at least two spiral modes with very different pitch angles. In this case, the mode driving non-axisymmetric behaviour in the solar neighborhood should be an open one.

REFERENCES

Hejlesen, F., Jorgensen, H., Peterson, J., Roncke, L.: 1972, IAU Coll. 17, "On stellar ages", G. Cayrel de Strobel and A.M. Delplace (eds.), Observatoire de Paris, p. XVII-1.
Oblak, E.: 1983, Astron. Astrophys. 123, p. 238.

SPIRAL STRUCTURE AND KINEMATICS OF HI AND HII IN EXTERNAL GALAXIES

R.J. Allen, P.D. Atherton, and R.P.J. Tilanus,
Kapteyn Astronomical Institute, University of Groningen,
P.O. Box 800, 9700 AV Groningen, The Netherlands.

ABSTRACT

Compression of the interstellar gas by a passing density-wave is thought to be responsible for triggering the presently observable star formation processes in the arms of spiral galaxies. We present new observations of the distribution and kinematics of the Hβ emission in the spiral arms of M83, obtained with the TAURUS imaging Fabry-Pérot system. These results, when combined with observations of the neutral and molecular components with sufficiently high resolution, should contribute to our understanding of the time sequence by which stars form out of the interstellar gas.

1. INTRODUCTION

Until recently our knowledge of the spiral structure and kinematics in external galaxies, and the basis on which we are able to construct and test models of Density-Wave-Induced Star Formation (DWISF), has been largely obtained from detailed HI mapping of a few nearby galaxies with radio-synthesis telescopes. The analyses by Rots and Shane (1975) and by Visser (1980a, b) of the distribution and motions of HI in M81 several years ago indicated the presence of streaming motions across spiral arms consistent with the predictions of Spiral Density-Wave Theory (SDWT). Some models of DWISF predict that the kinematics of the ionized gas may be different from the neutral gas (Bash & Visser, 1981) and there are observational indications that this is so (Bash 1983, Allen et al. 1983). These differences are of great interest for the study of the mechanisms and the time-scales for the formation of massive stars from the interstellar HI.

Improvements in sensitivity and angular resolution of radio synthesis telescopes and the advent of Imaging Fabry-Pérot Spectrometers in the visible (e.g. TAURUS) now enable a more comprehensive study of the star formation process in nearby galaxies.

H. van Woerden et al. (eds.), The Milky Way Galaxy, 275–279.

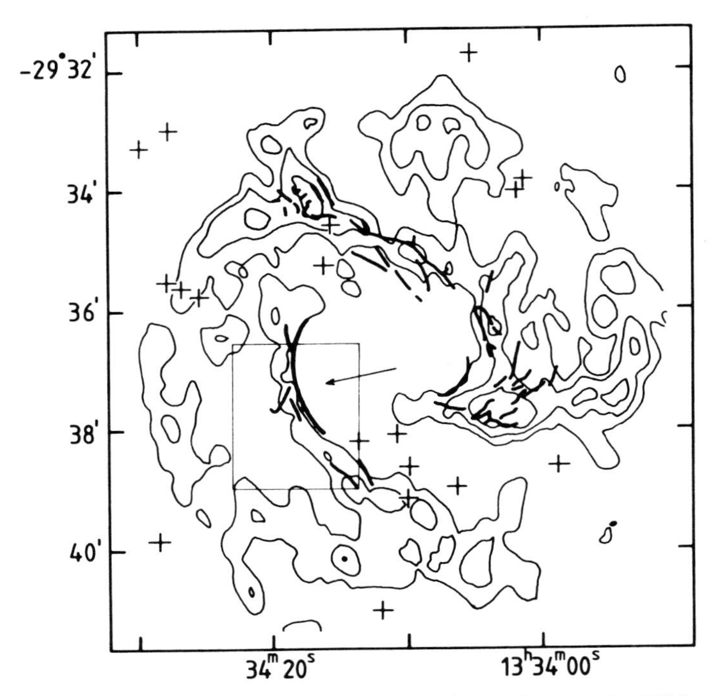

Figure 1 - The distribution of HI and dust lanes in M83. Contours are taken from the VLA-HI synthesis observations at 25" resolution by Ondrechen and van der Hulst. The dust lanes have been drawn from inspection of an optical photograph kindly provided by R.M. Humphreys. The crosses refer to a set of secondary standard stars whose positions have been determined by us to an accuracy of about 1". The arrow indicates the inner eastern arm which is the subject of this paper; the frame drawn here corresponds with the frames of Figures 2a and 2b.

We report here on a continuing study of the galaxy M83, combining radio continuum and 21 cm observations with optical Hβ emission. This study is aimed at determining the relative positions and kinematics of the neutral and ionized hydrogen.

2. DISCUSSION

In order that we might make an adequate observational test of the predictions of SDWT and DWISF, we need to choose a part of the galaxy in which we can be reasonably sure that this is the dominant mechanism for the formation of the massive stars. These stars become observable through the ionization which they produce on the neighbouring gas clouds. It seems likely that various other star formation processes will be superposed on DWISF in a particular galaxy (Elmegreen 1979) and this will tend to confuse the picture. Indeed, massive gas clouds will cause local gravitational perturbations to the underlying spiral

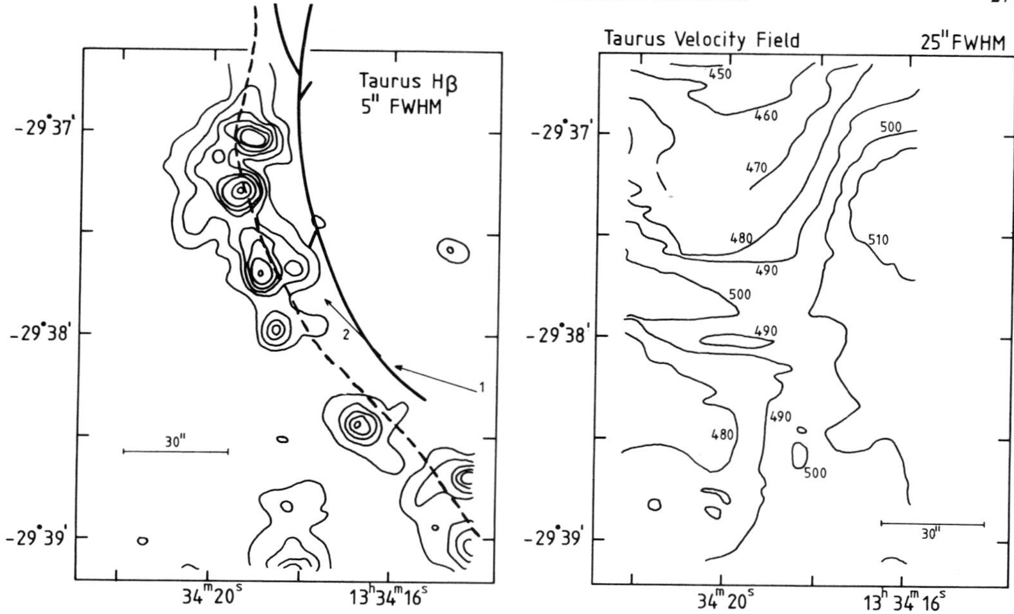

Figure 2
a) (left panel). The distribution of dust (thick solid line), Hβ emission (contours, 5" resolution), and the ridge line of the HI from the contours in Figure 1 (dashed line) are shown here for a section of the inner eastern spiral arm in M83. The horizontal bar at the lower left indicates 30", corresponding to 530 pc or 1280 pc along the major axis of M83 for assumed distances of 3.7 and 8.9 Mpc, respectively. The arrows marked 1 and 2 indicate schematically the pre- and post-shock gas flow. The Hβ (λ4861 Å) observations were obtained at the AAT using the TAURUS imaging Fabry-Pérot system (Atherton et al. 1982).
b) (right panel). Isovelocity contours (km s^{-1}, heliocentric) of the Hβ kinematics at 25" resolution in the area of Figure 2a.

potential of the old stellar population, and these pertubations could substantially reduce or enhance the strength of the shock locally in different parts of the galaxy. Furthermore one might expect that, once stars have begun to form, they might induce further star formation, through the interaction of strong stellar winds from OB associations with the surrounding material. Following Roberts (1969) and Shu (1984), we assume that the dust lane is the interstellar material that has been shocked and compressed by the locally changing spiral potential of the underlying old stellar population. This in turn is thought to induce the formation of giant molecular clouds, which may be launched from the shock as ballistic particles (Bash & Visser 1981), and wherein stars are formed which begin to radiate ionizing flux some 10^7years later, finally appearing as HII regions downstream of the shock. Secondary star formation processes may now be triggered (Elmegreen 1979) producing the chaotic structures observed in some parts of galaxies. In order to select against these secondary processes, we choose to study a part of the galaxy where the dust lane is long, clean and unfragmented. This is a matter of degree rather than division as seen in Fig. 1, where we superimpose a sketch of the dust-lane structure (from an optical photograph by R.M. Humphreys) on a VLA map of the HI. We interpret the long, unbroken dust lane on the inside of the eastern spiral arm as evidence that DWISF is predominant in this area, i.e., the large-scale organisation of the shock front is determined mainly by the spiral potential. Note that the western spiral arm appears considerably more chaotic both in the HI and in the appearance of the dust lanes.

In Figure 2a we show part of the inner eastern spiral arm in more detail, comparing the location of the Hβ emission (contours) with the position of the dust lane (solid line) and the ridge line (dashed) of the HI emission. From the picture of DWISF sketched above, we expect that the interstellar gas (mostly molecular in this region of the galaxy) streams into the shock (dust lane) as indicated for example by arrow #1 in Figure 2a. Upon leaving the shock (arrow #2) we might expect the spatial separation to follow the time sequence from molecular gas in the region of the dust lane through neutral (i.e. dissociated molecular) gas to ionized gas, and in a general sense this is clearly evident in the observations. The exact position of the HI ridge line is, however, uncertain owing to the relatively low resolution (25") of the VLA data presently available. Visser (1980b) has explained the shift of the HI ridge to the downstream side of the dust lane in M81 as arising from beam-smearing on an HI distribution which is intrinsically asymmetric on the two sides of the shock. The choice between these two different interpretations of the position shift between the HI and the dust lane requires new HI observations with an angular resolution of 10" or better.

The separation of dust lane and HII in Figure 2a is very clear. It seems unlikely that geometric projection effects could account for the lack of a clear separation in the other parts of M83. We suspect that these differences are due to variation in the potential, shock strength and secondary processes of star formation. Similar variations in separation can be seen in for example NGC 628, which is almost face-on.

In Figure 2b we show the velocity field of the ionized gas as determined from TAURUS observation at Hβ with a resolution of 25". These confirm in more detail our previous Hα results (Allen et al. 1983), indicating deviations in the kinematics of the ionized gas from those expected from SDWT for the neutral gas. At the seeing limit there is also evidence for much turbulence and disorder in the velocity field at the 5-10 km sec^{-1} level. Future work on M83 will include a more detailed comparison of the HI and HII kinematics.

ACKNOWLEDGEMENT

This work has been partially supported by NATO research grant no. 052.81. We are grateful to M. Ondrechen and J.M. van der Hulst for permission to use their unpublished VLA results, and to R.M. Humphreys for a superb photograph of M83. We have benefitted from discussions with W.W. Roberts and F.H. Shu, and with our colleagues at the Kapteyn Astronomical Institute.

REFERENCES

Allen, R.J., Atherton, P.D., Oosterloo, T.A., Taylor, K. 1983, in Internal Kinematics and Dynamics of Galaxies, ed. E. Athanassoula (Reidel, Dordrecht), IAU Symp. 100, p. 147
Atherton, P.D. Taylor, K., Pike, C.D., Harmer, C.F.W., Parker, N.M., Hook, R.N. 1982, M.N.R.A.S. 201, 661
Bash, F.N. 1983 (private communication)
Bash, F.N., Visser, H.C.D. 1981, Ap. J. 247, 488
Elmegreen, B. 1979, Ap. J. 231, 372
Roberts, W.W., 1969, Ap. J. 158, 123
Rots, A.H., and Shane, W.W. 1975, Astron. Astrophys. 45, 25
Shu, F.H. 1985, this volume
Visser, H.C.D. 1980a, Astron. Astrophys. 88, 149
Visser, H.C.D. 1980b, Astron. Astrophys. 88, 159.

DISCUSSION

W.H. Waller: Have you looked at the radio-continuum distribution in the same region?

Allen: We have a rough continuum map of M83. It shows a ridge along the dust lane, possibly indicating that the density of relativistic gas reaches a maximum there.

In M51, a much better VLA map by Van der Hulst, Kennicutt and others shows thermal emission (that is, HII regions) on the outside of the spiral arms, and nonthermal emission on the inside, in agreement with our findings for M83.

Top: R.J. Allen welcomes participants to the new home of the Kapteyn Institute. CFD

Bottom: At conference dinner, clockwise: Terzides, unidentified, Dame, Kathryn Mead, Leisawitz, Bash, Twarog, Kutner, Cohen, Jelena Milogradov LZ

A SYSTEMATIC STUDY OF M81

Frank Bash
Department of Astronomy, University of Texas
Austin, Texas

ABSTRACT

We describe a series of observational studies of M81 which are presently underway. We hope that these studies will allow improved understanding of this one, simple, density-wave spiral galaxy.

RADIO MAPS

Bash and Kaufman (Ohio State) have observed M81 using the VLA in the radio continuum at wavelengths of 6 cm and 20 cm using configurations B, C and D, and in the polarization mode. At this time we have only preliminary results. The strongest source in M81 is the VLBI source at the center which has a flux density ranging from ~50 to ~100 mJy/beam-area depending on the frequency and configuration. Regions which are very bright on the Hodge and Kennicutt (1983) Hα plates and the central part show up also on the 20 cm B-configuration maps and the 6 cm D-configuration maps while essentially only the center can be seen at 6 cm on the B-configuration.

Perhaps the most interesting of our continuum maps up to this point is that made using the C-configuration at 20 cm. Figure 1 shows that map. The faint radio emission along the arms between the HII regions can now be seen. Although it is saturated in Figure 1, the central part shows a source northeast of the center whose peak brightness is about 0.8 mJy/beam area. The VLA beam here is 12".7 x 12".5 and the brightness of the central source is ~90 mJy/beam area.

In addition to the continuum study, Hine (Texas) and Rots (VLA) are making a HI, 21-cm line map of M81 with the VLA's D and B configurations.

HII-REGION RADIAL VELOCITIES

Levreault, Leisawitz and Bash have observed 15 HII regions in M81 and measured their radial velocities to an accuracy of typically better than 10 km s^{-1}. These data are partly reduced and they should allow us to check Bash and Visser (1981), who predict non-circular velocities of up to 20 km s^{-1}.

H. van Woerden et al. (eds.), The Milky Way Galaxy, 281–282.

STELLAR-POPULATION SYNTHESIS

Hilton and Bash are studying the surface photometry, in the U, B, V and I bands, and spectrophotometry, with about 10 Å resolution, of the spiral arms of M81 and NGC2841. This differs from the pioneering work of Schweizer (1976) in that it uses standard photometric bands, adds a longer-wavelength filter and adds spectra. We desire to determine the stellar populations of the arms of M81 which are consistent with the observed photometry and spectroscopy, and to contrast those populations with the arms of NGC2841. NGC2841 is a classical chaotic, non-density-wave galaxy (Kormendy and Norman, 1979).

Much of this work has been supported by NSF grant AST-8116403.

REFERENCES

Bash, F. N., and Visser, H. C. D. 1981, Astrophys. J., 247, 488.
Hodge, P. W., and Kennicutt, R. C., Jr. 1983, Astrophys. J., 267, 563.
Kormendy, J. and Norman, C. 1979, Astrophys. J., 233, 539.
Schweizer, F. 1976, Astrophys. J. Suppl. 31, 313.

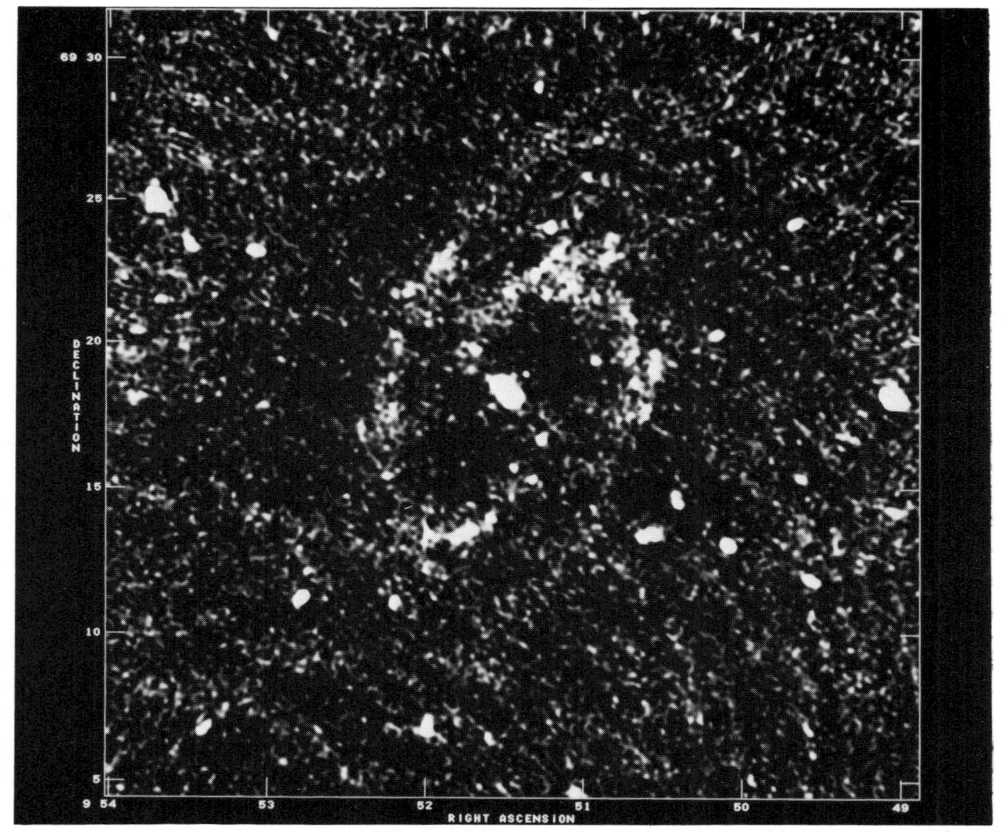

Fig. 1 - VLA radio map of M81 at 20 cm using the C-configuration. The intensity scale saturates at 1 mJy/beam-area; the noise is 60 μJy/beam-area.

DETERMINATION OF GALACTIC SPIRAL STRUCTURE AT RADIOFREQUENCIES

H. S. Liszt
National Radio Astronomy Observatory

ABSTRACT. We consider some results of attempts to trace the pattern of galactic spiral structure in HII regions, HI, and CO. There is really no adequate method available for solving this problem, a fact reflected in the lack of consensus regarding the "correct" spiral pattern. The newly-begun process of deriving galactic structure in CO seems to be recapitulating the history laid down by HI observers.

I. INTRODUCTION

We have at our disposal a variety of instruments capable of penetrating to the farthest reaches of the Milky Way, and use of this equipment for the purpose of deriving any large-scale galactic spiral structure is a major field of astronomical endeavor. The motivation for this work is strong and its goal a highly desirable one. Once having achieved it, we could combine very detailed observations of the physical state of the interstellar medium with a suitably detailed picture of the disk kinematics and dynamics to reveal the processes whereby the Galactic System evolves and is maintained over its lifetime.

Nonetheless, there exists a wide spectrum of views regarding how well we have done in deciphering galactic structure to date, or indeed, how well we shall be able to do in the future. The history of charting the galactic spiral pattern is a curious one, as structures have been found even when the observations employed were fallacious or when the spiral tracers "observed" do not exist in nature. One example of the latter phenomenon is the pattern derived for stellar rings (Figure 1) by Schmidt-Kaler and Isserstedt (see Schmidt-Kaler 1971), and another more recent example may be the "arm-like concentrations" found in weak-lined CO clouds in the outer Galaxy by Kutner and Mead (1981). Crampton (1971) has shown that stellar rings are not physical associations and Solomon, Stark, and Sanders (1983) have not been able to detect and confirm the vast majority of Kutner and Mead's sources.

H. van Woerden et al. (eds.), The Milky Way Galaxy, 283–300.

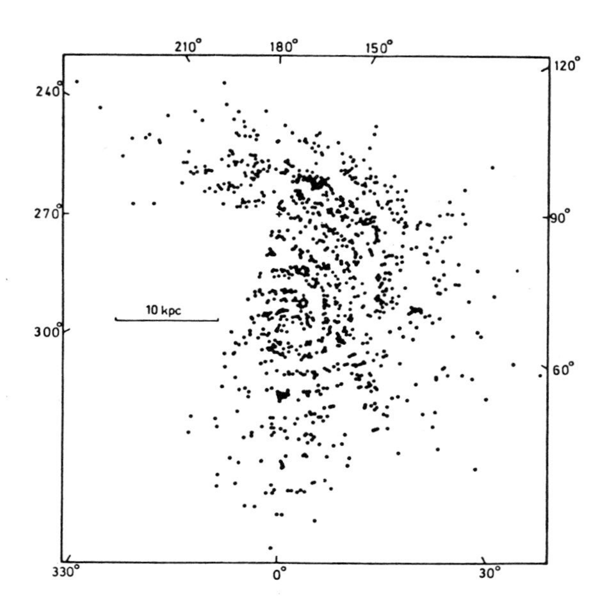

Figure 1. The spatial distribution of stellar rings from Schmidt-Kaler (1971). These objects are no longer believed to be real, but evince a clear spiral pattern.

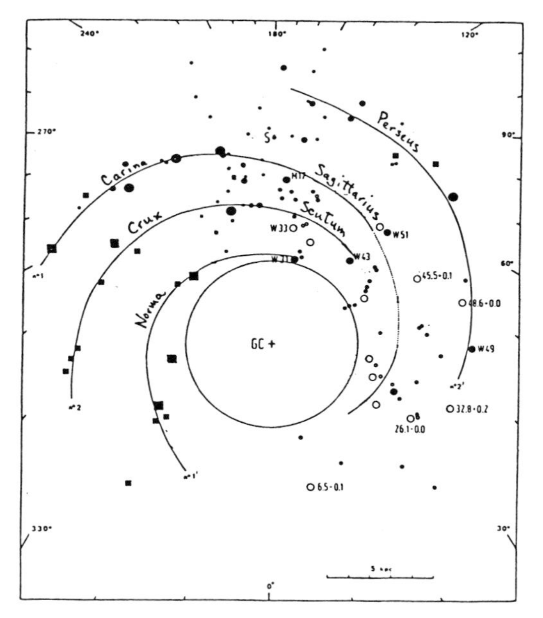

Figure 2. The spatial distribution of HII regions: data from Georgelin and Georgelin (1976) at longitudes above 60° and from Downes et al. (1980) elsewhere. The original spiral pattern drawn by the Georgelins is shown.

The problems inherent in tracing spiral patterns from within the galactic disk are not especially subtle, notwithstanding the fact that they are frequently ignored in greater or lesser degree. We lack perspective and often must settle for determination of a radial velocity and ambiguous kinematic distance when what we desire is an accurate heliocentric distance. All galactic tracers have an intrinsic velocity dispersion and they are often too widespread and confused in our observations to isolate the contribution of any given region or source. Only the mean circular-velocity field is known across the galactic disk, while significant perturbations of this motion occur frequently and especially in association with spiral arms. The purely kinematic patterns observed in HI and CO are sufficiently complicated that there is no consensus as to which loci in position-velocity space actually constitute single, connected features.

Here we summarize some of the many, many efforts which have been directed at deciphering galactic spiral structure. We concentrate on radiofrequency measurements and on tracers with some degree of kinematical information, neglecting the galactic continuum. We also neglect recent developments concerning the spectacular flaring, warping and corrugation of the outer galactic disk (Henderson, Jackson, and Kerr 1982, Kulkarni, Blitz, and Heiles 1982), and the peculiarities of the innermost regions (Burton and Liszt 1983), all of which will be discussed by others at this Symposium.

II. HII REGIONS

In the inner Milky Way, HII regions are especially useful spiral tracers because their kinematic-distance ambiguity may sometimes be resolved through use of absorption spectra, because they tag some molecular clouds by causing high CO-line temperatures, and (most importantly) because most everyone seems to expect them to show a clear grand design (but see below). Optically, of course, actual distances are available near the Sun and in the outer portions of the galactic disk.

Shown in Figure 2 is the inferred distribution of HII regions taken from Georgelin and Georgelin (1976:GG) at longitudes above 60° and from Downes et al. (1980:DWBW) elsewhere (a similar but more comprehensive diagram is given by Forbes 1983). The dataset of GG has been modified slightly by correcting several errors in resolving the kinematic-distance ambiguity as noted in the footnote to their paper and in the tables of Lockman (1979); as well, we have removed a few objects whose velocities are cited by Lockman as being highly unreliable. Plotted in the Figure are the original spiral patterns put forth by GG, and inspection (see also Figure 4 of DWBW) will show how substantially they must be altered to fit the newer northern dataset. No matter what pattern is fit to the data, very long segments of the supposed arms will be devoid of detected sources. As remarked by Forbes (1983), quite a few beads must have slipped off our galactic string.

Spiral features in Figure 2 have been labeled with their usual names, and it will be seen further on that the tangent points of the GG arms also appear in several other measures of large-scale galactic structure. One of these is significant perturbation of the observed maximum line-of-sight velocity. But this phenomenon, when itself taken as a signpost of spiral structure, implies that the pattern which has been derived is seriously defective. Essentially all the HII regions found at large distances are radio objects and have been placed in galactic perspective using the mean axisymmetric rotation curve. Unfortunately, this rather idealized function does not provide a detailed prescription for accurately locating any given region, and is especially inadequate near a spiral arm. All claims to the contrary aside, large-scale deviations from the mean pure circular motion will have a significant effect whenever we substitute kinematic for actual distances.

Even if large-scale perturbations did not exist, two further effects would need to be accounted for in dealing with radiofrequency sources. They are more properly located in a probabilistic manner, because their velocity dispersion of 5 km s^{-1} (Lockman 1979) and the line-of-sight velocity gradients due solely to rotation (varying between 0 and 20 km s^{-1} kpc^{-1} in the inner Galaxy) together imply typical distance errors (± one standard deviation) of about a kpc. It is also the case that the center of mass of an HII region-molecular cloud complex usually resides within the molecular material, which may have a velocity a few km s^{-1} different from that of any recombination lines. The sign of the difference probably depends on which side of the cloud first encountered any galactic-scale shock, HII regions on the closer side being more likely to be expanding toward us, and the overall effect is a systematic one depending on our particular viewing angle.

Problems encountered in using kinematic distances are taken seriously by Lockman (1979), who still concludes that the densest HII regions in the inner Galaxy lie in a two-armed spiral pattern (although ring models cannot be excluded entirely). The salient feature of these sources is their avoidance of certain regions of the longitude-velocity plane, particularly near the maximum expected line-of-sight velocity from 30°-50° and 310°-330° (a characteristic which is certainly not shared by the HII or CO discussed in Section III and IV). This aspect of the observations is insensitive to the assumed underlying kinematics, because it is so gross an effect and cannot be replicated by imposing a perturbed velocity field on an axisymmetric surface-density distribution which is too broad in galactocentric radius. It is, however, well fit by confining the HII regions to the two-armed spiral of Burton (1971) which specifies the Scutum and Sagittarius features in the North and their counterparts in Carina and Crux in the South. Kinematics similar to those of the HII regions are also exhibited by the 1720-MHz OH clouds mapped by Turner (1983).

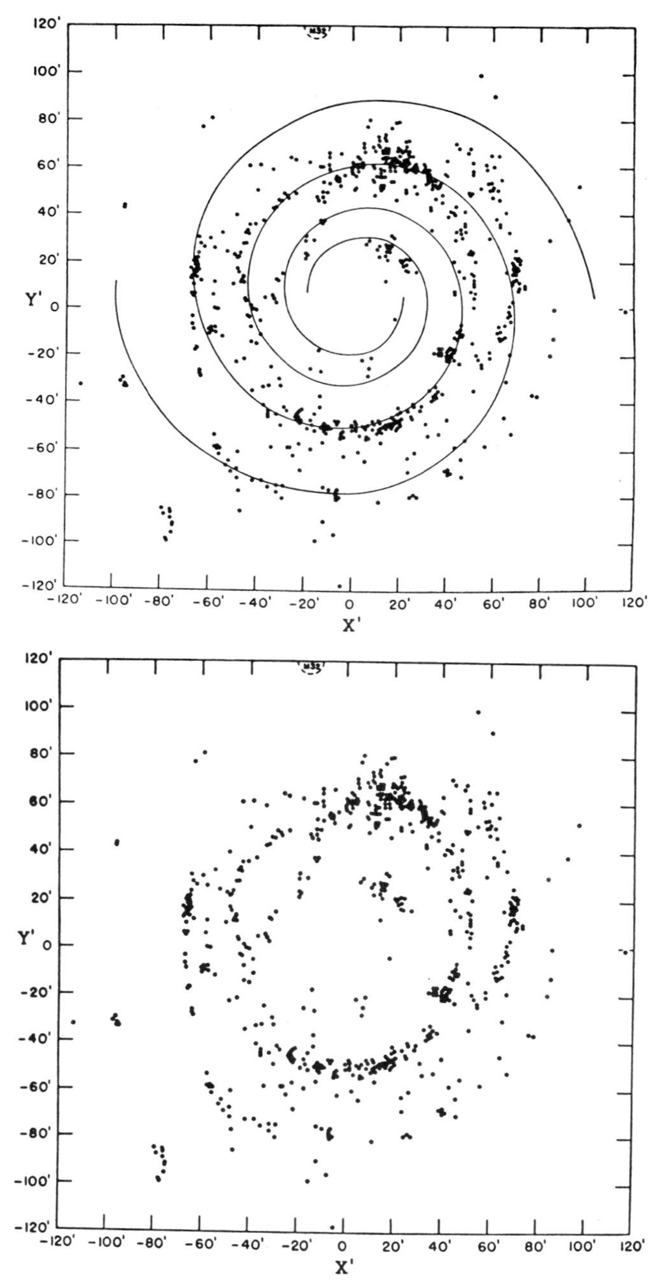

Figure 3. Deprojected positions of emission nebulae across the face of M31 (Figure 114 of Bok and Bok 1981), with and without a fitted logarithmic spiral. The Boks use this diagram to illustrate the difficulty of visual discrimination between spiral and ring distributions.

It is not actually necessary to exhibit an elegantly simple (and manifestly inconsistent) spiral as in Figure 2 in order to conclude that a definite pattern is present in the Galaxy. Much of the strongest evidence for such structure occurs in such a way that transformation from observational to Galaxy-centered spatial coordinates is difficult or impossible. One of the real embarrassments of Figure 2 is the proliferation of weaker sources near the Sun; most of these would not be detectable at large distances where the GG arms may appear to be well-defined. The usual excuse for such behaviour is the occurrence of a local arm or spur, and examples of the latter are often posited at larger distances to account for the presence of emission in regions which cannot be occupied by more major features. When too many of these minor features are present the overall grand design can become rather obscure.

How well should we have expected the HII regions to trace a grand design? Bok and Bok (1981) address this point, and their Figure 114 (due to Arp) is reproduced here as Figure 3. In those diagrams, the positions of 688 emission nebulae have been deprojected over the face of M31 and plotted with and without a fitted logarithmic spiral. The Boks stress that there is really no way to discern visually between ring structures and spiral arms, although the matter may clearly be forced by fitting one or the other distribution. This heuristic exercise must serve as a cautionary note for all the discussion here, as our perspective on the Milky Way is vastly inferior to that on Andromeda.

III. ATOMIC HYDROGEN

The most unambiguous and therefore strongest indicators of galactic spiral structure in the neutral gas species are perturbations of the circular velocity observed in the first and fourth longitude quadrants. These are shown for northern data (the HI survey of Westerhout 1976) in Figure 4, where the measured HI terminal velocity and the maximum projected line-of-sight velocity in the Burton-Gordon (1978) rotation curve are followed. The supposed tangent longitudes of the Scutum (30°) and Sagittarius (50°) features are accompanied by increases in velocity and the intervening or interarm region by a decrease of comparable magnitude; the perturbations occur over distances of order 800 pc across the line of sight. These fairly direct measures are usually (not always) taken as very strong constraints on possible spiral patterns, but our inability to observe similar effects in velocity away from the locus of sub-central points constitutes a serious obstacle in transforming the HI profile shapes into more "useful" information.

Another formidable obstacle is the longitude variation of integrated HI intensity, which exhibits only very minor deviations from the behaviour expected of a uniform axisymmetric gas distribution: in Figure 4 we show the results calculated for a constant density

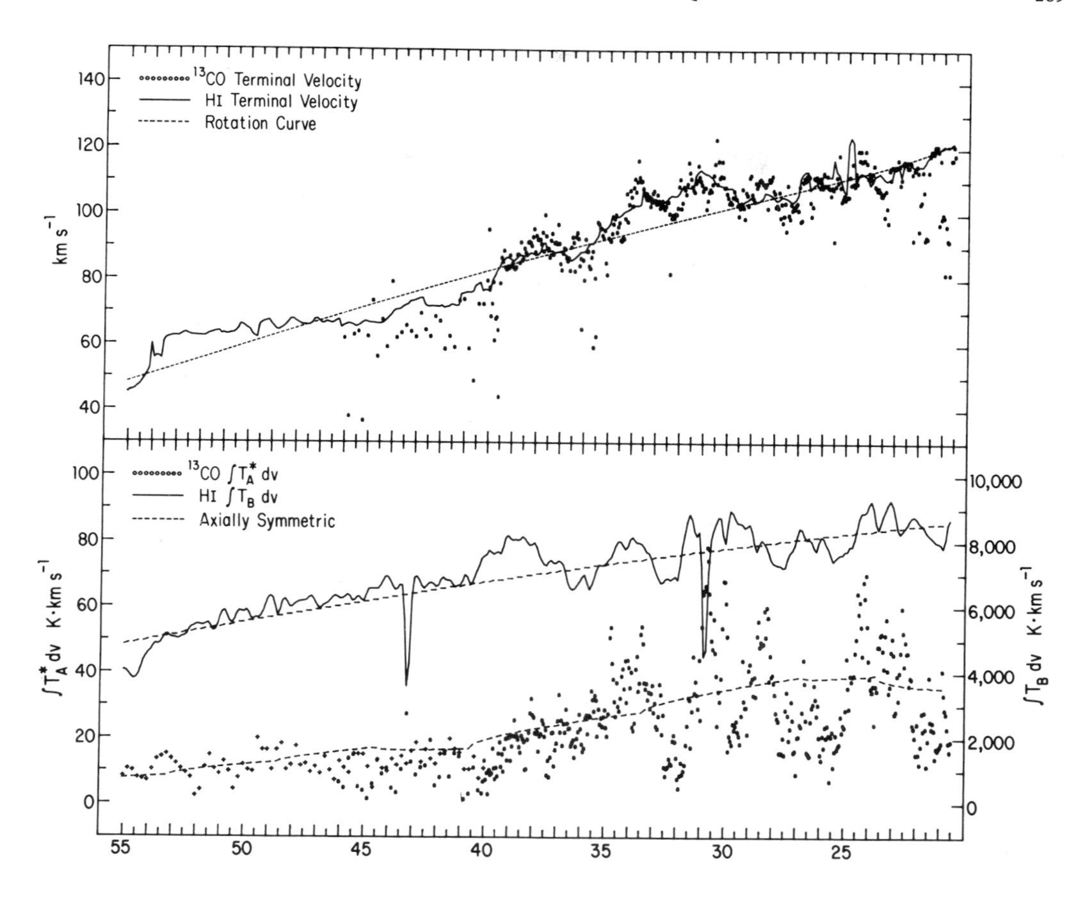

Figure 4. Upper panel: terminal velocities measured on HI (Westerhout 1976) and ^{13}CO (Liszt and Burton 1983) line profiles, and the maximum rotation velocity from Burton and Gordon's (1978) rotation curve. Lower panel: integrated intensities and the predictions of axisymmetric models. The HI model has constant density and temperature, the CO uses the usual intensity-abundance histogram derived from the data. The ^{13}CO data have been extended above 40° using scaled data of Burton and Gordon (1978).

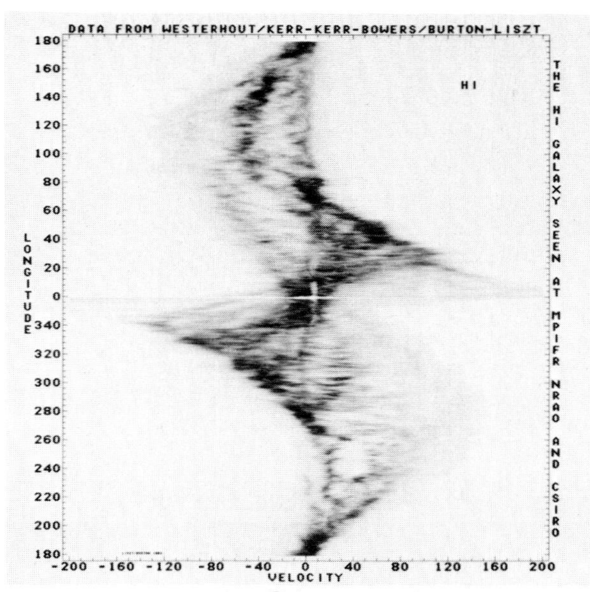

Figure 5. A grey-scale representation of HI profiles over the whole galactic equator with data of several observers, as indicated.

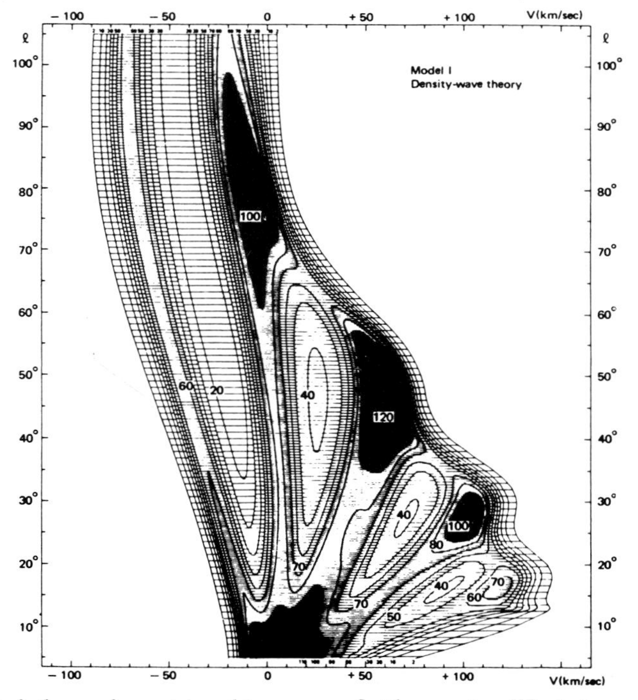

Figure 6. Model, schematic diagram of the main HI-intensity ridges in the first longitude quadrant from Burton (1974, 1971). The HI structures occur in CO as well.

0.38 H-nuclei cm^{-3} and spin temperature 125 K at R ⩾ 4 kpc. There is no observational peak corresponding to the tangent point of the Sagittarius Arm and very little for Scutum.

This is not to say that the HI is largely unstructured, for it certainly does display a high degree of ordering in ridges, loops, etc., as displayed in Figure 5, a presentation of the data over the whole galactic equator, and in Figure 6, where Burton's (1971, 1974) schematic model representation of the main intensity ridges is shown for the first quadrant. The model diagram was used to prove a fundamental point concerning HI intensity structures. They arise not from density enhancements at well-defined locations within an otherwise smooth distribution of gas in pure rotation, but from velocity perturbations such as we may observe more directly at the terminal velocity. The intensity enhancements do not represent added emission from extra material, but rather, emission that has been concentrated in certain velocity ranges at the expense of immediately adjacent portions of the spectrum. Portions of the spectrum corresponding to regions over which the line-of-sight velocity gradient is smaller--near the terminal velocity, and elsewhere if the gas motion is suitably perturbed--will have higher intensity than those in which it is large. Such an interpretation is of course entirely compatible with a nearly featureless run of integrated intensity with longitude. This circumstance was not self-evident when study of HI began, however, and its consequences for interpretation of CO (in which the same ridges appear) are still too often ignored.

We end this brief discussion of HI with a collage of some of the schematic spiral structures inferred to exist from HI observations, Figure 7 (the GG pattern and one other explained in Section V are included for comparison purposes). There is a remarkable variety of structure, especially considering that the observations are not in dispute at all, and surprisingly little consensus as to which aspects of the data should be weighted most strongly. The patterns shown represent only a few of those that might have been cited (see also Kerr 1970, Oort, Kerr, and Westerhout 1958, and Verschuur 1973), a veritable handbook of spiral anatomy occurring in the literature. There are few portions of either the longitude-velocity or inner galactic planes in which clear features have not been claimed to exist at some time.

IV. CARBON MONOXIDE

A. Integrated Intensities. While the kinematics of HI and CO are essentially identical, the run of their integrated intensities shows some important differences. The ^{13}CO data shown in Figure 4 (from Liszt and Burton 1983) have very strong peaks and troughs at several longitudes below 40°. Are these indicative of spiral arms? Only at 28°-31° can they be identified with intensity structures near the terminal velocity, while the rest occur in conjunction with emission concentrations occurring well below it (Figure 1 of Liszt and Burton

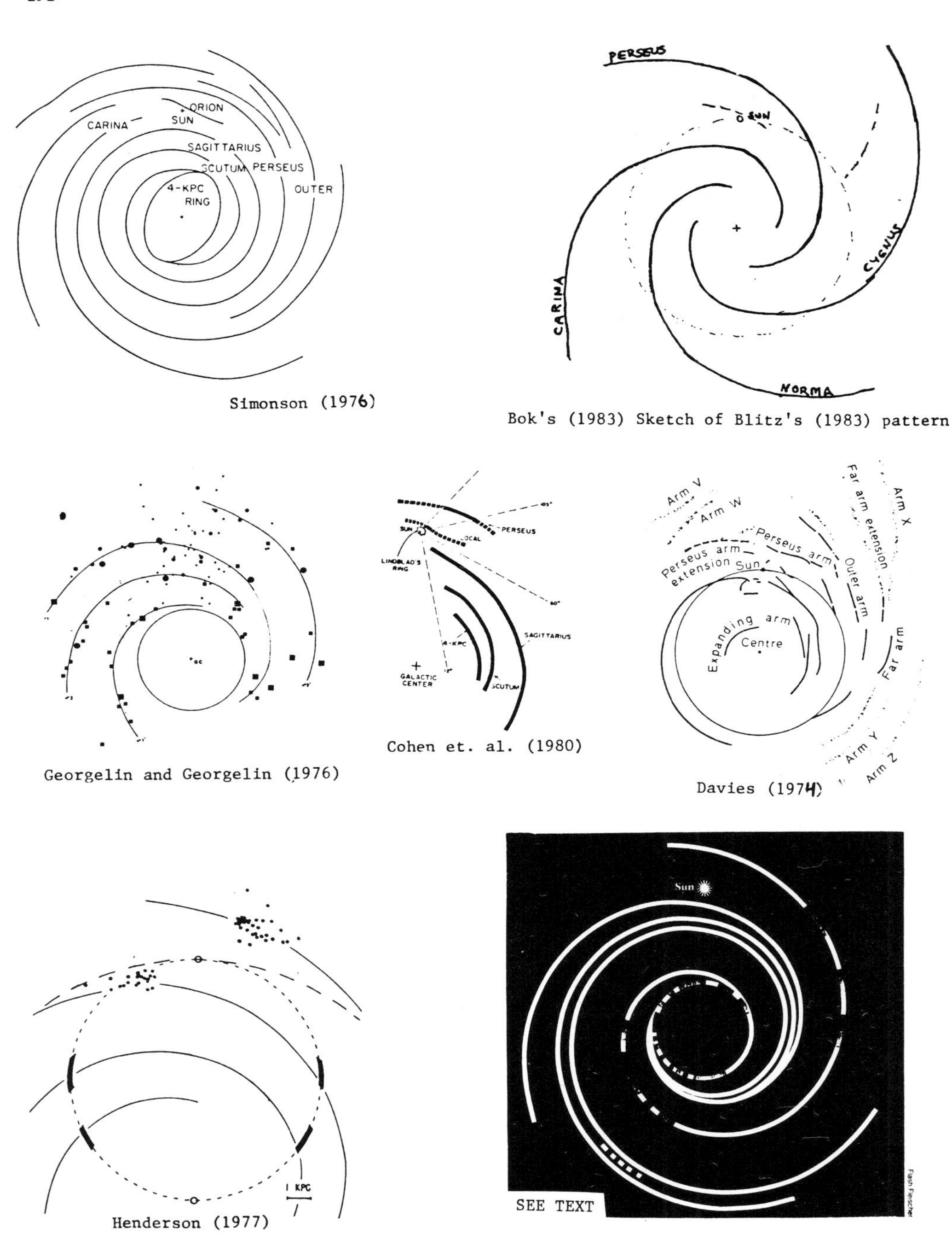

Figure 7. A few schematic spiral patterns inferred for the Milky Way. Except for the HII-region distribution (Georgelin and Georgelin 1976) and that in the lower right corner, all arise from consideration of essentially the same HI data.

1983 but see also Figure 8 here). What the intermediate-velocity features actually represent is not clear, but they are not readily related to tangent points of spiral arms. Indeed, the entire Sagittarius feature is as little evident in CO as in the HI.

The integrated CO intensity differs in another respect from that of HI. While no feature considered as a single cloud can by itself cause substantial variations in the total atomic-gas quantity, the largest molecular clouds probably do make a definite appearance in CO. Taking the characteristic sizes of the peaks in Scutum, about 0.7°, as indicating cloud surface areas $\leq 10^4$ pc^2, and using the usual very uncertain CO-intensity→H-column-density conversion factors, the inferred masses associated with the observed intensity structures are $2\text{-}7 \times 10^6$ M_{Sun}. The mass spectrum of interstellar clouds is frequently inferred to extend to or even beyond these values, and the larger clouds should be manifested in the run of integrated intensity.

B. Kinematics. The kinematics exhibited by HI and CO are actually quite similar, both at the terminal velocity and below it. The first point is made by Burton and Gordon (1978) and by Liszt and Burton (1983), but the full extent of the congruent behaviour in HI and CO is revealed most clearly in the more fully sampled CO data of Cohen *et al.* (1980:CCDT). CCDT noted that the intensity loops and ridges used to define spiral structures in the HI are also visible in CO (although the near portion of the Sagittarius feature is very weak, even after summing over latitude), and we have reproduced their Figure 2 in slightly modified form to stress this fact (Figure 8). Actually, their rendition of the "4-kpc arm" feature is not exactly correct (see Bania 1980 and Cohen and Davies 1976), but the claimed kinematic similarities are unassailable.

CCDT advanced the argument, reminiscent of early interpretations of HI, that the CO ridges and loops represent density-based enhancements of the molecular intensity and, further, that their appearance in the CO implies that most molecular clouds are confined to a few spiral arms: because of its lower velocity dispersion, increased clumpiness and higher arm-interarm contrast, the molecular-cloud ensemble might yield CO spectra which are less susceptible to the kinematic perturbations plaguing HI. Reader, take note. If the observed structure can only arise in the HI as the result of kinematic effects, and if this same structure appears in the CO, then it must be the case that the same kinematic effects are present in the CO. Indeed, Liszt and Burton (1981) show that introduction of a perturbed velocity field creates a substantial degree of intensity structure in CO emission from the molecular-cloud ensemble and induces a sizable apparent arm-interarm contrast when none actually exists. The discussion of this matter in Liszt and Burton (1981) is too long to bear repetition here. Instead, we list a few of the problems which arise when the galactic spiral is inferred too naively (these remarks are not strictly limited to interpretation of CO).

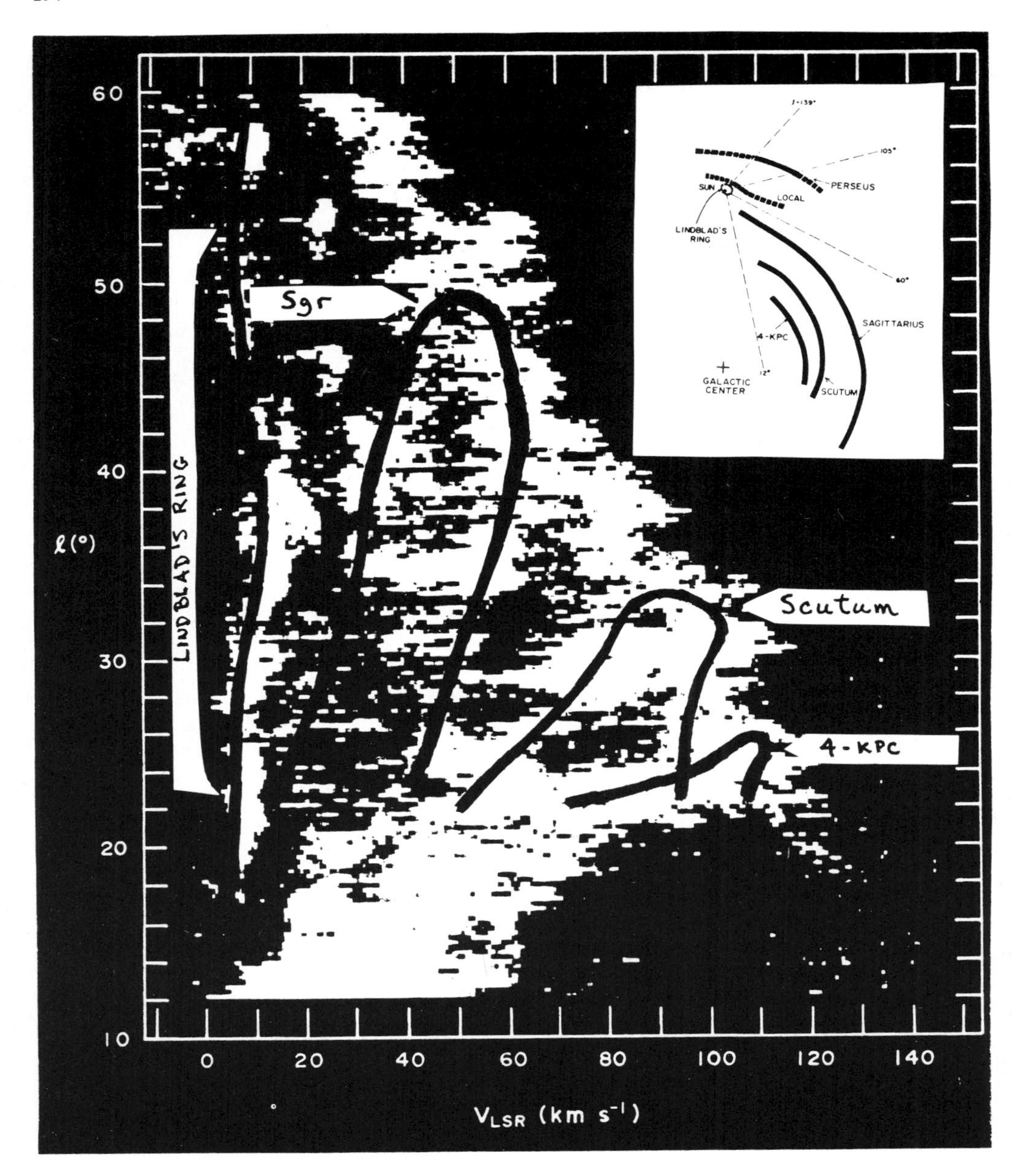

Figure 8. Grey-scale representation of ^{12}CO intensity, integrated over latitude, from Cohen et al. (1980). Major features are identified and sketched schematically in the insert.

(1) Neglect of velocity perturbations may cause the inferred spiral pitch angle to be too large, necessitating the supposition of more arms than are actually present. With non-circular or perturbed motions, some part of the velocity separation between front and back portions of a spiral pattern arises from streaming, etc. Without it the same velocity difference can arise only by placing the two segments at different galactocentric radii, and the arms therefore unwind too rapidly with spiral phase angle.

(2) The positions of the arm tangents will be mis-estimated because of possible offsets in spiral phase between maximum line-of-sight density and associated velocity perturbations.

(3) The inferred arm-interarm contrast will be too large, leading to the creation of spurs and other complications to account for emission which cannot be straightforwardly associated with the major arm pattern.

Interpretation of the molecular data is and will remain controversial. One clear feature of the northern data in all surveys is the unmistakable presence of the terminal-velocity intensity ridge at all longitudes above about 20°, continuing past the supposed Scutum-arm tangent at longitudes 30°-50° and past the Sagittarius feature over the remainder of the first quadrant into Cygnus. Such behaviour is not exhibited by the HII regions, as detailed earlier. This situation may be taken by one observer as evidence that a substantial portion of the molecular-cloud emission arises outside spiral arms and by another as indicative of where to place various spurs and other spiral anatomy, with all molecular emission confined to a relatively few well-defined features in space. Some ordering of the molecular-cloud ensemble is clearly necessary, but the extent to which one engages in this process is very much a matter of personal preference (supported to varying degrees by the actual data).

There is one further caveat regarding demonstration of spiral patterns in the molecular material. Estimates of the number of galactic molecular clouds range from 3000 to more than 30 000 (see Liszt and Burton 1981). The subset of these clouds with "well-determined", larger (> a few kpc) distances numbers at most a few dozen, as it is limited to those which can be shown to be physically associated with HII regions whose kinematic distance ambiguity is resolved. The reader is cautioned to view with suspicion diagrams like Figure 2 in which only a handful of clouds are placed in perspective, and to question the placement of even those sources!

V. CO NORTH VS. SOUTH AND OTHER REFLECTIONS

CO data are now available for the fourth longitude quadrant, and Robinson *et al.* (1983) have performed a naive decomposition of the combined north-south data into kinematic loops and their associated kinematic spiral arms. Their results are as different from those of CCDT as are many of the HI patterns from each other, even though no new northern data are introduced. Still, the geometrical parameters of their four-armed pattern are claimed to be in only mild disagreement with those of the Georgelins or of Henderson (1977). The spiral pattern which occurs in the lower right-hand corner of Figure 7 appears in the May issue of Science 83 magazine attributed to McCutcheon and collaborators, but bears no resemblance to the geometry discussed by Robinson et al. (1983), which has McCutcheon as a co-author.

Molecular mapping is just now beginning over about half the sky, and perhaps it is too early to demand that a coherent picture of the molecular clouds emerge. But it is not too early to demand, in general, that the utmost care be taken in interpretation of the data. Essentially the same HI data have now been decomposed literally dozens of times into loops and arms, and HII regions placed in galactic perspective, without achieving anything like the degree of consensus necessary for progress toward the ultimate goal of such work. But some lessons have been learned from this process and they should be heeded even when they complicate and make less straightforward our interpretational efforts. Kinematic effects cannot be ignored in any phase of the analysis; more objective methods must be developed to gauge the reality, importance, and connectedness of kinematic loops and other features; more rigorous tests must be made of the validity of decomposition of the observations into major structures. Perhaps our expectations should be tempered. In the meantime, the question of galactic spiral structure remains, as over the past 30-year history of HI, open.

REFERENCES

Bania, T. M.: 1980, Astrophys. J., 242, p. 95.

Bok, B. J.: 1983, Henry Norris Russell Lecture, Astrophys. J., 273, p. 411

Bok, B. J., and Bok, P. F.: 1981, THE MILKY WAY (Cambridge MA: Harvard University Press).

Burton, W. B.: 1971, Astron. Astrophys., 10, p. 76.

Burton, W. B.: 1974, in K. I. Kellermann and G. L. Verschuur (eds.), GALACTIC AND EXTRAGALACTIC RADIO ASTRONOMY (New York: Springer), p. 82.

Burton, W. B., and Gordon, M. A.: 1978, Astron. Astrophys., 63, p. 7.

Burton, W. B., and Liszt, H. S.: 1983, Astron. Astrophys. Supp., 52, p. 63.

Burton, W. B., and Shane, W. W.: 1970, in W. Becker and G. Contopoulos (eds.), THE SPIRAL STRUCTURE OF OUR GALAXY (Dordrecht: Reidel), p. 397.

Cohen, R. J., and Davies, R. D.: 1976, M.N.R.A.S., 175, p. 1.

Cohen, R. S., Cong, H., Dame, T. M., and Thaddeus, P.: 1980, Astrophys. J., 239, L53.

Crampton, D.: 1971, Pub. Dom. Astrophys. Obsy., 13, p. 427.

Davies, R. D.: 1974, in L. N. Mavridis (ed.), STARS AND THE MILKY WAY SYSTEM (New York: Springer), p. 124.
Downes, D., Wilson, T. W., Bieging, J., and Wink, J.: 1980, Astron. Astrophys. Supp., 40, p. 379.
Forbes, D.: 1983, in W.L.H. Shuter (ed.), KINEMATICS, DYNAMICS AND STRUCTURE OF THE MILKY WAY (Dordrecht: Reidel), p. 217.
Georgelin, Y. M., and Georgelin, Y. P.: 1976, Astron. Astrophys., 49, p. 57.
Henderson, A. P.: 1977, Astron. Astrophys., 58, p. 189.
Henderson, A. P., Jackson, P. D., and Kerr, F. J.: 1982, Astrophys. J., 263, p. 116.
Kerr, F. J.: 1970, in W. Becker and G. Contopoulos (eds.), THE SPIRAL STRUCTURE OF OUR GALAXY (Dordrecht: Reidel), p. 17.
Kulkarni, S. R., Blitz, L., and Heiles, C.: 1982, Astrophys. J., 252, p. 481.
Kutner, M. L., and Mead, K.: 1981, Astrophys. J. 249, p. L15.
Liszt, H. S., and Burton, W. B.: 1981, Astrophys. J., 243, p. 778.
Liszt, H. S., and Burton, W. B.: 1983, in W.L.H. Shuter (ed.), KINEMATICS, DYNAMICS AND STRUCTURE OF THE MILKY WAY (Dordrecht: Reidel), p. 135.
Lockman, F. J.: 1979, Astrophys. J., 232, p. 761.
Oort, J., Kerr, F. J., and Westerhout, G.: 1958, M.N.R.A.S., 118, p. 379.
Robinson, B. J., McCutcheon, W. H., Manchester, R. N., and Whiteoak, J. B.: 1983 in W. B. Burton and F. P. Israel (eds.), SURVEYS OF THE SOUTHERN GALAXY (Dordrecht: Reidel), p. 1.
Schmidt-Kaler, T.: 1971, in L. N. Mavridis (ed.), STRUCTURE AND EVOLUTION OF THE GALAXY (Dordrecht: Reidel), p. 85.
Simonson, S. C.: 1976, Astron. Astrophys., 46, p. 261.
Solomon, P. M., Stark, A. A., and Sanders, D. B.: 1983, Astrophys. J., 267, p. L29.
Turner, B. E.: 1983 in W.L.H. Shuter (ed.), KINEMATICS, DYNAMICS AND STRUCTURE OF THE MILKY WAY (Dordrecht: Reidel), p. 171.
Verschuur, G. L.: 1973, Astron. Astrophys., 27, p. 73.
Westerhout, G.: 1976, Maryland-Bonn Galactic 21-cm Survey (College Park: University of Maryland).

DISCUSSION

W.L.H. Shuter: Are the wiggles along the tangent lines really due to a spiral field, or could they be related or connected in some way to the scalloping at the outer edge of the disk?

Liszt: In the context of models that have streaming motions, or perturbations of the circular motions associated with real spiral arms, those bulges in the (run of) terminal velocity (with longitude) are taken as direct evidence for the existence of spiral arms. The terminal velocity is, however, quite uniform on small scales; it varies on scales of 800 pc or so across the line of sight.

H.C. van de Hulst: Since this is partly a historical meeting, I wish to point out that a number of your cautioning remarks can be found back almost literally in papers written around 1953.

Liszt: I am not saying my cautioning remarks are new. The unfortunate thing is that they are not always taken into account in CO papers written in 1983.

H. van Woerden: It is true that in some of the early papers attention was drawn to possible effects of non-circular motions. However, in these early papers such effects were generally considered to be minor. It was not until the late sixties (Burton 1966, Bull. astr. Inst. Netherl. 18, 247; Shane and Bieger-Smith 1966, Bull. astr. Inst. Netherl. 18, 263; Burton and Shane 1970, IAU Symp. 38, p. 397; Burton 1971, Astron. Astrophys. 10, 76; but see also Kerr 1962, Mon. Not. Roy. astr. Soc. 123, 327) that Burton and Shane demonstrated how seriously existing maps of the spiral structure of our Galaxy might have been distorted - or indeed counterfeited - by the effects of large-scale non-circular motions. (This comment was not made at the Symposium, but added later, for the sake of historical fairness - Editor.)

M.L. Kutner: You are using the lack of uniqueness of derived spiral patterns to discourage certain avenues of investigation. However, anyone who models anything complicated must learn to deal with non-uniqueness. Instead of looking for rules how to decompose the (l,V) plane, I suggest the procedure should be as follows: make a model of the Galaxy, including the kinematics, and then predict an (l,V) diagram.

Liszt: That is what I suggest in the end: do linear density waves, do nonlinear density waves, do two-armed spiral shocks, etc. - drive them, shear them, but at least start out with a model, and do not draw connecting lines in that (l,V) plane first.

A.A. Stark: My opinion is somewhat less pessimistic. We know there are external spiral galaxies in great variety. In our Galaxy we can identify arms and interarm regions - it may just be impossible to connect them up.

Liszt: I wish that were true. When it comes to a vote, I probably agree with your arms and interarms - but what do you do if I draw somebody else's pattern in this plane?

Stark: But surely, in the (l,V) diagram you can distinguish arm and interarm regions, and study objects in these regions, and forget about a global pattern.

Liszt: But what you call an arm, may not be somebody else's arm. And the philosophy in this game is: If you don't see what I see, you're blind. Of course, I agree that you have a valid way of defining arms.

And I agree with your interarm regions, they are ... between arms ... if such exist. But various observers have not come to the same conclusions - it was not as obvious to them what an interarm region was as it was to you, looking at the data.

Stark: There is one arm that everyone agrees on, and that is the 4-kpc arm, or 3-kpc arm, depending on what .... (Laughter).

Liszt: Right, there is one arm that everybody agrees on, and there may be spiral arms, hands, legs and feet. ----- In our Galaxy, the 4-kpc arm is one of the most enigmatic features, and it is associated with inner-Galaxy phenomena which probably do not propagate into the disk in their full glory.

F.J. Kerr: You said that we do not have guidelines. In fact, one guideline has been used by many people, namely they look for a very regular spiral pattern. However, this guideline must be wrong, as no other galaxies are so regular.

Liszt: We can learn from Dr. Elmegreen's review of external galaxies what it means to have a grand design: it may be grand, but it is not always as pretty as we like.

(The following remarks were made in the Discussion after the next paper, by B.G. Elmegreen, but their contents fit best here - Editor.)

F.J. Kerr to Elmegreen: You spoke of Gould's Belt-type complexes. A striking thing about Gould's Belt is its inclination to the galactic plane. We do not see such features in other galaxies.

B.G. Elmegreen: The shingles described by Schmidt-Kaler and House have a similar inclination. This phenomenon is not understood.

Kerr: But there is nothing similar in the CO known sofar.

Elmegreen: The classical Gould Belt is rather thin, with a large velocity dispersion. Its mass is too low for self-gravitation. It is unclear how it is held together, for it is much older than would follow from its width and velocity dispersion. The thinness and inclination may partly be a result of obscuration by dust.

R.S. Cohen: Liszt is not correct when he cites our 1980 paper as an example of yet another grand-design spiral model. The lines drawn in that paper are definitely not intended to be a grand-design model, they were drawn long ago by Burton on the basis of 21-cm data and were used in our paper simply to illustrate the clarity with which CO outlines previously identified HI features. One of the features in our diagram is the Cygnus Rift, a nearby naked-eye object. The inner Galaxy is much more complex, and I completely agree with Liszt that it is a mistake to fit logarithmic spirals naively. Nonetheless, I do think we see arm regions and interarm regions: there are regions rich in CO and others poor in CO. The confusion is how to connect things into a grand design, and I do not think we know how to do that yet.

B.G. Elmegreen: Part of the confusion about the Sagittarius Arm is a result of the clumping of CO clouds. Between the M16-M17 complex and the next clump in the Sgr Arm there is a huge gap. This is just due to the beady structure of spiral arms, which we can now study in our Galaxy, because we have distances to these clumps.

Liszt: Your spiral arcs are lines of almost constant velocity in the (l,V) plane, they are loops and ridges in the (l,V) plane placed in perspective in the galactic plane through an assumed rotation curve. And in fact I don't think many of those distances are really defensible - in many instances you have HII regions that are 1°.5 or 2° away, which you associate with a CO cloud to put it at a certain distance.

W.H.M. McCutcheon: One thing evident in the data of our Southern CO survey is the very clumpy emission along the run of terminal velocities. The holes are large and cannot be the result of non-circular velocities or streaming motions. The CO (l,V) diagram supports a gas distribution in large-scale features which undoubtedly consist of spurs and bifurcations, as well as segments of spirals. We do not want to claim a neat spiral pattern, but rather emphasize the large-scale, quasi-continuous features. An alternative view, expressed in a paper of which Liszt was co-author and suggesting that the emission can be accounted for by a random distribution of clouds, is not supported by our data.

At conference dinner, clockwise: Lynden-Bell, Burton, Jog, Mirabel, and Liszt looking diffidently at Ostriker LZ

GIANT CLOUDS AND STAR-FORMING REGIONS AS SPIRAL-ARM TRACERS

Bruce G. Elmegreen
Department of Astronomy, Columbia University, New York 10027

ABSTRACT

A variety of observations suggest that clouds of 10^6-10^7 $M_\odot$ and extended regions of star formation are the best tracers of spiral structure in the Milky Way.

OB associations such as Ori OB1, Per OB2, M16, and M17 have been used as spiral-arm tracers in the Milky Way since Morgan, Sharpless and Osterbrock (1952) and Morgan, Whitford and Code (1953) first delineated the nearby structure. The main spiral arms of our Galaxy are traced best, however, by larger regions of star formation (Mezger 1970). W42, W44, W47, and W51, for example, are among the giant HII regions that delineate the Sgr Arm. These giant regions are nearly invisible at optical wavelengths.

Studies of other galaxies indicate that the largest scales for cloud and star formation in the Milky Way should be between 1 and 4 kpc (Elmegreen and Elmegreen 1983, hereafter "EE"). Regions of this size should be the best spiral tracers. McGee and Milton (1964), for example, mapped 29 giant HI clouds (10^7 $M_\odot$) along 4 spiral arms in the outer part of the Milky Way. Such large HI-emission clumps were also observed by Burke, Turner and Tuve (1964), Kerr (1964), and others after the first 21-cm line surveys were available. Equally large structures are present in molecular emission. Wouterloot (1981) found 2 large (1-kpc) OH complexes in the Perseus Arm, and Dame et al. (1983) found 19 CO clouds with masses greater than 5×10^6 $M_\odot$ in the Sagittarius Arm. Other evidence for extremely large clouds comes from extinction surveys by Neckel (1967), Fitzgerald (1968) and Lucke (1978).

Clouds with masses in excess of 10^6 $M_\odot$ produce giant star complexes (300-pc scales) and conglomerates of clusters and OB associations. Such regions have been discussed by Efremov (1978), Shevchenko (1979) and others (see EE). A star complex typically contains one or more active sites of star formation, a few older clusters, and a large envelope of Cepheid variables and red supergiant stars. Such coherent large-scale

H. van Woerden et al. (eds.), The Milky Way Galaxy, 301–302.

star formation is probably responsible for the origin of comoving stellar groups (Eggen 1964). The most active conglomerate of star formation found locally is Gould's Belt. Similar regions elsewhere in the Galaxy have been discussed by Schmidt-Kaler and Schlosser (1973). Other examples are the giant clumps of A-type stars in the Carina Arm (Bok 1964; Maurice 1983).

A galaxy like ours produces new stars by forming enormous HI clouds, which then fragment and collapse into molecular clouds and clusters of star clusters (a family of 5 nearby clusters has been identified by Lyngå and Wramdemark 1983). The origin of the largest clouds is uncertain, but the cloud masses and internal motions suggest that they are weakly self-gravitating. They may be the result of Jeans-type instabilities in either the ambient interstellar medium (Elmegreen 1979; Cowie 1981; EE; Jog and Solomon 1983) or in large swept-up gas shells (Elmegreen 1982).

REFERENCES

Bok, B.J.: 1964, in IAU Symposium Nr. 20, ed. F.J. Kerr and A.W. Rodgers, Australian Acad. of Sci., p. 147
Burke, B.F., Turner, K.C., and Tuve, M.A.: 1964, in IAU Symposium Nr. 20, ed. F.J. Kerr and A.W. Rodgers, Australian Acad. of Sci., p. 131
Cowie, L.L.: 1981, Astrophys. J. 245, 66
Dame, T.M., Elmegreen, B.G., Cohen, R.S., and Thaddeus, P.: 1983, in prep. for Astrophys. J.
Efremov, Yu.N.: 1978, Sov. Astr. Lett. 4, 66
Eggen, O.J.: 1964, in IAU Symposium Nr. 20, ed. F.J. Kerr and A.W. Rodgers, Australian Acad. of Sci., p. 10
Elmegreen, B.G.: 1979, Astrophys. J. 231, 372
Elmegreen, B.G.: 1982, in Submillimeter Wave Astronomy, eds. J. Beckman and J. Phillips, University of Cambridge, Cambridge
Elmegreen, B.G., and Elmegreen, D.M.: 1983, Monthly Notices Roy. Astr. Soc. 203, 31, (EE)
Fitzgerald, M.P.: 1968, Astron. J. 73, 983
Jog, C.J., and Solomon, P.M.: 1983, preprint
Kerr, F.J.: 1964, in IAU Symposium Nr. 20, ed. F.J. Kerr and A.W. Rodgers, Australian Acad. of Sci., p. 81
Lucke, P.B.: 1978, Astron. Astrophys. 64, 367
Lyngå, G., and Wramdemark, S.: 1983, preprint
Maurice, E.: 1983, private communication
McGee, R.X., and Milton, J.A.: 1964, Austr. J. Phys. 17, 128
Mezger, P.G.: 1970, IAU Symposium Nr. 38, eds. W. Becker and G. Contopoulos, Reidel, Dordrecht, p. 107
Morgan, W.W., Sharpless, S., and Osterbrock, D.: 1952, Astron. J. 57, 3
Morgan, W.W., Whitford, A.E., and Code, A.D.: 1953, Astrophys. J. 118, 318
Neckel, T.: 1967, Landessternwarte Heidelberg-Königstuhl, Veröffentl. 19, 1
Schmidt-Kaler, Th., and Schlosser, W.: 1973, Astron. Astrophys. 25, 191
Schevchenko, V.S.: 1979, Sov. Astr. 23, 163
Wouterloot, J.: 1981, Ph. D. Dissertation, Leiden

THE LARGEST MOLECULAR COMPLEXES IN THE FIRST GALACTIC QUADRANT

T.M. Dame, B.G. Elmegreen, R.S. Cohen, and P. Thaddeus
Goddard Institute for Space Studies and Columbia University
New York City

The CO emission within a few kiloparsecs of the Sun is dominated by a small number of very large molecular complexes, including those associated with the Orion Nebula (Thaddeus 1982), M16 and M17 (Elmegreen, Lada, and Dickinson 1979), and NGC7538 (Cohen et al. 1980). These complexes have masses from several 10^5 to 10^6 $M_\odot$ and are generally very well-defined objects. They are also well endowed with HII regions, stellar clusters and associations, masers, and other Population-I objects whose distances can be measured. The complexes are thus valuable probes of the large-scale structure of the Galaxy.

We have used the Columbia University CO survey of the first galactic quadrant (Cohen et al. 1980, Dame 1983) to determine the locations and physical properties of the largest molecular complexes in the inner Galaxy. Masses for the complexes were determined from their velocity-integrated CO luminosities by assuming a proportionality between integrated ^{12}CO emission and H_2 column density: $N(H_2)/W(CO) = 2 \times 10^{20}$ cm^{-2} K^{-1} km^{-1} s (Lebrun et al. 1983). Distances to most of the complexes were determined kinematically, the distance ambiguity being resolved using a variety of methods; for most complexes several methods were applicable and gave consistent results. Within the range of our survey (ℓ=12°-60°) 19 complexes were detected with masses greater than 5×10^5 $M_\odot$. We estimate that roughly 100 such complexes exist in the Galaxy within the solar circle.

As figure 1 shows, the largest complexes delineate the Sagittarius Arm over more than 120° of galactocentric azimuth with remarkable clarity -- probably better in fact than any other Population-I tracer. The 15 large complexes which we identified in the arm have a total mass of 16×10^6 $M_\odot$ and an average spacing of about 1 kpc, comparable to the spacing on the regular strings of H II regions observed in many external spirals.

Although the complexes interior to the Sagittarius Arm do not reveal a very clear spiral pattern, this appears to be largely due to the fact that these complexes mainly cluster on the near side of the

H. van Woerden et al. (eds.), The Milky Way Galaxy, 303–304.

Galaxy. This apparent asymmetry may be partially due to the fact that distant clouds are more difficult to detect, but the clarity of the Sagittarius Arm out to a distance of 14 kpc suggests that this is not the complete explanation. It is worth noting that, if such an asymmetry exists in the so-called "molecular ring" region, then the total CO luminosity and subsequently molecular mass of the region has been overestimated by the axisymmetric models used to analyze CO-survey data (e.g., Cohen and Thaddeus 1977, Burton and Gordon 1978).

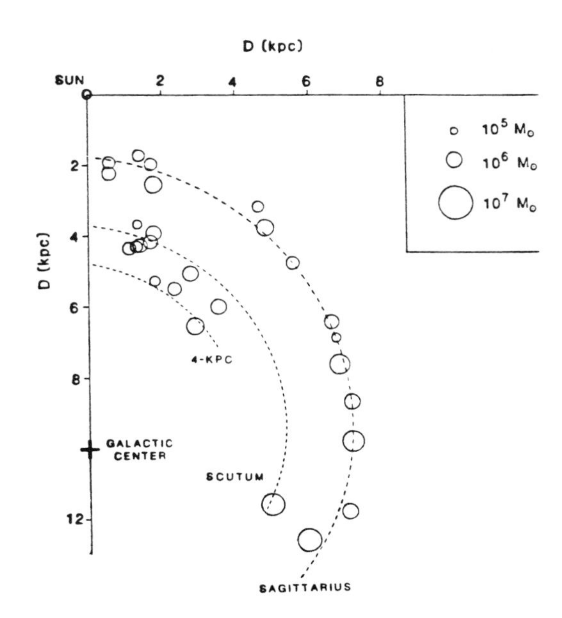

Figure 1. The locations in the galactic plane of the molecular complexes we have identified with masses > 10^5 $M_\odot$. The dashed lines are logarithmic spirals: the Sagittarius spiral is a least-squares fit with a pitch angle of 5°.3; the Scutum and 4-kpc spirals are not fitted, but are taken from the 21-cm analysis of Shane (1972).

REFERENCES

Burton, W.B., and Gordon, M.A.: 1978, Astron. Astrophys. 63, p. 7.
Cohen, R.S., and Thaddeus, P.: 1977, Ap.J. 217, p. L155.
Cohen, R.S., Cong, H., Dame, T.M., and Thaddeus P.: 1980, Ap.J. 239, p. L53.
Dame, T.M.: 1983, Ph.D. dissertation, Columbia University.
Elmegreen, B.G., Lada, C.J., and Dickinson, D.F.: 1979, Ap.J. 230, p. 415.
Lebrun, F., et al.: 1983, Ap.J., in press.
Thaddeus, P.: 1982, Ann. N.Y. Acad. Sci. 395, p. 9.
Shane, W.W.: 1972, Astron. Astrophys. 16, p. 118.

INTERPRETATION OF THE APPARENT ANOMALIES OF THE GALACTIC STRUCTURE

T. Jaakkola[1], N. Holsti[1], and P. Teerikorpi[2]
[1]Observatory and Astrophysics Laboratory,
University of Helsinki
[2]Observatory, University of Turku

In maps of the galactic structure based on the kinematical method (Fig. 1) several systematically heliocentric anomalies are found: 1. Assuming purely circular motion, the spiral arms are more tightly wound and the extent of neutral hydrogen is smaller in the northern galactic hemisphere than in the southern one. 2. With separate rotation curves for the north and the south, the arms become anomalously circular. 3. Consequently, there is a striking discrepancy with the stellar spiral structure. 4. There are long straight portions in the arms pointing towards the Sun. 5. There are abrupt knee-like features in the south. 6. Some arms seem to affect the structure of other, outer arms. 7. Conspicuously strong curvature of the arms is found in the north. 8. The HI-density is enhanced at symmetric longitudes on the far side. 9. With the northern rotation model HII-regions and HI avoid the southern tangential circle. 10. The Perseus Arm is displaced at $l = 180^{\circ}$.

These features are not random or small-scale fluctuations, nor can they be explained by a general expansion of the Galaxy or by a local outward motion. The systematic and heliocentric character of the anomalies indicates the presence of a non-Dopplerian, distance-dependent effect superposed on the kinematical line-shift data. A simple model involving a non-velocity redshift field within the Galaxy, enhanced within the spiral arms, is presented. The effect increases the positive velocity shifts and reduces the negative ones, with corresponding distortions in kinematically produced spatial maps.

Using this model and assuming that the spiral arms are not actually distorted, numerical simulation of the apparent spatial structure of the Galaxy reproduces all the anomalous features listed above (Fig. 2). The enhanced effect within the spiral arms causes the features 4, 5, and 6, and strengthens the other features. The continuous disk lineshift effect explains features 1 and 7-10. The reason for features 2 and 3, which have presented a major unsolved problem of galactic research, is an erroneous linking of a northern arm with an inner southern arm, as provoked at $l = 0^{\circ}$ by the distortion effect in the kinematical maps.

H. van Woerden et al. (eds.), The Milky Way Galaxy, 305–307.

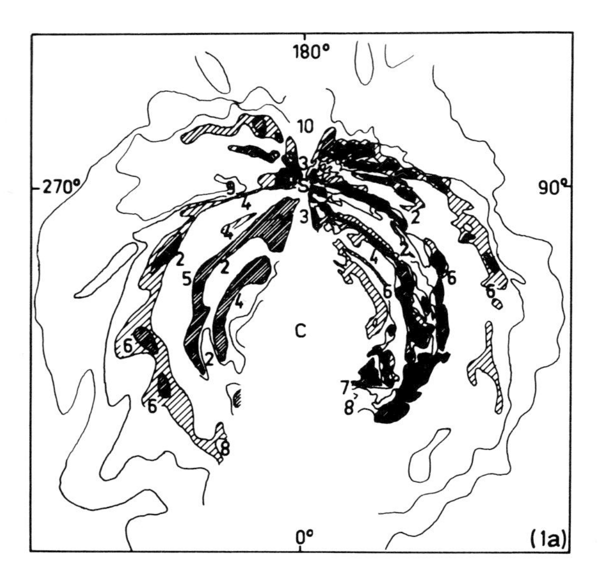

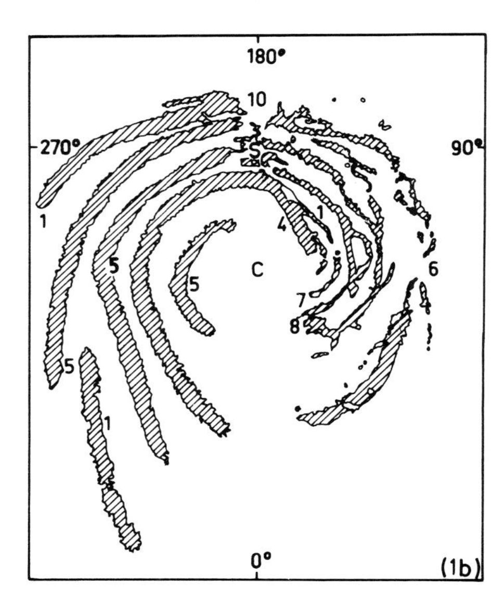

Figure 1. Schematic maps of the galactic neutral hydrogen redrawn from previously published maps. (a) From the classic map (Kerr and Westerhout, 1965) based on separate rotation models for the north and the south, (b) The map based on the northern rotation curve (from Weaver, 1975). Anomalous features discussed are indicated.

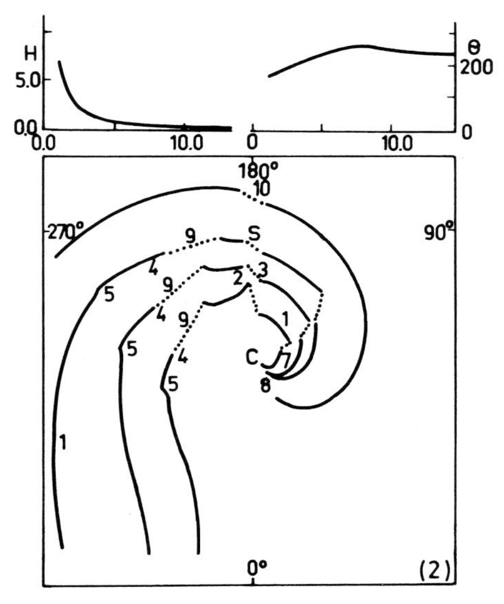

Figure 2. Numerical simulation of anomalies. The rotation model is shown in the upper right corner, the strength of the disk redshift (in km/ /(s kpc)) in the upper left corner; the additional redshift within the spiral arms is 4 km/(s kpc); the original undistorted arms are logarithmic spirals. Compare this figure with Fig. 1b.

The successful solution of ten very diverse structural anomalies demonstrates the existence of the non-velocity effect. Independent evidence is given by the data on a systematic redshift of the galactic nucleus (Jaakkola, 1978); by the data on the longitude-independent latitude--redshift relation (Moles and Jaakkola, 1977); and by the data on positive line-shift gradients across the disks of external galaxies (Jaakkola *et al.*, 1975). The strength of the effect within the Galaxy and in external galaxies is an order of magnitude higher than that of the cosmological redshift effect. Future observations of redshift gradients across the disks of galaxies will provide a definitive test of the present hypothesis.

By removing the non-velocity effect from the data it is quite possible to obtain a consistent, bilaterally and cylindrically symmetric picture of both the structure and the kinematics of the Galaxy.

REFERENCES

Jaakkola, T.: 1978, Sci. Inf. Astron. Council, Acad. Sci. USSR 45, 190.

Jaakkola, T., Teerikorpi, P., and Donner, K.J.: 1975, Astron. Astrophys. 40, 257.

Kerr, F.J., and Westerhout, G.: 1965, in "Galactic Structure", A. Blaauw and M. Schmidt, eds., Chicago Univ. Press, p. 167

Moles, M., and Jaakkola, T.: 1977, Astrophys. Space Sci. 48, 4.

Weaver, H.: 1975, Mercury 4, No. 6, 18.

DISCUSSION

F.J. Kerr: Are you suggesting any particular physical cause for your non-velocity lineshift?

Jaakkola: The cause is the same as for the cosmological redshift. Of course, if there is such an enhanced redshift field within the Milky Way, this indicates that the redshift is an interaction phenomenon, not a Dopplerian phenomenon.

Top: T.S. Jaakkola fills his glass, while Ria van Woerden considers her choice of wines at University President's reception. To her left: J.W. Pel. CFD

Bottom: Allen explains his views on star formation to Elmegreen. CFD

SECTION II.6

SMALL-SCALE STRUCTURE AND STAR FORMATION

Tuesday 31 May, 1725 - 1815
Wednesday 1 June, 0915 - 0945

Chairmen: F.J. Kerr and R.D. Davies

A. Blaauw introducing Symposium participants to the historical exhibition about Kapteyn and Van Rhijn CFD

SMALL-SCALE STRUCTURE AND MOTIONS IN THE INTERSTELLAR GAS

John M. Dickey
University of Minnesota

ABSTRACT: The interstellar medium shows turbulence whose velocity and density spectra are surprisingly like a Kolmogoroff law, in spite of the fact that the turbulence is generated mainly by microscopic explosions, rather than a cascade of energy from larger to smaller scales.

The question of how to describe the small-scale structure and motions in the interstellar medium leads to the question of turbulence. Although there are clearly random variations, it is less clear whether these fit a turbulence spectrum. The fundamental question is whether there is a cascade of kinetic energy from large-scale structures to smaller scales, just as there is in the classical Kolmogoroff problem of dissipation of energy in a viscous fluid. Some of the evidence in favor of such a cascade is compiled by Larson (1981), who draws a diagram similar to fig. 1. Here I take only observations of ^{13}CO, plotting the rms velocities (in this case taken from the line widths) versus the size of the region for many different clouds (points with names given). Recent work by Myers and Benson (1983) using NH_3 observations of very small structures in the Taurus and Ophiuchus clouds is indicated by the oval in the lower left of fig. 1. The random velocity distribution of diffuse clouds has been measured in several ways (reviewed by Crovisier 1978), this distribution is shown by the larger circle at the top right of fig. 1. Internal random velocities of diffuse clouds can be estimated knowing the spin temperature (Dickey et al. 1978, Crovisier 1981). These fall in the smaller circle on fig. 1. Finally recent results from the Columbia survey of CO in the galactic plane (Dame et al. 1983) show a fairly convincing correlation between size and random velocity for molecular clouds and giant-molecular-cloud+HII-region complexes which is indicated by the line of fig. 1. It is interesting that this line has quite a different slope from the data points compiled by Larson, which show velocity increasing roughly with size to the 0.38 power, curiously close to that found in a Kolmogoroff turbulence spectrum (0.33). This would not necessarily be expected, since the interstellar medium is anything but incompressible, and the motions are highly supersonic. On the other hand Dame et al. find slope one for their correlation. An easy way to understand Dame's relation is to use the virial theorem with

H. van Woerden et al. (eds.), The Milky Way Galaxy, 311–317.

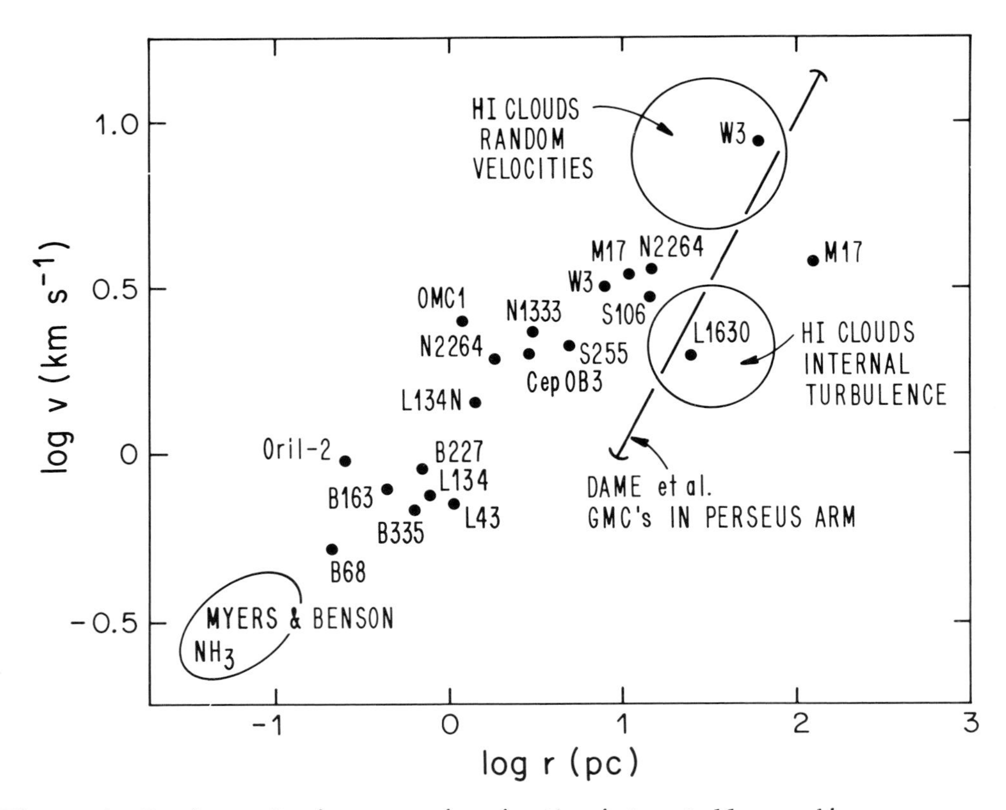

Figure 1. Random velocity vs. size in the interstellar medium.

constant density, which always gives a linear correlation between random velocity and size. The Columbia data suggest density $\langle n(H_2)\rangle \sim 8\ cm^{-3}$, two other densities (0.3 and 300 cm^{-3}) are shown as dashed lines on figure 2. It is interesting that giant HII regions in nearby spiral galaxies (Melnick 1977) also fit nicely with the correlation of Dame et al.; Larson on the other hand would have the density increase with decreasing size, as is shown by density tracers such as NH_3.

The source of the kinetic energy in these motions is not clear. One possibility is galactic differential rotation (eg. Hunter and Fleck 1982). For a flat rotation curve with A = 15 km s^{-1} kpc^{-1}, the shear due to differential rotation is illustrated by the dash-dotted line in the lower right of fig. 2. These velocities are not high enough to explain the random motions observed in the interstellar medium. For galactic rotation to cause the interstellar turbulence requires a moderator, such as spiral shocks or the magnetic field, which can concentrate the shear velocity over a large area and inject violent motions on small scales. An alternative is microscopic injection of kinetic energy by supernovae, as suggested long ago by Spitzer (e.g. 1978). Two possibilities for the expansion of a supernova remnant are traced on the top of fig. 2. These lines are obviously not random motions, they are the bulk motion of an expanding shell; but eventually they will be randomized by collisions

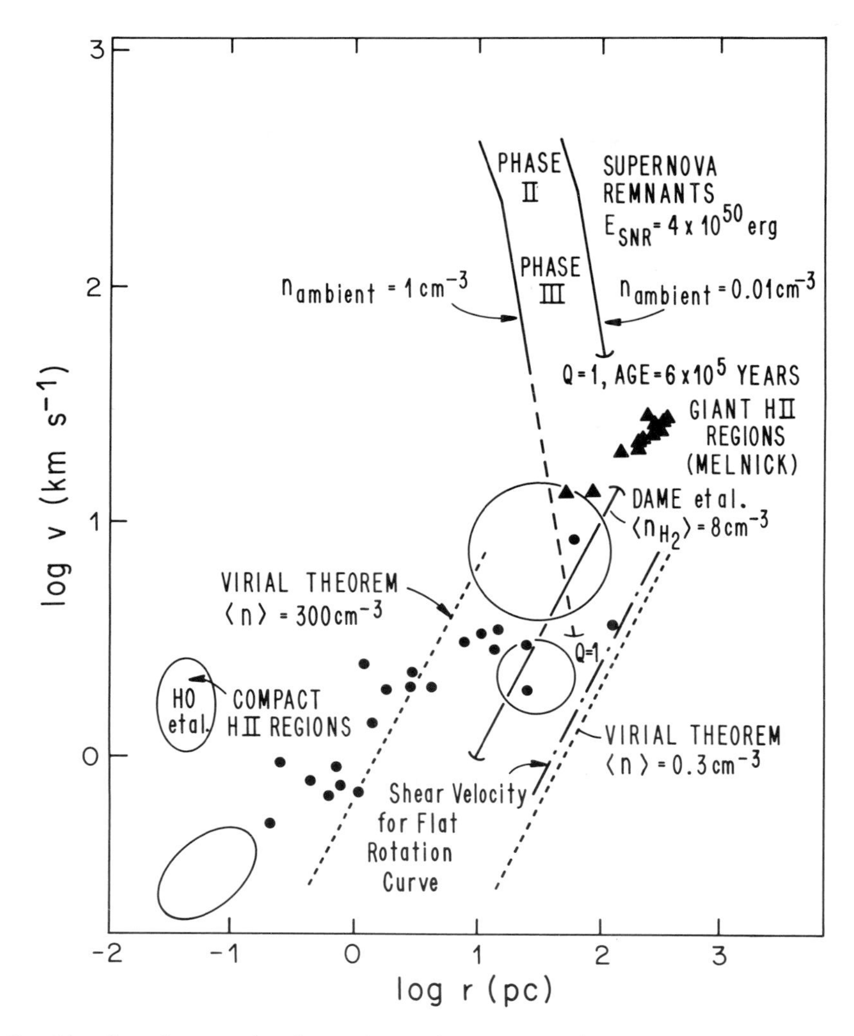

Figure 2. Simple theoretical explanations for fig. 1.

among nearby shells, just as rain drops falling on a pool of water cause at first ordered expanding rings which are soon converted to random motions by collisions with others. The age at which this randomization occurs is given by the porosity parameter Q (Cox and Smith 1974, McKee and Ostriker 1977); the velocity and size of the shell when Q = 1 predicts where on fig. 2 the turbulent motion is injected by this process. Unfortunately we do not understand how the system of clouds relaxes, since cloud collisions are so inelastic (Hausman 1981, Scalo and Pumphrey 1982). This problem is also central to understanding the cloud mass spectrum (Cowie 1980).

In dense molecular clouds other microscopic sources of turbulence are important. One theory which has abundant observational support is

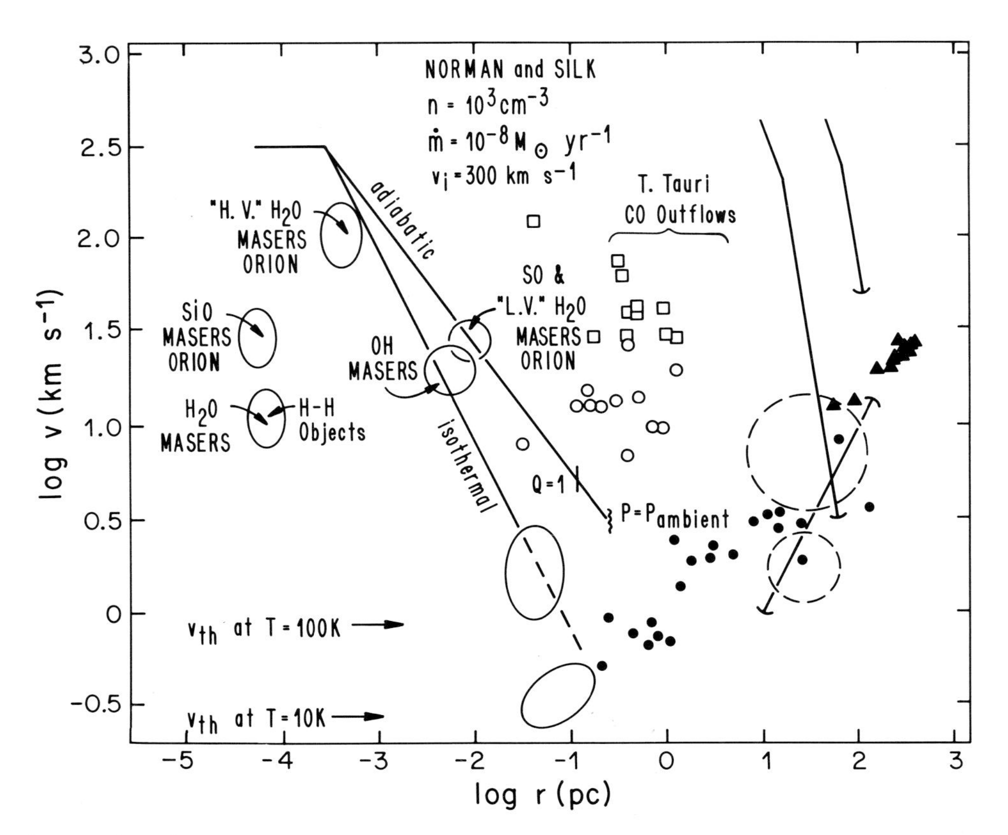

Figure 3. Figure 1 extended to star-formation regions.

that of Norman and Silk (1980), which discusses the effects of winds from pre-main-sequence stars on the dense clouds in which they form. This theory was originally motivated by Hα observations of T Tauri stars; recently there have been detections of such winds using many molecular tracers, especially CO. The CO detections are shown as circles (Edwards and Snell 1983) and crosses (Bally and Lada 1983) on fig. 3, which is once again the same velocity vs. size plot as above, but with a still smaller scale. Again these are not random velocities but ordered expansion which will be randomized by collisions with other shells. It is this collision process which causes the interstellar bullets which drive a rapid agglomeration process leading to star formation. Infrared observations of star-formation regions such as Chameleon (Hyland et al. 1982) and Ophiuchus (Wilking and Lada 1983) show just about the density of pre-main-sequence stars required by the Norman and Silk model (Beckwith et al. 1983). Theoretical predictions for the expansion of such T Tauri shells are illustrated on fig. 3. As for supernova shells, the expansion begins as ordered motion on the upper left, then converts to random motions somewhere near the lower right end of the diagonal lines, where $Q \sim 1$. It is interesting that the highest-velocity molecular outflows detected (e.g. Orion) are considerably stronger than predicted

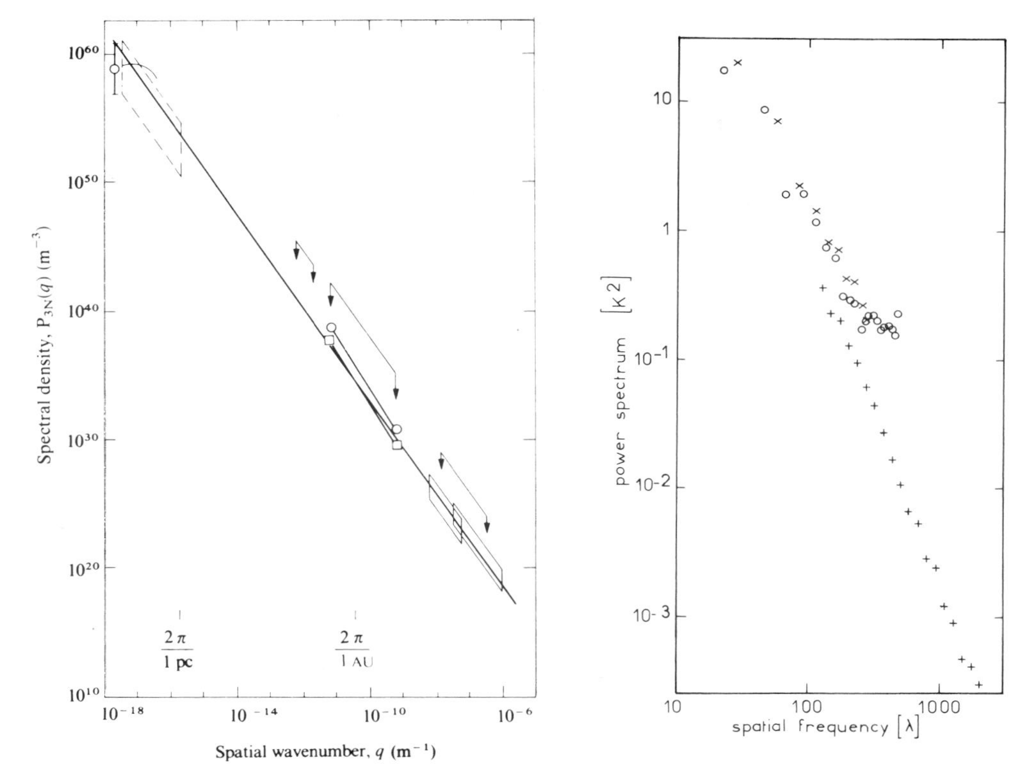

Figure 4. Recent measurements of the interstellar turbulence spectrum.

by ordinary T Tauri winds. Also shown on fig. 3 are observations of various masers which trace expanding shells around stars at smaller radii (eg. Genzel et al. 1982) and the recent aperture-synthesis study of SO in Orion (Plambeck et al. 1982). These scatter over various expansion velocities and sizes down to a few AU, but the evidence is fairly clear that stellar winds can generate the turbulent velocities needed to support dense molecular-cloud cores and globules.

This is the picture of small-scale motions which has been emerging over the last several years. There are problems with this picture. Figures like 1 - 3 illustrate inhomogeneities, they do not describe a statistically homogeneous process. It is not fair to compare the random motion measured in, say, the Kleinmann-Low region with the random motion measured over the entire Orion molecular cloud, and draw a spectrum between these points. The small-scale measurement should describe a region typical of any small volume chosen at random within the larger region. Obviously the OMC is not filled with K-L's! To construct a turbulence spectrum properly we should weight all points equally.

Two alternative means can be used to describe turbulent motions: the correlation function or the power spectrum. A nice example of both treatments applied to the motions of interstellar clouds is given by Kaplan (1966). A problem with Kaplan's analysis is that correlation in density (i.e. clouds) imposes a correlation in velocity. To study the turbulence spectrum we must first understand the density spatial power spectrum. This has been attempted for various phases of the interstellar medium, but with only preliminary results. One example, shown on the left on figure 4, is for the ionized gas which causes the pulsar scintillations (Armstrong et al. 1981). Various kinds of observations combine to suggest a power-law spectrum over a very broad range of sizes, from many parsecs to much less than an AU. The slope of this power law is surprisingly close to 11/3, which in this notation corresponds to the Kolmogoroff spectrum. This may have implications for the velocity spectrum as well, since the density and velocity spectra are often the same in astrophysical plasmas (Neugebauer et al. 1978). Another tracer to use to measure the power spectrum of small-scale fluctuations in the interstellar medium is the 21-cm emission. Studies of this have been done by Crovisier and Dickey (1983). On the right of fig. 4 is a spectrum from the former work, showing the spatial power spectrum as measured by four different telescopes. Here again we see a power law, with slope in the range 2.5 to 3, i.e. somewhat flatter than Kolmogoroff. More studies of this spectrum using other tracers of the interstellar medium (e.g. inside molecular clouds) would be of great interest.

REFERENCES

Armstrong, J.W., Cordes, J.M., and Rickett, B.J.: 1981, Nature 291, p. 561
Bally, J. and Lada, C.J.: 1983, Astrophys. J. 265, p. 824
Beckwith, S., Natta, A., and Salpeter, E.E.: 1983, Astrophys. J. 267, p. 596
Cowie, L.L.: 1980, Astrophys. J. 236, p. 868
Cox, D.P., and Smith, B.W.: 1974, Astrophys. J. Lett. 189, p. L105
Crovisier, J.: 1978, Astron. Astrophys. 70, p. 43
Crovisier, J.: 1981, Astron. Astrophys. 94, p. 162
Crovisier, J., and Dickey, J.M.: 1983, Astron. Astrophys. 122, p. 282
Dame, T.M., Elmegreen, B.G., Cohen, R.S., and Thaddeus, P.: 1983, preprint
Dickey, J.M., Salpeter, E.E., and Terzian, Y.: 1978, Astrophys. J. Suppl. 36, p. 77
Edwards, S., and Snell, R.L.: 1983, preprint (FCRAO report no. 206)
Genzel, R., Reid, M.J., Moran, J.M., Downes, D., and Ho, P.T.P.: 1982, Ann. N.Y. Acad. Sci. 395, p. 142
Hausman, M.A.: 1981, Astrophys. J. 245, p. 72
Hunter, J.H., and Fleck, R.C.: 1982, Astrophys. J. 256, p. 505
Hyland, A.R., Jones, T.J., and Mitchell, R.M.: 1982, M.N.R.A.S. 201, p. 1095
Kaplan, S.A.: 1966, "Interstellar Gas Dynamics" (Pergamon: Oxford), p. 116
Larson, R.B.: 1981, M.N.R.A.S. 194, p. 809
McKee, C.F., and Ostriker, J.P.: 1977, Astrophys. J. 218, 148
Melnick, J.: 1977, Astrophys. J. 213, p. 15
Myers, P.C., and Benson, P.J.: 1983, Astrophys. J. 226, p. 309

Norman, C., and Silk, J.: 1980, Astrophys. J. 238, p. 158
Plambeck, R.L., Wright, M.C.H., Welch, W.J., Bieging, J.H., Baud, B., Ho, P.T.P., and Vogel, S.N.: 1982, Astrophys. J. 259, p. 617
Scalo, J.M., and Pumphrey, W.A.: 1982, Astrophys. J. 258, p. L29
Spitzer, L. Jr.: 1978, "Physical Processes in the Interstellar Medium", (New York: Wiley), pp. 255-261
Wilking, B.A., and Lada, C.J.: 1983, preprint

DISCUSSION

B.G. Elmegreen: An important clue to the origin of the largest clouds, and to the initiation of star formation on the largest scales, comes from the observation that these clouds have the Jeans mass, and that their separations are the Jeans length, as measured in the ambient interstellar medium. Also, some of the hierarchy of scales you attribute to a turbulent cascade may instead be the result of self-gravitational fragmentation. This process may not operate down to scales as small as single stars, but it could account for the larger-scale clumpy structure seen in some clouds.

Dickey: You can play with the Jeans mass here, as Larson does, for example. Constant density gives lines parallel to that of Dame et al. in Figures 1-3; to follow both the Jeans mass and the virial theorem Larson changes the density as a function of size. I don't think we have very good observational indicators of the masses of individual clouds on small scales, so I have not talked about mass. But yes, if you assign masses to these clouds based on the virial theorem, you get an initial mass function. However, if you don't believe there is a cascade of turbulence down the virial-theorem lines, then I don't see how the "machine" runs.

Elmegreen: Well there are other ways to cascade, there is simple gravitational fragmentation.

Dickey: Yes, that is supposed to take over on the smallest scales. Gravitational fragmentation peels small clouds off the log(v)-log(r) line. That probably works, but the evidence to date is not entirely convincing.

Centre: Dickey between Iwanowska (left), G.D. van Albada and Liszt.
Front row: De Zeeuw, second row: Pismis and Wramdemark. Fourth row: Hermsen, Bloemen and Fujimoto. Further behind: Yuan and Terzides; Israel and Van der Laan; Jackson.

ATOMIC HYDROGEN TOWARDS 3C10

James S. Albinson
Netherlands Foundation for Radio Astronomy, Dwingeloo

New aperture-synthesis observations of HI in absorption towards 3C10 with high velocity resolution (0.6 km s^{-1}) and moderate angular resolution (1 arcmin) have been made with the Westerbork Synthesis Radio Telescope. These are an extension of the survey by Schwarz, Arnal & Goss (1980). A selection of the data has been studied in a preliminary way, covering the Perseus-Arm absorption feature (~ -50 km s^{-1} with respect to the Local Standard of Rest).

The absorption feature is sharply bounded in velocity, occurring between -58 km s^{-1} and -46 km s^{-1}. There are multiple components in velocity and in space across the face of the supernova remnant. The main component has a maximum optical depth of $\geq$ 3.5 and a FWHP of 2.0 km s^{-1}, centred on -48 km s^{-1}. Its angular size is ~ 6 arcmin by 3 arcmin, corresponding to ~ 4 pc by 2 pc at a distance of $2\frac{1}{2}$ kpc. This component covers the southeastern half of the remnant, the northwestern half is almost completely free of absorption - the boundary between the two regimes is very sharp ($\leq$ 1 arcmin). Part of the main absorption feature forms a thin filament right across the face of the remnant at a position angle of ~ 40°. Maps of HI emission made with the Cambridge Half-Mile Telescope (Albinson & Gull, 1982) show an emission filament ~ 1° in length lying to the northeast of the position of the remnant in a radial orientation. The radial velocity of the emission filament is ~ -49 km s^{-1}; it is suggested that the emission and absorption filaments are continuations of each other. The number density of HI in the main absorption feature is estimated to be ~300 to 1000 cm^{-3}.

This work will be published in full elsewhere in the near future (Albinson, Kalberla, Schwarz & Goss, in preparation).

ACKNOWLEDGEMENTS

I would like to thank the staff and students of the Kapteyn Laboratory for their generous assistance in operating the GIPSY system, which was heavily used in the course of this work. The Westerbork Synthesis

H. van Woerden et al. (eds.), The Milky Way Galaxy, 319–320.

Radio Telescope is operated by the Netherlands Foundation for Radio Astronomy, with the financial support of ZWO.

REFERENCES

Albinson, J.S., and Gull, S.F.: 1982, in "Regions of Recent Star Formation", eds. Roger, R.S. and Dewdney, P.E., Dordrecht: Reidel, pp. 193-199.

Schwarz, U.J., Arnal, E.M., and Goss, W.M.: 1980, M.N.R.A.S. 192, 67P.

Left to right: Van Driel, Bania, Crovisier, Pismis, Burton, Hu Fu-Xing and Mo Jing-Er
LZ

INTERFEROMETRIC OBSERVATIONS OF THE SMALL-SCALE STRUCTURE OF GALACTIC NEUTRAL HYDROGEN

J. Crovisier, Observatoire de Meudon, France
J.M. Dickey, University of Minnesota, USA

The small-scale structure of galactic neutral hydrogen may be statistically described by the spatial power spectrum of the 21-cm line. This latter may be readily observed by interferometer arrays since it is the squared modulus of the visibility function. We have observed the $l=52.^\circ5$, $b=0.^\circ0$ region with the Westerbork Synthesis Radio Telescope (Crovisier and Dickey, 1983). Brightness fluctuations of the 21-cm line were detected in this region on scales as small as 1.7 arcmin (corresponding to less than 5 pc). The Westerbork observations, combined with single-dish observations made at Nançay and Arecibo, allow determination of the spatial power spectrum over a dynamic range of about 10^6 in intensity. The spectrum follows roughly a power law with indices ~ -3 to -2. An interpretation in terms of the turbulence spectrum is proposed by Dickey (1985).

Another way to investigate galactic HI structure over very small scales is to measure, with an interferometer, 21-cm absorption profiles toward the components of double continuum background sources. The simple source structure allows an easy reduction by model fitting. Previous observations (Dickey, 1979) with the NRAO 3-element interferometer revealed only marginal optical-depth variations except toward the components of Cyg A. We recently made more sensitive observations with the WSRT toward 7 double extragalactic sources with angular separations ranging from 0.7 to 6 arcmin. Significant absorption differences were found between the components of each pair ($\Delta\tau$ = 0.05 to 0.3). Some of them correspond to linear scale-lengths as small as 0.2 pc, when the relevant clouds are placed at their kinematic distances. As an example, Figure 1 shows the spectra obtained toward 3C 69 ($l=136.^\circ26$, $b=-0.^\circ88$), whose components are separated by θ = 42 arcsec. These absorption differences may be attributed to optical-depth variations and/or velocity gradients within the HI clouds.

Interpretation of these interferometric observations will be done in the future in terms of HI distribution models.

H. van Woerden et al. (eds.), The Milky Way Galaxy, 321–322.

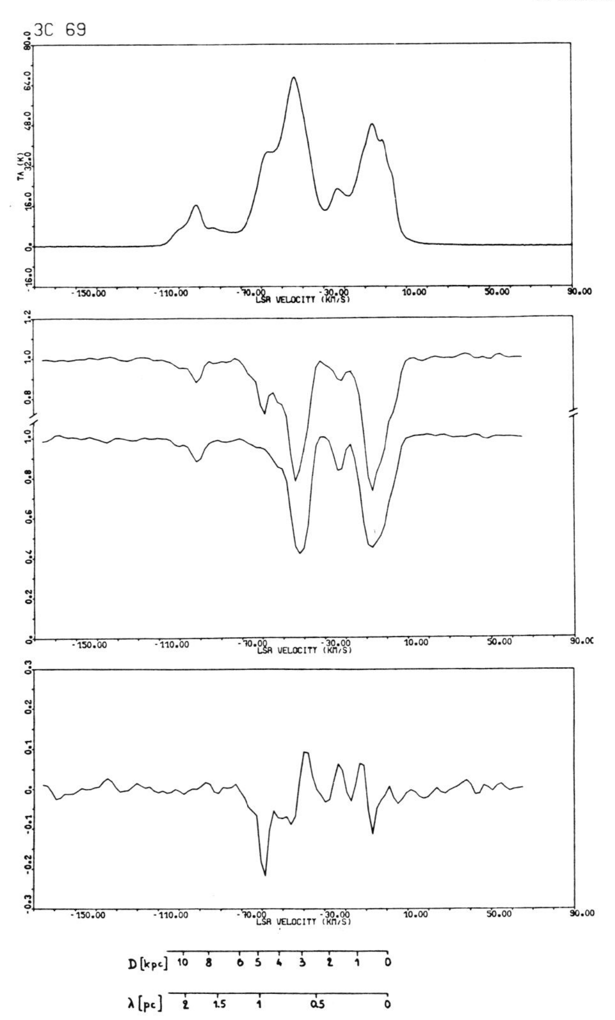

Figure 1. From top to bottom: a) the mean 21-cm emission spectrum in the 3C 69 direction observed with the Nançay radio telescope; b) the absorption profiles observed toward the south and north components of 3C 69 with the WSRT; c) the difference between the two absorption profiles; d) the relationship between velocity, kinematic distance D and linear separation $\lambda = D\theta$.

REFERENCES

Crovisier, J., and Dickey, J.M.: 1983, Astron. Astrophys., 122, 282
Dickey, J.M.: 1979, Astrophys. J. 233, 558
Dickey, J.M.: 1985, this volume

A 10-GHZ RADIO-CONTINUUM SURVEY OF THE GALACTIC-PLANE REGION AT THE NOBEYAMA RADIO OBSERVATORY - A COMPLEX REGION AT $\ell = 22° - 25°$

Y. Sofue*, H. Hirabayashi, K. Akabane, M. Inoue, T. Handa and N. Nakai
Nobeyama Radio Observatory, Tokyo Astronomical Observatory
*A. von Humboldt Fellow at MPI für Radioastronomie Bonn

ABSTRACT. Preliminary results of a 10-GHz radio-continuum survey of the galactic-plane region using the 45-m telescope at NRO are presented. An extensive study of a complex region at $22° \leq \ell \leq 25°$, $|b| \lesssim 1°$ has been made.

Figure 1 shows the distribution of brightness temperature at 10.2 GHz of the region. A comparative study with the Bonn 5-GHz survey (Altenhoff et al., 1978) has shown some remarkable features as below:

1. A Crab-like SNR?: Dashed circles in Fig. 1 show SNRs known so far (Milne, 1979). Among them the SNR G24.7+0.6 (arrowed) has an irregular shape. The peak-flux spectrum between 5 and 10 GHz is flat. These facts suggest that this object may be a new candidate for a Crab-like SNR.
2. HII Rings: Two apparently ring-like orientations of HII regions are found as indicated with full lines in Fig. 1. As the radial velocities of the HII regions in individual rings are similar to each other (80-100 km/s for the ring centred on G23.2+0.2 and 90 - 110 km/s for G24.6+0.0 (Downes et al., 1980)), there may be physical associations on rings or shells of diameter of about 100 pc. SNR-shock-enhanced star formation may be suggested for the formation of such HII rings.
3. Compact Nonthermal Sources: A comparison with the Bonn 5-GHz survey reveals a number of steep-spectrum compact sources of $S_{10GHz} = 0.1 - 0.3$ Jy or less (crosses in Fig. 1). From their distribution around the galactic plane (Fig. 2) the majority of the sources may be galactic. Possible origins of the compact sources are: (a) small-size, low-surface-brightness SNRs which do not satisfy the surface brightness-diameter relation; (b) very-short-period pulsars which have not been found because of the short period and high dispersion near the galactic plane; (c) active stars like cataclysmic variables; (d) background fluctuations due to irregularities of magnetic fields and cosmic-ray distributions.

REFERENCES

Altenhoff, W.J. et al.: 1978, Astron. Astrophys. Suppl. 35, 23
Milne, D.K.: 1979, Australian J. Phys. 32, 83
Downes, D. et al.: 1980, Astron. Astrophys. Suppl. 40, 379

H. van Woerden et al. (eds.), The Milky Way Galaxy, 323–324.

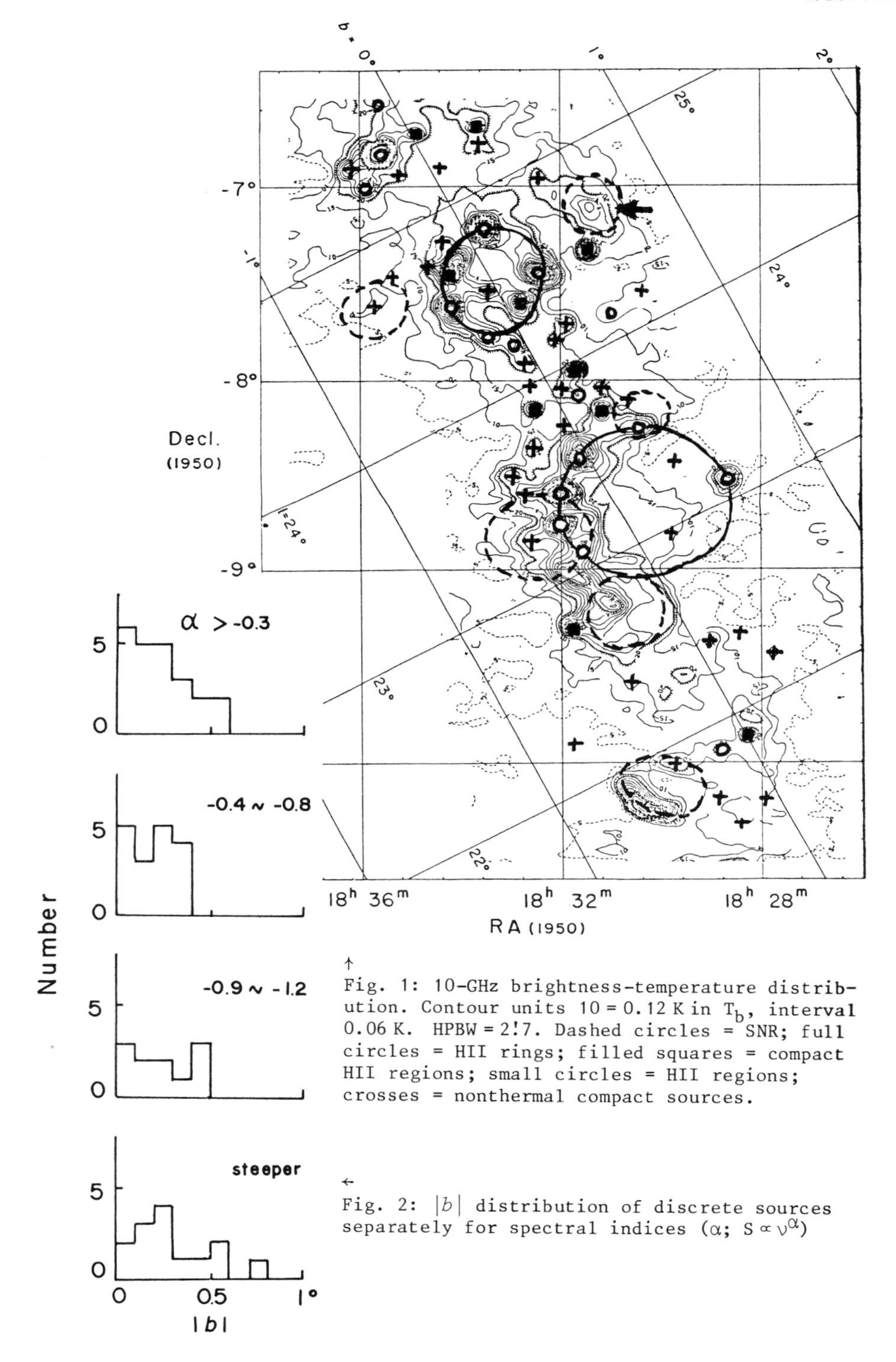

↑
Fig. 1: 10-GHz brightness-temperature distribution. Contour units 10 = 0.12 K in T_b, interval 0.06 K. HPBW = 2!7. Dashed circles = SNR; full circles = HII rings; filled squares = compact HII regions; small circles = HII regions; crosses = nonthermal compact sources.

←
Fig. 2: $|b|$ distribution of discrete sources separately for spectral indices (α; $S \propto \nu^{\alpha}$)

INTERSTELLAR SODIUM AND GALACTIC STRUCTURE. A HIGH-RESOLUTION SURVEY

E. Maurice, A. Ardeberg and H. Lindgren
European Southern Observatory - La Silla
Casilla 16317 Correo 9 Santiago de Chile

INTRODUCTION

Observation of absorption lines produced by interstellar gas is a straight-forward way to determine column densities and velocities along the line of sight of interstellar clouds. In practice, peculiar motions often mask galactic rotation and/or cause line blending. We have made a study of absorption lines of interstellar sodium covering a substantial part of the Galaxy at extremely high spectral resolution.

OBSERVATIONS AND RESULTS

We used the ESO Coudé Echelle Spectrometer with the 1.42-m Coudé Auxiliary Telescope and a Reticon silicon photodiode array. The resolution is 10^5 or 3 kms^{-1} at 5893 Å. At V=8 a s/n ratio of 150 is reached in 3 hours. We have made over 200 observations of 150 early-type stars with V from 0 to 9. We include also stars far from the galactic plane.

Almost all our stars have lines of interstellar NaI, also the very brightest ones. Less than 15% of the stars have spectra with less than two components, whereas close to 30% display 5 or more components. The presence of the two NaI D lines makes identifications rather clear-cut. Major features of interstellar sodium can be traced over large parts of the Galaxy. However, star-to-star variations are pronounced, also over small angular distances. It concerns line positions as well as equivalent widths. Some stars which are apparently close in space present line patterns of interstellar sodium which are rather difficult to match.

INTERSTELLAR SODIUM IN THE DIRECTION OF SCORPIUS OB 1

As an example we show in Figure 1 spectrograms of four supergiants in Sco OB 1. The strength and complexity of interstellar sodium may be compared to the fairly modest reddening (Schild et al., 1971). The largest sodium feature is similar for all four stars, with a "red" wing

H. van Woerden et al. (eds.), The Milky Way Galaxy, 325–327.

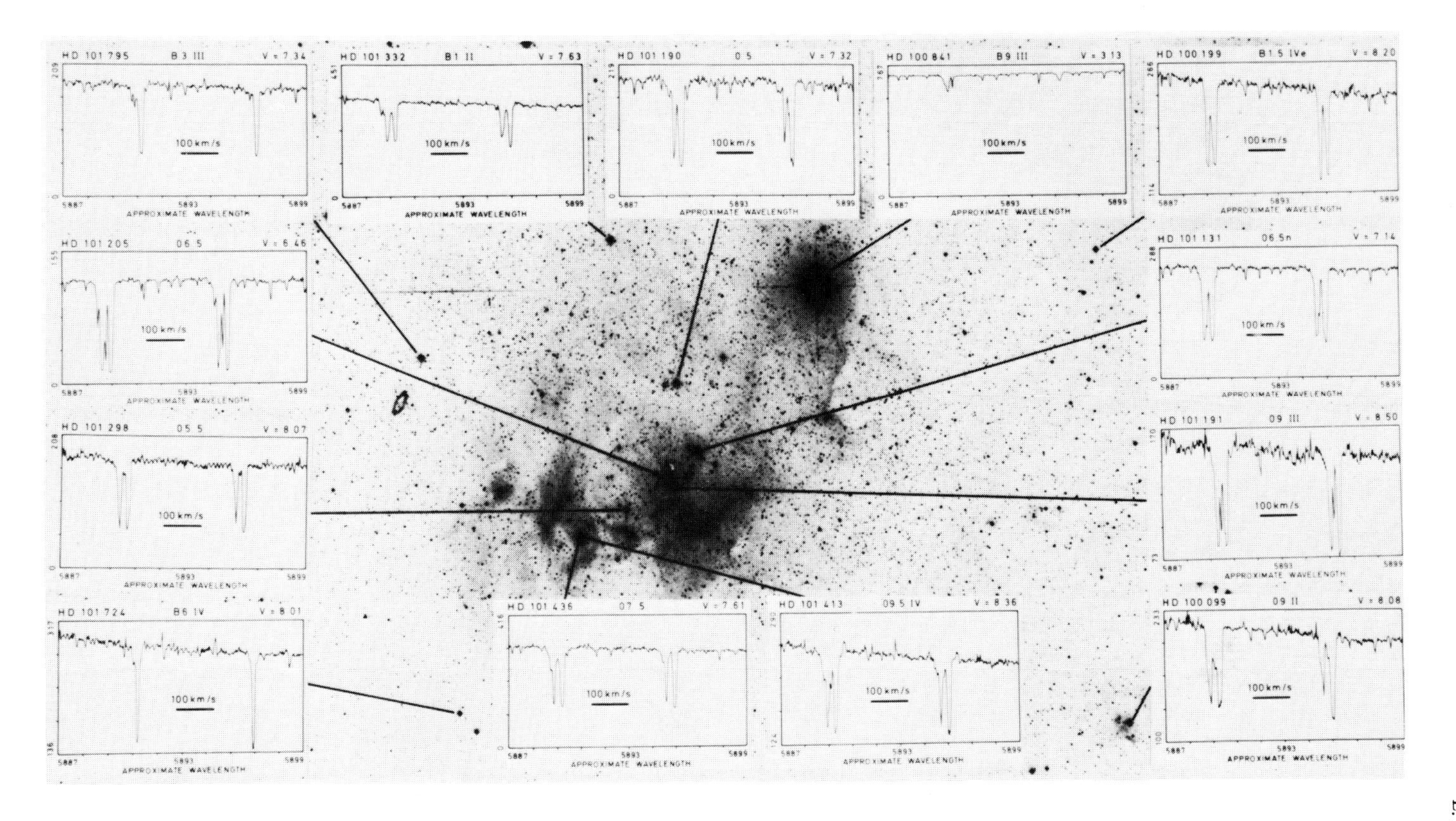

Fig. 2. Interstellar sodium in the direction of the Carina spiral arm

corresponding to velocities close to that of the Sun. Therefore, the major contribution seems to come from interstellar sodium close to the local spiral arm. The behaviour of the line components at "negative" radial velocities varies notably, as it does for the "blue" wing of the principal line feature. Preliminary estimates show the "blue" line components to fall between -25 and -55 kms^{-1} (LSR). These radial velocities are much more negative than expected from rotation models (Schmidt, 1965; Georgelin and Georgelin, 1976). At a much lower resolution, Rickard (1974) found the same effects for interstellar CaII.

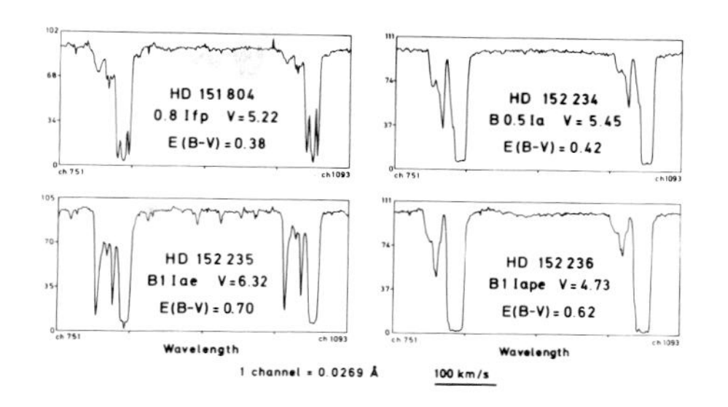

Fig. 1. Interstellar sodium in front of Scorpius OB 1

INTERSTELLAR SODIUM ALONG THE INNER SIDE OF THE CARINA SPIRAL FEATURE

A detailed study of interstellar sodium along the inner side of the Carina spiral feature covers a distance interval of four kpc with E(B-V) up to 0.38 (Ardeberg and Maurice, 1980). Tracings are given in Figure 2. The number of NaI line components varies from two for HD 100841 (λ Cen) to at least 7. Practically all sodium-line components fall at negative radial velocities (LSR), in good agreement with rotation models (Ardeberg and Maurice, 1981). For most stars the sodium feature is dominated by two strong components with a difference close to 20 kms^{-1}. The minor line components vary considerably in strength as well as in position.

REFERENCES

Ardeberg, A., and Maurice, E.: 1980, Astron. Astrophys. Suppl. 39, 325
Ardeberg, A., and Maurice, E.: 1981, Astron. Astrophys. 98, 9
Georgelin, Y.P., and Georgelin, Y.M.: 1976, Astron. Astrophys. 49, 57
Rickard, J.J.: 1974, Astron. Astrophys. 31, 47
Schild, R., Neugebauer, G., and Westphal, J.A.: 1971, Astron. J. 76, 237
Schmidt, M.: 1965, in "Stars and Stellar Systems V", A. Blaauw and M. Schmidt (eds.), 513

Dame (left) and Bronfman discussing their poster. Background: Leisawitz explains his poster to Iwanowska. At right: Higgs. CFD

MOLECULAR-CLOUD CLUSTERS AND CHAINS

D.B. Sanders, D.P. Clemens, N.Z. Scoville
Five College Radio Astronomy Observatory
University of Massachusetts, Amherst, MA

P.M. Solomon
Astronomy Program
State Univ. of New York, Stony Brook, N.Y.

We report some preliminary results from the Massachusetts - Stony Brook CO survey of the first galactic quadrant using the 14-meter millimeterwave telescope of the Five College Radio Astronomy Observatory. The survey contains approximately 50 000 observations spaced every 3 arcminutes in l and b between longitudes 0° and 90° and latitudes -1° and 1°. We have mapped emission from giant molecular clouds (GMC) which we identified, in earlier more limited strip surveys of the galactic plane (Solomon, Sanders and Scoville 1979; Sanders 1981), on size scales from a few parsecs to hundreds of parsecs, in order to determine the degree of clustering and organization into large-scale features. In addition to the characteristic size of 20 - 60 pc for individual GMC, we find clustering of clouds on a scale of from 100 to 300 pc.

Figure 1 (these Proceedings, page 3) shows approximately 40 percent of the survey data in the form of l,b strip maps between longitudes 20° and 50°, representing the integrated emission in 5 km s^{-1} velocity bins. Approximately 1500 emission features have been identified in this longitude range, and maps of a few hundred GMC known to be located on the near side of the tangent point have been analyzed. Several of the more prominent nearby clusters of clouds revealed by the survey can easily be seen at G35-0.6 (W44), v = 35-60 km s^{-1}; G49.5-0.4 (W51), v = 45-70 km s^{-1}; G42-0.4, v = 55-75 km s^{-1}; G28.3+0, v = 70-90 km s^{-1}; G24.5+0, v = 45-65 km s^{-1}. Clearly not all molecular clouds are contained in such clusters, but a full analysis of the degree of clustering and correlation lengths will be presented elsewhere.

Figure 2 (page 4) shows the relatively nearby cluster of GMC's near the supernova remnant W44. The cluster contains at least eight clouds with diameters larger than 20 pc and has a total diameter of approximately 110 pc. The cluster is nearly circular in projection and appears to have an ordered internal velocity structure, where the higher-velocity clouds are found at the largest radii and the central-velocity material is near the cluster center. The total velocity width is 25 km s^{-1}. We obtain a mass for the entire cluster of about 2×10^6 $M_\odot$. Larger clusters

H. van Woerden et al. (eds.), The Milky Way Galaxy, 329–330.

such as the 300-pc object near longitude 31° shown in Figure 3 (page 5) tend to be elongated in the plane; their width in the z direction is typically ≤150 pc, similar to the full width at half-maximum of the molecular disk as a whole.

Clouds on the far side of the tangent point typically subtend only one-tenth to one-twentieth the solid angle of nearby clouds at the same velocity, hence are more difficult to pick out from Figure 1. However, our 3-arcminute sampling, with a spatial resolution of 12 pc at a distance of 14 kpc, is adequate to easily resolve individual giant clouds and cloud clusters even at the most distant parts of the inner Galaxy. For example, one of the largest structures identified in the survey is the chain of molecular clouds nearly 2.2 kpc in length shown in Figure 3 (page 5). Over 70 distinct features with diameters larger than 12 pc can be identified, and at least 30 are GMC's with diameters larger than 20 pc. Three cloud clusters can be seen with sizes larger than 100 pc. At a distance of 14 kpc the entire chain can be characterized by a width perpendicular to the galactic plane of 100 pc (FWHM) with an H_2 mass of $2 \times 10^7\ M_\odot$, approximately 1 percent of the total mass of molecular hydrogen in the galactic disk. We note that the low-resolution GISS CO survey reported by Dame (these proceedings) has also identified some of the largest molecular clusters seen by us, but their report apparently has missed a large fraction of the strong emission, particularly from the far side of the tangent point which contains most of the area of the inner Galaxy. For example, the cloud cluster at l = 31° in Figure 3 is identified, but missed is the remaining two-thirds of the strong emission, including at least 25 clouds with diameters larger than 25 pc, each more massive than $10^5\ M_\odot$.

The cloud chain in Figure 3 is similar in length and number of clouds to two large, local galactic features: the Perseus Arm at longitudes 105° to 145°, and the M17-M8-NGC6334 cloud chain usually referred to as the Sagittarius-Carina spiral-arm segment. The origin of these molecular-cloud chains may be due to two-fluid gravitational instabilities as discussed by Jog and Solomon (1984) and Jog (these proceedings). They suggest that such instabilities "... may represent spiral-arm segments each of typical wavelength 2-3 kpc". In this view, many of the spiral features in our Galaxy are material arms, that is, randomly occurring sheared two-fluid gravitational instabilities.

REFERENCES

Jog, C. and Solomon, P.M. 1984, Astrophys. J. 276, in press

Sanders, D.B. 1981, Ph.D. Thesis, State University of New York at Stony Brook

Solomon, P.M., Sanders, D.B., and Scoville, N.Z. 1979, in "The Large-Scale Characteristics of the Galaxy", ed. W.B. Burton (Dordrecht: Reidel), IAU Symp. 84, pp. 35-52

THREE LARGE MOLECULAR COMPLEXES IN NORMA

L. Bronfman*, R.S. Cohen, and P. Thaddeus
Goddard Institute for Space Studies and Columbia University

H. Alvarez
Universidad de Chile, Santiago

A segment of the Milky Way in Norma, from $\ell = 327^o$ to 335°, $|b| \lesssim 1^o$, has been studied as part of the Columbia CO survey of the fourth galactic quadrant. Description of the entire survey is given by Cohen elsewhere in this volume.

In this region, just as in the corresponding part of the first galactic quadrant (Dame <u>et al</u>., 1983), the CO emission is dominated by large molecular complexes. These complexes are organized into three distant features apparently associated with three spiral-arm segments. Each feature extends over the entire longitude range, covering about 1 kpc (Figure 1). The natural division of the CO emission into three velocity ranges is clearly seen in the latitude-velocity diagram (Figure 2). For each velocity range we have identified the largest molecular complex. Each complex has a mass greater than 10^6 $M_\odot$. Complex 1 (Figure 3) is particularly massive ($M > 3 \times 10^6$ $M_\odot$), and may be the largest molecular complex in the Galaxy.

The positions of the three identified molecular complexes agree well with those of HII regions in the area. We find fifteen H 109α regions between $\ell = 331^o$ and 334° (Georgelin and Georgelin, 1976). Their mean velocities can be assembled into three groups, centered at $v = -86 \pm 5$ km/s, -66 ± 2 km/s, and -53 ± 2 km/s. The highest-velocity group is shown with Complex 1 in Figure 3. There are also two SNRs near $\ell = 332^o$, Milne 41 and 42 (Clark and Caswell, 1976). OH absorption at -88 km/s detected in their directions (Caswell and Haynes, 1975) indicates their possible association with Complex 1.

Once our survey is complete, we will extend this type of analysis in an attempt to identify all the massive molecular clouds in the fourth galactic quadrant.

---

*Permanent address: Departamento de Astronomía, Universidad de Chile, Casilla 36-D, Santiago de Chile.

H. van Woerden et al. (eds.), The Milky Way Galaxy, 331–332.

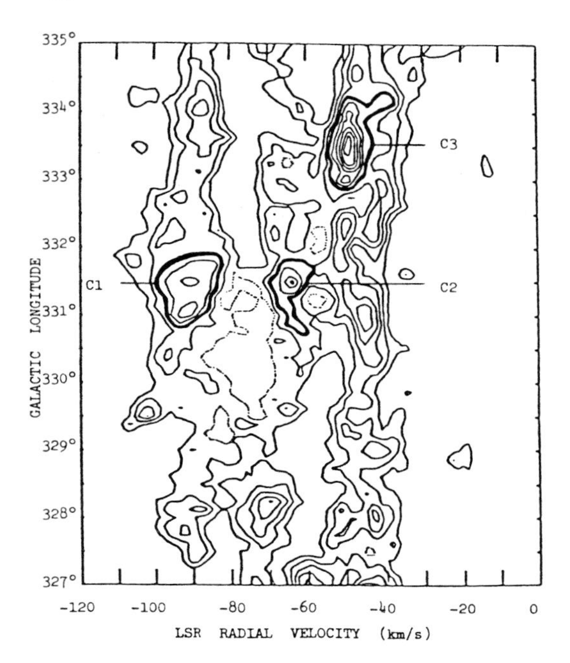

Figure 1. ℓ, v diagram obtained integrating the CO emission across the galactic plane. For each velocity range the largest molecular complex is enclosed by a darkened contour.

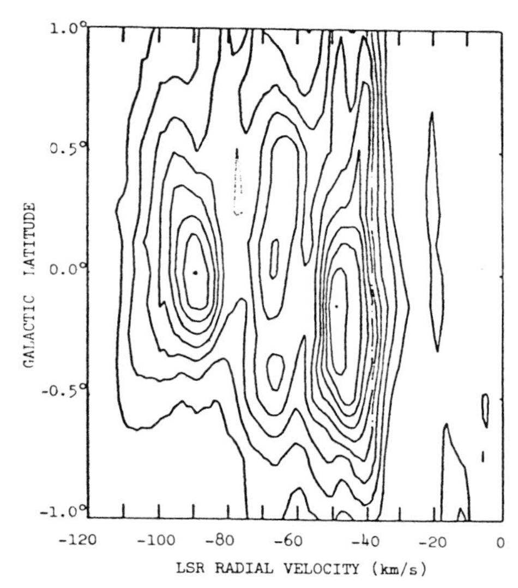

Figure 2. Latitude extent of the emission integrated over the whole longitude coverage. We can clearly distinguish three different velocity ranges.

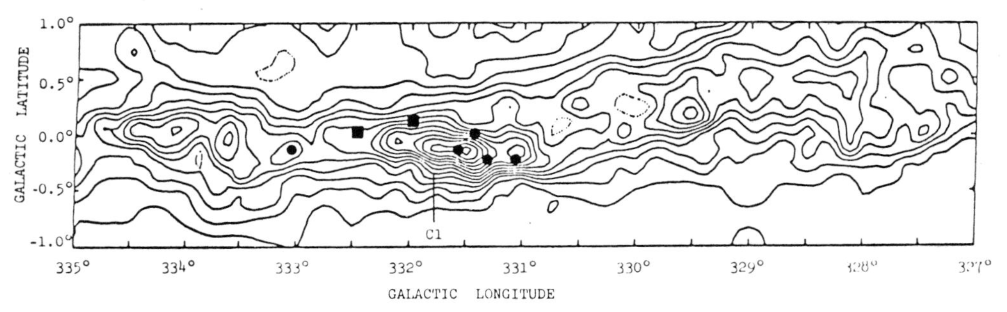

Figure 3. ℓ, v map of CO emission integrated over the highest velocity range ($-120 < v < -80$ km/s). Complex 1, in the figure, has a kinematical (near) distance of about 7 kpc. HII regions in the area are shown as filled circles, and SNRs as filled squares.

REFERENCES

Caswell, J.L., and Haynes, R.F.: 1975, Mon. Not. R. astr. Soc. 173, 649
Clark, D.H., and Caswell, J.L.: 1976, Mon. Not. R. astr. Soc. 174, 267
Dame, T.M., Elmegreen, B.G., Cohen, R.S., and Thaddeus, P.: 1983, Preprint
Georgelin, Y.M., and Georgelin, Y.P.: 1976, Astron. Astrophys. 49, 57

COMPARISON OF CO IN THE GALAXY AND THE MAGELLANIC CLOUDS

F.P. Israel, Th. de Graauw
Astronomy Division, ESA Space Science Department, Estec

H. van de Stadt, C.P. de Vries
Sterrewacht Sonnenborgh, Utrecht

We have used the Estec/Utrecht heterodyne submillimetre receiver and the ESO 3.6-m telescope at La Silla (Chile) to observe the CO(2-1) transition at 230 GHz (1.3 mm) in the Magellanic Clouds. We used a beam of 2 arcmin HPBW (corresponding to linear resolutions of 32 and 46 pc for LMC and SMC respectively), the system temperature was 2200 K (DSB) and the overall system efficiency was 0.55. In this paper we summarize the results, which are published in more detail elsewhere (cf. Israel et al., 1982, 1983; Israel, 1984).

In the LMC we sampled 22 positions, in the SMC 16 positions. The majority was near or coincident with bright HII regions and dark clouds. Four positions coincided with the known LMC OH masers and SMC H_2O masers. Five positions were taken as representative for the LMC Bar and three for the SMC Bar, i.e. they were not selected on the presence of HII regions, dark clouds etc. Detection statistics are overall 50 per cent for the LMC and 35 per cent for the SMC. All four maser positions were detected in CO. In the LMC, the region containing 30 Doradus and the bright HII regions and dark clouds to the south represents the greatest single concentration of molecular clouds. A strong CO signal of 2.6 K was detected in the direction of N159 which shows all the characteristics of a giant molecular cloud associated with a site of active star formation (Israel et al., 1982). The 30 Doradus complex itself is relatively poor in CO which may be due to the violent interaction of the HII region with its surroundings. In the SMC, molecular clouds are concentrated in the SW end of the Bar. The rest of the Bar, with the giant HII regions N66 and N76 in the NE, has a low detection rate (about 15 per cent) as has the LMC Bar. No CO was detected in the three Wing positions.

Thus, both LMC and SMC seem on the whole to have a lower CO cloud content than the Galaxy, especially when one allows for the fact that most observed positions were selected on the presence of objects that in the Galaxy usually are associated with giant molecular clouds. In addition, the CO content of the clouds detected is rather low. The strongest signals measured are of order 2.5 K, whereas a Galactic giant molecular cloud at Magellanic distances would be about two times stronger.

H. van Woerden et al. (eds.), The Milky Way Galaxy, 333–334.

The mean detected signal in the Clouds is only 1.5 K; all detections in the LMC and SMC added together show an area-integrated signal about as strong as expected from two to three Galactic GMC's, which is four to five times less than expected from an extrapolation of Galactic results. In view of the number of detections, this difference is statistically significant. Nevertheless, the detection pattern is similar to that seen in the Galaxy; CO is found near bright HII regions, dark clouds and molecular masers. In particular the concentrations of CO clouds south of 30 Doradus and in the SMC SW Bar correlate well with concentrations of dark clouds, neutral hydrogen, and young stars.

Elmegreen et al. (1980) have attempted to explain nondetection of CO in several Magellanic-type galaxies (but not including the LMC and SMC) by invoking the following explanations. 1. An underabundance of CO with respect to H_2. 2. A low level of cosmic-ray heating, resulting in low excitation temperatures of CO. 3. A relatively high rate of luminous-star formation, resulting in decreased molecular-cloud lifetimes. Local high star-formation rates may explain the difference at some positions (e.g. 30 Doradus), but most likely do not influence the majority of the observed positions. A low level of cosmic rays is also not a likely explanation (Israel et al., 1983), and an underabundance of CO with respect to H_2 would only show up at extremely high levels of depletion, when the CO line would become optically thin.

We suggest that the key to the difference is provided by the low dust-to-gas ratio in the Clouds; different dust properties are also suggested by low metal abundances and discrepant UV extinction laws. A lower dust content will cause a gas concentration in the Clouds to be more transparant to destructive UV radiation, leading to stronger CO dissociation. Because of its self-shielding properties, molecular hydrogen is not subject to dissociation. However, its formation is thought to take place primarily on dust-grain surfaces; it will thus be impaired. Compared to the Galaxy, a smaller fraction of the LMC and SMC total gas mass will therefore be in molecular form and a larger fraction in atomic form. Since the processes that lead to depletion of CO and H_2 are not the same, one would also expect the Galactic conversion factor of CO to H_2 not to apply to the Magellanic Clouds. Finally, the same reasoning suggests that the above also applies to different localities in galaxies with abundance gradients, such as the Galaxy.

REFERENCES

Elmegreen, B.G., Elmegreen, D.M., Morris, M.: 1980, Astrophys. J. 240, 455.

Israel, F.P., de Graauw, Th., van de Stadt, H., de Vries, C.P.: 1982, Astrophys. J. 262, 100

Israel, F.P., de Graauw, Th., van de Stadt, H., de Vries, C.P.: 1983, Astrophys. J., submitted

Israel, F.P.: 1984, in Proc. IAU Symposium No. 108, S. van den Bergh, and K.S. de Boer (eds.), 319

STAR FORMATION IN THE ORION ARM

A. Blaauw
Kapteyn Astronomical Institute
University of Groningen, The Netherlands

ABSTRACT

We assemble principal constituents of the morphology of the Orion Arm within 1500 pc as a starting point for the study of progression of star formation.

STRUCTURE FROM OB-ASSOCIATIONS

Although the Orion Arm (also referred to as the Local Feature or the Local Spur) is less prominent than major constituents of galactic spiral structure like the Perseus and Sagittarius Arms, it is of considerable interest for the study of the process of star formation. It is only here that, with current means, three-dimensional structure can be investigated in the relation between different star-forming regions and in the kinematic pattern of the young stellar components. Lack of photometric distance resolution is a major stumbling block for investigating such problems in the other galactic arms, let alone in extragalactic systems.

As a comparison with star-forming activity elsewhere, it is of some interest to note that the surface density (projected on the galactic plane) of luminous early-type stars in the Orion Arm is similar to the features of secondary importance in the Large Magellanic Cloud. For instance, the surface density of stars with M_V brighter than -4 in the string of subgroups numbered 1, 2, 5, 8, - - - - 66, 69, 107, 110 by Lucke and Hodge (1970) in their Fig. 1 is about the same as that of the Orion Arm. For further comparison, see also Fig. 4a to 4c of Schmidt-Kaler (1977).

The Orion Arm is a distinct feature in plots of the projection on the galactic plane of objects younger than 30 million years or selected by main-sequence spectral type B2 or earlier. Fig. 1 is such a plot, reproduced from Becker and Fenkart (1970). As observational information rapidly diminishes with increasing distance, we limit the present

H. van Woerden et al. (eds.), The Milky Way Galaxy, 335–342.

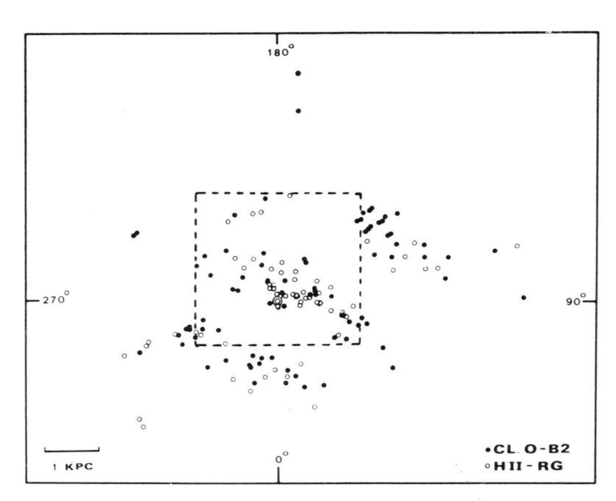

Figure 1. Positions of recently formed objects projected on the galactic plane according to Becker and Fenkart (1970). Dotted lines outline the region discussed in the present paper.

description to the section represented by the dotted square in Fig. 1, i.e. effectively to objects within 1500 pc. We shall see that even within this volume more data within reach of current observational means are required for a comprehensive analysis.

In Fig. 2a to 2c we present the spatial arrangement of the OB-associations, however with the omission of dubious cases like, for instance, Cas OB 14, Cyg OB7 and Vul OB4. We also omit, pending a more complete discussion, presentation of young clusters and of most of the loose groups of very young stars and of high-luminosity field stars; together these form a small fraction of the entire population of recently formed stars. We introduce a co-ordinate system s, t, z, adapted to the orientation of the Orion Arm, to be referred to as the OA system. The z co-ordinate has the conventional meaning; co-ordinate t is in the plane in the direction longitudes 60° - 240°, which is the rounded-off value for the direction of the ridge line of the Orion Arm in Fig. 1; hence the third co-ordinate, s, is roughly perpendicular to the OA. Accordingly, 30° is the chosen pitch angle for the OA. As this is considerably lower than the pitch angles found for spurs in extra-galactic spiral structure (see, for instance, Elmegreen 1980), we shall refrain from using the expression "spur" for the feature under study. Fig. 2a to 2c show the projections of the positions of the associations on the planes s,t, t,z, and z,s respectively.

Two of the main properties of the OB associations are represented in these diagrams. The size of the central dot is a measure of the star-forming activity. It is based on the number N, which is the sum of the number of OB stars brighter than $M_V = -5$ and the number of O-type stars; the latter ones are thus counted double if more luminous than -5. See the key at the bottom of Fig. 2a. In these counts we included the associated clusters. The circles around the central dots are roughly equal in diameter to the projected size of the association in the direction of galactic longitude.

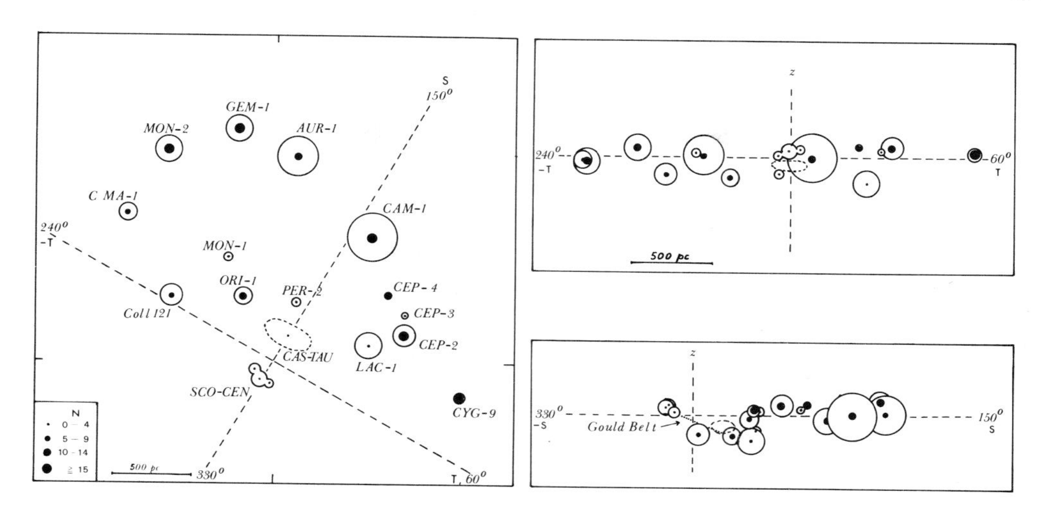

Figure 2. a (left): Position of OB-associations projected on the galactic plane. For meaning of the numbers N and for the Orion-Arm coordinate system s,t see the text; b (upper right): projected positions in the z,t plane; c (lower right): projected positions in the z,s plane.

From Fig. 2a to 2c we note some overall properties. The OA is very thin in the z-direction. The average distance of the associations from the plane (weighted according to the numbers N) is +3 pc and in absolute sense it is 50 pc. The thickness of the OA for the section under consideration here is less than 1/10 of its width measured along the co-ordinate s. At low s-values, the OA is mostly below the plane. We recognize the tilted feature known as the Gould Belt with Ori OB1, one of its constituents, farthest down below the plane; however note that at these s-co-ordinates also Coll 121 and Lac OB1 are well below the plane. Apart from this feature, no wavy or other systematic deviation from a plane structure is discernible in Fig. 2b and 2c. The largest central dots seem to show a preference for large s values, but it would be premature to suggest a systematic trend without more detailed data for Cam OB1, Aur OB1 and Gem OB1. Some of these may, with better data, be resolved into two or more associations and this is also the reason why caution should be exercised before concluding that large dimensions occur at largest s-co-ordinates.

Even within the limited distance range considered here, strong selection effects have to be kept in mind. Included in Fig. 2 is the Cas - Tau Association, a group in advanced state of desintegration containing no stars brighter than M_V = -5 and only 15 between luminosity -5 and -3 which moreover are spread over a relatively large volume; a group like this will not have been recognized at distances beyond a kiloparsec or even less. Associations with small membership at large distances are recognized only if concentrated in small volumes, like for instance Cep OB3 and Cep OB4.

MOLECULAR CLOUDS

Masses of the molecular clouds connected with the OB associations are presented in Fig. 3; there is strong evidence that they form the medium from which the associations originate according to the mechanism of sequential star formation. The numbers next to the positions of the associations give the estimated total mass of the cloud (unit 10000 solar masses), which sometimes includes several components. These masses lie in the range from several thousand to several hundred thousand solar masses, and as a rule far exceed the total mass represented by the association stars. There are no great differences between the linear sizes of the clouds, most of them have dimensions between 50 and 100 pc when measured in the direction of galactic longitude. For a summary of principal cloud properties, see Blitz (1980).

Surveys of molecular clouds (by means of CO or OH) have not yet covered the entire sky to such a degree of completeness that masses and sizes of the clouds of our sample are sufficiently known for statistical discussion. Intermediate and high galactic latitudes are only partly covered, and in some directions of small differential galactic rotation, particularly aorund 90°, it is very hard to separate clouds at different distances. One interesting question to explore will be, which OB associations notwithstanding their recent formation are free of associated clouds. A candidate seems to be Lac OB1, for which the high galactic latitude eliminates background complications. The reverse question, whether there are in this volume of space large molecular clouds free of an early stage of a stellar association, also deserves to be pursued further; positive evidence was found by Wouterloot (1981). A particularly intriguing problem is, to what extent the "hole" in the distribution of the associations at about one kiloparsec in the direction of 200° longitude is filled with still unignited clouds. An other intriguing problem concerns the thin filamentary links between the major cloud complexes, like for instance the one between Orion and Monoceros described by Thaddeus (1982).

AGES

Basic information for the study of the progression of star formation are the ages of the associations and, if applicable, the breakdown according to the ages of their subgroups. The ages, indicated in Fig. 4 (unit one million years), are all of essentially photometric or spectroscopic origin; we do not include here kinematic age estimates. In the most favourable cases - and this applies to the nearer, brighter, associations - multi-colour photometry, mostly in the ubvy-beta system, is available so that reddening-free colour-magnitude

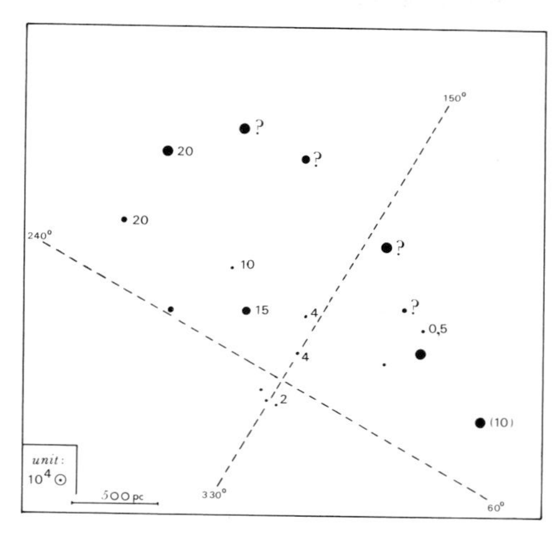

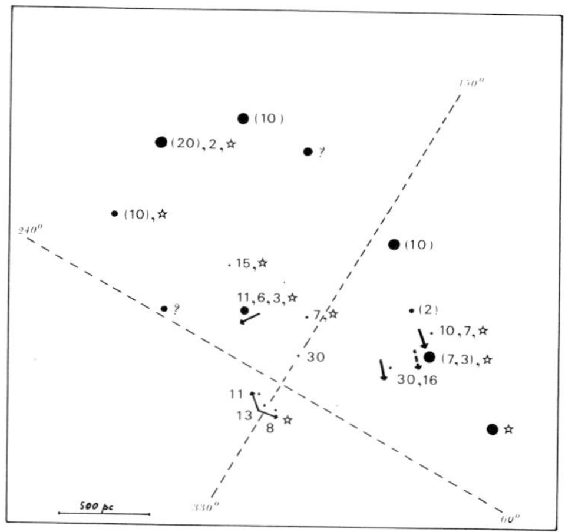

Figure 3 (left). Masses of the identified molecular clouds associated with the OB-associations (unit 10^4 solar masses).
Figure 4 (right). Photometric ages (unit 10^6 years) of the (subgroups of the) OB-associations. Arrows indicate the projected direction of the progression in the formation process within each association. Asterisks mark objects where star formation is still going on.
For identification of the names of the associations see Figure 2.

diagrams could be used. Sequences of numbers in Fig. 4 refer to the ages of subgroups, given in the order of decreasing age. The ages given here were made available to me by J. Brand and T. de Zeeuw of Leiden Observatory and are based on homogenized uvby photometry, carefully transformed into bolometric luminosity - log T_e diagrams. If no such age estimates could be made, less reliable photometric estimates were used, given in parentheses (f.i. based on UBV photometry), and where there is evidence that star formation is still going on this is marked by an asterisk.

The ages in Fig. 4 range from zero to about 30 million years. The oldest groups, Cas-Tau and Lac OB1, are spread very thinly in space and could only be recognized due to their proximity to the Sun or relatively high galactic latitude. Obviously identification of preceding generations of star formation, in the age range 30 to 60 million years, and their location in space will be of great importance for the study of the progression of star formation. One way to identify such groups is by means of space velocities accurate to a few km/sec. Hipparcos will be of basic importance for the nearest domain. For larger distances we will have to rely in addition to Hipparcos on accurate radial velocities in combination with structural features - not yet fully dissolved clustering - in the space distribution. With the data in Fig. 4 no overall gradient in the star formation can be established yet. This will require at least good multicolour photometry of the distant associatons.

Progression of star formation is, however, recognized within the individual associations, and we may ask whether the directions along which the sequential process proceeds reveal alignment or randomness. The available data, in Fig. 4, do not suggest the former. The arrows in Fig. 4 are the projected directions (perpendicular to the line of sight) of the progression from older to younger subgroups. Note that Sco-Cen deviates from the usual pattern in that the oldest subgroup seems to be Upper Cen, from which the sequential process proceeded in two nearly opposite directions: to Lower Cen and, via Upper Sco, to the Ophiuchus molecular cloud.

KINEMATICS

The state of motion of the assembly of OB associations and their associated molecular clouds can be described as, on the one hand, remarkable quietness for most parts of the domain surveyed and, on the other hand, the disturbance due to the expanding motions within the region of the Gould Belt. Using the well-observed radial velocities of the molecular clouds connected with the associations Cep OB3, Cep OB4, Mon OB1, Mon OB2, and C Ma OB1, for which moreover the photometric distances are well known, we find after elimination of the effects of differential galactic rotation and (standard) solar motion a residual radial velocity of only 3 km/sec. In the same way, we find for the OB associations themselves from ten objects a residual radial velocity of 4.5 km/sec only. Associations and connected clouds belonging to the Gould Belt system were not included in these figures. Within the Gould Belt we encounter deviations from this quiet pattern up to 10 km/sec (for Per OB2). Evidence for the remarkably quiet state of motion of the youngest stellar population beyond the Gould Belt has also been found from proper motions by Tsioumis and Fricke (1979).

RUN-AWAY STARS; STOCHASTIC IGNITION OF STAR FORMATION

Fig. 5 shows the position, projected on the galactic plane, of the well-established run-away O and B stars. Criteria were: a space velocity exceeding 40 km/sec and well-determined distance. A few stars were included on the basis of large radial velocity only. Marginal cases reported in the literature based on proper motions only were not included. Different symbols distinguish stars according to mass lower than, or exceeding, 20 solar masses. Dotted connecting lines mark those cases where the parent association is well identified. For the surroundings of the most remote associations, particularly Aur OB1, Gem OB1 and Mon OB2 (see Fig. 2), no well-established run-away stars are marked, reflecting the incompleteness of our data. We estimate that for the whole volume surveyed, only one half of all cases have been identified so far, even less for the low-mass category.

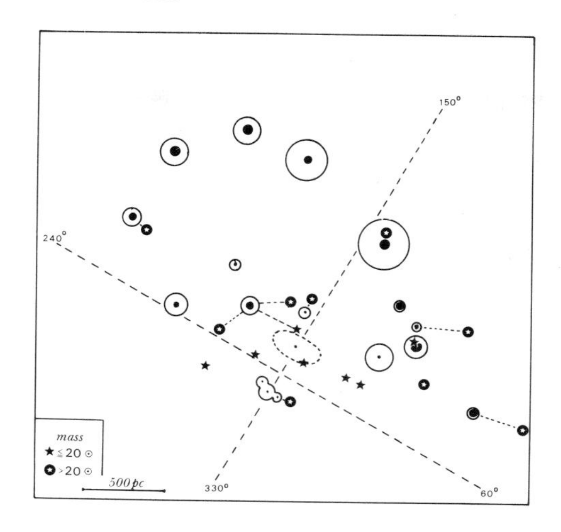

Figure 5. Projected positions of the known OB run-away stars. Dotted lines mark cases where the parent association could be traced back.

Two aspects of the study of run-away stars will deserve further study. One concerns their origin; accurate space velocities to be obtained with Hipparcos may allow identification of the origin for many more than are marked in the diagram, and hence the locations and epochs of past supernova events. An other aspect is their role in remote inducement of star formation. A provisional estimate of the chances of the run-away stars to turn into supernovae in the proximity of molecular clouds away from the parent association - having travelled up to several hundred parsecs - indicates that they may contribute significantly as a stochastic cause of the progression of star formation through the Orion Arm. For this purpose, again, more complete mapping of the molecular clouds, particularly at large distances from the galactic plane, will be essential.

REFERENCES

Becker, W., and Fenkart, R.: 1970, in IAU Symposium no. 38, "The Spiral Structure of our Galaxy", ed. W. Becker and G. Contopoulos, p. 205
Blitz, L.: 1980, in "Giant Molecular Clouds in the Galaxy", ed. P.M. Solomon and M.G. Edmunds, p. 1
Elmegreen, D.M.: 1980, Ap. J. 242, 528
Lucke, P.B., and Hodge, P.W.: 1970, Astron. J. 75, 171
Schmidt-Kaler, Th.: 1977, Astron. Astrophys. 54, 771
Thaddeus, P.: 1982, in "Symposium on the Orion Nebula to honor Henry Draper", ed. A.E. Glassgold, P.J. Huggins, E.L. Schucking, Ann. New York Acad. Sci. 395, p. 9
Tsioumis, A., and Fricke, W.: 1979, Astron. Astrophys. 75, 1
Wouterloot, J.: 1981, Thesis University of Leiden, ch. VI.

DISCUSSION

G.D. van Albada: There is one very clear gradient in your pictures, but that may very well be a selection effect. The distant OB associations appear to be larger on average than the nearby ones.

Blaauw: The distant, large circles in the diagram only indicate the presence of fairly large groups of young stars - we do not have sufficient data to discriminate subgroups or, perhaps, superpositions. The resolution at a distance of, say, 1500 pc is already very small for these objects. Hence the apparent size gradient may not be significant.

B.G. Elmegreen: The velocities of the Orion, Perseus and Sco-Cen OB Associations are directed away from the Sun, and are consistent with the idea that these regions are condensations inside the expanding Lindblad Ring. This region is an example of long-range propagating star formation, with the old Cas-Tau Association being the first generation, located in the ring centre, and the Orion, Perseus and Sco-Cen Associations being the second generation, located on the ring itself.

Blaauw: Indeed, in this region the motions are systematic. Analysis of the relative motions in the Gould Belt indicates a time scale of order 40×10^6 years, similar to that found for Lindblad's Feature A. I agree with the suggestion of propagating star formation, and the question is: To what extent does that occur also in other parts of the Local Spur?

B.G. Elmegreen: Regarding your statement that star formation began at about the same time all over the Local Spur, I wonder if this conclusion would change if you included the local galactic clusters, some of which have ages of 60 million years or more. Would OB associations of this large age still be recognized?

Blaauw: Clusters younger than 20 Myr are very scarce in the region. Clusters with ages of 30-60 Myr exist, but are not plotted in Figure 4. Still older clusters give a sytematically deviating pattern, shifted to the left. I have omitted the clusters, because the stellar content of clusters is small compared to that of associations.

A.I. Sargent: In your Figure 4 I should like to reverse one of the arrows indicating the direction in which star formation progresses. I refer to the feature which includes Cep OB2, Cep OB3 and Cep OB4. You have shown star formation proceeding in Cep OB3. In fact, there is a rather larger-scale progression: few stars are now forming and little molecular gas is left in the vicinity of Cep OB2; active star formation is continuing in the relatively small molecular cloud associated with Cep OB3; while around Cep OB4 is a fairly quiescent molecular cloud, $\sim 80 \times 60$ pc in size, in which we may expect future activity. Similar large-scale patterns have been noted by Elmegreen, for example, in the extended M17 complex, and I should expect that further investigation would reveal more such cases and improve our picture of large-scale star formation in the Galaxy.

A SURVEY OF MOLECULAR CLOUDS ASSOCIATED WITH YOUNG OPEN STAR CLUSTERS

D. Leisawitz and F. Bash
Department of Astronomy
The University of Texas
Austin, Texas, U.S.A.

A major study of the molecular gas surrounding young star clusters is underway. We are using the Columbia University 1.2-m millimeter-wave telescope to observe emission from the $J=1\rightarrow0$ rotation transition of ^{12}CO in the vicinities of 128 open star clusters. The survey region around each cluster is at least 10 cluster diameters in size, typically $\gtrsim$ 5 square degrees. Sensitivity is sufficient to detect lines as weak as 1 K over a range in velocity $\pm$ 83 km/s centered on the cluster velocity and with a velocity resolution of 0.65 km/s. Clusters in this sample have well-determined distances ranging from 1 to 5 kpc, and ages $\lesssim$ 100 million years (Myr).

Carbon-monoxide emission has already been mapped in regions surrounding 14 of these clusters. Preliminary results reveal a tendency for clusters younger than $\sim$ 20 Myr to show evidence of associated molecular clouds. Regions surrounding clusters older than $\sim$ 20 Myr, however, generally do not show CO emission at velocities coincident with those of the clusters. Contour maps of the velocity-integrated CO emission in five representative regions are shown in Figure 1.

The physical mechanism responsible for the removal of molecular gas from young clusters could be the propagation of a dissociation front into the molecular-cloud medium or acceleration of the clouds away from the clusters (e.g., by stellar winds or radiation pressure). A statistical analysis of a large sample of CO maps around cluster regions coupled with H I 21-cm observations, high-resolution observations of molecular line emission, and infrared maps of a small sample of representative regions is expected to reveal the nature of the mechanism. If the clouds are being destroyed, the molecular-cloud lifetime is of the order of 20 Myr.

H. van Woerden et al. (eds.), The Milky Way Galaxy, 343–344.

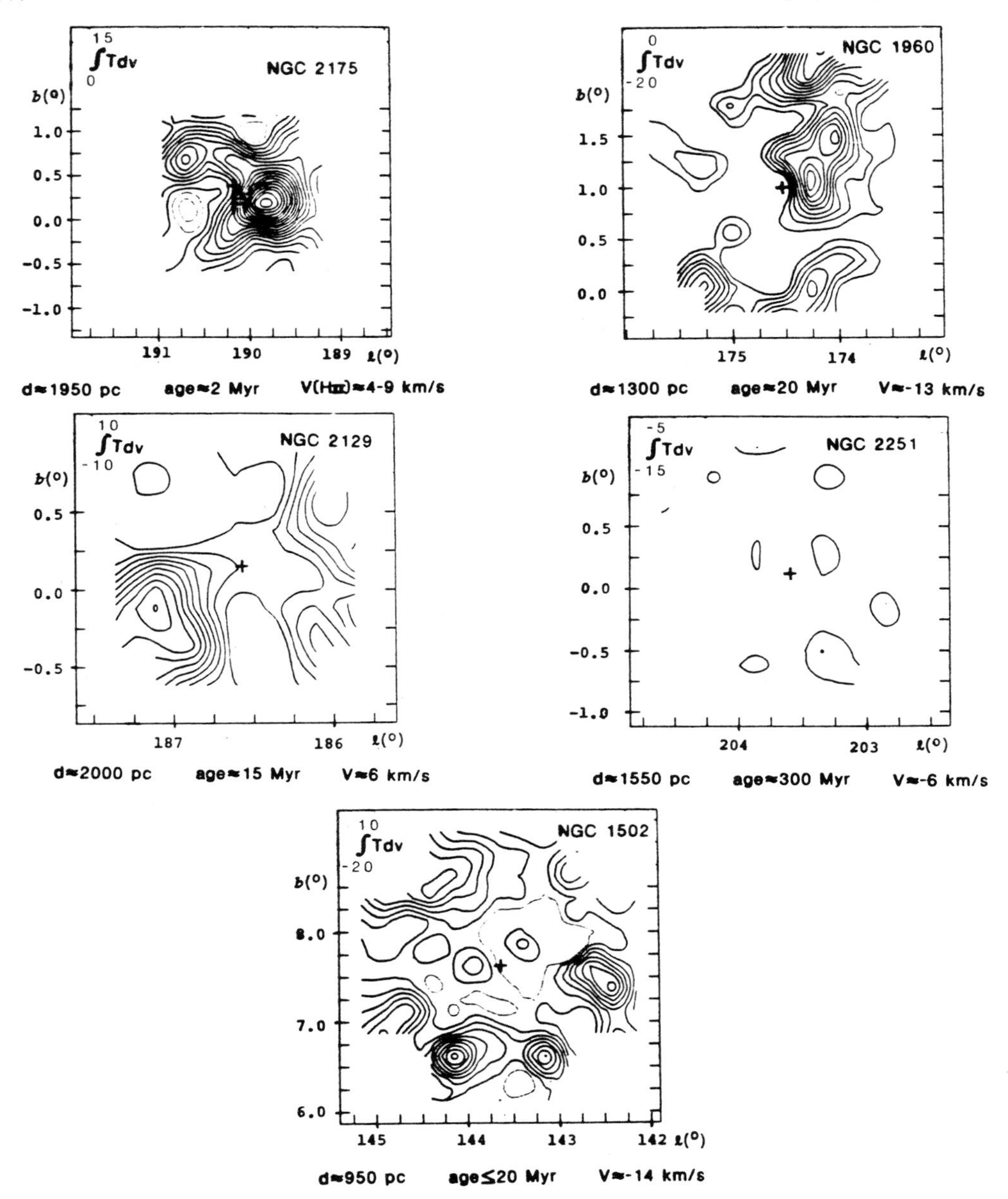

Figure 1. Velocity-integrated CO emission in regions around five open clusters. Units for integral limits are km/s. Cluster distances, ages, and velocities are indicated. Crosses mark cluster positions. All velocities are with respect to the local standard of rest. Contour levels are in steps of 1 K km/s and the lowest level is 1 K km/s, except around NGC 2175 where levels are separated by and begin at 2 K km/s.

A NEW STUDY OF THE NEUTRAL HYDROGEN IN GOULD'S BELT

T.J. Sodroski, F.J. Kerr, and R.P. Sinha
University of Maryland, College Park, Md. USA

The structure and kinematics of the neutral hydrogen associated with Gould's Belt have been studied using data of high velocity resolution and large latitude extent covering $\ell = 10^{\circ} - 350^{\circ}$. The data comprise the Berkeley Survey of Neutral Hydrogen (Weaver and Williams 1973, 1974), and an unpublished survey by Kerr, Bowers, Kerr, and Jackson (1983) using the 60-foot Parkes telescope. The latter is a fully-sampled survey of the region $\ell = 240^{\circ}$ to 350°, $b = -10^{\circ}$ to $+10^{\circ}$.

In order to separate the Gould's Belt feature from other galactic features, a low-pass velocity filter was applied to the data, thereby reducing features of intermediate and high velocity width while passing the Gould's Belt component and other narrow components. Final separation of the Gould's Belt feature was achieved by requiring the temperatures and velocities of the peaks constituting a single "structure" to vary continuously with longitude and latitude. This procedure was preferred over the more conventional gaussian analysis.

A contour map of the peak temperature of the Gould's Belt gas over the entire region studied was produced. In the northern hemisphere, the belt is well defined in the region $\ell = 150^{\circ} - 220^{\circ}$ (including the Perseus and Orion associations), but is extremely patchy outside this region. Also, the tilt of the gas to the galactic plane is significantly smaller than the recent estimates (Stothers and Frogel 1974) of the tilt of the stars in the belt. There is a large gap in the neutral hydrogen, situated near $\ell \sim 235^{\circ}$ and covering at least 25° in longitude. The gap may actually be over 50° in longitude, but a more extensive analysis of data from the southern hemisphere is needed to establish this.

The velocity-longitude relation of the gas was determined for $\ell = 10^{\circ} - 350^{\circ}$ by interpolating to the optical plane of the belt at each longitude. Figure 1 shows the interpolated values of velocity as a function of longitude. Also shown is a velocity-longitude relation for the gas (solid curve) derived by fitting a Fourier series to the

H. van Woerden et al. (eds.), The Milky Way Galaxy, 345–346.

observed velocities. It is intended to fit the velocities of Figure 1 with expanding-shell models of the style of Lindblad et al. (1973) and Olano (1982) to obtain new expansion parameters for Gould's Belt.

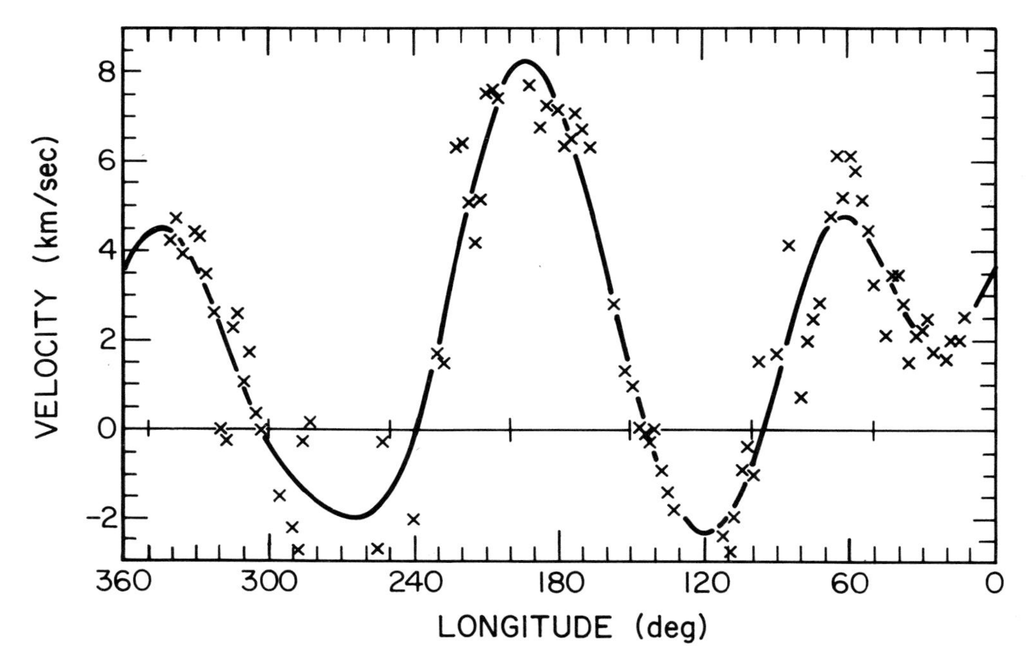

Figure 1: The observed values of velocity for the Gould's Belt gas as a function of longitude(x). Also shown is the "best fit" to the data using a Fourier series analysis (solid curve).

References

Kerr, F.J., Bowers, P.F., Kerr, M., Jackson, P.D.,: 1983, A survey of neutral hydrogen in the Southern Milky Way (In Preparation).

Lindblad, P.O., Grape, K., Sandquist, Aa., Schober, J.: 1973, Astron. Astrophys. 24, pp. 309-312.

Olano, C.A.: 1982, Astron. Astrophys. 112, pp. 195-208.

Stothers, F.M., Frogel, J.A.: 1974, Astron. J. 79, pp. 456-471.

Weaver, H.F., Williams, D.R.W.: 1973, Astron. Astrophys., Suppl. Ser. 8, pp. 1-516.

Weaver, H.F., Williams, D.R.W.: 1974, Astron. Astrophys., Suppl. Ser. 17, pp. 1-445.

SECTION II.7

THE GALACTIC NUCLEUS

Wednesday 1 June, 1115 - 1310

Chairman: B.F. Burke

Students of the Galactic Nucleus: Schwarz and Oort in the garden of Lauswolt.

LZ

THE GALACTIC NUCLEUS

J.H. Oort
Sterrewacht, Leiden, The Netherlands

ABSTRACT

The ionized gas within 2 parsec from the centre region is distributed in spiral-like features which appear to emanate from a nucleus. This nucleus is close to the ultra-compact radio source and the infrared source IRS 16, but does not exactly coincide with either. The spiral is strongly inclined to the galactic plane. The arms are clumpy. The spiral appears to rotate, but there appear to be also large radial motions.

The phenomena show some resemblance to what is observed on a very much larger scale in the nuclei of radio galaxies and quasars.

An exceptionally powerful, variable source of hard X- and gamma-rays, GCX, has been found in the direction of the centre. This may be the true galactic nucleus. Strong, variable emission of the 511-keV electron-positron annihilation line has also been observed to come from the central region.

The evidence for a central black hole is discussed. It is not compelling. The energy required for the strong infrared radiation emitted by the central few parsecs is probably coming from a large number of luminous stars, and due to a recent burst of star formation. The observed state of ionization indicates the burst to have occurred about one million years ago.

The distribution of the dense molecular clouds within 300 pc from the centre is very asymmetric: practically all lie at positive longitudes. Their motions differ radically from circular motions. The layer of clouds is tilted relative to the galactic plane. The so-called "+40-km/s cloud" lies in front of the nuclear spiral, and is therefore moving towards the centre. Different interpretations of the systematic motions are considered. The outspoken asymmetry excludes a static condition. If the "molecular ring" at R $\sim$ 150 parsec is actually an expanding feature, its origin might be sought in the same burst of star formation proposed to explain the state of ionization and the infrared radiation of the nuclear region.

H. van Woerden et al. (eds.), The Milky Way Galaxy, 349–365.

THE MINI-SPIRAL

An extremely interesting insight into the nucleus of the Milky Way System has recently been obtained from high-resolution observations at wavelengths between 2 and 20 cm. The region within ∿ 40", or 2 pc, from the centre is shown in Fig.s 1 and 2. Fig. 1, from an investigation by

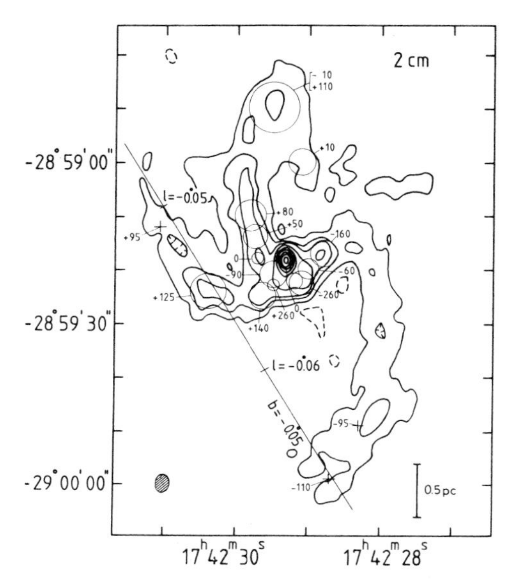

Fig. 1. Contour map at 2 cm of spiral structure in the galactic nucleus. Half-power beamwidth 2"x3" (Ekers *et al.* 1983). The circles indicate positions and diameters of [NeII] clouds; the numbers are velocities in km s^{-1}, determined by Lacy *et al.* (1980).

Ekers, van Gorkom, Schwarz & Goss (1983), shows the distribution at λ=2 cm measured with the VLA with a half-power beamwidth of 2"x3". Fig. 2 gives observations with the same instrument at 6 cm, and a still higher resolution (1"x2"), recently obtained by Lo and Claussen (private communication). Main features are a striking spiral-like pattern, with a strong central bar. The dense concentration near the centre in Fig. 1 is that around the compact radio source. The pictures suggest ejections of jets from the nucleus, principally in a nearly East-West direction, and turning into a North-South pattern at about 1/2 pc from the centre. It resembles the structures observed near the nuclei of radio galaxies and quasars. The structure is clumpy. The clumps are unstable, with half-lives of about 10^3 yr as estimated from their internal velocity. The average density in the spiral is about $4x10^3$ cm^{-3}. Its total mass is estimated at some tens of solar masses. Per year a mass of the order of $0.001M_\odot$ appears to have been ejected. The denser parts had been observed in the [NeII] line at 12.8 μm by Lacy *et al.* (1980) at Berkeley; these provide much information concerning the velocities in different parts of the pattern; part of their data are indicated in Fig. 1. High velocities (of +260 and -260 km s^{-1})

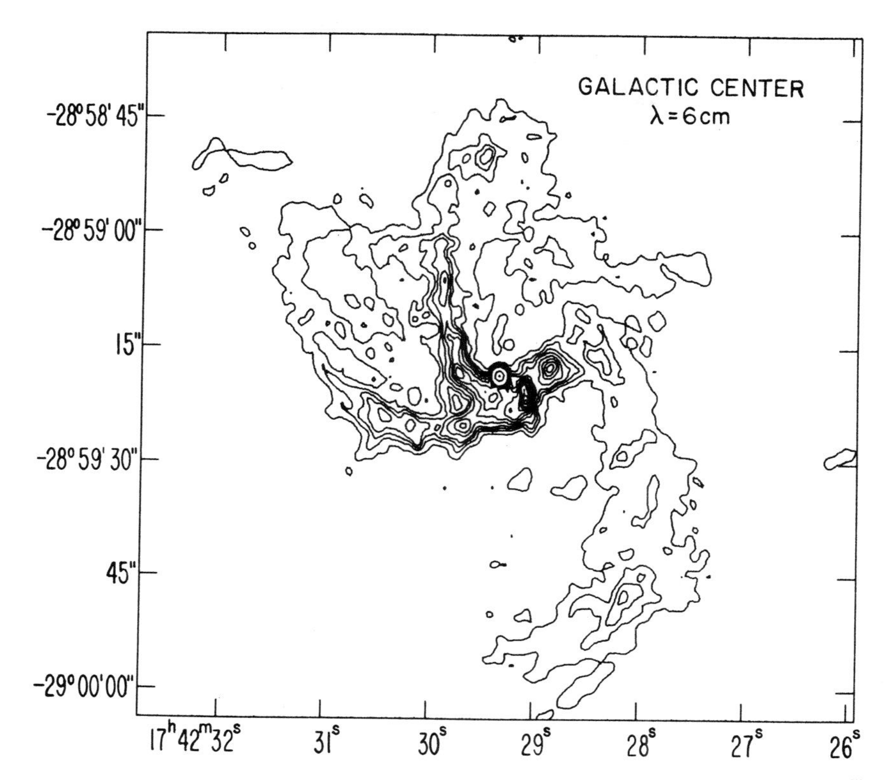

Fig. 2. Contour map at 6 cm. Resolution 1".1 x 1".3 (Courtesy of K.Y. Lo & M.J. Claussen).

occur in the immediate vicinity of the nucleus; at the ends of the "EW bar" they have dropped to +125 km s^{-1} on the E side and -160 km s^{-1} on the W side. In the N arm as well as in the NE arm the velocities are roughly +100 km s^{-1}, in the S arm -100 km s^{-1}; these velocities are in the same direction as the general rotation of the Galaxy, giving the impression that expelled gas is being swept along by gas already previously present in a rotating disk. Observations of hydrogen recombination lines at both Westerbork and the VLA (van Gorkom, Schwarz & Bregman 1985) have confirmed and extended the [NeII] velocities. If we are indeed witnessing ejection, its direction must evidently have made a large angle with the galactic plane. The hypothetical disk must therefore also have made a large angle with this plane.

The velocities do not furnish unambiguous evidence for outward motion. They can equally well be fitted to infalling clouds. But in order to come so close to the centre the clouds must have had extremely small initial transverse velocities, and it is a somewhat improbable accident that three clouds should be falling in simultaneously, two of which appearing to come from opposite directions. It is also difficult to see how the highly unstable clumps observed in the [NeII] lines could have formed in the infalling clouds.

Infall is strongly favoured by K.Y. Lo & M.J. Claussen, who are preparing an article on the subject. They suggest that a supernova exploded a few times 10^4 years ago and that clumps of swept-up gas are now falling back and are ionized by a concentration of OB stars near the centre.

Brown (1982), who first observed the spiral-like shapes (Brown et al. 1981), has proposed ejection from a rotating nozzle. Like the infall model this avoids the problem of having a pre-existing dense rotating disk.

The velocities observed at the positions nearest the centre (+260 km s^{-1} and -260 km s^{-1} respectively, at R $\sim$ 0.1 pc) furnish a rough indication of the mass near the centre required to obtain such infall velocities, or, alternatively, the mass required to explain the decrease of expulsion velocities from $\sim$ 260 km s^{-1} at 0.1 pc to $\sim$ 100 km s^{-1} at 1 pc. This is of the order of 10^6 $M_\odot$.

A CENTRAL BLACK HOLE?

It has variously been suggested that the bulk of this mass might be in a black hole. The evidence, however, is rather indirect, and cannot be considered as compelling. It rests principally, on the one hand, on observations of the systematic motion of the interstellar neutral hydrogen in the central 500 pc, and on the other hand, on the distribution of population II stars inferred from the near-infrared radiation, the latter combined with an $M/L_{2.2\mu}$ inferred from the nucleus of M31 (cf. Oort 1977). The observations give some evidence that the star distribution does not provide a sufficiently large mass to explain the motions, and that therefore an additional mass of the order of a few million solar masses may be required. But the arguments are not conclusive (cf. also Isaacman 1981, who has derived a value for the mass within R = 1 kpc from the distribution and motions of planetary nebulae).

It should be noted that a star concentration should have a density of 2 x 10^8 $M_\odot pc^{-3}$ to reach such a mass in a volume of 0.1 pc radius, while a density of 10^{10} hydrogen atoms per cm^3 would be needed for a gas clump of the same mass. Both values are improbably high. Whether a black hole at the centre would be observable depends among other factors on its present rate of accretion.

There are three peculiar objects which have been considered as possible candidates for the actual nucleus: (1) an ultra-compact radio source, (2) a sharp concentration of 2-μm radiation, called IRS 16, and (3) a source of X- and γ-rays called GCX. I shall deal with each of these in succession.

THE COMPACT RADIO SOURCE

VLBI observations have shown that the radio source is extremely small. The most recent measures at a wavelength of 1.3 cm, reported by K.Y. Lo, appear to have resolved it, and to have overcome the influence of interstellar scattering. These observations yield a diameter of 0".0035, or 5 x 10^{14} cm = 30 A.U. (Lo & Claussen, private communication). Previously Kellermann et al.(1977), observing at 3.8 cm, found a core

$\lesssim$ 0"001; this core has not been seen in later observations. The main source is variable with amplitudes of 20 to 40% on time scales ranging from days to years (Brown & Lo 1982). The variations are larger at 3.7 cm than at 11 cm; the 11-cm flux has increased systematically by a factor 1.5 in three years.

THE SOURCE IRS 16

This lies at the centre of a general concentration of 2- to 3-μm emission, and has been thought to be the core of the distribution of population II (cf. Bailey 1980). This is now doubted on the ground that its spectrum does not show the CO absorption typical for late-type giants; cf. Wollman *et al.*(1982) who conclude:"If it is a central condensation of stars, the population is abnormal". Recent observations of a 1500 km s^{-1} wide He emission line in this source, and the total absence of a similar H line (Hall *et al.* 1982) also indicate that something quite different is involved in IRS 16. Finally, IRS 16 does not coincide with the dynamical centre suggested by Fig.'s 1 and 2, but lies about 4", or 0.2 pc, NE of it. The ultra-compact radio source lies at a similar distance, but apparently does not coincide exactly with IRS 16; it lies 1"8 W of it (Ekers, private communication).

GCX

In the 0.5 - 4.5 keV energy range observations with the Einstein satellite at a resolution of 1' have located 12 sources within 0°5 from the centre, one of which coincides with Sgr A West. This is a relatively weak source, and has not varied in brightness between two observations 6 months apart (Watson *et al.* 1981). Observations with HEAO-1 in the 10-100 keV range have given particularly interesting information (Levine *et al.* 1979). The resolution, of $\sim$ 1°6, is much less than for the Einstein satellite, but a central source is found which in the higher half of this energy range surpasses all other sources in a field between ℓ -40° and $+40^{\circ}$, b -20° and $+20^{\circ}$ except for a transient source Nova Oph 1977. This central source is variable on a time scale of the order of half a year. Fig. 3, taken from Matteson (1982), illustrates the nature of the source in comparison to the other sources in the field. The maximum luminosity in the 10 keV to 10 MeV range is $\sim 3 \times 10^{38}$ erg s^{-1}, and at these energies is the largest of any known galactic source (Matteson 1982). Although the positional uncertainty is considerable ($\sim$ 0°5), the unique nature of this source, and its probable coincidence with the more accurately located Einstein X-ray source, make it likely that it is the galactic nucleus itself (cited from Matteson 1982). The source was observed in three periods separated by half-year intervals, and was found to vary considerably, in particular at the highest energies, indicating a diameter of 0.2 pc or less. The variation has been confirmed and extended by other observations, notably by the HEAO-3 satellite. At 300 keV variations by factors 4 to 8 have been found. At the still higher energies (10 MeV - 1GeV) observed by the COS-B satellite, with $\sim 1^{\circ}$ resolution, there is a source, 2CG359-00,

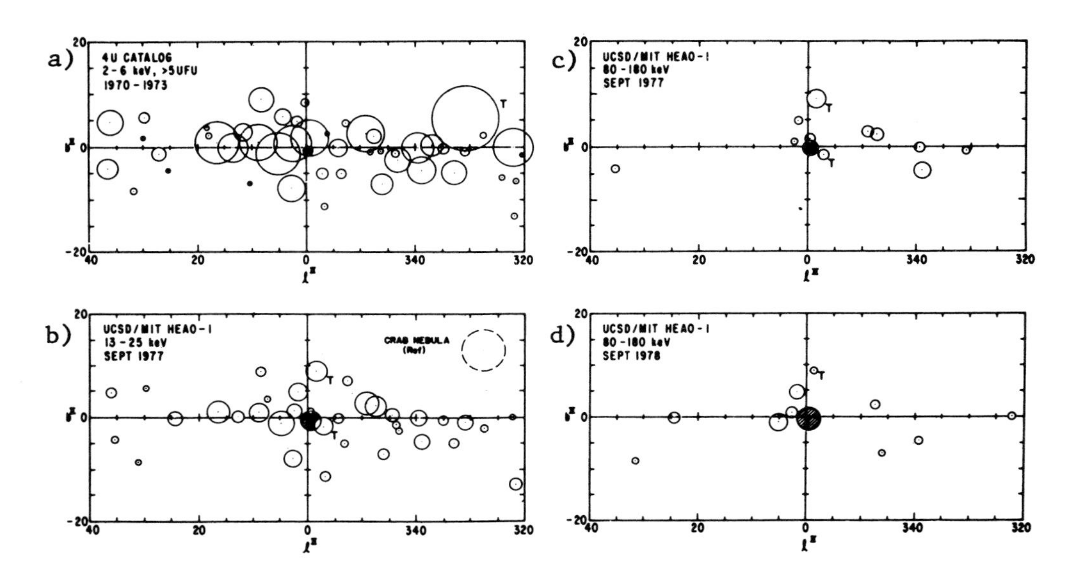

Fig. 3. X-ray maps of the galactic-centre region, with ℓ^{II}, b^{II} in degrees. The area of a circle is proportional to a source's luminosity per logarithmic energy band (Matteson 1982).

at 1° from the centre, but this is not unique in strength, and cannot yet with certainty be identified with the galactic nucleus.

It may be concluded from the above observations that the nucleus produces energetic particles corresponding with a maximum effective temperature of $\sim 10^{10}$ K, and varying in intensity in a period estimated to be of the order of 1 to 2 years.

THE POSITRON-ELECTRON ANNIHILATION LINE AT 511 keV

Beside continuum emission, wide-angle instruments have also detected line emission at a wavelength corresponding to 511 keV, coming from the general direction of the centre. Like the continuum the line emission is variable, and on a similar time scale. This proves that it probably comes from a single object with a diameter of the order of 0.2 pc or smaller. It is tempting to believe that it must in some way be connected with the nucleus itself. The line results from the annihilation of positrons. The narrowness of the line shows that the positrons must have been effectively cooled before recombination. The gas contained in the nuclear spiral could well provide the appropriate medium for this. According to Lingenfelter & Ramaty (1982) the positrons are most likely produced by photon-photon collisions in the vicinity of a compact source in the galactic centre. These authors conclude that the diameter of the positron source must be $\sim 3 \times 10^{8}$ cm and the mass of the black hole be $\sim 500\ M_{\odot}$. Blandford (1982) has, however, suggested that the positrons could be produced in the magnetosphere of a $\sim 10^6$ $M_\odot$ black hole, while Novikov et al. (preprint) have proposed interaction of a γ-ray beam from a massive black hole with an interstellar cloud as a possible mechanism. It is still unclear what relation, if any, exists between the variable X- and γ-ray source, the positron source and the minispiral of ionized gas.

ENERGETICS AND SOURCE OF IONIZATION IN THE CENTRAL FEW PARSECS

Lacy et al. (1980) have extensively discussed the state of ionization of the gas. This is considerably lower than in other known HII-regions, and indicates an ionizing source of about 30 000 K. Such a source could be provided by a burst of star formation about one million years ago.

For a 3' region centred on the galactic nucleus we have the following requirements:
From far-IR radiation and from grain temperature we conclude that $> 1.5 \times 10^{50}$ Ly_{cont} photons per sec are needed, while the state of ionization shows that the effective temperature of the ionizing stars must be less than 35 000°. From the presence of more than 7 red supergiants it is expected that there will be more than 80 blue supergiants. With a normal distribution in spectral type and luminosity these would give a luminosity exceeding 1.5×10^7 $L_\odot$, a Ly_{cont} flux of more than 1.9×10^{50} sec^{-1} and a Ly_{cont} temperature of $\sim$ 35 000 K.

Rieke & Lebofsky (1982) conclude: A star burst causing rapid star formation up to about 10^6 years ago can entirely explain the energetics in the inner few parsecs. The galactic centre is therefore similar to the star-burst powered nuclei of many spiral galaxies.

SUPERNOVA REMNANT CLOSE TO THE CENTRE

There is one more phenomenon which should be mentioned in connection with the region within 5 pc from the centre, viz. a shell-like structure of non-thermal radiation, which is clearly shown in Fig. 4 as an elliptical ring with a diameter of roughly 9 pc along the galactic plane and 6 pc perpendicular to it. This is most likely a supernova remnant. In earlier studies the brightest part on the Eastern side has been called Sgr A East. The centre lies at an apparent distance of roughly 50", or 2 pc, from the centre of the small spiral, and is probably also spatially close to it. There is no clear evidence that the shell has interacted with the gas in the spiral.

SUMMARY CONCERNING THE COMPACT NUCLEUS

The motions within 1 pc from the centre indicate the presence of a mass of at least one million solar masses which may plausibly be ascribed to a black hole, though the evidence is not yet convincing. A magnetized black hole of this order may also provide a mechanism for creating the large number of positrons which produce the strong, variable radiation in

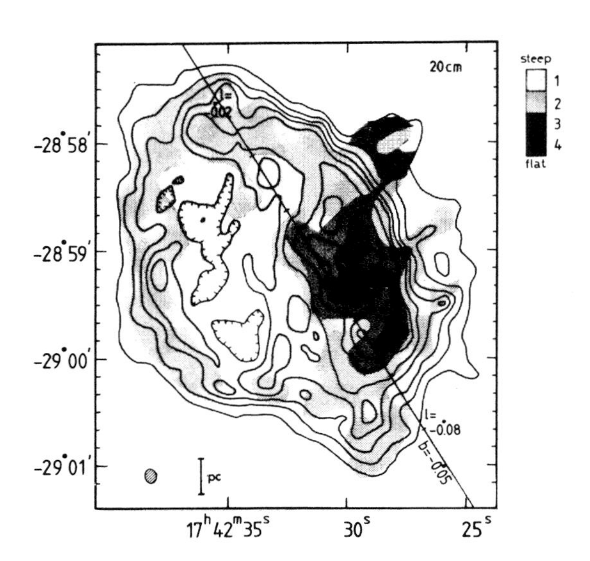

Fig. 4. Spectral index distribution superimposed on 20-cm map. Four intervals are given: 1. $\alpha < -1$; 2. $-1 < \alpha < -0.5$; 3. $-0.5 < \alpha < 0.0$; 4. $\alpha > 0.0$ (Ekers et al. 1983).

the 511 keV electron-positron annihilation line. A black hole may furthermore be needed for the production of the curious spiral features observed in the central 2 pc.

It is likely that the exceptionally strong and variable source of X- and γ-rays in the 50 keV to 10 MeV range called GCX coincides with the nucleus; this is corroborated by the close coincidence with Sgr A West of the accurately positioned, but less energetic point source observed in the 0.5 - 4.5 keV range.

Both the ultra-compact radio source and IRS 16 (whose positions appear to differ by 1".8) might coincide with the X-γ ray source; they are situated well within the uncertainty range of its position; but neither of them may qualify for the actual nucleus, because they lie about 4" away from what appears to be the dynamical centre of the spiral features.

THE WIDER SURROUNDINGS OF THE CENTRE

The General Mass Distribution

This can best be derived from the radiation in the near infrared, which may be assumed to come from population II stars. For the region within ∿ 1° from the centre the most relevant observations are those of Becklin & Neugebauer (1969) at 2.2μm. They have been discussed most extensively by Sanders & Lowinger (1972) and have been confronted with observations of HI velocities and with observations of the infrared and optical brightness in the nucleus of M31 to derive the mass distribution and the gravitational field (cf. also Oort 1977). The density varies approximately as $R^{-1.8}$. Recent observations by Matsumoto et al.(1982) with a resolution of 0°.5 and 0°.8 at 2.4 and 3.4μm have given a valuable extension to ∿ 10° from the centre. Their Fig. 1 shows the distribution at 2.4μm corrected for extinction. They also give a comparison of the rotation curve derived from these data with that derived from HI velocities.

HII REGIONS WITHIN 0°.25 (40 PC) FROM THE CENTRE

Fig. 5 shows the distribution (due to Pauls et al. 1976) of these regions. There are two peculiar circumstances: Firstly, all the ionized gas lies at positive longitudes. It forms a structure which has been

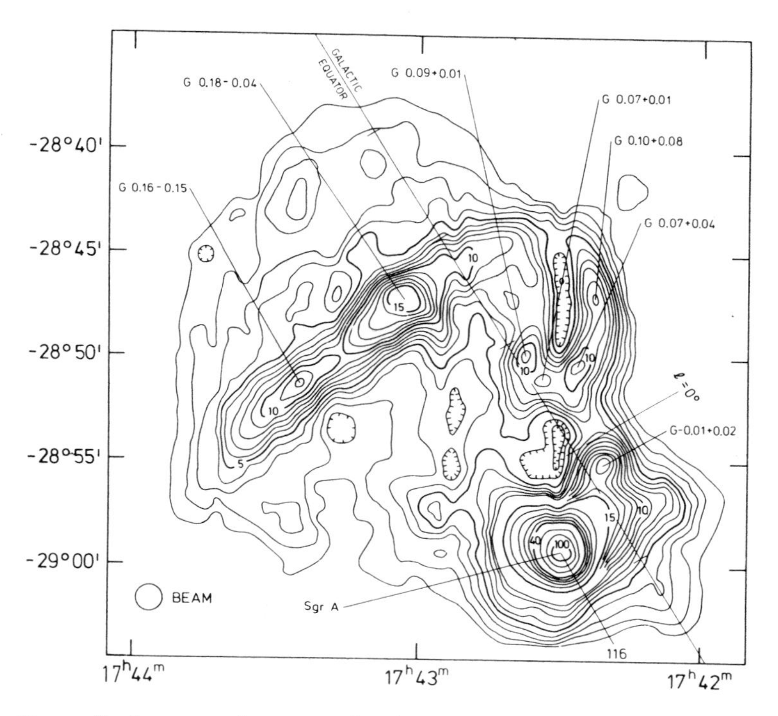

Fig. 5. Map of the continuum emission at 10 GHz within 15' of the centre, made with the Bonn 100-m telescope (Pauls et al. 1976).

called the Arc by the discoverers in Bonn. Secondly, the radial velocities of all clouds, as derived from hydrogen recombination lines, are remarkably small, ranging from -48 km s^{-1} at the longitude of the centre to -20 km s^{-1} at the sharp bend in the North, and smaller still along the North band of the arc. These velocities are much smaller than the circular velocities of 100-150 km s^{-1} at the same distance from the centre, inferred from the distribution of the infrared radiation. If the latter values for the circular velocity are correct the clouds clearly do not move in circular orbits. They might have been expelled from the nucleus and be near their apocentres. Westerbork observations have indicated that there is considerable fine-structure in these features.

MOLECULAR CLOUDS

Large-beam observations of the absorption lines of OH and H_2CO in Sgr A show two overpowering features, at ~ -130 and $\sim +40$ km s^{-1}, respectively; both are very wide, with half-intensity widths of ~ 70 km s^{-1} (Fig. 6). The first appears to be part of a large expanding feature extending to nearly 1° (150 pc) on either side of the centre, and has a counterpart of comparable mass on the far side of the centre, moving away from it at about 160 km s^{-1}, as inferred from observations of CO emission. Together, these features give the impression of a massive expanding ring; but other interpretations have also been suggested.

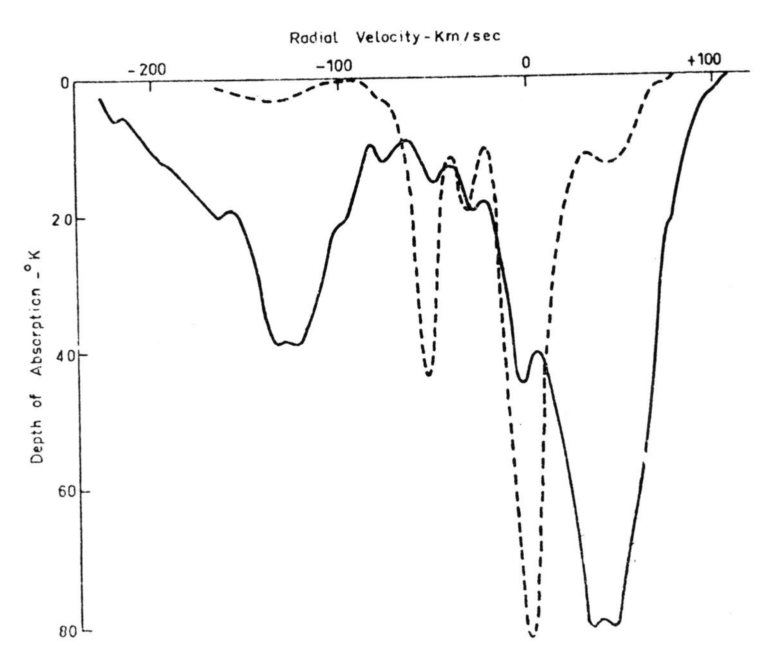

Fig. 6. Absorption profile of Sgr A observed with the Parkes telescope (Bolton et al. 1964). Full curve: OH; dashed curve: HI.

THE +40 KM S^{-1} CLOUD

The feature at +40 km s^{-1} is likewise part of an extended structure, as may be seen from Fig. 7 showing the HCN emission at 3.4 mm. Its main mass lies at positive longitudes, where it extends to $\sim +0\overset{\circ}{.}2$ (Fukui et al. 1977 and 1980). The cloud has a total mass estimated to be between 10^6 and 10^7 $M_{\odot}$. There has been considerable controversy concerning the question whether it moves inward towards the centre, as would be suggested by its

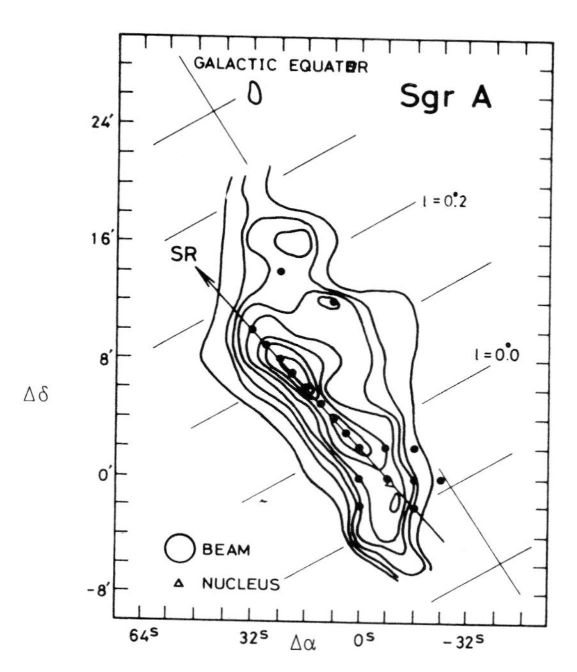

Fig. 7. The HCN emission from the "+40 to +50 km s^{-1} cloud" (Fukui et al. 1980; cf. also Fukui et al. 1977).

being seen in absorption against Sgr A; or whether it lies behind the centre and is seen in absorption against Sgr A East, which might lie beyond the centre, in which case it could be moving outward from the true centre. Recent VLA observations by Whiteoak et al. (1983) appear to have settled the problem. These observations with a beam of 3″.5 show H_2CO at velocities between +40 and +45 km s^{-1} in absorption against the small nuclear spiral discussed in the first part of this lecture. As the latter is presumably connected with the real nucleus, we must conclude that at least the part of the +40 km s^{-1} cloud covered by the VLA observations is moving inward towards the centre. A similar conclusion has earlier been stated by Sandqvist (1982). As the phenomena observed in the molecular clouds of the central region appear to require expulsion and subsequent falling back of these clouds, the observation of inward moving features should cause no surprise.

Most interesting information concerning in- and outward moving gas clouds has come from an investigation by Liszt et al. (1983), who have observed HI absorption against a region of about 3′ diameter around Sgr A West. The observations were made with the VLA at an angular resolution of 12″. The authors found absorption features over the range from -200 to +140 km s^{-1}. Only the gas at velocities larger than +150 km s^{-1} is unabsorbed and must lie beyond the centre.

An interesting detail is that, with their very small beam, Whiteoak et al. did not see any sign of absorption of the +40 km s^{-1} cloud against the compact radio source, although they did see the cloud in absorption against the small spiral at this position.

Must we conclude that the point source lies in front of the +40 km s^{-1} cloud? This is implausible, because the large size of the cloud would then place it at least 10 pc in front of the centre, while its lateral distance is only 0.1 pc or less. Liszt has drawn my attention to the extreme clumpiness of the H_2CO distribution in general, and that as a consequence the number of clumps of the +40 km s^{-1} cloud expected to lie directly in front of the point source might be sufficiently small for the line of sight to the point source not to cross any.

OTHER PHENOMENA OF THE GAS IN THE CENTRAL REGION

Let us now briefly consider some other phenomena observed in the distribution and motion of the gas in the nuclear region.

Three main peculiarities have to be explained: 1. the large radial components observed in the gas motions, 2. the striking asymmetry in the distribution of the molecular clouds (cf. Fig. 8), and 3. the fact that the high-velocity gas within the central 1-2 kpc lies in a tilted layer. The latter phenomenon has been studied most extensively by Liszt & Burton (1980), who concluded that between $\sim 1^{\circ}$ and $\sim 6^{\circ}$ from the centre the gas disk is tilted by as much as 24° to the galactic equator. In the inner few hundred parsecs the tilt is still present but is much less.

EXPANDING FEATURES?

I have mentioned the "expanding molecular ring" which, at a distance of 190 pc from the centre, appears to have an average radial component away from the centre of 150 km s^{-1}. If this gas has been expelled from the centre it has been decelerated by sweeping up gas from a rotating disk, as indicated by its having a transverse velocity of about 50 km s^{-1} in the direction of the general rotation. The mass of the ring is estimated at about 10^7 $M_{\odot}$.

Liszt & Burton (1980) have proposed a different interpretation of the feature. They proposed an alternative description of this as well as all features farther from the centre that had been ascribed to expanding gas. In their model the gas is supposed to move in central ellipses of considerable excentricity, with major axes in a direction making a considerable angle with the line of sight, as well as with the galactic plane. The model describes sufficiently well all observed velocity contours, but no satisfactory dynamical interpretation has yet been given.

The most straightforward interpretation of the molecular ring would still seem to be that it is ejected from the nucleus, although the mechanism causing the ejection is still obscure. The ejection should have taken place about a million years ago. It is possible that the star burst which has supposedly formed the late O stars required to explain the observed state of ionization of the nuclear gas clouds has been responsible for this ejection.

Güsten & Downes (preprint 1983) have recently drawn attention to formaldehyde at still higher velocity (~ -190 km s^{-1}) which is seen in absorption against Sgr A. They believe that it has been ejected. It extends

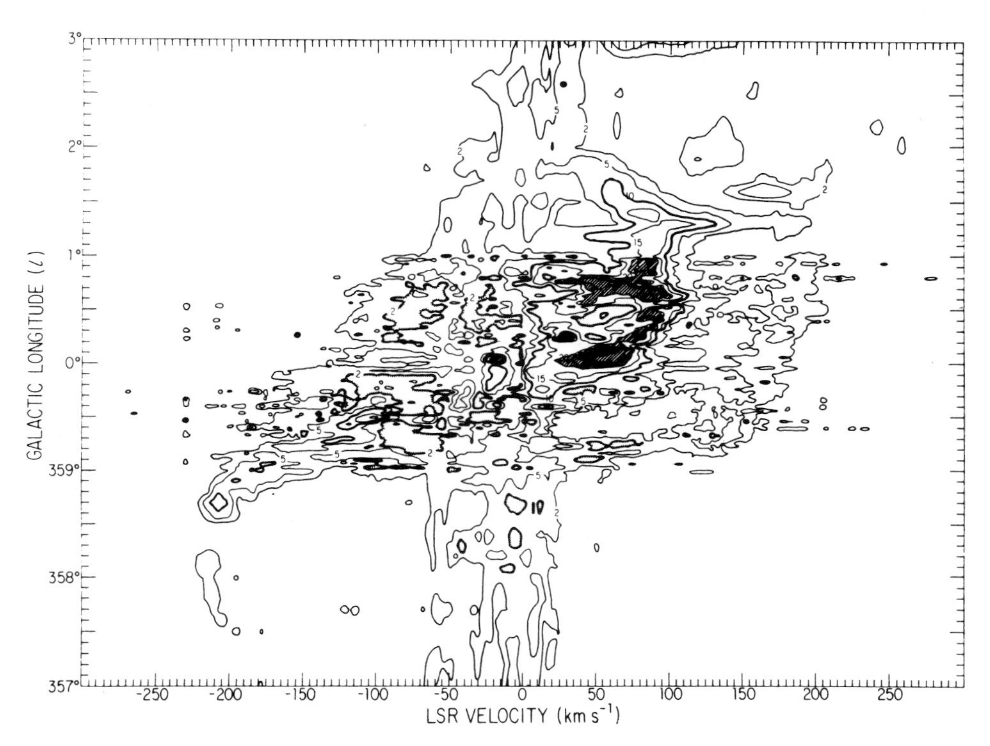

Fig. 8. Longitude-velocity map of ^{12}CO emission at b = 0°. From observations with the 11-m NRAO telescope on Kitt Peak (Bania 1977). The part between ℓ = 359° and ℓ = 1° is from a study by Liszt et al.(1977). Contour levels are drawn at T^* = 2, 5, 10, 15, 20, 25, 30 K. Heavy line: 10 K contour; hatched areas are above 20K; black is above 30K.

to at least 5 pc from the centre.

ASYMMETRY AND LOW ROTATION OF THE BULK OF THE MOLECULAR GAS

The distribution of the molecular gas as observed in the CO emission at 2.6 mm is extremely asymmetric, cf. Fig. 8. All dense clouds are situated at positive longitudes. The strongest concentration, Sgr B2, at ℓ = +0°.67 has been the subject of numerous investigations. It is a strong radio source, containing at least 7 giant compact HII regions, as well as some dense molecular and dust clouds; the CO clouds have a velocity range from +45 to +100 km s^{-1}. There is an apparent bridge at +70 to +90 km s^{-1} between the dense feature at ℓ = 0° (which is mostly due to the +40 km s^{-1} cloud) and Sgr B2. Some fairly dense features continue to +1°.5 longitude, beyond which there is a fairly rapid drop. The "+40 km s^{-1} cloud" itself lies also almost completely at positive longitudes.

A similar preference for positive longitude has been described above for the "Arc" of ionized gas near the centre.

The molecular gas at positive longitude exhibits a systematic motion in the same direction as the general rotation, but the velocity is considerably less than the circular velocities corresponding to the assumed gravitational field.

Clearly, the distribution and motions are widely different from equilibrium conditions. As the time of revolution in the region concerned is only a few million years, the disturbance which has caused the anomalous distribution and motions must have been quite recent. Its nature is unknown. In view of the large masses involved (of the order of 10^8 $M_\odot$) it must have been powerful.

A brief reference must be made to the anomalous motions of some large features at greater distances from the centre, the most massive of which is the 3-kpc arm, which at the longitude of the centre has an outward velocity of 53 km s^{-1}. There are half a dozen other extended features between ∿ 1 and 4 kpc from the centre which at $\ell = 0^o$ show outward motions between ∿ 100 and ∿ 150 km s^{-1}. (Cf. a summary by Oort (1977), based on discussions by van der Kruit (1970) and Cohen & Davies (1976)). All these may be interpreted as expanding structures pushed out by gas expelled from the centre, or, alternatively, as gas moving in excentric orbits in a bar-like potential field, as supposed by Liszt & Burton (1980). The expulsion theory would require a state of nuclear activity far greater than what is observed at present. The phenomena observed in Seyfert galaxies, and, in particular, those in the seemingly normal spiral NGC 4258, show that activity of the scope required does occur in spiral galaxies.

However, the most direct indication of violent activity - in a relatively nearby past - is shown by the molecular clouds within 250 pc from the centre, by the "Arc" within 40 pc, and by the minispiral of ionized gas within 1 pc from the centre.

The hard X- and γ-ray source at the centre shows that activity is continuing at the present time.

THE TILT OF THE GAS LAYER

This might or might not be connected with the same "disturbance" that caused the phenomena discussed above, but most likely this is mainly due to a triaxial distribution of the central mass. The nature of the motions to be expected in a triaxial gravitational field in the central region is being investigated extensively by W.A. Mulder and T. de Zeeuw in Leiden.

REFERENCES

Bailey, M.E., 1980, MNRAS 190, 217

Bania, T.M., 1977, Ap.J. 216, 381

Becklin, E.E., Neugebauer, G., 1969, Ap.J. Lett. 157, L 31

Blandford, R.D., 1982, in The Galactic Center, eds. G.R. Riegler & R.D. Blandford, p. 177

Bolton, J.G., Gardner, F.F., McGee, R.X., Robinson, B.J., 1964, Nature 204, 30

Brown, R.L., Johnston, K.J., Lo, K.Y., 1981, Ap.J. 250, 155
Brown, R.L., Lo, K.Y., 1982, Ap.J. 253, 108
Brown, R.L., 1982, Ap.J. 262, 110
Cohen, R.J., Davies, R.D., 1976, MNRAS 175, 1
Ekers, R.D., van Gorkom, J.H., Schwarz, U.J., Goss, W.M., 1983, Astron. Astrophys. 122, 143
Fukui et al, 1977, Publ. Astron. Soc. Japan 29, 643
Fukui, Y., Kaifu, N., Morimoto, M. Miyaji, T., 1980, Ap.J. 241, 147
Hall, D.N.B., Kleinmann, S.G., Scoville, N.Z., 1982, Ap.J. Lett. 260, L53
Isaacman, R., 1981, Astron. Astrophys. 95, 46
Kellermann, K.I., Shaffer, D.B., Clark, B.G., Geldzahler, B.J., 1977, Ap.J. Lett. 214, L61
Lacy, J.H., Townes, C.H., Geballe, T.R., Hollenbach, D.J., 1980, Ap.J. 241, 132
Levine, A. et al, 1979, B.A.A.S. 11, 429
Lingenfelter, R.E., Ramaty, R., 1982, in The Galactic Center, eds. G.R. Riegler & R.D. Blandford, p. 148
Liszt, H.S., Burton, W.B., Sanders, R.H., Scoville, N.Z., 1977, Ap.J. 213, 38
Liszt, H.S., Burton, W.B., 1980, Ap.J. 236, 779
Liszt, H.S., van der Hulst, J.M., Burton, W.B., Ondrechen, M.P., 1983. Submitted to Astron. Astrophys.
Matsumoto, T. et al, 1982, in The Galactic Center, eds. G.R. Riegler & R.D. Blandford, p. 48
Matteson, J.L., 1982, in The Galactic Center, eds. G.R. Riegler & R.D. Blandford, p. 109
Oort, J.H., 1977, Ann. Rev. Astron. Astrophys. 15, 295
Pauls, T., Downes, D., Mezger, P.G., Churchwell, E., 1976, Astron. Astrophys. 46, 407
Rieke, G.H., Lebofsky, M.J., 1982, in The Galactic Center, eds. G.R. Riegler & R.D. Blandford, p. 194
Sanders, R.H., Lowinger, T., 1972, AJ, 77, 292
Sandqvist, Aa, 1982, in The Galactic Center, eds. G.R. Riegler & R.D. Blandford, p. 12
Van der Kruit, P.C., 1970, Astron. Astrophys. 4, 462
Van Gorkom, J.H., Schwarz, U.J., Bregman, J.D., 1985. In preparation and these Proceedings, p. 371
Watson, M.G., Willingale, J.E., Grindlay, J.E., Hertz, P., 1981, Ap. J. 250, 142
Whiteoak, J.B., Gardner, F.F., Pankonin, V., 1983, MNRAS 202, 11P
Wollman, E.R., Smith, H.A., Larson, H.P., 1982, Ap. J. 258, 506

DISCUSSION

B.F. Burke (Chairman): We are used to subjects getting larger and larger. This one becomes smaller and smaller as we proceed.

Oort: Not quite: Lo has now found the limit and resolved the source.

A. Blaauw: A fundamental, almost philosophical, point is whether there is a "true Nucleus" of the Galaxy, if one goes to smaller and smaller scales. Ambartsumian and his associates think there is something of primordial, even fundamental, significance in the Nucleus. But is it not true that, if one goes to smaller and smaller dimensions, of the order of parsecs, the notion of "the Nucleus" vanishes? A gravitational pit there one could only notice if one orbited around it. Do you think there is more to it than this?

Oort: I can answer that briefly: a black hole.

Blaauw: Yes, but is that not a secondary phenomenon? A black hole may have formed in the course of the Galaxy's evolution, but it was not primordial in the formation of the Galaxy.

Oort: I do not know that at all. It may have formed in the course of time, but there are now several phenomena which one would like to explain with a black hole. One might say: The fundamental thing is the black hole.

H. van Woerden: It seems to me that the real question is whether there is a spike in the mass distribution at the Centre.

Oort: Yes, certainly - but one cannot see that well enough. The source IRS 16 has been suggested to represent the actual condensation in the general distribution of Population II stars. But it is doubtful whether it is really a central cluster. On a larger scale, one cannot define the centre well enough.

J.P. Ostriker: If the "central pit" in the galactic stellar density distribution has a size of parsecs, then perhaps a variety of interesting constituents can collect in this small region, including massive clusters, black holes dragged in by dynamical friction, and gas shed during normal stellar evolution. Could we not be witnessing the complex interaction amongst these constituents in "the pit", with no one of them being properly called the "true Centre" of the Galaxy?

R.H. Sanders: I wish to make a point about the central point source at the "true Nucleus". Backer and Sramek have measured the proper motion of this object at the VLA, and they find a proper motion consistent with the rotation of the Sun around the centre of the Galaxy. The error at present is about 100 km/s, but this will get better with time. If the object is a massive one, one would expect that is has settled to the "true Galactic Centre", and that is has no additional, peculiar motion in addition to the reflex of the Sun's rotation around the Centre.

H.S. Liszt: Even though we observe HI absorption over a very wide range of velocities, its behaviour is not at all chaotic and it can all be related to the molecular-cloud emission that we observe on arcminute scales. We do not understand it fully, but it is not chaotic.

Another point is that the +50 km/s HI cloud is seen in absorption against the central point source, and that is important to stress. Absorption by H_2CO may be absent, but in HI it is strong: 20%.

Oort: But your HI observations have a much wider beam than Whiteoak's 3.5 arcsec for H_2CO. His narrow beam shows the minispiral absorbed at +40 km/s, but there is no sign whatever of absorption against the compact source.

Liszt: The thermal and nonthermal fluxes are much more heavily tangled at 20 cm than at 6 cm, hence the sources are difficult to separate. We should probably reobserve the source with the VLA, but I think the measured optical depth of 0.2 is too high for that to arise from absorption of the nonthermal emission.

Oort: That would be sympathetic, because it is very difficult to imagine the point source to be so far in front of the Centre as would be required if it really lies outside the +40 km/s cloud.

Liszt: I agree, but I think - HI is an atom, it samples atomic gas, hence can exist at lower densities; H_2CO and other things are molecules, which require different conditions for their existence. Hence it is important to consider chemical as well as geometric and kinematic factors.

The Chairman: Lynden-Bell.

Oort: Don't ask too difficult questions! (Laughter.)

D. Lynden-Bell: How definite do you regard a) the positron annihilation line itself, and b) its variability?

Oort: To me the line itself seems very convincing, but I am not an X-ray specialist and cannot judge well enough. Its wavelength fits perfectly. As to its variability, the earlier suggestion by HEAO results has been confirmed by recent balloon observations: it has disappeared more or less, so it has varied from a fairly strong line to nothing at present.

Lynden-Bell: But it did not go through an intermediate phase: it was just on once, off twice, right?

C.J. Cesarsky: Since HEAO it has been observed twice from balloons.

The Chairman: We must bring this discussion to a close. Several people still want to be heard, but it is better to be unfair to several, and at least give duty to the speakers.

Oort and Mrs. Oort (right), with Mrs. Borgman, at the Kapteyn-Van Rhijn exhibition. Background: official academic portrait of Kapteyn CFD

FINE STRUCTURE OF MOLECULAR CLOUDS WITHIN 1 MINUTE OF ARC OF THE GALACTIC CENTER

Norio Kaifu, Junji Inatani, Tesuo Hasegawa and Masaki Morimoto
Nobeyama Radio Observatory, Tokyo Astronomical Observatory

We have observed the HCN(J=1-0) line in the vicinity of the galactic center with the 18" beam of the Nobeyama 45-m telescope. Profiles were taken at 53 points within 1 arcmin from the galactic nucleus [R.A. = 17h42m29.29s, Dec. = -28°59'17.6" (1950)] with a 10" grid (see figure 1). A SSB cooled-mixer receiver (T_{RX} = 600 K) and a wideband AOS (acousto-optical radiospectrometer) with 250 kHz resolution were used.

Figure 1 shows the brightness distributions in selected velocity ranges where the effects of absorption are not dominant. These maps show a remarkable change of cloud velocity from lower right (negative Δl, negative Δb) to upper left (positive Δl, positive Δb), apparently

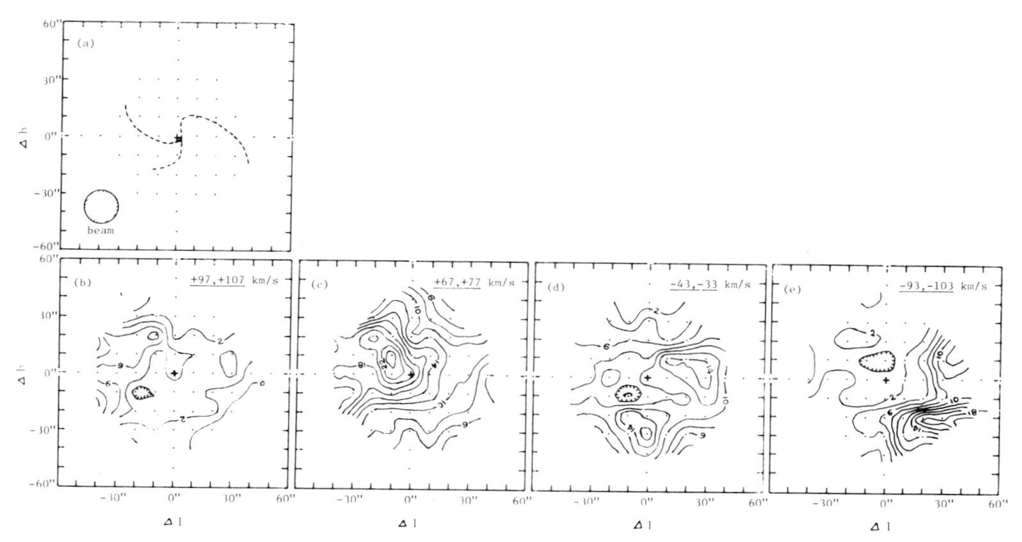

Figure 1. (a): Observed points and beam. The broken lines show the loci of ionized spirals taken from Ekers et al. (1983). (b)-(e): Brightness distributions in the HCN line, in selected velocity ranges. The cross indicates the position of the compact radio source.

H. van Woerden et al. (eds.), The Milky Way Galaxy, 367–369.

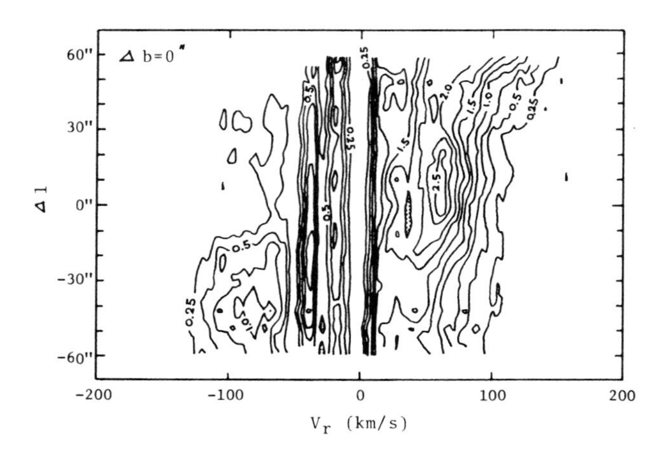

Figure 2. Longitude-velocity map of HCN.

indicating a rapid rotation of massive, cold molecular clouds around the galactic center. The axis of rotation is tilted about 40° from that of the Galaxy, and roughly coincides with that of the proposed "precessing jet" (Brown, 1982).

In figure 2 the negative-velocity feature seems to be separated into two components near the nucleus; on the other hand the positive-velocity feature reaches the position of the nucleus. This separation in the negative-velocity feature is probably caused by absorption against the central continuum sources. The existence of strong mm-wave continuum sources is indicated by deep absorption features with negative line temperatures in the profiles from these points. The longitude-velocity map (figure 2) also indicates that the negative-velocity feature is affected by absorption near the nucleus.

For this reason we put the negative-velocity cloud in front of the nucleus and the positive-velocity cloud, which does not show any sign of absorption, behind the nucleus. The expansion velocity derived from figure 2 is about 30 km s^{-1} for the negative-velocity side and somewhat lower than 60 km s^{-1} for the positive side. The rotation velocity is 80-90 km s^{-1}. Because of the limited observed area, the outer boundaries of the rotating clouds were not observed. But we can recognize corresponding features in the CO maps by Liszt et al. (1983), from which we have estimated the radius of this cloud to be 6 pc or somewhat smaller. The similar extent of far-IR radiation shown by Becklin et al. (1982) also supports the existence of massive molecular clouds surrounding the nucleus.

Comparing our results with the continuum mini-spiral (Brown et al. 1981, Ekers et al. 1983), we note that the masive cold clouds seem to have a correlation with the ionized gas. The arms of the mini-spiral seem to run along the edges of the molecular clouds. H110α measurements (Bregman and Schwarz 1982) show high positive velocities in the east arm which runs along the positive-velocity part of the molecular cloud, and also show high negative velocities in the west arm which runs along the negative-velocity molecular cloud. Taking into account that the mini-spiral has a thermal spectrum, we point out the possibility that the spiral arms are ionized regions distributed along the boundaries of the molecular clouds. The age of these expanding and rotating molecular clouds should be short, due to the tidal force of the central mass and due to the expanding motion. Thus we consider that these clouds may be remnants of recent activity in the nucleus.

REFERENCES

Becklin, E.E., Gatley, I., Werner, M.W.: 1982, Astrophys. J. 258, 134
Bregman, J.D., Schwarz, U.J.: 1982, Astron. Astrophys. 112, L6
Brown, R.L.: 1982, Astrophys. J. 262, 110
Brown, R.L., Johnston, K.J., Lo, K.Y.: 1981, Astrophys. J. 250, 155
Ekers, R.D., van Gorkom, J.H., Schwarz, U.J., Goss, W.M.: 1983, Astron. Astrophys. 122, 143
Liszt, H.S., van der Hulst, J.M., Burton, W.B., Ondrechen, M.P.: 1983, Astron. Astrophys. 126, 341

DISCUSSION

C.A. Norman: Where would you put the third component of the central spiral, which might be a jet, in your model?

Kaifu: We do not know its shape, but the rotation is not rigid. I think it is larger on the far side, but the change is slow.

J.H. Oort: Has any of this material been published?

Kaifu: Not yet. We hope to publish it soon.

B.F. Burke: Where will it be published?

Kaifu: We do not know yet.

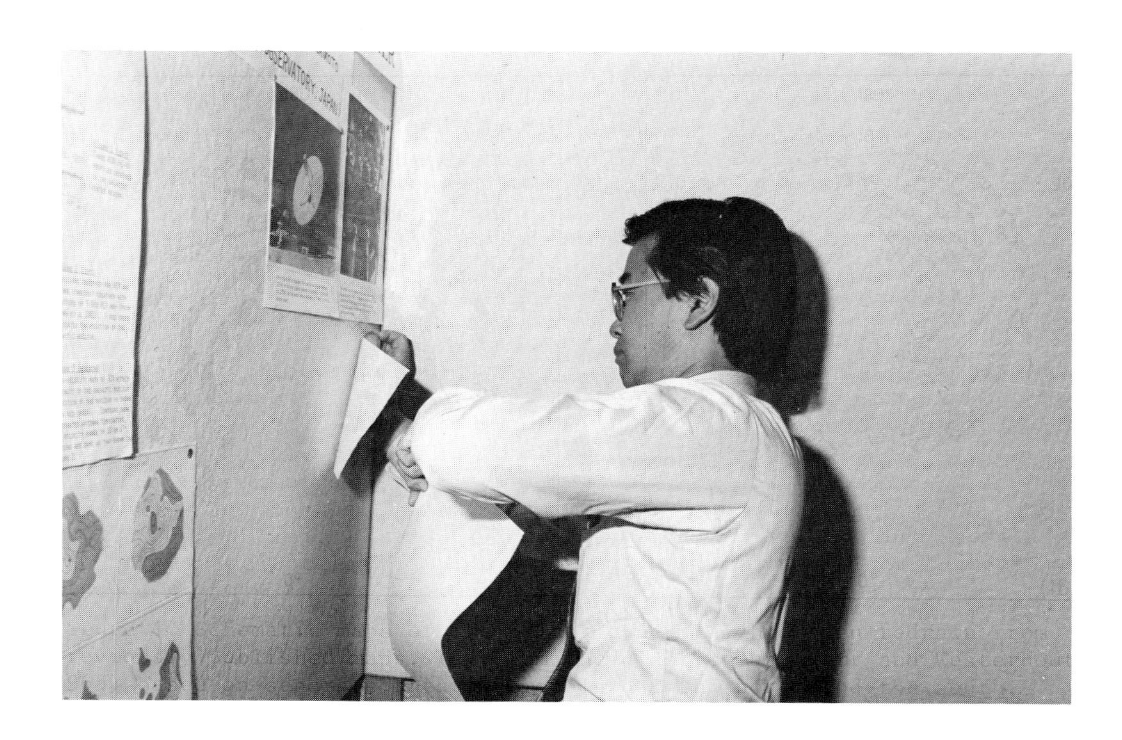

Top: Kaifu mounts his poster CFD
Bottom: At dinner, left to right: Katrin Särg, Mo Jing-Er, Fujimoto and Kaifu LZ

RECOMBINATION-LINE OBSERVATIONS OF THE GALACTIC CENTRE

J.H.van Gorkom[1], U.J. Schwarz[2], J.D. Bregman[3]
[1]National Radio Astronomy Observatory, Socorro
[2]Kapteyn Astronomical Institute, Groningen
[3]Netherlands Foundation for Radio Astronomy, Dwingeloo

ABSTRACT

Aperture-synthesis observations of the H76α and H110α recombination lines are presented for the inner (3pc) region of the Galactic Nucleus. The large line width measured with single dishes (Pauls et al., 1974) is caused by well-ordered large-scale motions of the ionized gas. The velocities of the NeII clumps (Lacy et al., 1980) fit well into our smooth velocity field and smooth intensity distribution. We suggest therefore that the cloud picture of the NeII gas is (at least partly) invalid.

1. INTRODUCTION

The presence of multiple, curved thermal features in the immediate vicinity of the Galactic Centre (Ekers et al., 1983) makes it possible to get kinematical information on the inner part of the Galaxy by recombination-line observations. Already single-dish observations of various recombination lines of hydrogen have shown (cf. Pauls et al., 1974) the lines are unusually wide, reflecting the existence of high velocities.

Fig. 4 in the paper by Oort (1984, this volume) shows the 20-cm continuum observations of Ekers et al. (1983), where one sees the spiral-shaped, thermal feature imbedded in a larger non-thermal shell. We present here results of the H76α line at 2 cm of this thermal 'spiral', observed with the VLA, and the H110α line observed at Westerbork. The results are then compared with the [NeII] observations by Lacy et al. (1980), and we give some contribution to the discussion of possible models to explain the results.

2. THE OBSERVATIONS

The VLA observations at 2 cm of the H76α line were made in January 1982, using the inner 18 antennas in the C configuration (0.1 to 1.7 km baselines). The continuum is shown in Fig. 1 of the preceding paper by Oort (1984). We fitted Gaussians to the spectra at each position in the

H. van Woerden et al. (eds.), The Milky Way Galaxy, 371–376.

maps, which were smoothed to a 4".5 × 4".5 resolution in order to improve the signal/noise ratio. The peak line intensity and velocity field are shown in Fig. 1a.

The H110α-line observations at 6 cm were made with the WSRT with a 3" × 20" resolution (Bregman and Schwarz, 1982). The spectra were also fitted with Gaussians and the results are shown in Fig. 1b.

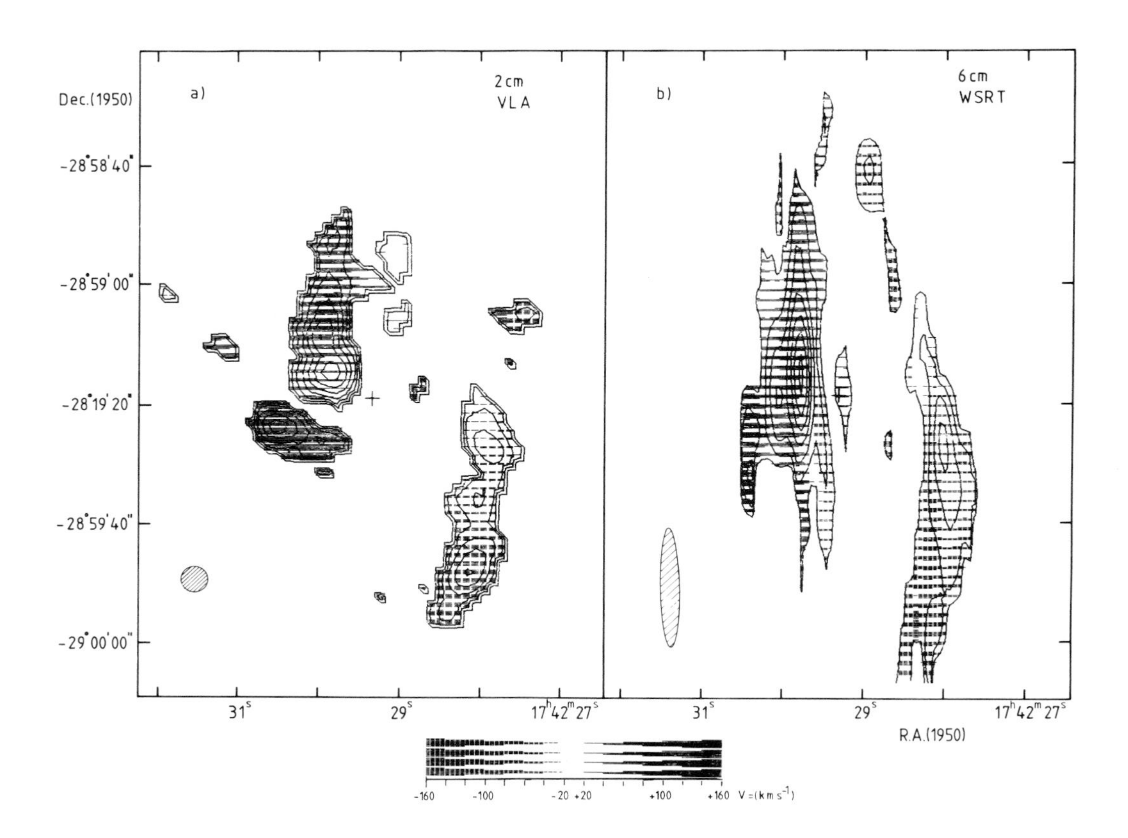

Fig. 1
a. VLA map of the H76α recombination line (2 cm). The contours give the peak intensity of the line (the contours are: 1.3, 2.6, ... K) whereas the raster gives the velocity, cf. the key at the bottom. The beam is 4.5 × 4.5 arcsec (shaded circle), the velocity resolution 32 km/s.

The large-scale structure and the regular velocity field of the main features are clearly seen.
b. WSRT map of the H110α recombination line (6 cm), as Fig. 1a. The contours correspond to 3.3, 5, 6.6, 8.3, 10 K. The beam is 3 × 20 arcsec (shaded ellipse), the velocity resolution 67 km/s.

The three elongated features all show a remarkable degree of coherence in velocity. The NS features appear to rotate around the Centre, with an axis close to that of the general galactic rotation. The EW features suggest that in- or outflow is occurring. The average line width of the H76α line is 40-50 km s^{-1}; at the bend in the N arm it goes up to 100 km s^{-1}, which is probably a blend of 2 features.

The electron temperature (Fig. 2) derived from the H76α results (the assumption of LTE will not affect the numbers significantly) shows the same degree of coherence and seems to go up close to the Centre. The fact that we do not see lines very close to the Centre, combined with the fact that [NeII] lines have been detected there, makes it likely that the electron temperature is greater than 12000K in the Centre.

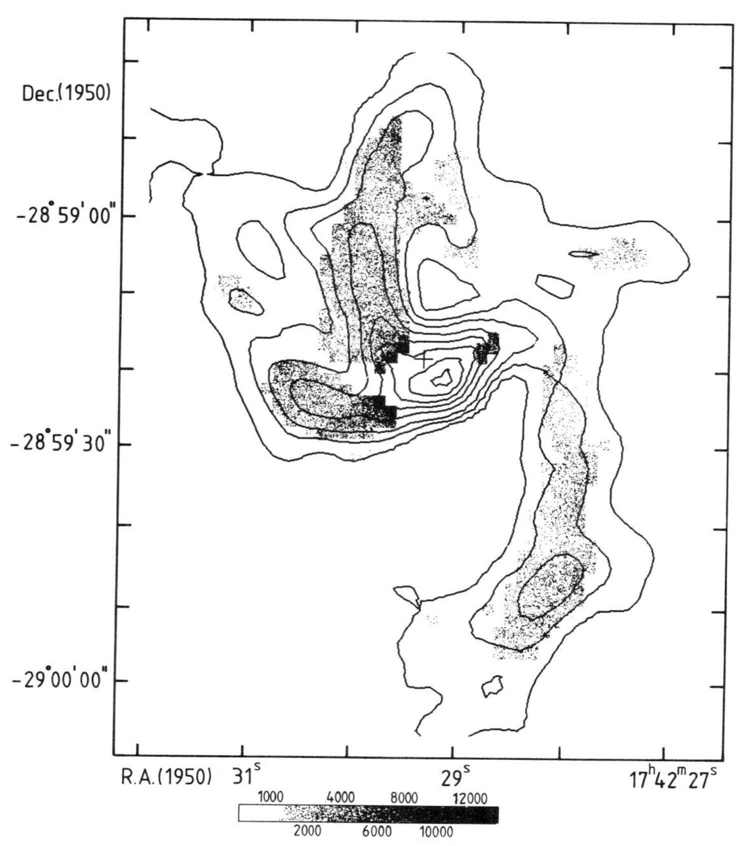

Fig. 2
2-cm continuum map, smoothed to 4.5 × 4.5 arcsec resolution. The contours are at 2.7, 13, 26, 39, ...K. The gray tone gives the electron temperature, derived from the line-to-continuum ratio assuming LTE. The cross gives the position of the non-thermal point-source, which was subtracted. At the centre where we detect no recombination line, the temperatures must be higher than 12000K.

3. COMPARISON WITH [NeII] LINE OBSERVATIONS.

In Fig. 3 we give the H76α results with intensity and velocity contours, together with the [NeII] concentrations as derived by Lacy et al. (1980). There is mostly an excellent agreement. The highest [NeII] velocities are found closest to the centre, where we cannot detect a recombination line.

The excellent agreement between the regular velocity field based on the recombination lines in the spiral features and the velocities of the [NeII] 'clumps' can hardly be fortuitous. Are the 'clumps' imbedded in the large-scale features, or is it possible to interpret the [NeII] results in a different way than in terms of 'clumps'? In any case this apparent morphological difference cannot be an effect of the limited resolution: the 2-cm continuum observations with the original 2 × 3 arcsec resolution, which is superior to the 3 × 3" resolution of the [NeII] observations, do also show the filamentary large-scale structure. In order to answer the above question, we tried to reanalyze the [NeII] data in a similar way as the recombination-line results, by fitting Gaussians to the spectra, within the regularly sampled region. At many positions, mainly in the Centre region, there was more than one Gaussian. But if we exclude components with velocities $|V| > 170$ kms^{-1}, we find one prominent component per position. Amplitude and mean velocity of these components are shown in Fig. 4. Clearly one sees the 'roots' of two of the three spiral features on the E side of the point source; these 'roots' are in perfect agreement with the recombination-line results. Also towards the W of the point source the distributions of [NeII] and its velocity are very smooth.

We therefore conclude that at least part of the [NeII] line originates in the same distributed gas as the recombination lines; possibly there are in addition small [NeII] features. The same conclusion was put forward by Bregman and Schwarz (1982).

4. DISCUSSION

A range of possible interpretations of these data has been discussed by Ekers et al. (1983), who favour tidal distortion of infalling material. We do not want to repeat that discussion here; see also the discussion by Oort in this Volume.

We would like to point out one thing here which is not emphasized by Ekers et al. (1983). The whole complex Sgr A East + West shows a remarkable similarity in the radio continuum to other supernova remnants. The filamentary structure in West and the extension of the N arm outside the shell looks just like the Crab. An obvious difference with other supernova remnants is the large amount of thermal gas and the presence of many infrared sources. A possible alternative interpretation is that a supernova has gone off near a dense, rotating molecular cloud.

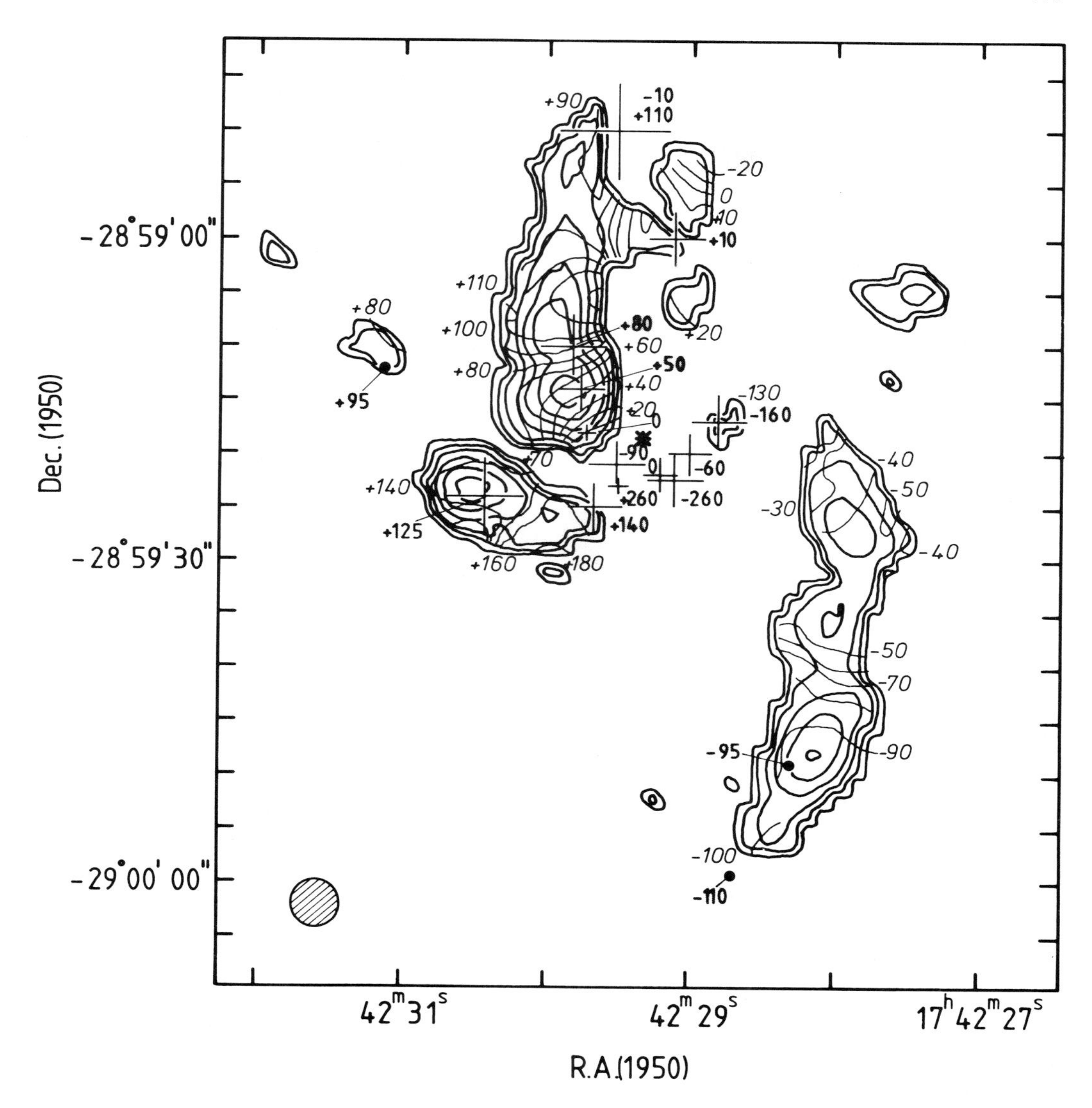

Fig. 3
Comparison of H76α and [NeII] results. Contours show peak intensity and velocity (slanted numbers) of the H76α line (cf. Fig. 1a); the concentration in the NW has a velocity of - 85 km s^{-1}. The crosses indicate positions and sizes of the [NeII] 'clumps', upright numbers (in shaded rectangle) velocities as found by Lacy et al. (1980). A few positions are added with our own estimates of velocities. The agreement between the [NeII] velocities and the recombination-line velocities is very good. The asterisk indicates the position of the point source.

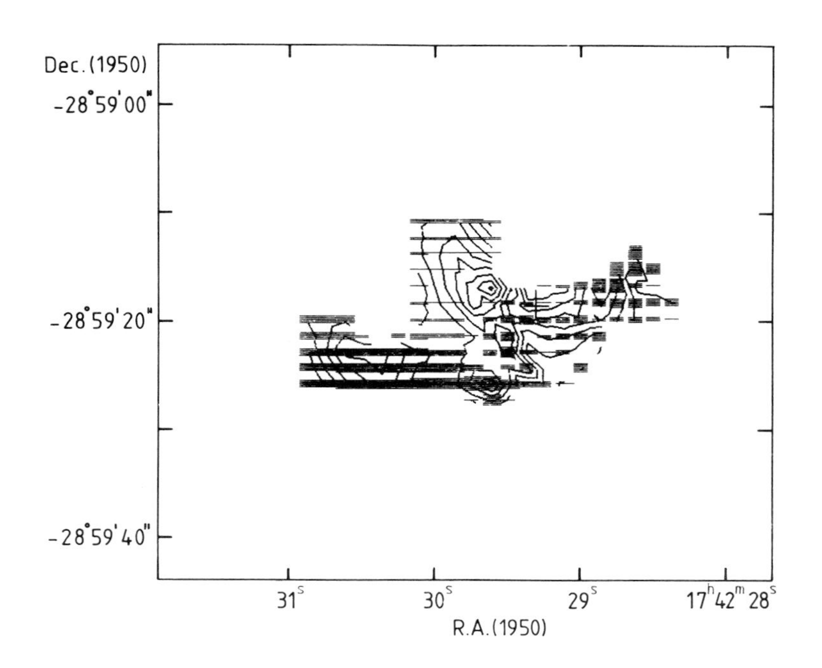

Fig. 4

Peak intensity and velocity of the [NeII] line. A Gaussian fitting was applied to the spectra. Components with velocities $|V| > 170$ km s^{-1} were left out, for key of the velocities see Fig. 1. The scale is the same as for Fig. 3.

REFERENCES:

Bregman, J.D., Schwarz, U.J., 1982, Astron. Astrophys. 112, L6

Brown, R.L., Lo, K,Y., 1982, Ap.J. 253, 108

Ekers, R.D., van Gorkom, J.H., Schwarz, U.J., Goss, W.M., 1983, Astron. Astrophys. 122, 143

Lacy, J.H., Townes, C.H., Geballe, T.R., Hollenbach, D.J., 1980, Ap. J. 241, 132

Oort, J.H., 1985, in The Milky Way Galaxy, eds. van Woerden H., Burton, W.B. and Allen, R.J., IAU Symp. 106, p. 349

Pauls, T., Mezger, P.G., Churchwell, E., 1974, Astron. Astrophys. 34, 327

DISCUSSION

R.H. Sanders: This southern arm is quite a bit fainter than the rest of the structure, and it seems to fit rather well into the general shape of the larger shell described by Oort in his review. Do you consider it possible that this southern arm is not connected to the three-armed structure closer to the Centre?

Schwarz: The continuum data may indicate a weak link between the southern arm and the large shell. In the northern arm there is no such relation.

ARE SPIRAL NUCLEI "ACTIVE" IN THE RADIO CONTINUUM?

Paul T. P. Ho
Harvard University

Jean L. Turner
University of California, Berkeley

ABSTRACT

Using the VLA, we obtained matched-array continuum observations at 6 and 2 cm. An angular resolution of $\sim$1" and an rms sensitivity of $\sim$0.05 mJy were achieved for a sample of 17 nearby spiral galaxies. Spectral-index maps derived for the nuclear regions reveal a mixture of thermal and nonthermal activity. The use of high angular resolution and high frequencies was the key to the success in detecting thermal activity. Conversion from thermal fluxes to total number of ionizing photons suggests that star formation is very active in some of these cores (inner 500 pc), with a rate typically 10 times greater than in our own nuclear region. A number of the nuclear regions appear to be dominated by extended nonthermal emission with a steep spectrum. Among these, some are closely associated with thermal emission and hence are consistent with supernova activity. However there are sources exhibiting aligned structures, suggesting possible connections with a central active nucleus. In any case, at our achieved sensitivity level 16 out of 17 galaxies were detected in the radio continuum.

DISCUSSIONS

A number of sensitive surveys have been made recently to study the radio continuum emission from nearby spiral galaxies, e.g. single-dish work of Klein and Emerson (1981), Gioia, Gregorini and Klein (1982), Berkhuijsen, Wielebinski and Beck (1983); interferometric studies of Israel (1980), Hummel (1981), vander Hulst, Crane and Keel (1981), and Condon et al. (1982). The current experiment which we report here concentrates on the inner nuclear region. The objectives are to (a) achieve high sensitivity ($\sim$0.05 mJy), (b) match angular resolution (1") to the expected sizescale of giant HII regions ($\sim$10 pc), (c) use the shortest wavelengths (6 and 2 cm) to enhance the thermal contribution. We used the B array at 6 cm and the C array at 2 cm in order to be sensitive to the same spatial structures at both wavelengths. Reliable spectral-index maps were thereby derived.

H. van Woerden et al. (eds.), The Milky Way Galaxy, 377–378.

Our sample consists of 17 spirals, at distances of 1-7 Mpc: NGC253, Maffei 2, IC342, NGC2403, NGC2903, NGC3031, NGC4236, NGC4258, NGC4736, NGC4826, NGC5194, NGC5236, NGC5457, NGC5474, NGC6946, M31 and M33. We detected emission in 16 cases; NGC5474 being the exception. The sources are predominantly weak, with 14 regions having peak fluxes per 1" beam of 0.2-7 mJy, while 2 regions (NGC253 and NGC3031) have peak fluxes $\geq$100 mJy.

The morphology of the continuum emission can be roughly classified into 3 distinct groups: (a) point-like compact source (e.g. NGC3031, NGC4736), (b) extended clumpy complex (e.g. IC342, NGC5236), (c) disk-like complex with aligned compact structures (e.g. NGC253, Maffei 2). From spectral-index maps, the smooth disk-like sources appear predominantly non-thermal while the clumpier complexes appear thermal. The point-like sources, which are unresolved with our angular resolution, tend to have rather flat spectra. These sources are most likely also non-thermal. For the brighter sources such as NGC3031 (M81), we can in fact conclude that they are non-thermal from brightness arguments and also from existing VLBI measurements (Bartel et al. 1982).

Preliminary results based on this technique of attempting to separate thermal and non-thermal emission from high-angular-resolution spectral-index maps have been reported for 3 galaxies (Turner and Ho, 1983). With the larger sample, we find that, as before, whenever thermal activity can be detected, the total number of ionizing photons correspond to vigorous star formation with $\sim 10^3$ OB stars. This is probably due to selection effects, in that weaker thermal activity cannot be detected at present sensitivity. Nevertheless, it appears that most of the nearby spirals are in fact active at the 1 mJy level in the inner 500-pc region, although the nature of the radio emission is not always clear-cut.

PTPH acknowledges support by a Henri Chrétien Award administered by the American Astronomical Society. JLT acknowledges support by an Amelia Earhart Fellowship, and NSF Grant AST 81-14717.

REFERENCES

Bartel, N. et al.: 1982, Astrophys. J. 262, 556.
Berkhuijsen, E.M., Wielebinski, R., Beck, R.: 1983, Astron. Astrophys. 117, 141
Condon, J.J., Condon, M.A., Gisler, G., Puschell, J.J.: 1982, Astrophys. J. 252, 102.
Gioia, I.M., Gregorini, L., Klein, U.: 1982, Astron. Astrophys. 116, 164.
Hummel, E.: 1981, Astron. Astrophys. 93, 93.
Israel, F.P.: 1980, Astron. Astrophys. 90, 246.
Klein, U., Emerson, D.T.: 1981, Astron. Astrophys. 94, 29.
Turner, J.L., Ho, P.T.P.: 1983, Astrophys. J. 268, L79.
van der Hulst, J.M., Crane, P.C., Keel, W.C.: 1981, Astron. J. 86, 1175.

PHENOMENA AT THE GALACTIC CENTRE - A MASSIVE BLACK HOLE?

Martin J. Rees
Institute of Astronomy
Madingley Road, Cambridge CB3 OHA

1. INTRODUCTION

About 10^7 $L_{\odot}$ of luminosity, mainly in ionizing flux and infrared radiation, emerges from the central pc^3 of our Galaxy. This exceeds the luminosity from the corresponding region of most nearby galaxies, though it is surpassed by M82 and NGC 253 (Reike and Lebofsky 1982), but perhaps involves nothing more exotic than a starburst 10^6 - 10^7 years ago. But the manifestations of activity at the Galactic Centre that are unambiguously non-thermal in character are at a much lower level: the γ-ray annihilation line ($\sim 10^{38}$ erg s^{-1}) and the compact radio source ($\sim 10^{34}$ erg s^{-1}). I shall comment briefly on these two phenomena, and also suggest an interpretation of the remarkable pseudo-spiral structures revealed by the NeII infrared and the radio-continuum maps. These phenomena relate to the old question (cf. Lynden-Bell and Rees 1971) of whether our Galaxy has ever experienced a more violent phase, leaving a massive collapsed remnant.

2. THE GAMMA-RAY SOURCE

The problems posed by the 0.511-Mev annihilation line have been reviewed by Lingenfelter and Ramaty (1983). The variability implies that the annihilations occur within a volume $\lesssim 10^{18}$ cm across, in gas clouds with density $\gtrsim 10$ particles cm^{-3}; the narrowness of the observed line implies that the gas must be cooler than $\sim 5 \times 10^{4}$ °K, with random internal motions $\lesssim 700$ km s^{-1}, but must nevertheless not be completely neutral (since, otherwise, the annihilation would occur via charge exchange, leading to a width $(\Delta\nu/\nu) \simeq (e^2/\hbar c)$). The line is not shifted from its expected 0.511 Mev energy: this implies that it cannot come from a compact region close to (say) a massive black hole - if there were such a hole of mass 10^6 M_6 solar masses, then the annihilation must occur $\gtrsim 10^{16}$ M_6 cm from it.

The source of the positrons may, of course, be much smaller than the annihilation volume. Irrespective of its size, this source has the

H. van Woerden et al. (eds.), The Milky Way Galaxy, 379–384.

rather surprising property that its apparent luminosity over the entire X-ray and γ-ray band - from a few kev up to $\sim$ 1 Gev photon energies - is no larger than the power going into the annihilation line (i.e. into the rest mass of the pairs). This could imply any of the following three possibilities.

(i) The pairs are produced at energies of only $\lesssim$ 1 Mev. This could happen if there were a central source emitting primarily γ-rays, which was sufficiently compact that $\gamma + \gamma \rightarrow e^+ + e^-$ occurred. The requisite dimensions, for a pair luminosity $\sim 10^{38}$ erg s^{-1}, are $\sim 10^9$ cm. This would be smaller than the gravitational radius of a black hole unless its mass were below a few thousand solar masses (Eichler and Phinney 1983).

(ii) The pairs could be produced with high Lorentz factors but could lose their energy (before annihilation) via synchrotron radiation in the form of soft (optical or UV) photons. Even if all the ionizing flux came from this central source - and none from stars - the pairs would still need to form with $\gamma \lesssim 10^3$.

(iii) Much more continuum luminosity is perhaps being beamed along other directions than our line of sight. An elaborate model along these lines - the "gamma gun" - has been developed by Kardashev and Novikov (1983).

3. THE COMPACT RADIO SOURCE

The small central radio component is known to have an unusual (rising) spectrum. If there were a central hole with $M_6 \gtrsim 1$, then the radio emission could come from plasma participating in a low-level accretion flow and gravitationally bound to the hole (Rees 1982): if the plasma were unable to cool on the inflow timescale, then the only substantial radiative output would come from relativistic electrons at $< 10^3$ Schwarzschild radii, where even the average ion or electron could have > 1 Mev of kinetic energy and the field would be $10 - 10^3$ G. If there were no central mass larger than $10^3 - 10^4$ $M_\odot$, then the radio emission could come from a wind (Reynolds and McKee 1980, Kardashev 1983). While the compact radio source does not unambiguously require a massive black hole, the presence of such an object, accreting at a low rate, would almost inevitably yield radio emission - the absence of a peculiar radio source at the Galactic Centre would therefore have been evidence against a massive black hole there.

4. ARE THE PSEUDO-SPIRAL FEATURES TRIGGERED BY TIDALLY DISRUPTED STARS?

The stellar disruption rate due to a central hole depends on the star density within the central pc ; it could be as high as $10^{-3} M_6^{4/3}$ yr^{-1}, though loss-cone depletion effects may reduce this. (Note also that this rate is calculated on the assumption that there is indeed a dense star cluster at the Galactic Centre, which would not be obligatory if the infrared and the inferred UV radiation were due to black-hole accretion.) The fate of disrupted stars has been much discussed (see Rees 1982 and references therein). The original binding energy of the disrupted star must be supplied by its orbital energy: on average, therefore, the debris must be bound to the hole by an energy $\sim v_*^2$ per unit mass, v_* being the escape velocity from the surface of the star. However, as Lacy et al. (1982) emphasize, some fraction of the debris can escape at $\gtrsim 10^3$ km s^{-1}. The reason for this is that at peribothron a star undergoing tidal disruption is moving at $V_T \simeq 3 \times 10^4 M_6^{1/3}$ km s^{-1}: it becomes somewhat compressed and elongated into a prolate shape (cf. Carter and Luminet 1982) and pressure gradients can impart to material on the leading side of the star an excess velocity δv over the parabolic orbital velocity which is a significant fraction of V_*. This corresponds to a large excess orbital energy - enough for the debris to escape on a hyperbolic orbit with terminal velocity $\sim \sqrt{(\delta v) v_*}$. Whenever a star is disrupted, some fraction of its "remains" will spray out in a fan or cone. The linear features revealed by the NeII infrared and by the radio continuum observations (Lacy et al. 1982; Lo and Claussen 1983; Oort 1985) are perhaps indirect manifestations of these ejecta.

The pseudo-spiral features each involve several solar masses of material, moving with speeds $\sim$ 100 km s^{-1}; they are therefore not themselves composed of the high-speed ejecta. Nevertheless, the formation of these features may have been triggered by ejecta from a disrupted star - this at least seems a tenable alternative to the infall hypothesis favoured by Lo and Claussen (1983).

Suppose that the region within 1 pc is pervaded by predominantly photoionized gas with density $\sim 10^3$ cm^{-3}. [The emission measure from this gas (which may be deficient in dust (Gatley 1982)) would be swamped by that from the $\sim 10^5$ cm^{-3} concentrations within it.] Suppose - as an illustrative example - that each stellar disruption leads to the expulsion of ~ 0.1 $M_\odot$ of material with velocities $\gtrsim 10^3$ km s^{-1} directed on a cone of solid angle $\sim 10^{-2}$ radius. By the time this debris has travelled $\sim$ 1 pc through gas of density $\sim 10^3$ cm^{-3} it will have swept up at least its own mass. The initial kinetic energy will thus have been turned into thermal energy, leaving an overpressured cone or swath of hot ($> 10^6$ ^oK) gas along the path of the ejecta. The clouds of density $\sim 10^5$ cm^{-3} which delineate the observed pseudo-spiral features would then be produced as this overpressured material expands sideways, causing a radiative shock. (Conceivably the broad HeII emission (Hall et al. 1982) could be the only direct manifestation of the fast moving ejecta itself).

In this interpretation, the central parsec of our Galaxy could (though it need not) be in a steady state, averaged over timescales $\gg 10^4$ yrs. Stellar disruptions every few thousand years (each leading to a $\sim 10^{44}$ erg s^{-1} flare lasting perhaps only a few years (Rees 1982)) would provide an energy input into the gas, clearing the central parsec of dust (Gatley 1982). The high-density clouds would then lie along the tracks of the most recent ejecta - those in which the overpressure has not been erased by cooling and by sideways expansion.

The apparent spirality might indicate some overall rotation in the ambient medium within the central parsec, as is obligatory in other interpretations of these features. If there were no such rotation, the debris from a disrupted star could in principle produce a curved track via an "aerodynamic lift" effect; the ejecta would generally not have a uniform momentum density per unit solid angle, so the shock where it interacts with the ambient medium would be oblique. (It would then of course be a coincidence, though only at the 25% level, that the three arms all curve the same way.)

5. CONCLUSIONS

The question of whether our Galactic Centre harbours a $\gtrsim 10^6\ M_\odot$ black hole - a question which bears on the relation between our Galaxy and the Seyfert galaxies - still cannot be settled definitely. Such a hole would account naturally for the compact radio component; it could also generate the $e^+ - e^-$ pairs inferred from the 0.511-Mev annihilation line, but this can be done more naturally by a $10^3 - 10^4\ M_\odot$ hole than by a larger one. If the pseudo-spiral features are triggered by directed expulsion of material rather than by infall, then tidal disruption of stars offers a possible explanation, and this would certainly require a mass $\gtrsim 10^6\ M_\odot$. It is still, however, open to sceptics to say that the best argument in favour of a massive black hole is that there is no firm evidence against it.

REFERENCES

Carter, B., and Luminet, J.P.: 1982, Nature 296, 211
Eichler, D., and Phinnery, E.S.: 1983, preprint
Gatley, I.: 1982, in Riegler and Blandford (1982) p. 25
Hall, D.N., Kleinmann, S.G., and Scoville, N.Z.: 1982, Astrophys. J. (Lett), 262, L53
Kardashev, N.S.: 1983, preprint
Kardashev, N.S., and Novikov, I.D.: 1983, in "Positron-Electron Pairs in Astrophysics", ed. M.L. Burns et al. (New York A.I.P. in press).
Lacy, J.H., Townes, C.H., and Hollenbach, D.J.: 1982, Astrophys. J. 262, 120
Lingenfelter, R.E., and Ramaty, R.: 1983, in "Positron-Electron Pairs in Astrophysics", ed. M.L. Burns et al. (New York, A.I.P. in press).
Lo, K.Y., and Claussen, M.J.: 1983, preprint
Lynden-Bell, D., and Rees, M.J.: 1971, M.N.R.A.S. 152, 461
Oort, J.H.: 1985, these proceedings
Rees, M.J.: 1982, in Riegler and Blandford (1982) p. 166
Reike, G.J., and Lebofsky, M.J.: 1982, in Riegler and Blandford (1982)p194
Reynolds, S.P., and McKee, C.F.: 1982, Astrophys. J. 239, 893
Riegler, G.R., and Blandford, R.D. (eds.): 1982, "The Galactic Center" (New York, A.I.P.)

DISCUSSION

B.F. Burke: What role is the compact radio source playing? Oort emphasized that it is not at the middle.

Rees: If it turns out not to be at the middle, then we have to say that either there is a large black hole giving the gamma rays and the radio source which is not at the dynamical centre of the spiral arms, or we have to say that the radio source is an irrelevance.

J.P. Ostriker: How many arms should the spiral have which is made by this star-disruption process?

Rees: If the spiral is due to ejecta, then naively one would expect just one arm per stellar flyby. If there is real evidence for symmetric ejection, then I prefer not to have this tidal-disruption mechanism, but to say that the jets are produced when a single star is being digested, and this gives us a thick-disk doughnut geometry. So if the evidence for symmetry is strong, then I think it would be best to say it is due to twin-jet productions of the kind we also discuss in the context of extragalactic nuclei. If there is no evidence for spatial symmetry - and there does not seem to be very much on the basis of the present data -, then each stellar flyby will give you one swath of ejecta.

DISCUSSION OF PAPER BY HO AND TURNER

T.M. Bania: In NGC 4736, is there a difference in spectral index between the central source and the outer ring?

Ho: The core source appears to have a flat spectrum between 2 and 6 cm. The outer ring is very faint, and we have not been able yet to determine its spectral index reliably. It may be thermal, while the central source is nonthermal.

E.M. Berkhuijsen: Why do you think so?

Ho: Preliminary, rough maps suggest that the more compact knots in the ring may be thermal.

Bania: I believe there is an Hα ring surrounding the nucleus of this galaxy with a radius of 1 arcmin.

F.P. Israel: In Sc spirals, the ratio of CO line strength to total radio continuum is almost always higher in the central regions (diameter of order 1') than in the disk. This supports your conclusion about strong star formation in the centres of spiral galaxies.

Ho: Our study attempts to quantify such statements. We are also pursuing the correlation between thermal radio continuum, CO emission and infrared radiation.

In University garden, at President's reception: (left to right) Hoskin, Higgs, Murray, Davies, Hodge, Elmegreen, Smith, Hilditch, N.N.
CFD

SECTION II.8

OUTSKIRTS AND ENVIRONMENT

Wednesday 1 June, 0945 - 1040

Chairman: R.D. Davies

Hugo van Woerden in his opening address as Chairman of the Scientific Organizing Committee CFD

HIGHLIGHTS OF HIGH-VELOCITY CLOUDS

Hugo van Woerden[1], Ulrich J. Schwarz[1] and Aad N.M. Hulsbosch[2]

[1] Kapteyn Institute, University of Groningen, Netherlands
[2] Astronomical Institute, Catholic University, Nijmegen, Netherlands

1. INTRODUCTION

Twenty years after their first discovery at Dwingeloo (Muller, Oort and Raimond 1963), the nature and origin of high-velocity clouds (HVCs) remain enigmatic. Yet, much important progress has been made in the study of their properties, and prospects are brightening that the problem of their distances, which holds the key to their understanding, may soon be solved.

The present paper reviews recent results, partly unpublished. For earlier reviews, we refer to Davies (1974), Giovanelli (1978, 1980b), Hulsbosch (1975, 1979a, b), Mirabel (1981a), Oort and Hulsbosch (1978), Van Woerden (1976, 1979) and Verschuur (1975).

The structure of this paper is as follows. In Section 2, we define the HVC phenomenon and review the sky distributions of HVCs and their velocities. In Section 3, we summarize attempts at interpretation and evaluate the merits of Bregman's Galactic-Fountain model. Section 4 discusses new large-scale surveys, Section 5 the velocity structure of the largest HVC complex. Section 6 reviews the small-scale structure of HVCs, highlighting new results on Chain A. Sections 7 and 8 summarize recent work on the Magellanic Stream and on the clouds with very high velocities. Section 9 discusses attempts to determine the distances of HVCs, and the related problem of their chemical composition. Section 10 draws a few conclusions.

2. THE PHENOMENA - AN OVERVIEW

The HVCs were discovered in the 21-cm HI line, and, with few exceptions (Section 9), they have never been observed in any other way.

No sharp, generally accepted definition exists. HVCs may be defined as HI clouds whose velocities are inconsistent with the standard model of the galactic HI distribution: a thin, flat disk in differential

H. van Woerden et al. (eds.), The Milky Way Galaxy, 387–408.

rotation. In using this definition, one must of course account for the well-known warp of the HI layer (see e.g. Henderson et al. 1982), for the random motions (velocity dispersion in one coordinate $\sim$ 7 km/s) and for local, systematic non-circular motions (e.g. Burton, 1966) in the interstellar medium. At low galactic latitudes, the range of radial velocities (with respect to the local standard of rest) expected from differential galactic rotation may be considerable; for instance, for a flat rotation curve with circular velocities Θ_o out to radius xR_o, where R_o is the Sun's distance from the Galactic Centre, it reaches in the first galactic quadrant at longitude ℓ from $V = \Theta_o (1/x - 1) \sin \ell$ to $V = \Theta_o (1 - \sin \ell)$, in the second quadrant from $V = \Theta_o (1/x - 1) \sin \ell$ to $V = 0$, with opposite values in the fourth and third quadrants. At latitudes $|b| > 15°$, however, the line of sight leaves the disk within about 1 kpc, and differential-galactic-rotation effects remain minor, facilitating the recognition of HVCs. In the analysis of HI surveys at $|b| > 15°$, it has become customary to call velocities $|V|$ between 20 and $\sim$ 80 km/s "intermediate velocities" (see e.g. Wesselius and Fejes 1973), and reserve the name "high-velocity clouds" for those with $|V| \gtrsim 80$ km/s.

The most recent compilation of HVC observations from various sources is that by Mirabel (1981a). Figure 1a (see page 6 of these Proceedings) shows the distribution of HVCs and of their velocities (coded in colour) over the sky. The map is based on surveys published up to 1980; however, later surveys - which we summarize in Section 4 - have not materially changed the picture.

The most striking feature of Figure 1a is the uneven distribution of both position and velocity over the sky. Negative velocities dominate at longitudes $30° < \ell < 210°$, positive velocities at $210° < \ell < 330°$. At first sight, this might be interpreted as an effect of galactic rotation, since the velocities shown are with respect to the local standard of rest (LSR). However, as emphasized by Oort and Hulsbosch (1978; also Hulsbosch 1979a), negative velocities also dominate near the galactic poles and in the Anticentre region, where galactic-rotation effects should be minor. Indeed, a plot of velocities corrected for the rotation of LSR about the Galactic Centre (Figure 2, taken from Giovanelli 1980b) also shows a preponderance of negative velocities. Although the observed radial velocities represent only one component of the space motions (cf. Kerr 1967), the dominance of negative values suggests an inflow of high-velocity gas into the Galaxy, in keeping with the early suggestions by Oort (1965, 1966, 1967, 1970) - or, at least, an inflow into its Disk (Bregman 1980, Mirabel 1981a). The inflow rate depends on assumptions as to the distance and thickness of HVCs, but most estimates are of order 1 $M_\odot$ per year (Oort 1967, Hulsbosch 1975, Mirabel 1981a).

Another major asymmetry is that HVCs appear much more numerous in the range $0° < \ell < 180°$ than at $180° < \ell < 360°$. A similar asymmetry exists in the intensities, but these are not shown in Figure 1a. In connection with the asymmetry in the distribution, it should be noted

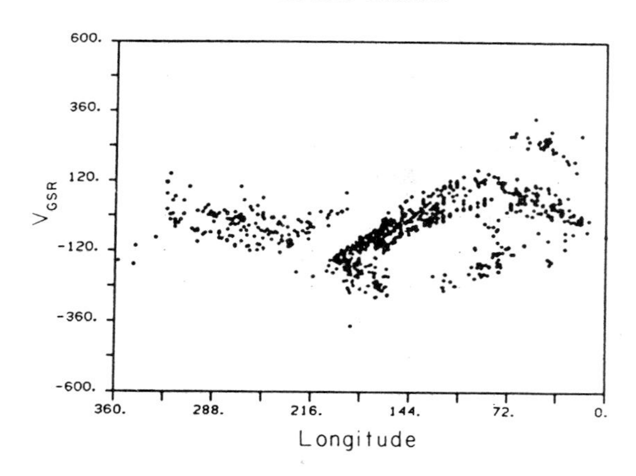

Figure 2. Velocities of HVCs with respect to the galactic standard of rest, V_{GSR} = V_{LSR} + 250 sin ℓ cos b km/s (Giovanelli 1980b).

that most of the longitude range 180° < ℓ < 360° lies in the southern celestial hemisphere, where surveys have been less complete and less sensitive than in the north. We return to this point in Section 4. However, the asymmetry is probably genuine. Obvious is the asymmetry between the positive and negative latitudes in this longitude range: the quadrant b > 0, 180° < ℓ < 360° contains many faint, scattered clouds - which are better shown in Figure 1 of Hulsbosch (1979a) -; the quadrant b < 0, 180° < ℓ < 360° is dominated by the gaseous envelope of the Magellanic Clouds, and by the Magellanic Stream: a long, narrow filament extending from the Clouds to the south galactic pole, and beyond to $\ell \sim 90°$, b $\sim$ -40°. We discuss the Magellanic Stream in Section 7.

Figure 1a shows several other elongated features. The best known is "Chain A", a narrow string of HVCs between ℓ = 132°, b = +23° and $\ell \sim 165°$, b $\sim$ +45°, which we shall discuss in detail in Section 6. "Complex C" is a broader feature between $\ell \sim 70°$, b $\sim$ +30° and $\ell \sim 140°$, b $\sim$ +55°, to which we return in Section 5. Long, narrow features also exist at southern latitudes in the Anticentre region (cf. Giovanelli 1980b, Figure 12). Each of these elongated features contains a number of condensations, "clouds", typically a few degrees in size (see Section 6).

One further asymmetry must be mentioned. Clouds with velocities $|V_{LSR}|$ > 200 km/s occur almost exclusively at **negative** velocities, in the quadrant b < 0, 0° < ℓ < 180°. Hulsbosch (1978a) pointed out that in this region the distribution of velocities relative to the **galactic** standard of rest, $V_{GSR} \equiv V_{LSR}$ + 250 sin ℓ cos b km/s, shows two distinct components, suggesting two separate categories of HVCs. Figure 3, taken from Giovanelli (1980b), shows that the distinction is also clear in a plot of V_{LSR} versus ℓ for HVCs in the southern galactic hemisphere. We discuss the "very-high-velocity clouds" (VHVCs) in Section 8.

Giovanelli (1980b) has estimated that about 10% of the sky is covered by HVCs. The masses involved depend on the distances. Hulsbosch (1979a) estimates that the sums of $M_{HI}d^{-2}$ over the whole sky are about

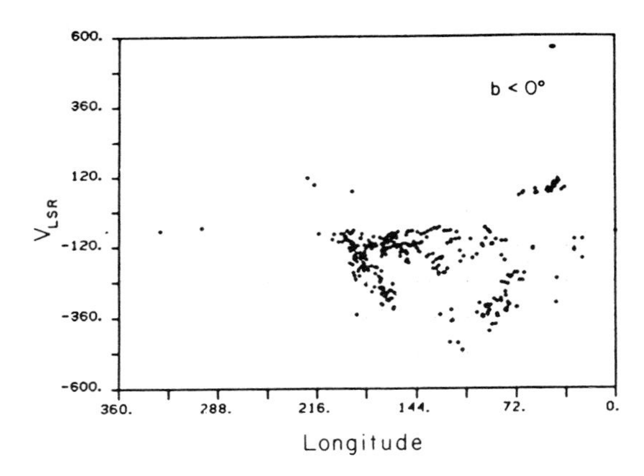

Figure 3. Velocities with respect to the local standard of rest, V_{LSR}, of HVCs in the southern galactic hemisphere (Giovanelli 1980b). Note the presence of two distinct components in the distribution; the clouds with more negative velocities are called "very-high-velocity clouds" (VHVCs, see Section 8).

10^5 $M_\odot$ kpc^{-2} and 0.5 x 10^4 $M_\odot$ kpc^{-2} for HVCs with negative and with positive velocities, respectively.

3. ATTEMPTS AT INTERPRETATION

A variety of models for the nature and origin of high-velocity clouds have been considered in the literature. We shall mention the more important models, with key references, but refrain from detailed discussion. Arguments pro and contra have been given in the reviews by Oort (1966, 1967), Davies (1974), Verschuur (1975), Giovanelli (1978), Oort (1978) and Oort and Hulsbosch (1978).

3.1. A general synopsis

A possible interpretation of HVCs as nearby supernova shells has been discussed in detail by Oort (1966) and by Verschuur (1971). Among the objections raised by Oort (1966, 1967, Oort and Hulsbosch 1978) are: the prohibitively high supernova rate required; the distances (> 1 kpc) estimated for a few HVCs; the straight, narrow shape of Chain A (cf. Section 6); the lack of correspondence between HVCs and the radio-continuum Loops (cf. also Verschuur 1975; however, Cohen (1981) suggests a relation between HVC 160-50-110 and the Cetus Arc, Loop II). Verschuur (1971) proposes a supernova model for the large HI complex with intermediate negative velocities at high northern latitudes; see also Wesselius and Fejes (1973) for discussion. Giovanelli (1980a) suggests that the Anticentre Stream between ℓ, b = 140°, -8° and 190°, -20° at V = -110 km/s may be due to a supernova explosion in the Perseus-Taurus region. Bruhweiler et al. (1980) have proposed that HI supershells of the type described by Weaver (1979) and Heiles (1979) form around evolving associations, and after breaking up into clouds may fall back as HVCs. In ultraviolet absorption lines, Cowie et al. (1979, 1981) have observed supershells of radii $\sim$ 100 pc and velocities $\sim$ 100 km/s around the Orion Association ("Orion's Cloak") and in Carina, but these have not been identified with HVCs.

Superexplosions in the Galactic Disk as a source of HVCs have been discussed by Oort (1966). His main objections were: the very high number of supernovae required, and the short lifetime of the structure in HVCs (Section 6) as compared with their long travel times from distant explosion sites.

Explosions in the Galactic Nucleus appear unsuitable (Oort 1966, 1978) because of the observed shortage of positive-velocity clouds, the lack of angular momentum caused by such ejecta, and again the short lifetime of HVC structure.

Condensations in a hot Galactic Corona have been briefly considered by Oort (1966, 1978), and in more detail by Bregman (1980) - see Section 3.2.

Oort (1965, 1966, 1967, 1969, 1970) has proposed that the HVCs are caused by an inflow of intergalactic gas into the Galaxy. The gas falls in at $\sim$ 500 km/s, but is decelerated in the Halo; the HVCs obtain their structure in the lower Halo. The total inflow would amount to $\sim$ 1 $M_\odot$ per year. The intergalactic gas density required is of order 10^{-4} cm^{-3}; streams of gas must come from various directions. The Halo must be replenished by explosions of supernovae in the Disk. Verschuur (1975) has sharply criticised this model. Davies (1974) points out that the positive-velocity clouds and the elongated strings remain unexplained. More recently, Oort and Hulsbosch (1978) have suggested that the long strings may have had their shapes before entering the Galaxy, possibly originating from old tidal debris in the Milky Way - Magellanic System. Holder (1980) discusses the thermal instabilities caused by radiative cooling of the infalling gas. Tenorio-Tagle (1981) has calculated the evolution of shocks caused by collisions of HVCs with the Galactic Disk.

Oort (1966, 1967) had considered, but rejected, a possible identification of the HVCs with satellites of our Galaxy. However, Kerr and Sullivan (1969) showed that the distribution of HVCs in ℓ, b and V could be well fitted by satellites with orbits having semi-major axes of 30-80 kpc, excentricities 0.5 to 0.8, and inclinations 40°-70°. The total mass required would be of order 10^8 $M_\odot$. The origin of these small satellites might lie in Galactic tidal action on the Magellanic Clouds during earlier passages (cf. the tidal models for the Magellanic Stream, Section 7). Similarly, Einasto et al. (1976) proposed to consider the HVCs as a system of companions in elliptical orbits about our Galaxy.

Following an early suggestion by Burke (1967), Verschuur (1969) showed that HVCs would be virially stable at distances of $\sim$ 400 kpc. With masses of order 10^9 - 10^{10} $M_\odot$ for the HVC complexes at such distances, they might then be considered protogalaxies. The failure of Lo and Sargent (1979) and of Haynes and Roberts (1979) to find intergalactic HI clouds - other than in the form of tidal debris - in several nearby groups of galaxies has made this hypothesis unlikely.

Whether the new, much fainter clouds with very high velocities (VHVCs, see Sections 2 and 8) might be intergalactic clouds in the Local Group, as suggested by Oort and Hulsbosch (1978), deserves further analysis. Eichler (1976) has shown that, at distances of order 1 Mpc, stock thermal energy would be insufficient to support HVCs against collapse. Support might be provided by turbulence, by star formation, or by the presence of substructure.

After Habing (1966) and Kepner (1970) showed that the outer arms of the Galaxy have faint extensions to a few kpc ($b \sim 10°-20°$) above the plane, Davies (1972, 1973) and Verschuur (1973a, b) proposed that the HVCs must be interpreted as features in a highly warped outer spiral structure. Hulsbosch and Oort (1973) accepted that a Galactic warp could explain the HVCs at the lower latitudes, but pointed out that no reasonable rotation of the warped disk could produce the velocities observed at $b \gtrsim 45°$ and in the Anticentre region. Davies (1972, 1973) ascribed these phenomena to tidal debris from an encounter between Galaxy and Magellanic Clouds, but Oort and Hulsbosch (1978) outline some dynamical objections to this suggestion.

It seems likely to us that several of the models mentioned above are actually represented among observed HVCs. Distance determinations will be required to prove this.

A new, detailed model, allowing prediction of the velocity field of HVCs, has been worked out by Bregman (1980). We discuss this in the next subsection.

3.2. Bregman's "Galactic Fountain of High-Velocity Clouds"

Bregman (1980) has proposed that HVCs form by condensation in a hot, dynamic Corona above (and below) the galactic plane, and has modelled this process by hydrodynamic calculations. Supernova-heated gas rises or bubbles up from the Disk, and flows outward into the Corona. There, radiative cooling will lead to thermal instabilities and to condensation of neutral clouds; these clouds fall ballistically back toward their point of origin. The cloud velocities are explained through conservation of angular momentum. Bregman obtains an optimum reproduction of the HVC velocity field in a model with $T_o \sim 1 \times 10^6$ K, $n_o \sim 1 \times 10^{-3}$ cm^{-3} at the base of the Corona; the coronal mass is then 7×10^7 $M_\odot$ and the HVC mass flux onto the Disk 2.4 $M_\odot$/year. Differential rotation in the Corona will stretch the density perturbations and lead to the elongated appearance of HVC groups.

Figure 1b (in colour, see page 6 of these Proceedings) shows the velocity field of HVCs predicted by Bregman (1980, his Figure 6) on the basis of his model E1. It displays the predicted relative numbers of clouds in several velocity intervals at various positions in the sky. These predictions (which are symmetric with respect to the galactic equator) may be compared with the **observed** velocities of HVCs as summarized in Figure 1a (also page 6). In general, Bregman's predictions fit the observed velocity field fairly well. In particular, the predominance of negative velocities around the galactic poles and throughout the longitude range $30° \leqq \ell \leqq 210°$ (i.e., **beyond** the Anticentre) is well represented: in Bregman's model, just as in real nature, the HVC velocity field is not symmetric about the meridional plane through Sun and Galactic Anticentre, but rather about a plane $\ell \sim 210°$. However, the observed **structural** asymmetry between the hemispheres $30° < \ell < 210°$ and $210° < \ell < 390°$ is not predicted by the model; nor are the observed asymmetries between the northern and southern galactic hemispheres. Furthermore, the observed velocity distributions appear to deviate systematically in a few respects from those predicted: 1) The model predicts both positive and negative high velocities in every region of sky, but there are very few regions where both are observed in nature. 2) In comparison with HVCs at $80 < |V| < 130$ km/s, those at higher velocities $130 < |V| < 220$ km/s are more frequently observed than predicted. 3) Finally, the observed clouds with extreme velocities $|V| > 220$ km/s (VHVCs, Section 8) are not predicted and - as emphasized by Mirabel (1981a) - it is unclear how a Galactic Fountain could produce the extreme observed asymmetry of VHVCs with respect to the Galactic Disk.

Thus, although the Galactic-Fountain model gives, in part, a fair representation of HVCs, important features remain unexplained. In particular, the very-high-velocity clouds may represent a separate category located outside the Fountain.

4. NEW LARGE-SCALE SURVEYS

The overview given in Section 2 and Figure 1a was based on quite incomplete information. So far, HVC surveys have either covered large areas of sky with a coarse grid, or mapped small areas in detail, or lacked sensitivity (notably the early surveys). In Section 6 we shall mention new small-scale surveys (of areas $\lesssim 30°$). The most complete published large-scale survey is that by Giovanelli (1980b) with the NRAO 91-meter telescope. It covers the range $-10° < \delta \lesssim +50°$ with a grid of 2-3 degrees in α and δ, and spans the velocity range $-900 < V(LSR) < +900$ km/s with a resolution of 22 km/s. Its detection limit is about 0.06 K, corresponding to 3×10^{18} atoms/cm^2. The results of this survey have been incorporated in Figure 1a, and in the discussion of Sections 2 and 3. Unfortunately, Giovanelli's survey is still quite incomplete in its sky coverage, and biased against small clouds: with its 10' beam, it samples less than 1 percent of the large area (almost 20 000 deg^2) covered.

Much more complete is the new survey by Hulsbosch with the Dwingeloo 25-m telescope. With a $0\overset{\circ}{.}6$ beam, it covers the sky from $\delta = -18°$ to $+90°$ with grid spacings $\Delta b = 1°$ and $\Delta\ell \cos b \approx 1°$, sampling 40 percent of the 27 000 deg^2 sky area covered. The velocity range spanned is $V_{LSR} = -1000$ to $+1000$ km/s, with 16 km/s resolution. The detection limit is 2×10^{18} atoms/cm^2 for resolved structures, but of course poorer for small clouds. Partial results, including the discovery of a large number of VHVCs, have been reported by Hulsbosch (1978) and incorporated in Figure 1a. A map of results in the region $0° < \ell < 200°$, $-70° < b < +70°$ - i.e., about 70 percent of the survey area - is contained in the contribution (Hulsbosch 1985) immediately following this review. The survey has not disclosed any new large-scale structures. However, many new VHVCs and a number of positive-velocity clouds have been found. Also, the survey provides much improved information on the structure and extent of many HVCs and HVC-complexes, and on their mutual relationships. As an example, we discuss in the next section the velocity structure of "Complex C". In view of its relatively dense sampling, the new survey will further provide good statistics of cloud sizes, masses, velocities etc. For further details, we refer to the paper by Hulsbosch (1985).

Another important, though more limited, survey is that by Mirabel and Morras (1984, and these Proceedings) in a wide region around the Galactic Centre. Their survey covers the region $320° < \ell < 50°$, $-90° < b < +40°$, $\delta > -44°$ with a 2° grid (2000 positions) and a detection limit $\sim 5 \times 10^{18}$ cm^{-2}. We discuss the results of this survey, which has yielded many new VHVCs, in Section 8. Since most of these VHVCs are small (often < 1°), and the beamwidth (0°.35) was small compared to the grid spacing (2°), many VHVCs may still have been missed.

At southern declinations beyond the limits of northern radio telescopes, major surveys have been made by Mathewson et al. (1974) and by Mathewson (1976). A complete survey of the sky south of $\delta = -30°$ was carried out by Cleary with the Parkes 18-meter dish (beam 0°.8); the grid spacing was 1° at $b > -25°$ and 2° x 1° at $b < -25°$, the velocity resolution 7 km/s, and the detection limit 0.3 K, or $\sim 10^{19}$ atoms/cm^2 for HVCs. Within the velocity range -148 to +300 km/s at $b > -25°$, and -230 to +218 km/s at $b < -25°$, this survey has not turned up any new HVCs (Cleary et al. 1979).

5. LARGE-SCALE VELOCITY STRUCTURE IN COMPLEX C

In Section 2 we have drawn attention to the prominence of elongated features ("strings") in the sky distribution of HVCs. The most striking strings, Chain A and the Magellanic Stream, will be discussed in Sections 6 and 7. Complex C is a broad, long feature running from $\ell \sim 70°$, $b \sim +25°$ to $\ell \sim 130°$, $b \sim +60°$. Velocities range from -80 to -200 km/s, with an average of $\sim$ -130 km/s. Intensities are generally low. The full extent of this complex, about 70° long and 20° wide, is shown in the map of Hulsbosch (1985). At lower ℓ and b, the complex connects (or, at least, becomes confused) with the "Outer Arm" in the warped HI disk, which reaches up to $b \sim +20°$ at $\ell \sim 70°$ (Habing 1966). In fact, Davies (1972, 1973) and Verschuur (1973) have advocated that complex C and other HVCs be viewed as very strongly warped parts of the outermost spiral structure, possibly so deformed by tidal action of the Magellanic Clouds.

Figure 4 (in colour, page 7 in these Proceedings) shows the velocities measured by Hulsbosch (1985) on a 1° grid in this Complex. Velocities between -90 and -110 km/s (red) occur along the high-b top of the Complex, and near its low-b, low-ℓ end. Velocities -110 > V > -130 km/s (orange) are found throughout the Complex, but especially along its long axis and in the northern half. Velocities -130 > V > -150 km/s (yellow) are seen mostly at lower b and higher ℓ, but also in patches elsewhere. Velocities -150 > V > -170 km/s (green) are observed along the axis of the Complex and in its southern half. Finally, the highest negative velocities (blue) are found in patches along the axis.

Clearly, the velocity field cannot be simply described by e.g. rotation about a long or a short axis. In many places the line profiles contain two peaks, suggesting two velocity systems overlapping on the sky. On the Davies-Verschuur model, these might represent two warped spiral arms. An alternative possibility is that we see a number of HVCs of modest size projected on a large complex. The material has not yet been analyzed in detail, and clarification may require observations at higher velocity resolution. Also, distance determinations of the overlapping structures would be of great interest.

6. SMALL-SCALE STRUCTURE AND ASSOCIATED MOTIONS

6.1. HVCs in general

The large, elongated complexes discussed in Section 2 in general consist of a number of discrete clouds, with sizes of a few degrees and velocity widths of 20-40 km/s, embedded in a tenuous medium. In fact, the clouds were known (1963) before the larger complexes, and gave the HVC phenomenon its name. In Chain A, the cloud structure is particularly strong (Giovanelli et al. 1973, and Figure 5), and large jumps in velocity occur between neighbouring clouds (Oort 1978). Similar jumps have been found in the Anticentre Complex, but these may be due to overlap of clouds - cf. the situation in Complex C (Section 5). The Magellanic Stream also consists of a number of clouds (Mathewson 1976), but its velocity structure appears to be smoother.

Since 1973, observations at high angular resolution (≦ 10', more recently even 1', see Section 6.2) have revealed much detailed structure within the clouds. Most HVCs are found to consist of a small, dense core and an extended, tenuous halo or envelope (Greisen and Cram 1976; Cram and Giovanelli 1976; Schwarz et al. 1976; for further references see Hulsbosch 1979a, b). The cores have angular sizes below 10 arcmin, often unresolved, velocity halfwidths of 5-10 km/s, and brightness temperatures up to 65 K - fully comparable with those measured in low-velocity, even low-latitude hydrogen! The envelopes have sizes of a few degrees, velocity halfwidths of order 25 km/s, and brightnesses of a few K or indeed much less. Also, the clouds often have very steep edges (Giovanelli et al. 1973, Hulsbosch 1978b), suggestive of shock fronts. Collisions of infalling intergalactic material with gas in the Disk or lower Halo may have led to these shocks, and by further compression to instabilities and core formation (Giovanelli and Haynes 1977, Oort and Hulsbosch 1978).

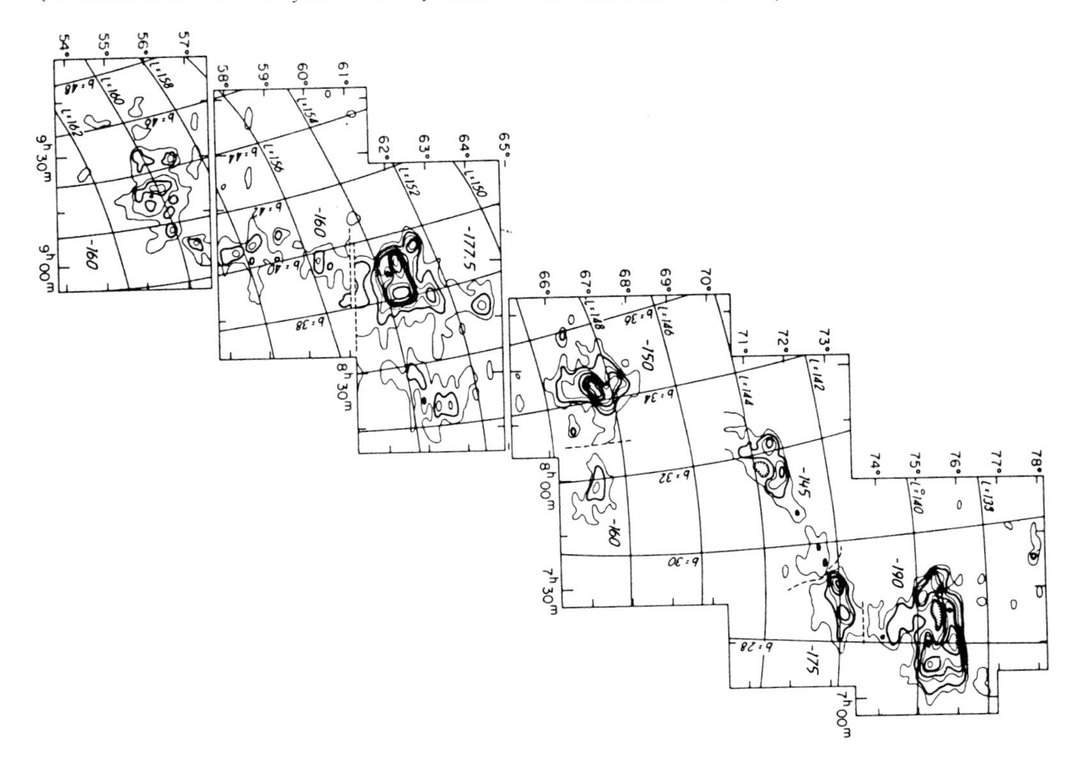

Figure 5. Cloud structure in part of Chain A (Giovanelli, Verschuur and Cram, 1973). The figure consists of a number of monochromatic contour maps, at velocities representative for the individual clouds; consequently, it does not show the full column-density distribution in Chain A. The velocity fields of the clouds at ℓ = 139°, b = +28° (AI) and at ℓ = 153°, b = +39° (AIV) are shown in Figures 7 and 8 (page 7 of these Proceedings).

While much of the above refers especially to Chain A, which has been studied in greatest detail (for a summary, see Oort 1978), it may well apply to HVCs in general. Cohen (1981) finds that HVC 160-50-110, a feature ≧ 25° in length, is similar to Chain A in some respects; and that it coincides with a low-velocity filament, suggesting collision of this HVC with Disk gas. Evidence for such collisions has also been reported by Mirabel (1982) and by Burton and Moore (1979). Core-envelope structure has also been found in several HVCs with **positive** velocities (Giovanelli and Haynes 1976, Cohen and Ruelas-Mayorga 1980, Morras and Bajaja 1983). However, the structure of VHVCs (Section 8) and of the Magellanic Stream (Section 7) appears to be different; this may indicate differences in origin and physical conditions, although one must watch out for possible differences in linear resolution caused by differences in distance.

The spin temperatures T_s of gas in HVCs can, in principle, be determined by measuring the absorption spectrum of a background radio source and comparing it with neighbouring emission spectra, measured at sufficiently high angular resolution. Payne et al. (1980) have very carefully applied the method to 8 sources located behind clouds with $|V_{LSR}| > 50$ km/s. However, the clouds detected in absorption are at low latitudes and would not normally be called high-velocity clouds. A measurement of T_s in cloud AI is discussed in Section 6.2.

6.2. New results in Chain A

The best-known string of HVCs is Chain A, which spans at least 33° from $\ell = 132°$, $b = +23°$ to $\ell = 161°$, $b = +46°$. Oort (1978) gives a detailed discussion of its structure, internal motions, probable distance, space motion and possible origins. Figure 5 shows the location and velocities of condensations in a major portion of the Chain. The core-envelope structure of these clouds has been discussed above.

Hulsbosch (1978b) has mapped the cloud HVC 132+23-211, at the low-latitude end of the string, with 9 arcmin and 1 km/s resolution. This cloud contains a number of very bright cores. Schwarz et al. (1976) mapped part of this cloud at Westerbork, with 2 arcmin resolution, and found 3 clumps of 3 arcmin size on a long, narrow ridge. Figure 6 (in colour, page 6 of these Proceedings) is a new Westerbork map, obtained at 1 arcmin resolution (Schwarz, unpublished); it shows a filament of 6 arcmin width and ≥40 arcmin length, containing several small but bright condensations. The filament runs along a line of constant galactic latitude, and is crossed by another, fainter structure. Velocity variations (coded in colour) are small in this field.

Schwarz and Oort (1981) have mapped a field in HVC 139+28-190 (cloud AI in figure 5) with 50" and 2 km/s resolution. The field falls on the intense ridge and steep slope found by Giovanelli and Haynes (1977) at the edge of cloud AI. Schwarz and Oort find much fine structure, down to 1 arcmin scale, and often of a filamentary nature. These small structures typically have velocity halfwidths ~ 5 km/s and column densities $\sim 1 \times 10^{20}$ atoms/cm^2, densities $\sim 100\ d^{-1}$ cm^{-3} and masses between 0.03 and 10 $d^{-2}\ M_{\odot}$, with d expressed in kpc. Their distribution in position and velocity (see Figure 7, page 7 of these Proceedings) is irregular, with a velocity dispersion of 7 km/s about the mean for the cloud. For the temperatures in 3 of the condensations both upper and lower limits can be given: kinetic temperature $T_k \lesssim 500$ K from the velocity widths, and spin temperature $T_s \gtrsim 25$ K from the lack of detected 21-cm absorption against background sources. The fine structure thus has $T \sim 100$ K, within a factor ~ 5, and it may be contained by the presence of the surrounding, more tenuous cloud of 2°

diameter, whose temperature can be estimated at 6000 K from the velocity dispersion. The doubling time scale for the cloud is 1 d Myr; for the filaments it would be an order of magnitude less, in the absence of pressure equilibrium with the cloud. A striking fact is that the filaments show no preferred direction of elongation, not even parallel to the steep slope. The orientation of the filament in Figure 6 might thus be a coincidence.

The chaotic structure of HVC 139+28-190 appears not to be unique: HVC 153+39-178 (Figure 8, page 7) shows similar filamentary features. We note in particular the velocity gradients across these filaments. The narrow filaments in Figures 7 and 8 are strongly suggestive of shock fronts, due to interaction of high-velocity gas with low-velocity material.

It is tantalizing that the physical conditions in these intriguing structures are still so poorly known, owing to the lack of distance information. Further detailed studies should be of great interest, but must be accompanied by strong efforts to obtain spin temperatures from 21-cm absorption and distances from optical absorption measurements (Section 9).

The high densities found invite the speculation that stars might be formed in the HVC cores or shock fronts. Dyson and Hartquist (1983) calculate that the internal motions within Chain A might be due to energization of the gas by young stars. So far, however, no direct evidence for such young stars has been found.

7. THE MAGELLANIC STREAM

This object will be extensively discussed at IAU Symposium 108 on the Magellanic Clouds (Van den Bergh and De Boer, 1984); hence, we shall review it only briefly here.

Originally known as the South Galactic Pole complex, this HVC was partly mapped by northern observers (see Wannier and Wrixon (1972) and references given there). Then Mathewson, Cleary and Murray (1974a, b) discovered its relationship to the Magellanic Clouds, named it the Magellanic Stream and mapped it more fully (see also Mathewson 1976). Since then, Haynes and Roberts (1979) have shown its extent and complexity in the Sculptor region, and Morras (1982) has observed a few, possibly related, HVCs on the opposite side of the Magellanic Clouds. McGee et al. (1983) have made sensitive measurements of 21-cm profile structure in the directions of Magellanic Clouds and Stream, compared these with ultraviolet absorption lines, and discussed implications for the haloes of Milky Way Galaxy and Magellanic Clouds.

Maps of the small-scale structure in the northern tip of the Stream have been made by Mirabel et al. (1979), Cohen (1982a), and Mirabel (1981a, c). The former observations show condensations of 0°.5 size, but at Arecibo (3!3 resolution) Mirabel (1981a, c) finds condensations of 5' size. In these various structures the velocity halfwidths are consistently ≧ 20 km/s; no narrow spectral components have so far been found. Towards its tip, the Stream appears to broaden, weaken and break up into clouds with velocity differences of order 30 km/s. Cohen (1982a) finds filamentary structures on scales from 10° to 0°.5, and velocity gradients both along the Stream (∿ 2 km/s per degree) and across it (∿ 6 km/s per degree); he claims that this favours a tidal interpretation.

Tidal models for the Stream, as debris from the Magellanic Clouds after a close encounter with the Galaxy, have been developed or advocated by Mathewson et al. (1974b), Fujimoto and Sofue (1976, 1977), Davies and Wright (1977), Lin and Lynden-Bell (1977, 1982), Fujimoto (1979), Kunkel (1979), and by Murai and Fujimoto (1980). The earlier models, with the Stream leading the Clouds, had trouble fitting its velocity field, but both Murai and Fujimoto (1980) and Lin and Lynden-Bell (1982) obtain good fits for a trailing Stream, caused through tidal action of a Milky Way Galaxy with a massive halo on the Clouds during a previous passage. Recently, Fujimoto and Murai (these Proceedings, Section III.1) have shown that a recent close encounter between the Large and Small Magellanic Clouds is required for a proper reproduction of the Stream.

Mathewson and Schwarz (1976) proposed to interpret the Stream as primordial gas clouds in the same orbit as the Magellanic Clouds. Similarly, Einasto et al. (1976) considered the Magellanic Stream as permanent members of the "Hypergalaxy" (the system of Galaxy and companions with total mass 1.2×10^{12} $M_{\odot}$), in elliptical orbits about the central Galaxy. Lo and Sargent (1979) consider the primordial interpretation unlikely, in view of their failure to detect inter-galactic HI clouds - other than tidal debris - in several nearby galaxy groups. Later, Mathewson et al. (1977, 1979) interpreted the Stream as the result of thermal instabilities in the wake of the Magellanic Clouds on their passage through a hot halo around our Galaxy; but Bregman (1979) raised doubts about this model. The non-tidal models have not been discussed in recent years, and a tidal origin of the Stream now appears well established.

8. CLOUDS WITH VERY HIGH VELOCITIES

A few HVCs with exceptionally high velocity, in the range -250 to -450 km/s, were discovered by Wright (1974) near M33, by Davies (1975) near M31, by Cohen and Davies (1975) and by Shostak (1977). Then Hulsbosch (1978a, 1984) and Giovanelli (1980b) found large numbers of very-high-velocity clouds (VHVCs) in their sensitive surveys of large areas of sky, and Giovanelli (1981) defined the VHVCs as a separate population in (ℓ, b, V). Most of these objects are small, at most 1 or 2° in diameter, and almost all of them lie in the galactic quadrant 10° < ℓ < 190°, b < 0°. Mirabel (1981b) has found several VHVCs in a survey of a region of 34° x 18° around ℓ ∿ 25°, b ∿ -20°; and Mirabel and Morras (1984) have recently found many more - see below.

Detailed maps of VHVCs have been made by Cohen (1982a, b), Cohen and Mirabel (1978, 1979), Giovanelli (1981), Mirabel (1981a, b) and Wright (1979 a, b). Most VHVCs appear to have simple structure, often with considerable velocity gradients indicating rotation. Two-component structure, with a small core having velocity halfwidth W ∿ 7 km/s and an envelope with W ∿ 25 km/s, has been found only in VHVCs 114-10-440 and 24-19-235 (Cohen and Mirabel 1979) and 17-25-230 (Mirabel 1981b). Steep edges suggesting shock fronts have not been observed. Thus, the structure of most VHVCs appears to differ from that in most northern (b > 0) HVCs at more modest velocities - or is this an effect of distance and resolution?

Two VHVCs of considerable size are known. Wright's (1974) cloud 128-33-400 near M33 extends over 7° x 5° and has much detailed structure (Wright 1979a). Its mass is 330 d(kpc)^2 $M_\odot$, or 2 x 10^8 $M_\odot$ if at the distance of M33. Although its edge is only 1° from that galaxy, Wright argues that it is too large to withstand the tides of M33, and too light to cause the warp in that galaxy, and hence it probably is an unrelated VHVC, seen close only in projection - unless it is a remnant of tidal interaction between M31 and M33.

The first VHVC discovered (by Meng and Kraus 1970, and Van Kuilenburg 1972, _avant la lettre_) is a large complex centred near ℓ = 165°, b = -45°. According to a detailed study by Cohen (1982b), it has ∿ 12° diameter and spans the velocity range -360 to -190 km/s. This object, with a total mass 3800 d^2 $M_\odot$, contains many clumps of ∿ 0°.5 diameter, density ∿ 3 d^{-1} cm^{-3} and mass ∿ 30 d^2 $M_\odot$, which should be investigated for smaller-scale structure. The velocity field is complex, with gradients both along and across the length of the object. Velocity halfwidths range from 18 to 54 km/s, with an average of 30 km/s. This object looks more like the northern HVCs than like the other VHVCs; its velocity, too, is not very different from that of Chain A.

The distribution of VHVCs in (ℓ, b, V) has been discussed by Giovanelli (1979, 1981). The pronounced difference of this distribution from that of other HVCs provides a clear distinction between the two categories. The small size of VHVCs and the proximity of a few early discoveries to M31 and M33 had suggested that the VHVCs might be intergalactic gas clouds in the Local Group (Hulsbosch 1978a). Whether this is compatible with the lack of detection of such clouds in other galaxy groups (Lo and Sargent 1979, Haynes and Roberts 1979) has not been fully assessed. However, Giovanelli (1981) finds that the (ℓ, b, V) distribution of VHVCs is quite different from that of galaxies in the Local Group, and is actually more similar to that of the Magellanic Stream. Rejecting both a local and a fountain-model interpretation on the basis of the observed velocities, Giovanelli (1981) proposes to explain the VHVCs as shreds of the Magellanic Stream, fragmented at its tip and now dispersed through a galactic quadrant. Similar suggestions have been made by Mirabel (1981a) and by Cohen (1982b). These suggestions appeared attractive as long as the Magellanic Stream was considered leading the Clouds, with a distance of ∿ 10 kpc at its tip - although the wide dispersion of fragments needed to be supported by orbit calculations. However, with the Magellanic Stream trailing as argued by Lin and Lynden-Bell (1982) and its tip at ∿ 60 kpc distance, the proposed explanation of VHVCs as Magellanic debris appears speculative. Giovanelli (1981) further suggests that collisions of (former) VHVCs with the Galactic Disk have decelerated them to HVCs with more modest velocities, and via shocks have caused the structures observed in these HVCs. This explanation of the HVCs in general meets severe difficulties (cf. Oort 1978, Oort and Hulsbosch 1978). Further objections to accretion of Magellanic Stream gas by the Galaxy have been raised by Bregman (1979).

In a new survey of a wide region around the Galactic Centre (cf. Section 4), Mirabel and Morras (1984, and these Proceedings) have found a large number of VHVCs. The great majority of these, about 85%, have negative velocities, also after correction for the rotation of LSR about the Galactic Centre (GC). Combining this finding with the fact that in the Anticentre (AC) region all VHVCs have $V < 0$, Mirabel and Morras conclude that the VHVCs fall towards the centre of the Galaxy. The small angular sizes observed in the GC region, smaller than in the AC region, support this view. The distances of VHVCs from the Centre may be of order 20 kpc, considering that two VHVCs (24-1.9-293 and 8.9-1.5-174) are observed at very low latitudes and yet have very high velocities, hence must probably still be outside the Disk. Mirabel and Morras (1984) estimate the inflow of gas with very high velocities - hence not yet decelerated - at about 0.2 $M_\odot$ per year.

9. DISTANCE AND CHEMICAL COMPOSITION

The problem of the distances of HVCs probably holds the key to their understanding. In derivations of size, density, mass, and various timescales the distance enters to the first or higher power. The distance of HVCs is also crucial in assessing their relationships to our Galaxy, its structural components such as spiral arms, stellar associations and supernova remnants, or to extragalactic objects.

The very definition of HVCs precludes the use of differential galactic rotation for their distance determination. Also, the claim by Giovanelli (1980b) and Mirabel (1981a), that one can roughly estimate the distances (nearby or distant) of HVC complexes or populations from the amplitude of variation of their LSR velocities with longitude, appears to us unfounded.

Until recently, no HVC had ever been detected outside the 21-cm line, and no convincing optical relationships had been found. In 35 hours of integration on Chain A, Giovanelli has not detected any CO or OH emission (Hulsbosch 1979a). Jura (1979) has pointed out that

Doppler-shifted Hα, due to scattering on dust grains, might be observable in HVCs, but no detection has been reported. In fact, the detailed infrared sky survey by IRAS should allow a careful search for correlations between the distributions of HVCs and dust in high latitudes. Dyson and Hartquist (1983) have suggested that OB stars might be formed in the dense cores of HVCs, but none have been found so far.

Attempts to determine the distance of HVC 131+1-200, which lies in the same direction as the galactic radio source 3C58, by measuring the absorption spectrum of this source at 21 cm wavelength, have given an ambiguous result (Hulsbosch 1975, Schwarz and Wesselius 1978). No absorption at the HVC's velocity was detected, hence either the HVC is behind 3C58 (which may be at 8 kpc distance) or its spin temperature is too high ($T_s > 200$ K).

Oort and Hulsbosch (1978), using various indirect arguments, estimate a distance of ∿ 2 kpc for Chain A. One of their arguments is the distance of intermediate-velocity clouds (IVCs), measured through the presence or absence of absorption lines of Ca^+ and Na in the spectra of early-type stars at high latitudes (cf. Münch and Zirin 1961). However, the relationship between IVCs and HVCs is uncertain, and may depend on assumptions as to the nature and origin of HVCs.

Indirect arguments have also been given by Cohen (1981) and by Watanabe (1981). Cohen estimates $z = -200 \pm 100$ pc for HVC 160-50-110, which appears to be interacting with low-velocity gas in the disk (cf. Section 6). Watanabe (1981), assuming that the absence or presence of elongated structures in HVCs is due to tidal effects, derives lower and upper limits to HVC distances. We believe this procedure to be invalid, since processes other than tides may play a decisive role in shaping the structures.

The most promising method to determine distances appears to be through measurement of optical interstellar absorption lines in the spectra of early-type stars. Kepner (1968) published a list of potential probes, and Oort and Hulsbosch (1978) have reviewed the early unsuccessful attempts to use them. The basic problem is that young B-type stars are very rare at large distances z from the galactic plane. Hobbs et al. (1982) have measured the Ly α absorption line in 10 distant, high-latitude OB stars, and compared the HI column densities so measured in absorption with those found in the 21-cm emission line in the same directions. They find that the ratio $N_{HI}(\mathrm{Ly}\ \alpha)/N_{HI}(21\mathrm{cm})$ increases with stellar $|z|$, and that little HI exists above $|z| = 2$ kpc; but this work carries no velocity information. In a very thorough study, Albert (1983) has measured Ti II, Ca II and Na I absorption lines in 9 pairs of distant (average $|z| = 1900$ pc) and nearer ($<|z|> = 170$pc) B-stars, and compared those with 21-cm emission profiles. He finds that gas with velocities $10 < |V| \lesssim 50$ km/s lies in the lower Halo, at $|z| > 170$ pc, but his spectra include no HVC lines.

Interstellar absorption at velocities of order 100 km/s has been measured in the high-ionization UV lines, with IUE, and also in the lower-ionization UV and visible lines, on lines of sight to globular clusters, to the Magellanic Clouds and to other extragalactic objects, by Savage and De Boer; by Songaila, Cowie and York; by Blades and Morton; and by others. These absorptions are identified as due to hot, largely ionized gas in the galactic halo, and no convincing identifications with HVCs measured in the 21-cm line have been reported. De Boer reviews this work in a separate invited paper at this Symposium.

The best attack appears to be to search for suitable probes seen projected on known HVCs. An absorption line at the HVC's velocity then shows that the HVC is in front of the probe and yields an upper limit to the HVC's distance. Failure to find absorption at the proper velocity allows several interpretations: 1) the HVC may be behind the probe; 2) the HI column density in the HVC may be too low in the direction of the probe; 3) the chemical composition and physical conditions in the HVC may be such that the column density of the optical absorbers (Ca^+, Na, or other) is too low for detection. A high-angular-resolution 21-cm observation in the direction of the probe can be used to test for item 2); a spectrum of a nearby extragalactic probe (quasar, Seyfert) can serve to test item 3). Several teams of observers are working on this program. Songaila (1981) has found Ca II K absorption, at the velocity of the Magellanic Stream (MS), in a background galaxy (Fairall 9, distance 135 Mpc) but not in the globular cluster NGC 2808 (distance 9 kpc). This leaves the MS distance uncertain by ±2 orders of magnitude, but it is consistent with the - generally assumed - association of the MS with the Magellanic Clouds. It further shows that the chemical composition of the MS is not primordial, that in fact the ratio of Ca II and HI line strengths is much **higher** than in low-velocity gas. Pettini and Boksenberg (unpublished) have detected Ca II K absorption in two galaxies behind HVC-Complex C, again indicating a high ratio of Ca II and HI line strengths. They have, however, so far not found any HVC absorption in RR Lyrae stars in this direction. These preliminary results suggest that Complex C may be beyond 10 kpc distance.

Distance determinations will be required for several HVC complexes of different character and structural properties. Blue horizontal-branch stars and blue Population II giants will be the best stellar probes, since their spectra are much freer of stellar lines than those of RR Lyrae stars. Deep surveys for such stars, down to $\sim$ 17th magnitude, will be required in order to cover distances up to 20 kpc - if HVCs are at such great distances.

10. CONCLUDING REMARKS

The name high-velocity cloud covers a great variety of phenomena. The discussion in the previous sections suggests that these require a variety of explanations.

An almost complete survey of HVCs north of −18° declination will now soon be available. A southern supplement is urgently needed. Such surveys can serve as a basis for statistical analysis of HVC properties, and for further small-scale studies. Aperture-synthesis maps are required to reveal the small-scale structures and the physical processes in HVCs.

Optical and ultraviolet absorption spectra may soon provide distances to several HVCs. As an immediate consequence, their physical properties, chemical composition, and relationships to our Galaxy and its constituents will then be clarified. Theories of the nature and origin of HVCs can then be put on a firmer basis.

ACKNOWLEDGEMENTS

The Dwingeloo HVC survey was carried out by Hulsbosch with partial support from ZWO, through the Netherlands Foundation for Astronomical Research (ASTRON). The Dwingeloo and Westerbork Radio Observatories are operated by the Netherlands Foundation for Radio Astronomy (RZM), with financial support from the Netherlands Organization for Pure Scientific Research (ZWO). We thank the RZM staff for their assistance, and N. Zuidema, G. Comello and B.P. Wakker for help with the figures.

REFERENCES

Albert, C.E. 1983, Astrophys. J. **272**, 509
Bregman, J.N. 1979, Astrophys. J. **229**, 514
Bregman, J.N. 1980, Astrophys. J. **236**, 577
Bruhweiler, F.R., Gull, T.R., Kafatos, M., Sofia, S. 1980, Astrophys. J. **238**, L27
Burke, B.F. 1967, in "Radio Astronomy and the Galactic System" (ed. H. van Woerden), IAU Symp. **31**, p. 299
Burton, W.B. 1966, Bull. Astron. Inst. Netherlands **18**, 247
Burton, W.B. and Moore, R.L. 1979, Astron. J. **84**, 189
Cleary, M.N., Heiles, C., Haslam, C.G.T. 1979, Astron. Astrophys Suppl. **36**, 95
Cohen, R.J. 1981, Mon. Not. Roy. Astron. Soc. **196**, 835
Cohen, R.J. 1982a, Mon. Not. Roy. Astron. Soc. **199**, 281
Cohen, R.J. 1982b, Mon. Not. Roy. Astron. Soc. **200**, 391
Cohen, R.J. and Davies, R.D. 1975, Mon. Not. Roy. Astron. Soc. **170**, 23P
Cohen, R.J. and Mirabel, I.F. 1978, Mon. Not. Roy. Astron. Soc. **182**, 395
Cohen, R.J. and Mirabel, I.F. 1979, Mon. Not. Roy. Astron. Soc. **186**, 217
Cohen, R.J. and Ruelas-Mayorga, R.A. 1980, Mon. Not. Roy. Astron. Soc. **193**, 583
Cowie, L.L., Hu, E.M., Taylor, W., York, D.G. 1981, Astrophys J. **250**, L25
Cowie, L.L., Songaila, A., York, D.G. 1979, Astrophys. J. **230**, 469
Cram, T.R. and Giovanelli, R. 1976, Astron. Astrophys. **48**, 39
Davies, R.D. 1972, Nature **237**, 88

Davies, R.D. 1973, Mon. Not. Roy. Astron. Soc. **160**, 381
Davies, R.D. 1974, in "Galactic Radio Astronomy" (eds. F.J. Kerr and S.C. Simonson), IAU Symp. **60**, p. 599
Davies, R.D. 1975, Mon. Not. Roy. Astron. Soc. **170**, 45P
Davies, R.D. and Wright, A.E. 1977, Mon. Not. Roy. Astron. Soc. **180**, 71
Dyson, J.E. and Hartquist, T.W. 1983, Mon. Not. Roy. Astron. Soc. **203**, 1233
Eichler, D. 1976, Astrophys. J. **208**, 694
Einasto, J., Haud, U., Jôeveer, M., Kaasik, A. 1976, Mon. Not. Roy. Astron. Soc. **177**, 357
Fujimoto, M. 1979, in "Large-scale characteristics of the Galaxy" (ed. W.B. Burton), IAU Symp. **84**, p. 557
Fujimoto, M. and Sofue, Y. 1976, Astron. Astrophys. **47**, 263
Fujimoto, M. and Sofue, Y. 1977, Astron. Astrophys. **61**, 199
Giovanelli, R. 1978, in "Structure and Properties of Nearby Galaxies" (eds. E.M. Berkhuijsen and R. Wielebinski), IAU Symp. **77**, p. 293
Giovanelli, R. 1979, in "Large-scale characteristics of the Galaxy" (ed. W.B. Burton), IAU Symp. **84**, p. 541
Giovanelli, R. 1980a, Astrophys J. **238**, 554
Giovanelli, R. 1980b, Astron. J. **85**, 1155
Giovanelli, R. 1981, Astron. J. **86**, 1468
Giovanelli, R. and Haynes, M.P. 1976, Mon. Not. Roy. Astron. Soc. **177**, 525
Giovanelli, R. and Haynes, M.P. 1977, Astron. Astrophys. **54**, 909
Giovanelli, R., Verschuur, G.L., Cram, T.R. 1973, Astron. Astrophys. Suppl. **12**, 209
Greisen, E.W. and Cram, T.R. 1976, Astrophys. J. **203**, L119
Habing, H.J. 1966, Bull. Astr. Inst. Netherlands **18**, 323
Haynes, M.P. and Roberts, M.S. 1979, Astrophys. J. **227**, 767
Heiles, C.E. 1979, Astrophys. J. **229**, 533
Henderson, A.P., Jackson, P.D., Kerr, F.J. 1982, Astrophys. J. **263**, 116
Hobbs, L.M., Morgan, W.W., Albert, C.E., Lockman, F.J. 1982, Astrophys. J. **263**, 690
Holder, R.D. 1980, Mon. Not. Roy. Astron. Soc. **191**, 417
Hulsbosch, A.N.M. 1975, Astron. Astrophys. **40**, 1
Hulsbosch, A.N.M. 1978a, Astron. Astrophys. **66**, L5
Hulsbosch, A.N.M. 1978b, Astron. Astrophys. Suppl. **33**, 383
Hulsbosch, A.N.M. 1979a, in "Large-scale characteristics of the Galaxy" (ed. W.B. Burton), IAU Symp. **84**, p. 525
Hulsbosch, A.N.M. 1979b, Trans. IAU **17A**, Part 3, p. 141
Hulsbosch, A.N.M. 1985, IAU Symposium 106, in press (Section II8)
Hulsbosch, A.N.M. and Oort, J.H. 1973, Astron. Astrophys. **22**, 153
Jura, M. 1979, Astrophys. J. **227**, 798
Kepner, M.E. 1968, Bull. Astr. Inst. Netherlands, **20**, 98
Kepner, M.E. 1970, Astron. Astrophys. 5, 444
Kerr, F.J. 1967, in "Radio Astronomy and the Galactic System" (ed. H. van Woerden), IAU Symp. **31**, p. 297
Kerr, F.J. and Sullivan, W.T. 1969, Astrophys. J. **158**, 115
Kunkel, W.E. 1979, Astrophys. J. **228**, 718
Lin, D.N.C. and Lynden-Bell, D. 1977, Mon. Not. Roy. Astron. Soc. **181**, 59

Lin, D.N.C. and Lynden-Bell, D. 1982, Mon. Not. Roy. Astron. Soc. **198**, 707
Lo, K.Y. and Sargent, W.L.W. 1979, Astrophys. J. **227**, 756
Mathewson, D.S. 1976, Royal Greenwich Obs. Bull. No. 182, p. 217
Mathewson, D.S., Cleary, M.N., Murray, J.D. 1974a, in "Galactic Radio Astronomy" (eds. F.J. Kerr and S.C. Simonson), IAU Symp. **60**, 617
Mathewson, D.S., Cleary, M.N., Murray, J.D. 1974b, Astrophys. J. **190**, 291
Mathewson, D.S., Ford, V.L., Schwarz, M.P., Murray, J.D. 1979, in "Large-scale characteristics of the Galaxy" (ed. W.B. Burton), IAU Symp. **84**, p. 547
Mathewson, D.S. and Schwarz, M.P. 1976, Mon. Not. Roy. Astron. Soc. **176**, 47P
Mathewson, D.S., Schwarz, M.P., Murray, J.D. 1977, Astrophys. J. **217**, L5
McGee, R.X., Newton, L.M., Morton, D.C. 1983, Mon. Not. Roy. Astron. Soc. **205**, 1191
Meng, S.Y. and Kraus, J.D. 1970, Astron. J. **75**, 535
Mirabel, I.F. 1981a, Rev. Mexicana Astron. Astrof. **6**, 245
Mirabel, I.F. 1981b, Astrophys. J. **247**, 97
Mirabel, I.F. 1981c, Astrophys. J. **250**, 528
Mirabel, I.F. 1982, Astrophys. J. **256**, 112
Mirabel, I.F., Cohen, R.J., Davies, R.D. 1979, Mon. Not. Roy. Astron. Soc. **186**, 433
Mirabel, I.F. and Morras, R. 1984, Astrophys. J., in press.
Morras, R. 1982, Astron. Astrophys. **115**, 249
Morras, R. and Bajaja, E. 1983, Astron. Astrophys. Suppl. **51**, 131
Muller, C.A., Oort, J.H., Raimond, E. 1963, C.R. Acad. Sci. Paris **257**, 1661
Münch, G. and Zirin, H. 1961, Astrophys. J. **133**, 11
Murai, T. and Fujimoto, M. 1980, Publ. Astr. Soc. Japan **32**, 581
Oort, J.H. 1965, Trans. IAU **12A**, 789
Oort, J.H. 1966, Bull. Astr. Inst. Netherlands **18**, 421
Oort, J.H. 1967, in "Radio Astronomy and the Galactic System" (ed. H. van Woerden), IAU Symp **31**, p. 279
Oort, J.H. 1969, Nature **224**, 1158
Oort, J.H. 1970, Astron. Astrophys. **7**, 381
Oort, J.H. 1978, in "Problems of Physics and Evolution of the Universe" (ed. L. Mirzoyan), Armenian Acad. Sci., Yerevan, p. 259
Oort, J.H. and Hulsbosch, A.N.M. 1978, in "Astronomical papers dedicated to Bengt Strömgren" (eds. A. Reiz and T. Andersen), Copenhagen Univ. Observatory, p. 409
Payne, H.E., Salpeter, E.E., Terzian, Y. 1980, Astrophys. J. **240**, 499
Schwarz, U.J. and Oort, J.H. 1981, Astron. Astrophys. **101**, 305
Schwarz, U.J., Sullivan, W.T., Hulsbosch, A.N.M. 1976, Astron. Astrophys. **52**, 133
Schwarz, U.J. and Wesselius, P.R. 1978, Astron. Astrophys. **64**, 97
Shostak, G.S. 1977, Astron. Astrophys. **54**, 919
Songaila, A. 1981, Astrophys. J. **243**, L19
Tenorio-Tagle, G. 1981, Astron. Astrophys. **94**, 338
Van den Bergh, S. and De Boer, K.S. (editors) 1984, "Structure and

Evolution of the Magellanic Clouds", IAU Symposium No. 108, Dordrecht, Reidel.
Van Kuilenburg, J. 1972, Astron. Astrophys. Suppl. **5**, 1
Van Woerden, H. 1976, Trans IAU **16A**, Part 3, p. 86
Van Woerden, H. 1979, Trans IAU **17A**, Part 3, p. 77
Verschuur, G.L. 1969, Astrophys. J. **156**, 771
Verschuur, G.L. 1971, Astron. J. **76**, 317
Verschuur, G.L. 1973a, Astron. Astrophys. **22**, 139
Verschuur, G.L. 1973b, Astron. Astrophys. **27**, 407
Verschuur, G.L. 1975, Ann. Rev. Astron. Astrophys. **13**, 257
Wannier, P. and Wrixon, G.T. 1972, Astrophys J. **173**, L119
Watanabe, T. 1981, Astron. J. **86**, 30
Weaver, H.F. 1979, in "Large-scale characteristics of the Galaxy" (ed. W.B. Burton), IAU Symp. **84**, p. 295
Wesselius, P.R. and Fejes, I. 1973, Astron. Astrophys. **24**, 15
Wright, M.C.H. 1974, Astron. Astrophys. **31**, 317
Wright, M.C.H. 1979a, Astrophys. J. **233**, 35
Wright, M.C.H. 1979b, Astrophys. J. **234**, 27

DISCUSSION

B.F. Burke: As a matter of principle, I object to presentation of anonymous data.

Van Woerden: What I showed is not anonymous. It was an unpublished spectrum of a background galaxy, with Complex C in absorption. I had to withhold the name of the galaxy, but the observation is by Pettini and Boksenberg, and Pettini has kindly allowed me to show this spectrum.

J.H. Oort: In his Green Bank survey Giovanelli found a remarkable thin feature at about -110 km/s extending over 50° in longitude between -10° and -20° galactic latitude. I have been puzzled by the nature of this object. Its width is no more than 3° to 4°.

Van Woerden: String A is a similarly intriguing, thin feature. I wonder whether these are shock fronts.

R.D. Davies: R.J. Cohen at Jodrell Bank has studied the feature mentioned by Oort, and suggested that its thinness was due to interaction between infalling gas and material in the upper Disk.

Oort: That may well be, but it extends over about 40 degrees!

Davies: He suggested that it might be very close to us.

Van Woerden: There is a misunderstanding. Giovanelli's Anticentre Stream runs from 140°-8° to 190°-20° (Giovanelli 1980a). Cohen's feature runs between 145°-50° and 180°-40° (Cohen 1981); it also has V = -110 km/s.

W. Iwanowska: Are the high velocities of clouds shown in your map (Figure 1a of your paper) corrected for the solar galactic-rotation component? If the clouds do not participate in the galactic rotation, many of them may show negative radial velocities.

Van Woerden: That is quite correct. Figure 1a shows velocities relative to the local standard of rest. Velocities relative to the galactic standard of rest (i.e. after correction for the Sun's rotation about the Galactic Centre) are shown in Figure 2 of my paper. In that figure, there is still a slight preponderance of negative velocities. In the interpretation one must, of course, take the effects of galactic rotation into account.

I.F. Mirabel: There is a clear preponderance of very-high-velocity clouds with inward motions in the Galactic Centre and Anticentre regions of the sky. Since the solar-motion component in these two directions is small, these findings are evidence for inflow of HI toward the Milky Way. Cf. my poster paper.

J.P. Ostriker: I think it very unlikely that there is a significant infall of mass into the Galaxy from large distances. Infall of $1 M_{\odot}$/yr corresponds to 10^{41} erg/s released gravitationally, which is two orders of magnitude more than the total X-ray luminosity of a similar galaxy like NGC 4565. Fountain models or other supernova-driven processes are acceptable.

Mirabel: The large infall velocities of up to 250 km/s of some HVCs are an indication that these clouds with extreme velocities are coming from great distances above the galactic plane. This kind of very-high-velocity clouds contribute a rate of accretion much smaller than 1 solar mass per year.

Ostriker: If the accretion rate is 2 orders of magnitude less, it would be consistent with the X-ray observations.

A DEEP SURVEY FOR HIGH-VELOCITY CLOUDS

Aad N.M. Hulsbosch
Astronomical Insitute of the University of Nijmegen
The Netherlands

The deep Dwingeloo survey for high-velocity clouds is now nearing completion. This survey consists of more than 28 000 positions in a 1° grid for $\delta>-18°$. The velocity coverage is −1000 to + 1000 km/s and the detection limit (4σ) in T_b about 0.05 K, corresponding to $\int T_a dv$ = 1.0 K km/s or 10 Jy km/s or 2.2 $r^2 M_\odot$ (r the distance in kpc) if we assume an average profile width of 25 km/s.

The preliminary results now available for $0°<l<200°$, $-70°<b<+70°$ are presented in the figure. Shown are the outer contours of all the larger features with velocities in excess of 100 km/s, as well as most of the small features which are observed at only one or two grid positions.

There is a clear difference between the northern and southern galactic hemisphere. In the northern hemisphere we can see an extended low-latitude feature at all longitudes, having velocities down to about −125 km/s around l=115°, which presumably originates in the Outer Arm. This feature has a rather sharp upper boundary at b=+25°, except near l=70° where a spur rises up and continues to l,b=140°,+55°. Embedded in this spur we find a complicated composition of features with velocities down to −200 km/s. At l,b=142°,+46° the spur contacts the upper region of the well-known string A which, at higher b, appears to be much more extended than was previously recognized. A few new complexes at l,b = 180°,+55° are also shown.

At negative latitudes the picture is dominated by the Magellanic Stream (l=90°, $b<-35°$). Furthermore, we find a complicated string running from l,b=155°,−45° to 185°,−10° (velocities −100 to −330 km/s), a complex around l,b=130°,0°, and a large number of small very-high-velocity clouds with velocities down to −465 km/s. Of these HVC 128-32-385 is the most extended one, but in general they are no larger than one or two grid points. The objects around l,b=38°,+6° also belong to this group.

The only positive velocities in the field shown are those at l,b = 20°,+5°, at 40°,−20°, and around l=195°. At greater longitudes (not included in the figure) many more positive-velocity features are found, mainly at declinations not much above −18°, but their total mass is at least an order of magnitude less than of the negative-velocity gas.

More detailed maps and a catalogue of all components found are in preparation.

H. van Woerden et al. (eds.), The Milky Way Galaxy, 409–410.

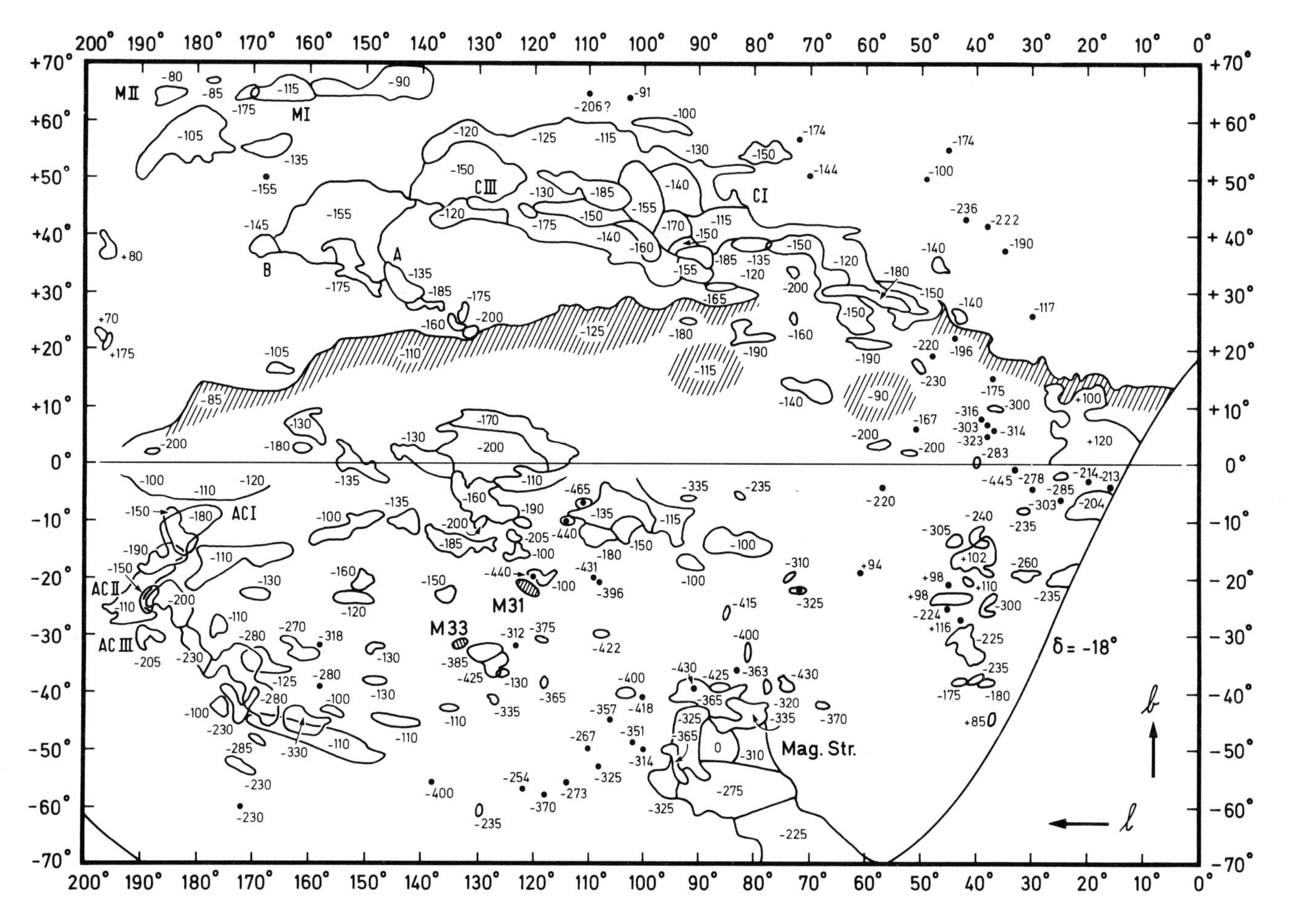

Figure 1. Preliminary results of a deep HVC survey (June 1983). Shown are outer contours of T_b = 0.05 K. Numbers give radial velocities (km s^{-1}) with respect to l.s.r.

A SURVEY FOR HIGH–VELOCITY CLOUDS IN THE INNER GALAXY

I.F. Mirabel and R. Morras
Department of Physics
University of Puerto Rico

The results from a search for high–velocity hydrogen around the direction to the galactic center are presented. About 2000 positions were surveyed with the 43–m radiotelescope of NRAO with a rms of 0.03 K on a velocity interval of −1000 to +1000 km s^{-1}.

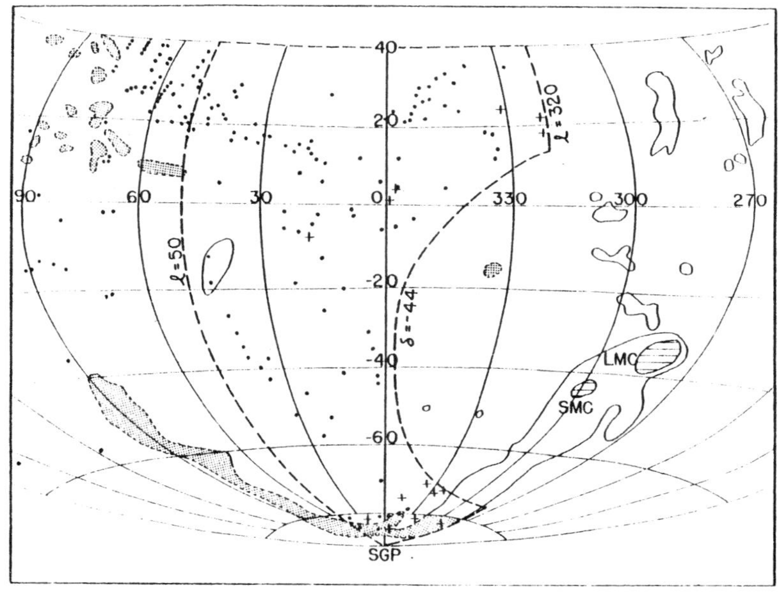

Figure 1. Distribution of HI with $|V_{LSR}|$ > 80 km s^{-1} in the inner Galaxy. The area surveyed by us is limited by the broken lines at $\ell=50^{o}$, $b=+40^{o}$ and $\delta=-44^{o}$, the southern limit of the observable sky with the 43–m telescope of NRAO. Data outside that area are from surveys by other authors (Mirabel, 1981 and references therein). Dots are clouds smaller than 3° with V_{LSR} < −80 km s^{-1}. Crosses are clouds smaller than 3° with V_{LSR} > +80 km s^{-1}. Contours with hatching are clouds greater than 3° with V_{LSR} < −80 km s^{-1}. Contours without hatching are clouds greater than 3° with V_{LSR} > +80 km s^{-1}.

H. van Woerden et al. (eds.), The Milky Way Galaxy, 411–412.

Figure 1 shows the presence of several streams or complexes. In the first place, **there appears** a stream that extends from $\ell=5^o$ $b=+5^o$ to $\ell=90^o$ $b=+40^o$. This stream has velocities in the range of -80 to -140 km s^{-1} and extends up to $\ell=120^o$ $b=+55^o$ (Giovanelli, 1980). On the other hand, there seems to be a stream of small and faint clouds with $V_{LSR} < -140$ km s^{-1} extending from $\ell=20^o$ $b=-55^o$ to $\ell=48^o$ $b=+26^o$. It is interesting that this stream could stretch out from the southern into the northern galactic hemisphere, across the galactic plane.

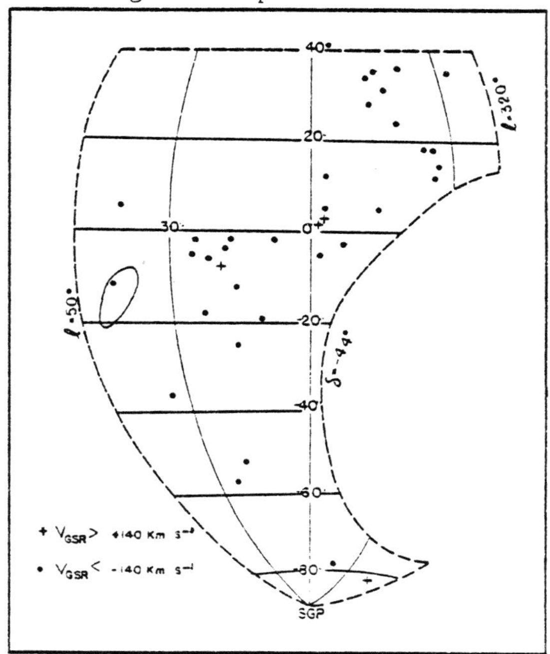

Figure 2. Distribution of clouds with very high galactocentric velocities ($|V_{GSR}| > 140$ km s^{-1}) in the region surveyed around the direction to the galactic center. More than 85% of the very-high-velocity detections are negative in sign. **There appears** a complex of clouds with very high negative velocities in the region $330^o<\ell<30^o$ $-40^o<b<+40^o$, with a striking void at $\ell>0^o$ $b>0^o$. There are clouds at angular distances smaller than 2° from the galactic plane (e.g. HVC 24.4-1.9-293).

Since similar extreme velocities of about -200 km s^{-1} are observed in the galactic center and anticenter, these results are strong evidence for a high-velocity inflow of neutral hydrogen toward the Milky Way. Assuming that the gas that is infalling with velocities greater than 140 km s^{-1} is at a distance of 20 kpc from the galactic center, we estimate a present net influx of 0.2 $M_\odot$ yr^{-1} toward the Galaxy.

This research was supported by NSF grant AST 80-13148.

REFERENCES

Giovanelli, R.: 1980, Astron. J. 85, 1155.
Mirabel, I.F.: 1981, Rev. Mexicana Astron. Astrof. 6, 245.

OBSERVATIONS OF HIGH-VELOCITY CLOUDS COLLIDING IN THE ANTICENTER

R. Morras and I.F. Mirabel
Department of Physics
University of Puerto Rico

High-velocity clouds that are colliding with Milky-Way material in the anticenter were observed in the 21-cm line of neutral hydrogen, using the Arecibo telescope with a system temperature of 40 K. We confirm the reported (Mirabel, 1982) positional and kinetic correlations between a high-velocity cloud that is infalling with a velocity of -200 km s^{-1} and a strong disturbance in the interstellar medium (see figure 1).

A region in the anticenter with large anomalous motions in the interstellar gas known as "Weaver's jet" (Weaver, 1974) was also observed. The events in this region show a striking resemblance with the phenomena observed in the region of the colliding cloud AC I. The obervations suggest that strong disturbances in the permitted-velocity gas that take place in the anticenter are the result of the impingement on the galactic disk of neutral-hydrogen high-velocity clouds (Burton and Moore, 1979).

We estimate that the infall of a single cloud with a mass of 10^4 to 10^5 solar masses deposits an energy of 5 x 10^{51} to 5 x 10^{52} ergs on a relatively small region of the Milky Way. This energy is several times the energy involved in the blast wave from a typical supernova. It is suggested that the on-going accretion of HI clouds must be an important source of energetic events in the interstellar medium, and may trigger large-scale structural peculiarities inside and outside the galactic disk.

This research was supported by NSF grant AST 80-13148.

REFERENCES

Burton, W.B. and Moore, R.L.: 1979, Astron. J. 84, 189.
Mirabel, I.F.: 1982, Astrophys. J. 256, 112.
Weaver, H.F.: 1974, in "Galactic Radio Astronomy", IAU Symp. 60, eds. F.J. Kerr and S.C. Simonson III, p. 573-583.

H. van Woerden et al. (eds.), The Milky Way Galaxy, 413–414.

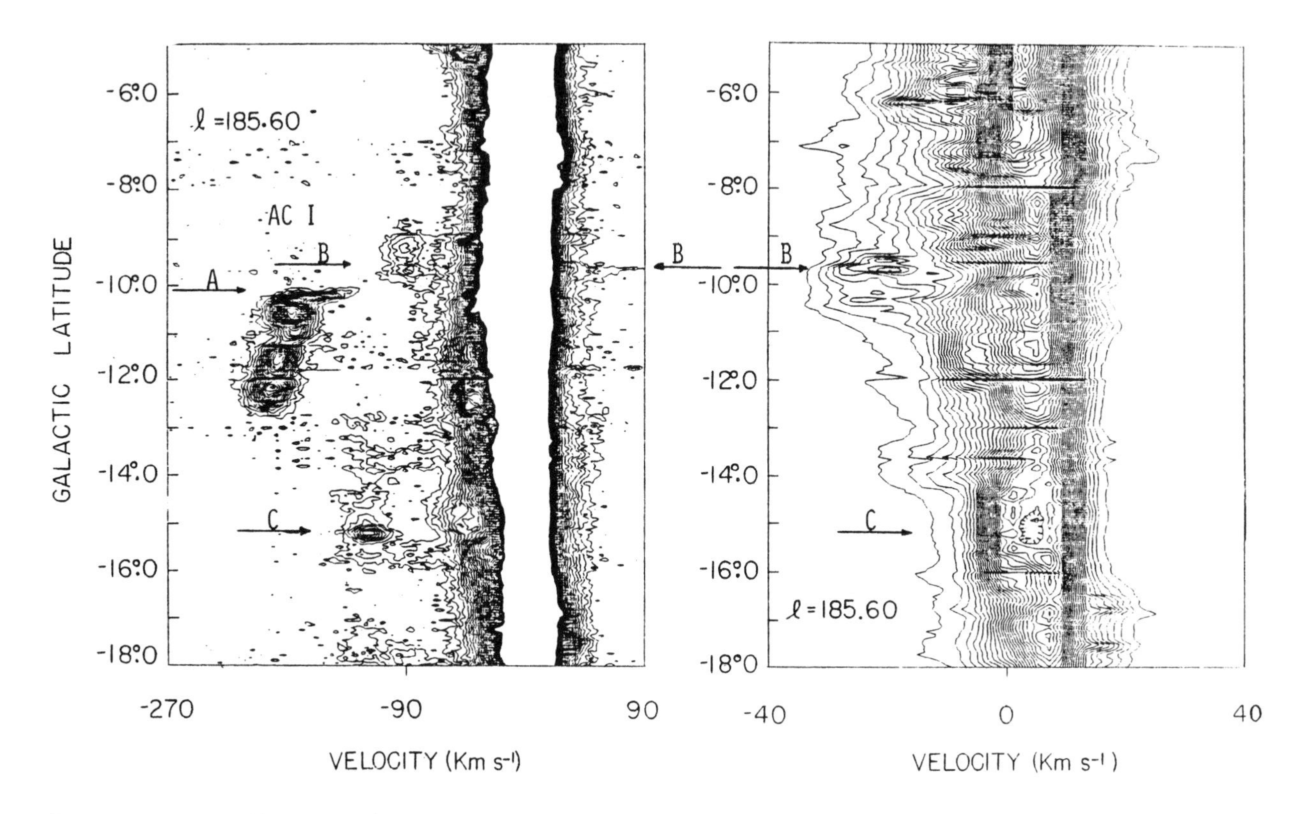

Figure 1. HI contour diagrams in the region of the infalling cloud with a velocity of -200 km s^{-1} that is colliding with Milky-Way material in the anticenter. Arrow A in the left panel points to a steep edge and abrupt deceleration of the high-velocity cloud AC I. In the left panel, at adjacent positions toward the galactic plane, there is gas that is being accelerated to velocities of -90 and +80 km s^{-1} (arrows B). These features are at the same position as the perturbation at V=-20 km s^{-1} and the relative decrease of gas density at V=+8 km s^{-1} indicated by arrow B in the right panel. The high-velocity cloud with V=-120 km s^{-1} pointed by arrow C in the left panel is at the same position as the "hole" at V=+8 km s^{-1} pointed by arrow C in the right panel.

THE MILKY WAY: A HALO, A CORONA, OR BOTH?

Klaas S. de Boer
Astronomisches Institut Tubingen, D-7400 Tübingen,F.R.Germany

The detection in absorption lines of gas clouds outside the galactic plane at high velocities by Münch and Zirin (1961), high velocities then defined as velocities differing by more than 20 km/s from the LSR, showed that the space outside the Milky-Way disk contains not just stars. Of course, from a continuity argument it had been all along clear that some transition zone had to exist between the dense (relatively speaking) gas of the Milky-Way plane and the vast (almost) emptiness of intergalactic space. The presence of these clouds requires a mechanism to prevent their evaporation, and Spitzer (1956) proposed that dilute hot gas had to exist outside the Milky-Way disk reaching, in his hydrostatic-equilibrium model, temperatures of a few million K at several tens of kpc. These high temperatures led him to name these gases the Galactic Corona. Observational confirmation of the abundance of these cool clouds came from the measurements of 21-cm HI emission, but no one-to-one correspondence with clouds detected in the visual did appear (Habing 1969). For the majority of the high-velocity (HV) clouds (Hulsbosch 1978) no distances are known, and all of those are believed to exist as a gaseous halo with the halo stars. Thus our Milky Way appears to have outside the disk: a halo, a gaseous halo, and a corona.

The nomenclature in the studies of the distant material is getting confused. In studying the gas, the location aspect has been emphasized using the word halo, the (high-temperature) gas aspect by using the word corona (De Boer and Savage 1982). In cosmic-ray and gamma-ray studies (Ginzburg 1978; Stecher 1978) also the spatial aspect is emphasized with the word halo. For models of the dynamics of the Milky Way the nuclear bulge, disk and halo are well-known entities. Einasto (1978 and references given there) demonstrated the need from dynamics for a large and very extended spheroidal mass, which he called corona. Later papers (Caldwell and Ostriker 1981; Rohlfs and Kreitschmann 1981; etc.) used the word corona as well. There is, however, an etymological objection against the use of the word corona in this case, since this hypothetical component is neither brilliant, nor always taken spherical. MASsive DArk Component (MASDAC) is more descriptive and is not in conflict with the older use of the word corona for hotter gases such as

H. van Woerden et al. (eds.), The Milky Way Galaxy, 415–420.

stellar corona, galactic corona (Spitzer 1956) and (less appropriate) the geocorona (Wegener 1911).

Cool gas in the Milky-Way halo has been detected on many lines of sight, using galactic as well as extragalactic background light sources. The original CaII-line studies of Münch and Zirin were extended by Blades (1981), and by Songaila and York (1980), but UV large-optical-depth absorption lines (e.g. CII, SiII, MgII and FeII) are very good probes for clouds in the halo (Savage and De Boer 1979, 1981 = SdB81; Pettini and West 1982; York et al. 1982; De Boer and Savage 1983 = dBS83). I want to stress that the CII lines seen in extragalactic sources show saturated absorption over wide velocity ranges, so there is "neutral" gas along vast portions of sightlines. Abundances are near solar (SdB81).

Distances of HV clouds are hardly known. Using galactic background light sources, one has a limit on the distance of the absorbing gas. Lines of sight to stars at z larger than 1 kpc and having HV clouds are rare: Münch and Zirin found 2, Pettini and West added 1, Songaila and York possibly 1 (M15), all with velocities differing less than 30 km/s from the LSR, and dBS83 found absorption at up to -100 km/s toward M13 (z=4.1 kpc). To extragalactic sources essentially always absorption by low-ionization-state material was detected in the Milky-Way halo (see York 1982 and dBS83 for references). This, in connection with the paucity of detections using galactic (so relatively nearby) background sources, I think means that many HV clouds exist at distances beyond 2 kpc.

Velocities are easy to measure in absorption lines. Insofar as the detections are toward high-latitude objects, the information supplements the 21-cm HV-cloud survey data of Giovanelli (1980). In particular, UV data from LMC and SMC stars extend the evident sinusoidal distribution of data points in a (l,v) plot into directions not extensively searched. Note that the HV clouds seen toward LMC and SMC stars now have been detected at 21 cm by McGee, Newton and Morton (1983). The large forbidden negative velocity found toward M13 led dBS83 to emphasize that the halo very likely rotates at lesser speed than the disk. This also would be more in line with dynamical models (Feitzinger and Kreitschmann 1982). An extended gas disk ("halo") as found by Lockman (1983) is compatible with e.g. an exponential decrease of velocity with height outside the first 1 kpc, with a scale height of perhaps 4 kpc. The limiting cases have been discussed recently by De Boer (1982).

Hot gas outside the disk shows its presence in the CIV absorption lines. The first detection, together with simple halo corotation, suggested an extent of CIV gas of up to z = 10 kpc (Savage and De Boer 1979). With the evidence for lesser rotational speed, the extent of the CIV gas as derived from velocities may be just a few kpc (dBS83). The scale height of CIV gas can be derived from plots of latitude-corrected CIV column densities versus z. There is no doubt that there is little, if any, CIV gas within 1 kpc from the Milky-Way disk (SdB81; Pettini

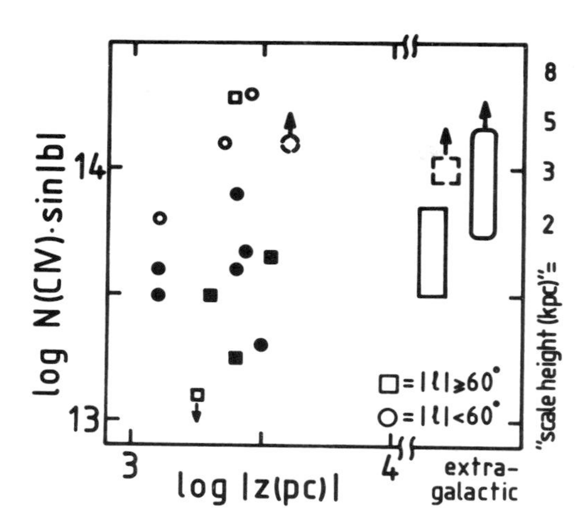

Figure 1. All CIV column densities toward high-latitude objects available in the literature are (latitude-corrected) plotted versus distance away from the Milky Way plane. The data are sorted twofold: filled symbols represent "accurate" data while open symbols the less secure determinations; the shape of the symbols refers to the general galactic directions of the light sources. Data for log z < 3.55 are from Pettini and West (1982). The entry at 3.6 is for M13 from dBS83, the tall box represents the range from the LMC star data (SdB81; Gondhalekar et al 1980), the rounded box the SMC star data (SdB81; Prevot et al 1980), and the small box 3C273 data (Ulrich et al 1980). Both SdB81 and Pettini and West have only upper limits below 1 kpc. The scale height of the CIV gas may be a few kpc. The entries for the "outer" longitudes of the Milky Way seem to suggest a smaller scale height than those for the "inner" directions.

and West 1982). Figure 1 shows that the scale height may be a few kpc. I think there is some indication that the scale height of the CIV gas is larger inside the solar circle than in more outward directions. If true indeed, this may have to do with larger star-formation and supernova rates in the Milky Way between R = 5 and 8 kpc.

The density of the gas seen in the halo can be roughly determined. The column densities (to LMC stars) in 21-cm gas and in CIV gas both are about $N(H)=10^{19}$ cm^{-2}. The clouds detected in front of the LMC appear in all UV spectra recorded, over 3° of the sky (SdB81). If such a cloud were at z = 2 kpc, its radial extent would be 100 pc. At 2 kpc the gas pressure is perhaps a factor 2 less than in the disk, at about 10^4 $K.cm^{-3}$. The HI cloud may be at 10^3 to 10^4 K, thus indicating a density of 10 to 1 cm^{-3}, and so a thickness of 0.3 to 3 pc. From the CIV absorption profile, as well as from the total column density, it followed (SdB81) that on the LMC line of sight $n(CIV)=10^{-8}$ cm^{-3} at a (revised, smaller) z of about 4 kpc. The lower limit to the gas density so becomes (solar abundance, all C in CIV) $n(H)=2\times10^{-5}$ cm^{-3}. Abundances are near normal on these lines of sight (SdB81; McGee, Newton and Morton 1983), but the ion fraction may be small (10% at most for collisional ionization), and so n(H) may be even 10^{-3} cm^{-3}. On the other hand, if indeed the CIV gas is at 10^5 K (Hartquist and Tallant 1981) and at about 4 kpc, where the gas pressure may be 300 $K.cm^{-3}$, the density of this gas phase would be 0.03 cm^{-3}. At very large distances hints to the gas density have been derived from the structure of filaments in the Magellanic-Stream gas. Mirabel, Cohen and Davies (1979) argue that these filaments require an outside pressure of roughly 400 $K.cm^{-3}$ at 50 kpc. With the maximum temperature of 3×10^6 K for bound coronal gas this implies $n(H)=10^{-4}$ cm^{-3}. And so the Milky Way (MW) halo-corona forms, also from the observational point of view, a smooth transition between MW disk and intergalactic space.

The corona could be part of a galactic fountain (Shapiro and Field 1976) and be supported by the hot gas forming the matrix of the Milky Way plane (Cox 1981). Actually, the star-formation rate, the supernova rate, the outflow from the disk and the density in the MW plane form a system which is a self-regulating closed loop (Cox). MHD waves from supernova explosions were proposed as a heating mechanism for the lower galactic halo (Hartquist and Tallant 1981), producing CIV but not requiring the existence of a million-degree corona. The dynamics of the coronal flow was modelled by Bregman (1980), allowing for the possible existence of a galactic wind from the nuclear bulge (Bregman 1981) and producing a net galactocentric flow of the returning cool halo clouds. A net galactocentric motion was noted by dBS83 in the Giovanelli (1980) high-velocity-cloud sample. The cooling volumes, I think, will fragment during descent into clouds. Is CIV the cooling phase of the galactic fountain? The 10^5 K derived from collisional ionization would have a lifetime (cooling) of about 500/n(H) yr (Chevalier 1981). Here I comment that continuous descent would replenish the CIV stage from the top of the corona, resulting in a much longer time where CIV can be detected. NV, with a somewhat longer lifetime, is seen only once (Fitzpatrick and Savage 1983), which has to do with the lesser optical depth of the NV lines (factor 5) and possibly also with smaller than solar abundance (SdB81).

Quasar and intervening-galaxy data may shed light on the actual extent of galaxy halos-coronas. Since the suggestion by Bahcall and Spitzer (1969) the speculations have been underpinned with the Milky-Way halo data. Savage and Jeske (1981) pointed out that the Milky-Way-halo absorbing column densities are similar to those in a particular (but perhaps representative) absorption-line system seen in a quasar spectrum. Hartquist and Snijders (1982) discussed quasar absorption-line statistics and find support for the intervening-corona hypothesis. Bregman and Glassgold (1982) did not see X-ray coronae around galaxies.

REFERENCES

Bahcall, J.N., Spitzer, L.: 1969, Ap. J. 156, L63.
Blades, J.C.: 1981, M.N.R.A.S. 196, 65p.
Bregman, J.N.: 1980, Ap. J. 236, 577.
Bregman, J.N.: 1981, in "The Phases of the Interstellar Medium", Ed. J.M. Dickey, p 191; N.R.A.O.
Bregman, J.N., Glassgold, A.E.: 1982, Ap. J. 263, 564.
Caldwell, J.A.R., Ostriker, J.P.: 1981, Ap. J. 251, 61.
Chevalier, R.A.: 1981, in "The Phases of the Interstellar Medium", Ed. J.M. Dickey, p 175; N.R.A.O.
Cox, D.P.: 1981, Ap. J. 245, 534.
de Boer, K.S.: 1982, in "Highlights of Astronomy" 6, Ed. R.M. West, p 657; Reidel.
de Boer, K.S., Savage, B.D.: 1982, Scientific American 247, no 2 (August).
de Boer, K.S., Savage, B.D.: 1983, Ap. J. 265, 210 (dBS83).

Einasto, J. 1978, in "The Large-Scale Characteristics of the Galaxy", IAU Symp. 84, Ed. W.B. Burton, p 451; Reidel.
Feitzinger, J.V., Kreitschmann, J.: 1982, Astron. Astrophys. 111, 255.
Fitzpatrick, E.L., Savage, B.D.: 1983, Ap. J. 267, 93.
Ginzburg, V.L.: 1978, in "The Large-Scale Characteristics of the Galaxy", IAU Symp. 84, Ed. W.B. Burton, p 485; Reidel.
Giovanelli, R.: 1980, A.J. 85, 1155.
Gondhalekar, P.M., Willis, A.J., Morgan, D.H., Nandy, K.: 1980, M.N.R.A.S. 193, 875.
Habing, H.J.: 1969, B.A.N. 20, 177.
Hartquist, T.W., Snijders, M.A.J.: 1982, Nature 299, 783.
Hartquist, T.W., Tallant, A.: 1981, M.N.R.A.S. 196, 527.
Hulsbosch, A.N.M.: 1978, in "The Large-Scale Characteristics of the Galaxy", IAU Symp. 84, Ed. W.B. Burton, p 525; Reidel.
Lockman, F.J.: 1983, in "Kinematics, Dynamics and Structure of the Milky Way", Ed. W.L.H. Shuter, p 303; Reidel.
McGee, R.X., Newton, L.M., Morton, D.C.: 1983, M.N.R.A.S. 205, 1191.
Mirabel, I.F., Cohen, R.J., Davies, R.D.: 1979, M.N.R.A.S. 186, 433.
Münch, G., Zirin, H.: 1961, Ap. J. 133, 11.
Pettini, M., West, K.A.: 1982, Ap. J. 260, 561.
Prévot, L., et al.: 1980, Astron. Astrophys. 90, L13.
Rohlfs, K., Kreitschmann, J.: 1981, Astrophys. Space Sci. 79, 289.
Savage, B.D., de Boer, K.S.: 1979, Ap. J. 230, L77.
Savage, B.D., de Boer, K.S.: 1981, Ap. J. 243, 460 (SdB81).
Savage, B.D., Jeske, N.A.: 1981, Ap. J. 244, 768.
Shapiro, P.R., Field, G.B.: 1976, Ap. J. 205, 762.
Songaila, A., York, D.G.: 1980, Ap. J. 242, 976.
Spitzer, L.: 1956, Ap. J. 124, 20.
Stecher, F.W.: 1978, in "The Large-Scale Characteristics of the Galaxy", IAU Symp. 84, Ed. W.B. Burton, p 475; Reidel.
Ulrich, M.H., et al.: 1980, M.N.R.A.S. 192, 561.
Wegener, A.: 1911, Physik. Zeitschr. XII, 170.
York, D.G.: 1982, Ann. Rev. Astron. Astrophys. 20, 221.
York, D.G., Blades, J.C., Cowie, L.L., Morton, D.C., Songaila, A., Wu, C.C.: 1982, Ap. J. 255, 467.

DISCUSSION

J.M. Dickey: Is there a strong correlation of column density with $|\text{cosec } b|$? Is there strong evidence for a disk-type distribution?

De Boer: The data are insufficient to make such a statement. We know too little.

H. van Woerden: What are the temperatures of the gas seen in the ultraviolet lines? Some HVC cores have such small internal motions that the temperatures there cannot be higher than a few hundred K.

De Boer: Temperatures are not really known. If CIV is in collisional ionization equilibrium, it would be 10^5K. The outer reaches of the corona probably are at a few million K and so the CIV is anywhere between that and your value. The CIV profiles, however, are smooth (see Savage and de Boer 1981), clearly indicating temperatures well over 1000 K. But during descent from far out, recombination may be faster than cooling.

Van Woerden: Unfortunately, the kinetic temperatures I mentioned are upper limits only. There are so far no direct measurements of spin temperatures, derived from comparisons of emission intensities with optical depths of absorption in the spectra of background radio-continuum sources. In fact, we only have lower limits to the spin temperatures.

J. Milogradov-Turin: Do you see asymmetry in the distribution of UV hot gas in galactic coordinates?

De Boer: There are too few lines of sight giving absorption-line information. Anyway, I doubt if an asymmetry would be present. One of van Woerden's slides shows Giovanelli's velocities at positive latitudes; it displays a sine wave structure. The sine form in such a diagram indicates that solar motion is a dominating factor. In one of my slides the run of data points does not go through zero at $\ell = 180°$, but 20° away from there. This indicates an average galactocentric motion of order 20-30 km/s in the HVCs (as defined in my paper).

Van Woerden: As to asymmetry in the distribution of high-velocity gas on the sky, I refer to figure 1 in my review paper. The **velocity** pattern shows, to zero order, mirror symmetry about the Anticentre, as expected from galactic rotation, or, more precisely, about $\ell \sim 210°$. But only to zero order; the pattern shows no symmetry in any detail. Also, the **density** distribution is highly asymmetric. This suggests to me that an extragalactic factor (infall of intergalactic gas?) also plays a role.

De Boer: The velocity asymmetry I pointed out is the same thing.

R.D. Davies: It is also visible in the spiral arms in that area.

SECTION II.9

COMPARISON OF ANDROMEDA AND MILKY WAY GALAXIES

Thursday 2 June, 0910 - 1100

Chairman: R. Sancisi

Above: Sancisi (centre left) talking with Blitz during boat trip. LZ

Below: Italians at dinner. Left to right: Phil Seiden, Antonella Natta, Renzo Sancisi, Stefano Casertano and Giuseppe Bertin. LZ

A COMPARISON OF THE ANDROMEDA AND MILKY WAY GALAXIES

Paul Hodge
Astronomy Department, University of Washington, Seattle, WA

ABSTRACT

A comparison of some of the basic properties of M31 and the Milky Way indicates that in almost every respect M31 is larger than the Galaxy. It is more luminous, redder, more massive, and of earlier Hubble type. A detailed comparison of the spiral structure, based on optical tracers, for comparable areas in the outer parts of each galaxy shows differences in the arm spacings, in density enhancement, and in pitch angle.

1. INTRODUCTION

The Andromeda Galaxy, M31, is the nearest spiral galaxy and the only giant spiral in the Local Group. Although not a close match to our Galaxy, it is nevertheless a useful galaxy for comparing with the Milky Way, as it is the only large spiral for which we can obtain detailed information of certain kinds. Recent work, especially at radio wavelengths, has made it one of the most thoroughly-observed galaxies beyond the Milky Way. The following review is not an exhaustive comparison of all features of M31 with those of the Galaxy, which would require vastly more time than is available, but rather is a discussion of some of the more challenging areas of similarity and of contrast between them. Insight into certain questions is certainly gained by their comparison, but it must also be confessed that, frustratingly, some problems that confound students of the Milky Way because of our location in the plane, similarly confound those who study M31 because of its nearly edge-on inclination angle.

2. INTEGRATED PROPERTIES

Table 1 compares some of the basic properties of Andromeda and the Milky Way, as best we can determine them so far. These are not easy parameters to measure and the table indicates the range found in the recent literature. For M31, the Hubble type is clearly Sb; Hubble him-

H. van Woerden et al. (eds.), The Milky Way Galaxy, 423–430.

Table 1. Comparisons of M31 and the Galaxy

Parameter	Galaxy	M31
Adopted distance (kpc)	8.5	765
Hubble type	Sc	Sb
M_B (face-on)	-20.5±.5	-21.2±.4
$(B-V)_T^o$	0.53±.04	0.74±.06
V(R = 8.5 kpc) (km/sec)	220±20	265±10
$\sigma_V(0)$ (spherical component) (km/sec)	130±10	160±30
D_o at 25.0 mag/arcsec^2 (kpc)	24±5	41±2
Effective diameter (D_e) (kpc)	11	16
D (from open-cluster system) (kpc)	40	62
D (from HII-region system) (kpc)	45	50
$D_{26.6}$ (kpc)	34	51
Mass ($M_\odot$)	?	?
M_H ($M_\odot$)	5×10^9	4×10^9

self used it as a typical example and there has been relatively little controversy on the subject, other than to argue about a possible weak bar-like structure in the central bulge (Lindblad, 1956; Sharov and Lyutyi, 1980). For our Galaxy, however, the Hubble type is not so clearly known. Of the eleven different methods recently used to gauge this parameter, two conclude that it is Sb, two that it is Sbc, and seven that it is Sc (Baade 1951, Arp 1964, 1965, Becker 1964, van den Bergh 1968, Georgelin and Georgelin 1976, de Vaucouleurs and Pence 1978, and Hodge 1983).

The relative sizes of M31 and the Galaxy can be examined anew. Table 1 summarizes five of the various measures of the diameters. No matter how the diameter is defined, that of M31 is consistently larger than that of the Galaxy by about 50%. For M31 the most distant open cluster, for example, is 139 arcmin from the center (Hodge 1979), which corresponds to 30.9 kpc, while for the Milky Way, Christian and Janes (1979) find Be 20 to be the most distant open cluster, with a distance of 20 ± 3 kpc. The objects are rather comparable. I calculate from Christian and Janes' (1979) CM diagram that the absolute magnitude of Be 20 is $M_V = -7.4 \pm 0.2$. At M31 this would be $m_V = 17.2$. The integrated brightness of M31's most distant cluster, C1, is estimated from KPNO 4-m plates to be $m_V = 17.9$.

The outermost luminous blue stars in the anticenter direction are estimated to lie at a distance of 20 kpc (Chromey 1978). For M31, Richter (1971) has catalogued aggregates of OB stars at $\sim$ 28 kpc.

The neutral-hydrogen diameters of the two galaxies are also in about this ratio. Emerson's (1976) outermost HI contours are at 32.4 kpc, for example, while Baker (1976) finds an HI cut-off for our Galaxy at about 25 kpc from the Galactic center.

When these data are combined with the various photometric measures of size (Table 1), we must conclude that the Milky Way is M31's inferior, being roughly two-thirds as large.

3. STRUCTURE

There are both similarities and striking differences between M31 and the Milky Way in structure. This discussion will concentrate in particular on the spiral structure, but mention should be made of other components as well. Both galaxies have compact nuclei, with that of M31 being somewhat less interesting (that is, less active). The galaxies both have central bulges made up largely of old, metal-rich stars, with that of the Galaxy being somewhat the smaller and bluer of the two (Sharov and Lyutyi 1980; Arp 1965; de Vaucouleurs and Pence 1978). The latter reference attempts to isolate the spheroidal component, assuming it to be structurally governed by an $r^{1/4}$ projected-density law, and finds the Galaxy's spheroid to have a blue absolute magnitude about 1.2 mag fainter than that of M31. The central bulge in both cases contains numerous examples of planetary nebulae (Ford and Jacoby 1978), SN remnants (Blair _et al_. 1981, Dennefeld and Kunth 1981, D'Odorico _et al_. 1980, Kumar 1976), novae (Rosino 1972), X-ray sources (van Speybroeck _et al_. 1979), and dust clouds (Hodge 1980).

The haloes of the two galaxies can also be compared. Both are surrounded by a nearly spherical and extremely extended system of globular clusters. Andromeda has almost three times as many catalogued globular clusters as the Milky Way. Its most distant globular lies at a projected distance of 130 kpc (Sargent _et al_. 1977), while our Galaxy's system appears to extend out to 116 kpc (Aaronson _et al_. 1983). Both also are surrounded by a thin population of dwarf elliptical galaxies (van den Bergh 1972, Hodge 1971).

It is on the disk component, especially the spiral-arm structure, however, that I wish to concentrate in this review. Most attempts to disentangle the spiral structure of M31 have used familiar, traditional "tracers": the OB stars, young clusters, HII regions, and HI, in much the same way the local spiral structure is sought in our Galaxy. The difference, of course, lies in the type of difficulty encountered: obscuration and distance uncertainties for the Milky Way, and a steep angle of inclination for M31. All agree that both galaxies _have_ spiral arms, but the literature is full of contradictory interpretations of the detailed spiral structure. For M31, arms are outlined by the counts of brightest stars (van den Bergh 1958, Reddish 1962), by population morphology (Baade 1963), by OB associations (van den Bergh 1964), from radio-continuum data (Pooley 1969, van der Kruit 1972, Berkhuijsen and Wielebinski 1974, Beck 1982), from HI surveys (Roberts 1966, Byrd 1978, Brinks 1985, Walterbos and Kennicutt 1985, Sofue and Kato 1981, and many others), from CO data (Stark 1985), using HII region surveys (Arp 1964, Simien _et al_. 1978), from ultraviolet images (Deharveng _et al_. 1980), from Cepheids (Efremov _et al_. 1981), from surface photometry (Hodge and

Kennicutt 1982), from dust lanes (Hodge 1980), and from young open clusters (Hodge 1979). Although the raw data available for these studies are fairly consistent, their interpretation has been very divergent. While most fits have been of a two-armed trailing spiral pattern, Simien et al. (1978) instead argued for a one-armed leading spiral, with some theoretical justification provided by the work of Kalnajs (1975). Such a contradiction seems to indicate that the situation has been, at least recently, even worse than in the case of our Galaxy.

The leading one-armed spiral pattern, based primarily on HII regions, fits rather poorly the young open clusters' positions. Although the fit within ∿ 50 arcmin of the center looks reasonable, in the outer parts, where the open clusters better define a pattern than the sparser HII regions, the fit is seriously deficient. The clearly-defined arm segments near the major axis at 70-90 arcmin clearly have a slope opposite to that of the model. Arp's two-armed spiral also does not fit these segments, but at least they are roughly parallel, with pitch angles of the same

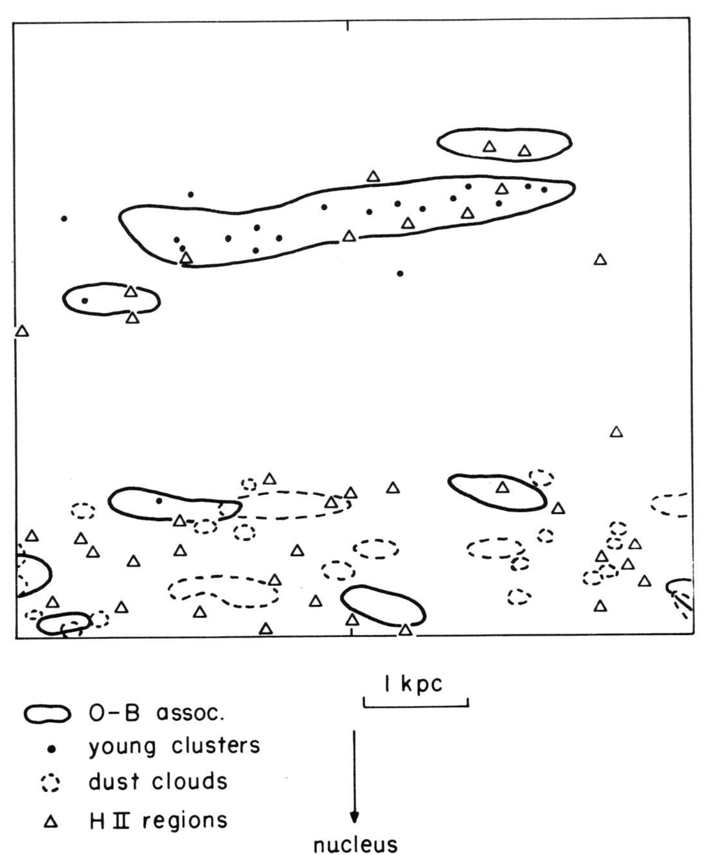

Figure 1. Map of part of the portion of M31 along the NE axis, rectified to face on, and extending from 7 to 14 kpc. Spiral-arm tracers are identified. Baade's arms N_3 and N_4 are conspicuous.

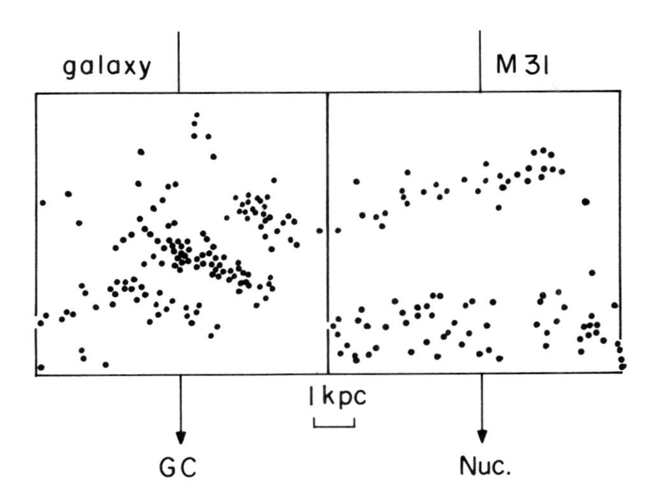

Figure 2. A comparison of the area in Figure 1 and a comparable area in the Milky Way galaxy, with the same kinds of tracers plotted.

magnitude and the same sign. It is clear that the arms of M31 are not to be fit to a simple, perfect logarithmic spiral pattern. Byrd's (1978) gravitationally-distorted spiral pattern looks much more promising. Perhaps the moral of this comparison is that students of the Milky Way's spiral structure should not be too discouraged if their data do not fit a perfect mathematical model, as M31 does not attain such perfection, nor do many other galaxies.

As a graphic demonstration of some of the similarities and differences in the spiral arms of the two galaxies, I have taken a section of the NE portion of M31 that corresponds in location to the solar neighborhood of the Milky Way. In Figure 1 I have plotted a rectified face-on map of the area in question, in which I have identified positions of young open clusters, dust clouds, OB associations, and HII regions, from the sources listed above. Two spiral-arm crossings are obvious. The inner has a pitch angle of perhaps 5°. They are separated by approximately 4 kpc at the position of the optical major axis, which vertically crosses the center of the diagram. All of the optical tracers agree remarkably well.

Figure 2 compares this map with a similar one for the Milky Way, plotted to the same scale and for the same distance from the nucleus (based on the summary diagram in Bok and Bok (1981)). Although the diagrams show a general similarity, there are three conspicuous differences: (1) the arms of M31 are much better defined, with interarm areas almost empty of tracers, (2) the M31 arms are at least twice as widely-spaced, and (3) the pitch angle for the M31 arms is much smaller. These facts, of course, have been known for many years; Fig. 2 merely supplies an especially graphic demonstration of these important differences.

There is not space to cover many other interesting comparisons that could be made, especially about the radial distributions of different components of the galaxies, rotational parameters, kinematics, warping of the plane, velocity dispersions, and so on. Much progress is being made in these fields. Questions do remain, however. I particularly point out that for both galaxies we still cannot answer the following rather basic questions:

1. How many arms are there?
2. What are the shapes of the arms?
3. What dominates the spiral pattern, generally and in detail?
4. What is the history of star formation in the plane and in the halo?
5. What is the total mass?
6. How is the mass distributed?

REFERENCES

Aaronson, M., Schommer, R., and Olszewski, E.: 1983, preprint
Arp, H.: 1964, Astrophys. J. 139, 1045
Arp, H.: 1965, Astrophys. J. 141, 43
Baade, W.: 1951, Publ. Obs. Univ. Michigan 10, 16
Baade, W.: 1963, Evolution of Stars and Galaxies (Harvard Univ. Press)
Baker, P.: 1976, Astron. Astrophys. 48, 163
Beck, R.: 1982, Astron. Astrophys. 106, 121
Becker, W.: 1964, Z. Astrophys. 58, 202
Berkhuijsen, E.M., and Wielebinski, R.: 1974, Astron. Astrophys. 34, 173
Blair, W.P. and Kirshner, R.P.: 1981, Astrophys. J. 247, 879
Bok, B.J., and Bok. P.F.: 1981, The Milky Way (Harvard Univ. Press)
Brinks, E. and Burton, W.B.: 1983, this volume
Byrd, G.: 1978, Astrophys. J. 226, 70
Christian, C., and Janes, K.: 1979, Astron. J. 84, 204
Chromey, F.: 1978, Astron. J. 83, 162
de Vaucouleurs, G., and Pence, W.D.: 1978, Astron. J. 83, 1163
Deharveng, J.M., Jakobsen, P., Milliard, B., and Laget, M.: 1980, Astron Astrophys. 88, 52
Dennefeld, M., and Kunth, D.: 1981, Astron. J. 86, 989
D'Odorico, S., Dopita, M., and Benvenuti, P.: 1980, Astron. Astrophys. Suppl. 40, 67
Efremov, Y.N., Ivanov, G.R., and Nikolov, N.S.: 1981, Astrophys. Space Sci. 75, 407
Emerson, D.T.: 1976, Monthly Not. Roy. Astron. Soc. 176, 321
Ford, H., and Jacoby, G.: 1978, Astrophys. J. Suppl. 38, 351
Georgelin, Y.M., and Georgelin, Y.P.: 1976, Astron. Astrophys. 49, 57
Hodge, P.: 1971, Ann. Rev. Astron. Astrophys. 9, 35
Hodge, P.: 1979, Astron. J. 84, 744
Hodge, P.: 1980, Astron. J. 85, 376
Hodge, P., and Kennicutt, R.: 1982, Astron. J. 87, 264
Hodge, P.: 1983, paper submitted

Kalnajs, A.: 1975, "La Dynamique des Galaxies Spirales", ed. L. Weliachew, Paris, CNRS No. 241
Kumar, C.K.: 1976, Pub. Astron. Soc. Pacific 88, 323
Lindblad, B.: 1956, Stockholms Obs. Ann. Bd. 19, No. 2
Pooley, G.: 1969, Monthly Not. Roy. Astron. Soc. 144, 101
Reddish, V.: 1962, Z. Astrophys. 56, 194
Richter, N.: 1971, Astron. Nachr. 292, 275
Roberts, M.S.: 1966, Astrophys. J. 144, 639
Rosino, L.: 1972, Astron. Astrophys. Suppl. 9, 347
Sargent, W.L.W., Kowal, C., Hartwick, D., and van den Bergh, S.: 1977, Astron. J. 82, 947
Sharov, A.S., and Lyutyi, V.B.: 1980, Astr. Zh. 57, 449
Simien, F., Athanassoula, E., Pellet, A., Monnet, G., Maucherat, A., and Courtès, G.: 1978, Astron. Astrophys. 67, 73
Stark, A.A.: 1985, this volume
Sofue, Y., and Kato, T.: 1981, Pub. Astron. Soc. Japan 33, 449
van den Bergh, S.: 1958, Pub. Astron. Soc. Pacific 70, 109
van den Bergh, S.: 1964, Astrophys. J. Suppl. 9, 65
van den Bergh, S.: 1968, Comm. D. Dunlap Obs. No. 195
van den Bergh, S.: 1972, Astrophys. J. 171, L31
van der Kruit, P.C.: 1972, Astrophys. Lett. 11, 173
van Speybroeck, L., Epstein, A., Forman, W., Giacconi, R., Jones, C., Liller, W., and Smarr, L.: 1979, Astrophys. J. 234, L49
Walterbos, R.A.M., and Kennicutt, R.: 1985, this volume

DISCUSSION

E. Athanassoula: The analyses of spiral structure in M31 mentioned by you (Baade + Arp; Simien et al.) try to fit one single logarithmic spiral to the whole galaxy. By relaxing this assumption and using a Fourier analysis which allows for several components, Considère and I (1982, Astron. Astrophys. 111, 28) find back the one-armed leading spiral first proposed by Kalnajs (1975) and Simien et al. (1978). However, in the outer regions (beyond 65' or 70') we find a two-armed trailing spiral with pitch angle 8°. It is thus no surprise that the extrapolation of the one-armed spiral does not fit the outer regions you showed.

H. van Woerden: Is your pitch angle of 5° in M31 based on an analysis of the run of arms all around the system?

Hodge: No, it is based on one segment only, and the value is 5° ± 5°.

P. Pismis: Your Table 1 shows that M31 is in many respects, including the arm spacing, larger than our Galaxy. Could this be due to an overestimate of the distance to the Andromeda Nebula or does a physical cause underlie the difference?

Hodge: I do not think such an overestimate of the distance is very likely. As to the spacing of the arms, the experts for our Galaxy do not seem to find compelling reasons to expect this to be equal to that in Andromeda. It seems preferable to consider the "arms" in our Galaxy as a complex set of portions of arms and spurs, rather than changing the distance scale.

W.L.H. Shuter: Your type estimate of Sc for our Galaxy is based on a scale of R_o = 10 kpc. What would happen if R = 8.5 kpc?

Hodge: It would make the Galaxy slightly earlier, but not much.

Hodge burns his lips at conference dinner, while Baud and Van der Laan discuss arrangements for after-dinner lecture. At right: Kormendy LZ

RADIAL DISTRIBUTIONS OF CONSTITUENTS IN M33, THE GALAXY, AND M51

Elly M. Berkhuijsen and Ulrich Klein
Max-Planck-Institut für Radioastronomie, Bonn, FRG

The radial distributions of the surface brightness or column density of thermal and nonthermal radio emission, far-infrared (FIR) emission, blue light, HI and CO in the Sc galaxies M33 and M51 are compared with the corresponding distributions in the Galaxy. Information on the variation of the absorption at Hα and on the variation of the abundance ratio O/H is also shown.

The data are presented in Figure 1. Table 1 gives radial scale lengths L of various disk components (NTH = nonthermal radio continuum, TH = thermal emission, L_B= blue surface brightness, $A_{H\alpha}$= absorption at Hα). The following similarities between the distributions of the <u>disk</u> components of the 3 galaxies may be noted:

a. L(NTH) > L(TH): this may be due to diffusion of relativistic electrons, if the electrons originate in the young population.
L(NTH) ≅ 2L(TH); L(NTH) - L(TH) is largest for M51, which has the strongest magnetic field of the 3 galaxies (Beck, 1983).

b. L(TH) < L(L_B): after correction of L(L_B) for absorption using A_B= 1.69 $A_{H\alpha}$, L(TH) ≅ 0.8 L(L_B). This suggests that the fraction of massive (ionizing) stars decreases outwards.

c. L($A_{H\alpha}$) ≅ 1.3 L(O/H): in M33 correction of L($A_{H\alpha}$) for the gradient in T_e (Berkhuijsen, 1983) yields L($A_{H\alpha}$) ≅ L(O/H). Thus the absorption at Hα may be linearly correlated with the element abundance.

d. L(CO) ≅ L(FIR) ≅ 0.5 L(O/H): the gradients in CO and FIR are equal, but they are much steeper than the gradients in O/H and $A_{H\alpha}$. This indicates that the radial distribution of the dust is less steep than that of CO or FIR.

An extensive discussion of this work wil be given elsewhere (Berkhuijsen, 1984; Klein, Wielebinski and Beck, 1984).

H. van Woerden et al. (eds.), The Milky Way Galaxy, 431–434.

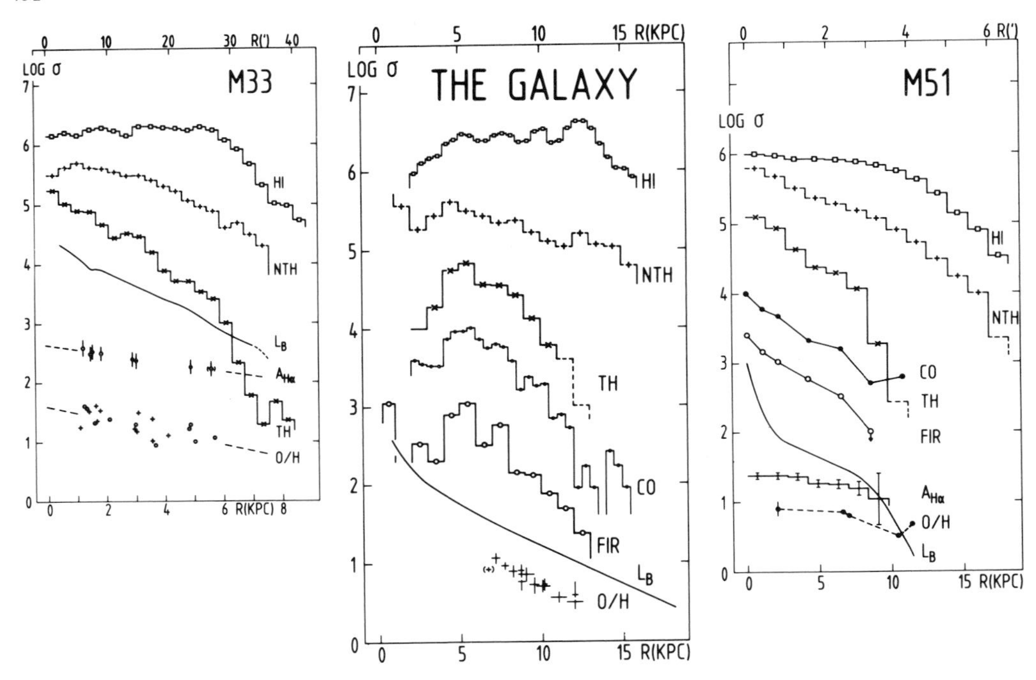

Figure 1. Radial distributions of constituents in M33, the Galaxy and M51. In each case the surface brightness σ is given on a relative scale. Data averaged in circular rings in the plane of the galaxy, centred on the nucleus, are shown as step curves.
M33: HI - column density of neutral hydrogen (Newton, 1980); NTH - surface brightness of nonthermal radio continuum emission at λ6.2 cm, obtained by subtracting the thermal emission from the total emission at λ6.2 cm (Berkhuijsen, 1983); TH - surface brightness of thermal emission at λ6.2 cm, derived from the catalogue of HII regions detected in Hα (Boulesteix et al., 1974). The absorption factor $A_{H\alpha}$ shown below and a calibration factor were applied; L_B- surface brightness of blue light (de Vaucouleurs, 1959); $A_{H\alpha}$- absorption factor of thermal emission between Hα and λ6.2 cm (Berkhuijsen, 1983), assuming $T_e = 10^4$K; O/H - abundance ratio by number, observed in HII regions (circles, Kwitter and Aller, 1981) and supernova remnants (plusses, Dopita et al., 1980).
The Galaxy: HI - column density of neutral hydrogen (Burton and Gordon, 1978); NTH - surface brightness of nonthermal radio continuum emission at λ74 cm (408 MHz), obtained by deconvolving the observed emission at b = +3°, 0° and -3° for all longitudes (Beuerman et al., 1983); TH - production rate of Lyc photons per unit area of compact HII regions and ELD HII regions (Güsten and Mezger, 1983); CO - column density of $^{12}C^{16}O$ (Gordon and Burton, 1976); FIR - surface brightness of far - infrared emission (Boissé et al., 1981); L_B- surface brightness of blue light, based on a two-component model (de Vaucouleurs and Pence, 1978); O/H - abundance ratio by number, observed in HII regions as compiled by Güsten and Mezger (1983). NTH and L_B refer to all longtitudes, other curves to northern longtitudes only.

M51: HI - column density of neutral hydrogen (Shane, 1975); NTH - surface brightness of nonthermal radio continuum emission at λ2.0 cm (Klein, 1981); TH - surface brightness of thermal emission at λ2.0 cm, derived from radio spectral-index maps (Klein, 1981); CO - column density of CO (Scoville and Young, 1983); FIR - surface brightness of far-infrared emission (Smith, 1982); L_B - surface brightness of blue light (Okamura et al., 1976); $A_{H\alpha}$- absorption factor of thermal emission between Hα and λ2.0 cm (Klein, 1981), derived from the Hα map of Tully (1974) and the radio map at λ2.0 cm assuming $T_e = 10^4$K; O/H - abundance ratio by number as compiled by Pagel and Edmunds (1981).

Table 1. Radial scale lengths L(kpc)

Constituent	M33	Galaxy	M51
HI	-27 ± 11	-132 ± 324	9.1 ± 0.8
NTH	2.0 ± 0.3	5.7 ± 0.5[a]	6.9 ± 0.6
TH	1.2 ± 0.1	2.3 ± 0.2	2.7 ± 0.2
CO	-	2.2 ± 0.2	3.9 ± 0.9
FIR	-	2.2 ± 0.2	3.6 ± 0.2
L_B	1.76 ± 0.04	4.1 ± 0.3[a]	4.6 ± 0.4
$A_{H\alpha}$	5.6 ± 0.4	-	9.8 ± 1.3
O/H	4.3 ± 1.0	4.2 ± 0.4	7.6 ± 1.6
Range in R (kpc)	1 $\lesssim$ R $\lesssim$ 6	5 < R < 12	2 $\lesssim$ R $\lesssim$ 10

[a] All longitudes; other L for northern longitudes only

REFERENCES

Beck, R.: 1983, in IAU Symp. 100, "Internal Kinematics and Dynamics of Galaxies", ed. E. Athanassoula, Reidel, Dordrecht, p. 159

Berkhuijsen, E.M.: 1983, submitted to Astron. Astrophys.

Berkhuijsen, E.M.: 1984, in preparation

Beuermann, K., Kanbach, G., Berkhuijsen, E.M.: 1983, preprint

Boissé, P., Gispert, R., Coron, N., Wijnbergen, J.J., Serra, G., Ryter, C., Puget, J.L.: 1981, Astron. Astrophys. 94, 265

Boulesteix, J., Courtès, G., Laval, A., Monnet, G., Petit, H.: 1974, Astron. Astrophys. 37, 33

Burton, W.B., Gordon, M.A.: 1978, Astron. Astrophys. 63, 7

de Vaucouleurs, G.: 1959, Astrophys. J. 130, 728

de Vaucouleurs, G., Pence, W.D.: 1978, Astron. J. 83, 1163

Dopita, M.A., D'Odorico, S., Benvenuti, P.: 1980, Astrophys. J. 236, 628

Gordon, M.A., Burton, W.B.: 1976, Astrophys. J. 208, 346

Güsten, R., Mezger, P.G.: 1983, to appear in "Vistas in Astronomy"

Klein, U.: 1981, Ph.D. Thesis, Wilhelms-Universität Bonn

Klein, U., Wielebinski, R., Beck, R.: 1984, in preparation

Kwitter, K.B., Aller, L.H.: 1981, Monthly Notices Roy. Astron. Soc. 195, 939
Newton, K.: 1980, Monthly Notices Roy. Astron. Soc. 190, 689
Okamura, S., Kanazawa, T., Kodaira, K.: 1976, Publ. Astron. Soc. Japan 28, 329
Pagel, B.E.J., Edmunds, M.G.: 1981, Ann. Rev. Astron. Astrophys. 19, 77
Scoville, N., Young, J.S.: 1983, Astrophys. J. 265, 148
Shane, W.W.: 1975, in "La Dynamique des Galaxies Spirales", ed. L. Weliachew, Paris, CNRS, p. 157
Smith, J.: 1982, Astrophys. J. 261, 463
Tully, R.B.: 1974, Astrophys. J. Suppl. 27, 415

Leiden students and alumni at dinner. Left to right: Walterbos, Hermsen, Bloemen, Van Driel (now at Groningen), Schwering, Lub. Foreground: Waller; background: Hu and Salukvadze. LZ

A COORDINATED RADIO AND OPTICAL SURVEY OF M31

R.A.M. Walterbos
Sterrewacht Leiden

R.C. Kennicutt
University of Minnesota

At Leiden we are obtaining coordinated radio, optical and infrared observations of the Andromeda galaxy, M31. Its proximity offers us a unique opportunity to study both the large-scale and small-scale structure of a galaxy which is similar in many respects to the Milky Way. The WSRT has been used to obtain high-resolution (24" x 36") maps of M31 in the HI line and 21- and 49-cm radio-continuum emission. Recently the radio data have been complemented with optical surface photometry in UBVR and Hα, using the Burrell Schmidt telescope at Kitt Peak and the Palomar Schmidt. Results from the HI and IRAS infrared observations are presented elsewhere. Here we present some preliminary results from the radio-continuum and optical surveys.

The figure shows a composite map of the 21-cm radio-continuum emission, made by combining 5 separate Westerbork fields. The emission is concentrated in a ring structure coincident with maxima in the distribution of the HI, HII regions, OB associations, far-infrared, and far-UV radiation. The continuum ring is partly resolved into individual sources at 21 cm; the spectral indices and optical counterparts indicate that both thermal and non-thermal sources are being detected. A separation of thermal and non-thermal emission is being attempted using both the 49- and 21-cm maps and the Hα-emission distribution. Both the thermal and non-thermal emission appear to be strongly correlated with the young stellar population, in agreement with recent results for other spiral galaxies, and with an earlier low-resolution survey by Berkhuijsen.

The UBVR and Hα plates are being scanned at Leiden using the ASTROSCAN reticon densitometer, and will be combined with published and unpublished photoelectric photometry to produce maps of the color distribution of the galaxy with a resolution of 4-8 seconds, and with an accuracy of order $\pm 0^{m}.1$. The data will provide direct information on the current and recent star formation, the properties of the interstellar dust, and the spiral structure of the galaxy. When combined with the radio-continuum and line data, we will be able to study the interactions between the interstellar medium and the young stellar population on scales of order 100 pc.

H. van Woerden et al. (eds.), The Milky Way Galaxy, 435–436.

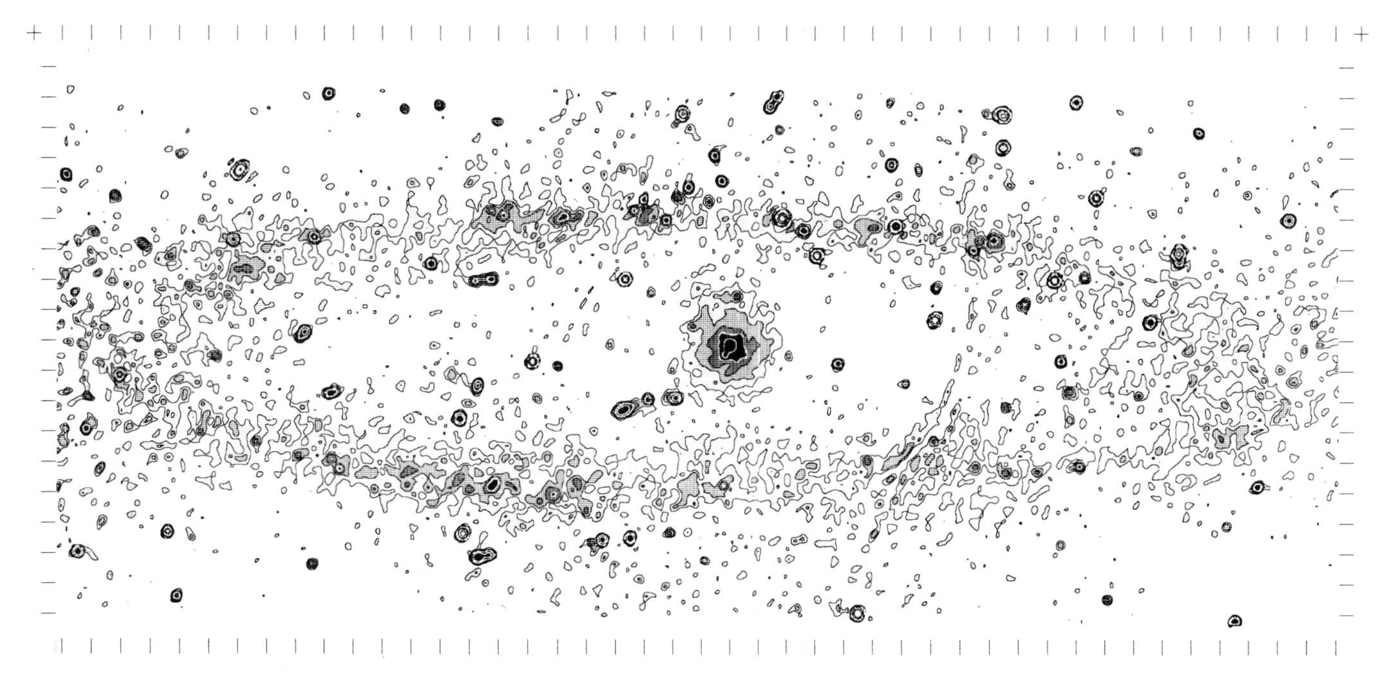

Map of the 21-cm radio-continuum emission of M31 made by combining five separate fields observed with WSRT. Short spacings taken from an unpublished 17.4-cm survey with the Effelsberg telescope are included for the large-scale structures. North East is to the left. The resolution is 35" x 35"; the distance between tick marks is 2!4. Contour levels are at 0.3, 0.6, 1, 1.6, 3.3, 10, 16, 40 K.

Most of the strong sources are far-extragalactic and not related to M31. This includes the double-lobed source close to the nucleus. The weaker sources in or close to the ring of emission generally coincide with HII regions or supernova remnants. The weak ring-like structure to the right of the nucleus is a residual grating ring due to 5C3.107, the only point source subtracted in this map. At the edges the noise increases due to the primary-beam correction.

DISTRIBUTION AND MOTIONS OF HI IN M31

E. Brinks and W.B. Burton
Sterrewacht, Leiden, The Netherlands

1. INTRODUCTION

It has become clear in the past few years that the distribution and kinematics of HI in M31 is far from simple. The new high-resolution survey made with the Westerbork SRT in the 21-cm line of atomic hydrogen by Brinks and Shane (1983) shows this dramatically. Along almost any line of sight through M31 two separate velocity systems are sampled. Based on a previous survey Shane (1978) and later Bajaja and Shane (1982) proposed that the extra component is due to warping of the plane of the galaxy into the direction of and crossing the line-of-sight. Roberts *et al.*(1978) and Whitehurst *et al.* (1978) emphasized that the observed profile structure ruled out confinement of the gas to a thin plane. Unwin (1983) reached a similar conclusion on the basis of his survey. The most complete model produced up until now which accounts for the two velocity systems is the one by Henderson (1979), based on the 100-m Effelsberg survey of M31 by Cram *et al.* (1980)

The new WSRT survey which was made at a resolution of $\Delta\alpha \times \Delta\delta \times \Delta V = 24'' \times 36'' \times 8.2$ km s^{-1} enables one to see both velocity systems well separated over large portions of the observed area. This separation is most prominent in position-velocity maps made parallel to the major axis of M31. This can be seen in Figure 3a, which shows a map made parallel to the major axis $4\overset{'}{.}8$ to the West. This map was produced after smoothing the original data to a resolution of $\Delta\alpha \times \Delta\delta = 48'' \times 72''$. A complete set of such position-velocity maps is given by Brinks and Shane (1983).

2. THE MODEL

Based on the high-resolution HI data a new model was constructed (Brinks and Burton, in preparation). This model is an extension of the model derived for the inner part of our Galaxy by Burton and Liszt (1978). Only the main characteristics of the model will be discussed here. Figure 1 shows the geometry of the model in a plane which is perpendicular to

H. van Woerden et al. (eds.), The Milky Way Galaxy, 437–442.

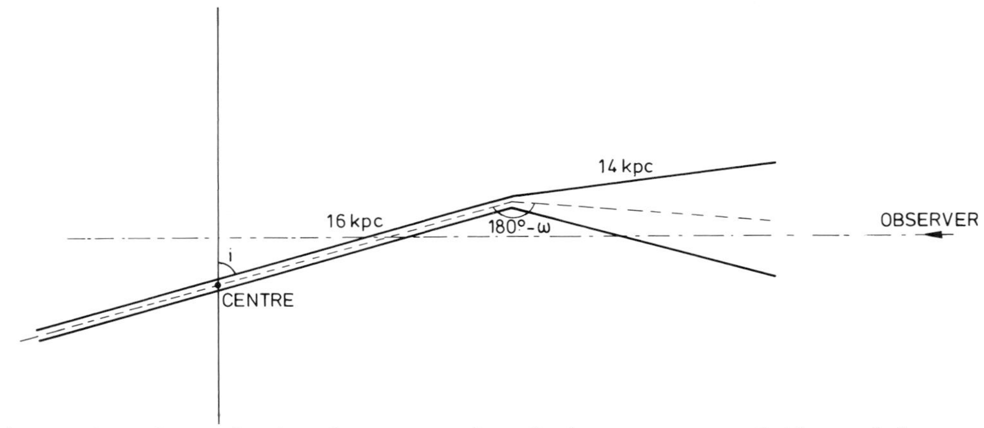

Figure 1: Schematic drawing on scale of the geometry of the model described in the text. The drawing shows the model in the plane which contains the observer and which is perpendicular to the major axis of M31.

the major axis of M31 and which contains the observer. The model starts out with a disk which is inclined at an angle of $i = 74^{\circ}$ (0° is face-on).

Because many features in the spectra are caused by kinematic rather than by gas-density or temperature effects, we produced synthetic spectra for each line of sight and sorted them to simulate the observed position-velocity diagrams. No convolution with the beam of the observing instrument was applied. The parameters which are needed to specify the geometry and the characteristics of the gas are summarized in Table 1. The model parameters for the gas temperature and density are those employed by Burton and Liszt (1978). The value for the velocity dispersion was deliberately chosen to be lower than the standard 10 km s^{-1} to reduce smearing effects in the synthetic spectra. The HI volume density is kept constant, except for the inner 5 kpc where the lack of gas is simulated by reducing the volume density to 5% of the value used for the rest of the disk.

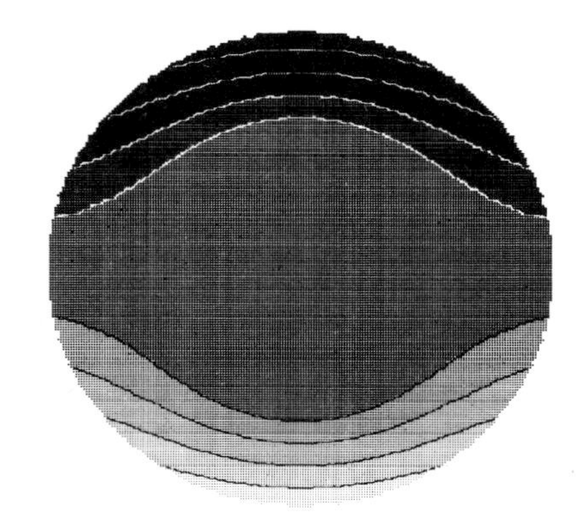

Figure 2: Map showing the deviation of the warp from the mean plane in a face-on view of the model. Contour levels are at -4(1)4 kpc; the dark gray levels correspond to the near side. The diameter of the picture is 60 kpc.

Table 1: Model parameters

Maximum rotational velocity	V_{rot}	$= 260$ km s^{-1}
Region of flat rotation	R	≥ 5.2 kpc
Temperature of the gas	T	$= 120$ K
Velocity dispersion	σ	$= 4$ km s^{-1}

Disk:			Warp:		
Inclination	i	$= 74^o$	Angle	ω	$= 20^o$
Scale height	h_d	$= 0.33$ kpc	Scale height	h_w	$= 0.3$-3.0 kpc
Density	n_d	$= 0.33$ at cm^{-3}	Density	n_w	$= 0.33$-0.0 at cm^{-3}
Radius	R_d	$= 16$ kpc	Extent	R_w	$= 16$ kpc to 30 kpc

To keep the model simple and to keep the number of free parameters to a minimum, we assume that the warp has a linear dependence of height from the plane of the (extended) unwarped disk along any radius in the galaxy, and that it makes an angle $\omega = 20^o$ with the disk at the azimuth, measured in the disk, where the warp reaches its maximum deviation from the plane. The degree of warping, ω, varies as the sine of azimuth. In the model discussed here the line of nodes of the warp runs parallel to the major axis of M31 in such a manner that it bends towards the observer on the near side of the galaxy and away from the observer on the far side, i.e. getting more edge-on. Figure 2 is a face-on view of the model, showing the deviation of the mean plane of the warp with respect to the plane of the disk. We assume cylindrical rotation throughout the disk and the warp. The rotation curve is flat at a circular velocity of 260 km s^{-1} beyond 5.2 kpc, rising linearly from 0 km s^{-1} at the center to the maximum velocity.

3. RESULTS

The most obvious attribute of the warped-galaxy model is its ability - recognized already in the earlier work referenced above - to reproduce the multiply-peaked structures characterizing the observed profiles. Most lines of sight through M31 evidently systematically transverse at least two separate regions of the galaxy. Figure 3 and 4 show both the observed and modelled situation for cuts respectively parallel and perpendicular to the major axis. In Figure 3 the rather linear feature running diagonally across the map corresponds to the warped gas; the other feature is the inner disk. We note that all intensity enhancements in the model are caused by the details of the projections involved; no density structure was put into the model. The model intensity enhancements resemble in a general way the intensity enhancements observed.

In Figure 4 the pedestal features on either side of the major-axis crossing correspond to the warped gas; the inverted-vee shaped feature corresponds to the inner-galaxy disk. During the first stages of the modelling, we constrained the HI scale to be the same, 300 pc, for both the inner disk and the warped part. We found in that case that the broad, diffuse nature of the pedestal corresponding to low-intensity emission in Figure 4a more than 20' from the major axis could not be reproduced.

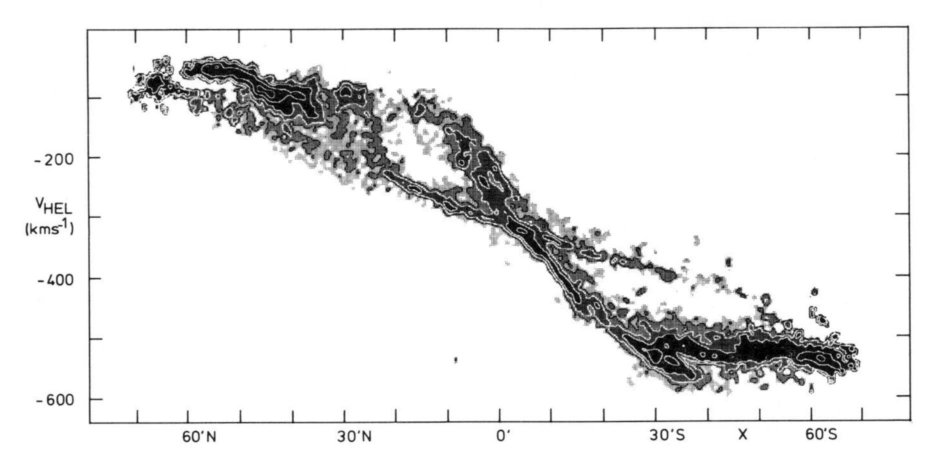

Figure 3a: Position-velocity map made along a line 4'.8 West and parallel to the major axis. Contour levels are at 2.5 , 5 , 10 , and 25 K.

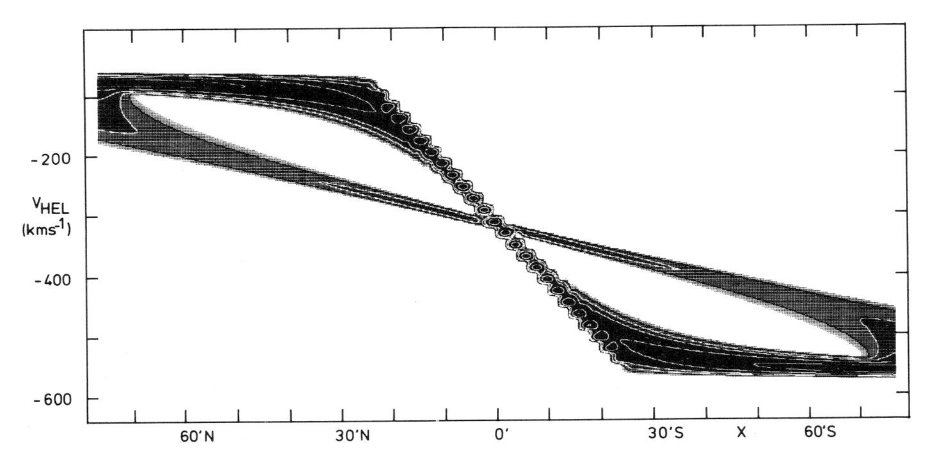

Figure 3b: Model position-velocity map. Same specifications as for Figure 3a.

On the other hand, a flaring warp, in which the scale height of the warped gas increases linearly with radius by about a factor of 10 between the initiation of the warp and the edge of the galaxy does result in synthetic spectra similar in this respect to the ones observed. Also, the low-density HI envelope around M31 seen in a map of the integrated HI brightness (Brinks and Shane, 1983) can be understood in this way. In the model the surface density of the HI in the flaring warp decreases linearly from the value it has in the conventional disk at its initiation to zero at its outer edge.

The principal refinement which we expect to make to the preliminary model shown here involves allowing the line of nodes of the warp to be

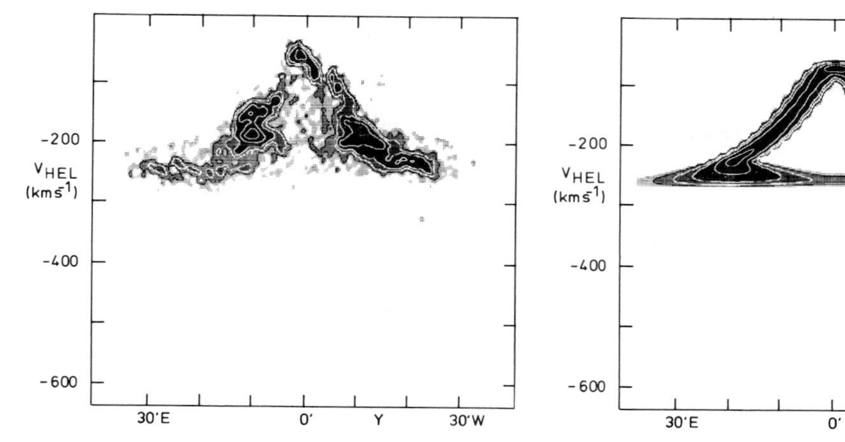

Figure 4a: Position-velocity map made along a line 28'.8 North and parallel to the minor axis. Contour levels are at 2.5, 5, 10, and 25 K.

Figure 4b: Model position-velocity map. Same specifications as for Figure 4a.

oriented other than perpendicular to the line of sight. We are investigating if a twist of this orientation may account for the asymmetries in the position-velocity maps with respect to the major and minor axes shown by the observations. Evidence shown by Emerson and Newton (1978) suggests that the warp *is* twisted with respect to the line of sight. The manner in which our model reproduces the observed spectra leads us to favor the warped-galaxy interpretation of the double-peaked position-velocity maps over the interpretation suggested by Byrd (1977) based on the possible gravitational influence of M32 on the disk of M31. Although Byrd showed by computer simulation of such an interaction that multiply-peaked maps could result, only a limited portion of the disk is distorted. Consequently the predicted effects differ much more dramatically on opposite sides of the axes than is observed. In general, the model poorly represents the observed velocity field.

We note that although the model is an *ad hoc* one in the sense that it lacks dynamical justification, some confidence in the plausibility of the model may be found in its similarity to the well-established situation in our own Galaxy. The aspects of the M31 model which are most important to getting a reasonable fit to the data include the sinusoidal variation with azimuth of the warp, the flaring nature of the gas layer in the warp, the large extent of the warp, the general run of HI surface density, and the flat (or at least not rapidly falling) rotation curve. These are all aspects of the morphology of the outer gas layer in our Galaxy (see e.g. Henderson *et al.*, 1982, and Kulkarni *et al.*, 1982) and of other galaxies (Sancisi, 1983). We will make a detailed comparison of the characteristics of the M31 and the Milky Way warps in our paper currently being prepared; we will also address the relative orientation of the two warps.

REFERENCES

Bajaja, E., Shane, W.W.: 1982, Astron. Astrophys. Suppl. 49, 745.
Brinks, E., Shane, W.W.: 1983, Astron. Astrophys. Suppl. (submitted).
Burton, W.B., Liszt, H.S.: 1978, Astrophys. J. 225, 815.
Byrd, G.G.: 1977, Astrophys. J. 218, 86.
Cram, T.R., Roberts, M.S., Whitehurst, R.N.: 1980, Astron. Astrophys. Suppl. 40, 215.
Emerson, D.T., Newton, K.: 1978, in "Structure and Properties of Nearby Galaxies" (IAU Symp. No. 77), eds. E.M. Berkhuijsen and R. Wielebinski (Reidel, Dordrecht), p. 183
Henderson, A.P.: 1979, Astron. Astrophys. 75, 311.
Henderson, A.P., Jackson, P.D., Kerr, F.J.: 1982, Astrophys. J. 263, 116.
Kulkarni, S.R., Blitz, L., Heiles, C.: 1982, Astrophys. J. 259, L63.
Roberts, M.S., Whitehurst, R.N., Cram, T.R.: 1978, in "Structure and Properties of Nearby Galaxies" (IAU Symp. No. 77), eds. E.M. Berkhuijsen and R. Wielebinski (Reidel, Dordrecht), p. 169.
Sancisi, R.: 1983, in "Internal Kinematics and Dynamics of Galaxies" (IAU Symp. No. 100), ed. E. Athanassoula (Reidel, Dordrecht). p. 55.
Shane, W.W.: 1978, in "Structure and Properties of Nearby Galaxies" (IAU Symp. No. 77), eds. E.M. Berkhuijsen and R. Wielebinski (Reidel, Dordrecht), p. 180.
Unwin, S.C.: 1983, Monthly Notices Roy. Astron. Soc. (preprint).
Whitehurst, R.N., Roberts, M.S., Cram, T.R.: 1978, in "Structure and Properties of Nearby Galaxies" (IAU Symp. No. 77), eds. E.M. Berkhuijsen and R. Wielebinski (Reidel, Dordrecht), p. 175.

DISCUSSION

J.P. Ostriker: Can you comment on the character of the rotation curve?

Brinks: As far as I can tell, it is flat at 260 km/s out to R ~25 kpc.

D. Lynden-Bell: That is quite different from Emerson's result.

Brinks: Yes, but Emerson did not take the warp of the disk into account; the warp explains most of the difference.

I.F. Mirabel: Is there any correlation in position between "holes" in the HI disk and shells going out of the disk? And what is the range of kinetic energies of the HI shells in M31?

Brinks: The average diameter of the holes is about 300 pc; this is smaller than the average of Heiles' supershells. The kinetic energies of the shells are 10^{50} -10^{51} erg, again smaller than Heiles' ~10^{53} erg.
We have not yet been able to discriminate between holes in the HI disk at z = 0 and holes which are offset from the mean HI layer.

J.M. Dickey: Can you estimate the filling factor of neutral hydrogen?

Brinks: On the basis of an analysis of the holes, I estimate the filling factor to be somewhere between 1 and 5%.

THE PRODUCTION OF A 16-MM FILM OF M31

G. Seth Shostak
Kapteyn Astronomical Institute

Eli Brinks
Leiden Observatory

INTRODUCTION

The high-resolution HI survey of M31 (Brinks, this volume) was made at a resolution of $\Delta\alpha \times \Delta\delta \times \Delta V$ = 24" × 36" × 8.2 km s^{-1}. These measures comprise a data cube of 147 channel maps, each 1024 by 1024 pixels in size, and separated in velocity by 4.1 km s^{-1}. Interpretation of these data is hindered by the inability to quickly and easily peruse the maps. It is especially hard to see features which may be continuous along the velocity axis or along a single coordinate axis.

For these reasons we chose to make a record of the data set on 16-mm film. By viewing gray-scale representations of the radio maps at high speed (typically 12 frames per second), any continuity in the data would become manifest as a result of persistence of vision. Further, such a film would allow easy dissemination of the data to others.

HARDWARE DESCRIPTION

All data reduction was performed at Leiden Observatory (Brinks and Shane, 1983). The processed data were then brought to Groningen where they were displayed and photographed using the GIPSY image processing system (Shostak and Allen, 1980).

The GIPSY facility consists of an International Imaging Systems Model 70 image processor connected to a DEC PDP 11/70 host computer. The Westerbork data, compressed to maps of 512 x 512 or 256 x 256 pixels in size, were transferred from tape to disk for rapid access. Several PDP RPO 5 (88 Mbyte) disk packs were used for this purpose. Maps on disk could be quickly loaded into one of the Model 70's six 512 x 512 x 8-bit deep frame buffers for display. The analog video signal from the display controllers was also output to a video disk recorder, which had the capacity to store up to 300 monochrome images. The playback speed of the recorder was selectable from one frame per four seconds up to standard video rate (25 frames per second).

H. van Woerden et al. (eds.), The Milky Way Galaxy, 443–444.

For reasons of speed in production of the film, images were first loaded onto the (analog) video disk, and then photographed in real time upon playback. A once-per-frame synchronization pulse was output from the disk and used to drive a synchronous motor mounted on a Bolex film camera. Since exposure times were short (0.02 sec), a reasonably sensitive emulsion was demanded. We used Eastman type 7250 high-speed reversal color film, daylight type, which has an ASA index of 250.

DISPLAY CONSIDERATIONS

All maps had to be numerically scaled to the 8-bit range of the display memories. Since the dynamic range of the data was approximately 20 dB, this presented no problem. However, because of the transfer characteristics of the monitors, the range of displayed brightnesses was not 256:1 but rather 256^{γ}:1, where $\gamma \simeq 2.4$. Consequently, special software was employed to reduce image contrast by effectively taking the $(1/\gamma)$-th root of frame buffer values before display. The final printed film shows the full dynamic range of the data and has a resolution comparable to the original images.

FILMED SEQUENCES

The film primarily consists of three sequences, each of which shows the data cube along one of its principal axes. A sequence is repeated four to five times at a relatively high frame rate and then three times at a lower rate. The recordings are in black and white, as we feel it is easier to translate gray-scale levels into intensity levels, and because it appears less "noisy".

The first sequence shows the channel maps at full resolution, with a velocity range from -618.1 to -16.2 km s^{-1} in steps of 4.1 km s^{-1}, i.e. at half the velocity resolution. The second series consists of position-velocity maps, made along a line perpendicular to the major axis. The position-velocity maps were made after the channel maps had been smoothed to twice the original spatial resolution in order to enhance the extended low-level emission. The film shows this sequence of position-velocity maps, each separated by 0.6 arcmin, starting about 70 arcmin north of the nucleus. The last sequence comprises position-velocity maps made parallel to the major axis, also at 0.6 arcmin intervals, running from east to west of the nucleus.

The duration of the film is about 8 minutes. Copies may be obtained at cost (about 250 Dutch guilders) from E. Brinks.

REFERENCES

Brinks, E., Shane, W.W.: 1983, Astron. Astrophys. Suppl. (submitted)
Shostak, G.S., Allen, R.J.: 1980, in "ESO Workshop on Two-Dimensional Photometry" (ed. P. Crane and K. Kjär)

DISTRIBUTION AND MOTION OF CO IN M31

Antony A. Stark
Bell Laboratories, Holmdel, NJ 07733 USA

ABSTRACT

M31 is among the few galaxies whose solid angle is more than 100 times the beam area of currently operational millimeter-wave telescopes. Roughly 10% of its area has been mapped in the $J=1\rightarrow0$ line of CO with the 7-m telescope at Bell Laboratories. The data lie on the major and minor axes and in a filled 20' by 15' field. The average CO emissivity is less than 1/5 that of the Milky Way Galaxy, and the inferred H_2 column densities are less than 10% of the HI column densities at the same positions. Thus M31 has a predominantly atomic interstellar medium, whereas our Galaxy seems to have a predominantly molecular interstellar medium. The molecular emission is strongly concentrated in spiral-arm segments, with greater than 25:1 contrast between arm and interarm regions. Overall, the CO emission is well correlated with the HI emission, but shows greater concentration to the spiral arms. There is no evidence for a spatial separation of the HI and CO arms to within our resolution ($\sim$1'): all the spiral-arm tracers are spatially coincident. Almost all the gas on the minor axis exhibits non-circular velocities of about 20 km s^{-1}. There are systematic streaming motions as large as 80 km s^{-1} associated with spiral arms.

INTRODUCTION

Detection of spiral structure in a galaxy requires spatial resolution finer than about 500 pc. Since operational millimeter-wave telescopes have beam sizes $\gtrsim$30", the presence or absence of spiral structure in CO emission can be seen only within the Local Group. Combes et al. (1977a, b) have shown that the CO line is detectable only towards dust lanes in M31; Boulanger, Stark and Combes (1981) made a filled 20-point map to show that CO and HI are well correlated, the CO having a somewhat greater contrast and concentration to the spiral arms. Observations of ^{13}CO towards a northern spiral-arm region (Encrenaz et al. 1979) showed ratios $T_A^*(^{12}CO)/T_A^*(^{13}CO) \approx 10$, roughly twice the average value for the Milky Way Galaxy, MWG (Stark, Penzias and Beckman 1983).

H. van Woerden et al. (eds.), The Milky Way Galaxy, 445–450.

More than 200 CO spectra of M31 have been obtained so far with the 7-m telescope at Bell Laboratories (Stark, Frerking and Linke 1979; Stark, Linke and Frerking 1981; Linke 1981): these data include a filled 20'x15' field on the SW major axis and a strip map of the minor axis.

A FILLED MAP OF A SPIRAL-ARM REGION IN M31

The existence of spiral structure in the CO line is shown clearly in Figure 1. There is > 25:1 contrast between the brightest and faintest map points - the point (-34, 6.5) shows no emission at the 5-mK rms noise level in 1-MHz filters. Projected onto the disk of M31, the telescope-beam half-power circle is an ellipse 200pc x 900pc in size. The beams located on the spiral arms have on the order of 10 giant molecular clouds in them, while the beams on the interarm regions have none. The CO emission is well correlated with all spiral-arm tracers: HI, HII regions, and dust. There are 35 Baade and Arp (1964) HII regions within the bounds of Figure 1, and only 5 lie outside the 1 K km s^{-1} contour where a random distribution would put $\sim$20.

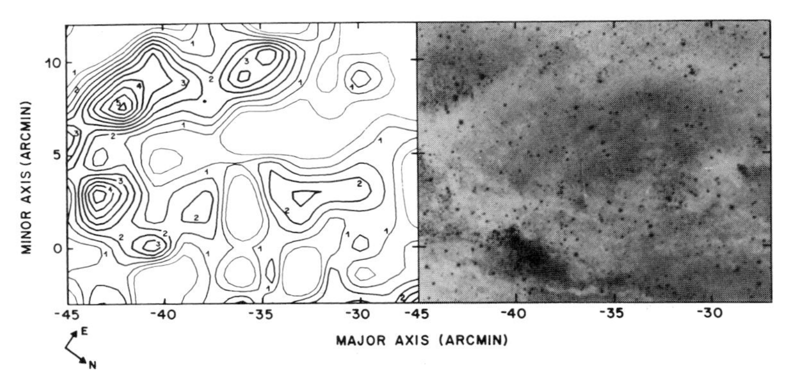

Figure 1. (Left) Integrated ^{12}CO brightness of a field in M31, sampled on a 1.5' square grid (Stark, Linke and Frerking 1981). Contours are labeled in K km s^{-1}. The coordinate system is that of Baade and Arp (1964). (Right) Blue photograph of the same region, showing the correlation of dust and CO.

CO ON THE MINOR AXIS OF M31

This data is summarized in Figure 2. The distribution of CO brightness with radius is very different from the Galaxy: there is no detectable central feature in M31 at a level which is a factor of 300 less than the CO surface brightness of the MWG central region (e.g. Sanders, Solomon and Scoville 1983). The radius of M31's "molecular ring" is nearly twice as large as in the MWG. A comparison of surface brightnesses corrected for inclination shows M31 to be weaker by a factor of $\sim$ 5 compared to the MWG (Stark, Linke and Frerking 1981), and weaker by a factor of 10 to 20 with respect to some other galaxies studied in CO (cf. Morris and Rickard 1982). M31 is not, however, an unusually CO-faint galaxy - it is, in fact, typical of Sb spirals of similar luminosity class. This apparent discrepancy results from a selection effect: those galaxies that are well-observed in CO are a special class of objects with high CO surface brightness, chosen because of limits imposed by instrumental sensitivity.

The velocities shown in Figure 2 have some peculiar features. Almost none of the CO emission on the minor axis is at the systemic velocity of -300 km s^{-1} (e.g. Rubin and D'Odorico 1969). The intensity-weighted average of the CO velocities on the minor axis is -321 ± 5 km s^{-1} indicating a "sloshing" motion of the whole disk. This discrepancy between the gas motion and the systemic velocity was seen in HI by Burke, Turner and Tuve (1963). It is not the typical signature of velocity perturbations by a bar, since both sides of the disk have streaming motions of the same sign. The spiral arm at the -5' position is moving inwards, towards the nucleus of M31, with a radial streaming motion of 80 km s^{-1}. Inward motion cannot be the direct effect of an explosion in the center. The FWHM line widths are typically 25 km s^{-1}, much larger than any single molecular cloud. This width results from the superposition of several clouds, spread out in velocity by a one-dimensional rms cloud-to-cloud velocity dispersion of about 8 km s^{-1} (Stark 1979; Blitz 1981).

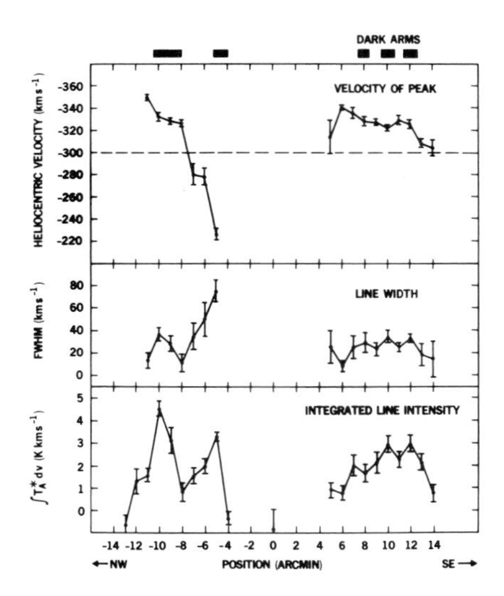

Figure 2. Parameters of CO J=1→0 lines on the minor axis of M31 (Stark, Frerking and Linke 1979). Positions of dust lanes are shown above the frame. (Top) Velocity of the line peak: the systemic velocity is indicated by the dashed line at -300 km s^{-1}. (Middle) Line width (Full width at half-maximum). (Bottom) Line intensity integrated over all velocities. One arcminute corresponds to a radial distance of $\approx$ 1 kpc. There is no detectable emission between the -4' and +4' positions.

SUMMARY

I) The CO distribution in M31 has a hole in the center - there is little or no emission from the inner 4 kpc. The CO brightness peaks at R ~ 10 kpc.

II) The emission is strongly concentrated to spiral arms, with 25:1 contrast. All spiral-arm tracers are well correlated with CO.

III) M31 has a predominantly atomic ISM, whereas the Galaxy's molecular component is greater than or roughly equal to the atomic component.

IV) The observed line widths imply a one-dimensional root-mean-square cloud-to-cloud velocity dispersion $\sigma_v \approx 8$ km s^{-1}.

V) One arm is streaming inwards at 80 km s^{-1}.

VI) The ISM in the disk of M31 is "sloshing" by 20 km s^{-1} with respect to the systemic velocity.

REFERENCES

Baade, W., and Arp. H.: 1964, Ap.J. 139, 1027

Blitz, L.: 1981, in Extragalactic Molecules, ed. L. Blitz and M. Kutner (NRAO Publ. Div.: Green Bank W.Va.) pp. 93-100

Boulanger, F., Stark, A.A., and Combes, F.: 1981, Astron. Astrophys. 93, L1

Burke, B.F., Turner, K.C., and Tuve, M.A.: 1963, Annual Report, Department of Terrestrial Magnetism, Carnegie Institution (Washington, D.C.), p. 289

Combes, F., Encrenaz, P.J., Lucas, R., and Weliachew, L.: 1977a, Astron. Astrophys. 55, 311

Combes, F., Encrenaz, P.J., Lucas, R., and Weliachew, L.: 1977b, Astron. Astrophys. 61, L7

Encrenaz, P.J., Stark. A.A., Combes, F., and Wilson, R.W.: 1979, Astron. Astrophys. 78, L1

Linke, R.A.: 1981, in Extragalactic Molecules, ed. L. Blitz and M. Kutner (NRAO Publ. Div.: Green Bank W.Va.), pp. 87-92

Morris, M., and Rickard, L.J: 1982, Ann. Rev. Astron. Astrophys. 20, 517

Rubin, V.C., and D'Odorico, S.: 1969, Astron. Astrophys. 2, 484

Sanders, D.B., Solomon, P.M., and Scoville, N.Z.: 1983, Ap.J. in press

Stark, A.A.: 1979, Galactic Kinematics of Molecular Clouds (thesis, Princeton University), pp. 12-14

Stark, A.A., Frerking, M.A., and Linke, R.A.: 1979, Bull. Am. Astr. Soc. 11, 415

Stark, A.A., Linke, R.A., and Frerking, M.A.: 1981, Bull. Am. Astr. Soc. 13, 535

Stark, A.A., Penzias, A.A., and Beckman, P.: 1983, in "Surveys of the Southern Galaxy", ed. W.B. Burton and F.P. Israel, Dordrecht: Reidel, pp. 189-194

DISCUSSION

B.G. Elmegreen: What are the masses of the 4 large CO clouds in the dusty regions?

Stark: I have no ^{13}CO data, so I hesitate to answer, but they are of order 10^6 $M_\odot$, hence comparable to complexes of giant molecular clouds.

T.S. Jaakkola: In your figures, as in those of Brinks, there are features indicating more positive velocities on the far side than in the symmetric position on the near side. It was shown several years ago (T.S. Jaakkola, P. Teerikorpi, K.J. Donner, 1975, Astron. Astrophys. 40, 257) that this is a general effect seen in the iso-velocity data of galaxies (18 positive cases out of 25 available at that time). I do not believe in a general expansion of the disks of spiral galaxies ten times faster than in the Metagalaxy. Rather, there is a systematic effect superposed on the lineshifts due to radial velocities.

Stark: The largest negative velocities occur on the side nearest to us.

H. van Woerden: Your cloud-to-cloud velocity dispersion of 8 km/s, is it the dispersion of cloud motions around the mean, or rather an r.m.s. difference between 2 clouds?

Stark: It is the one-dimensional velocity dispersion (r.m.s.) around the mean.

A. Blaauw: How much of that is due to differential rotation in the system?

Stark: The contribution of differential rotation to the linewidths is small, and has been taken into account in calculating the dispersion.

J.P. Ostriker: In comparing our Galaxy and possibly similar galaxies to M31, you find they do not quite agree and an additional parameter may be needed. Another solution, especially after hearing Hodge's review, is that our Galaxy is really an Sc (and then it agrees rather nicely), rather than an Sbc.

Stark: My point there is that the selection effects in extragalactic CO observations are still very great, hence it is too early to judge the relationships between morphological type and other properties.

C.J. Cesarsky: You give an arm/interarm contrast for CO. If you convert from CO to H_2, and take the HI into account, what arm/interarm contrast do you get then?

Stark: The arm/interarm contrast is entirely dominated by the HI, since the molecular constituent is always less than one-tenth of the total mass, even in the arms, unlike in the central parts of our Galaxy.

J.S. Young: The low CO abundance in M31 (so that the surface density of HI is greater than that of H_2) is very important for the question of possible confinement of the molecular clouds to spiral arms. In the inner regions of high-luminosity Sc galaxies, where $H_2 >> HI$ (e.g. IC 342, NGC 6946 and M51), one cannot confine the dominant component of the interstellar gas into small areas. In M51, the CO studies at 20" - 45" resolution by Rydbeck, by A. Sargent, and by Scoville and myself show arm/interam contrasts of factors of 2-3, and that only in the outer parts where the HI and H_2 surface densities become comparable.

With regard to the Hubble type of our Galaxy, its smaller size compared to M31 as indicated by Hodge, combined with its lower V_{max}, suggests that our Galaxy has a lower luminosity than M31. If our Galaxy is an Sc, it must be a high-luminosity one (according to the rotation curves of Rubin, Ford and Thonnard); but it does not have as much gas as the high-luminosity Sc's (as I showed on Monday), so the Milky Way is more likely to be an intermediate-luminosity Sb.

Stark: As to M51, I prefer to wait till the Nobeyama telescope maps this galaxy - then I'll believe the story. The arm/interarm contrast in our Galaxy is about 3:1 in CO emissivity and more than 10:1 in the density of giant molecular clouds, even in regions where molecular hydrogen dominates.

B.F. Burke: Considering the high arm/interarm contrast in CO, where would the old CO go to die?

Stark: I suspect it blows up, as we saw for the local spur yesterday. It probably gets destroyed after 30 million years.

LARGE-SCALE MAPS OF M31 AT MIDDLE AND FAR INFRARED WAVELENGTHS

H.J. Habing
Sterrewacht, Leiden
(on behalf of the IRAS science team)

The satellite IRAS (see Neugebauer et al. 1984; see also Gautier and Hauser, this volume) was in operation from January 27 until November 22, 1983. Its main goals were to carry out a survey of the sky at 12, 25, 60, and 100 micron and to make observations around selected positions achieving higher sensitivity or better angular resolution than provided in the survey. Among the very first targets of these pointed observations was the Andromeda galaxy, M31, because it was considered of prime importance and was about to leave the observing window. Here we report briefly on crude maps made from the first measurements. A more elaborate presentation may be found in Habing et al. (1984) and a publication of all the data on M31 is planned for the future.

We obtained two adjacent maps of 3 by 3 degrees each by making raster scans with the full set of 59 active detectors. Each detector has a width of approximately 5 arcmin and a height that depends on the wavelength band: from 0.8 arcmin at 12 micron to 3 arcmin at 100 micron. The scan direction practically coincided with the minor axis. Therefore the angular resolution of the maps is (almost) the same in the direction of the major axis, but in the direction of the minor axis it degrades progressively with wavelength. In Figure 1 we show a combination of three maps.

The following are the main conclusions drawn so far:

1. The total luminosity of M31 between 12 and 100 micron is 1.2×10^9 $L_\odot$. This value has an uncertainty of at least a factor of two because of remaining uncertainties in the calibration and because of possible baseline effects. In addition M31 emits perhaps the same amount of radiation beyond 100 micron. We thus estimate that M31 emits at most 5×10^9 $L_\odot$ of infrared radiation. Compared to the total bolometric luminosity of M31 it is seen that less than 10 percent of the stellar light is re-emitted in the infrared. This is a small percentage compared to what is found in most other spiral galaxies, especially those of later type (Sbc, Sc, Sd) (see de Jong et al., 1984; Soifer et al., 1984). It is unknown whether M31 is a poor emitter for its type, that of an Sb galaxy.

H. van Woerden et al. (eds.), The Milky Way Galaxy, 451–456.

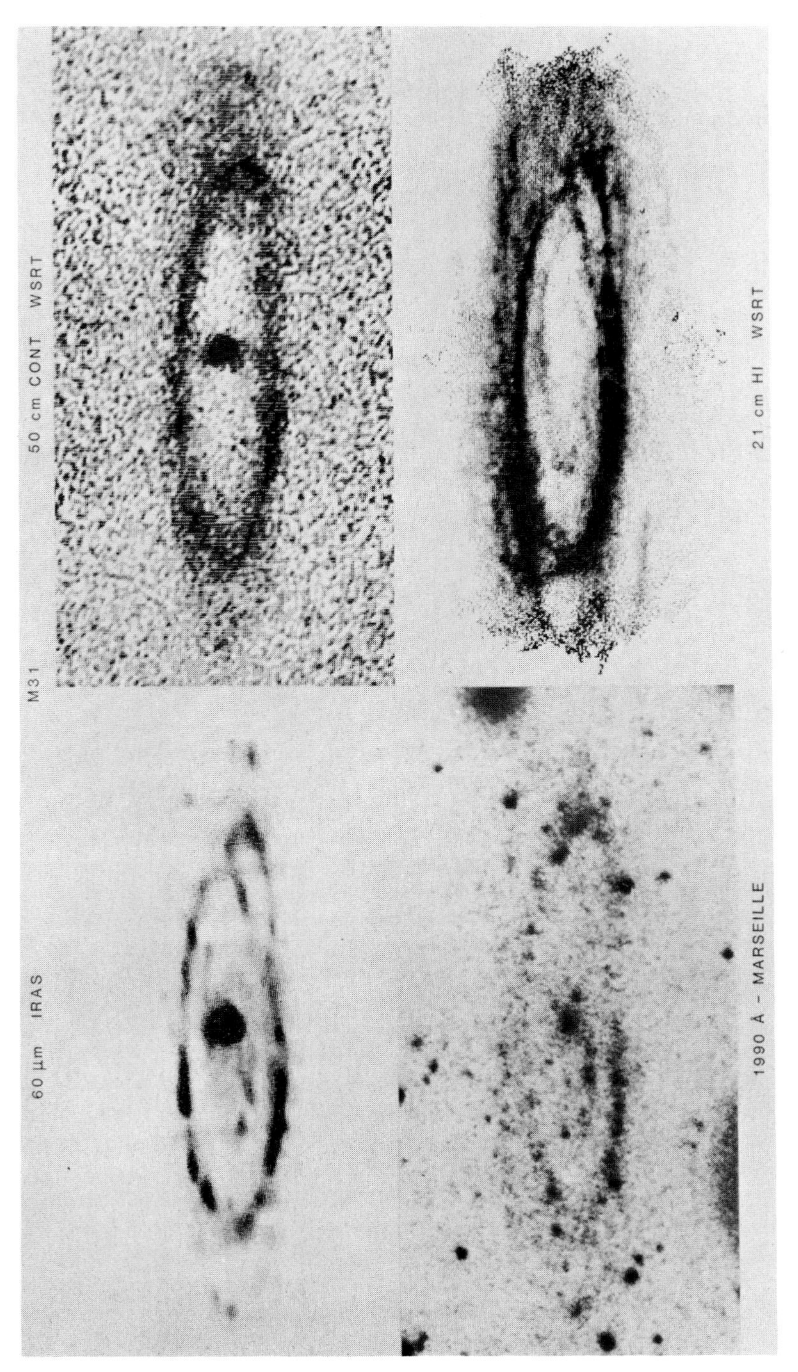

Figure 2. Comparison of the IRAS infrared emission from M31 with emission in the UV (Deharveng et al.) in the 50-cm radio continuum (Bystedt et al.), and in the 21-cm line (Brinks and Shane).

2. At each of the four wavelengths the emission comes mostly from a ring of about 45 arcmin (major axis) diameter. This ring is also seen at several other wavelengths (see Figure 2) and it apparently represents the part of M31 that is actively forming new stars: the same ring can be seen in a UV map at 100 nanometer (Deharveng *et al.*, 1980), in the 21cm line of HI (Unwin, 1980a, 1980b; Brinks and Shane, 1984) and in the non-thermal radio continuum at 50 cm (Bystedt, Brinks, de Bruyn, Israel, Schwering and Shane, in preparation). Also the CO emission is very probably concentrated in this ring (Boulanger *et al.*, 1981). The distribution of HII regions in M31 as given by Pellet *et al.* (1978) closely resembles the distribution of the IR emission. The same is true for the 21cm line emission. On optical photographs the prominent dust lanes coincide very precisely with the IR emission in all four wavelength bands. These various coincidences convince us that the IR emission is indeed from dust heated by embedded, probably young stars. It is conceivable that the emission at the longest wavelength is, in part, originating from dust that is less intimately associated with the star-forming regions but that is part of the more diffuse interstellar medium. This seems to be the case in our Galaxy (e.g. Mezger *et al.*, 1982) and could equally apply to M31, as far as our maps are concerned.

Although the ring is the most dominant feature in our maps, some IR emission is associated with dust lanes inside the ring and with a few HII regions outside of the ring on the major axis.

3. An unexpected feature of all maps is the emission from the central part of M31. The centre is quite prominent in the maps, but it contributes less than two percent to the total IR emission. Interstellar matter in the centre of M31 is only known in the form of dark clouds (Baade, 1963). An upper limit of 10^6 $M_\odot$ has been set for the total cloud mass by Gallagher and Hunter (1981). HI and CO emission has not yet been detected. The IR emission from the centre has a higher ratio of 60 to 100 micron flux density than that of the ring and the emitting dust at the centre will thus be somewhat hotter. The grain properties in the far IR are still rather uncertain, but adopting values recommended by Hildebrand (1983) we estimate that the dust has a temperature of 34K and a total mass of 3000 $M_\odot$. For a normal gas-to-dust ratio of 100 the total amount of interstellar matter is comparable to the value quoted above. The temperature of 34K is rather high, when compared with e.g. the grain temperature in the solar surroundings(see e.g. Mezger *et al.*, 1982). Most certainly this can be attributed to the much stronger radiation field at wavelengths longward of 300 nanometer, produced by the large number of late-type giants in the bulge of M31. The ultraviolet radiation field there is comparable in energy density with that in the solar surroundings and thus insufficient to heat the grains.

4. We have detected IR emission from the two close companions, M32 and NGC 205. Both are elliptical galaxies but the nature of their IR emission is radically different. M32 is strongest at the shortest wave-

length, very weak at 25 micron and not at all detectable at the longest wavelengths. The amount of radiation is compatible with emission in the photospheres of late-type stars. NGC 205 is not detected at the shortest wavelengths, is weak at 60 micron and is a 2-Jansky source at 100 micron. NGC 205 is known to possess blue stars and star formation is probably taking place (see e.g. Deharveng et al., 1982). The character of the IR emission supports this conclusion.

ACKNOWLEDGEMENT. I thank Piet Schwering for massaging the M31 maps and producing the colour picture. The Infrared Astronomical Satellite IRAS was developed and has been operated by the Netherlands Agency for Aerospace Programs (NIVR), the U.S. National Aeronautics and Space Administration (NASA), and the U.K. Science and Engineering Research Council (SERC).

REFERENCES

Baade, W.: 1963, Evolution of Stars and Galaxies (edited by C. Payne Gaposchkin), Harvard University Press, Cambridge, Mass.
Boulanger, F., Stark, A.A., Combes, F.: 1981, Astron. Astrophys. 93, L1
Brinks, E., Shane, W.W.: 1984, Astron. Astrophys. Suppl. Ser. (in press)
Deharveng, J.M., Jakobsen, P., Milliard, B., Laset, M.: 1980, Astron. Astrophys. 88, 52
Deharveng, J.M., Joubert, M., Monnet, G., Donas, J.: 1982, Astron. Astrophys. 106, 16
De Jong, T., Clegg, P.E., Soifer, B.T., Rowan-Robinson, M., Habing, H., Houck, J.R., Aumann, H.H., Raimond, E.: 1984, Astrophys. J. Letters (March 1)
Gallagher, J.S., Hunter, D.A.: 1981, Astron. J. 86, 1312
Habing, H.J., Miley, G.K., Young, E., Baud, B., Boggess, N., Clegg, P., de Jong, T., Harris, S., Raimond, E., Rowan-Robinson, M., Soifer, B.T.: 1984, Astrophys. J. Letters (March 1)
Hildebrand, R.: 1983, Quart. J.R.A.S. 24, 267
Mezger, P.G., Mathis, J.S., Panagia, N.: 1982, Astron. Astrophys. 105, 372
Neugebauer, G., Habing, H., van Duinen, R.J., Aumann, H.H., Baud, B., Beichman, C.A., Beintema, D.A., Boggess, N., Clegg, P.E., de Jong, T., Emerson, J.P., Gautier, T.N., Gillett, F.C., Harris, S., Hauser, M.G., Houck, J.R., Jennings, R.E., Low, F.J., Marsden, P.L., Miley, G.K., Olnon, F.M., Pottasch, S.R., Raimond, E., Rowan-Robinson, M., Soifer, B.T., Walker, R.G., Wesselius, P., Young, E.: 1984, Astrophys J. Letters (March 1)
Pellet, A., Astier, N., Viale, A., Courtes, G., Maucherat, A., Monnet, G., Simien, F.: 1978, Astron. Astrophys. Suppl. Ser. 31, 439
Soifer, B.T., Rowan-Robinson, M., Houck, J.R., de Jong, T., Neugebauer, G., Aumann, H.H., Beichman, C.A., Boggess, N., Clegg, P.E., Emerson, J.P., Gillett, F.C., Habing, H.J., Hauser, M.G., Low, F., Miley, G.K. Young, E.: 1984, Astrophys. J. Letters (March 1)
Unwin, S.C.: 1980a, M.N.R.A.S. 190, 551
Unwin, S.C.: 1980b, M.N.R.A.S. 192, 243

DISCUSSION

A.A. Stark: IRAS is, as you said, a sensitive detector of dust. There is a feature in the data, a line of dust that connects the nucleus of M31 to the ring of gas in the disk, in position angle roughly 90°. This feature also appears in Einstein X-ray and soft-UV maps of M31. It may be that this object is the projection of the major axis of the bar in M31. It is possible to fit the photometry (including the tilt in the isophotes of the bulge) with a triaxial model of the bulge of M31, where the (3-dimensional) major axis of the bulge projects onto that feature.

Stark, between Elaine Sadler and Reid, in discussion with Burton

LZ

P.W. Hodge: Is the infrared flux for NGC 205 consistent with the optically-measured mass of dust?

Habing: I do not know. The flux is a twentieth of that received from the nucleus of M31; but the dust in NGC 205 is cooler, so its amount should be relatively greater.

Brinks (left) and Bash on the Frisian lakes. Background: Hu (left) and conference secretaries Joke Nunnink, Marijke van der Laan and Ineke Rouwé LZ

PART III

DYNAMICS AND EVOLUTION

Local organizers Willy Bosman (standing) of Groningen University Congress Office and Jan de Boer, between secretary Ineke Rouwé and Eli Brinks (left). Behind De Boer: LOC Chairman Allen, between Lin (left) and Paul; continuing around the table: Rydbeck, Johansson, D.M. and B.G. Elmegreen.
LZ

SECTION III.1

MILKY WAY, MAGELLANIC SYSTEM AND LOCAL GROUP

Thursday 2 June, 1220 - 1320

Chairman: T.S. van Albada

.bove: Oort and Van Albada. In background, to their left: D.B. Sanders, unidentified, Liszt, Lub and Higgs; at right: Terzides and Alladin. Foreground, from left: Van Driel, Mrs. Oort, Kormendy, Fujimoto, Okuda, Seiden, Brink, Carignan. LZ
Below: Mieke Oort (right) and Anneke van Albada LZ

SLIPPERY EVIDENCE ON MASSES IN THE LOCAL GROUP

D. Lynden-Bell
Institute of Astronomy, Cambridge, U.K.

Once it is agreed that not all mass gives a significant contribution to light or any other emission, then one must rely on the dynamics of the visible objects to determine the total gravity field. It is clearly impossible to do this without subsidiary hypotheses. Here we shall assume that all members of the Local Group began together in the Big Bang, and that their dynamics have been governed by their mutual gravitational interactions since the system first achieved a size of some 200 kpc. This must have been some 10^9 years after the Big Bang.

The light of the Local Group is dominated by that of M31 and the Milky Way, so we shall assume that they and their haloes dominate and these masses will in turn determine the dynamics of the Group. Our aim will be to determine their masses.

Objects close to either of the big two will typically have orbits with periods much less than 10^{10} years, so that they will have been in and out more than once. There are nine satellites of the Milky Way that fall into this category and at least seven satellites of M31. One may use their separations from their primary galaxy and their velocities to determine the masses of the two primary galaxies as far out as the satellite systems extend. Such estimates assume the satellites are randomly phased in their orbits which are randomly oriented to the line of sight (Lynden-Bell 1983). Some subsidiary data come from satellites whose light distributions show evidence of tidal limitation by the primary. As we shall see in the next section, the data are not yet sufficient to do a good job on M31, while our special observing position leads to an ambiguous result for our satellites.

However, of greater interest than the discussion of the short-period orbits is the discussion of those with periods greater than the current age of the Universe. For these orbits we are not only interested in the amplitudes of the motions but also in their phases, because we have the boundary condition that all the objects must have been together at the Big Bang. The simplest example of this constraint was given in the timing argument of Kahn & Woltjer (1959). They assume that the present

H. van Woerden et al. (eds.), The Milky Way Galaxy, 461–469.

velocity of approach of M31 has been caused by our mutual attraction, which first slowed our relative expansion in the Big Bang and then reversed it. This picture gives us a relationship between the sum of the masses M, the time since the Big Bang t, the relative velocity of the two galaxies v and their separation r. To sufficient accuracy (3%) this relationship is

$$\left(\frac{GM}{r^3}\right)^{\frac{1}{2}} t + 0.85\ vt/r = 2^{-2/3}\pi = 1.11\ . \qquad (1)$$

For $r = 700$ kpc, $t = (1.5\pm0.5)\times10^{10}$ yr, and $v = -123$ km/s we have $M = (4.3\mp1.0)\times10^{12}\ M_\odot$. This result is increased if significant transverse motion exists in the Milky Way - M31 binary. Such a transverse motion was formerly found, but with Sandage's new determination of Local Group members we find that our Galaxy's motion through <u>them</u> is towards Andromeda to within better than the accuracy of determination. v depends on the heliocentric velocity of Andromeda, for which we used -301 km/s, and the circular velocity of the LSR for which we used 220 km/s. Recent radio data may favour -310 km/s for M31 which would require even more mass. Alternatively, if our circular velocity is 250 km/s, then M is reduced to $(3.4\mp0.9)\times10^{12}\ M_\odot$.

The largest uncertainty is in the time since the Big Bang. If we knew the distances to more distant members of the Local Group more accurately, we could in fact determine that time as well, because the distance and velocity of the third member from the barycentre also obeys relation 1 in the approximation in which M31 and the Galaxy are treated as a single heavy mass. With two equations like (1) both M and t may be solved for. My attempts to do this using the small galaxy Wolf-Lundmark-Melotte have yielded ages between 16 and 10 billion years, depending on whether WLM is at 1.6 Mpc or 1.3 Mpc distance (Lynden-Bell 1981).

A crucial assumption of all methods based on Local Group dynamics is that M31 and the Galaxy are truly dynamically related. If the masses are <u>insufficient</u> to bind us, then as we proceed into the past, M31 will continue to recede and other galaxies will come between us. Eventually there will be such a mass between that we will all be dragged back together into the Big Bang. It is <u>not obvious</u> that the relative motion of M31 and the Galaxy must have been reversed by <u>our</u> gravity rather than by the intervention of others.

THE M31 SUBGROUP

M31, M32, NGC205, 185, 147, and the dwarfs I, II & III are members. Other more doubtful candidates are M33, IC10, Pisces and IC1613 - the last is 40° away from M31 but at a similar distance. Figure 1 shows $\log 3(\Delta v_r)^2$ plotted against distance from M31. The former should equal $\log V_c^2$ on average because of the theorem that in time average around any orbit $\langle \underline{v}^2 \rangle = -\langle \underline{r}.\underline{\nabla}\psi \rangle = \langle V_c{}^2 \rangle$. For this reason the rotation curve

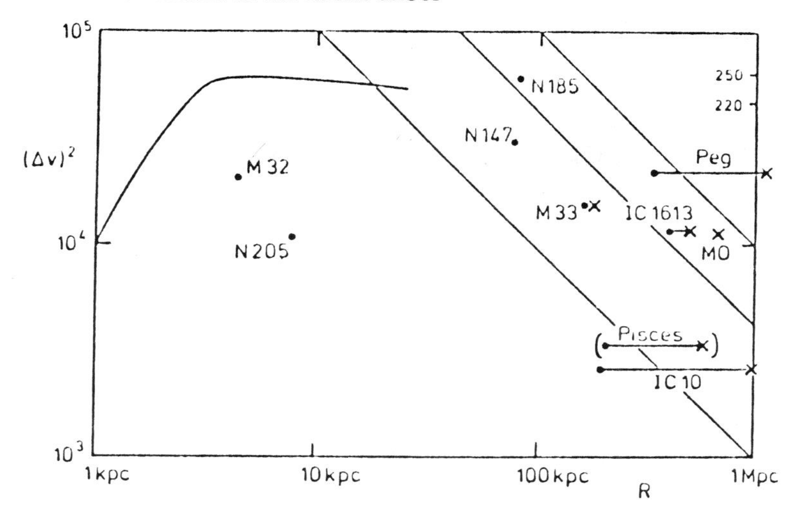

FIGURE 1: Possible Andromeda Satellites. Sloping lines are 2.3, 10 and 23 x 10^{11} $M_\odot$.

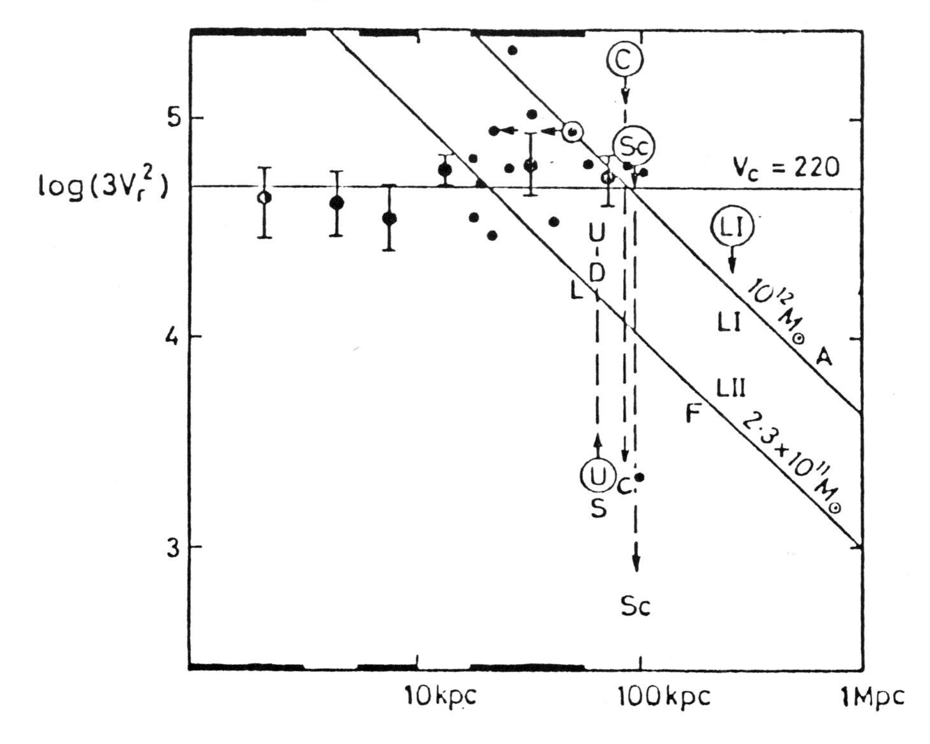

FIGURE 2: Satellites of the Galaxy (obsolete values ringed) and means of globular clusters (with error bars); distant globulars are also plotted individually.

of M31 taken from Newton & Emerson is also plotted. It looks as though the rotation curve probably stays high out to nearly 100 kpc and that a total mass appropriate for M31 is of the order of 1 or 2 x 10^{12} $M_\odot$.

THE MILKY WAY SUBGROUP

Large masses have been obtained by several workers trying to get precise gravitational models for the Magellanic Stream, but such modelling depends crucially on the idea that only gravitational forces are involved. Relaxing this constraint would probably allow too many new parameters to give any definitive result. Davies & Wright (1977) early showed that even within gravitational models a heavy mass for the Galaxy was not the only possible way of getting the high velocities in the Magellanic Stream. If we want more definite assurance that the Galaxy is heavy, we should turn to the observations of globular clusters and satellite galaxies, Figure 2. At first glance it appears that the flat rotation curve of the Galaxy must continue out to 100 kpc. However, there are disturbing reasons to treat this result with caution. Firstly, not all the globular-cluster data are accurate and some clusters still have their distances changed by factors of 2 or 3! In our problem clusters of large v^2r are important. The majority lie below a line corresponding to 2.3 x 10^{11} $M_\odot$. The scatter of points above and to the right of that could be sent there by poor velocities, poor distances or a suitable orientation of the velocity vector so that $v^2 \cong v_r{}^2$ rather than being three times as large. Note it is the distant, faint halo globulars with low metal abundance and weak lines in their spectra that determine the outer extension of the Galaxy. It is these that will have the greatest distance and velocity errors. A hint that this may indeed be so comes from the dwarf-spheroidal galaxies, where the method of measuring the velocities off Carbon-star band heads has greatly improved the data over the last two years, particularly by the hands of Richer & Westerlund (1983), and Aaronson, Olszewski & Hodge (1983). Whereas Draco has remained with the velocity as determined off red giants by Hartwick and Sargent (1978), there have been large changes in the velocities of Sculptor and Ursa Minor. Very recently Lynden-Bell, Cannon and Godwin (1983) redetermined Carina's velocity and showed that, far from being the Galaxy's fastest-moving satellite, it was in fact one of the slowest. The heliocentric velocity is not 450±100 but 240±10 km/s, and this changes the Galactocentric velocity from 235 to 24! Of our 9 satellites, 7 now have well-determined velocities: LMC, SMC, Draco, Ursa Minor, Sculptor, Carina and Fornax. The velocities of Leo I and Leo II are currently less secure, mainly because the Carbon stars are so faint in these distant systems.

On the hypothesis that the motions are randomly oriented to the line of sight, we can determine the rms circular velocity at the distance of these satellites (52 to 220 kpc) to be

$$V_c = \langle 3v_r{}^2 \rangle^{\frac{1}{2}} = 106 \pm 19 \text{ km/s}.$$

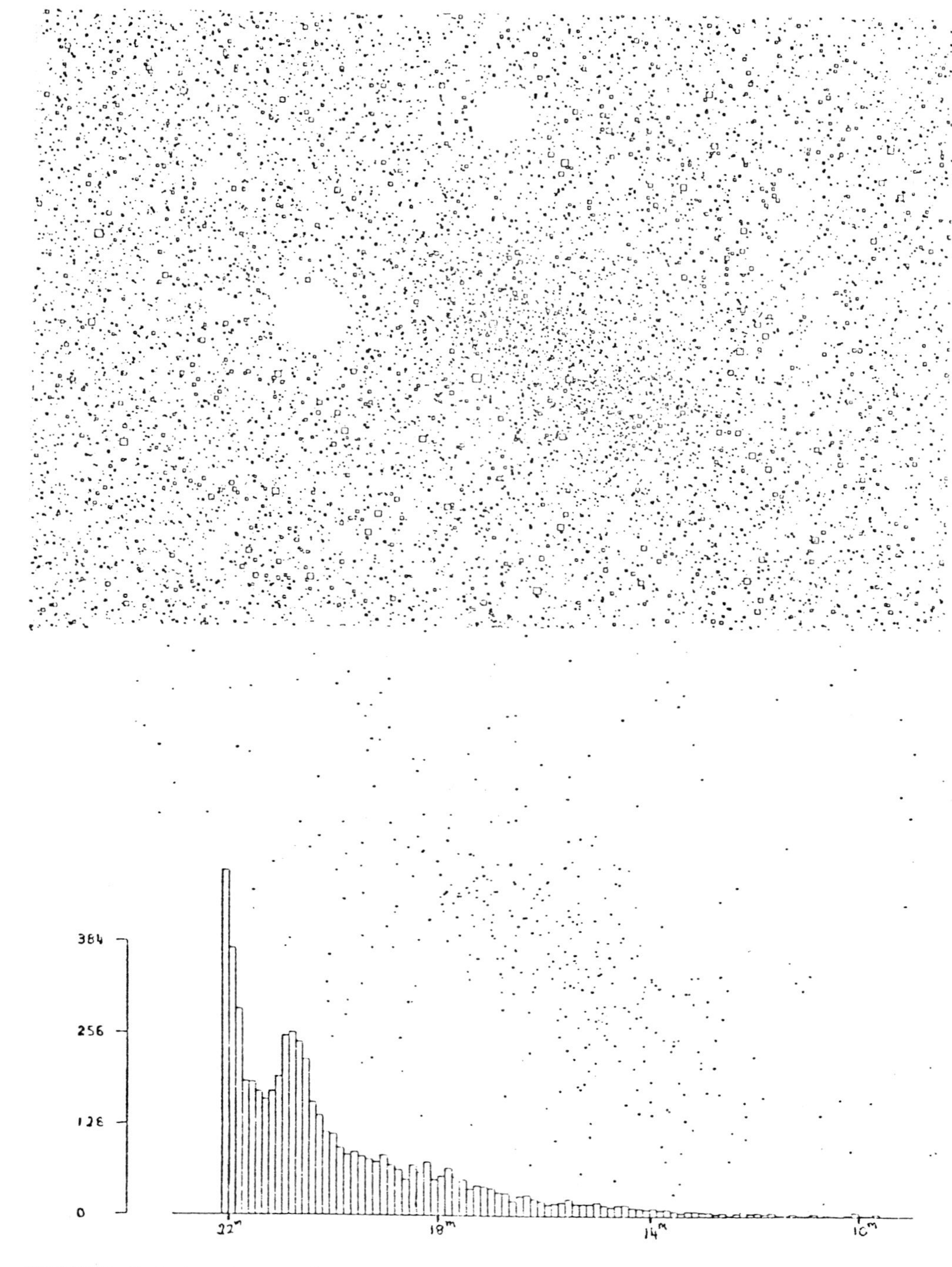

FIGURES 3 & 4: Ursa Minor dwarf. Processed by Kibblewhite's A.P.M. Group, Cambridge.

This low value suggests that a point-mass approximation to the orbits might be appropriate. For such orbits the radial velocity is given by

$$v_r^2 r = GM\, e^2 \sin^2\phi\, /(1+e\cos\phi),$$

or time-averaging over the orbits

$$\langle v_r^2 r\rangle = \tfrac{1}{2}\, GM \langle e^2\rangle .$$

Now Jeans (1928) showed that, if the velocities are isotropically distributed, then e is distributed like 2e de between 0 and 1. We therefore expect $\langle e^2\rangle = \frac{1}{2}$ and

$$GM = 4 \langle (v_r^2 - \varepsilon^2) r\rangle .$$

We have replaced v_r^2 by $v_r^2-\varepsilon^2$ to allow for measurement errors in velocities, ε, which are significant for Leo I and Leo II. Applying this formula to the nine satellites of the Galaxy (which lie between 52 and 220 kpc from the centre) we find

$$M = (2.6 \pm 0.8)\times 10^{11}\ M_\odot$$

A result which, if taken at face value, implies no heavy halo. As implied by our title, this result should not be taken uncritically. $\langle e^2\rangle = \frac{1}{2}$ implies a typical $e \sim 0.7$ or a typical radial excursion by a factor 6. Most of the satellite galaxies could not withstand the tide of the Galaxy if they were brought 6 times closer. Thus it is not unlikely that the orbits of our distant satellites are just those that allow those satellites to survive. The more nearby circular orbits ($e<\frac{1}{2}$) will never have subjected them to tides capable of tearing them apart. If in place of Jeans's distribution we say that the eccentricity distribution is truncated at $e = \frac{1}{2}$, then we have a distribution 8e de and a $\langle e^2\rangle$ of 1/8 in place of $\frac{1}{2}$. This gives a mass of four times our former estimate, consistent with a heavy halo which extends to 100 kpc.

The great importance of tides in the above argument has led us to look more carefully at the distribution of faint stars in Ursa Minor on a deep IIIaJ Schmidt plate, kindly provided by the Palomar Observatory. Although Ursa Minor is never easy to see on an original plate, it shows up well on a plot of all faint stars on the plate, Figure 3. We may study its extent by taking only the blue horizontal—branch stars which give a remarkably clean picture, Figure 4. At the bottom we superpose the luminosity function. All this work is the product of the A.P.M. unit at Cambridge. In Figure 3 Ursa Minor has a sharp NW edge and is elongated at $53^\circ \pm 3^\circ$ with $b/a = 0.47$, values in good agreement with those of Hodge (1964). The orientation is almost precisely along the Magellanic Stream, and it lies as accurately in the stream as the SMC, to which it is antipodal in the galactocentric sky. Since Draco is also antipodal to the LMC and oriented along the Magellanic Stream, the idea of a tidal origin of all dwarf spheroidal galaxies should be taken seriously. Sculptor, Leo I, Leo II and Fornax lie too precisely in a galactocentric great circle for chance alignment. If they were made from debris torn off Fornax in some long-dead Fornax stream, we have a natural explanation of this great circle.

DWARF SPHEROIDALS AND MISSING MASS

If the missing mass were heavy neutrinos with a rest mass of 10-30 eV/c^2, then they will have moved collisionlessly since they were created in the Big Bang. Their phase-space density will be the same as it was initially, but on coarse graining it may decrease. Such arguments show that it is difficult to account for galaxies' heavy haloes with such particles, and it would be impossible to account for missing mass in the weakly bound dwarf spheroidals. Aaronson's work on Draco (Aaronson 1983) is especially interesting in this regard, as it may indicate the need for missing mass there, in which case we would have an example of abundant dark matter that is not neutrinos. However, the reliability of Aaronson's result can be called in question, because different Carbon stars have dredged up different amounts of Carbon from their interiors. Thus correlation techniques may give slightly different velocities for Carbon stars whose spectra differ, even if the true velocities are the same. It is possible that Aaronson's apparent velocity dispersion arises from such a situation, in which case the estimate of the mass and missing mass of Draco would be meaningless. An easier test may be made with Fornax, where the one planetary nebula and the five globular clusters have velocity differences within their errors (Cohen 1983). There is no case for Fornax having a mass-to-light ratio greater than that of a normal stellar population. This must cast some doubt on the Draco result.

ENHANCED TIDES AND HEAVY HALOES

By giving galaxies more mass the heavy-halo hypothesis clearly enhances tides between them, but there is also an indirect effect that makes tides about three times greater still. For example, the tide of M31 will distort our halo, and our distorted halo then helps to distort our disk. Thus the tide imposed by M31 and its halo will be magnified by the distortion of our halo. To be definite, we take a static model with an initially spherical halo with density $\rho = A\, r^{-\alpha}$ and a free boundary at r_h. We shall assume that the halo is a polytrope of index n with $\alpha = 2n/(n-1)$ and follows the same equation of state during perturbation. Let the tide-producing body be of mass M and at distance R. The hydrostatic equation for the perturbed potential ψ reads

$$\nabla\left(\psi - \frac{dp}{d\rho}\frac{\delta\rho}{\rho}\right) = 0\ ,$$

from which one deduces that $\delta\rho = \rho\, \dfrac{d\rho}{dp}\, \psi$.

Inserting this perturbation in Poisson's equation and taking ψ and ρ to have the P_2 tidal form, we have

$$\nabla^2\psi = r^{-2}\, d/dr(r^2 d\psi/dr) - 6\psi/r^2 = -4\pi G\rho\, d\rho/dp\ \psi$$

If we write $\psi_\ell = d\psi/d\ln r$ and evaluate $G\rho\, d\rho/dp$ from the unperturbed state, this equation becomes

$$\psi_{\ell\ell} + \psi_\ell - [6 - N]\psi = 0 \quad \text{where } N = \alpha(3-\alpha)$$

hence $\psi = C(r/r_h)^\beta\, P_2$,

where β is the positive root of $\beta^2 + \beta - (6 - N) = 0$

which is $\beta = -\frac{1}{2} + (\frac{25}{4} - N)^{\frac{1}{2}}$.

Outside the halo $\psi = [C_1(r_h/r)^{-3} + GM\, r^2/R^3]\, P_2$.

To evaluate C we must use the boundary conditions at $r = r_h$. Continuity of ψ gives $C = C_1 + GM\, r_h^2/R^3$ (2). The free boundary condition becomes

$$\beta C/r_h = -3C_1/r_h + 2GM\, r_h/R^3 + 4\pi G\rho\xi_r/P_2. \qquad (3)$$

Here ξ_r is the radial displacement at the boundary. ξ_r can be evaluated from the condition that on the displaced boundary there is no pressure change, which gives on the unperturbed boundary $r = r_h$ the condition

$$\delta p = -\rho\ \underline{\xi}\cdot\underline{g}\ ,$$

where $\underline{g}$ is the acceleration due to gravity at r_h. Evaluating these expressions we find

$$4\pi\, G_\rho\ \xi_r = (3-\alpha)\ \psi/r_h.$$

Hence from (2) and (3) $\quad C = [5/(\alpha+\beta)]\ GM\, r_h^2/R^3$.

Thus

$$E = \frac{\text{Total potential change}}{\text{imposed potential}} = \frac{\psi}{GM\, r^2\ P_2/R^3} = \frac{5}{(\alpha+\beta)}\left(\frac{r}{r_h}\right)^{\beta-2},$$

where the halo density is $Ar^{-\alpha}$ and $\beta = -\frac{1}{2} + [\frac{25}{4} - \alpha(3-\alpha)]^{\frac{1}{2}}$.

For $\alpha = 2$, $\beta = 1.56$ and $E = 1.4(\frac{r}{r_h})^{-0.44}$. For $r_h/r \sim 10$ this enhancement factor is 3.9. However, even with such a factor, Andromeda's tide remains a negligible effect out to well beyond the observed warp. The above calculation cannot describe the dynamic tide of the Magellanic Clouds which orbit within our dark halo.

REFERENCES

Aaronson, M.: 1983, Astrophys. J. 266, L11.
Aaronson, M., Olszewski, E.W., & Hodge, P.W.: 1983. Astrophys. J. 267, 271.

Cohen, J.G.: 1983, Astrophys. J. 270, L41.
Davies, R.D., & Wright, A.E.: 1977, Monthly Notices Roy. Astron. Soc. 180, 71.
Hartwick, F.D.A., & Sargent, W.L.W.: 1978, Astrophys. J. 220, 453.
Hodge, P.W.: 1964, Astron. J. 69, 438.
Jeans, J.H.: 1928, Astronomy & Cosmology, p. 287, Cambridge Univ. Press.
Kahn, F.D., & Woltjer, L.: 1959, Astrophys. J. 130, 705.
Lynden-Bell, D.: 1981, Observatory 101, 111.
Lynden-Bell, D.: 1982, Observatory 102, 202.
Lynden-Bell, D.: 1983, in "Kinematics, Dynamics and Structure of the Milky Way", ed. W.L.H. Shuter, Dordrecht: Reidel, p. 349.
Lynden-Bell, D., Cannon, R.D., & Godwin, P.J.: 1983, Monthly Notices Roy. Astron. Soc. 204, 87 p.
Richer, H.B., Westerlund, B.E.: 1983, Astrophys. J. 264, 114.

DISCUSSION

J.P. Ostriker: Have you reanalyzed the satellite data, assuming circular orbits? And if so, what is the conclusion?

Lynden-Bell: I know the exact answer to that: the mass of the Galaxy, assuming circular orbits, is infinite since we do discover actual velocity shifts.

Ostriker: It would be infinite anyway, because we are on an other side of the Galaxy.

Lynden-Bell: Yes, I agree, but the velocities do not correlate with position on the sky in the way one would like them to.

Above: Oort, Mrs. Oort, Lynden-Bell, Van der Laan and Hartwick. LZ
Below: Mrs. Blaauw and Alladin enjoy University President's supper, while Schmidt and Ostriker have another debate. CFD

A COLLISION BETWEEN THE LARGE AND SMALL MAGELLANIC CLOUDS 2×10^8 YEARS AGO

M. Fujimoto and T. Murai
Department of Physics, Nagoya University, Chikusa, Nagoya 464

ABSTRACT

A number of orbits are obtained for the Large and Small Magellanic Clouds (LMC and SMC) revolving around a model Galaxy with a massive halo. It is suggested that the SMC approached the LMC as close as 3 to 7 kpc about 200 million years ago, if these clouds have been in a binary state for the past 10^{10} years, and the Magellanic Stream (MS) is due to the gravitational interaction among the triple system of the Galaxy, LMC, and SMC.

1. PAST BINARY ORBITS OF THE LARGE AND SMALL MAGELLANIC CLOUDS

We search for the past binary orbit of the LMC and SMC in the gravitational potential ϕ_G of the flat rotation curve of constant velocity V_G,

$$\phi_G = - V_G^2 \ln r \, , \tag{1}$$

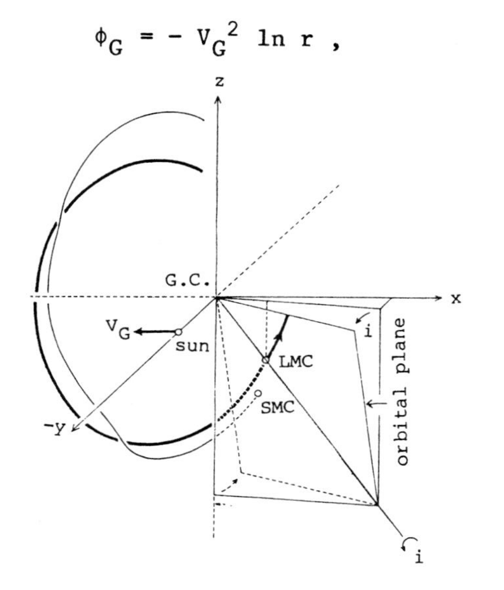

Figure 1. The geometrical relationship of the Galaxy, LMC, and SMC. The Galaxy rotates with the constant velocity V_G. The orbital plane of the Magellanic Clouds is inclined by the angle i.

H. van Woerden et al. (eds.), The Milky Way Galaxy, 471–475.

where r is the distance from the Galactic Centre (see Rubin et al. 1978; Blitz 1979). The geometrical relationship among the Galaxy, LMC and SMC, and some necessary parameters to be employed are given in Figure 1 and Tables 1 and 2. The quantities with suffixes G, L, and S refer to those associated with the Galaxy, LMC and SMC, respectively. When the superscript ^{0} is attached to the position vectors, $\mathfrak{r}_L$ and $\mathfrak{r}_S$, and the velocities, $\mathfrak{v}_L$ and $\mathfrak{v}_S$, they represent the respective values at t=0 (at present).

Table 1. Observed positions x, y, z and line-of-sight velocities of the LMC and SMC for various assumed values of $r_\odot$ and V_G*.

Position	$r_\odot$ = 10 kpc $x^0_{L,S}$	$y^0_{L,S}$	$z^0_{L,S}$ (kpc)	$r_\odot$ = 9 kpc $x^0_{L,S}$	$y^0_{L,S}$	$z^0_{L,S}$ (kpc)
LMC	42.9	-2.4	-28.3	42.9	-1.4	-28.3
SMC	37.4	14.3	-44.6	37.4	15.3	-44.6

	V_G = 230 km s^{-1}	250 km s^{-1}	280 km s^{-1}
	Line-of-sight velocity (km s^{-1})		
LMC	70	54	29
SMC	14	2	-16

* $r_\odot$ is the galactocentric distance of the Sun, V_G the velocity of the assumed flat rotation curve.

Table 2. Assumed masses for the LMC and SMC ($M_\odot$)

	V_G = 230 km s^{-1}	250 km s^{-1}	280 km s^{-1}
LMC	2×10^{10}	6.1×10^9	1.6×10^{10}
SMC	2×10^9	1.5×10^9	0.4×10^{10}

We integrate backward the equations of motion of the LMC and SMC,

$$d^2\mathfrak{r}_L/dt^2 = \partial\phi(\mathfrak{r}_L)/\partial\mathfrak{r}_L + \partial\phi(\mathfrak{r}_L - \mathfrak{r}_S)/\partial\mathfrak{r}_L + \mathfrak{F}_L , \quad (2)$$

and

$$d^2\mathfrak{r}_S/dt^2 = \partial\phi(\mathfrak{r}_S)/\partial\mathfrak{r}_S + \partial\phi(\mathfrak{r}_S - \mathfrak{r}_L)/\partial\mathfrak{r}_S + \mathfrak{F}_S , \quad (3)$$

with the initial (present) positions, $\mathfrak{r}^0_{L,S}$, given in Table 1 and various assumed velocities, $\mathfrak{v}^0_{L,S}$, whose line-of-sight components must be, however, identical with the observed values in Table 1. The dynamical frictions (Chandrasekhar 1942) on the LMC and SMC, $\mathfrak{F}_{L,S}$, caused by the dark halo (Tremaine 1976; Murai and Fujimoto 1980) are included in the calculation.

Figures 2a to 2e show some of our binary orbits in the form of $|\mathfrak{r}_L - \mathfrak{r}_S|$ ($\lesssim$D), enduring for the past 10^{10} years, where D denotes the perigalactic distance of the LMC orbit. We find easily that $|\mathbf{r}_L - \mathbf{r}_S|$ became temporarily as small as a few kpc at $t=-2\times10^8$ years, so the SMC made a close encounter with the LMC about 200 million years ago.

2. A VERY POSSIBLE COLLISION BETWEEN THE LMC AND SMC OF 2×10^8 YEARS AGO

We have chosen systematically a set of initial velocity values $\mathbf{v}_S^0$ for integrating equations (2) and (3). The result can be summarized in the form of a "capture window" in $\mathbf{v}_S^0$-space; inside this window the initial (present) velocity guarantees the bound state of the SMC and the LMC for 10^{10} years (Fujimoto and Sofue 1976, 1977; Murai and Fujimoto 1980). Five binary orbits are given in figures 2a to 2e in order to show in how many cases the last collision occurred in our computed orbits for $\mathbf{v}_S^0$ in the capture window. In fact only rarely are realized such orbits as in Figure 2e where the last collision did not occur.

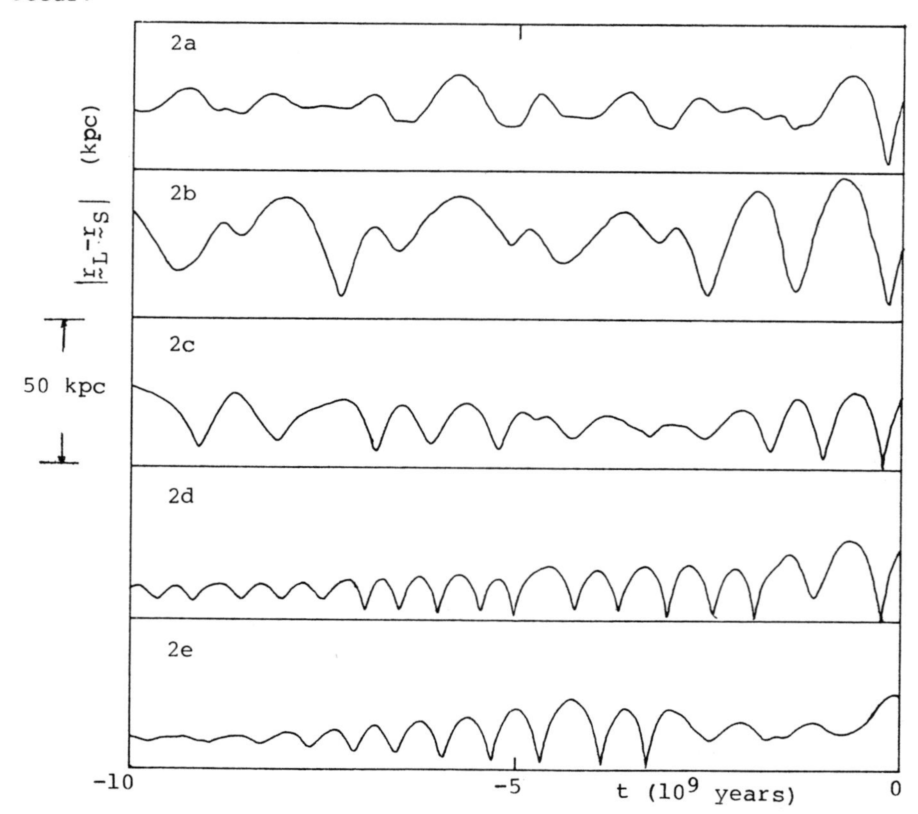

Figures 2a to 2e. The time variation of $|\mathbf{r}_L - \mathbf{r}_S|$ for various initial velocities. The collision is found at $t = -2\times10^8$ years (2a to 2d). The main parameters are: $M_L = 2\times10^{10}\ M_\odot$, $M_S = M_L\times10^{-1}$, $i = 0^0$, $D = 50$ kpc and $V_G = 250$ km s^{-1}.

3. THE MAGELLANIC STREAM AND THE BINARY ORBIT OF THE LMC AND SMC

We now examine whether these binary orbits may be responsible for the Magellanic Stream (MS) in the time-dependent gravitational potential due to the Galaxy, LMC and SMC. We conduct the particle simulation developed by Toomre and Toomre (1972), and show two typical results in Figures 3 and 4 as projected onto the plane of sky and superimposed on the MS. It is found that the binary orbit which went through the last collision can reproduce the MS (Figure 3), whereas that without the collision cannot do it (Figure 4). Since this tendency holds for other parameters in Tables 1 and 2 (except for V_G= 280 km s^{-1}), we consider that our collision of 200 million years ago is a very possible event between the LMC and SMC.

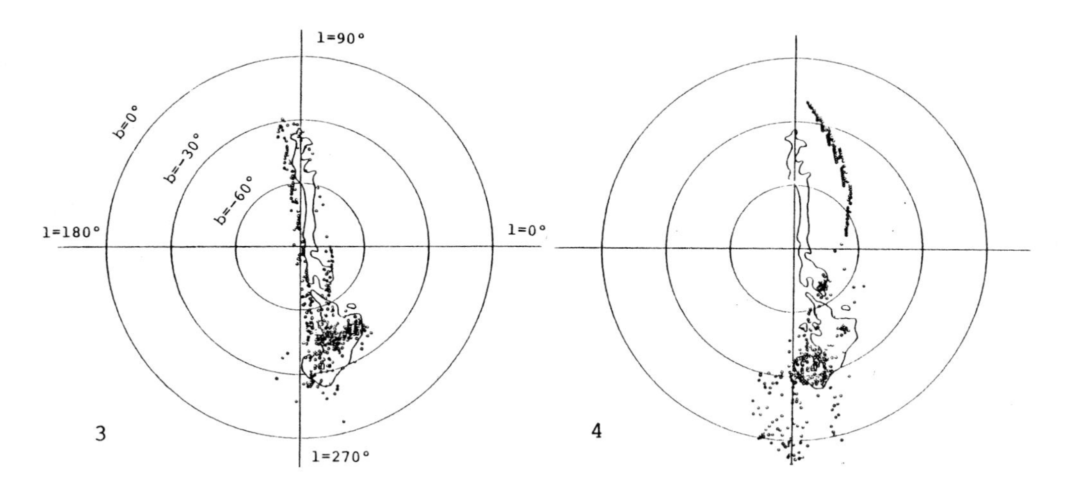

Figures 3 and 4. The Magellanic Stream (outline, Mathewson et al. 1974), and the particle distributions for the orbits in Figures 2a and 2e.

Finally we refer to the spatial distributions of the neutral hydrogen gas (Hindman et al. 1963), the plane of optical polarization of bright stars (Mathewson and Ford 1970; Schmidt 1969, 1976), and Hα emission (Johnson et al. 1982) in the Magellanic Clouds. They are observed along the LMC-SMC line, as if the SMC has stretched the common magnetized hydrogen gas as it departs from the LMC after the collision.

REFERENCES

Blitz, L.: 1979, Astrophys. J. Letters 231, L115
Chandrasekhar, S.: 1942, "Principles of Stellar Dynamics" (Dover, New York)
Fujimoto, M. and Sofue, Y.: 1976, Astron. Astrophys., 47, 263
Fujimoto, M. and Sofue, Y.: 1977, Astron. Astrophys., 61, 199

Hindman, J.V., Kerr, F.M. and McGee, R.X.: 1963, Australian J. Phys., 16, 570
Johnson, R.G., Meaburn, J. and Osman, A.M.J.: 1982, Mon. Not. Roy. Astr. Soc., 198, 985
Mathewson, D.S., Cleary, M.N., and Murray, J.D.: 1974, Astrophys. J., 190, 291
Mathewson, D.S. and Ford, V.L.: 1970, Astrophys. J. Letters, 160, L43
Murai, T. and Fujimoto, M.: 1980, Publ. Astron. Soc. Japan, 32, 581
Rubin, V.C., Ford, W.K., Jr., and Thonnard, N.: 1978, Astrophys. J. Letters, 225, L107
Schmidt, Th.: 1969, Astron. Astrophys., 6, 294
Schmidt, Th.: 1976, Astron. Astrophys. Suppl., 24, 357
Toomre, A. and Toomre, J.: 1972, Astrophys. J., 178, 623
Tremaine, S.D.: 1976, Astrophys. J., 203, 72

DISCUSSION

H. van Woerden: Does your work imply that the Magellanic Stream is a tidal tail formed in the collision of the Large and Small Magellanic Clouds? If so, why is the Magellanic Stream at right angles to the galactic plane?

Fujimoto: No, the Magellanic Stream is due to the tidal interaction between the LMC and SMC in the past $\sim 2 \times 10^9$ years. The collision is a very recent dynamical event, in which the Magellanic Clouds would be much perturbed. The Magellanic Stream may be in the orbital plane of the Magellanic Clouds, which is perpendicular to the galactic plane, or the orbital plane is perpendicular to the line joining the Sun and the Galactic Centre.

On the lakes - front to back and left to right: Van Driel and Schwering, Judy Young and Seiden, Chanda Jog and Fujimoto, Anneke van Albada and Okuda, Mieke Oort, Kormendy, Oort (standing). LZ

TIDAL INTERACTIONS BETWEEN THE GALAXY AND THE MAGELLANIC CLOUDS

N. Ramamani, T. Meinya Singh and Saleh Mohammed Alladin
Centre of Advanced Study in Astronomy, Osmania University,
Hyderabad - 500 007 India

The merging time and the disruption time in a binary galaxy system are analytically obtained under the Adiabatic Approximation (AA). Applications are made to the Galaxy-LMC (Large Magellanic Cloud) pair.

Estimates for the time of merging of the Magellanic Clouds with the Galaxy were made earlier by Tremaine (1976) using the standard dynamical-friction formula, and by Alladin and Parthasarathy (1978) under the Impulsive Approximation (IA). In the latter case, it is assumed that the speeds of the stars in the galaxies may be neglected in comparison with the orbital speed of the pair. Although this assumption is good when the relative motion of the galaxies is hyperbolic, it becomes worse as the orbital motion becomes slower. In the present paper, we have considered the other extreme case wherein the motion of the galaxies is neglected (AA). In the case of Galaxy-LMC pair, we find that even the outermost star in LMC has an angular speed, ω_s, larger than the orbital angular speed, ω_g, of the pair assuming circular orbits. Hence the approximation $\omega_s >> \omega_g$ (AA) is better than $\omega_s >> \omega_g$ (IA).

Using the geometry given in Avner and King (1967), who used this approximation earlier in their study of the galactic warp, the velocity increment of a star is obtained by integrating the tidal force on it over a period of the star. Analytical expressions are derived from it for the average increase in the binding energy/unit mass $dU(R_h)/dt$ at the median radius, R_h, of a galaxy and the average decrease in the orbital energy of the pair dE/dt, assuming that the two galaxies are spherically symmetric, non-penetrating, the stars move in circular orbits and have a circularly symmetric distribution of velocity vectors. The first term in the tidal force which contributes to $dU(R_h)/dt$ in the case of IA, contributes nothing to it secularly. However, the subsequent term does contribute a small amount.

We obtain from the dominant non-zero term in the tidal force, the disruption and merging times:

H. van Woerden et al. (eds.), The Milky Way Galaxy, 477–478.

$$t_d = \frac{|U|}{dU(R_h)/dt} = \frac{0.0136}{\sqrt{G}} \frac{M^{1.5}}{M_1^2} \frac{D^8}{R_h^{6.5}}$$

$$t_m = \frac{E}{dE/dt} = \frac{0.048}{\sqrt{G}} D^7 \left[\frac{M_1}{M} \int_0^R \frac{r^{5.5}}{\sqrt{M(r)}} \frac{dM}{dr} dr + \frac{M}{M_1} \int_0^{R_1} \frac{r_1^{5.5}}{\sqrt{M_1(r_1)}} \frac{dM_1}{dr_1} dr_1 \right]^{-1}$$

where M_1 and M are masses of the perturbing and the test galaxies, D the separation of the galaxies assumed constant, r the distance of the star from the galactic centre, and U is the binding energy/unit mass. Thus the times of merging and disruption increase rapidly with increasing D, as expected from the numerical work of Lin and Tremaine (1982) and T.R. Bontekoe (private communication).

We take the mass of LMC as 5×10^9 $M_\odot$ within 5 kpc (Feitzinger, 1980) and the mass distribution as that of a polytrope n=2. We obtain for the Galaxy-LMC pair, with D=55 kpc, $t_m \sim 6\times10^9$ yrs with mass models of Rohlfs and Kreitschmann (1981), and Ostriker and Caldwell (1979) truncated at 50 kpc, for the Galaxy. Schmidt's (1965) model for the Galaxy which does not have such a massive halo gives $t_m \sim 6\times10^{10}$ yrs. The merging rate at this distance is considerably faster than the disruption rates of the galaxies. The ratio of the tidal force to the main force indicates that the outermost parts of the Galaxy will be more affected than the outermost parts of the LMC.

A comparison of the results obtained in the present treatment with those obtained under IA leads to the conclusion that as the orbital motion of the galaxies becomes slower, the density ratio of the galaxies becomes more important than their mass ratio in determining the effects of mutual disruption. That a small satellite galaxy can appreciably influence the structure of a big, extended galaxy is also being suggested by the fact that galaxies with spectacular spiral structure are generally accompanied by companions (Toomre, 1981).

REFERENCES

Alladin, S.M. and Parthasarathy, M.: 1978, M.N.R.A.S. 184, pp 871-883
Avner, E.S. and King, I.R.: 1967, Astron. J. 72, pp 650-662
Feitzinger, J.V.: 1980, Space Sci. Rev. 27, pp 35-105
Lin, D.N.C. and Tremaine, S.D.: 1982, Astrophys. J. 264, pp 364-372
Ostriker, J.P. and Caldwell, J.A.R.: 1979, IAU Symp. 84, pp 441-450
Rohlfs, K. and Kreitschmann, J.: 1981, Astrophys. and Space Sci., 79, pp 289-319
Schmidt, M.: 1965, "Galactic Structure" (eds. A. Blaauw and M. Schmidt), Univ. of Chicago, pp 513-530
Toomre, A.: 1981, "The Structure and Evolution of Normal Galaxies" (ed. S.M. Fall and D. Lynden-Bell), Univ. of Cambridge, pp 111-136
Tremaine, S.D.: 1976, Astrophys. J. 203, pp 72-74

SECTION III.2

DYNAMICS OF THE DISK

Thursday 2 June, 1440 - 1620

Chairman: C.A. Norman

Norman dons a tie at conference dinner. To his left: Secretary Marijke van der Laan and L. Bronfman

LZ

DYNAMICAL EVOLUTION OF THE GALACTIC DISK

Roland Wielen and Burkhard Fuchs
Institut für Astronomie und Astrophysik,
Technische Universität Berlin, Germany

ABSTRACT

After some general remarks on the dynamical evolution of the galactic disk, we review mechanisms which may affect the velocities of disk stars: stochastic heating, deflections, adiabatic cooling or heating. We compare the observed velocities of nearby disk stars with theoretical predictions based on the diffusion of stellar orbits.

1. INTRODUCTION

In this paper we discuss some aspects of the dynamical evolution of the axisymmetric part of the stellar galactic disk. We shall not consider here the evolution of modes in the disk, such as bars or spiral density waves.

Direct observational evidence for a dynamical evolution of a galactic disk is provided by the increase of the velocity dispersion $\sigma(\tau)$ of disk stars with age τ, observed for nearby disk stars, and by the age-dependence of the z distribution of disk stars, observed both for our Galaxy and for external edge-on galaxies. The observable age-dependence at least reflects the unobservable time-dependence of properties of galactic disks.

There is also some indirect observational and theoretical evidence for the dynamical evolution of the galactic disk: (1) Infall of gas: The galactic disk has very probably been formed within the galactic halo and corona by infall of gas. The dynamical effects of such an infall depend strongly on the time-scale of the infall. If the gaseous disk was formed very rapidly, then this would have caused a violent initial phase in the evolution of the disk, while the later dynamical evolution would not be affected. A slow formation of the disk by a rather steady infall, say over 10^{10} years, would produce important effects all the time (Gunn 1982). Estimates on the rate of infall can be obtained from observed high-velocity clouds or from the interpretation of the chemical evolu-

H. van Woerden et al. (eds.), The Milky Way Galaxy, 481–490.

tion of the disk. (2) Star formation alone can produce changes in the stellar and gaseous mass fractions, and such variations may drive a dynamical evolution of the disk, especially in the z direction. (3) Instabilities in the galactic disk, such as bars, spiral density waves, warps, produce often dynamical effects in the underlying disk: an increase of the random velocity dispersion of disk stars ('heating' of the disk), a redistribution of matter (especially of gas by bars), a change of the distribution of angular momentum in the disk by the non-axisymmetric perturbations. (4) The influence of the environment of the galactic disk on the dynamical evolution is difficult to assess: The mean gravitational field of the luminous halo and the dark corona is probably rather constant in time. If, however, the corona contains a very large number of massive black holes, then the heating of the stellar disk by such penetrating objects would be important (Ostriker 1983). Tidal effects of passing galaxies are of minor importance for the evolution of the inner parts of the disk, probably also for the solar neighbourhood.

2. RELAXATION IN THE GALACTIC DISK

The random peculiar velocities of disk stars may have been affected by the following mechanisms: stochastic heating, deflections, and adiabatic cooling or heating.

2.1 Stochastic heating

Irregularities in the galactic gravitational field produce a steady increase of the velocity dispersion σ of disk stars with the age τ of the stars. Such an increase is well observed for nearby stars (Wielen 1974, 1977).

The basic physical process responsible for the observed stochastic heating of disk stars has not been identified with certainty up-to-now: (1) Encounters between stars are known to be completely inefficient (e.g. Chandrasekhar 1960). (2) Encounters of stars with massive interstellar clouds have first been investigated by Spitzer and Schwarzschild (1951, 1953). However, even the observed giant molecular clouds seem to be insufficient for explaining the observed heating of nearby disk stars (see e.g. Lacey 1983, Villumsen 1983), mainly because these clouds are too rare at the distance of the Sun from the galactic center. (3) If the dark corona of the Galaxy mainly consists of massive black holes, with individual masses of about 10^6 $M_\odot$, this would explain the observed heating of disk stars and especially the observed age-dependence of $\sigma(\tau)$, namely $\sigma(\tau) \propto \sqrt{\tau}$ for old disk stars (Ostriker 1983). (4) One of the most promising mechanisms for the heating of the disk is the fluctuating gravitational field provided by transient instabilities in the disk, such as local Jeans instabilities, wavelets, transient spiral arms (see e.g. Carlberg and Sellwood 1983).

Since the basic physical mechanism for the heating of the disk is not well known at present, the phenomenological description of the heating process by the theory of orbital diffusion seems to be rather adequate (Wielen 1977, Wielen and Fuchs 1983). In this theory, the heating of the disk is basically described by a diffusion process in velocity space, and the diffusion coefficient D is empirically determined from the observed age-dependence of the velocity dispersion of nearby stars (Wielen 1977). Some results of the theory will be discussed in detail in Section 3.

2.2 Deflections

Deflections are random changes in the direction of the velocity vector of a star. Similar to the stochastic heating, deflections may be caused by irregularities in the galactic gravitational field, e.g. due to clouds (Lacey 1983), spiral arms etc.. The overall importance of deflections is that the deflections may transfer energy (and energy changes) between the motions of a star perpendicular and parallel to the galactic plane. Therefore, the axial ratios of the velocity ellipsoid of disk stars may be governed by deflections (Lacey 1983) rather than by the heating mechanism. In Chandrasekhar's notation (1960), the time-scale for heating is the relaxation time T_E, while the relaxation time T_D is the time-scale for deflections. If the velocity dispersion of clouds is much smaller than that of the field stars, then we find for encounters of stars with clouds that T_D is much smaller than T_E. This means that deflections due to certain irregularities in the galactic gravitational field may be of primary importance for the axial ratio of the velocity ellipsoid even if the heating effect of the same irregularities is nearly negligible. Giant molecular clouds lead to a deflection time-scale T_D of about 10^9 years for nearby stars, slightly longer than the observed heating time-scale, but probably of some importance for the axial ratio of the velocity ellipsoid.

2.3 Adiabatic heating or cooling

Slow changes in the regular gravitational field of the galactic disk produce adiabatic changes in the velocities of the stars. This adiabatic heating or cooling of the disk is probably strongest in the z direction, perpendicular to the galactic plane, because the disk is nearly self-gravitating in z, but is heavily supported in the radial direction by the combined gravitational field of halo and corona.

We shall discuss here two typical examples of adiabatic changes of a galactic disk which is assumed to be essentially self-gravitating in the z direction: (1) Adiabatic cooling due to stochastic heating: The stochastic heating increases primarily the velocity dispersion σ_W of the disk stars. The higher velocity dispersion σ_W leads to a larger thickness H of the disk in the z direction, thereby lowering the z force K_z of the disk. The adiabatic decrease of K_z causes a corresponding decrease in the W motion of the disk stars, thus decreasing σ_W. In total, the effect of the stochastic heating in σ_W is partially compensated by the adiabatic cooling (see also Section 3). (2) Adiabatic heating due to infall: Infall increases the surface density μ of the disk. Then the thickness H of the disk decreases for a given velocity dispersion σ_W, and the force K_z increases both because of the higher mass and the smaller thickness of the disk. The W motions of the stars increase adiabatically due to the increase in K_z. Hence, the infall leads finally to an adiabatic heating of the disk stars, i.e. an increase of σ_W.

While the adiabatic heating and cooling discussed above affect primarily the z motions of the stars, i.e. the velocity dispersion σ_W, deflections may transfer this energy change from the z direction partially to the motions parallel to the galactic plane, i.e. the velocity dispersions σ_U and σ_V may be changed too.

3. DIFFUSION OF STELLAR ORBITS

For an introduction into the theory of the diffusion of stellar orbits, we refer to our earlier papers (Wielen 1977, Wielen and Fuchs 1983). We shall use here the same notations as Wielen and Fuchs (1983). In that paper, we have already presented a solution of the appropriate Fokker-Planck equation for a 'standard case' using the following assumptions: (1) constant isotropic diffusion coefficient D_0; (2) no radial variation of the distribution function in phase space, $f(U,V,W,z,t)$; (3) constant star formation rate $S(t_f)$; (4) linear K_z force, $\ddot{z} = -\omega_z^2 z$; (5) no adiabatic cooling in z, i.e. a time-independent K_z force; (6) no infall. Our theoretical results for the standard case are already in good agreement with the observed velocity distribution of nearby disk stars, represented by 317 McCormick K+M dwarfs in Gliese's catalogue (Gliese 1969, Jahreiß 1974). We shall now discuss some modifications of the standard case.

The introduction of a more realistic, non-linear K_z force, which agrees essentially with that derived by Oort (1965), produces only minor changes in the predicted velocity distribution of disk stars at $z = 0$ (Fig. 1). The predicted values of the velocity dispersions σ_U, σ_V and σ_W as functions of the height z are slightly smaller than for the standard case (Fig. 2). The predicted density $\rho(z)$, normalized at $z = 0$, has stronger wings at higher z (Fig. 2).

We shall now take into account the adiabatic cooling in the z direction. For harmonic oscillations of stars in the z direction, the adiabatic invariant I_z is given by

$$I_z = E_z(t)/\omega_z(t) = \text{const.} \quad , \qquad (1)$$

where E_z is the energy of a star in the z motion and ω_z is the frequency of oscillation in z. For non-linear K_z forces, the form of the adiabatic invariant is more complicated. For simplicity, we use Eq. 1 even for such non-linear K_z forces, since the majority of stars still moves in the quasi-linear regime, and we evaluate ω_z always at $z = 0$.

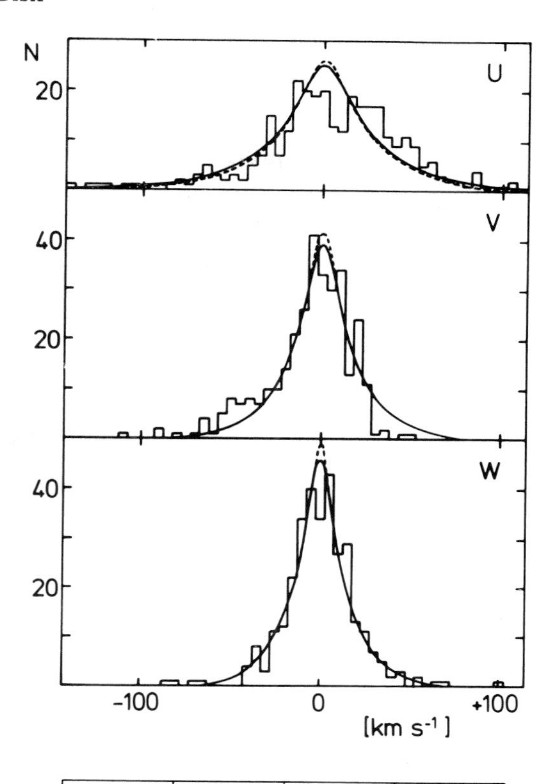

Figure 1. Velocity distribution of nearby disk stars at z=0. Observations: histograms. Linear K_z : solid curves. Non-linear K_z : dashed curves.

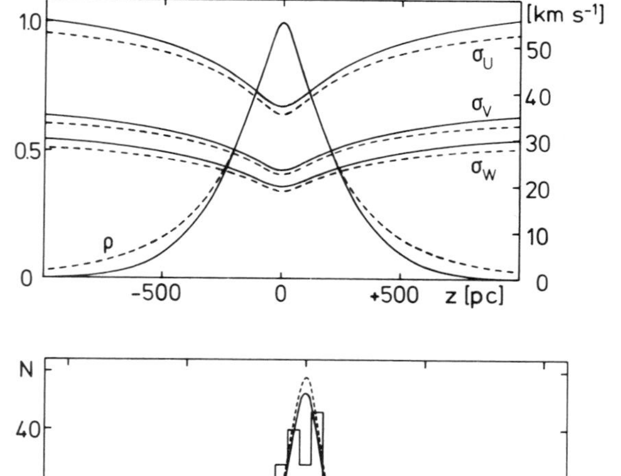

Figure 2. Predicted z-dependence of the overall velocity dispersions σ_U, σ_V, σ_W, and of the overall space density ρ. Linear K_z: solid curve. Non-linear K_z: dashed curve.

Figure 3. Distribution of W velocities of nearby disk stars at z=0. Observations: histogram. Standard case: solid curve. With adiabatic cooling: dashed curve.

The Fokker-Planck equation can be approximately solved for a linear K_z force with $\omega_z(t)$ depending slowly on the time t, by a transformation after which I_z (Eq. 1) occurs as one of the independent variables. The resulting distribution function f for a generation of stars formed at time t_f is essentially the same as obtained for the standard case (Eq. 14 of Wielen and Fuchs). We have only to replace the velocity dispersion $\sigma_W(\tau)$ by

$$\sigma_W^2(t,t_f) = \omega_z(t) \left(\sigma_{W,0}^2 \omega_z^{-1}(t_f) + \alpha_W D_0 \int_{t_f}^{t} \omega_z^{-1}(t')dt' \right) \quad , \quad (2)$$

with $\alpha_W = 1/2$. Differentiating Eq. 2 with respect to t gives

$$d\sigma_W^2/dt = \alpha_W D_0 + \sigma_W^2 \, (d\omega_z/dt)/\omega_z \quad . \quad (3)$$

The second term on the right-hand side of Eq. 3 can be derived directly from Eq. 1 by noting that the energy in the z direction, averaged over a stellar generation, $< E_z >$, is proportional to σ_W^2.

If we assume that the disk is self-gravitating in the z direction, we can derive $\omega_z(t)$ from known quantities, by evaluating the Poisson equation at z = 0:

$$\omega_z^2 = 4\pi G\rho(z=0) \quad . \quad (4)$$

For a linear K_z force and a single generation of stars, the density ρ at z = 0 can be derived by integrating Eq. 14 of Wielen and Fuchs (1983) over the velocities U, V, W. The result is

$$\rho(z=0) = (2\pi)^{-1/2} \mu\omega_z/\sigma_W \quad . \quad (5)$$

Combining Eqs. (4) and (5), we obtain

$$\omega_z(t) = (8\pi)^{1/2} G\mu/\sigma_W(t) \quad . \quad (6)$$

The same result, with a slightly different numerical constant $(2^{1/2}\pi)$, is found for a self-gravitating isothermal stellar disk which obeys Eqs. (19) and (20) of Wielen and Fuchs (1983) and

$$\rho(z=0) = \mu/(2H) \quad . \quad (7)$$

From Eq. 6, we derive, independent of the value of the numerical constant,

$$(d\omega_z/dt)/\omega_z = -(d\sigma_W/dt)/\sigma_W \quad . \quad (8)$$

Inserting Eq. 8 into Eq. 3, we obtain

$$d\sigma_W^2/dt = (2/3)\alpha_W D_0 \quad . \quad (9)$$

The adiabatic cooling reduces therefore the effect of the diffusion in W by factor of 2/3 for a single generation of stars. Hence we find in this case

$$\sigma_W^2(\tau) = \sigma_{W,0}^2 + (1/3)D_0\tau \quad . \quad (10)$$

For a disk made up of many stellar generations, we have to integrate Eq. 5 over all generations, i.e. over t_f with an appropriate star formation rate $S(t_f)$, in order to derive the total density $\rho(t)$ at z = 0.

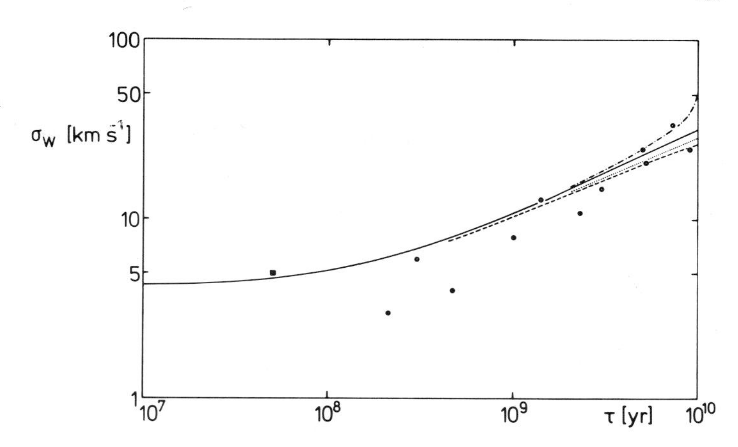

Figure 4. Velocity dispersion σ_W, averaged over z, as a function of age τ. Observations: symbols. Standard case: solid curve. With adiabatic cooling: dashed curve. With infall: dotted (low rate) and dash-dotted (high rate).

Then we have to find $\omega_z(t)$ from Eq. 4 and finally to use Eq. 2 for obtaining the new velocity dispersion σ_W for each stellar generation. In order to be more realistic, we have also added a gaseous component to the disk. The velocity dispersion of the gas is kept constant. Stars are formed out of the gas according to a chosen star formation rate, thereby decreasing the amount of gas in the disk. The results of such a calculation are shown in Figs. 3 and 4. The adiabatic cooling in z has only minor effects for both the velocity distribution in W at z = 0 (Fig. 3) and the age-dependence of the velocity dispersion σ_W (Fig. 4). As long as we do not invoke deflections, the distribution of the velocity components U and V, integrated over z, and the age-dependence of the velocity dispersions $\sigma_U(\tau)$ and $\sigma_V(\tau)$, averaged over z, are not affected by the adiabatic cooling in z (nor by a non-linear K_z force).

We shall now consider the effect of infall. To simulate the infall, we add gas to the disk according to a constant infall rate. We have used either a 'low rate' of about 2 $M_\odot/(10^9$ years pc$^2)$ or a 'high rate' of about 7 $M_\odot/(10^9$ years pc$^2)$ over 10^{10} years. The present total disk surface density μ adopted is about 70 $M_\odot$/pc^2. Besides the adiabatic heating due to the infall, we have also included the adiabatic cooling caused by the stochastic heating and the depletion of the gas by star formation. For both infall rates, the resulting velocity distribution in W at z = 0 is so close to the standard case that the difference would be hardly visible in Fig. 3. The age-dependence of the velocity dispersion $\sigma_W(\tau)$ is also not dramatically affected by infall (Fig. 4). Only in the case of the high infall rate, the heating of the older stars is slightly higher than the observed values.

We conclude that the simple standard case of stellar diffusion gives already a rather good agreement between theory and observations for both the velocity distribution at z = 0 and the age-dependence of the velocity dispersion of nearby disk stars. This indicates that the stochastic heating is the main dynamical mechanism, while other processes like adiabatic cooling and infall are dynamically only of secondary importance.

The isotropic diffusion provides a simple explanation for the observed ratio of the velocity dispersion σ_W, perpendicular to the galactic plane, to the dispersions parallel to the plane, σ_U and σ_V (Wielen 1977). The observed values for the McCormick K+M dwarfs in Gliese's Catalogue, as the most representative sample of nearby stars, averaged over z, are σ_U = 47 km/s, σ_V = 29 km/s, σ_W = 25 km/s. If we normalize the predicted dispersions always so that $\sigma_U^2 + \sigma_V^2$ is equal to the observed value, then the following values for σ_W are predicted: (1) Isotropic diffusion without adiabatic cooling (standard case) predicts σ_W = 25 km/s, in perfect agreement with the observations. (2) Adiabatic cooling lowers the predicted value slightly to 22 km/s for the case of a constant star-formation rate (not to 20 km/s as would be expected from the factor 2/3 in Eq. 9, which is valid only if all the stars have the same age and if there is no gas). (3) Deflections by clouds (Lacey 1983) would predict σ_W = 33 km/s, if other sources of stochastic heating and the adiabatic cooling are neglected. (4) In reality, all three effects, namely stochastic heating, deflections and adiabatic cooling, will jointly determine σ_W. The predicted value of $\sigma_W/\sqrt{(\sigma_U^2 + \sigma_V^2)}$ would then depend strongly on the direction-dependence of the stochastic heating and of the deflections and on the relative time-scales of the heating and of the deflections. One should remember here that the main sources of stochastic heating and of deflections may be quite different ones. We conclude that there is probably no significant discrepancy between the observed and predicted axial ratio of the velocity ellipsoid of common nearby stars, if all the uncertainties are taken into account.

REFERENCES

Carlberg, R.G., Sellwood, J.A.: 1983, IAU Symposium No. 100, p. 127.
Chandrasekhar, S.: 1960, Principles of Stellar Dynamics, Dover Publ., New York.
Gliese, W.: 1969, Veröffentl. Astron. Rechen-Inst. Heidelberg No. 22.
Gunn, J.E.: 1982, in Astrophysical Cosmology, eds. H.A. Brück, G.V. Coyne, M.S. Longair, Pontificia Academia Scientarium, Citta del Vaticano, p. 233.
Jahreiss, H.: 1974, Diss. Naturwiss. Gesamt-Fak. Univ. Heidelberg.
Lacey, C.G.: 1983, Monthly Not. Roy. Astron. Soc. (in press)
Oort, J.H.: 1965, Stars and Stellar Systems 5, 455.
Ostriker, J.P.: 1983, private communication.
Spitzer, L., Schwarzschild, M.: 1951, Astrophys. J. 114, 385.
Spitzer, L., Schwarzschild, M.: 1953, Astrophys. J. 118, 106.
Villumsen, J.V.: 1983, Astrophys. J. (in press).
Wielen, R.: 1974, IAU Highlights of Astronomy 3, 395.
Wielen, R., :1977, Astron. Astrophys. 60, 263.
Wielen, R., Fuchs, B.: 1983, in Kinematics, Dynamics and Structure of the Milky Way, ed. W.L.H. Shuter, D. Reidel Publ. Co., Dordrecht, p. 81.

DISCUSSION

F.H. Shu: Toomre's Q parameter as a stability index is based on an analysis which assumes only a single Schwarzschild distribution. If you have a superposition of Schwarzschild distributions, then the Q parameter loses its precise original meaning. In particular, a formal value >1 may not guarantee even axisymmetric stability, because the population of stars with low velocity dispersion contribute most importantly to the instability mechanism.

Wielen: I agree completely with you. One has to figure out the meaning of Q for a sequence of generations of stars.

C.C. Lin: In my review I shall show that there may be a variety of reasons for Q to exceed 1.

On another matter: how does your rate of growth of velocity dispersions compare with that obtained by Carlberg and Sellwood through numerical modelling? Do they find more rapid evolution? Does their evolution follow the same linear trend?

Wielen: In principle they find similar trends, although their results show a large scatter. Villumsen will discuss the diffusion by molecular clouds, and he also finds agreement.

R.G. Carlberg: The value of Q will depend on the assumed surface density of the disk. You have used a value of 70 $M_\odot/pc^2$, while Schmidt in his review (section II.1) gave 48 $M_\odot/pc^2$.

Wielen: Schmidt uses a value of 50 $M_\odot/pc^2$ in his model. The count of known stars and gas amounts to only 40 $M_\odot/pc^2$. The 70 $M_\odot/pc^2$ adopted by me agrees with the estimate of the expected surface density in Schmidt's review, which is based on the local determination of K_z by Oort, after allowance for a contribution of 10 $M_\odot/pc^2$ by the dark corona. This estimate, then, includes "hidden mass". It would imply a value of Q greater than 2.

L. Blitz: You argued that results for the diffusion calculations do not change very much, as long as the star formation rate is less than a factor ~5 greater than it is now. But I wish to point out that in the past the star formation rate in the disk may indeed have been much higher. We know now that most stars, essentially all stars probably, form in molecular clouds. Since stars with mass less than 1 $M_\odot$ essentially lock up all that mass for a Hubble time, one can argue that a Hubble time or half that time ago, there may have been 5 or even 10 times as much molecular material in the Galaxy as there is now. If the star formation rate goes as the mass of molecular gas, the effects on the diffusion calculations might be strong. Also, when one looks at the diffusion by molecular clouds in the solar neighbourhood, one should consider not the current surface density of such clouds, but rather a time average over a substantial period.

Wielen: The observed increase of the velocity dispersion with age over the last 10^8 years is mainly determined by the present surface density of molecular clouds in the solar neighbourhood. Hence the discrepancy discussed remains at least for the present situation. A variation of the star formation rate does not affect the relation between velocity dispersion and age, but the integrated velocity distribution.

J.P. Ostriker: Wielen has demonstrated that the velocity dispersion increases as $t^{\frac{1}{2}}$, in contradiction to expectations based on existing physical theories of diffusion. This led Lacey, Schmidt and me to look at the unlikely possibility of diffusion due to interaction of disk stars with a dark halo comprised of massive black holes. We find that the correct diffusion rate is obtained if the mass of such holes is 2×10^6 $M_\odot$, and the predicted axial ratio of the velocity ellipsoid is also correct. The major byproduct of such a halo would be the existence at the centre of the Galaxy of one or two black holes dragged into that region by dynamical friction.

M. Iye: The existence of massive black holes (10^3 $M_\odot \lesssim M \lesssim 10^6$ $M_\odot$) in the halo might be observationally confirmed, if they do exist, by searching for gravitational-lensing effects on images of stars caused by such black holes. One way to do this would be monitoring the movement and light variation of stellar images in globular clusters with a spatial resolution of 0.1 arc second or better for a period of tens of years.

Carlberg: How confident are you as to the power of time in the heating rate of stars?

Wielen: That depends mainly on the ages of the stars used, and I don't think that they can be very wrong. I would say that the linear dependence of σ^2 on time is quite established. Even if the diffusion coefficient D varies with velocity, as e.g. proposed by Spitzer and Schwarzschild ($D \propto 1/v$), one can correct that by assuming the constant to be time-dependent. The irregularities in the gravitational field may indeed have been bigger in the past.

EVOLUTION OF THE VERTICAL STRUCTURE OF GALACTIC DISKS

Jens Verner Villumsen
Institute for Advanced Study
Princeton, New Jersey, U.S.A.

Numerical simulations of the evolution of the vertical structure of galactic disks have been performed. The physical mechanism for the evolution is the scattering of stars off Giant Molecular Clouds (GMCs) as proposed by Spitzer and Schwarzschild (1951). A model galaxy consists of a fixed, nearly isothermal halo plus an axisymmetric, thin exponential disk consisting of 1000 stars. A population of GMCs is embedded in the disk. The stars interact with each other via a self-consistent axisymmetric field determined from an expansion in spherical harmonics to twelfth order. The stars scatter off the GMCs that are modelled as soft particles. The equations of motion of the stars and the GMCs are integrated directly to high accuracy. Adiabatic cooling is therefore included implicitly. In order to avoid axisymmetric instabilities, the stellar component is initially relatively hot in the plane of the disk. Nineteen simulations were performed with varying parameters to check the consistency of the results.

The results can be summarized as follows. At a given position in the disk the scale height and z-velocity dispersion grow approximately as the square-root of time. The short axis of the velocity ellipsoid points in the vertical direction and the axial ratio is typically 0.55. The vertical density distribution is well fitted by an isothermal sech^2 solution. The z-velocity distribution is well fitted by a Gaussian distribution. These results are consistent with observations of the structure of the solar neighbourhood (Wielen 1977), but differ strongly from the analytical results by Lacey (1983). Further, the scale height can be constant with radius, if the GMC distribution is more concentrated towards the galactic center than the stellar distribution. The effective mass of GMCs can be enhanced by wakes set up in the stellar component. The scattering efficiency depends strongly on the masses of the GMCs, and the resultant scale height will scale with the square of the GMC masses. This makes it difficult to estimate the scale height induced by an observed GMC population.

Details of this work will be reported elsewhere (Villumsen 1983).

H. van Woerden et al. (eds.), The Milky Way Galaxy, 491–492.

REFERENCES

Lacey, C.G.: 1983, M.N.R.A.S., preprint
Spitzer, L., and Schwarzschild, M.: 1951, Astrophys. J. 114, p. 385
Villumsen, J.V.: 1983, Astrophys. J., 274, p. 632
Wielen, R.: 1977, Astron. Astrophys. 60, p. 263

DISCUSSION

R.J. Allen: You have told us what happens to the stars. What happens to the molecular clouds?

Villumsen: By construction nothing happens to them. They run around happily orbiting the Galaxy, unchanged throughout the simulation. The computer program is made that way. I have tried both infinite lifetimes for the molecular clouds, and finite lifetimes with random dissolution and reconstitution; for the same radial distribution it made no perceptible difference.

S.M. Fall: What were the initial conditions, such as Q and shape of velocity ellipsoid, for the simulations?

Villumsen: The initial Q was 1.1, so that the disk was self-supporting and axisymmetrically stable. The ratio of tangential and radial velocity dispersions was taken as 3:4, based on the Oort constants. Another thing to do would be to study the evolution of a cold population with an initial velocity dispersion of a few km/s, rather than a hot population.

R. Güsten: How do you treat the molecular clouds?

Villumsen: As Plummer models. The potential is $U(r) = -GM\ (r^2+\varepsilon^2)^{-1/2}$, where ε is an assigned size of the clouds. For clouds of 10^6 solar masses, I chose $\varepsilon = 50$ pc. I tried different values of ε, but it made no difference.

HEATING OF STELLAR DISKS BY MASSIVE GAS CLOUDS

Cedric G. Lacey
Institute of Astronomy, Cambridge, England

I have analytically calculated the evolution of the three components of velocity dispersion of the stars in a galactic disk when these are scattered by massive gas clouds, in a generalization of Spitzer & Schwarzschild's (1953) calculation. The principal assumptions made are: (i) The stellar orbits obey the epicyclic approximation. (ii) The gas clouds are massive, long-lived and on circular orbits. (iii) The typical star-cloud encounter time is short compared to the orbital time. (iv) The evolution due to encounters is treated as a diffusion process.

The effect on the stars of encounters with clouds is then given by the standard expressions for the diffusion coefficients $\langle \Delta V \rangle$ and $\langle (\Delta V)^2 \rangle$ (Chandrasekhar 1960). These are used to derive expressions for the rates of change of the epicyclic energies $E_e = \frac{1}{2}(u^2 + \beta^2 v^2)$ and $E_z = \frac{1}{2}(w^2 + \nu^2 z^2)$, where $\beta = 2\Omega/\kappa$ and Ω, κ, ν are the frequencies of circular motion and of horizontal and vertical epicyclic oscillations. The expressions are then averaged over the epicyclic phases and over the stellar distribution function, which is assumed to be always approximately isothermal. Expressed in terms of the independent velocity dispersions σ_u and σ_w, the results are

$$d\sigma_u^2/dt \simeq 2G^2 N_c M_c^2 \ln\Lambda \; K(\alpha,\beta)/(\sigma_u (h_s^2 + h_c^2)^{\frac{1}{2}}) \qquad (1)$$

$$d\sigma_w^2/dt \simeq 2G^2 N_c M_c^2 \ln\Lambda \; L(\alpha,\beta)/(\sigma_u (h_s^2 + h_c^2)^{\frac{1}{2}}) \qquad (2)$$

where $\alpha = \sigma_w/\sigma_u$, N_c and M_c are the number of clouds per unit area and their mass, h_s and h_c are the Gaussian scale-heights of the stars and gas, and $K(\alpha,\beta)$ and $L(\alpha,\beta)$ are dimensionless integrals over the epicyclic phases.

The evolution derived from these equations is in two distinct stages:

(i) Transient Relaxation: The shape of the velocity ellipsoid relaxes to a steady final state with (see also Fig. 1)

$$\sigma_u : \sigma_v : \sigma_w = 1 : 1/\beta : \alpha_s(\beta) \qquad (3)$$

H. van Woerden et al. (eds.), The Milky Way Galaxy, 493–495.

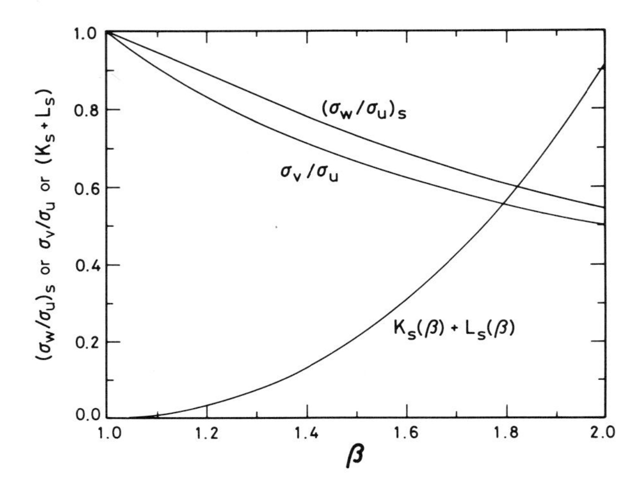

Figure 1. Dependence of velocity-dispersion ratios σ_w/σ_u and σ_v/σ_u and heating rate (as given by $K_s(\beta) + L_s(\beta)$) on rotation-curve shape in steady-heating phase.

(ii) Steady Heating: The total velocity dispersion σ increases steadily on a longer timescale. For $h_s \gtrsim h_c$:

$$d\sigma^2/dt \propto N_c M_c^2 \nu/\sigma^2 \tag{4}$$

Comparison with Observations: (i) Equation (3) predicts $\sigma_u > \sigma_w > \sigma_v$, whereas $\sigma_u > \sigma_v > \sigma_w$ is observed (e.g. $\sigma_u : \sigma_v : \sigma_w = 1 : 0.59 : 0.52$ according to Wielen (1974)). (ii) Equation (4) predicts $\sigma \sim t^{1/4}$ for $N_c M_c^2$ constant, compared to the observed dependence $t^{1/3}$ or $t^{1/2}$ (Wielen 1974), but this is not conclusive since N_c was probably larger in the past. (iii) Using observational determinations for cloud masses and number densities, I compute $\sigma \simeq (10\text{-}40)$ km s^{-1} for the oldest disk stars compared to $\sigma \simeq (60\text{-}80)$ km s^{-1} observed (Wielen 1974). (iv) If one assumes $N_c \propto \mu_D$, M_c = constant, the disk scaleheight is predicted to vary as $h_D \propto \mu_D^{-1/5}$, where μ_D is the disk surface density. This is consistent with the approximately constant scaleheights observed by van der Kruit & Searle (1981) in edge-on galaxies.

In conclusion, (iv) is consistent with the observations, (ii) and (iii) may be, but (i) (the axial ratios of the velocity ellipsoid) is discrepant. This may either mean that some of the assumptions of the theory must be modified, or that a different physical mechanism, such as scattering by spiral density waves or by massive black holes in the galactic halo, dominates the heating. A fuller account of this work appears in Lacey (1984).

REFERENCES

Chandrasekhar, S. 1960, "Principles of Stellar Dynamics" (Dover).
Lacey, C.G. 1984, Mon. Not. R. astron. Soc., in press.
Spitzer, L. and Schwarzschild, M. 1953, Astrophys. J. 118, p. 306.
van der Kruit, P.C. and Searle, L. 1981, Astron. Astrophys. 95, p. 105.
Wielen, R. 1974, Highlights of Astronomy 3 (Dordrecht: Reidel), p. 395.

DISCUSSION

L. Blitz: You use a different value for the velocity dispersion of giant molecular clouds than Villumsen?

Lacey: I assume that the velocity dispersion of the clouds can be neglected compared to that of the stars. Since the observed value is only 3-4 km/s in one coordinate, I think that this is a better approximation to the true situation than that of Villumsen, who assumes a much larger value for the cloud velocity dispersion.

J.V. Villumsen: I do not remember my exact value, but it was taken equal for stars and molecular clouds, and I took $Q = 1.1$.

Blitz, C.A. Norman: So your dispersion would be about 15-20 km/s.

J.H. Oort: I wonder how uncertain the observed value of 60-80 km/s for the velocity dispersion of the oldest disk stars is. These stars have been strongly selected according to proper motion, and the "observed value" may well be too high.

Lacey: This value comes from the oldest-age group in the McCormick stars, ranked in age by Wielen according to their Ca II emission. It is higher than the average for the old disk as a whole.

Oort: It might be good to look into that again sometime.

J.P. Ostriker: I think the disagreement between prediction and observation is really very severe, because for 50 km/s dispersion you need more massive molecular clouds than most people find.

Lacey: The value 40 km/s for the dispersion of the oldest disk stars corresponds to a mass-weighted mean mass $M_c = 10^6\ M_\odot$ and a surface density $N_c M_c = 5\ M_\odot\ pc^{-2}$, as is found by Sanders, Scoville and Solomon (1983) for instance. To obtain a dispersion of 60-80 km/s one would require the past average value of $N_c M_c^2$ to have been (5-15) times its present value.

Ostriker: OK - and you should really compare in your model the observed and predicted values of σ^4, because that is how the diffusion coefficient goes, and then you'd see a big difference between the predicted and observed diffusion coefficient .... (Laughter)

At dinner, clockwise around table: Mrs. Mayor, ?, Schmidt, Ewine van Dishoeck, De Zeeuw, Lacey, Mayor LZ

STELLAR VELOCITY DISTRIBUTION IN THE PRESENCE OF SCATTERING MASSIVE CLOUDS

W. Renz
Department of Theoretical Physics
RWTH Aachen, Templergraben 55, 5100 Aachen, FRG

The purpose of this contribution is to show

1. a systematic perturbation-theoretical method of solving multivariate Fokker-Planck equations (FP eq.) as they appear in galactic dynamics.
2. that it is not only important to consider the cooling of the stellar disk by birth of new stars but also to include stellar death processes in the equation. This leads to different stellar velocity distributions.
3. that the input of new stars into circular orbits as it is done in recent numerical work (Sellwood and Carlberg, 1983) leads not only to the cooling effect wanted but also to an additional damping effect on spiral waves.

The FP approach has already been proposed by several authors (Spitzer and Schwarzschild, 1951; Wielen and Fuchs, 1983) as a useful method to calculate the velocity distribution of stars which undergo the influence of random gravitational forces. Nevertheless people used the Langevin approach in connection with a Gaussian velocity distribution, which may be a quite good approximation in certain cases but is generally inconsistent (Lacey, 1984). For simplicity we assume a constant diagonal diffusion matrix $\underline{\underline{D}}$ (Wielen, 1977). As in our Galaxy the diffusion process is known to be very slow in comparison with the rotation period, we can choose $D/\kappa|\underline{\sigma}^2|$ as the small expansion parameter. κ means the epicyclic frequency and $\underline{\sigma}^2$ the main velocity dispersions of the disk stars. The FP eq. is transformed on action angle variables. By introducing the epicyclic approximation one can reobtain the FP eq. (Wielen and Fuchs, 1983) and also the solution given there as the 1st order solution in our perturbation expansion. We give also 2nd order corrections (Renz, 1983).

In order to model stellar birth and death processes we introduce an additional source term into the FP eq. From an empirical initial stellar mass spectrum and the main-sequence lifetime (Zinnecker, 1981) we can calculate that part of stellar mass which is still left as stars after time t (Figure 1). It turns out that about 50% of the initial mass gets lost in the relevant range of dynamical evolution of the Galaxy. For the further analytical calculations we fit the remaining mass by

(*) $M(t) = \{0.2 + 0.5\exp(-\gamma . t)\} . M(10^7 \text{ yr})$ and $\gamma^{-1} = 1.2 \times 10^9$ yr.

H. van Woerden et al. (eds.), The Milky Way Galaxy, 497–498.

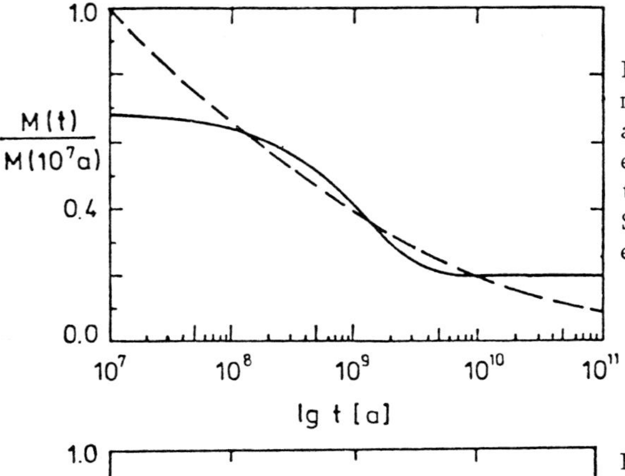

Figure 1. Dashed line:remaining mass M(t) of one generation of stars after time t, derived from an empirical initial mass function and the main-sequence lifetime. Solid line: the approximation of eq. (*) for the same quantity.

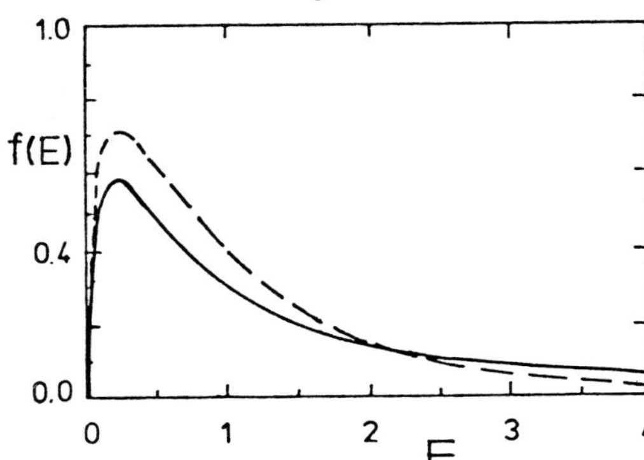

Figure 2. The normalized distribution of epicyclic energies E. The dashed line shows the distribution with only stellar birth processes. The solid line shows the quasi-stationary distribution by taking also stellar death processes into account.

As result we get a quasi-stationary velocity distribution on a time scale of 10^9 yr, which changes slowly only within 10^{10} yr due to the 20% of stars with nearly infinite lifetime (Figure 2). The damping effect can be seen from hydrodynamical equations which derive from the modified FP eq. Details will be presented elsewhere.

REFERENCES

Lacey, C.G.: 1984, M.N.R.A.S., in press
Renz, W.: 1983, Ph. D. Thesis, RWTH Aachen
Sellwood, J.A., Carlberg, R.G.: 1983, preprint, to appear in M.N.R.A.S.
Spitzer, L., Schwarzschild, M.: 1951, Astrophys.J. 114, 385
Wielen, R.: 1977, Astron. Astrophys. 60, 263
Wielen, R., Fuchs, B.: 1983, in "Kinematics, Dynamics and Structure of the Milky Way", ed. W.L.H. Shuter, Dordrecht: Reidel, p. 81
Zinnecker, H.: 1981, Ph.D. Thesis, MPI Garching

WARPS AND HEAVY HALOS

Linda S. Sparke
Institute for Advanced Study, Princeton, N.J. 08540, U.S.A.

The outer gas disk of our Galaxy (and many others) is warped, bending away from the plane defined by the inner disk. The bend begins just outside the solar circle; gas at longitudes $\ell \approx 80^o$ reaches highest above the plane, while material at $\ell \approx 260^o$ lies below it. Short-wavelength ripples are superposed. The orbit of a free particle inclined to the galactic disk precesses at a rate which depends on galactocentric radius; warped structures will tend to do the same, winding the warp into a tight spiral. The large-scale galactic warp has no sense of spirality, and is not obviously of recent origin - why then has it survived?

If the galactic disk had a discrete mode of vertical oscillation, it could vibrate like a drum under its own gravity. Any forces which perturb the disk would excite the bending mode; once a warp had been set up it would persist for many rotation periods. Unfortunately an isolated self-gravitating disk has no discrete bending modes (Hunter and Toomre 1969) if the density falls smoothly to zero at the edge; any initial warp is lost within a couple of revolutions. The galactic rotation curve suggests that over half its mass may not lie in the disk, but form a massive unseen halo. This paper discusses how an axisymmetric (but not spherical) heavy halo affects the warping modes of a galactic disk.

The disk (of surface density falling exponentially with radius within a sharp edge) is approximated as a system of gravitating rings, subject to the halo potential. The halo density ρ_H depends on spherical radius R and polar angle θ as

$$\rho_H = \frac{\rho_o}{R^2 + R_c{}^2} \left[1 - \varepsilon(R) P_2(\cos\theta)\right]; \qquad (R < R_t)$$

$\varepsilon > 0$ in an oblate halo. Modes are sought with the form

$$z(r,\phi,t) + h(r) e^{i(\omega t - \phi)}$$

in cylindrical polar coordinates (r,ϕ,z); material at any given radius

H. van Woerden et al. (eds.), The Milky Way Galaxy, 499–502.

lies above the mean plane on one side of the disk and below that plane on the other, while the warped shape precesses at an angular rate ω.

When the halo is <u>spherical</u> or <u>oblate</u> with $\varepsilon(R)$ increasing no faster than linearly with R, the shape of the most slowly precessing warped mode is found to depend on where the sharp disk edge is taken to lie. The precession rate tends to zero as the disk extends further out. As the edge is made smooth rather than sharp, the frequencies of different modes come together to form a continuum. This is exactly what Hunter and Toomre found for the isolated disk; a smooth-edged disk has no discrete warping modes, but supports a continuum of dispersive waves which carry away the energy of any initially imposed warp.

If the ellipticity ε rises more rapidly (for example, quadratically) with radius, and the disk is 2 or 3 times smaller than the halo radius R_t, a qualitatively different result is found. Ths disk has at least one discrete mode, which is not sensitive to the edge; its shape and precession rate stay constant as the disk is extended. But as the disk radius approaches R_t, the mode behaviour reverts to that in a uniformly oblate halo; discrete modes are found only if the disk has a sharp edge. Disks in <u>prolate</u> halos appear to have at least one discrete warping mode even if the disk edge is smooth, so that the system can sustain a long-lived warp.

These results can be understood using the WKB (short-wavelength) theory of bending waves. If the precession rate Ω_p of a free particle in the halo <u>decreases</u> outward at the disk edge, then a discrete warping mode exists independently of the details of the edge; otherwise, a smooth-edged disk has a continuum of bending waves. If the halo is uniformly prolate, the (positive) precession rate falls at all radii; if it becomes rapidly more oblate, the (negative) precession rate may reach a minimum before rising towards zero at large radii. In these cases the Galaxy has an inner region where Ω_p is falling, and a smooth-edged disk can have discrete warping modes. By contrast, a uniformly oblate halo yields no region of falling precession rate, and only a sharp-edged disk can keep a long-lived warp.

So, if our Galactic disk is in a normal mode of warping, the massive unseen halo is either a) <u>increasingly oblate at large radii</u>, or b) <u>prolate</u>. The warped outer disk then crosses the plane of the inner disk at a constant azimuthal angle at all radii (the warp has no sense of spirality). The figure precession (which can in principle be measured from kinematic data) is in the same direction as the circular motion if the halo is prolate, and retrograde if the halo is oblate. The precession is much slower than the disk rotation if the halo is not far from round.

A full account of this work has been submitted for publication in the Astrophysical Journal.

REFERENCE

Hunter, C., and Toomre, A.: 1969, Astrophys. J. 155, 747

DISCUSSION

J.P. Ostriker: I would be interested in any comparison you might wish to make between your proposed explanation and that of R.H. Sanders. I believe that he found that a warp would persist for a long time (but not for ever) in regions where the disk-to-halo mass ratio was small.

Sparke: Sanders considered the warp at distances of 30-35 kpc, where the disk has become quite light. The Galactic warp is observed out to 18-20 kpc radius (with the Sun at 8-10 kpc), and would require a lot of heavy halo out there. Toomre argued at Besançon (IAU Symposium 100, page 177) that the required halo mass would be implausibly large. Of course all these arguments are about factors 2 or 3.

H. van Woerden: Would our halo be made prolate or flattened by tidal interaction with the Andromeda Nebula?

Sparke: I suspect not the halo, but I am not sure. I do not know why haloes should have the shape they have.

D. Lynden-Bell: A galaxy with an extensive heavy halo is more susceptible to tides than a galaxy without one (cf. my review paper, in Section III.1 in this volume). The tide distorts the halo, which distorts the visible parts of the galaxy. For a galaxy whose halo extends a factor of 10 beyond the visible disk, this enhancement of an imposed tide is by a factor of about 3.

R. Güsten: We know from the recent CO surveys that in the inner Galaxy there is a distinct ripple in the position of the mean CO-layer relative to the galactic plane. Is this an expected (second-order?) response to the galactic warp phenomenon?

Sparke: Probably not. I think it is more likely that the gas layer in the inner Galaxy has been perturbed by star formation. (In the outer Galaxy, the mean plane of the CO follows that of the neutral hydrogen.)

D. Lynden-Bell: There is an important observational criterion by which one may distinguish steady modes from transient, propagating ones. A steady mode must have its nodal line along a diameter, whereas in a propagating disturbance the nodal line will have a spiral shape. Is there an observational answer to this?

F.J. Kerr: Yes, the warp in our Galaxy does have a radial form in the observations. It has always been a puzzle, in fact, that the phase (or azimuth) of the warp is roughly constant with radius, while the structure lies in a field of differential rotation.

Lynden-Bell: Is that true in all galaxies, or just in ours?

Kerr: I think it is true in all galaxies.

C.A. Norman (Chairman): Can we have some discussion of that point?

SILENCE!

Norman: No one willing? What about Renzo?

T.S. Jaakkola: This is a question to the theorists: if the rotation of galaxies is indeed the relic of a primordial vortex, as the round-the-corner explanation is usually given, how can it still be so extremely virile even within massive haloes, as seen in the rapid rotation of spiral galaxies out to their very outskirts??

SILENCE!

Bernard Burke (left), Frank Kerr (centre), Lloyd Higgs and Bill Shuter have a beer during Wednesday excursion.
Burke (1957, Astron. J. **62**, 90) and Kerr (1957, Astron. J. **62**, 93) were the first to discuss the warp in the Galactic hydrogen layer. GSS

3-DIMENSIONAL PARTICLE SIMULATION OF VERTICAL OSCILLATIONS IN GAS DISCS

T.C. Johns and A.H. Nelson
Department of Applied Mathematics and Astronomy,
University College, P.O. Box 78, Cardiff CF1 1XL. UK

The problem of the persistence of warping in the outer regions of many galactic HI discs is now well known. Such a warp is observed in the Milky Way, where further deviations from a flat plane are seen in the form of a systematic tilt of the gas disc near the centre, with a short-wavelength (approximately 2 kpc) corrugation at intermediate radius.

A kinematical model for warped HI discs suggests that in the absence of external influences such features should rapidly wind up by differential precession and dissipate for a realistic rotation law. However it was argued by Nelson (1981) that in reality the kinematical tilted-ring model may be an oversimplification, and a more complete analysis of the vertical wave modes available to the galactic gas disc including the pressure term revealed the existence of two distinct classes of wave mode, the so-called "fast" and "slow" corrugation waves. In practice we expect only the lowest-order modes to be dynamically important. The fast mode winds up rapidly over the whole disc being closely related to the kinematical model, but Nelson demonstrated that the retrograde slow wave has the property that the line of nodes winds up slowly or not at all in the central and outer regions, giving a persistent central tilt and outer warp.

Since the properties of the slow and fast wave modes were derived using a local approximation, it is desirable to test the global properties of such waves by numerical simulation. To this end a 3-dimensional quasi-particle code has been developed, based on a non-self-gravitating version of the quasi-particle hydrodynamics technique due to Lucy (1977) and Gingold and Monaghan (1977). In the problem of interest the only interparticle forces involved are short-range pseudo-pressure forces, and consequently it has been found to be convenient to introduce a rectangular mesh of 63x63x7 cells and make use of an indirect particle-mesh scheme for evaluation of pressure forces. This enables a large number of particles to be handled efficiently - approximately 30 000 in the simulations reported.

The gravitational potential used in the calculations was of the form

H. van Woerden et al. (eds.), The Milky Way Galaxy, 503–504.

$$\phi(\underset{\sim}{r}) = \phi_1(r) + \phi_2(r,z)$$

where the rotation law for the models was

$$\Omega(r) = \Omega_c(1+(r/a)^p)^{-q}, \quad q = (1+2^{-p})/p, \ p=4 \text{ for Model 1 (Figs 1,2)}$$

and $\Omega(r) = \Omega_c(1+r^2/a^2)^{-3/4}$ for Model 2 (Fig 3).

The vertical potential is modelled as

$$\phi_2(r,z) = \tfrac{1}{2}\nu(r)z^2$$

with $\nu(r) = k\Omega^2(2a)\exp(1-r/2a)$, $k = 16.4$ respectively for Models 1,2;

$\Omega_c-1 = 1.22\times10^7$, 1.82×10^7 yr respectively for the two models and $a = 6$ kpc.

Figures 1 to 3 below show the wavefronts corresponding to the crossing of the galactic midplane for models in which the gas disc was initially tilted at an angle of about 2 degrees with respect to the plane of the potential. In all figures the sense of rotation is anti-clockwise and shaded areas indicate gas above the midplane.

In Figures 1 and 2 the initial velocity was chosen so that the gas was in approximate centrifugal equilibrium. In Figure 1 the z-velocity was chosen so that initially the gas was moving in the plane of the tilted disc; in Figure 2 the z-velocity was reversed. Note the different sense of the wind-up in the two cases and the reversal that occurs in the centre in Figure 2. In Figure 3 the velocity field was similar to that in Figure 2, but with an additional asymmetric perturbation in the plane of the disc, which caused large-scale deviations from circular motion and shocks in the gas disc. In this case the wavefront does not wind up significantly in the centre, perhaps evidence for the presence of a retrograde slow wave mode.

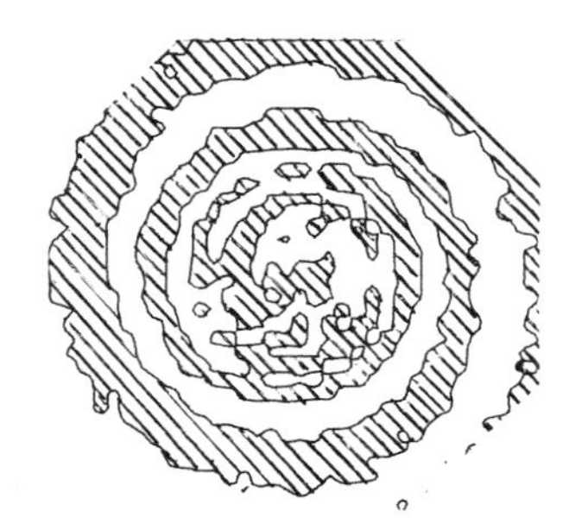
Figure 1.

Figure 2.

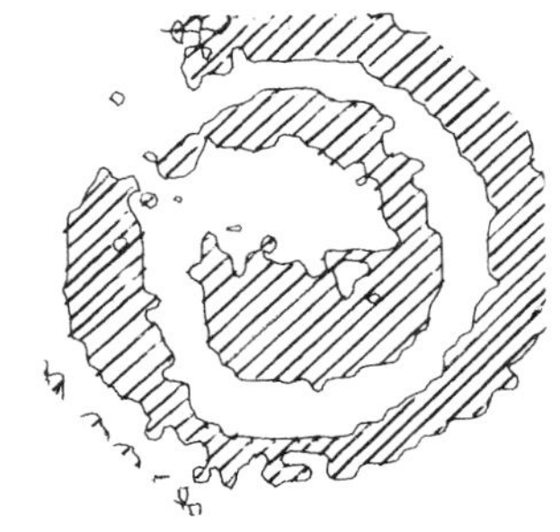
Figure 3.

REFERENCES

Gingold, R.A., Monaghan, J.J.: Monthly Notices Roy. Astron Soc. 181, 375.
Lucy, L.B.: 1977, Astronomical J. 82, 1013.
Nelson, A.H.: 1981, Monthly Notices Roy. Astron. Soc. 196, 557

TWO-FLUID GRAVITATIONAL INSTABILITIES IN A GALACTIC DISK

Chanda J. Jog
Princeton University Observatory, Princeton, New Jersey 08544

We formulate and solve the hydrodynamic equations describing an azimuthally symmetric galactic disk as a two-fluid system. The stars and the gas are treated as two different isothermal fluids of different velocity dispersions ($C_s \gg C_g$), which interact gravitationally with each other. The disk is supported by rotation and random motion. The formulation of the equations closely follows the one-fluid treatment by Toomre (1964). We solve the linearized perturbation equations by the method of modes, and study the stability of the galactic disk against the growth of axisymmetric two-fluid gravitational instabilities.

We find that even when both the fluids in a two-fluid system are separately stable, the joint two-fluid system, due to the gravitational interaction between the two fluids, may be unstable. The ratio of the gas contribution to the stellar contribution towards the formation of two-fluid instabilities is substantially greater than μ_g/μ_s, the ratio of their respective surface densities--this is due to the lower gas velocity dispersion as compared to the stellar velocity dispersion ($C_g \ll C_s$). The two contributions are comparable when the gas fraction (μ_g/μ_s) is only $\simeq 0.10 - 0.25$. Therefore, the galactic disk is a meaningful two-fluid system even when the gas constitutes only 10%-20% of the total surface density. Figure 1 contains plots of ω^2=(angular frequency)^2 vs. λ^{-1} =(wavelength)$^{-1}$ for the two-fluid perturbation, at different gas fractions. The values used for κ, the epicyclic frequency, and μ_t, the total disk surface density, represent the solar neighbourhood (Caldwell and Ostriker 1981). As a result of the increasing gas fraction, the most unstable mode grows faster and $\Delta\lambda$, the range of unstable wavelengths, increases. The wavelength and the time of growth of a typical two-fluid instability in the inner Galaxy, for μ_g/μ_s = 0.1-0.2, are ~ 2-3 kpc and ~ 2-4 × 10^7 years respectively, and each of these instabilities contains gas of mass ~ 4 × 10^7 -10^8 $M_\odot$.

The existence of even a small fraction of the total disk surface density in a cold fluid (that is, the gas) makes it much harder to stabilize the entire two-fluid disk. $(C_{s,min})_{2-f}$, the critical stellar velocity dispersion for a two-fluid disk, is an increasing function of

H. van Woerden et al. (eds.), The Milky Way Galaxy, 505–507.

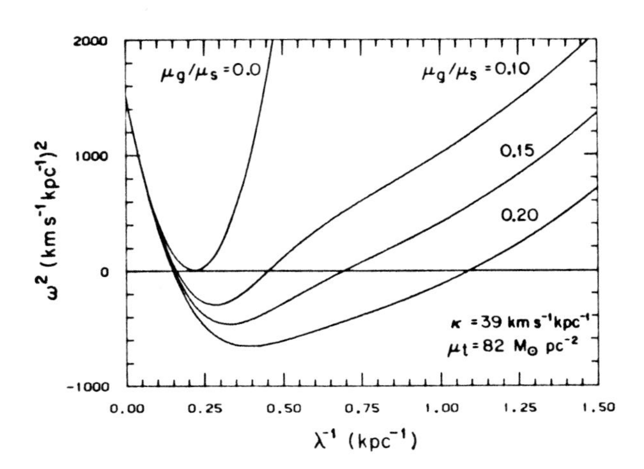

Figure 1. Influence of the gas content on the two-fluid gravitational instabilities.

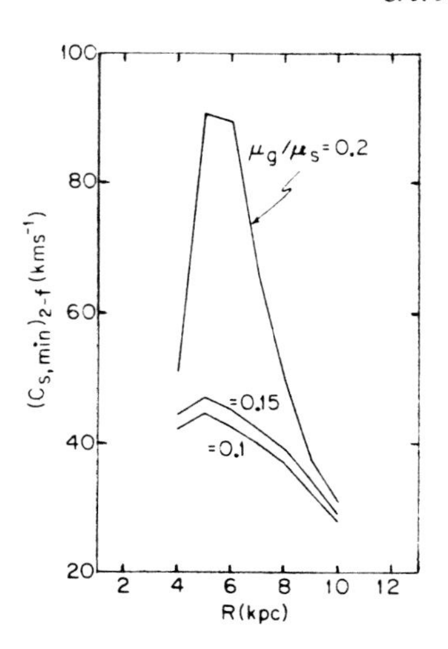

Figure 2. Results for $(C_{s,min})_{2-f}$ vs. R in the galactic disk.

μ_g/μ_s and μ_t/κ. In the Galaxy, $(C_{s,min})_{2-f}$ as a function of R, the galactocentric radius, peaks when μ_t/κ peaks--that is, at R ~ 5-7 kpc (see Figure 2); two-fluid instabilities are most likely to occur in this region. This region does coincide with the peak in the observed molecular cloud distribution in the Galaxy (see e.g., Scoville and Solomon 1975).

At the higher effective gas density resulting from the growth of a two-fluid instability, the gas may become unstable--even when originally the gas by itself is stable. The wavelength of such a typical (induced) gas instability in the inner Galaxy is 400-500 pc and it contains gas of mass ~ 1-2 × 10^7 $M_\odot$; this may be identified with a cluster of molecular clouds.

The above two-fluid analysis is applicable to any general disk galaxy consisting of stars and gas. The details of this analysis are given in Jog (1982) and Jog and Solomon (1984a,b).

REFERENCES

Caldwell, J.A.R. and Ostriker, J.P.: 1981, Ap. J., 251, 61.
Jog, C. J.: 1982, Ph.D. Thesis, State University of New York at Stony Brook.
Jog, C. J. and Solomon, P. M.: 1984a,b, Ap. J., in press.
Scoville, N.Z. and Solomon, P.M.: 1975, Ap. J. (Lett.), 199, L105.
Toomre, A.: 1964, Ap. J., 139, 1217.

DISCUSSION

M. Iye: How do your results compare with those of others, for example by Kato (1972, Publ. Astron. Soc. Japan 24), who studied a similar problem for a two-component system, but treated the stellar component not as a fluid but as a collisionless system?

Jog: I am not aware of his work.

F. Shu: I made similar calculations in 1968, and I do not believe I was the first (Lin and Shu, 1968, Brandeis Lectures in Astrophysics; Lin, Yuan, and Shu, 1969, Ap. J. 155, 721). Without thickness corrections, the calculations preceded 1966. We always made the assumption that the combined star-gas system was stable.

Jog: As far as I am aware, this is the first time that rotation as well as random motion have been taken into account in a two-fluid calculation. Lynden-Bell derived in 1967 the criterion for instability, but I think he did not consider the rotation.

Lynden-Bell: That is probably correct. I believe Toomre had a preprint which he never published.

Colin Norman (foreground) chairs at dinner table of conference secretariat, counter-clockwise: Ineke Rouwé, Jan de Boer, Eli Brinks (guest), Joke Nunnink, Marijke van der Laan, Leonard Bronfman LZ

COLLECTIVE PHENOMENA IN A MULTI-COMPONENT GRAVITATING SYSTEM

A.M. Fridman[1], A.G. Morozov[2], J. Palous[3], A. Piskunov[1], V.L. Polyachenko[1]
[1] Astronomical Council, USSR Academy of Sciences, Moscow
[2] Volgograd State University, Volgograd, USSR
[3] Astronomical Institute, Czechoslovak Academy of Sciences, Praha, CSSR

ABSTRACT

The paper aims at a demonstration of the principal differences between the oscillation spectra of multi-component systems and a one-component medium. The character of the mutual motions of components appears then to be of importance. Three cases are considered: (1) motionless components in the inertial frame of reference; (2) inertial subsystems moving at constant relative velocities; (3) a rotating n-component system. The oscillation spectra in these three cases have qualitative differences between each other and when compared with those of resting or rotating one-component systems.

(1) It is shown that in the motionless, n-component homogeneous gravitating medium, along with sound (Jeans) oscillations being synphasic, there is a new class of "asynphasic" oscillations having much in parallel with the Langmuir plasma oscillations and with optical oscillations of a crystal lattice. In contrast with the sound oscillations, the new class remains oscillatory at arbitrarily large wavelengths, which alters the conventional point of view that the gravitational condensation of all long-wave perturbations is inevitable. The theorem on the stability of a motionless n-component gravitating medium is formulated. It states that the Jeans instability may develop only on the synphasic branch, the remaing (n-1) different oscillation branches being asynphasic.

(2) For a system consisting of n components moving with constant relative velocities, a theorem on the number of beam instabilities as function of stationary parameters of the components is formulated. The maximum number of instabilities in such a system is n, of which (n-1) are beam instabilities and one a Jeans instability.

(3) For an n-component rotating gravitating disk, the maximum number of unstable regions in the space of perturbation wave vectors is also n. Moreover, the relative rotational motions of the components may

H. van Woerden et al. (eds.), The Milky Way Galaxy, 509–510.

be completely absent, i.e. all the n unstable regions of the k-space will belong only to one unstable Jeans branch. Consequently, the rotation removes the degeneration of case (1), splitting the Jeans "level" into n "sublevels". This once again demonstrates the analogy between rotation and the effect of a magnetic field.

REFERENCES TO RELATED WORK

Morozov, A.G., 1981, Astron. Zhurnal Letters, 7, 9-13.
Polyachenko, V.L., Fridman, A.M., 1981a, Zh. Exp. Th. Phys. (Moscow), 81, 13.
Polyachenko, V.L., Fridman, A.M., 1981b, Astron. Zhurnal Letters, 7, 136.

SECTION III.3

SPIRAL STRUCTURE AND STAR FORMATION

Thursday 2 June, 1645 - 1850

Chairman: D. Lynden-Bell

Above: Conclave of density-wave theorists; left to right: Shu, Bertin, Lin, Yuan. Background: Alladin and G.D. van Albada. CFD
Below: Confrontation with observers. Foreground: Cohen, Lin, Yuan, with Salukvadze in front. In background: Dame, D.M. Elmegreen, Oort, Ostriker, Kormendy, Van der Laan, Elmegreen, Wouterloot, Kutner, Alladin, Mead. LZ

FORMATION AND MAINTENANCE OF SPIRAL STRUCTURE IN GALAXIES

C. C. Lin and G. Bertin*
Massachusetts Institute of Technology
Cambridge, MA. USA
*Permanently at Scuola Normale Superiore, Pisa, Italia

ABSTRACT. The formation and maintenance of spiral structure in galaxies is discussed in terms of spiral modes with a general perception consistent with the hypothesis of quasi-stationary spiral structure. The latter is explained in some detail to give it the original proper perspective and to contrast it with recent studies of non-stationary spiral structures. We again emphasize the possible coexistence of fast-evolving spiral features and slow-evolving spiral grand designs. Spiral modes are described in terms of three categories of propagating waves. The reason is given why isolated non-barred galaxies with spiral grand design are found to be mostly two-armed. Some suggestions are made for future research.

I. INTRODUCTION

Research work on the formation and maintenance of large-scale spiral structure has been pursued along several lines of approach. In this paper, we shall attempt to clarify the relationship among the various approaches by presenting a coherent description of one of them and comparing the conclusions for this approach with those obtained from other alternative approaches on a few crucial issues.

The basic mathematical formalism adopted is the theory of linear spiral modes; these modes are in turn described in terms of steady wave trains. To connect our theoretical analysis with observations, we continue to adopt the hypothesis of quasi-stationary spiral structure (QSSS hypothesis), whose precise nature will be explained in some detail.

To give a proper perspective to our discussions, we again quote from Oort (1962) for his statement of the problem:

> "In systems with strong differential rotation, such as is found in all nonbarred spirals, spiral features are quite natural. Every structural irregularity is likely to be

H. van Woerden et al. (eds.), The Milky Way Galaxy, 513–532.

> drawn out into a part of a spiral. But this is not the phenomenon we must consider. We must consider a spiral structure extending over the whole galaxy, from the nucleus to its outermost part, and consisting of two arms starting from diametrically opposite points. Although this structure is often hopelessly irregular and broken up, the general form of the large-scale phenomenon can be recognized in many nebulae."

A few comments will now be made to clarify the above statements.

Galaxies may or may not exhibit a grand design. But it is the grand design that challenges the theorists and the observers alike. The theorists should be able to explain why it is possible for grand designs to exist, in spite of the winding dilemma. The observers should find out how the various features in the grand design are related to each other. The theorists should again explain why. Indeed, such efforts to try to explain the grand-design structure lead us to find out more about the astrophysical processes that prevail in the interstellar medium and in the process of star formation. The key step that made such investigations possible is the adoption of the hypothesis of quasi-stationary spiral structure (QSSS) as a working hypothesis in a semi-empirical context. (See Lin, 1971, pp. 89, 91.)

Some unnecessary controversy has arisen around this hypothesis in recent years. There are at least the following two reasons for this controversy: (1) The hypothesis has sometimes been misunderstood and given a role beyond that of a working hypothesis. This confusion can be easily cleared away by referring to the original literature, as is done in this paper. (2) There have been studies of dynamical mechanisms which suggest the possibility of spiral patterns in rather rapid evolution, not quasi-stationary structures. At the meeting in Besançon last year, only one aspect of the latter issue was addressed. In this paper, we shall attempt to deal with the issues more fully. Before we go on with further details, we should perhaps make a few remarks on the general perception.

From laboratory experiments on hydrodynamic stability and turbulence and from the study of weather patterns, -- both of which can be observed to evolve in time, -- we know that some flow patterns are quasi-stationary while others are transient. There are advantages in focussing our attention on regular quasi-stationary structures. In the study of galaxies, the adoption of the QSSS hypothesis has enabled us to pursue the study of a number of astrophysical phenomena and astrophysical processes (see Section V for a short list of examples). We therefore first focus our attention on the study of quasi-stationary phenomena, and leave the study of transient processes to separate investigations. Clearly, the studies of quasi-stationary and transient phenomena are complementary rather than competitive, unless it can be shown that quasi-stationary structures cannot possibly exist or that they can exist only under exceptional circumstances. Observational evidence gives us assurance that this is not so (cf. paper by D.M. Elmegreen at this conference). At the same time, we shall devote a part of this paper (Section II.3) to the discussion of studies which apparently yield transient impressions and point out why they do not, in fact, conflict with the possible existence of quasi-stationary spiral features.

Actually, this issue was already quite carefully examined by Bertil Lindblad (1963) twenty years ago, when he discussed the implications of some pioneering computational studies by Per Olof Lindblad. Specifically, his paper was entitled "On the possibility of quasi-stationary spiral structure in galaxies", and this possibility was recognized despite indications from the computational studies that the spiral structure may be more likely to be "quasi-periodic" (as described in Lindblad's paper), or regenerative (as described in the paper by Goldreich and Lynden-Bell, 1965).

Besides these general issues, two further points should be noted in the above quotation from Oort. (1) Attention is directed towards non-barred spirals as being the more challenging. (2) Even these galaxies preferentially have two-armed structures. In Section IV, we shall show why this is the case even though we know that there are isolated galaxies with other types of symmetry (cf. Iye et al., 1982 for a specific example). As it turned out, Landau damping of waves at inner Lindblad resonance holds the key; multiple-armed spiral structures are more likely to encounter inner Lindblad resonance as already noted in the early nineteen-sixties. Without inner Lindblad resonance, multiple-armed spiral features may indeed overwhelm two-armed spirals (see Section IV.4).

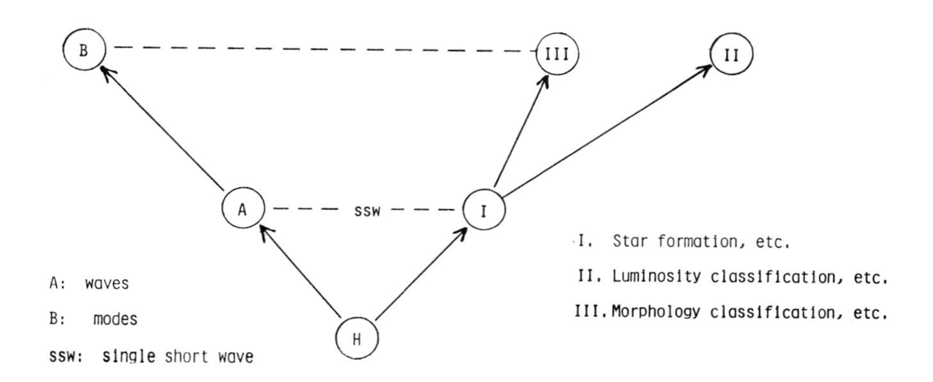

H: QSSS as a working hypothesis in a semi-empirical approach. Existence of a grand design, Hubble classification and winding dilemma are key issues.

Fig. 1. The hypothesis of quasi-stationary spiral structure.

II. THE HYPOTHESIS OF QUASI-STATIONARY SPIRAL STRUCTURE

1. The Semi-Empirical Approach. As a working hypothesis in a semi-empirical approach, its correctness is primarily to be judged by the inferences made from it. We require consistency both with observations and with internal dynamical mechanisms. As mentioned above, the QSSS hypothesis has led to a number of successful applications to observed phenomena, a partial list of which will be given in Section V. We call special attention to the application to the process of star formation and to the possibility of providing the basis for the luminosity and the morphology classifications of galaxies. On the other hand, we also derive support from dynamical studies, which will be given in some detail in this paper.

The research efforts may, therefore, be represented in terms of a picture of "mountain climbing", cf. Toomre (1977) and Tremaine (1983). Both these authors appear to emphasize a deductive approach from first principles. From our experience with the much simpler system of homogeneous fluids, we know that it would be quite difficult to "deduce" the existence of regular patterns in the full non-linear theory. Rather, we may attempt to describe such patterns by first accepting their plausible existence on the basis of experiments (see examples of regular and irregular patterns given by Figures 128, 131, 137, 158, 177 in van Dyke's Album, 1982). We then attempt to reach a better understanding of the underlying mechanisms that distinguish irregular flow patterns from regular grand designs. These mechanisms are, however, not always easy to describe.

The QSSS hypothesis receives support from the theory of linear spiral modes which has been quite successfully developed in recent years, since the coexistence of a small number of such modes, slowly growing from a certain basic state, provides a natural basis for the eventual existence of a quasi-stationary spiral structure. However, as is well known in the studies of hydrodynamic stability and turbulence, the calculation of linear modes "merely marks the beginning of the dynamical studies", and the same comment applies to the study of galactic spirals (Lin and Lau, 1979, p. 107). The existence of linear modes does not imply that the spiral structures are always necessarily permanent in shape. "The complicated spiral structure of galaxies indicates the coexistence of material arms and density waves -- and, indeed, of the possible coexistence of several wave patterns. These features [in general] influence but do not destroy one another." (Lin, 1971, p. 91.)

The basic time rate in all dynamical studies is of course the shear rate or the epicyclic frequency. However, linear modes in general evolve at a much slower rate. [Indeed, if the time scale for a feed-back cycle in a linear mode were comparable to that of its evolution, the "mode" would hardly be a useful concept from a physical point of view.]

Experience in hydrodynamical studies shows that the observed nonlinear states which are qualitatively similar to slowly evolving linear modes usually also evolve slowly. Thus, the study of linear modes gives a powerful method, perhaps the most convenient, to approach the problems of formation and maintenance of spiral structure on a global scale, since they do represent the intrinsic characteristics of the dynamical system. Thus, even the effect of external excitation may be described in terms of these modes. As an analogy, we may recall that essentially a slowly decaying quasi-stationary mode is excited through the ringing of a church bell.

2. The Single Short Spiral Wave. Here we wish to call attention to one fact which has not always been put in proper perspective. The phenomena listed in Section V have mostly been described by the use of a single short spiral wave. Such a short wave requires a feedback process (must be "replenished", see Toomre, 1969) for its long-term existence as a part of a modal formulation (Shu, 1968). The question is whether this feedback system can be observed. This will be discussed in the section on linear modes (Sections III and IV) where the feedback mechanism will be examined in some detail.

3. Nonstationary Impressions. The plausible existence of quasi-stationary spiral structures can be questioned if all studies suggest that the dynamical processes always imply rapid evolution of spiral structures. Thus, the following three issues deserve our special attention:

a. Rapid dynamical evolution is often found in N-body simulations;
b. strong amplification is associated with a "swinging" process;
c. strong "instability" seems to be present even when the disk is stabilized against axisymmetrical disturbances ($Q \geq 1$).

The first issue will be the main topic of this subsection. The third issue will be taken up in Section III. The second issue, discussed at Besançon (Lin, 1983b), will be briefly reviewed here. As an alternative to the time-dependent "swinging" approach (Toomre, 1981), Lin referred to Drury's work (1980), which does not involve a rapid evolution. It is therefore compatible with growing modes, even though both approaches may still be labelled "swing amplification", in the sense that they both describe the transformation of a leading spiral into an amplified trailing spiral. Bertin (1983b) presented still another analysis of leading-trailing wave amplification in terms of steady wave trains satisfying the local dispersion relationship. Again no transient behavior is implied, i.e., time evolution is not required. Rapid time evolution does occur in certain contexts; for example, as a result of certain special initial conditions (cf. Toomre, 1981). We shall take up this topic again in Sections III and IV in the context of modes.

Let us now turn to the first item. It is well known that there are apparent conflicts between results obtained through the use of numerical simulation and those obtained through analytical methods. There are perhaps three reasons for this. (1) The requirement on the number of particles is highly demanding when simulation of phase-mixing is needed. Often there are simply not enough particles in the phase-space. The simulation of $N_0 \sim 10^6 - 10^8$ particles by a single super-particle severely distorts the distribution function in velocity space. (2) The small number and the large mass of each superparticle magnifies the possible fluctuations of surface density and gravitational field by a factor of the order of $N_0^{1/2}$, which could be as large as $10^3 - 10^4$, thus greatly accelerating the evolutionary processes. These difficulties are not easy to remove. (3) Sometimes artificial physical processes are introduced to control effects such as the rapid rise of velocity dispersion. This could lead us to even worse situations, where it is no longer clear what physical system is being simulated. (The computation may become like a computer game which can be intellectually stimulating and even suggestive, but whose results cannot be accepted at face value. Proper interpretation is needed.)

All these ideas and discussions are not new. Computer experiments in plasma physics have been performed for at least two decades (see an excellent extensive review by Dawson, 1983). Millions of particles have been used, but experts are still demanding more in order to get a proper simulation of processes involving resonances, where phase mixing and Landau damping are important.

4. Some Additional Remarks. In addition to these studies giving nonstationary impressions, there is also the subtle mathematical point of the existence of an infinite number of modes with a continuous spectrum. This issue was examined by Lin and Bertin (1981), and found not to be serious.

In the spirit of the QSSS hypothesis, we therefore adopt the modal approach, because it is mathematically sound and it is convenient for the description of comparatively regular grand designs, including their evolution. The complete description of the evolutionary process must, of course, be based on the use of a nonlinear theory which is yet to be developed. The general nature of such developments can be partly visualized by examining existing theories for some simpler systems. For homogeneous fluids, the mathematical theory can be sketched in the following manner. Any dynamical variable $\psi(\underline{x},t)$ may be represented in the form

$$\psi(\underline{x},t) = \sum_n A_n(t)\phi_n(x)$$

in terms of a complete set of eigenfunctions $\phi_n(\underline{x})$. [Integration is implied by Σ when a continuous spectrum exists.] In the linear theory, each $A_n(t)$ is exponential in time t. In the full non-linear theory, the set of variables $\{A_n(t)\}$ satisfies nonlinear differential equations which describe the process of evolution. Extensive investigations along this line have been carried out, especially in meteorological studies. Galaxies are, of course, complicated systems with many components. Non-linear effects in the gaseous component have been shown to be likely to play an important role.

III. DYNAMICAL MECHANISMS: WAVES AND MODES

For the study of dynamical mechanisms, we shall focus our attention on linear processes, with one important exception. This is the issue of angular momentum transfer associated with the spiral patterns. Bertin (1983a) has shown that the angular momentum transfer associated with normal spiral modes is not going to produce a significant impact on the basic state over a period of several billion years even when the amplitude of the disturbance is about one-quarter of the basic distribution. Thus, we need not be unduly concerned with any fast evolution from this mechanism.

Linear modes can be described in terms of waves propagating in opposite directions, together with mechanisms of feedback from the central regions and over-reflection of these waves near the corotation circle. In earlier studies, we recognized two categories of waves: long waves and short waves. With leading and trailing configurations in each case, there are altogether four types of waves. Detailed studies based on the Lau-Bertin (1978) dispersion relationship now lead us to the recognition of a third category of waves: the open waves, again with possibly leading and trailing configurations. Open waves are more open than long waves, but propagate like short waves.

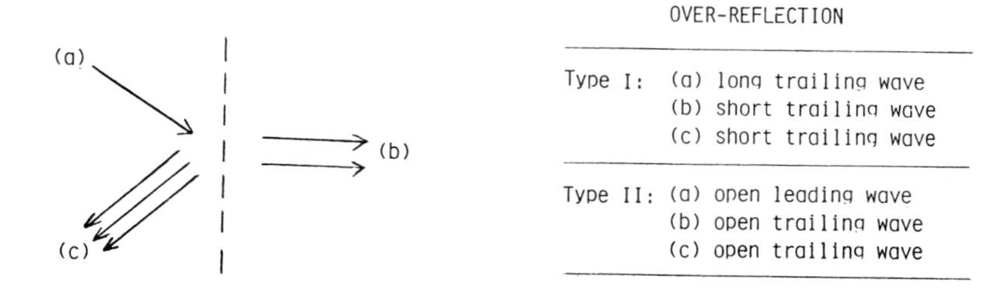

(a)
(b)
(c)
OVER-REFLECTION
Type I: (a) long trailing wave
(b) short trailing wave
(c) short trailing wave
Type II: (a) open leading wave
(b) open trailing wave
(c) open trailing wave

Fig. 2. Two types of over-reflection

In this new perspective, the two principal types of processes of over-reflection near corotation are shown in Fig. 2, which is amended from Fig. 1 of Lin's Besançon paper. The term over-reflection is used to describe a process in which the reflected wave contains more energy than the incoming wave. Over-reflection of Type I is the WASER (Mark, 1976) where only trailing waves are involved; that of Type II corresponds to the "swing amplification" (Toomre, 1981), in the sense that leading waves are converted into amplified trailing waves. Open waves are especially important for ranges of dynamical parameters for which there is only one real solution. For those cases, the distinction between open waves and short waves becomes less apparent.

The Lau-Bertin dispersion relationship is a cubic, with two parameters J and Q (see their paper for definitions of symbols). For the treatment of all three waves, it is therefore convenient to introduce the (J,Q) plane. The same format can be used for the study of the factor of over-reflection after the value of the shear parameter s is specified. In the case of constant linear velocity of circular motion, the shear parameter s equals unity.

Let us now briefly outline some of the principal results which differ from previous conclusions.

1. Waves in the (J,Q) Plane. First, the cubic dispersion relationship leads to the following overall outlook. The two real solutions representing the long/short waves can exist only if

$$JQ^2 \leq 4\sqrt{3}/9.$$

The third real solution represents an open wave, which is often too long to fit into the disk geometry, and should be excluded. However, this is not always the case. We may have to deal with all three categories of waves under certain circumstances (usually for J not too small). On the other hand, for

$$JQ^2 > 4\sqrt{3}/9 \quad ,$$

there is only one real solution, i.e., the open wave. The solutions which represent long/short waves have become complex conjugate pairs. It is in this range of parameters that the "swing" process is usually discussed.

Thus, in general, neither the long/short wave cycle nor the swing cycle represents the complete picture. We shall present examples of calculated spiral patterns in which the feedback processes include all three categories of waves (cf. Fig. 6 in Section IV).

2. Over-Reflection in Moderate Amounts. The second point is that one can identify a range of parameters in the (J,Q) plane such that the amplification factor (over-reflection) is moderate, whether the over-reflection is of Type I or of Type II (see Fig. 2). Moderate over-reflection is typified by an energy ratio of two (and an amplitude ratio $\alpha = \sqrt{2}$). The curve for this condition has been calculated by two methods. The system of curves shown in Fig. 3 is obtained from the approach used by Bertin in his Besançon paper, for different values of the shear parameter $s = - d\ln\Omega/d\ln r$. We also

checked Drury's calculations of the amplification factor for the case $s = 1$ by using an equation equivalent to Eq. (35) in Goldreich and Tremaine (1978). (Robert Thurstans did the computer work. Drury used a different, though equivalent, procedure.) The results are shown in Fig. 4 as curves of constant over-reflection (logarithmic scale). The two figures may be compared for $s = 1$, $\ln \alpha = 0.35$ [curve (b) in Fig. 4 versus curve marked $s = 1$ in Fig. 3]. The numerical values are not identical because in Bertin's formulation the Weber equation is used, just as in Mark's original WASER formalism. The two approaches are indeed two different approximations, and there is no good reason to decide that one approximation is better than the other. Fortunately, they support each other within the proper limits of these approximations.

3. The (J,Q) Parameters for Moderate Over-Reflection. In the new perspective just discussed, one is led to emphasize the role of the parameter J (or J_1, its value for $m = 1$) in the determination of the stability characteristics of the spiral modes. As shown before (cf. Lin and Lau, 1979, Bertin, 1980), tightly wound spirals with moderate rates of growth are associated with small or moderate values of the J-parameter, and with a value of the Q-parameter close to unity. Such modes are maintained by a specific feedback mechanism.

Let us leave the feedback mechanism aside for the moment and focus our attention on the over-reflection factor at corotation. If one follows the curve of "moderate" over-reflection factor, $\alpha = \sqrt{2}$, in Fig. 4, one finds that $Q \simeq 1$ only if $J \leq 0.5$ (Axisymmetry implies $J = 0$). In contrast, for $J = 1$, very high over-reflection is reached at $(J, Q) = (1,1)$; the factor becomes moderate ($\alpha = \sqrt{2}$) only when Q is raised to about $\sqrt{3}$. Thus, the condition $Q = 1$ does not always have special significance in all galactic disks. It is essentially the condition to be expected in galaxies with relatively low disk mass (low J), typified by normal spirals of types Sa and Sb.

There is another reason for expecting to observe higher values of Q associated with stars. This is the presence of a small amount of gas, which may cause a significant reduction of the effective Q-parameter (see Lin and Shu, 1966). For the specific combination of stellar and gaseous disks considered, marginal stability requires that

$$Q_* \simeq 1 + 2(\sigma_g/\sigma_*) \quad .$$

Thus, for 30% gas-to-star mass ratio (which is not large when disk thickness is taken into account), we find that $Q_* = 1.6$. Combining this with the factor $\sqrt{3}$ discussed above, we see that $Q_* = 2.7$ would not be, in some cases, an unreasonable value. Of course, this calculation gives only an indication of the type of situation to be expected and the result should not yet be directly compared with observations (cf. Kormendy's paper at this meeting).

With respect to observational studies, one should note that the usual definition of Q is quite easily influenced by the presence of a

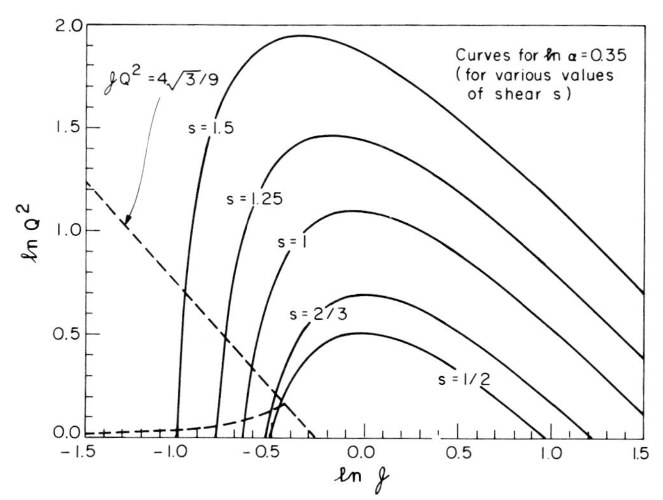

Fig. 3. Curves for a given coefficient of over-reflection ($\alpha = \sqrt{2}$) in the (J,Q) plane for various values of the shear parameter s.

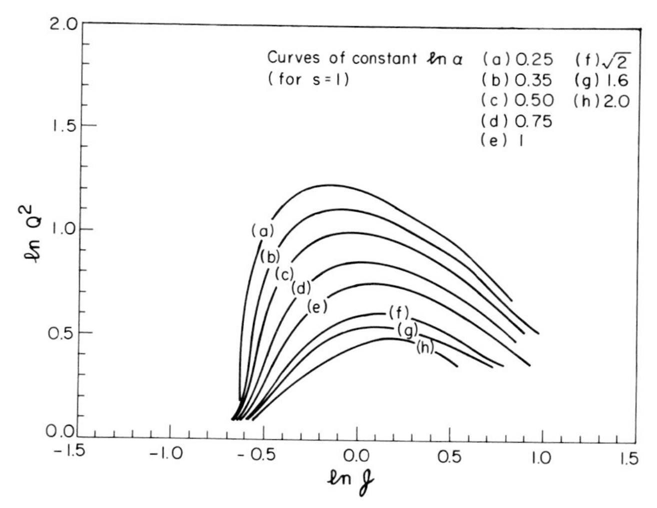

Fig. 4. Curves for various values of the coefficient of over-reflection α in the (J,Q) plane for a given value of the shear parameter (s=1).

small amount of stars with high velocity dispersion. This is the "long-tail" problem common to many statistical distributions. After all, the classical Cauchy distribution implies an infinite value for Q, if one applies the same definition as that for quasi-Maxwellian distributions. "Contamination" from other stellar components (e.g., halo stars) has to be removed. Basically, it may be wise to refine the definition of the stability parameter or even to introduce a new parameter that emphasizes the role of the responsive stars with low dispersion velocities.

IV. DYNAMICAL MECHANISMS: SPIRAL MODES AND MORPHOLOGICAL CLASSIFICATION

1. A Process of Self-Regulation. Many modal calculations have been done in the fluid model. The process of Landau damping at inner Lindblad resonance is therefore not incorporated in the calculations. An additional step must be taken to check the result for possible resonance effect. Such a step will make it apparent that the number of unstable spiral modes is expected to be reduced, because the feedback mechanism may be seriously impaired by the Landau damping process. (See subsection 4 below for further details.)

If we combine these considerations with the above discussions of the parameters J and Q, we are led to the following scenario for the self-regulation of the instability of spiral modes. Suppose that the galactic system, to begin with, has $J \simeq 1$ (for $m = 3$, say), and $Q = 1$. The excitation of spiral waves and modes, possibly violent, will tend to cause the stellar dispersion speed to increase until the disk is "heated up" to such high values of the Q parameter and such a form for its distribution that only a small number of spiral modes may eventually remain moderately unstable. This reduced instability also restricts the increase in the Q-parameter, and the final spiral structure may be expected to be quasi-stationary when nonlinear processes, involving both stars and gas, are included. [In this final state, the over-reflection factor may also be expected to be relatively moderate, but not necessarily very small.] It is also possible that the galaxy will evolve into a state where the dispersion speeds are so large that only a gentle oval distribution of stars results. Such a mass distribution would drive the Population I objects into a bar-like distribution, as shown by Roberts, Huntley and van Albada (1979).

The process of self-regulation just discussed is crucial in the context of the QSSS hypothesis, which can be invalidated by the presence of too many unstable spiral modes. We shall discuss it further in connection with some examples below. Having developed a scenario for the realization of quasi-stationary spiral structures, we are ready to venture to relate our theoretical investigations to one of the major objectives: the search for a dynamical basis for morphological classification. For this purpose, we must first be able to produce modes of various morphological types in the context of the linear theory. Based on experience with hydrodynamical experiments, especially those with rotating fluids, we may then infer that the grand designs in the final state are approximately simulated by the superposition of modes obtained from the linear theory.

2. Modes of Various Morphological Types. In previous publications, (see Lin and Bertin, 1981; Haass, Bertin and Lin, 1982) we have shown how one may use linear modes to simulate the various morphological types of the galaxies by judiciously prescribing three distribution functions: (a) the rotation curve, (b) the mobile component of the disk mass (including the effect of disk thickness) and (c) the distribution of dispersion speed of stars (or the acoustic velocity in the gaseous model). These different morphological types are often associated with different feedback mechanisms. For example, it is clear that the (r), (s) types relate to different degrees of propagation of waves through the center of the galaxies. The time is now ripe for relating the morphological type to dynamical mechanisms and modal calculations. Some examples were published by Haass, Bertin and Lin; some additional examples will now be cited.

Bertin (1983b)contrasted an open mode with a tightly wound spiral mode obtained from the original asymptotic theory. Specifically, their propagation diagrams were compared. The open mode shown by Bertin has an inner turning point for wave propagation. This implies a "reverse swing" of the wave and a rather long feedback cycle. It thus has only a moderate growth rate, smaller than that of the normal mode shown, even though the over-reflection factor is quite high, similar to that of the normal mode. Indeed, this open mode is not represented in Fig. 2 in Norman's Besançon paper (1983). The avoidance of violently unstable modes can be achieved through an increase of the Q-parameter.

For each of the two types of modes discussed by Bertin, we may assign a point in the (J, Q) plane according to the values of these parameters at the corotation circle. In Fig. 5, which shows a portion of Fig. 3, these points are located in the general areas marked I and II, corresponding respectively to the two types of over-reflection. We now note that there is a triangular region in the (J, Q) plane, marked III, corresponding to a gradual transition from one type of feedback to another.

To demonstrate this transition, a sequence of modes has been calculated for which the conditions at the corotation circle vary from Region I to Region II, traversing through Region III. The propagation diagrams show clearly that, in the cases shown in Fig. 6, it is the most open of the three waves that carries the primary signal across and around the corotation circle. The dashed lines show the real part of the complex conjugate pair of comparatively shorter waves. Their roles are, however, not insignificant. Thus, all three categories of waves are present, albeit some in modified forms. The over-reflection process and the feedback process are thus quite complicated; both types are involved. But all of these modes show moderate growth rates. Thus, the presence of over-reflection of Type II does not necessarily imply violent instability. For details, we refer to future publications.

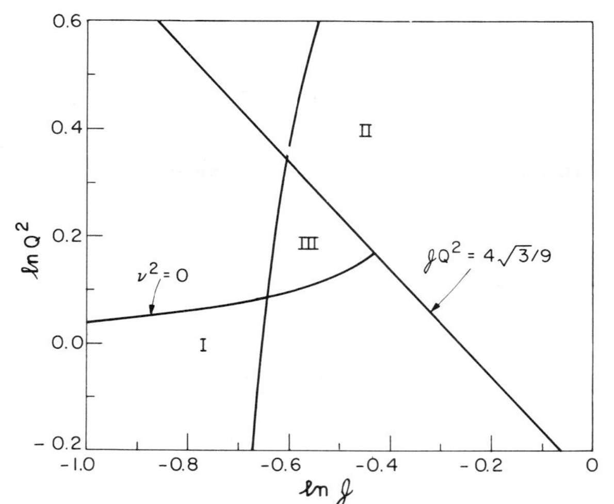

Fig. 5. Typical areas in the (J,Q) plane where over-reflections of Types I and II are important. These are respectively labelled I and II. In domain III, all six waves may be active. The nearly vertical curve is that for $\alpha = \sqrt{2}$ for the open waves. Unstable modes with moderate growth rates have been calculated for all three domains.

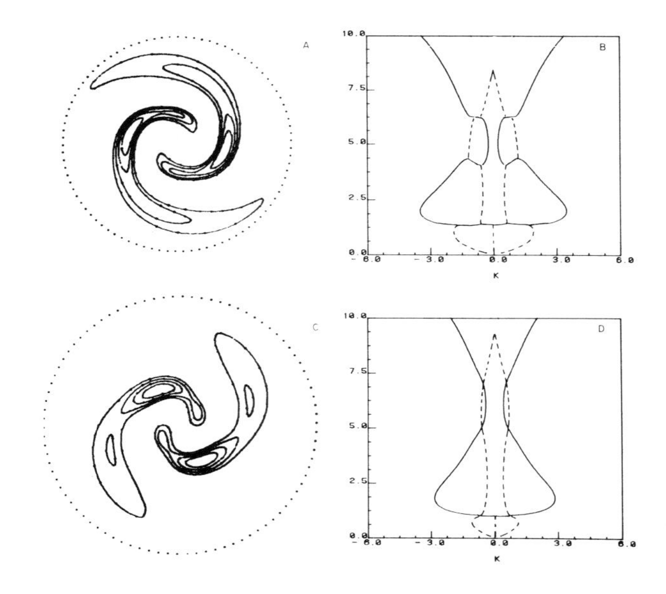

Fig. 6. Spiral patterns and propagation diagrams for two cases, A and C, where both types of over-reflection are active. Note that near co-rotation the open waves, which are real solutions represented by the solid lines, are the most important. For conditions far away from the co-rotation circle, the short waves are the important ones represented by solid lines. In Case A, (ℓn J, ℓn Q^2) = (-.511,+.191) near Region I; in Case C, (ℓn J, ℓn Q^2) = (-.371,+.365) in Region II.

3. Are Feedback Processes Observable? The short wave is assigned a predominant role in much of the application theory. What can be expected from the feedback process from an observational point of view? There is no general answer to such a question, but the following comments may be helpful.

In the case of normal spirals, the long trailing wave, which is responsible for the feedback process, produces a modulation on the short trailing wave. (See Bertin et al., 1977, p. 4728, and reference to Mark's earlier work.) For empirical purposes, it simply appears that there is a change of intensity of the spiral field with radial distance from the center. The simplicity of the situation results in part from the fact that in the amplified modes, the over-reflection process gives the short trailing spiral a considerable advantage. For a real galaxy, one can construct a scenario compatible with an over-reflection process of the same type by including the damping of such short waves by the shock wave in the gaseous component.

In the case of open modes such as that discussed by Bertin, the situation is somewhat different. The open waves lead to something like an oval distortion in the central regions, which exhibits itself through the gaseous response to a bar-like forcing. Again, the general impression may be that of only one trailing spiral extending out of a bar.

4. Why Two-Armed Modes? We now turn to the following important issue: why do we not see many three-armed grand designs in galaxies even though the example shown by Haass at the Besançon meeting exhibits higher growth for multiple-armed modes? The key to this puzzle lies with Landau damping at inner Lindblad resonance, which is not included in the fluid code used by Haass (cf. III.3, IV.1 above). In the example shown by Haass, the rotation curve has a nearly constant angular velocity at the center ($\Omega \simeq \Omega_0$, somewhat like that of M33). Thus, the curve for $\Omega - \kappa/2$ has very low values at the center, but the curve for $\Omega - \kappa/3$ has a peak value of $\Omega_0/3$ at the center. Thus, if a three-armed mode has a relatively large corotation circle (and hence a low Ω_p), it would tend to suffer from inner Lindblad resonance. Note that the three-armed modes shown by Haass (1983b) have a rather small corotation circle, since he has deliberately chosen the Q-distribution to have a very small scale, smaller than the scale for the velocity distribution, (V_{max}/Ω_0). In real galaxies, the Q-distribution may have a much larger scale. In such cases, one may regard the corotation circles as being pushed outwards as the scale for the Q-distribution expands. Lindblad resonance follows, and the mode is expected to be damped in the stellar model (even though this is not apparent in the gaseous model calculated). Two-armed modes, being relatively free from Lindblad resonance, thus become predominant. Notice, however, that low frequency two-armed modes would also suffer from inner Lindblad resonance, and consequently the number of unstable two-armed modes can become quite small.

It is obvious that the above line of reasoning applies to four-armed spirals with stronger justification. One-armed spirals do not suffer from inner Lindblad resonance, but they usually have lower growth rates than two-armed spirals. Observationally, their presence together with two-armed spirals is often not as easily detectable. But the well-known asymmetrical motion of the dust lanes in M31 might indeed be due to the presence of a one-armed structure. However, it is not yet entirely clear why galaxies like NGC4254 (Iye et al., 1982) apparently show only a spiral structure with odd harmonics.

V. CONCLUDING REMARKS

We have surveyed the study of dynamical mechanisms from a semi-empirical perspective. The following is a short list of the successful applications of the theory.

A Short List of Some Physical Phenomena Studied

1. Possible dynamical basis for the Hubble classification
2. Plausible dynamical basis for the luminosity classification (Roberts, Roberts and Shu 1975)
3. Process of star formation
4. Dust lanes: distribution
5. Synchrotron emission (e.g., in M51): distribution
6. Distribution and motion of atomic hydrogen (e.g., in the Galaxy and in M81; including the modal theory of Visser and Haass)

On the theoretical side, the following points appear to be worthy of our attention.

1. We should recognize the possible coexistence of (a) spiral grand designs which evolve rather slowly, and (b) less regular spiral features which evolve more rapidly, generally on the dynamical time scale of κ^{-1}, Ω^{-1} or $(s\Omega)^{-1}$. The hypothesis of quasi-stationary spiral structure is thus re-affirmed to provide a basis for the dynamical approach to the morphological classification of galaxies (cf. Fig. 1).

2. The mathematical model adopted and the mathematical formalism used are crucial to the plausible demonstration of the above statements. While the use of wave packets tends to give impressions that favor the less regular spiral features, the equivalent method of using steady wave trains and modes emphasizes the possible existence of long lasting grand designs. Numerical experiments can be misleading unless Landau damping is properly simulated.

3. It is important to recognize different behaviors of the spiral waves and patterns for different physical models and for different ranges of the dynamical parameters. Dynamical processes tend to re-adjust the distribution of the three

basic parameters listed above (Section IV.2), especially the dispersion velocity of stellar motions. This process of self-regulation, aided by the presence of the gaseous component, may be crucial to the quasi-stationary existence of grand designs of various types (cf. Sections IV.1 and IV.2).

Looking into the future, we see that the study of the morphology of galaxies continues to require the collaboration of observers and theorists. In the dynamical approach to the morphological classification of galaxies, we adopt a logical framework based on our understanding of the dynamical mechanisms and on the spiral appearance to be expected under various dynamical situations. We then compare these conclusions with the morphological classification based on observations. Preliminary work (e.g., by Haass, Bertin and Lin, 1982) shows that this approach is highly promising.

Other promising lines of research include the analysis of observed spiral structures into Fourier components and logarithmic spirals, the study of bars and rings, especially in SBa galaxies, and the study of the dispersion of stellar velocities in external galaxies. Some of these studies are being carried out by various researchers. Because of limitation of space, comments on these topics have to be omitted from this paper.

It is perhaps unfortunate that inadequate appreciation of the theoretical points discussed above has led to considerable confusion in the literature. It is hoped that the present paper helps to clear away much of the confusion and to focus our attention on the substantive issues of attempting to explain the observed phenomena such as the different morphology of the galaxies.

VI. THE MILKY WAY

At this conference, various aspects of the structure of the Milky Way have been discussed, including the successful modelling of the outgoing motion of the 3-kpc arm by Yuan as a flow driven by a rotating bar. The resultant picture is quite similar to that appearing in the inner parts of NGC5364. For this paper, we shall restrict ourselves to some short comments and a few references.

We recall (see Lin and Yuan 1978, Lin 1983a) that there have been rather extensive theoretical examinations of the spiral structure in the solar neighborhood from both the observational and the theoretical points of view. Consistency with the two-armed pattern originally suggested by Lin and Shu (1967) has been found. But there is coexistence of other spiral features together with the two-armed pattern, which is found to be compatible with spiral modes calculated by Haass (1983 a). The Carina arm may be an extra feature associated with the condition $|\nu| = 1/2$ (Shu, Milione and Roberts, 1973). It is not surprising that multiple-armed features can exist in the outer parts of the Galaxy in addition to a dominant two-armed structure in the more massive inner part. It is also not surprising that an additional one-armed spiral structure can lead to asymmetries of the kind that appears to be observed in the fourth quadrant. However, the long stretches of spiral arms observed present an overall picture that shows a predominantly two-armed pattern (see Burton, 1976). Finally, it should be emphasized that the structure of individual spiral features such as the high-velocity stream outside of the Sagittarius arm and other streaming motions (see Burton and Shane, 1970) played an important role in the development of the density-wave theory by giving observational support to the basic concept, -- a support that remains valid today.

ACKNOWLEDGEMENT

We wish to thank Jon Haass for many helpful discussions and for permitting us to quote his unpublished results on the sequence of models that are calculated for the demonstration of the various types of over-reflection.

REFERENCES

Bertin, G.: 1980, Phys. Reports 61 (1), pp. 1-69 (217 references)
Bertin, G.: 1983a, Astron. Astrophys. 127, pp. 145-148
Bertin, G.: 1983b, IAU Symposium No. 100, ed. E. Athanassoula (Reidel), pp. 119-120
Bertin, G. et al.: 1977, Proc. Nat. Acad. Sci. 74, pp. 4726-4729
Burton, W.B.: 1976, Ann. Rev. Astron. Astrophys. 14, pp. 275-306
Burton, W.B., and Shane, W.W.: 1971, IAU Symposium 38, eds. W. Becker and G. Contopoulos (Reidel), pp. 397-414
Dawson, J.M.: 1983, Rev. Modern Phys. 55, pp. 403-447
Drury, L. O'C.: 1980, Monthly Notices Roy. Astron. Soc. 193, pp. 337-343
Goldreich, P., and Lynden-Bell, D.: 1965, Monthly Notices Roy. Astron. Soc. 130, 97, pp. 125
Goldreich, P., and Tremaine, S.: 1978, Ap. J. 222, pp. 850-858
Haass, J.: 1983a, in 'Kinematics, Dynamics and Structure of the Milky Way', ed. W.L.H. Shuter (Reidel), pp. 283-287
Haass, J.: 1983b, IAU Symposium No. 100, ed. E. Athanassoula (Reidel), pp. 121-124
Haass, J., Bertin, G., and Lin, C.C.: 1982, Proc. Nat. Acad. Sci. USA 79, pp. 3908-3912
Iye, M., Okamura, S., Hamabe, M., and Watanabe, M.: 1982, Astrophys. J. 256, pp. 103-111
Lau, Y.Y., and Bertin, G.: 1978, Ap. J. 226, pp. 508-520
Lin, C.C.: 1971, in 'Highlights of Astronomy', ed. C. de Jager (International Astron. Union, Reidel, Dordrecht, Netherlands), Vol. 2, pp. 88-121
Lin, C.C.: 1983a, in 'Kinematics, Dynamics and Structure of the Milky Way', ed. W.L.H. Shuter (Reidel), pp. 277-281

Lin, C.C.: 1983b, IAU Symposium No. 100, ed. E. Athanassoula (Reidel), pp. 117-118
Lin, C.C., and Bertin, G.: 1981, in 'Plasma Astrophysics', eds. T.D. Guyenne and G. Levy (European Space Agency, Noordwijk, Netherlands), SP-161, pp. 191-205
Lin, C.C., and Lau, Y.Y.: 1979, Studies in Appl. Math. 60, pp. 97-163
Lin, C.C., and Shu, F.H.: 1966, Proc. Nat. Acad. Sci. 55, pp. 229-234
Lin, C.C., and Shu, F.H.: 1967, IAU Symposium No. 31, ed. H. van Woerden (Academic Press, London), pp. 313-317
Lin, C.C., and Yuan, C.: 1978, 'Astronomical Papers dedicated to Bengt Strömgren', eds. A. Reiz and T. Andersen, published by Copenhagen University Observatory, pp. 369-386
Lindblad, B.: 1963, Stockholm Observ. Ann. 22, No. 5
Mark, J.W.-K.: 1976, Astrophys. J. 205, pp. 363-378
Norman, C.: 1983, IAU Symposium No. 100, ed. E. Athanassoula (Reidel), pp. 163-174
Oort, J.: 1962, in 'Interstellar Matter in Galaxies', ed. L. Woltjer (Benjamin, New York), pp. 234-244
Roberts, W.W., Huntley, J.M., and van Albada, G.D.: 1979, Astrophys. J. 233, pp. 67-84
Roberts, W.W., Roberts, M.S., and Shu, F.H.: 1975, Astrophys. J. 196, pp. 381-405
Shu, F.H.: 1968, Ph. D. Dissertation, Harvard University
Shu, F.H., Milione, V., and Roberts, W.W.: 1973, Astrophys. J. 183, pp. 819-841
Toomre, A.: 1969, Astrophys. J. 158, pp. 899-913
Toomre, A.: 1977, Ann. Rev. Astron. Astrophys. 15, pp. 437-478
Toomre, A.: 1981, in 'Structure and Evolution of Normal Galaxies', eds. S.M. Fall and D. Lynden-Bell (Cambridge University Press), pp. 111-136
Tremaine, S.: 1083, IAU Symposium No. 100, ed. E. Athanassoula (Reidel), pp. 411-416
Visser, H.C.D.: 1978, The dynamics of the spiral galaxy M81. Ph. D. thesis, University of Groningen, Netherlands
Van Dyke, M.: 1982, An Album of Fluid Motion (The Parabolic Press)

DISCUSSION

F.H. Shu: In this conference many people have commented on the destabilizing contribution of interstellar gas. I would like to remind people that interstellar gas can also have an opposite effect, namely to saturate growing normal modes by dissipative effects. A. Kalnajs pointed out in 1972 that galactic shocks introduce a phase shift between the response of the interstellar gas and the background stellar density wave that drives the gas. This tends to damp the wave; and W.W. Roberts and I suggested that this might turn out to be a good thing, if an instability mechanism lies behind spiral structure. The various instability mechanisms subsequently worked out and described in Lin's talk all have the following features in common: an amplifier (at corotation) plus feedback (via long, leading, or open waves). Thus, consider

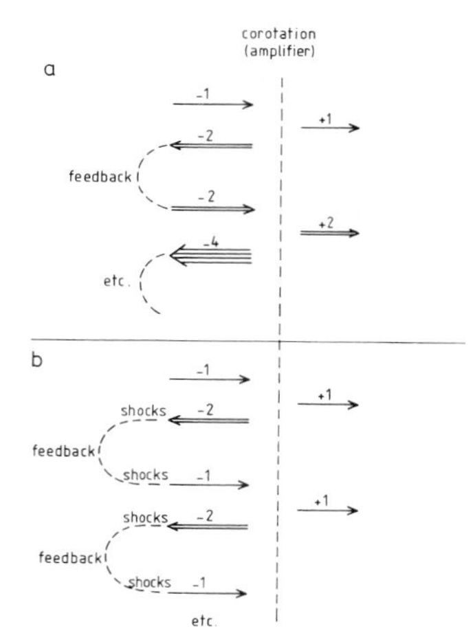

Figure D1 - a) Rapid growth of wave amplitude by amplification at corotation, followed by dissipationless feedback, leading to instability.
b) Quasi-stationary situation: zero growth of wave amplitude; amplification at corotation is followed by dissipative feedback, caused by galactic shocks.

(Figure D1a) the simplest generic example: propagation to corotation of a wave carrying -1 unit of angular momentum, transmission of +1, over-reflection of -2 (to conserve angular momentum), feedback of -2 (if there are no losses), transmission of +2, over-reflection of -4, etc. Thus, the square of the wave amplitude would grow by a factor of 2 per cycle, and may be in danger of growing to catastrophically high values.

Consider, however, how dissipation of the waves by galactic shocks changes this picture (Figure D1b). Start again with a wave approaching corotation and carrying -1 unit of angular momentum. There is transmission of +1, over-reflection of -2, followed now by shock dissipation to cut down the reflected wave as it propagates. By the time the wave is fed back, it might again have only -1 when it comes back to corotation, starting the cycle over again in nearly the same configuration as before. The resulting pattern would be quasi-stationary with evolution only on a slow timescale.

Why should such a quasi-stationary situation obtain? I think it is automatic. Because density waves are dispersive, they do not steepen into shocks until they reach a finite critical amplitude. Until the gaseous density wave reaches this amplitude, the overall wave (stars plus gas) would grow in time. Once the critical amplitude is exceeded significantly, strong shocks develop that halt the growth of the wave, giving a unique solution of finite amplitude.

This consideration may be important in a complementary context. Suppose that in some galaxy there is not enough gas to hold the mode amplitudes to fairly small values. The mode(s) may then grow catastrophically and heat the stellar disk enough to suppress all instabilities. Could something like this have happened to the "flocculent spirals"?

W. Renz: In this connection, I wish to refer to my non-linear calculation of the amplitude to which waves will be limited by damping due to shocks (W. Renz 1982, in "Applications of modern dynamics ...", ed. Szebehely, Reidel, Dordrecht, p. 356). Assuming, with Kalnajs (1972, Astrophys. Letters 11, p. 41), a relaxation time of about 10^9 years and taking amplification at corotation into account, I found the amplitude of a quasi-stationary wave to be about 5%.

P. Pismis: You stated at the beginning of your talk, Dr. Lin, that there would be little problem in having a driving mechanism for spiral arms in a barred galaxy, as the existence of the bar would give rise to such a mechanism. However, a good many barred spirals have a small, well-defined and tight spiral within the nucleus in the middle of the bar. Good examples are NGC 1097 and 4314. How would the existence of such features agree with the scheme of the density-wave theory?

Lin: I must not give the impression that everything is deductive. In the text of my paper I emphasize the semi-empirical approach of the density-wave theory. I would assume that the structures you describe may perhaps also be quasi-stationary. One has to look for the dynamical process that is responsible.

G.D. van Albada: At least in some cases the nuclear spirals observed in barred galaxies may be an artifact produced by the projection of a curved dust lane on the luminous nuclear region. This conjecture is based on the comparison of recent numerical results with e.g. the pictures shown in the Hubble Atlas.

J.V. Feitzinger: A crucial point is the time scale of growth rates.

Lin: Indeed, that is a very important point. We believe that the growth rates are moderate. Wielen's study of the increase of stellar velocity dispersions in the Milky Way points to the same conclusion. The time scale for these changes may indeed be too short in the numerical experiments (for reasons given in Section II.3 of the text).

D. Lynden-Bell: Sellwood has made thorough comparisons of the modes determined analytically by Toomre and by Kalnajs with his own detailed numerical calculations. He went as far as 10^5 bodies, without getting accurate growth rates, when starting with Poisson noise in phase space. However when he started with a special smooth procedure, he managed to get almost exact agreement. This may serve as a warning to those in the audience who make numerical simulations: it is hard work to get exact agreement with analytical results.

Lin: Was Lindblad resonance involved in that case?

D. Lynden-Bell: I must ask Sellwood about that.

J.A. Sellwood (answer supplied after Symposium, on request by Editor): No inner Lindblad resonances were involved in these models. The apparent growth rates were too high, because noise interfered with the measurements at early times. The amplitudes also soon became non-linear.

Lin: I would say that both approaches have their own limitations. The analytical theory certainly is now restricted to the linear case; it is a difficult step to go on to non-linear analysis. As to numerical simulations the plasma physicists, with their decades of experience, are very cautious about involving resonances.

TWO-DIMENSIONAL CALCULATIONS OF TIGHTLY WOUND SPIRAL SHOCKS

A.H. Nelson and T. Johns
Applied Mathematics, University College, Cardiff, U.K.

M. Tosa
Tohoku University, Sendai, Japan

Recently it has been possible to perform computer calculations with sufficiently high resolution to obtain the response of the gas in galaxies to a tightly wound spiral potential due to a stellar density wave. Previously for pitch angles $\sim 10^{o}$ the problem had to be simplified using the tightly wound approximation, with gradients parallel to the arms neglected compared to perpendicular gradients, yielding a quasi-1-dimensional calculation. This paper reports the early results of a 2-dimensional calculation, and compares them with the 1-dimensional results.

The calculation was performed on an ICL Distributed Array Processor at QMC, London, allowing a polar grid of dimensions 63x126 in r and θ. The gas was initially in circular motion with uniform density, and a two-armed trailing spiral potential was switched on slowly over a few mean rotation periods. The response was then followed for up to 10 such periods, and the figure shows the response at a time of 3×10^{9} years for a pitch angle of 20°, and a perturbing potential amplitude of 3%. The figure shows the density profile in azimuth across a single arm at varying galactic radii, with the arm unwound for clarity. The position of corotation at 22 kpc is indicated by the arrow. Inside corotation there is an inner shock on the left edge of the arm (rotation is from left to right), while outside corotation there is an outer shock on the right hand edge as expected. In between, at around corotation, the density response is hardly diminished and the width of the arm is as narrow as it is away from corotation.

These results are in contrast to previous quasi-1-dimensional results, in which the density amplitude decreases markedly at corotation and the width of the peak widens to a broad bump (see Nelson and Matsuda, 1977); while away from corotation the density peak is very narrow compared to that shown here. It seems therefore that the 1-dimensional tightly wound approximation has a tendency to artificially broaden the density peak at corotation, and to make it sharper elsewhere.

H. van Woerden et al. (eds.), The Milky Way Galaxy, 533–534.

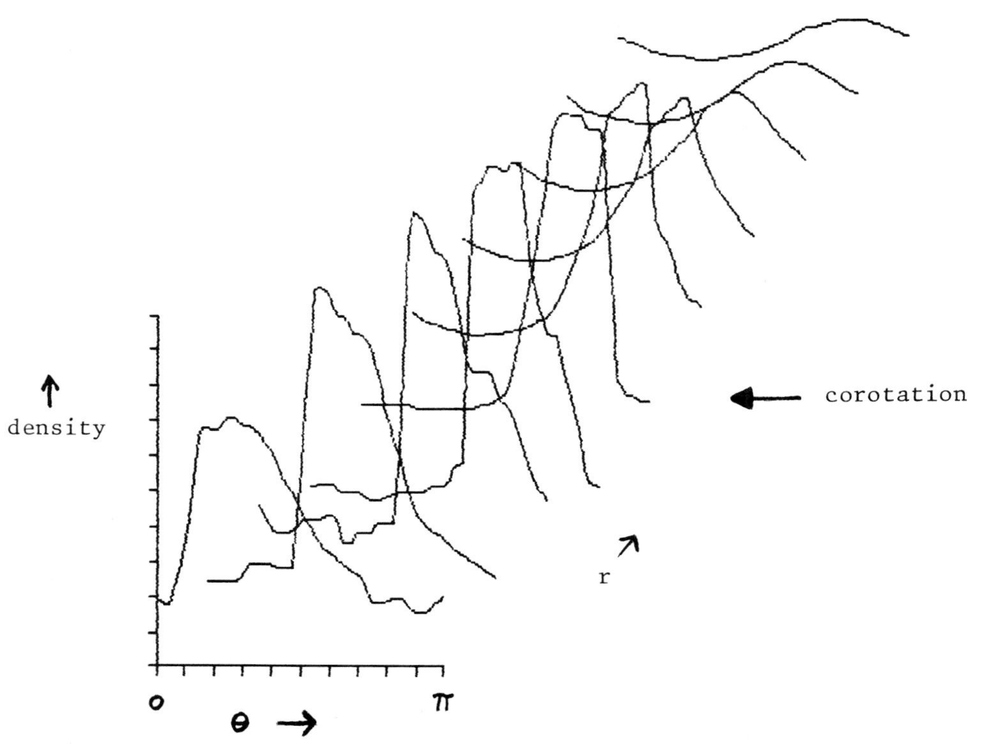

Figure 1. Density profiles in azimuth at various radii. Only one spiral arm is shown, and it has been unwound for clarity. The profiles start at a radius of 7 kpc and go outwards in steps of 3.5 kpc to 38.5 kpc.

REFERENCE

Nelson, A.H. and Matsuda, T.: 1977, M.N.R.A.S. 179, 663

OSCILLATIONS OF ROTATING GAS DISKS : *p*-modes, *g*-modes, and *r*-modes

Masanori Iye
Institute of Astronomy, Cambridge CB3 OHA, U.K.

Short-wavelength adiabatic oscillations of an infinitesimally thin, rotating and self-gravitating gas disk in dynamical equilibrium are studied. The problem is formulated in terms of the theory of oscillations of gas spheres (*i.e.* stars) in order to help galactic and stellar seismologists towards their mutual understanding of their rather similar subjects.

It is natural to expect that the oscillations of a rotating gas disk (*e.g.* a galactic gas disk) share a good deal of physical similarity with those of a rotating gas sphere (*i.e.* a star). Unfortunately, however, density-wave theoreticians and stellar seismologists have not always communicated easily, partly because they haven't shared a common physical language. Schutz and Verdaguer (1983) and Verdaguer (1983) were the first, to the author's knowledge, who tried to fill in this gap. They studied the oscillations of *isentropic* gas disks using the theory of oscillations of stars (*e.g.* Unno et al. 1979).

The present paper gives the outline of a more general discussion of adiabatic oscillations of *non-isentropic* gas disks. Full details will be found in Iye (1983).

The linearized equations of continuity, of motion (in two directions), and of adiabatic perturbation show that there are four modes in general. One can indeed derive a quartic dispersion relation for them (see Iye 1983). These modes can be classified, in the short-wavelength limit, into three types according to their main restoring forces for the perturbation, *i.e.* pressure force, effective gravity (or buoyancy) and Coriolis force. The self-gravity of perturbations makes the oscillations less stable, or even unstable, for certain perturbations of longer wavelengths. This destabilizing force is, however, not significant in the short-wavelength limit.

It is shown that non-isentropic disks have a pair of *pressure* modes ($\pm p$-modes) for which the restoring force is two-dimensional pressure force and a pair of *gravity* modes ($\pm g$-modes) which are restored by

H. van Woerden et al. (eds.), The Milky Way Galaxy, 535–536.

effective gravity. All of these waves drift with the local rotational flow. The sign of each pair denotes the direction of wave propagation with respect to the local rotating frame. In the short-wavelength limit, $|kr| \gg 1$, the frequency of oscillation ω measured in the corotating frame will be

$$\omega = \pm c\, k \qquad \text{for } {}^{\pm}p\text{-modes} \qquad (1)$$

and

$$\omega = \pm N\, k_t / k \qquad \text{for } {}^{\pm}g\text{-modes} \qquad (2)$$

respectively, where k and $k_t (=m/r)$ are the total and tangential wave number, c is the sound speed, and N is the Brunt-Väisälä frequency defined by $N = (-gA)^{-\frac{1}{2}}$ with g and A being the effective gravity and the Schwarzschild discriminant. The discriminant A for convective stability is defined by $A = d\ln \mu_o/dr - 1/\Gamma_1\, d\ln P_o/dr$, where μ_o is surface density, P_o is two-dimensional pressure of the disk and Γ_1 is the adiabatic exponent.

Since $N = 0$ for isentropic disks, they have no g-modes. In the limit of $A \to 0$, the frequencies of ${}^{\pm}g$-modes merge at $\omega = 0$ in the first order of $(kr)^{-1}$. However, it is found that in the next order of approximation $(kr)^{-2}$, the ${}^{+}g$-mode behaves as another type of oscillation called *Rossby* mode (r-mode) while the ${}^{-}g$-mode remains strictly neutral (ω=0). This r-mode is restored by Coriolis force. Its frequency is

$$\omega = \zeta\, F\, k_t \,/\, k^2 \qquad \text{for } r\text{-mode}, \qquad (3)$$

where ζ is the vorticity, F is defined by $F = d\ln(\zeta/\mu_o)/dr$. F is the inverse of the scale length of a quantity ζ/μ_o, which Lynden-Bell and Katz (1981) identified as an essential conserved quantity for isentropic disks. Isentropic disks, therefore, have a pair of pressure modes and a Rossby mode. One may regard the trivial neutral mode as the fourth mode.

All of these p-, g-, and r-modes are stable in the short-wavelength limit. For perturbations with longer wavelengths, the couplings among these modes become essential and the effect of self-gravity of perturbations may become important.

The standard technique widely used in the study of oscillations of individual stars is proved to be useful for studying the oscillations of gas disks.

REFERENCES

Iye, M.: 1983, *M.N.R.A.S.* (in press).
Lynden-Bell, D. and Katz, J.: 1981, *Proc. R. Soc. London*, A378, pp.179-205.
Schutz, B.F. and Verdaguer, E.: 1983, *M.N.R.A.S.*, 202, pp.881-901.
Unno, W., Osaki, Y., Ando, H., and Shibahashi, H.* 1979, *Nonradial Oscillations of Stars*, University of Tokyo Press.
Verdaguer, E.: 1983, *M.N.R.A.S.*, 202, pp.903-925.

THE STABILIZING EFFECTS OF HALOES AND SPIRAL STRUCTURE

H. Robe
Institut d'Astrophysique, Université de Liège
B 4200, Belgium

The equilibrium and the stability of homogeneous gaseous and uniformly rotating MacLaurin spheroids imbedded in a rigid, homogeneous and spherical stellar halo are considered. Such systems may be useful as crude models of disk-halo galaxies. Explicitly, we have determined the modes of oscillations stemming from the fundamental radial mode of the homogeneous compressible sphere and its non radial "f" (or Kelvin) modes belonging to the second ($\ell=2$) and third ($\ell=3$) spherical harmonics, together with the first "pressure" and gravity modes p_1 and g_1 belonging to the first spherical harmonics ($\ell=1$). It has been assumed that the oscillations are adiabatic with a ratio of specific heats $\gamma=5/3$, and several sequences of spheroids corresponding to various values of the halo mass have been examined.

Two interesting results emerge from this study. We first find that the halo does not always exert a stabilizing effect on dynamical instability. Indeed, the gravity mode g_1 becomes stable for a value of the meridional excentricity e of the spheroid greater in the presence of the halo than without it, and this value increases with the ratio q of the halo mass to the spheroid mass (fig. 1).

On the other hand, the dynamical instabilities exhibited by the "two-arms" mode (i.e. corresponding to the spheroidal harmonics $\ell=2$ with the $\exp(i2\phi)$ azimuthal dependence) and the "three-arms" mode ($\ell=3$ and $\exp(i3\phi)$) are successively suppressed when the ratio q reaches rather low values: $q > q_2 \approx 1.2$ for $\ell=2$, and $q > q_3 \approx 2.1$ for $\ell=3$. This agrees with previous expectations (fig. 2). However, it is worth noting that when $q > q^* \approx 0.2$, those instabilities reverse order: for $q > q^* \approx 0.2$, the $\ell=3$ instability occurs first and even more, for $q_2 < q < q_3$, only this instability manifests itself. This suggests that the occurrence of three-armed spiral structures might be due to the presence of a halo of high enough mass.

H. van Woerden et al. (eds.), The Milky Way Galaxy, 537–538.

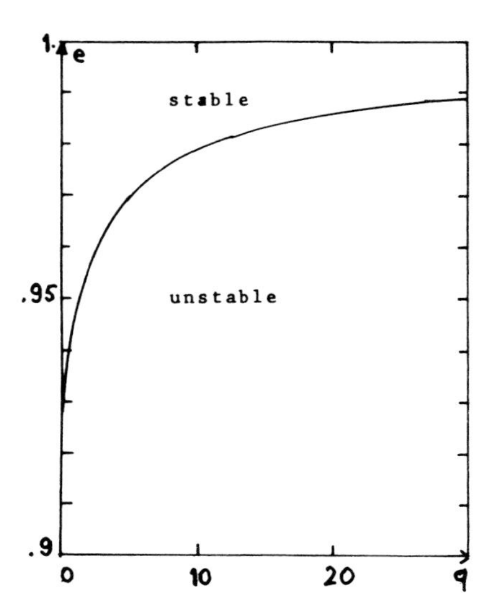

Fig. 1. Spheroid excentricity e at dynamical stability as a function of the ratio q of the halo mass to the spheroid mass, for the gravity mode g_1 with $\ell=1$.

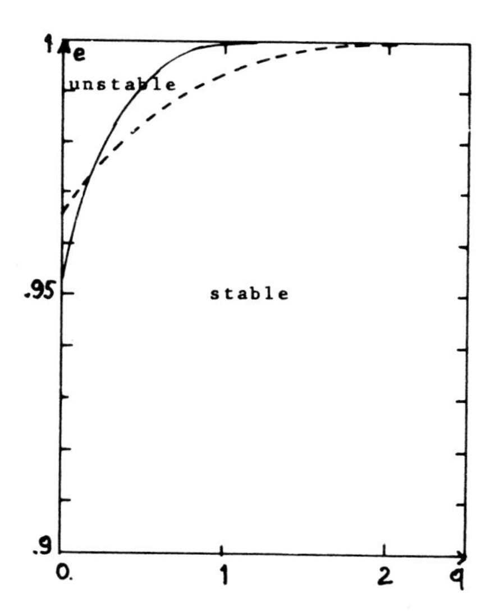

Fig. 2. Spheroid excentricity at dynamical instability as a function of q. The full and dashed lines correspond to the Kelvin modes $\ell=2$ and $\ell=3$, respectively.

NEW N-BODY EXPERIMENTS ON THE SPIRAL STRUCTURE OF GALAXIES

K.O. Thielheim and H. Wolff
University of Kiel, F.R.G.

N-body experiments with stellar disks like those performed by Hohl (1971) and Sellwood (1981) show regular, global, two-armed spiral structures associated with bar formation. The question is whether the spiral is caused by the bar, and if this is true, how the generating mechanism works.

To answer this question we have performed three types of experiments (Thielheim and Wolff 1984), in which the motion of some 10 000 stars in a model disk galaxy under the influence of their mutual gravitational interaction is followed by means of a two-dimensional N-body code. This code differs from others mainly in that the gravitational field of the stars is calculated by expanding their distribution into a biorthogonal system of surface mass density and potential functions, the so-called Hankel-Laguerre functions (Clutton-Brock 1972). These functions are particularly suited to the problem and thus only few are needed to approximate the large-scale mass distribution.

In a first series of experiments we have studied the evolution of unstable disks. These experiments show that spiral structure develops during the formation of a central bar. This process is accompanied by a mass redistribution away from corotation and by radial transfer of angular momentum that leads to the "heating" of the disk.

In a second series of experiments we study the response to oval distortions. Evolved disks are symmetrized azimuthally to produce stable axisymmetric disks. These stable disks are perturbed by an imposed oval distortion. These response experiments show that spiral patterns very similar to those in unperturbed, self-gravitating disks can be generated by bar forcing.

To study the influence of self-gravity, a third series of experiments is performed. The response experiments are repeated with disks of non-interacting stars, i.e., without self-gravity. The response patterns are virtually identical. Hence, the self-gravity of the disk is not required for the response mechanism.

H. van Woerden et al. (eds.), The Milky Way Galaxy, 539–540.

The results of our experiments are summarized schematically in Figure 1. N-body models with halo masses not exceeding approximately half of the total galactic mass show global instabilities that lead to the formation of bars that extend to the corotation radius. The growth of the bar amplitude during the bar instability destroys the symmetry of the stellar equations of motion and produces a quasi-stationary spiral-shaped response outside of the corotation radius. This mechanism does not require the self-gravity of the stars that participate in the spiral wave. The gravitational field of the spiral, however, exerts a torque on the bar and thus transports angular momentum from the bar region in outward direction to the region of the outer Lindblad resonance, around which the spiral is located. The extraction of angular momentum from the stellar orbits in the bar region leads to an increasing eccentricity of these orbits and thus to a reinforcement of the bar, which therefore continues to grow. This step closes the positive-feedback cycle.

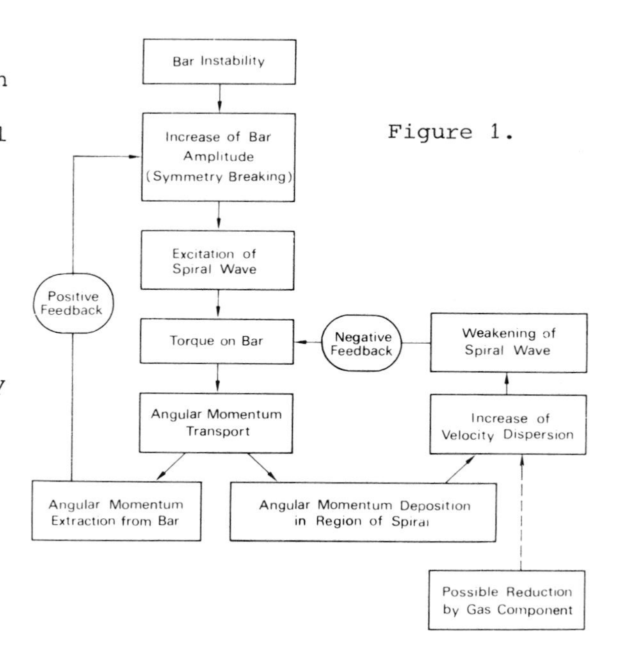

Figure 1.

On the other hand, the angular momentum drained from the bar is deposited around the outer Lindblad resonance and is also increasing the orbital deviations from circular motion which are partially transformed into random motion. Thus the velocity dispersion of the stars in this region is increased and the amplitude of the spiral wave is decreased. Consequently, the torque exerted on the bar is reduced. This step closes the negative-feedback cycle. Since the deposition of angular momentum near the outer Lindblad resonance is a cumulative effect, the negative-feedback cycle eventually dominates and the spiral structure vanishes, leaving the bar rotating without further change. This saturation can be avoided or at least slowed down by interaction with the interstellar gas through star formation which is "cooling" the disk.

REFERENCES

Clutton-Brock,M. 1972, Ap. Space Sci. 16, 101
Hohl,F. 1971, Ap. J. 168, 343
Sellwood,J.A. 1981, Astr. Ap. 99, 362
Thielheim,K.O. and Wolff, H. 1984, Ap. J., in press

VELOCITY DISPERSION AND THE STABILITY OF GALACTIC DISKS

John Kormendy
Dominion Astrophysical Observatory
Victoria B.C., Canada

ABSTRACT

This paper estimates the effect of the stellar velocity dispersion on local and global instabilities in galaxy disks. Measurements of rotation velocities and velocity dispersions are illustrated for the disks of NGC 488 (Illingworth and Kormendy, in preparation), NGC 936 and NGC 1553. These are used to derive the Toomre stability parameter Q, i.e. the ratio of the observed dispersion to that needed for marginal stability against local modes. In both NGC 488 and 1553, Q = 2.5 - 4, depending on the assumed mass-to-light ratio. These values larger than 2 imply that the stellar disks are stable, although the gas in NGC 488 may remain unstable. This is consistent with the observation that both galaxies largely lack coherent spiral structure: NGC 1553 is an SO and NGC 488 a flocculent spiral. The SBO galaxy NGC 936 is more extreme: Q = 5 (estimated error is a factor of two). This is consistent with Sellwood's suggestion that a bar heats the disk until it is too hot to support spiral structure. In contrast to the above galaxies, Toomre has shown that our Galaxy has Q = 1.6 ± 0.5 near the Sun. It is therefore responsive enough to make spiral structure, as observed.

We also explore the stellar kinematics of the disk of NGC 1553 as a function of radius. The exceptionally strong lens component in this galaxy is very hot. At half of the radius of the lens the velocity dispersion appears to be large enough to have an effect even on global stability.

H. van Woerden et al. (eds.), The Milky Way Galaxy, 541.

Roland Wielen (left) and Hugo van Woerden discuss scientific program during excursion, with Ute Wielen and Ria van Woerden (right) listening. Background, left to right: Bruce Elmegreen, John Kormendy, Garth Illingworth, Eli Brinks. LZ

PERIODIC ORBITS RELEVANT FOR THE FORMATION OF INNER RINGS IN BARRED GALAXIES

M. Michalodimitrakis and Ch. Terzides
Department of Astronomy and Theoretical Mechanics
Thessaloniki, Greece

The study of orbits of a test particle in the gravitational field of a model barred galaxy is a first step toward the understanding of the origin of the morphological characterstics observed in real barred galaxies. In this paper we confine our attention to the inner rings. Inner rings are a very common characteristic of barred galaxies. They are narrow, round or slightly elongated along the bar (with typical axial ratios from 0.7 to near 1.0), and of the same size as the bar. A first step to test the old hypothesis that inner rings consist of stars trapped near stable periodic orbits would be a study of particle trapping around periodic orbits encircling the bar. Such a study is contained in the work of several authors (Danby 1965, de Vaucouleurs and Freeman 1972, Michalodimitrakis 1975, Contopoulos and Papayannopoulos 1980, Athanassoula et al. 1983). In the above works the stability of periodic orbits was studied with respect to perturbations which lie on the plane of motion $z = 0$ (planar stability). To ensure the possibility of formation of rings, a study of stability with respect to perturbations perpendicular to the plane of motion (vertical stability) is necessary. In this paper we investigate the properties of periodic orbits which we believe to be relevant for the inner-ring problem using a sufficiently general model for the galaxy and sets of values for the parameters which cover a wide range of different possible cases. We also study the stability, planar and vertical, with respect to large perturbations in order to estimate the extent of particle trapping. A detailed numerical investigation of three-dimensional periodic orbits will be given in a future paper.

We find that: a) for a wide range of values of the dimensionless parameters stars can be trapped in a significant three-dimensional region around the direct and retrograde periodic orbits in the vicinity of the bar. Such trapping may be responsible for the formation of inner rings. b) the main contribution to the particle trapping is due, in general, to the family x_2 of direct periodic orbits which extends between the two inner Lindblad resonances.

H. van Woerden et al. (eds.), The Milky Way Galaxy, 543–544.

A comparison of our results with those of other authors shows that the identification of the families which are relevant for the inner-ring problem as well as the relative significance of these families crucially depends on the model used and on the range of values of the parameters involved.

We must note that our work, as well as the works of the above-mentioned authors, offers no proof that a visible inner ring is formed from the orbits we propose. Such a proof is a difficult task which requires additional investigation.

REFERENCES

Athanassoula, E., Bienaymé, O., Martinet, D. and Pfenniger, D.: 1983, Astron. Astrophys. 127, 349
Contopoulos, G., Papayannopoulos, Th.: 1980, Astron. Astrophys. 92, 33
Danby, J.M.A.: 1965, Astron. J. 70, 501
De Vaucouleurs, G., Freeman, K.C.: 1972, Vistas in Astronomy 14, 153
Michalodimitrakis, M.: 1975, Astrophys. Space Sci. 33, 421 and 37, 131

Terzides discusses his poster with Lynden-Bell (left) and Casertano. CFD

ON THE 3-KPC ARM: WAVES EXCITED AT THE RESONANCE BY AN OVAL DISTORTION IN THE CENTRAL REGION

Chi Yuan
NASA Ames Research Center, Moffet Field, CA 94035 U.S.A.
On leave from the City University of the City College of New York

A relatively minor oval distortion in the central region of the Galaxy, turning at a representative angular pattern speed, can excite outgoing waves at the outer Lindblad resonance of that pattern speed. Associated with the density crest of these waves is fast-expanding gas flow. The physical basis of this phenomenon can be understood through a linear analysis. However, to explain the observed expanding velocity in the "3-kpc arm", the non-linear theory must be used. In our calculations an oval distortion turning at 118 km s^{-1} kpc^{-1} with a perturbation of 10% of the mean gravitational field at the outer Lindblad resonance (located at 3 kpc in the present case) can generate an outgoing velocity of 53 km s^{-1} at the first density crest of the wave (located at 3.6 kpc).

The mechanism which is responsible for the expanding motion can be understood as resonance excitation through a linear analysis. Within the short-wave approximation of the linear theory, the amplitude of the wave, say, for the radial velocity, is governed by a second-order inhomogeneous ordinary differential equation. This equation has a turning point at the outer Lindblad resonance such that its solution decays algebraically towards the center and has a wave-like behaviour outside the resonance. Together with the phase term, the solution is represented by two tightly-wound trailing spiral waves in the rotating frame of the oval distortion. The density and the radial velocity are in phase with each other, so the density crest is always associated with the maximum outmoving velocity. The amplitude of the wave for the density in the absence of viscous damping remains fairly constant as it propagates, although the forcing term decreases like a quadruple moment from the center. This suggests that the fast-turning oval distortion pumps its energy and angular momentum into the disk system mainly by means of exciting waves at the resonance regions, and that this excess of energy and angular momentum is carried away by the gas medium in the form of compression waves.

An assessment of the importance of viscous effects for these waves can be made by calculating a kinematic viscosity associated with a mean-free-path for cloud-cloud collisions, ℓ, and the random speed of HI

H. van Woerden et al. (eds.), The Milky Way Galaxy, 545–546.

clouds, a. A viscosity of 1 kpc km s^{-1} obtains for $\ell = 0.1$ kpc and $a = 10$ km s^{-1}. The e-folding distance for this viscosity is about 1 kpc. Thus one expects that the energy dissipation and angular-momentum deposition will occur within a short distance beyond resonance and will therefore not interfere with the outer spiral waves.

The resonance-excitation mechanism described above is so effective that an oval field of merely 1% of the mean gravitational field at the resonance would produce a disturbance which well exceeds the limit of the linear theory. We must therefore use the non-linear theory in order to explain the "3-kpc arm" phenomenon. Since we are only interested in a narrow region in the neighbourhood of the resonance, a transformation may be introduced to reduce a two-dimensional steady problem to a one-dimensional unsteady but periodic problem. Some accuracy is lost in this transformation, but the general non-linear nature is preserved. The non-linear solution has the same wave character as the linear solution. With a 10% oval field, the first density crest reaches a peak of 4.6 (from an unperturbed density equal to 1) and the corresponding radial velocity is 53 km s^{-1}. The mesh size in the numerical calculation is taken such that the numerical viscosity involved is comparable to the kinematic viscosity estimated above. The results are depicted in the figures 1 and 2.

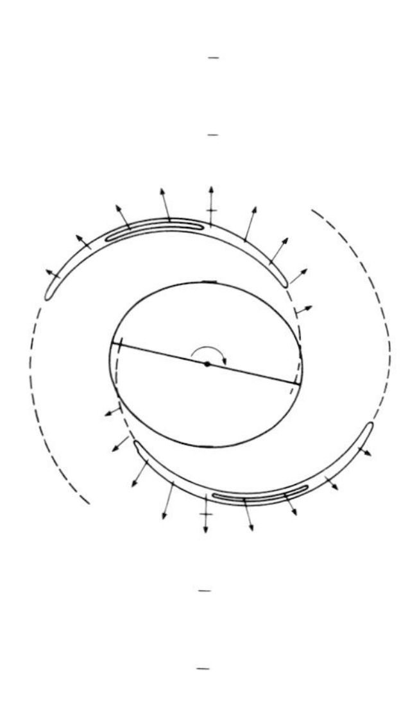

Figure 1. Density Contour of the 3-kpc arm. Arrows indicate radial expansion velocity.

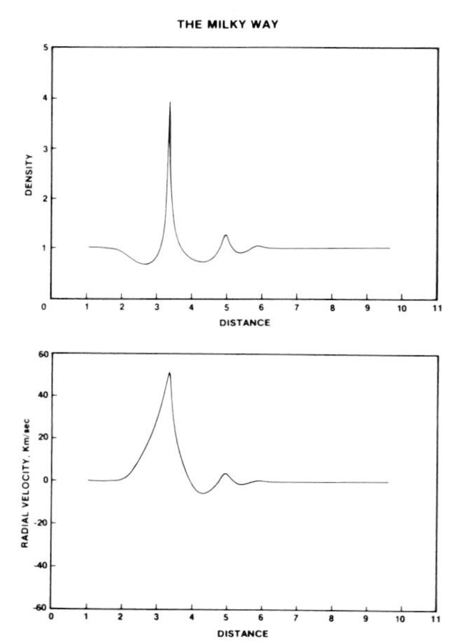

Figure 2. Distributions of density (above) and radial velocity (below) in the ray towards the Sun.

A BARRED GALAXY: THE INSIDE VIEWPOINT

G.D. van Albada
Kapteyn Astronomical Institute
Groningen, the Netherlands

ABSTRACT: The observed gas distribution and kinematics in the central 4 kpc of our Galaxy show many aspects reminiscent of barred spirals. In this note I describe a collection of (l,v) maps obtained from a gas flow model of a barred spiral.

The central 4 kpc of our Galaxy shows a number of features that may be very similar to features commonly observed in barred spirals. The most striking of these are: a) the ring of HII regions around R = 4 kpc, similar to the inner ring in SB(r) or SAB(r) galaxies; b) the general lack of gas inside this ring down to R = 1 kpc; c) the inner molecular ring and nuclear disk, corresponding to a nuclear ring of HII regions; d) the large observed deviations from circular rotation. Where important details, such as the 3-kpc arm, should fit in is, however, not entirely clear.

The very different vantage point from which we observe our own Galaxy makes a more detailed comparison very difficult. Comparison with detailed model calculations of the gas flow appears to be a promising approach to the problem. Computations are now available that reproduce most of the observed features of SBbc-type galaxies (G.D. van Albada, 1984). The gas distribution and gas streamlines obtained for one such model are shown in Figure 1.

The model displayed was obtained for a bar with an axial ratio of 2.5:1 and a maximum ratio of tangential to radial forcing (Qtmax; Sanders and Tubbs, 1980) of 24%. A number of features stand out clearly in this model, namely 1) the straight, narrow regions of high gas density, presumably corresponding to the dust lanes in the bar and ending in a central region of enhanced gas density, and 2) crossing a region generally lacking in gas, two streams that show similarities to the "feathers" often observed in bars. Due to inadequacies in the current mass model, the central gas disk is rather too large and interferes with the formation of more realistic feathers. Yet we can already obtain some insight from this model into the way the various observed

H. van Woerden et al. (eds.), The Milky Way Galaxy, 547–549.

features could arise. The identifications of the nuclear disk and that of the empty region outside it are obvious. The 3-kpc arm is somewhat more problematic, but also of greater diagnostic value. If we identify it with a feather, the bar must make an angle of about 40° with the line of sight, with the near end at positive longitude. The lack of an exact counterpart at positive velocities is not surprising, in view of the asymmetries generally observed in galaxies, while several features can be found that would correspond to the +135 km/s arm.

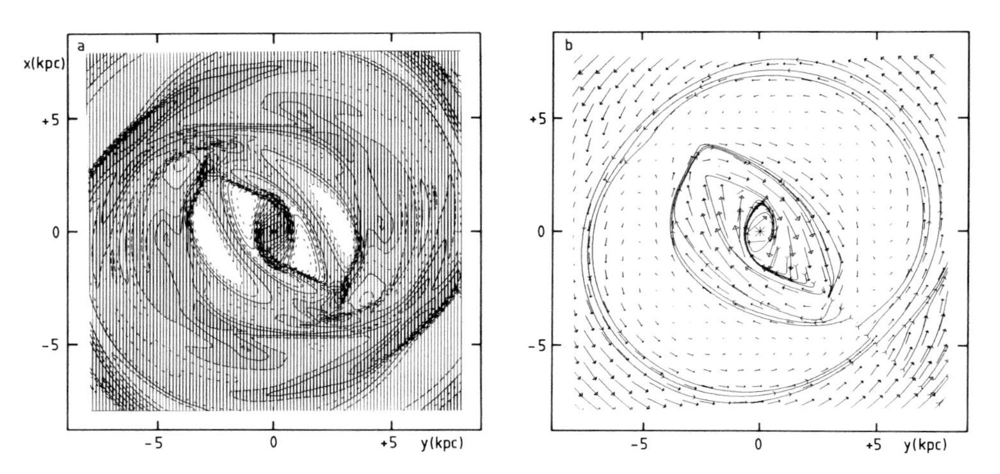

Figure 1. a: The central 8 kpc of the gas density distribution in a barred-galaxy model. b: The gas stream-lines in the rotating frame of the bar. The bar is at a position angle of +45° and has a semi-major axis of 5 kpc.

Using the computed gas distribution as input (and assuming that it represents one species, such as HI), it is possible to produce artificial observations. In Figure 2 we present (l,v) maps as they would be observed at a distance of 12 kpc (for scale purposes) from the centre, in the plane of the model in Figure 1. These (l,v) maps may be compared with the HI observations published by e.g. Burton (1970), or the CO observations published by e.g. Bania (1980). Many features similar to those observed in our Galaxy are indeed reproduced. The (l,v) map in Figure 2d especially appears to be a rather good model. However, in this case the feather corresponding to the 3-kpc arm would be at the far side of the nucleus. From Figure 2b, which corresponds most closely to the expected geometry, it is clear that a smaller, more rapidly rotating nuclear disk is required in order to obtain a more acceptable model.

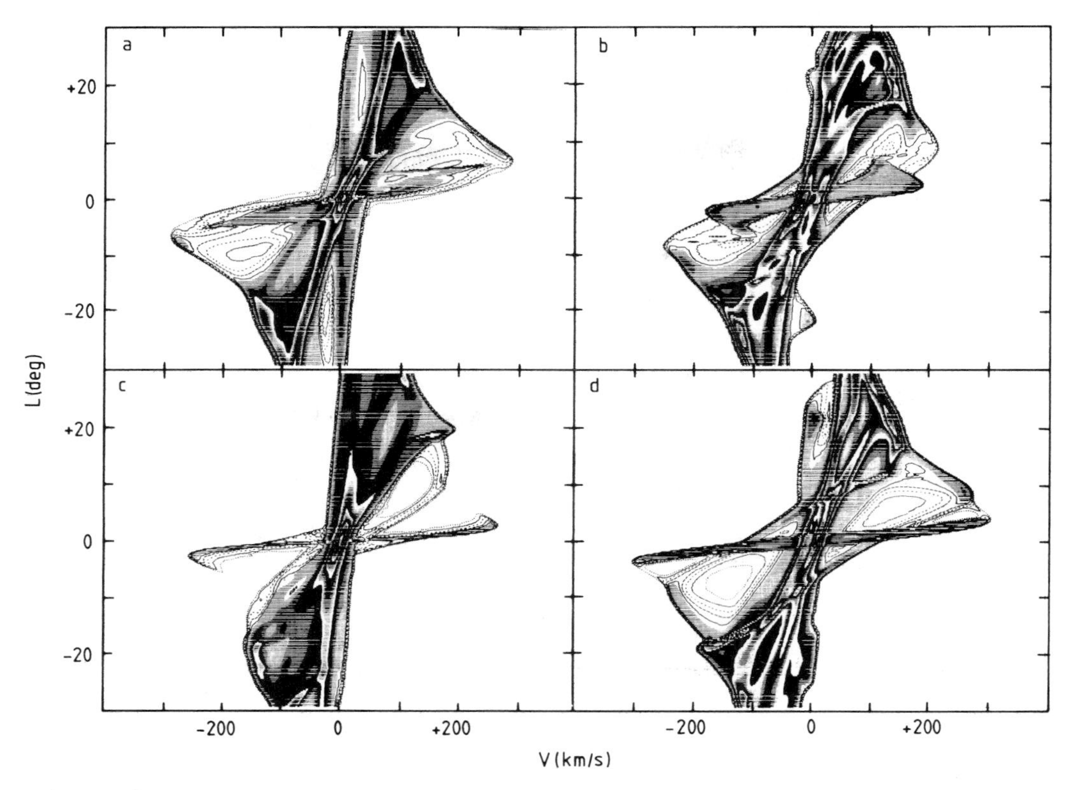

Figure 2. a: An (l,v) map such as would be observed along the bar from a distance of 12 kpc in the model galaxy of Figure 1.
b: Same as a, but 45° counter-clockwise relative to the bar.
c: Same as b, but perpendicular to the bar.
d: Same as b, but 135° counter-clockwise with respect to the bar.

REFERENCES

Albada, G.D. van, 1984, in preparation
Bania, T.M., 1980, Astrophys. J. 242, 95
Burton, W.B., 1970, Astron. Astrophys. Suppl. 2, 261
Sanders, R.H., Tubbs, A.D., 1980, Astrophys. J. 235, 803.

Van de Hulst (left) studies posters of G.D. van Albada (right) and Yuan.
Background: Hodge CFD

STOCHASTIC STAR FORMATION AND SPIRAL STRUCTURE

Philip E. Seiden
IBM Thomas J. Watson Research Center
Yorktown Heights, New York 10598

INTRODUCTION

Most approaches to explaining the long-range order of the spiral arms in galaxies assume that it is induced by the long-range gravitational interaction. However, it is well-known in many fields of physics that long-range order may be induced by short-range interactions. A typical example is magnetism, where the exchange interaction between magnetic spins has a range of only 10 ångströms, yet a bar magnet can be made as large as one likes. Stochastic self-propagating star formation (SSPSF) starts from the point of view of a short-range interaction and examines the spiral structure arising from it (Seiden and Gerola 1982). We assume that the energetic processes of massive stars, stellar winds, ionization-front shocks and supernova shocks, in an OB association or open cluster can induce the creation of a new molecular cloud from cold interstellar atomic hydrogen. In turn this new molecular cloud will begin to form stars that will allow the process to repeat, creating a chain reaction. The differential rotation existing in a spiral galaxy will stretch the aggregation of recently created stars into spiral features.

This process belongs to a class of phenomena called percolation and exhibits the phase transition characteristic of this kind of system (Schulman and Seiden 1982). It is only necessary that the short-range interaction be strong enough to cross the phase-transition point; its value after that is immaterial since, as shown below, the strong feedback between the gas and stars will control the process. In the rest of this paper I discuss only the statistical mechanics of the percolating process and not the details of the interaction itself.

THE MODEL

SSPSF is easily investigated by means of a computer simulation. We create a polar array of an appropriate number of rings and divide each ring into cells so that each cell in the array has the same area. The rings are allowed to rotate following a desired rotation curve. Three

H. van Woerden et al. (eds.), The Milky Way Galaxy, 551–558.

arrays are included, one of stars, a second of atomic hydrogen and a third of molecular hydrogen.

The process proceeds as follows. If there is a young cluster created in a cell at time t, it can create a new molecular cloud in any one of its neighbouring cells at time t+1 with a probability

$$P_{eff} = P\,D_a, \tag{1}$$

where P is a constant and D_a is the density of atomic hydrogen. The molecular gas returns to its atomic form with a gas recycling time, τ. The recycling time is not just the lifetime of some observed feature of a molecular cloud, but it is the length of time it takes for the gas to return to the cold atomic-hydrogen state suitable for shocking and repeating the process.

The feedback between the stars and the gas provides a close control of the amount of atomic hydrogen in the system (Seiden 1983). Figure 1 shows results for a model in which the total gas density was chosen to drop off exponentially with radius. The figure shows that, as long as the total gas density is large enough, the atomic hydrogen density is pinned at a constant value. This result is independent of the recycling time, as long as it is great enough to provide effective feedback.

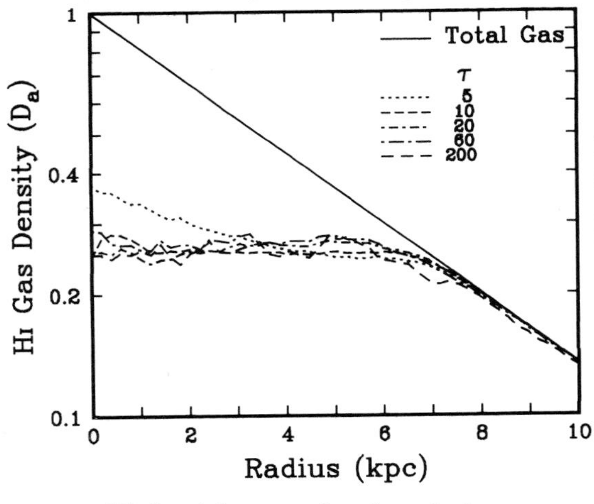

Figure 1. Atomic gas density for models with an exponential total gas disk. The scale of the ordinate is normalized to unity at R=0.

With this result in mind we can rewrite equation (1) as follows

$$P_{eff} = D_a/D_c, \tag{2}$$

where D_c is a critical gas density at which cloud formation is assured (the largest possible value of equations (1) and (2) is unity, greater values are truncated to unity). We can evaluate D_c from observations; Burton and Gordon (1978) found that the atomic hydrogen in our Galaxy is roughly flat over a wide range of radii with a value of about 0.35 H cm^{-3}. Simulations show that feedback pins the effective probability at about 0.3 (Seiden 1983). Therefore, D_c = 1 H cm^{-3}.

Since the atomic hydrogen is constant over the galactic disk, the molecular hydrogen will be given by the total gas content minus a constant. We don't know what the total gas content of a galaxy is in general, but we can argue that from the nature of the rotation curve it might be close to 1/r. Firstly, a flat rotation curve for a disk implies that the mass density of the disk is 1/r. For a spherical distribution it will be $1/r^2$, but if the disk is formed by the accretion of gas from the spherical distribution, the resulting gas disk will again be 1/r. Observational evidence seems to indicate that the star-formation rate in the Galaxy has not changed appreciably over the lifetime of the disk, so the gas distribution may not have appreciably changed either. In any event the essential point can be illustrated with a 1/r total gas distribution. This is shown in Figure 2: the 1/r line shows no trace of the rough exponential behaviour usually seen for both the luminosity of a galaxy and its molecular hydrogen (Young and Scoville 1982).

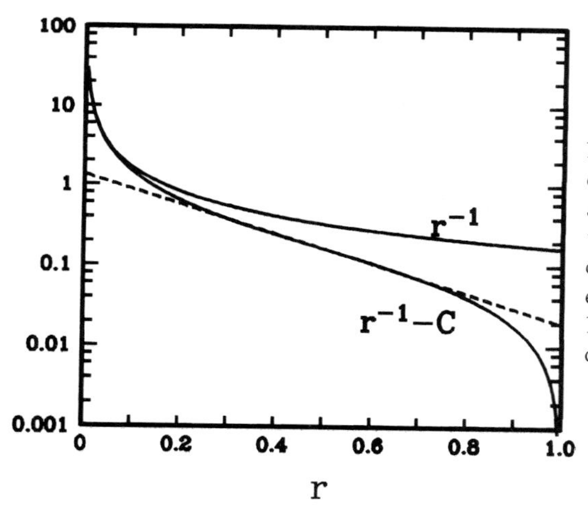

Figure 2. Origin of the exponential disk. The upper curve is 1/r; when a constant (C) is subtracted from this curve, the lower curve is obtained. The good fit to an exponential (superposed straight line) extends over a large fraction of the disk.

However, when a constant is subtracted from this curve, a clean straight-line section appears - the so-called exponential disk (Freeman 1970). See Seiden, Schulman and Elmegreen (1984) for a more extensive discussion.

THE MILKY-WAY SIMULATION

The SSPSF model described above will now be applied to a simulation of the Milky Way. The simulation was done with the following parameters. The array contains 90 rings and has a radius of 18 kpc so that the cell size is 200 pc. The position of the Sun is at 8 kpc. The rotation curve is rigid to 3 kpc and is then flat with a value of 220 km/sec. $D_c = 1\ cm^{-3}$ as above and the total gas density is chosen to vary as 1/r with a value of 0.3 cm^{-3} at 18 kpc. Inside of 3 kpc, where the rotation curve is rigid, the gas density is taken as flat, independent of radius. The time step is 10^7 years and the gas regeneration time is 2×10^8 years.

Figure 3 shows typical radial dependencies for this simulation. The atomic hydrogen is roughly flat as expected, and the molecular hydrogen falls off roughly exponentially from 3 kpc to about 14 kpc. The blue luminosity shows the same approximate exponential behaviour. The more rapid fall-off at larger radii is due to the phase-transition nature of the process. Such a fall-off has been observed by van der Kruit and Searle (1982) for edge-on disk galaxies.

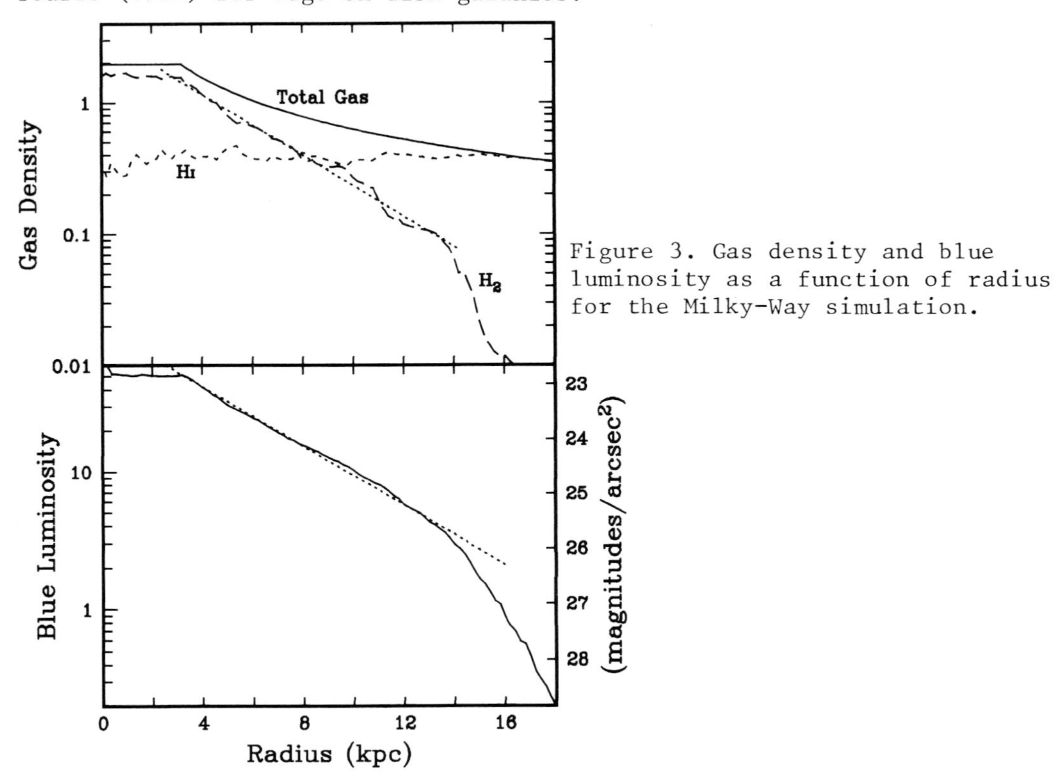

Figure 3. Gas density and blue luminosity as a function of radius for the Milky-Way simulation.

We can compare our results to observations in the following manner. First, we normalize the surface brightness by choosing r_{25} = 11.5 kpc as given by de Vaucouleurs and Pence (1978). Table I shows the values we get for a number of other parameters, along with the observed values.

TABLE I

	calculated	observed	
μ_e (mag/arcsec^2)	23.43	23.47	(de Vaucouleurs and Pençe 1978)
r_e (kpc)	6.1	5.68,5.98	(de Vaucouleurs and Pence 1978)
$r_{Holmberg}$ (kpc)	15.1	17.05	(de Vaucouleurs and Pence 1978)
r_D (kpc)	4.06	3.38,3.56	(de Vaucouleurs and Pence 1978)
r_{max} (kpc)	16.2,16.8	16-20	(Chromey 1978)
M(HI) (10^9 $M_\odot$)	0.8 z_{100}	1.66	(Baker and Burton 1975)
Arm/Interarm contrast:			
HI	3.6-5.2	4	(Kulkarni _et al_. 1982)
H_2	10-30	10-20	(Stark 1983)

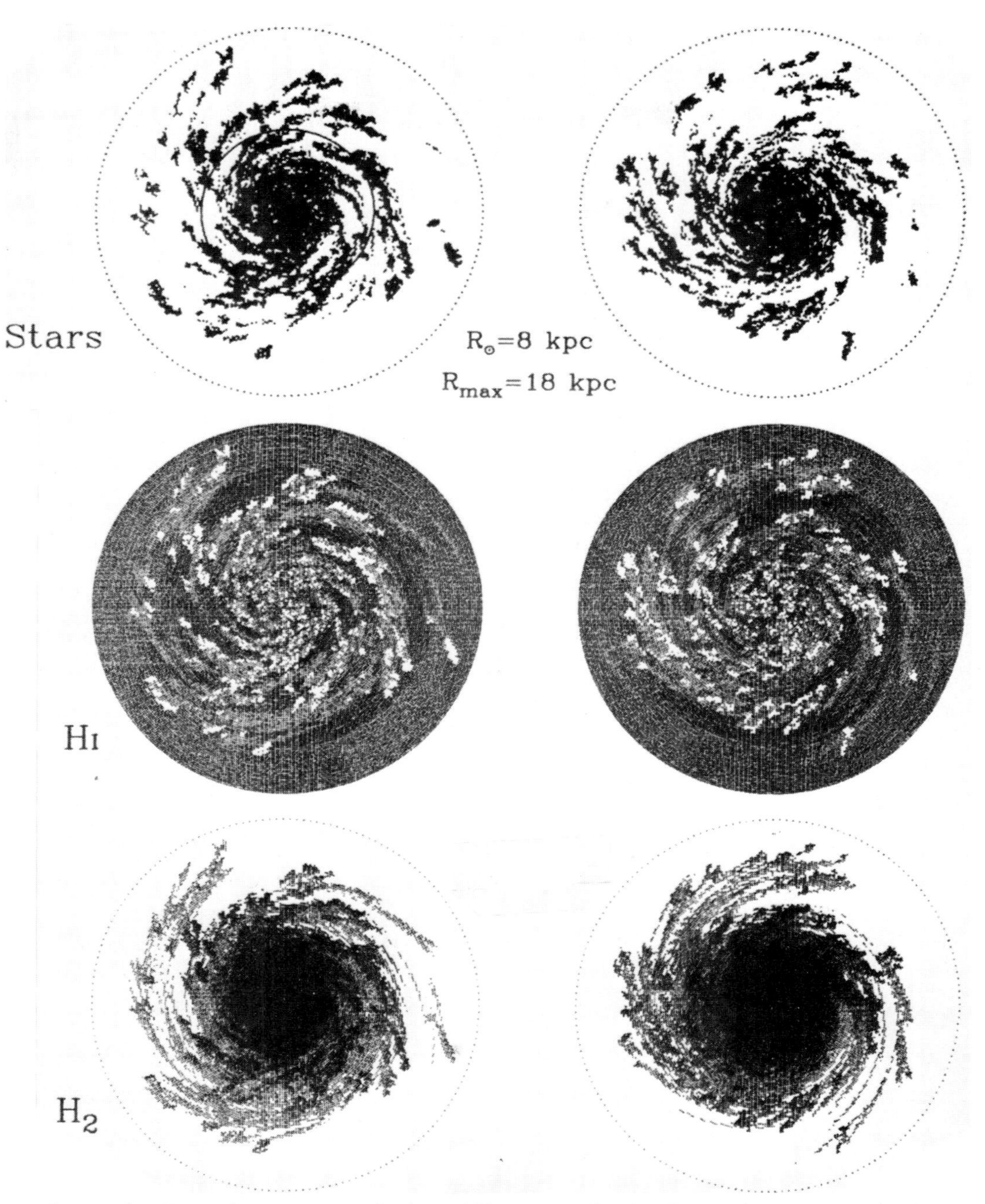

Figure 4. Two time steps of the Milky-Way simulation. The outer circles are the limits of the array. The inner circle in the upper left-hand frame is the radius of the Sun.

One major difference we find from the de Vaucouleurs and Pence results is in the value for the Holmberg radius. Theirs is considerably bigger since they chose an exponential disk all the way out. The rapid fall-off of Figure 3 gives a smaller value for the Holmberg radius. The parameter r_{max} is the van der Kruit and Searle parameter for the cut-off of the disk. We obtain it in two different ways. First, analysis shows that it is just $4r_D$ (Seiden, Schulman and Elmegreen 1984). Second, we can do a fitting procedure to the data similar to van der Kruit and Searle (1982). The observed value for our Galaxy comes from the radial decrease of Chromey's (1978) blue-star counts for the anticenter. The atomic hydrogen is the amount within a 12-kpc radius, z_{100} is the scale height in units of 100 pc. The observed value is the Baker and Burton (1975) result as corrected by de Vaucouleurs and Pence (1978) for 8 kpc.

Figure 4 shows the stars, atomic hydrogen and molecular hydrogen for two timesteps. The radius of the Sun is shown by the inner circle on the top left-hand picture. The spiral arms can be clearly seen in all three pictures. The stars and molecular hydrogen show the same structure but with different contrast. However, the atomic hydrogen is roughly a negative image of the stars and molecular hydrogen. The clearest arms are shown by a depletion of atomic hydrogen; nevertheless, one can still trace the arms by the ridges of dense atomic hydrogen, but these are displaced from the locus of the stellar and molecular-hydrogen arms by a few hundred parsecs. One should also note that there is atomic hydrogen surrounding each molecular cloud and this has not been included in Figure 4.

One feature that this simulation does not reproduce properly is the hole in the gas inside 3 kpc. This may be a three-dimensional effect, since in the inner region of the Galaxy the simple two-dimensional disk approximation may break down.

This one exception aside, the SSPSF model gives a reasonable description of the Milky Way with few adjustable constants. In fact the only constant that is really a free parameter unconstrained by observation is the gas recycling time.

REFERENCES

Baker, P.L., and Burton, W.B.: 1975, Astrophys. J. 198, p. 281.
Burton, W.B., and Gordon, M.A.: 1978, Astron. and Astrophys. 63, p. 7.
Chromey, F.R.: 1978, Astron. J. 83, p. 162.
de Vaucouleurs, G., and Pence, W.D.: 1978, Astron. J. 83, p. 1163.
Freeman, K.C.: 1970, Astrophys. J. 160, p. 811.
Kulkarni, S.R., Blitz, L., and Heiles, C.: 1982, Astrophys. J. 259, p. L63.
Seiden, P.E.: 1983, Astrophys. J. 266, p. 555.
Seiden, P.E., and Gerola, H.: 1982, Fund. Cosmic Phys. 7, p. 241.
Seiden, P.E., Schulman, L.S., and Elmegreen, D.B.: 1984, Astrophys. J. (to be published).
Schulman, L.S., and Seiden, P.E.: 1982, J. Stat. Phys. 27, p. 83.
Stark, A.A.: 1983, in "Kinematics, Dynamics and Structure of the Milky Way", W.L.H. Shuter (ed.), Reidel, p. 127.
van der Kruit, P.C., and Searle, L.: 1982, Astron. Astrophys. 95, p. 105.
Young, J.S., and Scoville, N.: 1982, Astrophys. J. 258, p. 467.

DISCUSSION

J.P. Ostriker: Several people have noted that cloud-cloud collisions will produce viscous transport of mass and angular momentum on a relevant time scale within a radius of a few kpc of the center of our Galaxy. If you had included dynamics: collisions and viscosity, you probably would have produced a central hole.

F.H. Shu: Could you list for us all the free parameters of your theory, what physical processes they are supposed to represent, and how you choose the numerical values for these parameters in your simulations?

Seiden: I can afford to be somewhat cagey about the physical processes, because I am doing statistical mechanics of the array of molecular clouds and clusters being formed. Similarly, in statistical mechanics you do not have to know the molecular orbitals of the atoms scattering off each other; you only have to know that they scatter off each other so that they reach equilibrium. In this case all I have to say is that a process exists whereby massive stars can create new clouds. I do not have to specify the details of that process in order to make calculations and obtain results, because in statistico-mechanical language I am doing the macroscopic theory of the phenomenon, not the microscopic theory (and "microscopic" here means 200 pc). The free parameters are: rotation speed, cell size, molecular recycling time, and gas density distribution. Their values are given in the paper.

J.H. Oort: Can you make a barred spiral?

Seiden: No, a barred spiral is undoubtedly a dynamical effect. There is no dynamics in these calculations, there is no gravity (other than the rotation curve which comes from gravity).

U.J. Schwarz: Have you compared the morphology of your Galaxy with real galaxies? They make the impression of a mixture of irregular galaxies with spiral patches, a sort of fireworks.

Seiden: There are lots of galaxies which look like that in the Hubble Atlas.

R.S. Cohen: As far as we can tell in our own Galaxy, the molecular and atomic arm features agree very closely.

Seiden: To better than 200 pc?

Cohen: Well, there are all these dangers of course in the (l,V) diagram. But as far as we can read off the diagram, they agree exactly.

Seiden: There should be a shift of a few hundred parsecs - that's all.

W. Renz: From your film I got the impression that the star-formation rate is rather constant in time. On the other hand, the poster paper by Feitzinger at this Symposium shows that for a certain choice of model parameters one can obtain a periodically variable star-formation rate. Have you performed numerical simulations about this, and what oscillation periods do you find?

Seiden: The original calculation was a collaborative effort by Feitzinger, Schulman and myself. The oscillations increase with the value of , the lifetime of the molecular clouds. For high enough values of , star formation occurs in pulses, with almost-zero values in between. Whether those values of are realistic or not, is an open question at this time. The oscillation period is related to , it is not quite linear in but it is of the same order.

A.A. Stark: A point of nomenclature. What you and others usually call the lifetime of molecular clouds, would be more properly called the "molecular recycling time".

Seiden: That's a wonderful word. I did not think of it, but I will use it from now on. (It is used indeed in the printed paper - Editor.)

Stark: I know, we people in AT&T are always competing with IBM, but ...

Seiden: Hahahaha. Yes, but being the only industrials around, we should collaborate here.

M.L. Kutner: It seems that this recycling time is too long to allow a strong contrast between arms and interarm regions as we observe in CO.

Seiden: No - it depends on what you are talking about. I have not made a complete analysis of all the contrasts that are available in these models, I just did a few to have some numbers here. But the highest contrast occurs around those cells that had clouds formed in them fairly recently. So the lifetime is not really important: the contrast is between cells that have not formed clouds for well over a timescale , and cells that have just formed clouds a few timesteps ago.

R. Beck: There is increasing evidence that the process of star formation is influenced by the surrounding interstellar magnetic field. Thus it seems important to consider the field strength and structure in your theory of star-formation propagation.

Seiden: That has been tried by Chiang and Elmegreen. They used an anisotropic, stimulated probability in the old-star model without gas. They found an effect in galaxies that do not have large rotation velocities, no large shears - for instance the Magellanic Clouds, or galaxies with even more rigid rotation than those. For large disk galaxies there was no effect: the strong rotation shear dominated everything.

SPIRAL STRUCTURE IN GALAXIES: LARGE-SCALE STOCHASTIC SELF-ORGANIZATION OF INTERSTELLAR MATTER AND YOUNG STARS

J.V. Feitzinger
Ruhr-University, Bochum, BRD

Galaxies are dissipative systems, and the spatial and time structure of the interstellar medium and young stars is governed by reaction-diffusion equations. The coherent galactic oscillations of star formation self-organized in spiral waves, previously detected by numerical simulations (Seiden, Schulman, Feitzinger, 1982) can be analytically described by the concept of a limit cycle. Analytical work on self-propagating stochastic star formation is also done by Kaufman (1979), Shore (1981, 1982) and Cowie and Rybicki (1982).

The star-formation rate from processes induced by massive stars and the fraction of the interstellar medium involved in these processes are related (Seiden 1983). The stochastic process of forming new stars through the compression of the interstellar medium by shock waves is coupled with galactic differential rotation. This produces the spiral structure, since the spiral arms are caused by waves of propagating star formation in a shearing disk.

The basic idea is that the interstellar medium does not respond immediately to its stellar environment. Instead there is a delay. The rate of exchange of the star-forming phase of the ISM is not a function of the present stellar population N(t), but of a past population $N(t-t_d)$, where t_d is the delay time:

$$dN(t)/dt = f(N(t-t_d))$$

The delay time may be the time it takes a molecular cloud to form stars, or the cooling time of the hot gas. If we apply the ideas behind logistic growth to the delay differential equations, then such a delay can cause the different populations to oscillate around an equilibrium situation. The description of an interaction between interstellar material (I) and star (N) results in predator-prey situations and Lotka-Volterra differential equations. This model can be extended in the frame of a three-component medium: Molecular clouds (I_M), hot gas (I_H) and new stars (N). Then a system of three reaction-diffusion equations governs the spatial and time structure of the interstellar medium and young stars:

H. van Woerden et al. (eds.), The Milky Way Galaxy, 559–560.

$$dI_M/dt = I_M(a-bI_M-cN) + D_M \nabla^2 I_M$$

$$dI_H/dt = I_H(-kN+aI_M+d) + D_H \nabla^2 I_H$$

$$dN/dt = N(-k+\lambda I_M) + D_N \nabla^2 N$$

The constants a, b, c, λ, k describe the different interaction and evolution processes between the three components; ∇^2 is the Laplace operator and D are the diffusion constants. The isoclines along which $dI/dN = 0$ are curves, separating regions in which a population is increasing from regions in which the same population is decreasing. The solution curves oscillate around an equilibrium value, yielding a periodic oscillation of the population. This is a limit cycle. In this framework spiral arms are dissipative structures emerging as a spatio-temporal order in the sense of Prigogine (Nicolis and Prigogine, 1977) and the synergetic concept of Haken (1978). The steady state and the oscillating mode depend on the values of parameters, i.e. on the physics of the interstellar medium.

One consequence of the limit cycle is the existence of travelling waves (trigger waves). The space behaviour of such waves can be followed using the kinematical or geometrical-optical method (Whitham 1960, 1974; Hunter 1973; Cowie and Rybicki 1982). The method involves determining the rays along which waves are propagated. Such rays run together to form an envelope or caustic. The caustics delineate the most prominent features of the pattern.

REFERENCES

Cowie, L.L., Rybicki, G.B.: 1982, Ap.J. 260, 504
Haken, H.: 1978, Synergetics, Springer Verlag, Berlin
Hunter, C.: 1973, Ap.J. 181, 685
Kaufman, M.: 1979, Ap.J. 232, 707 and 717
Nicolis, N., Prigogine, I.: 1977, Self-Organization in Nonequilibrium Systems, Wiley Interscience Pub., New York
Seiden, P.E., Schulman, L., Feitzinger, J.V.: 1982, Ap.J. 253, 91
Seiden, P.E.: 1983, Ap.J. 266, 555
Shore, S.N.: 1981, Ap.J. 249, 93
Shore, S.N.: 1982, Ap.J. 265, 202
Whitham, G.B.: 1960, J. Fluid Mech. 9, 347
Whitham, G.B.: 1974, Linear and Nonlinear Waves, Wiley Interscience Pub., New York

STAR FORMATION IN MOLECULAR CLOUDS

Frank H. Shu
Astronomy Department, University of California, Berkeley

ABSTRACT. We examine how star formation occurs in the Galaxy and come to the following conclusions. (1) The distribution of newly-born stars in the Galaxy depends on the origin of giant-molecular-cloud complexes. For individual complexes, we favor the mechanism of Parker's instability behind galactic shocks. The production of "supercomplexes" may require the mediation of Jeans instability in the interstellar gas. (2) Magnetic fields help to support the clumps of molecular gas making up a complex against gravitational collapse. On a timescale of 10^7 years, these fields slip by ambipolar diffusion relative to the neutral gas, leading to the formation of dense cloud cores. This timescale is the expected spread in ages of stars born in any clump. (3) When the cores undergo gravitational collapse, they usually give rise to low-mass stars on a timescale of 10^5 years. (4) What shuts off the accretion flow and determines the mass of a new star is the onset of a powerful stellar wind. The ultimate source of energy for driving this wind in low-mass stars is the release of the energy of differential rotation acquired during the protostellar phase of evolution. The release is triggered by the entire protostar being driven convectively unstable when deuterium burning turns on.

INTRODUCTION

The problem of star formation in spiral galaxies attracts a diverse group of workers, because the phenomenon spans at least twelve orders of magnitude in length scale. In this paper, I shall comment on a selected list of physical events that allow a galaxy of size 10^{23} cm to give birth to a star of size 10^{11} cm.

THE ORIGIN OF MOLECULAR-CLOUD COMPLEXES

It is widely believed that the bulk of star formation today takes place in giant-molecular-cloud complexes. The empirical evidence is consistent with the interpretation that, once we have a molecular cloud of sufficient mass, no external inducement is needed to yield stars.

H. van Woerden et al. (eds.), The Milky Way Galaxy, 561–566.

All we need to do is wait (on the order of 10^7 years according to the estimates of Mathewson, van der Kruit, and Brouw 1972). Moreover, once OB stars appear, the destruction of a molecular cloud also seems secure. Again, we just have to wait (on the order of 3×10^7 years according to the estimates of Bash, Green, and Peters 1977, and Blaauw 1985).

The first problem of star formation reduces, therefore, to the origin of giant molecular clouds. In particular, why should they aggregate along long spiral features, many kiloparsecs in length (Dame et al. 1985, Stark 1985)? And why should they gather in "supercomplexes" (complexes of complexes), tens of millions of $M_\odot$ in mass (Allen and Goss 1979, Elmegreen and Elmegreen 1983)?

These questions pose severe difficulties for the picture of stochastic self-propagating star formation (SSPSF). The issue, as Seiden (1983) has himself emphasized recently, is not how to induce star formation, but how to induce molecular-cloud formation. What is the PHYSICS in SSPSF that allows a giant molecular cloud here to induce the formation of a giant molecular cloud there, a few hundred parsecs away, much less many kiloparsecs away?

In density-wave theory, the answer is simple: the concentration of giant molecular clouds along long spiral fronts is organized by galactic shocks (Fujimoto 1966, Roberts 1969). In particular, even if molecular clouds of 10^5-10^6 $M_\odot$ do not exist in the interarm regions, they can form by the triggering of Parker's (1966) instability behind galactic shocks (Mouschovias, Shu, and Woodward 1974; Blitz and Shu 1980; Giz and Shu 1983). This proposal is well known; here, let me simply reiterate two points. First, no one has found any plausible way to prevent the instability, either by geometrical arrangement (Parker 1967), or by differential rotation (Shu 1974), or by tangled magnetic fields (Zweibel and Kulsrud 1975). Second, the buckling of the field lines and the escape of cosmic-ray particles to the elevated portions yield a natural account for the thin and thick disks in the nonthermal radio-continuum emission of external galaxies (Mathewson, van der Kruit, and Brouw 1972; Beck and Reich 1985).

The agglomeration of molecular clouds into "supercomplexes" also seems to have a convenient explanation within the context of density-wave theory. Length scales of roughly a kiloparsec or more and mass scales of 10^7-10^8 $M_\odot$ appear automatically, if the interstellar gas behind galactic shocks is Jeans-unstable (Elmegreen 1979, Cowie 1981, Jog and Solomon 1983, Elmegreen and Elmegreen 1983).

THE ORIGIN OF MOLECULAR-CLOUD CORES

Observed at high enough angular resolution, completely mapped giant molecular clouds break up into many clumps, each of which may contain several thousand $M_\odot$ and which move randomly with respect to each other at roughly the virial speeds appropriate to the complex (Sargent 1977; Blitz 1978; Solomon, Scoville, and Sanders 1979). The clumps themselves are probably supported against their internal gravity by a combination of magnetic fields (Mouschovias 1976) and turbulence (Larson 1981). The turbulent velocity fields can often be attributed to driving by stellar winds from newly-formed stars. The youngest of these stars are often found in the densest portions (the cores) of a clump (see the review of Wynn-Williams 1982).

Since gravitational contraction has never been documented for any molecular-cloud complex, or even for a single clump, it is tempting to speculate that star formation requires only the collapse of the dense cores. The second problem of star formation reduces, therefore, to the question of the origin of molecular cloud cores. This question, I believe, also has a simple answer. Once we have a molecular clump, supported against its self-gravity at least in part by magnetic fields, the production of dense cores is inexorable. We only have to wait.

Wait for what? Wait for the magnetic field to leak out by ambipolar diffusion (Mestel and Spitzer 1956, Nakano 1981, Mouschovias 1981). This leakage is inevitable, because magnetic fields can provide only indirect support of the neutral gas. It is the ions and electrons which are tied to the field lines and feel their stresses; they transmit the stresses to the neutrals via frictional coupling (through ion-neutral collisions), but this friction requires there to be slip of ions (and field) relative to the neutrals. As the field slips out and the medium becomes less elastic, the level of turbulence which can be sustained presumably also drops. Thus, the self-gravity of a clump tends automatically to produce concentrated cores where the neutrals have pulled past the magnetic field embedded in the less dense background of the envelope. Simple one-dimensional calculations give the timescale of core formation as roughly 10^7 years (see fig. 5 of Shu 1983). Preliminary analysis of more realistic cloud geometries and field configurations (Lizano-Soberon and Shu 1984) suggest that the cores try to acquire $1/r^2$ density profiles (singular isothermal spheres).

This picture for the formation of molecular-cloud cores is attractive from three observational viewpoints. First, radio studies of ammonia in molecular-cloud cores find many quiet cases where the line shapes are consistent with the cores being in hydrostatic equilibrium (at an "equivalent temperature" that includes some "subsonic turbulence") or, at most, in an early stage of gravitational collapse (Myers and Benson 1983). Although Myers and Benson analyzed their data in terms of <u>bounded</u> isothermal spheres (see fig. 3 of their paper), in fact the correlation they find for the average density divided by the "equivalent temperature" plotted versus size is consistent with all their cores being singular isothermal spheres (where $\bar{n}/T_{eq} \propto 1/R^2$). Second, a timescale of core formation of roughly 10^7 years would explain the spread in the ages of the T Tauri stars which have recently formed from the same regions (Cohen and Kuhi 1979). Third, the gravitational collapse of singular isothermal spheres (Shu 1977) with the "equivalent temperature" characteristic of Myers and Benson's observations would lead to a typical protostellar accretion rate $\dot{M} = 10^{-5}\ M_\odot/yr$, a value which Stahler, Shu, and Taam (1980) advocate as required to explain the locations of T Tauri stars in the Hertzsprung-Russell diagrams constructed by Cohen and Kuhi (1979, see also Stahler 1983).

THE ORIGIN OF STELLAR MASSES

If the above interpretation is correct, then we need to revise conventional ideas concerning how the masses of forming stars arise. It does not happen because the accretion process runs out of gas; there is no clean separation of the mass which forms the core and the mass of the envelope around it. The masses derived from the ammonia observations refer to a certain sensitivity level; going to lower density contours would almost certainly produce larger masses. And if we consider the molecular clump as a whole, then we have much more gas than is needed to form any conventional star. Why does the protostellar buildup process usually halt after only a solar mass, or less, of gas has been used up? Observations suggest the reversal of the accretion flow by the onset of a powerful stellar wind.

What is the basic energy source for driving this wind? From studies of the collapse of a model for a rotating molecular-cloud core plus its envelope (Terebey, Shu, and Cassen 1983), we find that a significant fraction of the gravitational binding energy of a protostar may be stored in the form of differential rotation of the protostar. If this store of mechanical energy can be tapped with reasonable efficiency and on a timescale short in comparison with the accretion timescale to drive a stellar wind, then there would be ample power to reverse the accretion flow.

What triggers the sudden release of stored rotational energy? In a low-mass star, it may be the dynamo action initiated when the differentially rotating star is driven convectively unstable by the onset of deuterium burning (see fig. 6 of Stahler, Shu, and Taam 1980 for a calculation of the latter process in a nonrotating context). The strong magnetic fields generated and twisted in this fashion would buoy up to the surface, ultimately driving the activity that astronomers have long associated with young stellar objects of low mass (Herbig 1962; Kuhi 1964; Strom, Strom, and Grasdalen 1975). It is interesting to note that, when T Tauri stars first appear as visible objects (after the onset of a wind that has reversed the accretion flow and swept clear the surrounding gas and dust), they are indeed completely convective (Cohen and Kuhi 1979), and their "birthline" lies close to the locus for the onset of deuterium burning (see Stahler 1983 for a different interpretation).

From this viewpoint, a T Tauri star ends with the mass that it does, because that is the mass it acquires (at an accretion rate of roughly 10^{-5} $M_{\odot}$/yr) before its interior temperature rises high enough to ignite deuterium. Differences in the masses of T Tauri stars then occur, because this condition is reached at different stages in the accretion flows of collapsing molecular-cloud cores of different "equivalent temperatures," angular rotation rates, magnetic fluxes, etc.

The above scenario cannot explain the formation of high-mass stars. It is quite possible that the production of high-mass stars requires exceptional circumstances (say, the additional compression provided by clump-clump collisions or other violent events), and therefore constitutes a separate mode of star formation from that of low-mass stars.

REFERENCES

Allen, R. J., and Goss, W. M.: 1979, Astron. Astrophys. Suppl., 36, p. 135.
Bash, F. N., Green, E., and Peters, W. L.: 1977, Astrophys. J., 217, p. 464.
Beck, R., and Reich, W.: 1985, this volume.
Blaauw, A.: 1985, this volume.
Blitz, L.: 1978, Ph.D. Thesis, Columbia University, published as NASA Tech. Memo. 79708.
Blitz, L., and Shu, F. H.: 1980, Astrophys. J., 238, p. 148.
Cohen, M., and Kuhi, L. V.: 1979, Astrophys. J. Suppl., 41, p. 743.
Cowie, L. L.: 1981, Astrophys. J., 245, p. 66.
Dame, T. M., Elmegreen, B. G., Cohen, R. S., and Thaddeus, P.: 1985, this volume.
Elmegreen, B. G.: 1979, Astrophys. J., 231, 372.
Elmegreen, B. G., and Elmegreen, D. M.: 1983, M.N.R.A.S., 203, p. 31.
Fujimoto, M.: 1966, in IAU Symp. No. 29, p. 453.
Giz, A., and Shu, F. H.: 1983, in preparation.
Jog, C. J., and Solomon, P. M.: 1983, Astrophys. J., in press.
Herbig, G. 1962, Adv. Astron. Astrophys., 1, p. 47.
Kuhi, L. V.: 1964, Astrophys. J., 140, p. 1409.
Larson, R. B.: 1981, M.N.R.A.S., 194, p. 809.
Lizano-Soberon, S., and Shu, F. H.: 1984, in preparation.
Mathewson, D. S., van der Kruit, P. C., and Brouw, W. N.: 1972, Astron. Astrophys., 17, p. 468.
Mestel, L., and Spitzer, L.: 1956, M.N.R.A.S., 116, 503.
Mouschovias, T. Ch.: 1976, Astrophys. J., 207, p. 141.
Mouschovias, T. Ch.: 1981, in IAU Symp. No. 93, p. 27.
Mouschovias, T. Ch., Shu, F. H., and Woodward, P. R.: 1974, Astron. Astrophys., 33, 73.
Myers, P. C., and Benson, P. J.: 1983, Astrophys. J., 266, p. 309.
Nakano, T: 1981, Prog. Theor. Phys. Suppl., No. 70, p. 54.
Parker, E. N.: 1966, Astrophys. J., 145, p. 811.
Parker, E. N.: 1967, Astrophys. J., 149, p. 535.
Roberts, W. W.: 1969, Astrophys. J., 158, p. 123.
Sargent, A. I.: 1977, Astrophys. J., 218, p. 736.
Seiden, P. E.: 1983, Astrophys. J., 266, p. 555.
Shu, F. H.: 1974, Astron. Astrophys., 33, p. 55.
Shu, F. H.: 1977, Astrophys. J., 214, p. 488.
Shu, F. H.: 1983, Astrophys. J., 273, p. 202.
Solomon, P. M., Scoville, N. Z., and Sanders, D. B.: 1979, Astrophys. J. Lett., 232, p. L89.
Stahler, S. W.: 1983, Astrophys. J., in press.

Stahler, S. W., Shu, F. H., and Taam, R. E.: 1980, Astrophys. J., 241, p. 637.
Stark, A. A.: 1985, this volume.
Strom, S. E., Strom, K. M., and Grasdalen, G. L.: 1975, Ann. Rev. Astron. Astrophys., 13, p. 187.
Terebey, S., Shu, F. H., and Cassen, P.: 1983, in preparation.
Wynn-Williams, C. G.: 1982, Ann. Rev. Astron. Astrophys., 20, p. 597.
Zweibel, E., and Kulsrud, R. M.: 1975, Astrophys. J., 201, p. 63.

DISCUSSION

B.G. Elmegreen: I believe that the physics of long-range propagating star formation is beginning to be understood in some detail. One mechanism is for an OB association to pressurize the surrounding interstellar medium, thereby causing a large shell to be swept up. If this shell moves fast enough when the pressure goes away and the shell enters the snowplough phase, then the shell will be able to collapse gravitationally along its periphery before it erodes. The growth of gravitational instabilities in expanding, sheared shells is being studied now. The preliminary results explain well the positions, ages and velocities of the large star-formation sites (the Orion, Perseus and Sco-Cen associations) that occur along the periphery of the expanding Lindblad ring.

Shu: I hope these physical calculations are fed back into the computer simulations.

V. Radhakrishnan: In this picture there is no connection whatever with magnetic fields associated with stars, and with the fields in the original gas from which the stars condensed.

Shu: That's right.

B.F. Burke: Garcia Barreto has recently determined the local magnetic-field values at several places in a star-forming region. The Zeeman effect in the OH-maser complex W3OH was measured for 6 different Zeeman pairs with secure identification. The spots are scattered over the entire complex, a linear spread of about 2000 AU, and all values of B are remarkably similar, ranging from 5.2 to 6.7 milligauss, all pointing in the same direction. This shows that the field is well-ordered, with a magnetic pressure entirely comparable to the local gas pressure.

Shu: That's fine. The magnetic fields, if they are to play a role, must be dynamically significant.

STAR FORMATION IN A DENSITY-WAVE-DOMINATED, CLOUDY INTERSTELLAR MEDIUM

W. W. Roberts, Jr. and M. A. Hausman
University of Virginia, Charlottesville, Virginia

We present a model of the interstellar medium in spiral galaxies. The ISM is simulated by a system of particles, representing gas clouds. It is a probabilistic N-body system in which "cloud" particles orbit ballistically, undergo dissipative collisions with other clouds, and experience velocity-boosting interactions with expanding supernova remnants (SNRs). Associations of protostars may form in clouds following collisions or SNR interactions (thus allowing sequential star formation); these associations take finite times before becoming active and undergoing their own SN events (Roberts and Hausman, 1983; Hausman and Roberts, 1983).

In the presence of a spiral-perturbed galactic gravitational field, the cloud distribution responds with a strong global density wave (galactic shock), with clouds concentrated in spiral arms. The appearance of large cloud-particle clumps in the spiral arms of the computations, due in part to the self-gravitation-mimicking effect of inelastic collisions, suggests how easily real clouds can agglomerate into giant complexes in spiral arms.

Our results are found to be remarkably independent of the cloud system's collisional mean free path. Figure 1 shows the cloud distributions at steady state in two cases whose mean free paths differ by a factor of five. The global coherencies of the two cases are not that dissimilar, although case E (longer mean free path) exhibits a somewhat weaker, more ragged spiral pattern.

The young star population in our model exhibits a coherent spiral pattern in all cases for which cloud collisions are an important star-formation mechanism. In contrast, a case, for which sequential (SSPSF) star formation dominates, is unsteady and shows only transient spiral structure. All cases, except this last one, reach steady state within 200 Myr and remain stable through the duration of the computations - 10^9 years in some cases. At the shock, the region of peak cloud velocity dispersion leads the locus of peak cloud density. Although continuum and "cloud-fluid" models can also approximately reproduce the global structure

H. van Woerden et al. (eds.), The Milky Way Galaxy, 567–569.

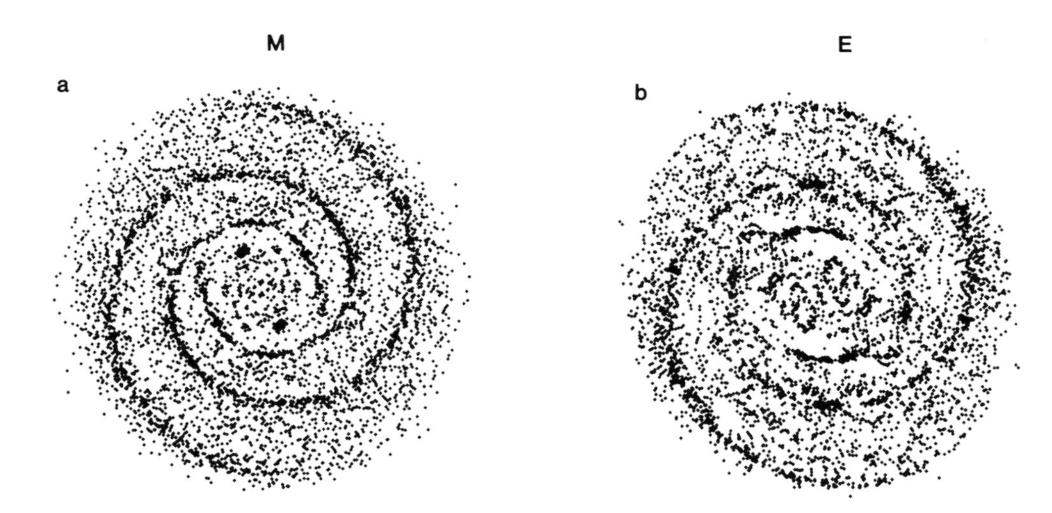

Figure 1. Global spiral structure exhibited by the gas-cloud distributions in two cases of greatly different collisional mean free path λ. Left panel - case M, λ = 200 pc. Right panel - case E, λ = 1000 pc.

of a cloudy ISM, they do not represent local effects (scales of a few hundred pc) as well.

An analytical theory based on fluid-dynamical equations is developed and compared to our N-body model of the ISM in spiral galaxies. When steady-state is assumed, the equations may be spatially averaged to give a mean galaxy-wide star-formation rate. The poorly-known physics of real star-formation processes is incorporated into two parameters which measure the relative importance of the cloud-collision and supernovae-sparking mechanisms. Despite the small-scale stochastic variations and the large-scale density inhomogeneities of the particulate, numerical model, its star formation rate is well described by the fluid-dynamical theory.

ACKNOWLEDGEMENTS

This work was supported in part by the National Science Foundation under grants AST-7909935 and AST-8204256.

REFERENCES

Hausman, M. A., and Roberts, W. W., 1983, Ap. J. (in preparation).
Roberts, W. W., and Hausman, M. A., 1983, Ap. J. (submitted for publication).

DISCUSSION

J.P. Ostriker: Your clouds will lose their velocities on the time-scale for cloud-cloud collisions, unless you have a mechanism for accelerating them, and you lose all your small clouds by making big clouds on the same time-scale. Since the cloud-cloud collision time scale is very short compared with the duration of your simulation, the exact way you put in more energy and more clouds must be important to the final results.

Roberts: That is a very good observation. In fact we have energy coming in through supernova explosions; so the clouds are dissipating, but an energy balance is maintained by the interaction of the supernovae with the clouds. New clouds are not made; we keep the same number of clouds in a given run throughout.

B.G. Elmegreen: How does the size of the clumps produced in your calculation depend on the number of clouds in the simulation? If you had a very large number of clouds, comparable to the expected number in a galaxy, would the resultant clump size be smaller than it is in the present calculations?

Roberts: If we had made our calculations with a much larger number of clouds than the 20000 adopted, and with a correspondingly smaller cloud cross-section (to still preserve the collisional frequency and collisional mean free path estimated for real clouds in the interstellar medium), then clumps on much smaller scales might well have resulted in the simulations. Future simulations with greater numbers of clouds should be able to address this question more adequately.

Heading for reception by President of Groningen University, right to left: Lyngå, Wramdemark, Roberts, Mrs. Oort, Oort and G.D. van Albada (hidden), Milogradov-Turin, Mrs. Schwarz, Hu, Twarog CFD

SPIRAL TRACERS AND PRESTELLAR INCUBATION PERIODS IN A CLOUDY INTERSTELLAR MEDIUM

M. A. Hausman and W. W. Roberts, Jr.
University of Virginia, Charlottesville, Virginia

We present further results of our model of the interstellar medium in spiral galaxies (Roberts and Hausman, 1983; Hausman and Roberts, 1983). The ISM is simulated by a system of particles, representing gas clouds, which orbit ballistically in a spiral-perturbed galactic gravitational field, collide inelastically with one another, and receive velocity impulses from expanding supernova remnants. Star formation may be triggered in the clouds by either collisions or SNR interactions.

The global morphology of bright, young stellar associations is influenced by our choice of delay times, i.e. the time between the star-birth triggering event and the period of high luminosity or SN explosions, after which associations are assumed to dim and lose their identity as spiral tracers. Short maximum-delay times (20 Myr or less) result in the young stellar associations appearing in coherent spiral arms. Longer maximum-delay times (50-100 Myr) wash out spiral patterns: the interarm regions become more substantially populated by young associations while the arms are less continuous, being formed by high-luminosity segments separated by gaps of low bright-star density. These long delay times allow the associations, whose birthsites are still concentrated in spiral arms, to drift long distances before they dim significantly. Such long delay times might be realistic if the most common bright stars in associations have long main-sequence lifetimes, or if the associations require long incubation periods between the star-birth triggering event and the time of peak luminosity.

The possibility that clouds may be sites of sequential star formation to different degrees is examined by varying our model clouds' refractory times, i.e. the period a cloud, which forms stars, must wait before it is again susceptible to further star formation. A short refractory time allows clouds the opportunity of forming stars repeatedly, whereas a very long refractory time more closely simulates the complete disruption of a cloud and the reconstitution of the same amount of gas into another cloud at an uncorrelated position elsewhere in the galaxy. We find that shortening the mean refractory time, although it increases the overall star-formation rate, has relatively little effect on the

H. van Woerden et al. (eds.), The Milky Way Galaxy, 571–572.

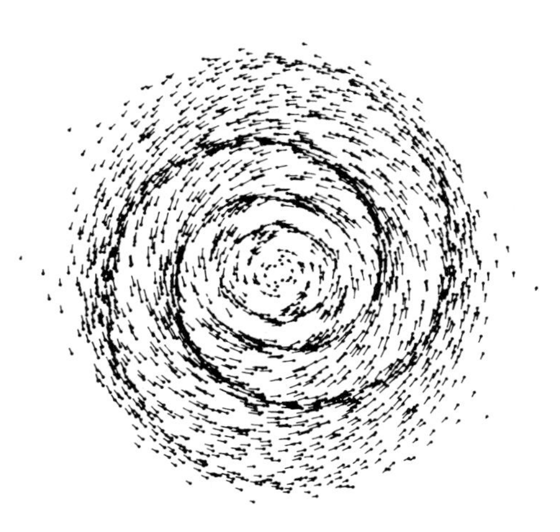

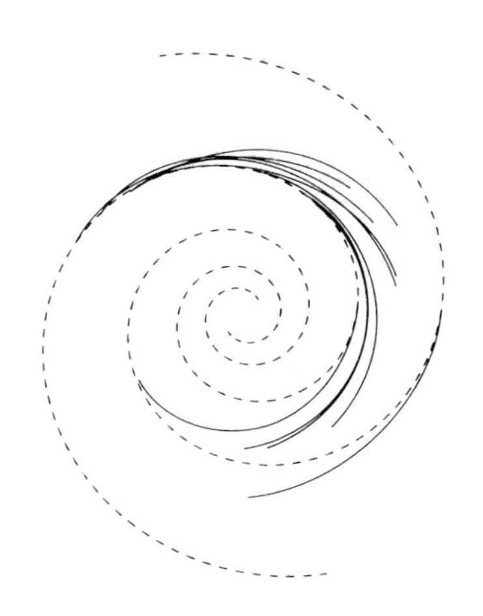

Figure 1. Left panel - Instantaneous-velocity vectors of 1000 representative clouds (dots). Vector lengths are the distances the clouds travel in 4 Myr (in the corotating frame). Right panel - Orbits, from age 0 to 100 Myr, of 10 stellar associations formed in the spiral arms.

coherence of spiral structure, slightly increasing the frequency of interarm spurs. Short refractory times slightly favor SNR-sparked over collisionally-induced star formation, which may explain the greater frequency of spurs.

The mean velocity field of clouds (Figure 1, left panel) shows oval streamlines, very similar to continuum gas-dynamical calculations, although individual orbit-segments are ballistic. Newly formed stellar associations leave the cloud-density peak at higher than post-shock velocities and do not recross the "shock" region. This result is illustrated in the right panel of Figure 1; ten representative associations which are formed in the arms are followed over 100 Myr orbits. This result is in contrast to earlier ballistic-particle models.

By varying choices of delay times, refractory times, and cloud mean free paths within physically plausible limits, we may reproduce spiral galaxies with a wide range of morphological appearances.

ACKNOWLEDGEMENTS

This work was supported in part by the National Science Foundation under grants AST-7909935 and AST-8204256.

REFERENCES

Hausman, M. A., and Roberts, W. W., 1983, Ap. J. (in preparation).
Roberts, W. W., and Hausman, M. A., 1983, Ap. J. (submitted for publication).

SECTION III.4

CHEMICAL COMPOSITION AND EVOLUTION OF THE DISK

Friday 3 June, 0900 - 1020

Chair: C. Cesarsky

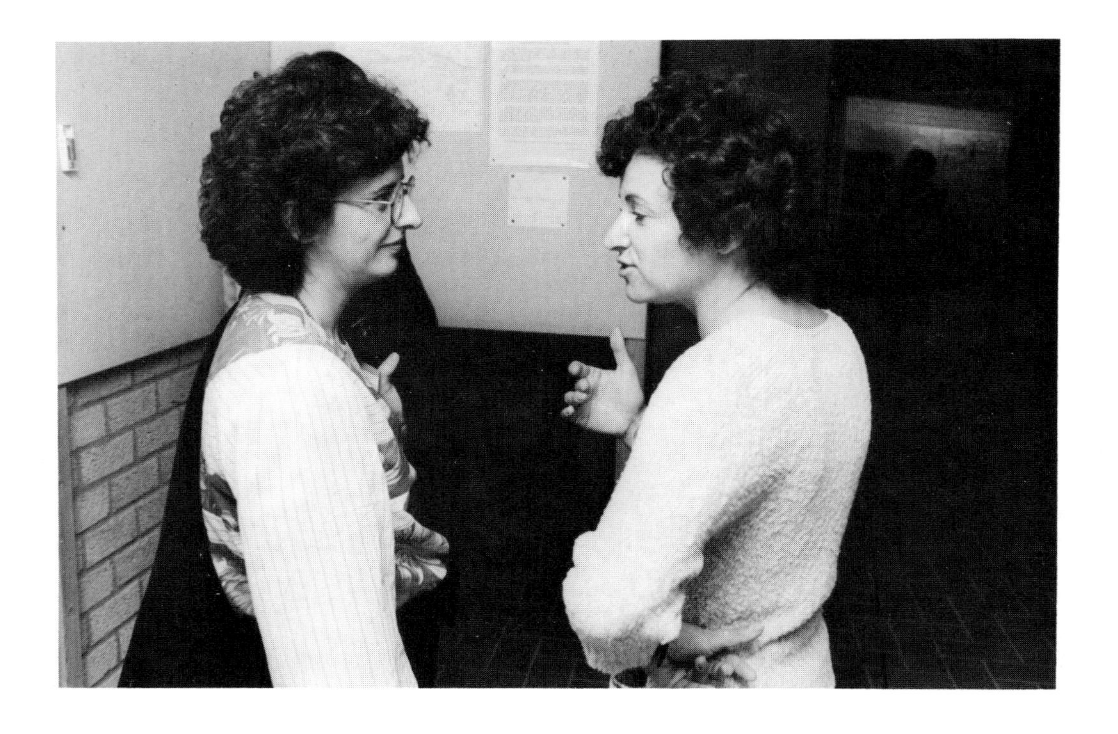

Catherine Cesarsky (right) and Antonella Natta CFD

CHEMICAL COMPOSITION OF INTERSTELLAR MATERIAL

S.R. Pottasch
Kapteyn Astronomical Institute
Groningen, The Netherlands

SUMMARY

Abundances in interstellar clouds, as determined from interstellar absorption lines, are discussed first, including abundances in 'abnormal' (high-velocity) clouds. HII-region abundances are then discussed and compared to results from the interstellar clouds. The present status of an abundance gradient as determined from HII regions is given. Abundances in planetary nebulae are then given for various categories of nebulae, and compared to HII regions. Finally a short status report on abundances near the galactic center is given.

1. ABUNDANCES DETERMINED FROM ANALYSIS OF INTERSTELLAR ABSORPTION LINES

In general, the interstellar absorption lines yield information on the chemical composition of the interstellar gas only for regions within several kiloparsecs of the Sun. There are several reasons for this. First, the stars in which the absorption lines are measured are weaker as they are more distant. Especially close to the galactic plane, where extinction becomes very important, it is impossible to find suitable stars which are very distant. Second, and more fundamental, the analysis of the lines becomes very difficult when the stars are distant. This is because the velocity of the absorbing material has a greater variation (in most directions) along the line of sight to distant objects, and the uncertainty in the velocity dispersion is reflected in a large uncertainty in the derived abundance. The errors become greater as the line becomes more saturated. Thus the best-known abundances are determined in short lines of sight with little material. Sometimes transitions with very low oscillator strengths (f-values) can be measured, which circumvents the above problem.

1.1. Abundances within 1 kpc of the Sun ($|z| < 100$ pc)

The abundances in the direction of about ten early-type stars have been carefully and completely studied, while another hundred directions have been studied as well, but somewhat less completely. The results for two well-studied directions are given in Table 1. The first is the

H. van Woerden et al. (eds.), The Milky Way Galaxy, 575–584.

TABLE 1
Abundances (log X/H + 12) toward α Vir and ζ Oph

Element	Abundance α Vir	Abundance ζ Oph	Element	Abundance α Vir	Abundance ζ Oph
C	9.0 ±.3:	8.4±.2	Mg	6.8±.1	6.1±.2
N	7.66±.1	7.8±.3	Si	6.7±.2	6.3±.3
O	8.6 ±.1	8.7±.1	Fe	6.4±.2	5.4±.2
Ar	6.5 ±.1	5.9±.2	Mn	4.8±.2	4.1±.1
S	7.5 ±.05	7.0±.3	Ca	3.8±.1	2.7±.1
Zn		4.2±.1	Cl	5.8±.2	5.0±.2
Al		4.6±.3			

direction toward α Vir (York and Kinahan, 1979). This star is relatively close (88 pc), and has a low column density of neutral hydrogen (10^{19} cm^{-2}) and a small extinction (E_{B-V} = 0.03). The other direction is toward ζ Oph (Morton, 1975; Lugger et al., 1978; Jenkins and Shaya, 1979; de Boer, 1981). This well-studied star is only two or three times more distant, but the column density of material in this direction is much higher. The neutral-hydrogen column density is about 5×10^{20} cm^{-2}, while the combined atomic and molecular hydrogen is about 1.4×10^{21} cm^{-2}. The extinction (E_{B-V} = 0.32) is also substantially higher. It can be seen from Table 1 that for some of the elements listed, the abundance in the line of sight to α Vir is higher than that toward ζ Oph. When combined with data from other lines of sight, the following summary can be made of the abundances:

(1) Some elements, especially S, Zn, N, O and probably C, show only a small variation from one line of sight to another. These same elements show no correlation between their abundance (relative to H) and the extinction E_{B-V} or the hydrogen column density in the line of sight.

(2) Other elements, especially Si, Mg, Fe, Mn and Ca show substantially larger variations from one line of sight to another. These variations are correlated with E_{B-V} and the hydrogen column density in the line of sight, such that the abundance is lower when E_{B-V} and the hydrogen column density are higher. This effect is most strongly seen in Ca. A line of sight has been found, toward HD 147889, a star deeply embedded in the ρ Oph dark cloud, indicating an upper limit of log Ca/H + 12 = 1.2 (Snow et al., 1983).

1.2. Element depletion and its interpretation

The abundances may be compared to some standard values, for which the solar abundances are very convenient. The solar values are listed in Table 2 for elements which are of interest. Two values are usually listed for each element. The first is taken from Ross and Aller (1976) and is an average of photospheric and coronal abundances. The second is taken from Parkinson (1977) or Pottasch (1967) and represents only the coronal value. The purpose of the second column is to show that for some elements, notably oxygen, some discrepancy exists in the solar values. The value for neon is only measured in the corona, and it sometimes is

TABLE 2
Selected Solar Abundances (log X/H + 12)

Element	Corona + Photosphere	Corona	Element	Corona + Photosphere	Corona
C	8.68	8.7	Al	6.52	6.3
N	7.95	7.8	Mg	7.60	7.65
O	8.84	8.45	Si	7.64	7.65
Ne	7.57 (8.04)	7.65	Fe	7.50	7.65
Ar	6.0 (6.56)		Mn	5.42	
S	7.2	7.3	Ca	6.35	
Zn	4.45		Cl	5.5	

related to the photospheric values by using the coronal O/Ne ratio together with the photospheric oxygen value.

If the abundances found from the interstellar absorption lines are compared with the solar values, the following results are obtained:
1) The abundances of Si, Mg, Fe, Mn and Ca are almost always lower than the solar values, usually by factors varying from 10 to 10^3. These elements are thus usually depleted with respect to the Sun.
2) The elements S and Zn have abundances very similar to their solar value.
3) The elements N, O and C are more controversial and have been studied extensively in very recent years. As can be seen by comparing these elements in Tables 1 and 2, the interstellar abundance is close to, or perhaps slightly less than, solar.

The controversy centers about whether the C,N,O abundances are significantly less than solar. In spite of the care taken in the analysis of the observations, the errors in the abundances in individual objects remain as high as 50%. When lines of low f-value are not available, the error can be substantially larger. Carbon is probably the most difficult case, since the low f-value line of the most abundant ion, C^+, is rarely seen. The most reliable abundances of carbon come from a study of C^0, coupled with a rather large correction for the ionization equilibrium by Jenkins et al. (1983). These authors conclude "that most stars are consistent with no depletion of gas-phase carbon, except for a number of stars in the Scorpius region". Ferlet (1981) who studied nitrogen, and de Boer (1981) in a study of oxygen, also concluded that there was no compelling evidence for depletion of these elements. On the other hand York et al. (1983), studying N and O, conclude that evidence for a small depletion is present. These authors do not take into account the accuracy of the solar abundance.

The depletions of Si, Mg, Fe, Mn and Ca are believed to be caused not by their absence from the interstellar medium, but by their presence in the form of dust which has condensed out of the gas. Apart from the known presence of dust in the interstellar medium, evidence for such a formation is found in a rough correlation between the condensation temperature and the amount of the depletion (Field, 1974). It is generally believed that there is not enough material in the elements listed above to provide all the mass of the dust found in the interstellar medium (Spitzer, 1978). Only carbon and oxygen are thought to be abundant enough to supply the extra mass to the dust. This is an added reason for the importance of a possible depletion of these elements in the gas phase.

1.3. Variation of element depletion further from the plane

Measurements of some selected lines have been made to determine if depletions change with the distance $|z|$ from the galactic plane. Jenkins (1983) studied the ratios of lines of FeII and SiII to SII. He finds that the ratio of Fe/S is about 3×10^{-2} within 500 pc of the plane (similar to the line of sight to ζ Oph) and increases by about of factor of 2 at 1000 to 1500 pc. He finds no significant change in the Si/S or Al/S ratios. Albert (1982) studied the titanium abundance in the interstellar medium. He finds that the Ti/H ratio varies by about two orders of magnitude, with depletions with respect to the solar abundance from 2×10^{-3} to 2×10^{-1}. For high-latitude stars the lowest depletion (about 10^{-1}) is found.

A lower depletion at high latitudes is also found for Ca and may be true of other elements as well.

1.4. 'Abnormal'-velocity-cloud abundances

It has long been recognized that, when the absolute value of the velocity of an interstellar cloud is greater than 20 km s^{-1} with respect to the local standard of rest, the line ratios vary in a systematic way. Routly and Spitzer (1952) showed that the ratio of the column densities Na^0/Ca^+ decreased strongly with increasing velocity. The effect is quite large in this case, the variation in the ratio is at least three orders of magnitude. A possible explanation of this effect suggested by Routly and Spitzer is a change in the depletion of one of these elements.

Since then several studies have been made of the abundances in these 'higher-velocity' clouds. They confirm that for the higher velocities the depletions are less than for the lower velocities, and in fact there appears to be a rough correlation between the depletion and the velocity. The precise values of the column densities are often difficult to obtain for two reasons. First, the lines are usually quite weak and sometimes confused with other components. Second, the hydrogen column density to be used for comparison often cannot be determined for the high-velocity component, since the Lyman-α absorption is always dominated by the low-velocity component. The depletions are often measured with respect to sulfur which, as discussed above, may be expected to have a near-solar abundance.

The results of an abundance analysis in the high-velocity component ($v = -75$ km s^{-1}) of HD 175754 are shown in Table 3 (Pottasch et al., 1980). It can be seen that practically all abundances which could be

TABLE 3
Abundances (log X/H + 12) in the high-velocity component in HD 175754

Element	Abundance	Element	Abundance
C	8.7	S	7.2
O	8.7	Fe	6.8
Mg	7.5	Ca	5.9
Si	7.5	Na	6.3
Al	6.2		

measured are solar, with the exception of iron and calcium, which are depleted by a factor of 5 and 3, respectively.

These 'high'-velocity clouds may have different origins. It seems clear that those observed in the spectra of stars within the Carina Nebula (Walborn and Hesser, 1982), in which components of several hundred km s^{-1} are observed, are due to large motions very close to the nebula itself. These motions are probably derived from the energy of the many hot stars within the nebula. On the other hand, the similarly high velocities found in the region of Vela (Jenkins, Silk and Wallerstein, 1976) probably have their origin in the supernova explosion which occurred there. It is likely that other 'high'-velocity clouds derive their energy, directly or indirectly, from similar events.

2. RESULTS FROM HII REGIONS

The greatest source of error in determining the abundances in HII regions has traditionally been the determination of the electron temperature. This is because an accurate value is necessary: a 40-percent change in the electron temperature can change the abundance by an order of magnitude. Optically the electron temperature is usually obtained by measuring the intensities of [OIII] λ 4363 Å or [NII] λ 5755 Å, both very weak lines which are detectable only in relatively hot ($\geqslant$7000 K) and bright HII regions.

This dependence on optical spectra to determine electron temperatures has disappeared in recent years. Using radio recombination lines, electron temperatures can be determined with accuracies of a few percent (Shaver, 1980). These lines are easily detected in cool, highly reddened or obscured and relatively faint HII regions, so that the radio and optical methods are complementary.

Accurate electron temperatures are now available for almost 100 HII regions from close to the galactic center to about 15 kpc from the center (Shaver et al., 1983; Wink et al., 1983). These temperatures may be used in conjunction with the optical spectra, which are available for about 33 HII regions (Shaver et al., 1983 and references cited there), to determine the abundances of some of the lighter elements as a function of radial position in the Galaxy (from 5.9 to 13.7 kpc from the centre). The average values of abundance for five elements from the

TABLE 4
Abundances (log X/H +12) in the Orion Nebula and in the average HII region in the solar neighbourhood

Element	Orion	Average HII region
C	8.5	
N	7.5	7.6
O	8.5	8.7
Ne	7.7	7.8
Ar	6.4	6.4
S	6.93	7.1
Fe	5.8	
Cl	5.0	

nearby HII regions are given in Table 4. The scatter in individual values is about 50%. Also shown in the table are the abundances in the Orion Nebula. For those elements not discussed by Shaver et al., the abundances were calculated using the same electron temperature (T_e = 8700 K). The carbon abundance is based on IUE measurements in the ultraviolet. The iron abundance is more uncertain than the others, because the electron collision cross-sections (Ω) are poorly known; the abundance given is based on an average value Ω = 1 for all observed transitions (Olthof and Pottasch, 1975). Despite this uncertainty, it is clear that iron is at least an order of magnitude less abundant in Orion than in the Sun, and this is true of most HII regions as well. It is rather remarkable that the abundances in Orion are in almost complete agreement with those in the line of sight to ζ Oph. If the latter are depleted because of condensation onto dust particles, this same dust must exist in Orion.

Table 5 gives the abundances of oxygen and nitrogen as a function of distance from the galactic center (assuming the Sun is 10 kpc from the center). The abundance gradient is beyond any doubt. The scatter in the results for individual HII regions can be entirely accounted for by the remaining errors in measurement (especially in T_e). The gradients for N and O are essentially the same and it appears that Ar, and possibly Ne (despite the uncertain conclusion of Shaver et al.), also have the same gradients. Sulfur seems to be an exception, with a gradient less than half that of the other elements.

3. ABUNDANCES IN PLANETARY NEBULAE

The problem of determining the electron temperature from optical line ratios is not as difficult for planetary nebulae as it is for HII regions because the electron temperature is usually considerably higher in planetary nebulae, and the surface brightness is also higher so that spectra are more easily obtained. The biggest problem until very recently has been the rather large corrections which had to be made for unseen ions. Nitrogen and sulfur have presented the greatest problem. In the optical wavelength region only NII lines are seen, and in the ionization

TABLE 5
Variation of HII-region abundances with distance from Galactic Center

Galactocentric Distance	log O/H + 12	log N/H + 12
6 kpc	9.2±.2	8.1±.2
8	9.0 "	7.9 "
10	8.7 "	7.6 "
12	8.5 "	7.4 "
14	8.3	7.2

equilibrium which exists in many nebulae less than 1% of the nitrogen is in this stage of ionization. Thus the correction factors were often large as were the consequent uncertainties in abundance. A similar situation existed for sulfur and several other ions. Carbon was only represented by very weak recombination lines, which were difficult to interpret in terms of abundance.

The situation has considerably improved at present. The IUE ultraviolet measurements have made it possible to observe NIII, NIV and NV, as well as CII, CIII and CIV. Infrared measurements of SIV and SIII have made the sulfur abundance much more reliable. The most difficult problem at present is to take into account the effects of density and temperature gradients in the nebulae.

Results have been summarized by many different authors (see Kaler, 1983; Aller and Czyzak, 1983; Pottasch, 1983 and the references given therein). The median abundance for 'average' nebulae is shown in column 2 of Table 6. By 'average' nebulae those nearby (within 3 kpc) nebulae are meant which do not fall into one of the other categories to be listed. The first seven elements in the table are the non-depleted elements. Variations for these elements ususally do not exceed a factor of two from nebula to nebula. Within this factor the abundances strongly resemble the abundances in nearby HII regions and the Sun. The abundances of Si, Fe, Ca and Mg are less known and probably have greater variation from one nebula to the other. In this respect they probably resemble HII regions and interstellar clouds, the elements being substantially depleted with respect to the Sun.

In column 3 of Table 6 the abundances in the so-called 'Type I' nebulae are listed. These are nebulae with very hot exciting stars (often the star is not seen), are very close to the galactic plane and often have a very filamentary structure (see Peimbert and Torres-Peimbert, 1983). The exciting stars are probably more massive than for the average nebulae. The abundances of He and N are considerably higher than for the 'average' nebulae, N being an order of magnitude higher. These abundance changes probably occurred within the star in the course of evolution to the planetary-nebula stage. The other elements are slightly more abundant than their counterparts in the 'average' nebulae which, if it is significant, may reflect the fact that these are younger objects formed out of somewhat enriched material.

TABLE 6
Abundances in Planetary Nebulae

Element	Average Nearby Nebulae	Type I Nebulae	Halo Nebulae	Galactic Center Nebulae
He	11.02	11.13	11.03	
N	7.80	8.70	7.5	8.9
C	8.60	8.70	8.1	
O	8.61	8.81	8.0	9.5
Ne	7.91	8.10	7.0	
Ar	6.32	6.50	4.6	7.4:
S	6.83	6.92	5.9	7.4:
Si	6.6			
Fe	6.2			
Ca	5.1			

The fourth column of the table gives the abundances in 'halo' nebulae. These are nebulae distinguished by a high velocity perpendicular to the galactic plane or membership in a globular cluster. They are thought to originate from stars as old as globular clusters. Three such objects are known and a fourth has recently been discovered. The abundances in these nebulae are substantially less than in the 'average' nearby nebulae, with the exception of He and possibly N. Especially Ar is underabundant by two orders of magnitude. These abundances must reflect the original abundance at the time of formation of the central stars.

4. ABUNDANCES NEAR THE GALACTIC CENTER

Little is known concerning the abundances near the center. Because of the large extinction near the center, interstellar lines cannot be measured there. The same is true of optical spectra of HII regions. Planetary nebulae near the center can be measured. Because of the high but uncertain value of the extinction, abundances are difficult to determine unless the spectra are accurately measured and the radio continuum is also known. Webster (1976) concluded that the abundances were consistent with being solar for a sample of nebulae she studied, but the conclusion must be considered uncertain because of the above. The results of a single nebula studied by Price (1981) are shown in the last column of Table 6. In this nebula the O, and possibly Ar and N abundances are an order of magnitude greater than in the nearby nebulae, while S is at least a factor of 2 more abundant. Further observations are needed.

The decrease of electron temperature of HII regions toward the galactic center is probably a consequence of the increasing abundance of the elements responsible for the cooling. Since the temperature continues to decrease until about 3 kpc from the center, it is likely that the abundance increases at least to this radius. No further electron-temperature measurements are available closer to the center, except within 300 pc of the center. The temperature of these HII regions varies between 5000 K and 7000 K so that the temperature does not by itself argue for a high abundance. Argon and sulfur lines in the far infrared have been observed from Sgr A (Lester et al., 1981; Herter et al., 1983) which indicate an Ar and S abundance about 5 and 3 times that observed in Orion. In summary, there is an indication that the abundances may be considerably higher near the galactic center, but the present observational material is not decisive.

REFERENCES

Albert, C.E. 1982, Astrophys. J. 256, L9
Aller, L.H., Czyzak, S.J. 1983, Astrophys. J. Suppl. 51, 211
De Boer, K.S. 1981, Astrophys. J. 244, 848
Ferlet, R. 1981, Astron. Astrophys. 98, L1
Field, G.B. 1974, Astrophys. J. 187, 453
Herter, T., Briotta, D.A. Jr., Gull, G.E., Shure, M.A., Houck, J.R. 1983, Astrophys. J. 267, L37
Jenkins, E.B. 1983, in "Kinematics, Dynamics and Structure of the Milky Way", ed. W.L.H. Shuter (Dordrecht: Reidel), p. 21
Jenkins, E.B., Jura, M., Loewenstein, M. 1983, Astrophys. J. 270, 88
Jenkins, E.B., Shaya, E.J. 1979, Astrophys. J. 231, 55
Jenkins, E.B., Silk, J., Wallerstein, G. 1976, Astrophys. J. 209, L87
Kaler, J.B. 1983, in "Planetary Nebulae", ed. D.R. Flower, I.A.U. Symp. 103, p. 245
Lester, D.F., Bregman, J.D., Witteborn, F.C., Rank, D.M., Dinerstein, H.L. 1981, Astrophys. J. 248, 524
Lugger, P.M., York, D.G., Blanchard, R., Morton, D.C. 1978, Astrophys. J. 224, 1059
Morton, D.C. 1975, Astrophys. J. 197, 85
Olthof, H. and Pottasch, S.R. 1975, Astron. Astrophys. 43, 291
Parkinson, J.H. 1977, Astron. Astrophys. 57, 185
Peimbert, M. and Torres-Peimbert, S. 1983, in "Planetary Nebulae", ed. D.R. Flower, IAU Symp. 103, p. 233
Pottasch, S.R. 1967, Bull. Astron. Inst. Netherl. 19, 113
Pottasch, S.R., Wesselius, P.R., Arnal, E.M. 1980, Proc. Second Eur. I.U.E. Conference, p. 13 (ESA SP-157)
Price, C.M. 1981, Astrophys. J. 247, 540
Ross, J.E., Aller, L.H. 1976, Science 191, 1223
Routly, P.M., Spitzer, L. 1952, Astrophys. J. 115, 227
Shaver, P.A. 1980, Astron. Astrophys. 91, 279
Shaver, P.A., McGee, R.X., Newton, L.M., Danks, A.C., Pottasch, S.R. 1983, Mon. Not. Roy. Astron. Soc. 204, 53
Snow, T.P., Timothy, J.G., Seab, C.G. 1983, Astrophys. J. 265, L67
Spitzer, L. Jr. 1978, "Physical Processes in the Interstellar Medium" (New York: Wiley)
Walborn, N.R., Hesser, J.E. 1982, Astrophys. J. 252, 156
Webster, B.L. 1976, Mon. Not. Roy. Astron. Soc. 174, 513
Wink, J.E., Wilson, T.L., Bieging, J.H. 1983, Astron. Astrophys. 127, 211
Wink, D.G., Kinahan, B. 1979, Astrophys. J. 228, 127

York, D.G., Spitzer, L., Bohlin, R.C., Hill, J., Jenkins, E.B., Savage, B.D., Snow, T.P. 1983, Astrophys. J. 266, L55

DISCUSSION

C.J. Cesarsky: What is the current thinking about abundances and grains in HII regions?

Pottasch: Grains probably exist in HII regions and in planetary nebulae. They seem to have a rather long lifetime. The presence of grains can be seen from their infrared emission, not easily from extinction in HII regions. The composition of the grains may be inferred from the underabundance of certain elements both in planetary nebulae as well as in the general interstellar medium. It is likely that the grains consist mainly of iron, magnesium and silicon, possibly in oxide or carbide form.

Cesarsky: What about abundance determinations from X-ray spectroscopy for hot regions? I understand they would be difficult, because such regions are often out of equilibrium. But eventually we should be able to obtain abundances for regions with less depletion.

Pottasch: The physical parameters involved in this emission are well-known, so some day with better observations one will be able to determine the composition.

INTERSTELLAR NITROGEN ISOTOPE RATIOS

R. Güsten
Max-Planck-Institut für Radioastronomie, Bonn

H. Ungerechts
I. Physikalisches Institut der Universität Köln

We report first results of a systematic study of the galactic interstellar nitrogen isotope ratios, based upon measurements of the (J,K) = (1,1) and (2,2) lines of the $^{14}NH_3$ and $^{15}NH_3$ molecules. We find significant deviations of the $[^{14}N]/[^{15}N]$ abundance ratio from the solar-system value (∿270), with the most extreme enrichment in the galactic-center region, where $[^{14}N]/[^{15}N]$ ∿ 1500. This value is consistent with a nitrogen nuclear-synthesis scheme in which ^{14}N is the main product of normal secondary CNO-burning, whereas ^{15}N is depleted by the same process.

Although nitrogen is one of the most abundant elements, research on its interstellar isotope abundances is still rudimentary. The processes by which the ^{14}N and ^{15}N nuclei are synthesized are rather well identified: ^{14}N is the product of 'normal' (cold) CNO-burning and ^{15}N is formed in the hot CN-cycle. On the other hand, the production efficiencies of the various stellar sites (i.e. where their nucleosynthesis and liberation into the interstellar medium occur) are uncertain, and predictions of the nitrogen yields by stellar-evolution models require an observational calibration (see Güsten and Mezger, 1983, for a recent compilation). In particular it is not known whether the production of ^{14}N requires a pre-enrichment of the star with, e.g., ^{12}C seed nuclei from a previous generation of stars, or whether its synthesis can take place in one stellar generation by CNO-burning in self-enriched giant envelopes (Renzini and Voli, 1981); in short, whether ^{14}N is a 'secondary' or 'primary' product of nucleo-synthesis (see, e.g., Pagel and Edmunds, 1981).

Up to now, determinations of the interstellar nitrogen abundance ratios have been based upon double isotope ratios of hydrocyanic acid $[H^{12}C^{15}N]/[H^{13}C^{14}N]$ (Wannier et al., 1981). These have the advantage of avoiding the problems caused by the strong saturation in the main isotope lines, but they suffer in accuracy from the required independent determination of the carbon isotope ratio. Moreover, in the very massive clouds close to the galactic center, line saturation becomes important even for $H^{13}C^{14}N$, and causes the nitrogen isotope ratio to be underestimated. For these reasons we have begun a systematic study of the single isotope ratio of the (J,K) = (1,1) and (2,2) lines of ammonia, $^{14}NH_3$ and $^{15}NH_3$, using the 100-m telescope of the MPIfR at

H. van Woerden et al. (eds.), The Milky Way Galaxy, 585–586.

Effelsberg. The rare-isotope lines were first detected in 1977 in the Orion molecular cloud (Wilson and Pauls, 1979).

In this short report we present the first results for two molecular clouds close to the galactic center as well as for DR21(OH), which is situated in the local galactic disk. We find that the gas in DR21(OH) has $[^{14}N]/[^{15}N]$ ∿400, and hence that the local disk has been only moderately enriched since the time when the solar system was formed 4.5×10^{9}yr ago. The ratio we find for the galactic-center clouds, ∿1500, is significantly larger than the lower limit (>500) which was given by Wannier et al. (1981), and means an enhancement by a factor of ∿5 over the solar-system abundance ratio (∿270).

Table 1: Solar System and Interstellar Abundances

Isotope	Solar System[5]	Galactic Disk 9-11 kpc	Galactic Disk 4-6 kpc	Galactic Center <0.3 kpc		IRC10216[8]
$[^{16}O/H]$		$\sim5\cdot10^{-4}$	$\sim12\cdot10^{-4}$	$(5-10)10^{-4}$	[1,2]	
$[^{14}N/H]$		$\sim4\cdot10^{-5}$	$(\sim12\cdot10^{-5})$	?	[1,3,4]	
$[^{12}C]/[^{13}C]$	89	70-75	∿50	∿20	[6]	40(±10)
$[^{14}N]/[^{15}N]$	270	∿400		∿1500	[7]	∿3000

References:

[1] Shaver et al. (1983, Mon.Not.Roy.Astr.Soc. 204, 53)
[2] Mezger et al. (1979, Astron. Astrophys. 80, L3)
[3] Lester et al. (1983, Astrophys. J., preprint)
[4] Binette et al. (1982, Astron. Astrophys. 115, 315)
[5] Cameron (1980, in "Nuclear Astrophysics", ed. C. Barnes et al.)
[6] Henkel et al. (1982, Astron. Astrophys. 109, 344)
[7] this paper
[8] Wannier et al. (1981, Astrophys. J. 247, 522)

In table I we compare our results for nitrogen with determinations of the abundance gradients of oxygen and nitrogen as well as the ratio of the carbon isotopes. The general trend in these data is consistent with the assumption that the material in the inner Galaxy is in a chemically more evolved state. For the carbon data this is most evident for the immediate galactic-center region. Nitrogen, however, does not follow the carbon isotope enrichment, as we would expect if both ^{14}N and ^{12}C were produced by primary nucleosynthesis. On the contrary, the strong $[^{14}N]/[^{15}N]$ enhancement in the galactic center suggests that:
(1) the bulk of the ^{14}N in the galactic center is of secondary origin,
(2) the ^{14}N enhancement which occurs mainly in low-mass stars is accompanied by strong ^{15}N depletion, as is both expected from nucleosynthesis theory and observed in the circumstellar shell ejected by the evolved giant star IRC10216 (compare last column of table I).

In a highly evolved stellar population such as the central bulge, the material ejected by low-mass stars is a major contribution to the enrichment of the ISM by astrated matter, and hence a large $[^{14}N]/[^{15}N]$ ratio is to be expected. A more quantitative analysis of our data in terms of a chemical-evolution model will be given in a future paper.

REFERENCES

Güsten, R., Mezger, P.G. 1983, Vistas in Astronomy, in press
Pagel, B.E.J., Edmunds, M.G. 1981, Ann. Rev. Astron. Astrophys. 19, 77
Renzini, A., Voli, M. 1981, Astron. Astrophys. 94, 175
Wannier, P.G., Linke, R.A., Penzias, A.A. 1981, Astrophys. J. 247, 522
Wilson, T.L., Pauls, T. 1979, Astron. Astrophys. 73, L10

THE CHEMICAL EVOLUTION OF THE GALAXY

Bruce A. Twarog
University of Kansas

1. INTRODUCTION

Over the last few years, our picture of the chemical evolution of the Galaxy has changed substantially. These changes are of interest because chemical evolution provides a common point of contact for most astrophysical processes of importance to galaxy evolution. By astrophysical processes we mean star formation, stellar nucleosynthesis, gas dynamics, etc. An understanding of galactic chemical evolution would allow us to place constraints on all of these topics simultaneously. This property, however, is a double-edge sword because, with so many variables involved, unique solutions to problems in chemical evolution are almost impossible.

The review is restricted to those areas, both observational and theoretical, where a consensus of opinion appears to be emerging on some of the major components of chemical evolution, particularly those which deviate from the simple, closed models. By necessity, many important topics will be ignored while others will reflect the particular bias of the reviewer. Detailed discussions can be found in a number of excellent reviews which have appeared in the last few years by Tinsley (1980a), Pagel and Edmunds (1981), and Mould (1982).

As a reference point, [Fe/H] refers to the logarithmic iron abundance relative to the Sun on a scale where 47 Tuc and M71 have [Fe/H] = -0.8, and -0.6, respectively.

2. THE SIMPLE, CLOSED MODEL

The classical starting point for discussions of chemical evolution is the simple, closed model. In this picture, the Galaxy is regarded initially as a closed box filled with gas of primordial composition, i.e., pure H and He. As the Galaxy ages, the gas is turned into stars, forming with a constant mass function. As a generation of stars evolves and dies, processed material is returned to the interstellar medium, while a significant fraction of the mass remains locked up in low-mass

H. van Woerden et al. (eds.), The Milky Way Galaxy, 587–596.

stars and stellar remnants. A critical parameter is the yield, y, the mass fraction of enriched material returned to the interstellar medium (ISM) relative to the mass locked up in low-mass and dead stars. The total yield for an element depends upon the mass function and the elemental yields as a function of stellar mass. The beauty of the model lies in the fact that the chemical history can be described by one equation

$$Z = y \ln \mu^{-1} \tag{1}$$

where Z is the metallicity of the gas, y is the yield, and μ is the gas fraction, the ratio of gas mass to total mass. If the initial mass function (IMF) is constant, the metallicity is dependent solely upon the gas fraction, independent of the star formation rate (SFR). It is assumed that the yield is recycled instantaneously and is well mixed. For twenty years, the classical G-dwarf problem (Schmidt 1963), a paucity of low-metallicity stars in the solar neighborhood, has been a dominant constraint on our picture of galactic evolution, with a variety of ad hoc assumptions presented to explain the failure of the simple model. Those features which now appear justified by observational evidence and their role in galaxy evolution will be discussed next.

3. COMPLEX, OPEN MODELS

The deviations from simplicity arise from the failure of a combination of assumptions. The specific areas of modification are:

3.1 The Continuity Question

The evolution of the Galaxy can be divided into two distinct phases, the halo and the disk; the conditions for star formation, the mass function, the degree of homogeneity, and the gas flows during these two phases are totally distinct. The metallicity distribution of the halo is very nicely matched by a simple, closed model as shown in Fig. 5 of Bond (1981), where the gas is initially enriched to [Fe/H] $\simeq$ -2.6, and the yield is constant at $y = 0.018\ Fe_\odot$, where $Fe_\odot$ refers to the mass fraction of iron in the Sun. The characteristics of the solar neighborhood cannot be matched by a simple model and the halo yield for iron is forty times lower than that for the disk. Though the lower halo yield is consistent with the fragmentary-protocloud model of Searle (1977), the difference could result from a mass function weighted toward low-mass objects ($m < 0.1\ M_\odot$), a mass function deficient in Fe-producing stars of intermediate mass ($1\text{-}10 M_\odot$), and/or elemental stellar yields which are composition dependent. The only deficiency of the fragmentary protocloud model is that the disk must form out of the relatively unenriched gas lost by the clouds. Though theoretical models with infall of zero-metals gas indicate that the metallicity of such gas will rise rapidly to levels consistent with old-disk abundances, there should still exist an intermediate population of old disk stars with metallicities overlapping those of the halo. Though decisive information is lacking, the thick-disk component of the Galaxy found by Gilmore and Reid (1983) could represent this transition population. The boundary line between these two phases is tentatively adopted as [Fe/H] = -0.90 ± 0.2.

3.2 SFR History and the Age-Metallicity Relation

Despite the prompt initial enrichment phase provided by the halo (Ostriker and Thuan 1975), the simple model for the disk still fails. The age-metallicity data of Mayor (1976) and Twarog (1980) both indicate an approximately linear increase in the mean metallicity of the gas in the disk, Z_g, over most of its lifetime with a possible leveling off over the last 5×10^9 years. (See Fig. 2 of Pagel 1981) When combined with the metallicity distribution for the disk (Pagel and Patchett 1975), one can exclude models in which the present SFR is a small fraction of the average past SFR, i.e. less than 15%. From a variety of techniques, the best estimates for this parameter are constrained such that $0.5 \leq \langle SFR \rangle / SFR_o \leq 3$ (Mayor and Martinet 1977; Miller and Scalo 1979; Twarog 1980), where subscript zero refers to current values. Attempts at further restricting the SFR history through the use of the G-dwarf distribution seem unjustified. The detailed structure of the distribution is dominated by photometric and statistical uncertainties caused by the small sample (133 stars), the use of UBV photometry (McClure and Tinsley 1976), and velocity corrections.

The reasons for believing that the SFR in the past was significantly greater than the present are tied to two assumptions of the models which appear to be highly questionable: (1) no gas flows in or out of the system and (2) the SFR is proportional to the gas density to some power greater than or equal to one. The first of these will be discussed next while the second will be returned to later.

3.3 Infall

The value of using infall to explain some of the deviations from the predictions of the simple model has been recognized for some time (Larson 1972; Lynden-Bell 1975). The surprising result of the more recent discussions (Twarog 1980; Chiosi and Matteucci 1982; Vader and deJong 1981; Lacey and Fall 1983) is that, despite a wide range of model parameters, the need for a constant, moderate rise in metallicity combined with the current stellar and gas density inevitably requires an inflow of gas at a rate which is a significant fraction of the SFR, approximately 1/3 to 1/2. It must be emphasized that the crucial parameter dominating the chemical history is the ratio of infall to star formation. The need for infall on a galaxy-wide scale is consistent with the observation that most spirals have surprisingly short gas depletion timescales based on gas content and SFR estimates. (Larson, Tinsley, and Caldwell 1980; Kennicutt 1983) The nagging uncertainty in this picture of harmony is that not only do we not know the source of the infalling gas, but no conclusive evidence exists that there is extensive infalling gas. (See van Woerden et al., this volume, for more information.)

3.4 The Initial Mass Function (IMF)

Of all the assumptions used in chemical-evolution models, that of a constant IMF and/or yield is most often greeted with skepticism. The elemental production can vary with time because of a change in the slope of the mass function and/or the elemental yields as a function of stellar mass can vary as the composition changes. Unfortunately, our understanding of star formation and stellar nucleosynthesis is so poor that plausible arguments can be made supporting almost any direction of variability.

If one accepts the general premise that the mean metal abundance of the disk increases with time (irrespective of the rate), the best test of a variable IMF is a plot of an elemental abundance ratio with [Fe/H], where the dominant source of one element is the high-mass stars, e.g. O, while the second element comes from low-mass stars, e.g. Fe. If the yield ratio is constant, the abundance ratio is constant, independent of the SFR history. Fig. 2 of Twarog and Wheeler (1982) shows [O/Fe] vs. [Fe/H] for the disk based on elemental abundances of dwarfs. The halo shows an overabundance of O relative to Fe equivalent to [O/Fe] $\simeq$ +0.5. The transition between halo and disk is apparent at [Fe/H] $\simeq$ -0.9 $\pm$ 0.2. At first glance, it would appear that not only is the O/Fe production rate lower in the disk but variable as well. This is not the case. The disk data can be well represented by a model in which the yield ratio is constant but lower than in the halo. The variation in [O/Fe] is due to the initially high O abundance in the gas of the disk which gradually weakens due to disk production.

4. BEYOND THE SOLAR NEIGHBORHOOD

Our picture of disk evolution above is based solely on information obtained from stars within a few hundred parsecs of the Sun. As one moves outside this small sphere, the quantity and quality of the information relevant to chemical evolution declines dramatically and little of significance can be said about the temporal evolution of the system. However the data on the spatial distribution of the evolutionary parameters within a galaxy can be used to place crude constraints on the possible sources of different chemical histories among galaxies, and thereby may shed some light on the crucial factors controlling galaxy evolution.

4.1 Abundance Gradients

The spatial abundance distribution within the disk has been studied using a variety of objects, Cepheids (Harris (1981), G and K giants (Janes 1979), and supergiants (Luck 1982). A comprehensive discussion of the abundance gradient based on HII-region abundances can be found in the recent study of Shaver et al. (1983). The gradients observed for specific elements are listed in Pottasch, this volume. For those elements exhibiting significant gradients, the trend of increasing abundance with decreasing galactocentric distance is typical of that found in unbarred spirals (see e.g. McCall 1982). Extrapolation to the nuclear bulge indicates a metallicity approximately a factor of three above solar, consistent with recent claims based on studies of long-period variables (Wood and Bessell 1983) that the bulge contains a young population of super-metal-rich stars.

In a simple model, higher metallicity implies a greater degree of processing, i.e. a higher SFR per unit gas density. However, once the constraints of the closed model are removed, variation of any number of parameters can lead to higher current metallicities. In light of our picture of the solar neighborhood, it is felt that the dominant factor in the formation of an abundance gradient is the radial variation of the timescale for star formation relative to that for infall as exemplified by the dynamical collapse models of Tinsley and Larson (1978).

Crucial tests of this solution are the size of the gradient within the old disk, and the range in mean metallicity among spiral galaxies, particularly the early types. While it is possible to make the instantaneous metallicity of the disk gas abnormally high relative to the yield in an infall model with a high SFR relative to infall, the mean metallicity of the stars over the lifetime of the system can never be significantly higher than the yield, unless the infalling gas is metal-rich. The existence of an old, metal-rich population within the bulge may require: (1) radial gas flow of low angular momentum infall (Mayor and Vigroux 1981); (2) radial variations in the IMF (Garmony, Conti, and Chiosi 1982); or (3) a radial abundance gradient within a thick-disk population of the inner halo (Zinn 1980).

While the assumption of a higher infall rate nearer the galactic center rests on theoretical arguments regarding free-fall collapse timescales, a gradient in the SFR is supported by direct observation of high mass star-formation tracers, supernova remnants (Guibert, Lequeux and Viallefond 1978), and Lyc photons (Mezger 1978). The higher SFR is often produced in chemical-evolution models through the assumption that the SFR is some power-law function of the gas density or gas mass, a parameterization first used by Schmidt (1959). Numerous attempts to discover n, the power of the relation, have been inconclusive, implying that star formation is a multistep process dependent upon a number of factors in addition to the gas density. At least one factor appears to be the ability to make molecular clouds (see Young, this volume), which occurs at a higher rate near the galactic nucleus in most star-formation models. The observed decoupling of CO and HI distributions in our Galaxy and others would explain why a definite correlation between the SFR and HI has failed to emerge (Kennicutt 1983) and why some gas-rich galaxies show so little recent star formation (Schommer and Bothun 1983).

Beyond some critical distance where the SFR is low and the infall is a significant contributor over a long timescale, both models for star formation would predict that molecular-cloud formation declines sharply, possibly reverting to an evolutionary burst history similar to that in the halo. Such a transition between the two phases could explain the apparent steepening of the Fe-abundance gradient toward the galactic anticenter found by Janes (1979), while the O gradient maintains a constant slope. An ideal test of this picture and any chemical evolution model is provided by the irregular galaxies.

4.2 Irregulars and Chemical Evolution

For some time now, irregular galaxies have been extolled as unique objects for study because of their ability to fit the predictions of the simple closed model, i.e. Z is linearly correlated with $\ln \mu^{-1}$. This claim appears to be unjustified for two reasons: (a) Given the large uncertainties in μ and Z, the data for irregulars are not uniquely fit by a simple closed model. Over a wide range in μ, infall models can very closely mimic simple closed models in the Z - μ plane. In Figure 1a is plotted the O abundance vs. $\ln \mu^{-1}$ for the solar-neighborhood infall model of Twarog and Wheeler (1982). The straight line is the prediction for a simple closed model within the disk with the same O yield as the infall model; (b) μ is normally calculated using dynamical estimates of the total mass (e.g. Lequeux et al. 1979). The assumption that all the mass is involved in chemical evolution is definitely unjustified in the solar neighborhood (Tinsley 1980b), where at least 1/3 to 1/2 of the disk material is in the form of dark matter. Whether or not a comparable problem exists in irregulars is unknown. The effect of adding such unseen matter to the total mass estimate is seen in Figure 1b, where 25% of the mass is in the form of dark matter uninvolved in chemical evolution. The effect is to steepen the slope of the relation, giving too large a yield.

Two recent studies of importance are those of Hunter, Gallagher, and Rautenkranz (1982) and McCall (1983). Both studies of HII regions in irregular and late-type spirals attempted fits to simple closed models, using gas fractions estimated with total masses based on assumed M/L ratios rather than dynamical estimates. Contrary to earlier studies, Hunter et al. found the irregulars as a group failed to match the predictions of the simple closed model using localized or galaxy-wide gas fractions. McCall found that HII regions in galaxies of type Scd or later fit the simple, closed model with an O yield of 0.0004, while earlier types deviated significantly from the prediction. If correct, this may imply that irregulars are not simple, isolated systems, but one extreme in a continuum of galaxy evolution influenced by gas flows.

5. SUMMARY AND SPECULATION

The topics of this talk have been those areas where supposedly our picture of chemical evolution of the Galaxy has changed for the better in recent years. However, two points cannot be overemphasized: (a) those areas of astrophysical importance to galaxy evolution where it can be said that our understanding is satisfactory, are meager, if not nonexistent, and (b) the degree to which we can claim to understand the chemical evolution of other galaxies is always significantly less than the degree to which we understand the chemical evolution of our own Galaxy. In light of (b), let me close by summarizing what our picture of

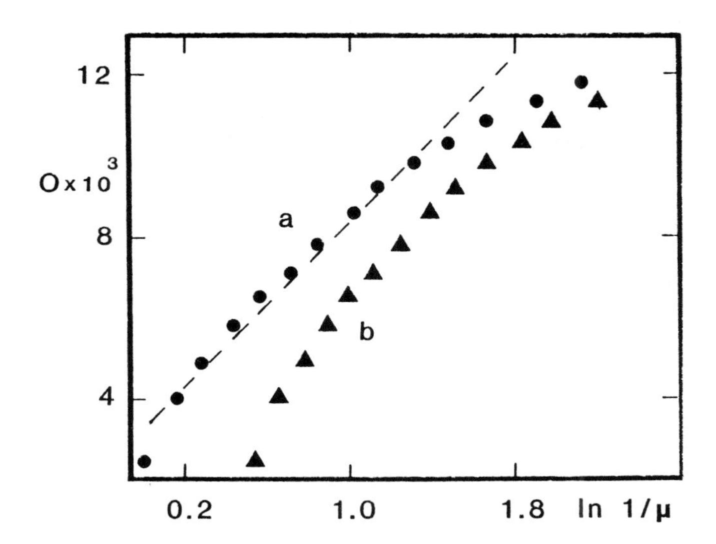

Fig. 1. Oxygen history for the solar neighborhood with (a) infall and (b) dark matter included.

the Galactic evolution may be telling us about chemical evolution of galaxies in general. Though the following statements are totally speculative, they are not original but represent a hybrid of ideas which have been stated or implied by a number of authors in the literature.

(1) Chemical evolution in spiral and irregular galaxies can be broken down into two distinct components: (a) a burst component similar to Searle's fragmented-protocloud model, characterized by large abundance inhomogeneities, reduced absolute yield but high O/Fe relative to the solar neighborhood, and little coherent spatial structure, and exemplified by the halo, the outermost regions of spirals, and irregular galaxies; (b) a continuous component, characterized by continuous star formation, a homogeneous ISM, yields typical of those in the solar neighborhood, and well-defined spiral structure as exemplified by early-type spirals, the solar neighborhood, and the inner regions of late-type spirals.

(2) All spirals and irregulars experience gas infall. The morphological trends along the Hubble sequence and radially within spirals have the same origin, the variation in the timescale for star formation relative to that for infall, with early-type galaxies and spiral nuclei having the highest rate of star formation relative to infall.

(3) The transition between the two phases among galaxies and within a galaxy is sharp and represents the point where molecular-cloud formation no longer occurs efficiently. Where this transition occurs is controlled by the mass distribution and the gas density which, in turn, is dominated by the variation of the star-formation to infall timescales within galaxies.

It is a pleasure to acknowledge the travel support kindly provided by the Kansas University Endowment Association and the University of Kansas Faculty Scholarly Travel Fund.

REFERENCES

Bond, H.E.: 1981, Astrophys. J. 248, 606
Chiosi, C. and Matteucci, F.: 1982, Astron. Astrophys. 110, 54
Garmany, C.C., Conti, P.S. and Chiosi, C.: 1982, Astrophys. J. 263, 777
Gilmore, G. and Reid, N.: 1983, Monthly Notices Roy. Astron. Soc. 202, 1025
Guibert, J., Lequeux, J., and Viallefond, F.: 1978, Astron. Astrophys. 68, 1
Harris, H.C.: 1981, Astron. J. 86, 707
Hunter, D.A., Gallagher, J.S., Rautenkranz, D.: 1982, Astrophys. J. Suppl. 49, 53
Janes, K.A.: 1979, Astrophys. J. Suppl. 39, 135
Kennicutt, R.: 1983, preprint
Lacey, C.G. and Fall, S.M.: 1983, Monthly Notices Roy. Astron. Soc., in press
Larson, R.B.: 1972, Nature Phys. Sci. 235, 7
Larson, R.B., Tinsley, B.M. and Caldwell, C.N.: 1980, Astrophys. J. 237, 692
Lequeux, J., Peimbert, M., Rayo, J.F., Serrano, A., Torres-Peimbert, S.: 1979, Astron. Astrophys. 80, 155
Luck, R.E.: 1982, Astrophys. J. 256, 177
Lynden-Bell, D.: 1975, Vistas in Astronomy 19, 299
Mayor, M.: 1976, Astron. Astrophys. 48, 301
Mayor, M., and Martinet, L.: 1977, Astron. Astrophys. 55, 221
Mayor, M. and Vigroux, L.: 1981, Astron. Astrophys. 98, 1
McCall, M.: 1982, Univ. of Texas, Publ. in Astronomy, No. 20
McClure, R.D. and Tinsley, B.M.: 1976, Astrophys. J. 208, 480
Mezger, P.G.: 1978, Astron. Astrophys. 70, 565
Miller, G.E. and Scalo, J.M.: 1979, Astrophys. J. Suppl. 41, 513
Mould, J.R.: 1982, Ann. Rev. Astron. Astrophys. 20, 91
Ostriker, J.P. and Thuan, T.X.: 1975, Astrophys. J. 202, 353
Pagel, B.E.J.: 1981, in The Structure and Evolution of Normal Galaxies, S.M. Fall and D. Lynden-Bell (eds.), Cambridge Univ. Press, 211
Pagel, B.E.J. and Edmunds, M.G.: 1981, Ann.Rev.Astron.Astrophys. 19, 77
Pagel, B.E.J. and Patchett, B.E.: 1975, Monthly Notices Roy. Astron. Soc. 172, 13
Schmidt, M.: 1959, Astrophys. J. 129, 243
Schmidt, M.: 1963, Astrophys. J. 137, 758
Schommer, R.A. and Bothun, G.D.: 1983, Astrophys. J. 88, 577
Searle, L.: 1977, in The Evolution of Galaxies and Stellar Populations, B.M. Tinsley and R.B. Larson (eds.), Yale Univ. Obs., p. 219
Shaver, P.A., McGee, R.X., Newton, L.M., Danks, A.C., and Pottasch, S.R.: 1983, preprint
Tinsley, B.M.: 1980a, Fund. of Cosmic Phys. 5, 287
Tinsley, B.M.: 1980b, Astron. Astrophys. 82, 246
Tinsley, B.M. and Larson, R.B.: 1978, Astrophys. J. 221, 559
Twarog, B.A.: 1980, Astrophys. J. 242, 242
Twarog, B.A. and Wheeler, J.C.: 1982, Astrophys. J. 261, 636
Vader, J.P., de Jong, T.: 1981, Astron. Astrophys. 100, 124
Wood, P.R. and Bessell, M.S.: 1983, Astrophys. J. 265, 748
Zinn, R.: 1980, Astrophys. J. 241, 602

DISCUSSION

J.P. Ostriker: Two questions: 1) Could the infall of metal-poor material not be supplied by evolving halo giants? 2) And conversely, could mass loss of metal-rich material (e.g. from type-I supernovae and runaway-star supernovae) leaving the Galaxy in a galactic wind remove the need for infall altogether?

Twarog: 1) Clearly you cannot have the gas come from too far out, for the reason you mentioned the other day: it would be inconsistent with the X-ray observations (see Discussion after paper by H. van Woerden in Section II.8). So, infalling material would have to come from the halo, within a constrained distance. Whether one can obtain it from halo stars, depends on the mass-loss rate of stars and on the star density in the halo.

Ostriker: In the model I made with Thuan several years ago, the early metal enrichment of the disk was from halo stars, and the late infall was hydrogen gas from halo stars - and both seemed to work out.

Twarog: 2) As to your second question, one could mimic a change in metallicity with time with such a model, but the star formation history also points to the need for infall and would be harder to reproduce in such an approach.

J. Milogradov-Turin: What do you mean when you say "halo" in the solar neighbourhood? Is it halo stars or the interstellar medium?

Twarog: What I call "halo", is stars with metallicity less than -1. This gets back to the matter of the thick disk (Section 3 of my paper). In Searle's picture of fragmenting proto-clouds, the gas falling into the disk has a range in metallicity from -2 to -1. So there should be disk-kinematic stars which have abundances like the halo. Hence the boundary between halo and disk is quite fuzzy.

R. Güsten: I disagree with two of your conclusions, namely the constant star-formation rate and the need for a high present-day infall rate.
1) The conclusion on the constancy of the star-formation rate (SFR) may be an overinterpretation of the stellar age-metallicity relation. Fits to the latter may also be obtained for the more reasonable assumptions of an exponentially decreasing infall rate and a e.g. linear dependence of the SFR on the available gas mass (see, for example, Vader and de Jong, 1981; Lacey and Fall, this meeting; Güsten and Mezger, 1983).
2) If you relax the instantaneous-recycling approximation, there is a natural flattening in the abundance gradients during late evolutionary phases, when evolved low-mass stars start diluting the ISM with 'metal'-poor ejecta.

Twarog: 1) The models do allow production of the age-metallicity relation using a star-formation rate which depends on the gas density via

some power law. However, this is irrelevant because all the observational tests to determine the exponent in this power law have been inconclusive: numbers between 0 and 3 have been found. And this is only to be expected, since star formation is far too complex a process to be described only by a simple function of the gas density. It has to be a combined function of gas density, molecular-cloud formation, composition changes, gas temperature, etc. Hence, even though parametrization may work in the models, I do not think it is relevant in terms of being physically correct. The constancy of the star formation rate is derived independently of the age-metallicity relation.
2) As to your second comment, I refer to the text of my paper.

V. Radhakrishnan: According to some supernova theories, a certain fraction of stars blow up completely, without leaving a neutron star behind. What can be said about the likelihood of such theories from the observed abundances?

Twarog: I had a point about that in my paper. Stellar-nucleosynthesis predictions for the abundances of O, Fe, Mg, Ne, etc. cannot be matched with continuous mass functions in which all stars above a certain mass contribute to chemical evolution. The only things you can do are to cut off the mass function, or to have it discontinuous. This may mean that we do not know enough about stellar nucleosynthesis to match the observations. There is no consistent theoretical prediction for stellar nucleosynthesis yields which will match the observations.

D. Lynden-Bell: Do the outermost HII regions in the galactic disk have the same O/Fe ratio as the halo or not?

Twarog: Unfortunately, there are hardly any data on this. One piece of information that we now have is C-abundance data from IUE on the Magellanic Clouds. C and Fe are made in the same stars, so their ratio is always 1. Dufour and Shields find that oxygen in the Magellanic Clouds is overabundant, relative to carbon, by about the same amount as in the halo: 0.3-0.7 dex.

Lynden-Bell: Does that indicate, then, that it is not really a matter of disk versus halo, but rather that the metal abundance is determined by the star-formation process?

Twarog: There are two modes of star formation, each with its own initial mass function. I used the terms "halo" and "disk", because these are the obvious examples of the two modes.

G.D. van Albada: How does your finding of a constant IMF tally with Kahn's and Viallefond's findings that at least the heavy end of the IMF strongly depends on the metallicity?

Twarog: Clearly it does not, but the observational tests for variation of IMF with metal abundance seem to be inconclusive so far.

KINEMATICAL AND CHEMICAL EVOLUTION OF THE GALACTIC DISK

Cedric G. Lacey and S. Michael Fall
Institute of Astronomy, Cambridge, England

We have calculated models for the kinematical and chemical evolution of the Galactic disk that include the infall of metal-free gas and the stochastic acceleration of disk stars. Our models are similar in some respects to those of Chiosi (1980) and Vader & de Jong (1981), but the results are compared with a greater variety of observational data. A complete description of this work will appear shortly in Monthly Notices of the Royal Astronomical Society.

The evolution of the stars, gas and heavy elements is treated in the approximation of instantaneous recycling, with a constant yield y and returned fraction R, and with no exchange of material between different galactocentric radii. The rate of infall of gas per unit area is assumed to vary as

$$f(r, t) \propto \exp(-\alpha r - t/t_f) \quad (1)$$

so as to give an exponential profile for the surface density. Following Schmidt (1959), the star-formation rate is assumed to vary as a power of the volume density of the gas; thus

$$(1 - R)\,\psi = C(\mu_g/H_g)^n\,H_g \quad (2)$$

where μ_g is the surface density of the gas layer, H_g is its half-thickness and ψ is the star-formation rate per unit area. We calculated models with n = 1, 3/2, 2 and adjusted the parameters t_f and C.

The stars are born with a small velocity dispersion comparable to that of the gas and are then accelerated by the irregular gravitational field of "clumps" in the disk. The evolution of the vertical velocity dispersion σ_s is represented by a phenomenological equation of the form proposed by Wielen (1977), with the additional assumption that the coefficient of diffusion in velocity space has the same time-dependence as the star-formation rate, thus:

$$d\sigma_s{}^q/dt = \beta d\mu_s/dt\ , \quad (3)$$

H. van Woerden et al. (eds.), The Milky Way Galaxy, 597–599.

Table 1. Input parameters for best-fit models

n	t_f	C	q	β	σ_{SO}
1	5.5	0.5	2	14	2
3/2	3.5	1.1	2	17	2
2	2.0	2.1	5/2	85	3

Units : t_f in Gy, C in $(M_\odot pc^{-3})^{1-n}$ Gy^{-1}
β in $(km\ s^{-1})^q (M_\odot pc^{-2})^{-1}$, σ_{SO} in $km\ s^{-1}$

Table 2. Output parameters for best-fit models

n	y	$\bar{\psi}/\psi_1$	$(1-R)\psi_1$	f_1
1	0.015	2.1	2.9	1.9
3/2	0.014	3.9	1.6	0.8
2	0.012	8.6	0.7	0.1

Units : ψ_1 & f_1 in $M_\odot$ pc^{-2} Gy^{-1}

where μ_S is the surface density of the stars and q and β are parameters to be determined. This is reasonable if the perturbing clumps are giant molecular clouds and perhaps also if they are spiral wavelets.

Models for the solar neighbourhood were compared with observations of the present gas density and the age-metallicity, number-metallicity and velocity dispersion-age relations for disk stars. For each value of n there is a best-fit model providing a good fit to all of these data. The input parameters of these models are given in Table 1 and some of the output parameters in Table 2 (in which the subscript "1" denotes the present value and a bar the time-average value). The n = 2 model is probably ruled out by its large value of $\bar{\psi}/\psi_1$, but the n = 1 and n = 3/2 models are consistent with the observational constraint $0.5 \lesssim \bar{\psi}/\psi_1 \lesssim 4$.

We also calculated the variation of properties with galactocentric radius for models with the same values of the parameters as fit the solar neighbourhood best. These models reproduce the approximate constancy of the stellar scale height with radius observed in edge-on galaxies, and the n = 1 and n = 3/2 models were consistent with the observed radial variation of gas density and star-formation rate. However, none of the models gave an abundance gradient as large as that observed. We are at present working on models with radial gas flows in an attempt to resolve this problem.

REFERENCES

Chiosi, C.: 1980, Astron. Astrophys. 83, p. 206
Schmidt, M.: 1959, Astrophys. J. 129, p. 243
Vader, J.P., and de Jong, T.: 1981, Astron. Astrophys. 100, p. 124
Wielen, R.: 1977, Astron. Astrophys. 60, p. 263

DISCUSSION

J.V. Villumsen: Your best value for q in equation (3) is about 2, while your analytical calculation of scattering (your paper in section III.2 of this Symposium) gives you q = 4. How do you reconcile those values?

Lacey: The value q = 2 was obtained by fitting the predictions of the model, including the phenomenological equation (3), to Wielen's (1977) observational data on the relation between vertical velocity dispersion and age. The mechanism of scattering by clouds with negligible random motions, which predicts q = 4 (c.f. eqn (4) of my paper in Sect. III.2), is thus not consistent with Wielen's data unless the product $N_c M_c^2$ has declined with time more rapidly than the star-formation rate.

R.W. Hilditch: A general observational remark. There is now plenty of evidence for the existence of main-sequence B, A stars and G, K giants with apparently normal, solar-type abundances at several kiloparsecs from the galactic plane. I refer to the work of Tobin and Kilkenny; Rodgers, Harding and Sadler; Hill, Hilditch and Barnes; and Hartkopf and Yoss.

T.M. Bania : I would like to comment on the possible uncertainty in our understanding of the evolution of low-mass stars, which contribute substantially to the chemical evolution of the ISM when their envelopes become fully convective at the base of the red-giant branch. Recently Bob Rood, Tom Wilson and myself have surveyed the Galaxy to determine the He abundance in giant HII regions. The most striking result is that the He abundance in W3, which is located beyond the solar circle, is at least seven times the proto-solar value whereas the He abundance in M17, an inner-Galaxy source, is solar or less. This is exactly opposite in sense to the observed C and He being produced in low-mass stars and slowly accumulating during the main-sequence phase into a zone of enhanced abundance near the mass fraction 0.3 - 0.4. If the C gets mixed to the surface and expelled, so should the He, and these species ought to show the same radial abundance gradient in the Galaxy. Thus something funny is going on in these sources, and our standard assumptions regarding the rate of enrichment from low-mass (and high-mass?) stars should be examined critically.

In the Aula of Groningen University, A. Blaauw presents descendants of J.C. Kapteyn with copies of "Sterrenkijken bekeken", a booklet about the history of Groningen astronomy. CFD

SECTION III.5

THE OLD POPULATION

Friday 3 June, 1045 - 1140

Chairman: K.C. Freeman

Freeman (right) and Allen talking over beer. Background: Fujimoto and Paris Pişmiş. GSS

THE FORMATION AND EARLY EVOLUTION OF THE MILKY WAY

S. Michael Fall

Institute of Astronomy, Cambridge, England

I. INTRODUCTION

In broad outline, the traditional picture for the formation of the Milky Way can be summarized as follows. The proto-galaxy consisted of a slowly rotating cloud of metal-free gas that cooled by bremsstrahlung and recombination radiation. As the internal pressure of the gas decreased, it collapsed in stages with smaller dimensions, faster rotation velocities and flatter shapes until it reached centrifugal support in a fundamental plane. At the same time, the gas was progressively depleted by the formation of stars and enriched with heavy elements by the ejecta from previous generations. The result is a general correlation between the kinematic properties, chemical compositions and relative ages of the stellar populations within the Galaxy. This picture was formulated at the Vatican symposium by Oort (1958) and others and was elaborated by Eggen, Lynden-Bell & Sandage (1962), Sandage, Freeman & Stokes (1970), Gott & Thuan (1976), Larson (1976) and others. Much of the recent work on galaxy formation has been an attempt to extend these ideas to a more comprehensive picture that includes large quantities of dark matter. The purpose of this article is to review several topics concerning the collapse phase in the evolution of the Milky Way.

In the following discussion, the Galaxy is assumed to consist of three main components: an exponential disc with a scale-radius of about 4 kpc, a de Vaucouleurs spheroid with an effective radius of about 3 kpc and an isothermal halo with much larger dimensions. This description, which relies heavily on extragalactic studies, is consistent with the available star counts and kinematic data for the Milky Way (Bahcall & Soneira 1980, Caldwell & Ostriker 1981). The traditional members of population II are provisionally assigned to the spheroid, although a distinction between globular clusters and field stars may be useful for some purposes. Suitable candidates for the dark matter in the halo include stellar remnants, low-mass stars and a variety of elementary particles. Following the arguments of White & Rees (1978), Fall & Efstathiou (1980), Silk & Norman (1981), Faber (1982), Gunn (1982) and

H. van Woerden et al. (eds.), The Milky Way Galaxy, 603–610.

others, the dark matter is assumed to be pre-galactic and the spheroid and disc are assumed to have formed from residual gas that collapsed within this arena. The halo itself may have formed by hierarchical clustering or as part of a pancake, depending on whether the primordial spectrum of density perturbations had a cutoff below or above galactic scales.

II ROTATION OF THE PROTO-GALAXY

The radial extent of the collapse depends critically on the rotation of the proto-galaxy. This can be quantified in terms of the dimensionless spin parameter $\lambda \equiv J|E|^{\frac{1}{2}}G^{-1}M^{-5/2}$ where J, E and M are respectively the total angular momentum, energy and mass of the system and G is the gravitational constant. It is instructive first to consider the problem without a massive halo. In this case, a reasonable model for the proto-galaxy just prior to collapse is a sphere with uniform density and negligible kinetic energy. In terms of the maximum radius r_t at turnaround, the specific angular momentum is $J/M = (5GMr_t/3)^{\frac{1}{2}}\lambda$ and the free-fall time is $\tau_{ff} = \pi(r_t^3/8GM)^{\frac{1}{2}}$ for $\lambda \ll 1$. A self-gravitating exponential disc with a scale radius α^{-1} has $J/M = 1.11(GM/\alpha)^{\frac{1}{2}}$ and a maximum circular velocity $v_c = 0.65(G\alpha M)^{\frac{1}{2}}$ at the radius $2.2\alpha^{-1}$ (Freeman 1970). Thus, conservation of J/M during the collapse implies

$$\alpha r_t = 0.74\lambda^{-2} \qquad \text{and} \qquad \alpha v_c \tau_{ff} = 0.46\lambda^{-3} \tag{1}$$

These results are not affected by any internal redistribution of angular momentum that may occur during or after the formation of the disc.

The relation between the initial spin and radius of the proto-galaxy without a halo is plotted as the upper curve in Fig. 1 and labelled with several values of the free-fall time for $\alpha^{-1} = 4$ kpc and $v_c = 220$ kms^{-1}. If the Galaxy formed by hierarchical clustering, it would have been endowed with some rotation by the tidal torques of neighbouring objects before it collapsed (Peebles 1969). The dashed lines in Fig. 1 show the 10, 50 and 90 percentile points in the distribution of spins generated by cosmological N-body simulations with Poisson initial conditions (Efstathiou & Jones 1979). For the median value, $\lambda \approx 0.065$, the initial radius of the proto-galaxy is $r_t \approx 700$ kpc and the corresponding free-fall time is $\tau_{ff} \approx 3 \times 10^{10}$yr, neither of which is acceptable. Unfortunately the distributions of spins for different initial conditions are not yet known with certainty and there are no reliable predictions for the pancake picture. As Fig. 1 shows, only a small range near $\lambda \approx 0.15$ is compatible with a spread of less than a few billion years in the ages of globular clusters.

The situation is changed dramatically by the presence of a massive halo. If this is modelled as a singular isothermal sphere out to some truncation radius r_t, then the specific angular momentum is $J/M = \sqrt{2}v_c r_t \lambda$ and the free-fall time is $\tau_{ff} = (\pi/2)^{\frac{1}{2}} (r_t/v_c)$ for $\lambda \ll 1$. As

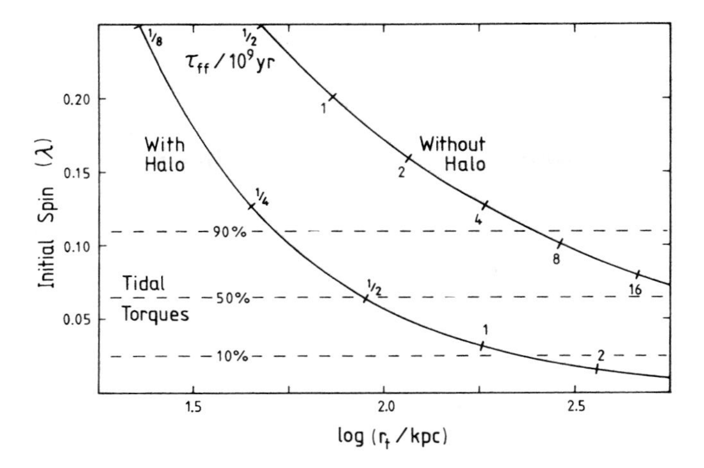

Fig. 1. Initial spin against initial radius for the proto-galaxy. The upper curve is from eqn. (1), the lower curve is from eqn. (2) and the numbers are free-fall times in units of 10^9 yr.

before, v_c denotes the velocity of test particles in circular orbits, which is generally smaller than the average rotation velocity of the dark matter. An exponential disc with negligible self-gravity in this halo has $J/M = 2v_c/\alpha$, which implies

$$\alpha r_t = \sqrt{2}\,\lambda^{-1} \qquad \text{and} \qquad \alpha v_c \tau_{ff} = \sqrt{\pi}\,\lambda^{-1} \tag{2}$$

if both components have the same specific angular momenta. These results are plotted as the lower curve in Fig. 1 and they agree with more detailed calculations that include the gravitational field of the disc (Fall & Efstathiou 1980). For $\lambda \approx 0.065$, the initial radius is $r_t \approx 90$ kpc and the corresponding free-fall time is $\tau_{ff} \approx 5 \times 10^8$ yr, both of which are quite acceptable. In fact, they are similar to the values deduced by Eggen et al. (1962) from the motions of old stars in the solar neighbourhood.

III. FORMATION OF SUBSTRUCTURE IN THE SPHEROID

Some substructure may have been present before the collapse, especially if the Galaxy formed by hierarchical clustering, but much of it may have developed during the collapse. Gunn (1980) has pointed out that the minimum mass for gravitational instability behind isothermal shocks in the spheroid is comparable with the masses of present-day globular clusters. The following arguments suggest that such objects might form under fairly general conditions. Any irregularities in the flow are likely to heat some of the gas up to the virial temperature of the halo

$$T_h = \mu_h v_c^2/2k \approx 1.8 \times 10^6 \text{ K} \tag{3}$$

where $\mu_h \approx 0.6$ amu is the mean molecular weight for an ionized mixture of hydrogen and helium and k is Boltzmann's constant. Since the dark matter is assumed to provide most of the gravitational acceleration, the free-fall time is $\tau_{ff} = (\pi/2)^{\frac{1}{2}}(r/v_c)$ from a radial position r in the halo. At a density ρ_h, the hot gas will radiate energy on a time-scale $\tau_{cool} = 3\mu_h kT_h/2\rho_h\Lambda(T_h)$, where Λ is the usual cooling function. It will remain hot if the condition $\tau_{ff} \lesssim \tau_{cool}$ is satisfied, which implies

$$\rho_h \lesssim 1.7\mu_h^{\frac{1}{2}}(kT_h)^{3/2}/r\Lambda(T_h) \approx 1.6 \times 10^{-3}(r/\mathrm{kpc})^{-1}M_\odot \mathrm{pc}^{-3} \quad (4)$$

for $\Lambda(T_h) \approx 2 \times 10^{-23}$ erg cm^3 s^{-1}.

Any gas that is more dense than the limit specified by eqn. (4) will cool rapidly and then be compressed to even higher densities by the hot gas. In this sense, the medium is thermally unstable and is likely to produce cold clouds at the temperature $T_c \approx 1.0 \times 10^4$K where the cooling function drops precipitously; they will be nearly neutral with a mean molecular weight $\mu_c \approx 1.2$ amu. In pressure balance, the densities in the hot and cold phases are related by

$$\rho_c/\rho_h = (\mu_c/\mu_h)(T_h/T_c) \approx 360, \quad (5)$$

irrespective of position in the halo. A compressed cloud is gravitationally unstable if its mass exceeds

$$M_{crit} = 1.18(\rho_h/\rho_c)^2(kT_h/\mu_h G)^{3/2}\rho_h^{-\frac{1}{2}} \gtrsim 3.2 \times 10^6(r/\mathrm{kpc})^{\frac{1}{2}}M_\odot, \quad (6)$$

where the inequality follows from eqns. (4)and (5). At the effective radius of the spheroid, $r \approx 3$ kpc, the internal density of the clouds is $\rho_c \lesssim 0.2\ M_\odot \mathrm{pc}^{-3}$ and their critical mass is $M_{crit} \gtrsim 5 \times 10^6\ M_\odot$. In comparison with globular clusters, the first prediction is low and the second is high, but they are both correct to within an order of magnitude. Moreover, the velocity dispersion in the cold gas, about 8 kms^{-1} in one dimension, is similar to that in globular clusters.

The free-fall time of a cold cloud, when approximated as a sphere of uniform density, is $\tau_{ff} = (3\pi/32G\rho_c)^{\frac{1}{2}}$ and its cooling time is $\tau_{cool} = 3\mu_c kT_c/2\rho_c\Lambda(T_c)$; thus

$$\tau_{cool}/\tau_{ff} = 2.8\mu_c kT_c(G/\rho_c)^{\frac{1}{2}}/\Lambda(T_c) \quad (7)$$

$$\gtrsim 3.2(r/\mathrm{kpc})^{\frac{1}{2}}[\Lambda(T_c)/10^{-28}\ \mathrm{erg\ cm^3 s^{-1}}]^{-1}.$$

At temperatures below 1×10^4K, the cooling function is $\Lambda(T_c) \lesssim 4 \times 10^{-28}$ erg cm^{3}s^{-1} for a fractional ionization $x = 0$ and a metallicity $Z \lesssim 0.01Z_\odot$ or for $x = 0.1$ and $Z \lesssim 0.001Z_\odot$ (Dalgarno & McCray 1972). When the metallicity is this low, τ_{cool} exceeds τ_{ff} and any clouds with masses greater than M_{crit} will contract quasistatically at roughly constant temperature. Since the Jeans mass within such a cloud varies as $\rho_c^{-\frac{1}{2}}$, it will fragment into a bound collection of smaller objects and perhaps ultimately into stars. If the condition $\tau_{cool}/\tau_{ff} \gtrsim 1$ is not satisfied and if the clouds are not heated by some means, they will cool below 1×10^4K before they collapse. In this case, the resulting substructures will have higher densites and smaller masses than those given above. It appears likely, however, that radiation from the hot gas and star formation within the cold clouds will keep their temperatures near 1×10^4K even when the metallicity is high and the cooling times are short. This possibility will be discussed in detail elsewhere (Fall & Rees, in preparation).

IV. DISRUPTION OF SUBSTRUCTURE IN THE SPHEROID

Several processes could play a role in the disruption of substructure in the spheroid. The tidal field of the halo insures that the internal density of any stellar or gaseous object exceeds

$$\rho_{lim} = (3/4\pi G)[\omega^2 + (v_c/r)^2] \gtrsim 2.7(r/kpc)^{-2} M_\odot pc^{-3}, \qquad (8)$$

where ω is the angular velocity and r is the radial position with respect to the Galactic centre. Even if a bound object is not tidally limited at the time of formation, it may become so near the pericentre of its orbit. If the mean density of the object is below ρ_{lim} it will lose some mass, and if the central density is below ρ_{lim} it will be completely destroyed. The debris of tidal disruption are therefore more likely to have radial orbits than the substructures that manage to survive this process. A comparison of eqns. (4) and (5) with eqn. (8) shows that the cold clouds produced by a thermal instability are particularly vulnerable and those with $r \lesssim 5$ kpc might even be prevented from collapsing if their central densities are low. As a result of tidal limitation, the clouds that do survive will have lower masses, higher mean densities and a greater resemblance to present-day globular clusters.

Another process that might disrupt some substructures while they are still mainly gaseous is star formation (Gunn 1980). The heat supplied by a few supernovae is enough to unbind a cloud with a mass of order $10^6 M_\odot$ and a velocity dispersion of order 10 kms^{-1}. If the metallicity is very low, this energy cannot be radiated efficiently at the densities of interest until the temperature reaches 1×10^4 K. Any clouds with much lower temperatures are therefore likely to become unbound whereas those near 1×10^4 K will almost certainly survive. However, the relevance of star formation is doubtful in the context of a thermal instability, because when the metallicity is low enough to permit the disruption of clouds with temperatures below 1×10^4 K, it is probably also low enough to prevent their formation. Any substructures that survive the relatively rapid effects of tidal limitation and star formation might be disrupted on much longer time-scales by several other processes (Fall & Rees 1977 and references therein). These include dynamical friction against the halo and the evaporation of stars by internal relaxation and external impulses. The effects that such processes have on present-day globular clusters are controversial, but they may have helped to narrow the range of surviving substructures.

The degree to which the spheroid consists of disrupted substructure might be revealed by any differences between globular clusters and field stars. The space distributions of both kinds of objects can be fitted by density profiles with similar effective radii, but obscuration prevents a detailed comparison in our Galaxy (de Vaucouleurs & Pence 1978). In some elliptical galaxies, however, the distribution of globular clusters is slightly more extended than the diffuse stellar component (Forte, Strom & Strom 1981). The velocity ellipsoid of metal-poor RR Lyrae variables in the solar neighbourhood, which may be typical of field stars in the spheroid, is elongated in the radial direction (Woolley 1978). In contrast, the velocity ellipsoid of globular clusters in our Galaxy appears to be nearly isotropic (Frenk & White 1980). These results are qualitatively consistent with the notion that substructures on radial orbits were preferentially disrupted and shed more of their mass than those on circular orbits. It would be interesting to refine this suggestion and make a quantitative comparison with the observations.

V CHEMICAL ENRICHMENT IN THE SPHEROID

The natural starting point for any discussion of chemical enrichment is the so-called simple model. It postulates that the initial composition of the gas is metal-free, the system is closed and well-mixed, the stellar mass function is constant and the recycling of enriched gas is instantaneous. In this case, the metallicity Z is related to the fraction of mass in gas μ by the familiar expression

$$Z = -y \,\ell n \mu , \tag{9}$$

where y is the nuclear yield. If the enrichment is stopped suddenly by the complete removal of gas with metallicity Z_m, then the fraction of stars $f(Z)dZ$ with metallicities between Z and $Z + dZ$ is given by

$$f(Z) = y^{-1} \exp(-Z/y)/[1 - \exp(-Z_m/y)] \tag{10}$$

for $Z < Z_m$ and $f(Z) = 0$ for $Z > Z_m$. This distribution is entirely the result of nucleosynthesis at different times rather than different places, and it could not be expected to apply if the enrichment occurred in isolated substructures. The main justification for applying the simple model to the spheroid is that the gas and stars might behave as a co-moving system during a free-fall phase of the collapse.

The observed distribution of metallicites for globular clusters beyond the solar circle and high-velocity subdwarfs in the solar neighbourhood can be fitted by an exponential function (Hartwick 1976, Searle & Zinn 1978). There is some evidence for a deficiency of objects with $Z \lesssim 0.003Z_\odot$, but an estimate of their frequency depends on how the samples are selected and calibrated (Bond 1981, Hartwick 1983). Even so, this sets a firm upper limit on any pre-galactic enrichment by stars from the hypothetical population III. For $Z \gtrsim 0.003Z_\odot$, the observed distribution is reproduced by eqn.(10) with the parameters $y \approx 0.02Z_{\odot}$ and $Z_m \approx 0.1Z_{\odot}$. The first is an order of magnitude smaller than the yield for the disc and the second implies, through eqn. (9), that nearly all of the gas was converted into stars during the formation of the spheroid. Unfortunately, the conditions that determine the stellar mass function in different environments are known too poorly to say whether the inferred value of y is a problem for the simple model. The depletion of gas in the spheroid is compatible with the formation of the disc by late infall, but it would be a problem if both components formed during a rapid collapse.

As an alternative to the simple model, Hartwick (1976) introduced a model in which the rate of gas removal is a constant c times the rate of star formation. In this case eqns. (9) and (10) are still valid, if μ is interpreted as the ratio of the mass in gas to the initial mass of the system and the nuclear yield y is replaced by the effective yield

$$y_{eff} = y(1 - R)/(1 - R + c), \tag{11}$$

where R is the fraction of mass returned to the gas by each generation of stars. With the conventional parameters for the disc, $y \approx 0.8Z_{\odot}$ and $R \approx 0.2$, the fit to f(Z) for the outer parts of the spheroid implies $c \approx 30$. Since the removal of this much gas from the proto-galaxy seems unlikely, Hartwick argues that it was transferred from the spheroid to the disc while they were in the process of formation. The fact that c is roughly equal to the ratio of the masses of the two components is consistent with this interpretation. Although the model is plausible, it does not explain why the rate that gas accumulates in the disc should be directly proportional to the rate that stars form in the spheroid. Clearly our understanding of chemical enrichment during the formation of the Galaxy is still rudimentary.

ACKNOWLEDGEMENTS

I thank Ray Carlberg, Ken Freeman, Cedric Lacey and Martin Rees for stimulating discussions of the material in this article.

REFERENCES

Bahcall, J.N. & Soneira, R.M.: 1980, Astrophys. J. Suppl. 44, 73.
Bond, H.E.: 1981, Astrophys. J. 248, 606.
Caldwell, J.A.R. & Ostriker, J.P.: 1981, Astrophys. J. 251, 61.
Dalgarno, A. & McCray, R.A.: 1972, Ann. Rev. Astron. Astrophys. 10, 375.
de Vaucouleurs, G. & Pence, W.D.: 1978, Astron. J. 83, 1163.
Efstathiou, G. & Jones, B.J.T.: 1979, Mon. Not. Roy. Astr. Soc. 186, 133.
Eggen, O.J., Lynden-Bell, D. & Sandage, A.R.: 1962, Astrophys. J. 136, 748.
Faber, S.M.: 1982, in "Astrophysical Cosmology", eds. Bruck, H.A., Coyne, G.V. & Longair, M.S., Pontificia Academia Scientiarum, p. 191.
Fall, S.M. & Efstathiou, G.: 1980, Mon. Not. Roy. Astr. Soc. 193, 189.
Fall, S.M. & Rees, M.J.: 1977, Mon. Not. Roy. Astr. Soc. 181, 37P.

Forte, J.C., Strom, S.E. & Strom, K.M.: 1981, Astrophys. J. 245, L9.
Freeman, K.C.: 1970, Astrophys. J. 160, 811.
Frenk, C.S. & White, S.D.M.: 1980, Mon. Not. Roy. Astr. Soc. 193, 295.
Gott, J.R. & Thuan, T.X.: 1976, Astrophys. J. 204, 649.
Gunn, J.E.: 1980, in "Globular Clusters", eds. Hanes, D. & Madore, B., Cambridge University Press, p. 301.
Gunn, J.E.: 1982, in "Astrophysical Cosmology", eds. Bruck, H.A., Coyne, G.V. & Longair, M.S., Pontificia Academia Scientiarum, p. 233.
Hartwick, F.D.A.: 1976, Astrophys. J. 209, 418.
Hartwick, F.D.A.: 1983, Mem. Soc. Astr. Italiana 54(1), 51.
Larson, R.B.: 1976, Mon. Not. Roy. Astr. Soc. 176, 31.
Oort, J.H.: 1958, in "Stellar Populations", ed. O'Connell, D.J.K., North Holland, Amsterdam, p. 415.
Peebles, P.J.E.: 1969, Astrophys. J. 155, 393.
Sandage, A., Freeman, K.C. & Stokes, N.R.: 1970, Astrophys. J. 160, 831.
Searle, L. & Zinn, R.: 1978, Astrophys. J. 225, 357.
Silk, J. & Norman, C.: 1981, Astrophys. J. 247, 59.
White, S.D.M. & Rees, M.J.: 1978, Mon. Not. Roy. Astr. Soc. 183, 341.
Woolley, R.: 1978, Mon. Not. Roy. Astr. Soc. 184, 311.

DISCUSSION

F.H. Shu: In your discussion of the disruption of substructure in the Galaxy, it is not clear to me that the processes would all lead to complete destruction. Would there be any remnants of objects that might have resided outside the part of parameter space where we now find the globular clusters?

Fall: That is very hard to say. It could just be that they become dense and that we do not see them. The argument is very crude, only to orders of magnitude. It is not even clear, for example, that the evaporation of stars from a globular cluster actually leads to its total disintegration.

T.S. Jaakkola: How can the disk angular momentum be constant in time, as in your model? Should not there be friction between the disk and the massive corona? To me, the fast rotation of spirals indicates that rotation is not a relic effect, but rather, that galaxies are "rotationally active".

Fall: Suppose the halo of the Galaxy were strongly triaxial, then it probably would exert some torque on the disk material, and could act to spin it up or down. The assumption made here is that this is a small effect, hence the disk angular momentum is approximately constant.

SUPPOSED HISTORY OF OUR GALAXY

Wilhelmina Iwanowska
Institute of Astronomy
N. Copernicus University, Torun, Poland

Looking back with modern instrumentation like that at Westerbork to possible progenitors of galaxies, we find at increasing redshifts active galaxies, radio galaxies, BL Lac objects and quasars. With improving instrumental sensitivity and resolving power, energetic activity is discovered in these objects: ejecta of millions of solar masses, mostly bipolar, were thrown out from the central bodies to hundreds of kiloparsecs at nearly luminal speeds. The rich variety of their structures and energetics is illustrated in a review paper by Miley (1980). Like miniatures of these phenomena, ejections of 10-100 $M_\odot$ with speeds of 10-100 km s^{-1} appear in star-forming regions of our Galaxy.

The powerful "engines" working in galactic explosions are still a mystery. Gravitational sources of energy accumulated in black holes through processes of accretion are considered. However, it is not easy to produce them in an expanding universe on a proper scale of mass, energy and time. It seems more natural to link the ubiquitous explosive phenomena to the general expansion of the Universe: to see the Big Bang as a long-lasting process of gradual explosive fragmentation of the primeval dense matter. In a sequence of "minor bangs" super-clusters and clusters of galaxies would be formed with large voids extending between them. Galaxies could gradually eject protoclusters of stars, beginning with far-reaching massive globulars, ending on small close open clusters and associations. Star-forming regions would still contain explosive seeds of primeval dense matter. Stars themselves could give birth to their planets, and these to their satellites in a similar process.

Thirty years ago Ambartsumian (see e.g. 1980) expressed the idea that expansion and fragmentation are leading processes in the present phase of cosmic evolution. He and his collaborators have found support for this idea in extensive observational studies of non-stationary phenomena in very young stars and active galactic nuclei.

The proposed imaginative scenario does not fit the conventional

H. van Woerden et al. (eds.), The Milky Way Galaxy, 611–612.

model or models of a hot Big-Bang. It calls also for different solutions of some basic problems. To give some examples, let us consider the spiral structure of galaxies. The string-like, mostly bipolar ejecta from quasars and galaxies, directed close along their symmetry axes, are probably aligned with the magnetic axes. Differential rotation, precessional and tidal interactions with the environment may reorient and wind up the arms to the shape of a spiral galaxy. It should be remembered that Pismis has proposed years ago and discussed in a series of papers (see e.g. 1979) a similar mechanism for spiral-structure formation. Strongly inclined, bar-like molecular features, in the bulge of our Galaxy, as well as a small spiral recently discovered in recombination-line studies of the nucleus (Oort, 1985) are examples of different orientations of substructures.

Concerning the chemical evolution of our and other galaxies there is an elaborate theory of nucleosynthesis in stars. Yet, we have to remember that the experimental test of the H-burning rate in the Sun yields a flow of neutrinos considerably lower than expected. There are also difficulties with the theory of formation of heavy elements which are present everywhere, in young and old objects. It seems worth-while to remember some alternative ideas concerning these problems. Mayer and Teller years ago (1950) suggested that the present chemical composition of cosmic matter is a result of two competitive processes: light elements in which lighter, proton-rich isotopes are more abundant were formed through nucleosynthesis; heavy elements with heavier, neutron-rich isotopes in preponderance could be formed through fission of a primordial, superdense "poly-neutron fluid". Could not a chain of nuclear fissions be active in primeval dense matter in the cores of quasars, active galactic nuclei and star-forming regions during their explosive phases?

In the frame of the proposed scenario one may also think of many other difficult or unsolved problems, like the flat rotation curves and dark matter, the origin of cosmic rays or the excessive X-ray emission - in order to encourage the cosmologists to work on models as close to the current observational results as possible.

REFERENCES

Ambartsumian, V.A.: 1980, Ann. Rev. Astron. Astrophys. 18, 1
Mayer, G.M. and Teller, E.: 1950, Rev. Mod. Phys. 22, 203
Miley, G.K.: 1980, Ann. Rev. Astron. Astrophys. 18, 165
Oort, J.H.: 1985, these Proceedings, section II.7
Pismis, P.: 1979, in I.A.U. Symposium 84, "Large-scale characteristics of the Galaxy", ed. W.B. Burton, Dordrecht: Reidel, p. 145.

SECULAR EVOLUTION IN GALAXIES

A. May[1], C.A. Norman[2,3] and T.S. van Albada[1]
[1] Kapteyn Astronomical Institute, Groningen, the Netherlands
[2] Sterrewacht, Leiden, The Netherlands
[3] Institute of Astronomy, Cambridge, U.K.

We have adapted the N-body code of Van Albada (1982) to study the secular evolution of a hot collisionless stellar component (E galaxy or galactic bulge) due to slow changes in another component of the same galaxy. Our equilibrium starting model is a non-rotating triaxial ellipsoid with axial ratios 1.3:1.4:2.0; the effects of the "other component" are then simulated by various simple means.

First, we consider the growth of a massive black hole or dense nucleus (represented by a point mass) at the centre of the galaxy, to a maximum mass $M_{p,max}$. We find an interesting large-scale structural effect: the system becomes rounder, out to radii more than 30 times the radius of influence GM_p/σ_o^2. This effect is greater for larger values of $M_{p,max}$; if $M_{p,max} = 10^{10}\ M_\odot = 0.05\ M_{gal}$, the observed change is one ellipticity subclass (E3.4 to E2.4). The observational statistic that radio galaxies appear rounder than normal E galaxies (Hummel 1980) may thus be evidence for massive black holes in active nuclei.

Secondly, we consider a torque applied slowly about the short axis, to mimic the effect on a galactic bulge of a rotating bar in the disc component. We find that, besides making the bulge rotate, the torque causes the system to assume a box or peanut shape in side view (see Fig. 1), like that observed in the bulges of the Milky Way (Hayakawa et al. 1981) and of some external galaxies (Kormendy and Illingworth 1982).

The key to both these effects may lie in the behaviour of individual stellar orbits under adiabatic changes (cf. Norman 1984): the first may be related to the loss of box-orbits, the second to the loss of long-axis tubes. Further work is in progress on these problems, as well as on the effect on a bulge component of the slow growth of a rigid disc potential.

H. van Woerden et al. (eds.), The Milky Way Galaxy, 613–614.

REFERENCES

van Albada, T.S.: 1982, Mon. Not. Roy. Astr. Soc. 201, 939
Hayakawa, S. et al.: 1981, Astron. Astrophys. 100, 116
Hummel, E.: 1980, Thesis, University of Groningen, pp. 151-152
Kormendy, J. and Illingworth, G.:1982, Astrophys. J. 256, 460
Norman, C.A.: 1984, in Proc. Moriond Conference on Galaxies, ed. J. Audouze and J. Tran Thanh Van, D. Reidel, pp. 327-336

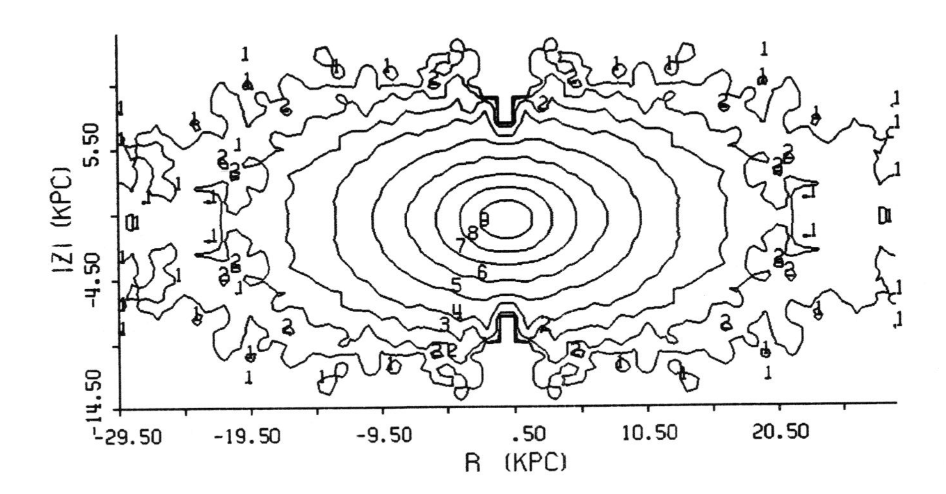

Fig. 1. Mean density in meridional (R, |z|) plane for torque model, after torque is switched off (averaged over 14 crossing times). Contour spacing is logarithmic (0.5 dex), in arbitrary units. Figure is reflected about R and |z| axes for clarity.

GAS DYNAMICS AND DISK-GALAXY EVOLUTION

R.G. Carlberg
University of Toronto

I. INTRODUCTION

Gas dynamics plays a vital role in the formation and ongoing evolution of a disk galaxy through its ability to have dissipative collisions. In this paper the simplest possible gas physics which provides a description of a gaseous component is incorporated into N-body experiments to investigate the formation of a disk galaxy.

II. DISSIPATIVE COLLAPSE IN A DARK HALO

It is necessary to maintain star formation over a Hubble time for a viable model of a disk galaxy, a problem identified in the pioneering models of Larson (1976). If all the gas collapses into a thin disk, the density is so high that star formation uses the gas up in a few rotation periods. The view of this paper is that the disk of the galaxy usually contains only a fraction of the gas present in the system, the rest being dispersed in the halo. The danger of ejecting gas from the disk is that it will never return, but help comes from the dark halo which provides a deep potential well. Dark halos also provide a theoretically attractive dynamical arena in which to form a disk galaxy (White and Rees 1978, Fall and Efstathiou 1980). In this paper a physical simulation for the formation of a disk galaxy is constructed, using the extreme assumption that the whole unit was present at the initial instant, with no pre-galactic conversion of gas to stars.

As the starting condition for the model we take a static dark halo with $\rho(r) \sim r^{-2}$. All computational units are dimensionless, with the outer radius of the protogalaxy being $r = 1$, containing a total mass of 1. The halo is filled with 10000 gas clouds having a combined mass equal to 10% of the static halo mass. The initial velocity distribution is an equilibrium isotropic velocity ellipsoid with a flat rotation curve such that $v/\sigma = 0.2$. In the absence of collisions these clouds and any stars present orbit under the combined gravity of the halo and their own self-gravity, neglecting nonaxisymmetric forces.

H. van Woerden et al. (eds.), The Milky Way Galaxy, 615–619.

The clouds collide to change direction, lose energy and form stars. The collision rate per cloud is modelled following Larson (1976), as $An_c^p\sigma$; where A is an adjustable parameter giving the effective cross-section, the rate of collisions increases linearly with the velocity dispersion σ of the gas, and p determines how the cross-section varies with the local density of gas clouds n_c. After a number of tests and comparison with models of elliptical galaxies the value of $p = \frac{1}{2}$ was chosen, although values in the range $1/3 < p < 2/3$ seemed to be suitable. An important point here is that the various phases of the interstellar medium present are assumed to be closely coupled, i.e. that the hot gas cannot slip by the cool gas out of which stars form. This would seem reasonable on the basis of the theory of McKee and Ostriker (1977), but in our galaxy molecular clouds are in a very thin layer, whereas there is evidence of hot gas in a much thicker layer suggesting that at some level the coupling breaks down.

When clouds collide, their random velocity about the centre-of-mass velocity is reduced by a factor f (<1), and randomly redirected. At the time of a collision there is a probability ε_* that a star will form from the gas cloud. When a star does form, it immediately ejects a metal-enriched gas cloud, with a random velocity S. The magnitude of S is estimated from the injection of 10^{51} ergs per supernova, and one SN per 100 $M_\odot$ of stars formed. Even at a 10% efficiency of conversion of the SN energy to thermal and kinetic energy this corresponds to a velocity of several hundred kilometers per second, i.e. comparable to the velocity dispersion in the dark halo.

The two major parameters describing the model then are the cross-section parameter A and the energy-ejection parameter S. These parameters can be constrained to about an order of magnitude from simple considerations. In order that any disk at all develop, the mean free path of gas clouds must be much smaller than the size of the system, and at least as short as the thickness of the disk. If S is taken to be so small as to be negligible, then the gas simply settles into a plane in one or two orbital times and turns into stars. A mean free path larger than the system size produces a very slowly growing ellipsoidal galaxy, and a very short mean free path in the halo quickly turns all the gas clouds into stars, more or less with their initial orbits, once again making an ellipsoidal galaxy. Both of these ellipsoids are flattened by rotation. If $S \approx 1$ and the mean free path is adjusted to be about equal to the thickness of the disk, then the gas is constantly ejected out of the disk into the halo, where the density is so low that no stars can form until the gas returns to the disk region. This is similar to the fountain advocated by Bregman (1980). Thus in order to get a reasonable disk galaxy, the parameters are determined to within an order of magnitude, all have straightforward physical interpretations, and most are open to observational investigation. The particular model to be discussed below has A = 10, S = 2, f = 0.95, and $\varepsilon_* = .01$. A scaling to galactic dimensions would typically take a total mass inside of 100 kpc as $10^{12}M_\odot$, the time unit is then 5×10^8 years.

The distribution of stars in the model at t=5.5 is shown in an edge-on view in Figure 1, and the galaxy clearly has a reasonably thin disk imbedded in a spheroidal component. At this time 15% of the mass is still in the form of gas, and star formation is continuing at a rate of 2.0% of the original gas mass per dynamical time, or with the suggested scaling, 4 $M_{\odot}$ per year.

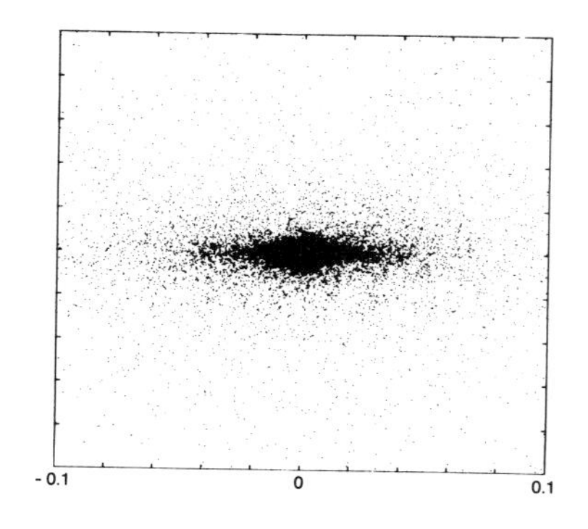

Figure 1. Edge-on view of stars in the final model.

In Figure 2 the metallicity (normalized to the yield) of stars forming in the halo, and in the "solar neighbourhood" ($|z| < .1$, $.05 < r < .1$), is shown as a function of the time of formation. Most of the halo stars are formed at early times, with a mean metallicity of around log (Z/y) = -1, with a very large scatter. The continuation to very high metallicities in the halo is only marginally consistent with the data for our Galaxy, and may indicate that one should include a cutoff minimum density for star formation.

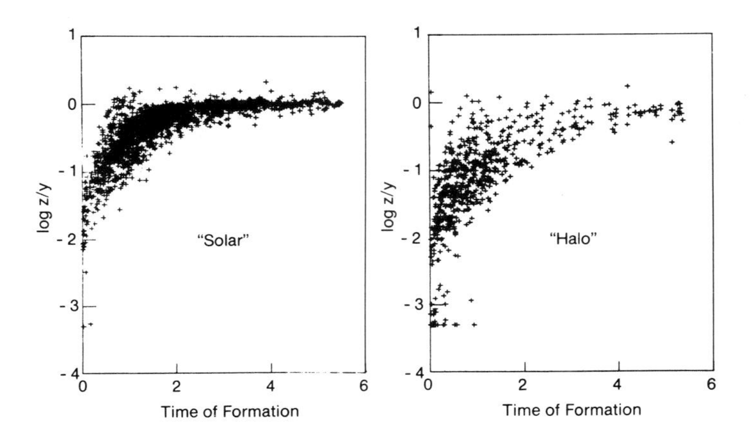

Figure 2. Age-metallicity relation in two regions.

Figure 3 shows the metallicity of stars in the spheroid of the galaxy as a function of radius (the gap near .01 is an artificial selection effect). It should be noted that there is very little gradient of abundance beyond the end of the disk at r = .1, but a very large scatter is present, as in the observations of outer-halo objects by Searle and Zinn (1978) and Zinn (1980). The reason for the lack of

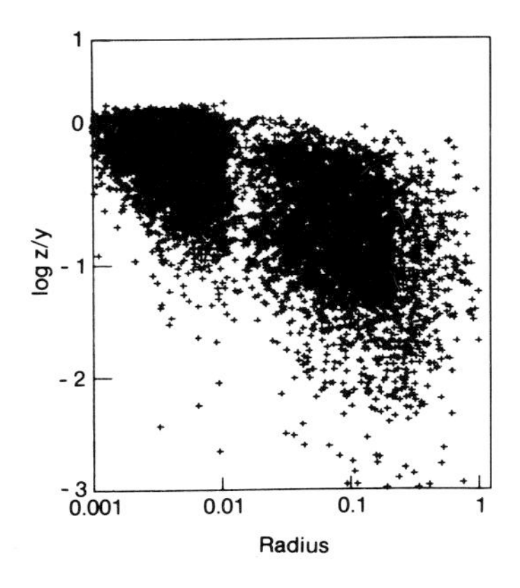

Figure 3. Spheroid radial metallicity distribution.

a gradient in the outer halo is twofold. The bulk of star formation occurs near the disk where the density is high, but the mean free path is long enough that the metals produced mix back up into the halo. There is a significant abundance gradient in the gas but also a considerable scatter about the trend. The halo objects are mostly formed between the start of the experiment and two dynamical times, which encompasses a large range of gas metallicities in the halo. Objects formed at the same time do not reflect the abundance gradient in the halo gas, but once the range of times is added together, and the orbits start mixing the stars around the galaxy, essentially no gradient is discernible.

The model has the large-scale characteristics of a disk galaxy, with clearly differentiated disk and spheroidal components, and star formation which extends over 20 rotation periods measured at the half-mass radius of the disk. However it is not possible to have the gas settle into a very thin plane and simultaneously keep the star-formation rate fairly low. The root of the difficulty may be in the assumption that the hot gas ejected from stars cannot slip past the cooler clouds out of which stars form,requiring a more detailed treatment of the interstellar medium.

III. SUMMARY

Gas-dynamical experiments modelling the formation of a disk galaxy in a dark halo are capable of producing an acceptable object under a limited range of conditions. Formation of a thin disk requires considerable dissipation without star formation. Supernovae and other stellar events eject gas from the disk back into the halo, regulating the gas content of the disk so that star formation is extended over many dynamical times. On the basis of these experiments an extensive dark halo is a necessary component of an actively star-forming disk galaxy.

REFERENCES

Bregman, J.N. 1980, Ap. J., 236, p. 557.
Fall, S.M. and Efstathiou, G. 1980, M.N.R.A.S., 193, p. 189.
Larson, R.B. 1976, M.N.R.A.S., 176, p. 31.
McKee, C.F. and Ostriker, J.P. 1977, Ap. J., 218, p. 148.
Searle, L. and Zinn, R. 1978, Ap. J., 225, p. 357.
White, S.D.M. and Rees, M.J. 1978, M.N.R.A.S., 183, p. 341.
Zinn, R. 1980, Ap. J., 241, p. 602.

DISCUSSION

K.C. Freeman: In your metallicity-radius diagram, you had a number of very metal-weak stars near the centre. Should we take that seriously?

Carlberg: Yes, in that the low-metallicity stars are simply the first stars formed in the bulge region. The overall high mean metallicity is a result of the large amount of infall of metal-enriched gas from elsewhere in the protogalaxy. One notes that the presence of RR Lyrae stars in the bulge is an indicator of the presence of a low-metallicity component.

Gathering for dinner at Lauswolt. In foreground, left to right by rows: Mrs. Wielen; Wielen; Denoyelle, Maurice; Mrs. Pismis (hidden); Van der Laan, Oort, Gingerich. LZ

PART IV

LIFE IN THE GALAXY

Wednesday 1 June, 2200 - 2230

After-dinner lecture by G.S. Shostak

Chairman: H. van der Laan

G. Seth Shostak and Karen Claffey, now Mrs. Shostak LZ

LIFE IN THE GALAXY?

G.S. Shostak
Kapteyn Astronomical Institute
University of Groningen, The Netherlands

1. INTRODUCTION

It is a decided honor to have the opportunity to address as distinguished a group as is present this evening, and I am particularly grateful for the very complimentary introductory remarks of Harry van der Laan. To paraphrase Einstein's comment on a dissimilar occasion, such a lovely preamble entails a risk that the packaging may prove better than the meat.

As you know, this year has seen the establishment of a special I.A.U. Commission (No. 51) to deal with the subject of exobiology and the Search for Extra Terrestrial Intelligence (SETI). It is, therefore, all the more appropriate that we give at least some attention to the question of biological activity in the Galaxy.

However, in keeping with the present setting and post-prandial atmosphere, my utterances will be more diverting than comprehensive, and this discourse should perhaps be likened to that most civilized of after dinner drinks, the digestive: capable of dissipating the effects of overindulgence and distinguished by its bad taste.

2. PURPOSE OF THE TALK

Our subject, then, is extraterrestrial life in the Galaxy. We believe in its existence, or don't, largely as a philosophical matter, since tenable arguments can be made both ways. As with many scientific questions, there is an unconscious tendency to array evidence on the basis of what we feel must be the correct answer.

One of the most compelling arguments for alien life, largely because of its generality, is being uncomfortable with uniqueness. It seems that every time man has thought that he was in the centre of things -- be he Ptolemy or Kapteyn -- the facts have proven otherwise. There are hundreds of millions of stars which are cousin to our Sun in

H. van Woerden et al. (eds.), The Milky Way Galaxy, 623–632.

the Milky Way. If they all have inhabited planets, then it is a virtual certainty that at this moment (whatever that may mean) a symposium on the Milky Way Galaxy is in session on a distant world. Thus, even this conference is not unique. (I remark, however, that we must extend our horizons to more distant galaxies to ensure a reasonable chance of finding a talk similar to my own taking place now.)

Indeed, for most of recorded history, it has been assumed that the Universe was teeming with life. The last century produced several schemes to signal potential neighbours, including a plan to set Siberia ablaze or, less environmentally destructive, to use large mirrors to brand our initials onto the tender surface of Mars. The aliens are a constant subject of science fiction, alternately interested in teaching us how to avoid nuclear war, and obsessed with the necessity to flatten Tokyo.

But the subject of life in space has heated up substantially since the Second World War, largely because there is now the chance that philosophizing and speculation may finally yield to experimental proof.

It's one thing to sit in a parlour discussing the possible sphericity of the Earth, and quite another to borrow three small ships from Spain and sail the Atlantic.

Tonight's discussion, then, is predicated on the fact that meaningful experiment is possible, and the first part of the talk will deal with that subject. Later, having pondered the most propitious techniques for uncovering the aliens, we will consider whether one should bother. (In deference to those who believe the aliens have already landed, a facility is provided on stage for any extraterrestrials who may wish to phone home [the speaker indicates a nearby telephone, provisioned with an upward-leading cable]. In light of the currently distressed economic situation in the Netherlands, I must ask that he call collect.)

3. FINDING THE EXTRATERRESTRIALS

In ferreting out the aliens, one might consider searches of both an active and a passive nature. Of the active quests, just looking around is the most patently obvious technique.

I will begin with "where they're not". One place they're not is the Earth. In fact, these stunning photos of our planet taken from but a few hundred miles away show how difficult it is to recognize that we're here.

The Moon is also a biological wasteland: geologically, organically, and, I might note, culturally dead.

Venus is Earth's sister planet, but is just too hot for life. The surface is a sterile, rocky desert, with temperatures of 400 C. As further dissuasion to settlement, the clouds contain, amongst other things, sulphuric and hydrochloric acid.

Thus we come to Mars, not only the scene of the teeming inhabitants of science-fiction fame, but also of the most sensitive scientific experiments to find life beyond Earth. Unfortunately, I have no time to describe the ingenious and thorough tests for biological activity made by the two Viking landers. The negative results of these probes could be questioned by the sceptical, but to quote one researcher: "it's possible that the experiments did indicate life, but it's also possible that the rocks seen at landing sites are actually living organisms which look like rocks." Despite centuries of speculative literature, Mars appears distressingly dead.

Unlikely as it may sound, Jupiter could be our last hope for company in the Solar System. It's conceivable that deep in the 1000 km thick atmosphere of this planet, a sort of microbal life floats in the churning mists. Here, in a never-never land between the Jovian cloud-tops above and a gloomy, bottomless black sea of liquid hydrogen below,

Earthlings watch as alien does push-ups in "This Island Earth".

these postulated suspended life forms would enjoy temperatures not much different from that of this room. This is a long shot for life, of course, but a space probe has been planned.

Taking the hard-nosed view, our Solar System appears lifeless beyond Earth, although as a final thought we should note Papagiannis' suggestion that extraterrestrials could be purloining our raw materials by secretly mining the asteroid belt. (As a personal aside, I do not begrudge the aliens our asteroids, but would draw the line upon the sudden disappearance of, say, Neptune.)

The "active" searches used to reconnoiter our Solar System lose their appeal once we move into the realm of the stars. This is because of the enormous distance-scale differences between the local planets and the not-so-local stars. Even so, the ploy has been tried: Carl Sagan's Pioneer plaque is moving into space at a few tens of kilometers per second. Still, it will be about a hundred thousand years before it has gone as far as the nearest stars. Even assuming the message is retrieved and correctly decoded (and who knows? Perhaps they'll interpret the nude couple as a map, and assume the inhabitants of our planet resemble sea urchins), we might have the wretched luck to make our presence known to an aggressive culture which will destroy civilization to obtain our chlorophyll or other commodity. In this view, Carl Sagan may be destined for eternal fame as the man responsible for the obliteration of Earth.

But a hundred thousand years is a long time, and Carl isn't worried: "Après moi, le déluge". Even his more audacious effort, the beaming of a radio signal to a globular star cluster, a communication travelling at the speed of the light, cannot possibly elicit a reply for about 500 centuries. In other words, active searches are truly implausible; if we wish to find the extraterrestrials within our lifetime, we must play the part of eavesdropper.

It is largely because our eavesdropping technology has suddenly come of age that SETI has developed respectability. After all, astronomers always demand observational proofs, and proof is now possible. If we make the (in our view) not unreasonable assumption that aliens use electromagnetic waves to communicate, it soon becomes obvious that for solid engineering reasons we should be listening for signals at decimetric wavelengths.

To date, several dozen searches have been conducted using radio telescopes on an occasional basis to weed out narrow-band signals emanating from, mostly, nearby stars. Other search objects have included neighbouring galaxies and the centre of the Milky Way. Most of this work has been done on an ad hoc basis, using equipment designed for astronomy, although specialized receivers and occasionally even an entire antenna are made available.

You will not be surprised when I tell you that no conclusive evidence for intelligent signals has been found to date (unless you are

Underside of the 1000-foot diameter Arecibo Telescope.

committed to the idea of a conspiracy of secrecy between radio astronomers and the US Air Force, as some are). Does this failure to hear the aliens mean anything? Probably not. The experiments done until now could generally detect transmitter powers of 100 MW or more, and the stars searched are still numbered in the hundreds only. Frank Drake has talked about the "cosmic haystack" having 8 dimensions: 3 spatial, 1 temporal, 1 frequency, 2 polarizations and 1 transmitter power. If you like to think in such terms, then we have investigated about 10^{-17} of the haystack for the Milky Way. This may, in view of later remarks, be unnecessarily pessimistic.

Radio searches have got most of the publicity, but there are other ways to find aliens, or at least life. As a first step, we might deal with the assumption that life originates on planets. Gatewood and others have stressed that, in fact, no fully reliable detection of planets around other stars has yet been made. Improved astrometric techniques and the advent of the Space Telescope may soon change this situation, but in the meantime we cannot fully deny the advertising claim that ET was 3 million lightyears from home. Maybe that's the nearest other planetary system, and it's beyond Andromeda. All the more reason to admire ET's tenacity in coming here to play with our kids.

Assuming that we do find other planets, if we could somehow measure their optical spectra then life might reveal itself in the planet's atmospheric composition. Oxygen, for example, is abundant in the air only because of the biomass. Methane in the atmosphere derives mainly from the flatulence of cows. A nice advantage of this technique is that it is capable of detecting life which isn't intelligent. While no inferences should be made, it is undoubtedly possible to detect northern Holland at considerable distance, given the locally high bovine density.

Still, despite the attractiveness of optical observation, radio remains the "technology of choice" for SETI, and several large projects have been proposed, if not built. Cyclops, for example, was a thoroughly engineered plan to decorate untold square kilometers of American desert with attractive white parabolas. Failing that, this very, very large array could be erected on the far side of the moon; a site not only out of sight, but also out of the direct influence of terrestrial interference.

But wait a minute. Before we risk Senator Proxmire's wrath by funding large SETI projects, perhaps it is worthwhile to ask whether searching for the aliens really makes sense. This brings us to the second subject....

4. WHO ARE THEY?

How many aliens can we expect, anyway? To begin to answer the question, we have to have some understanding of what kind of extraterrestrials we're talking about. Scientists generally are excruciatingly conservative in postulating alien life, and are inclined to assume that they are built the way we are built. Carbon, the stuff of pencil leads and diamonds, is singularly adept at hooking up with other atoms to form complex molecules, including those of our bodies, given a fluid environment and reasonable temperatures. (As chemists and sci-fi readers know, silicon also has this property, although to a far lesser extent: it tends to get locked up in stable, un-lifelike configurations such as quartz.) We assume, then, more out of ignorance than not, that the aliens are also constructed out of carbon compounds, and consequently that they, too, require a nice watery, not-too-cold and not-too-hot planet.

Now, I hope you will suffer through the following argument, because it has become a critical question in SETI circles. I will present the gist of the matter in a series of steps.

- Our Galaxy -- our home -- has at least one hundred million stars which are similar to the Sun.

- We don't know how many of these stars have planets, but for the sake of argument let's guess that one in a hundred do, each system consisting of ten planets.

- How many planets are like Earth, with water, air and brisk temperatures such as we enjoy in Holland? Who knows, but let's guess one in a hundred.

That already means one hundred thousand "earths" are floating around the Milky Way. And because our own Solar System is still relatively young, most of these other earths have had billions of extra years to develop life.

We have bravely estimated, then, that the number of civilizations in our Galaxy could be as high as one hundred thousand.

NOW.....

Any advanced civilization will invent medicine, and as a result will soon swamp the planet with a burgeoning populace. Orbiting space stations, each accommodating the population of Groningen, will be built. Contrary to our likely first reaction, it's probable that life in a spinning tube can be made quite pleasant. Children will be born, raised, and will die in this artificial environment without thinking twice about it. "Tube be, or not tube be" will not be a question.

With very little effort, such a space station can be sent off to another star. The stars are far, and the time of travel will be measured in hundreds or even thousands of years. But that's OK. After reaching someplace interesting, these "colonists" can be expected to take a thousand-year break for R and R, establishing their culture, increasing their population, and availing themselves of the facilities.

Then they, too, will send off some colonists, who will spend a millennium getting to the next star. And so forth. The colony will slowly spread, like coral in the sea.

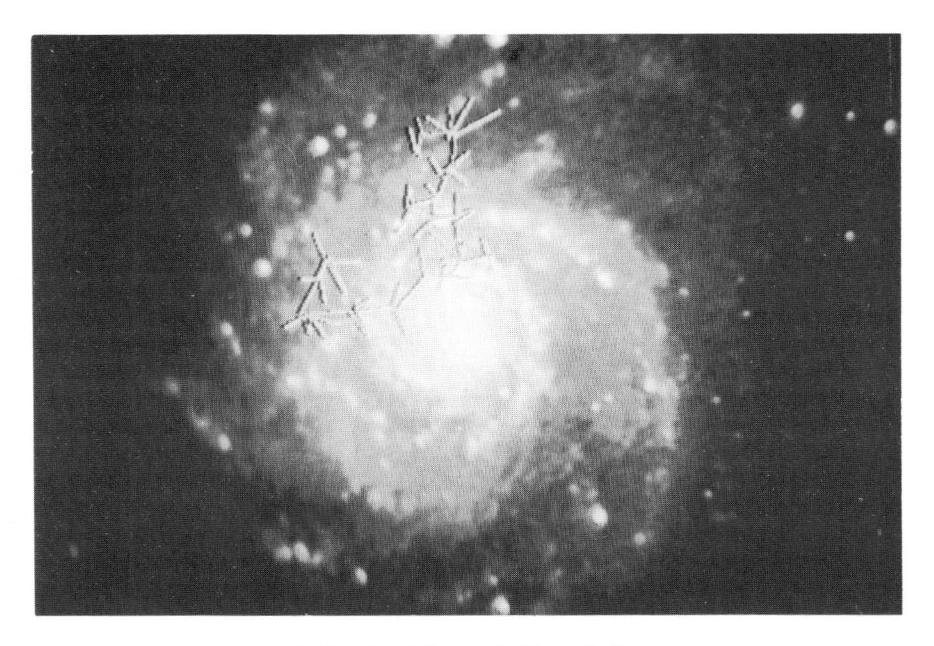

Colonization of the Galaxy.

Now the point is, they will run out of new stars in ten million years, which sounds like a lot, but is only the blink of an eye in the history of the Galaxy.

You may wish to liken the phenomenon to the Spaniards in the 16th century: once Columbus made his first voyage, the whole Caribbean and Pacific were visited in but a matter of decades. There were Spaniards everywhere.

THE BIG CONCLUSION, then, is that there has been plenty of time for an ambitious group of aliens to colonize the ENTIRE Galaxy. Maybe most aliens are passive, content to contemplate their antennae, and are not interested in space flight. Maybe many blow themselves up with nuclear weapons. Maybe, as Newman and Sagan have said, the colonists forget where they've come from, and waste a lot of time retracing their steps. But only ONE civilization has to do it. Most Europeans in the year 1500 weren't interested in sailing across the Atlantic, but it is enough that one person was.

Thus, the aliens should have been everywhere, including our own Solar System..... Now, the followers of van Däniken believe they have been here, and the UFO crowd thinks they still are. But the hard-headed, irrefutable evidence is missing. No one has as much as an ash-tray from an alien spaceship. Fact A of the SETI business is that we seem to be alone, and we shouldn't be.

Enrico Fermi summarized the problem in one sentence: "Where Are They?"

We must have made a mistake somewhere. What is it?

- The obvious possibility is that we really are alone in the Galaxy. Hart has pointed out the complexity of the nucleotides which make up DNA, and he figures that the random combining of atoms in the primordial soup of a distant earth will take 10^{32} billion years before dishing up something interesting. In his view, not only are we alone in the Milky Way, but probably in all the Universe we can see. An eminent Briton has put it another way: the chance of forming life in this manner is the same as assembling a Rolls Royce by sending a tornado through a junk yard. (I will resist all temptations to make the obvious remarks concerning British workmanship.)

- Frank Drake figures space travel is just too darn expensive. A rocket trip for 100 people will take the energy budget of the entire U.S. for hundreds of years. I can only respond to this argument by saying "more power to you", and note that Columbus' trip was also expensive by contemporary standards. Furthermore, as others have pointed out, an impending supernova would be strong incentive to emigrate, no matter what the cost.

- The Zoo Hypothesis has been offered as an appealing explanation for our apporent isolation: The aliens are leaving us alone because we amuse them, and we may even be seen as a useful exhibit. While I find this a "cagy" theory, it does seem a bit anthropocentric.

- Perhaps we don't know enough physics yet to recognize the aliens. Or maybe we're in a cultural backwater, like central Borneo, only occasionally visited. Von Hoerner has suggested that the only stable civilizations are those that are passive. But again, only one group has to be a bit adventurous.

5. A FINAL THOUGHT

Perhaps we're missing the boat. Possibly we should be compared to dinosaurs who sit around and try to consider the breed of dinosaurs which populate other worlds. We are but a stage in evolution, and although man may no longer be evolving as an individual (thanks to medicine), we may still be in the process of producing the next generation of species. Organic intelligence is fragile, fallible and barbarous. Human intelligence is a tool evolved for survival in a hostile world. But the environment is changing, and intelligence (like great size) may have limits to its survival value. We may disappear, but our progeny -- the machines -- may survive. (And what I mean by machines includes not only the Apple CCCIX, but also bio-engineered humans.)

Technical organic life -- mankind of the last hundred years, in other words -- is but a brief flash in the long night of Earth's existence. If that is also true on other worlds, then the chances of contact are small; neither we nor they are around long enough to arrange communication. But an intelligent machine can in principle overcome the limitations of hostile environments and biological mortality. It was suggested that such machines might leave their home planets to travel to the Galactic Center, where the action is, and where they could hook up with other devices, a kind of interspecies mating thankfully denied humans. In this view, we are on Earth only to give birth to the machines, and it is they that dominate the intelligence in the Galaxy. They may have no more interest in visiting us than we have in visiting the ant colonies in our garden. And their activities may be of a kind which make their detection difficult.

Thus, do not disparage the bus which brings you back to your hotels this evening: that lowly mechanism may be the progenitor of all that's enlightened in the Galaxy.

PART V

SUMMARY AND OUTLOOK

Friday 3 June, 1140 - 1230

Chairman: K.C. Freeman

Oort at the telescope - looking ahead to the next symposium?
LZ

THE MILKY WAY: SUMMARY AND OUTLOOK

Jeremiah P. Ostriker
Princeton University Observatory
Princeton, New Jersey 08544, USA

HISTORICAL

First let me review the historical discussions presented during our symposium: the papers by Paul, Gingerich, Hoskin and Smith. I was greatly impressed by the power of abstract human thought in its confrontation with resistant reality. On the one hand we see again and again extraordinary prescience, where abstract beliefs based on little or no empirical evidence--like the island-universe hypothesis--turn out to be, in their essentials, true. Clearly, we often know more than we know that we know. On the other hand, there are repeated instances of resistance to the most obvious truth due to ingrained beliefs. These may be termed conspiracies of silence. Van Rhijn and Shapley agreed about few things. But one of them was that there was no significant absorption of light in the Galaxy. Yet the most conspicuous feature of the night sky is the Milky Way, and the second most conspicuous feature is the dark rift through its middle. What looks to the most untutored eye like a "sandwich" was modeled as an oblate spheroid. These eminent scientists must have known about the rift, but somehow wished it away in their analyses. I find that very curious. Other examples from earlier times abound. We all know that the Crab supernova was seen from many parts of the globe but, though it was bright enough to be detected by the unaided eye in daylight, its existence was never--so far as we know--recorded in Europe. It did not fit in with the scheme of things, so it was not seen.

There is absolutely no reason that I know of to believe that this process has ceased. Thus even now there must be many important facts staring us in the face in a blatant and unequivocal way, which we are refusing to recognize, but may soon find to be of central importance. I found myself wondering "What are these facts?". Conversely, from the historians it is clear that much of what is believed to be scientifically accurate by the most precise and sober minds at one time, is thought (or "known") to be incorrect at a later date. Realization of this fact was humbling. And again, there is absolutely no reason to believe that the process has stopped. Which of our beliefs will seem childish at a later date? I have my candidates here. I think the learned

H. van Woerden et al. (eds.), The Milky Way Galaxy, 635–639.

discussions which we have all endured and some of us even inflicted on others concerning adiabatic vs isothermal perturbations in the early Universe, may be based on theories which will seem hopelessly naive to later scientists. Yet we also learned that not everything which was believed earlier was wrong--the picture of our Solar System has changed in detail but not in essence since the early part of the century. And much of the early progress concerning stars in the solar neighborhood--positions, motions, distances--has withstood the test of time quite impressively. The trick is to tell which material is in which category.

ASTRONOMICAL PAPERS

Now let me turn to the scientific presentations. First the large-scale galactic mass and light models. There seems, for better or worse, to be a developing consensus around models based on a few physical components. Convergence of the mass models of Schmidt, Einasto, myself and J. Caldwell, and others is perhaps an encouraging sign. Certainly the similarities are far greater than the differences. Among surprising features common to this work is that all seem to want, if not absolutely need, a significant dark-halo component ($\gtrsim$ 1/3 of total) within the solar circle.

The most important and presently uncertain data used by the model makers are of two kinds: classical-local and new-distant. In the first category the revived interest in star counts is impressive. The work of observing groups around Strömgren, Gilmore, Hilditch and others, combined with analytical studies such as those by Bahcall, should do two things. First, they will allow us to tighten up the embarrassingly loose constraint on the local surface density, the Oort limit. I think everyone would agree that the total mass within ±1 kpc of the plane, locally, is certainly more than 40 and less than 100 $M_\odot/pc^2$ and probably between 60 and 80 $M_\odot/pc^2$, but we should be able to do better and perhaps soon will. Also our optical studies will tell us if we need other components such as thick discs in the models, and will help to determine properties of other components.

The other really weak point in the models concerns the rotation curve exterior to the Sun where, classically, the 21-cm work gave little information. Evidence presented by Chini (radio) and Freeman (optical) at this conference seems to require that for $R_0 < R < 3R_0$ the rotation curve is flat or rising, consistent with older work by Blitz and others.

The 21-cm work performed here in the Netherlands, largely through the efforts and brilliance of Jan Oort, revolutionized galactic astronomy. After the initial burst of activity there was a pause, but now molecular CO, IR, X- and γ-ray windows have been opened on the Galaxy in the last 10 years and very rapid progress is again to be expected. The data are pouring in now and it will take scientists having the synthesizing power of Oort or Kapteyn to make sense of it all. My own feeling is that we are in the very early stages of such studies.

Direct observations of gradients in various components, from the gas (Burton) to fascinating COS B results concerning cosmic γ rays (Hermsen) to distant tracers such as M supergiants and H II regions, are increasingly reliable and important and sould ultimately (but not for the foreseeable future) allow us to test in detail our ideas on galactic evolution with information difficult to obtain in other systems. This subject was treated on the last day, and moderately good agreement between theory and observation already exists. Here I want to stress the obvious point that we are gaining in power enormously as we are able to view the Galaxy at wavelengths to which it is largely transparent.

It is logical to turn at this point from the large-scale studies to one particularly interesting part of the Galaxy, and especially to Jan Oort's exciting tour through the Nucleus. There is little I can add by way of review but it is perhaps worthwhile to note a few points that seem to me specially significant. Both the compact radio source and the hard X-ray source now are seen to be unique in the Galaxy. Are they the same thing, and, if so, what? The questions were sharpened. The discovery of the apparently spiral-like features at several wavelengths, especially through recombination-line work, allowed one to make a plausible interpretation of the hitherto baffling Ne^+ data, but I remain as confused as ever. I would not argue against a 10^6 $M_\odot$ black hole at present, but neither would I argue for one. We still do not know enough. I was left with one very simple question. If we could strip away all of the exciting stuff from this region and only look at the old stars, as we can in Andromeda, what would we see? What is the "core radius" of the old stars: is it $\leqslant$ 1 pc, 10 pc, or 100 pc? How big is the potential well, the room within which all these dramatic events are occurring?

Let me now leave the Galaxy and turn to comparison with other galaxies, to the fine papers by Young, Beck and Reich, Elmegreen, and the comparisons with our big brother M31 by Hodge, Brinks and Stark, which were very illuminating. We are becoming mature and are developing a realistic view of where we stand in the world of galaxies. We are no longer the only, or even the biggest, but a rather ordinary-sized Sc spiral. I must admit that I found myself a little disappointed; couldn't we at least be as big as NGC 4565? The regularities that are beginning to be found are impressive. But in most extragalactic (non-optical) studies the resolution available at present just barely allows us to note such features as spiral arms or central holes, and it is clear that detailed study of other galaxies is a field in its infancy. Optical analyses, not presented here, are also making great strides at present.

The situation concerning the mass of the Local Group has not changed in an essential way in over a decade. The paper by Lynden-Bell confirmed that we know or firmly believe it should be measured in units of 10^{12}, not 10^{11}, solar masses, but we have little clue as to where the mass is and no knowledge at all as to what it is. The best current candidates, moderately massive ($\sim 10^{6.5}$ $M_\odot$) black holes, strange particles (massive neutrinos) and low-mass stars, are all possible but none particularly likely.

The problem of spiral structure I found to be very depressing, not, as one might expect, due to its resistance to theoretical attack but, on the contrary, because of our successes. I shall explain my feeling by relating a dream. Following the pioneering work of Lin and collaborators, Toomre, and others, I believe that we really do understand this phenomenon. In fact I couldn't restrain myself during a recess from explaining to Frank Shu how I really understood the origin and maintenance of spiral structure. I proceeded to stitch together a theory. Of course it had gravity, but it also, of necessity, included the fact that star formation produces star formation--necessary since that is seen to be true. And above all, it required gas dynamics as an essential ingredient. Gas, dissipational, will always cool (in absence of supernovae) until the system is driven unstable and then, as Shu had mentioned during the discussion yesterday, dissipation will limit the instability as well. I commented that I am actually working on this problem with Len Cowie and Scott Tremaine. Frank Shu's answer to our proposed model was "Of course, we all agree, everyone in the field knows all that". Somewhat depressed, I went home for an early nap, in which I dreamt that I was blessed with an army of graduate students each one more eager, intelligent and industrious than the last. Then with this army, and unlimited access to computer time, all the relevant physics was put into the computer model. It took a year to get the bugs out but then, when the first run was complete, voilà! Out come pictures representing at every wavelength real spiral galaxies. By suitable adjustment of knobs every type of galaxy can be made. What a happy dream.

Success!? Perfect understanding of the phenomena. Peace and understanding will reign.

Of course I did not understand the pictures in the incredible detail provided by this wonderful code, so I gave them over to the observers to study. Some found trailing two-arm logarithmic spirals, other one-arm leading spirals. The theoreticians were not better. Each argued, with great logic, that the input of a particular bit of physics was essential to success. And each argued that other elements I had used were really trivial, inessential. The dream ends in chaos with cries and shouts, grown men fighting over control of overhead projectors, etc. I leave it to each individual to draw whatever conclusion he or she may wish from my dream. Mine was that we, in fact, collectively do understand this problem but that the solution is an unappetizingly complicated mess, and the will to simplify what is intrinsically complicated shall drive us to exhaustion. A depressing conclusion.

Slightly more hopeful to me were the theoretical papers of Wielen and Lacey, addressing the far simpler problem of the growth of the velocity dispersion of disc stars. Here we do not know the answer, but there is every expectation that we will understand the answer when we find it. My own bet would be on item (3) of Wielen's list, puffing-up of the disc due to spiral, tidal and other instabilities and perturbations. Carlberg's simulations seem to show this nicely.

Before going on to the future, I will give you one answer to the question I asked earlier of what it is that is in front of our eyes but which we are refusing to notice. My candidate answer is bars. M31 clearly has a bar. I think that our Galaxy has a bar. LMC has a bar. Ken Freeman tells me that SMC and NGC 205 have bars. In almost all members of the Local Group of galaxies bars are seen. Yet this fact never entered any discussion presented here. Even when C.C. Lin described Yuan's work on bar-driven spirals, the spiral was seen in the picture but there was no bar. Could this be a severe omission? Could formation of and effects due to bars be somehow vital, not trivial features? Perhaps.

THE FUTURE

I have already mentioned several of the areas where I believe progress is to be expected. The collection of observations made at many wavelengths (to which the Galaxy is transparent) and coordination of observations of many different types of object is likely to yield rapid progress. But collection of data by itself does not provide understanding, and some hard and some inspired thought will be necessary to produce an intellectually coherent picture. On the theoretical front the computer looms up over us, casting a bigger and bigger shadow. Our giant and exponentially growing helper has become indispensible. With this brute it seems we can do anything, and if not now then soon, with the next factor 10^1 or 10^2 in computer power. But we haven't yet come to terms with the monster, and there may be a contest down the road to see who is to be master.

Let us look further ahead, beyond the immediate and even the foreseeable battles. If I may change my metaphor, will there always be more to learn or will the galactic fishing hole eventually be "fished out"? With all the big problems solved, will we have to fish longer and longer to catch solutions to smaller and smaller problems? Logically that seems possible, even likely. But history teaches us otherwise. There was a synthesis reached with publication of Stars and Stellar Systems, Vol. V, edited by Blaauw and Schmidt, two gentlemen eminently in the tradition of Kapteyn's science. It is interesting to think how far back that book was published--1965, almost 20 years ago. And students reading that book (which I still assign to my classes) might be excused if they thought that we only had to nail down the Oort constants, get better galactic tracers, etc. to understand the Galaxy.

The two biggest current puzzles of galactic astronomy did not even appear, I believe, in that volume. These are: what is the nature, distribution and amount of dark matter, and what is happening in the galactic center? If new problems of that significance continue to arise at this rate, we will not soon fish out the galactic fishing hole, and I hope we can look to our Dutch brethren to continue to supply us with the world's most expert anglers-- in the Kapteyn tradition, ingenious, persistent and precise.

CONCLUDING GENERAL DISCUSSION

H.M. Maitzen: Wasn't the existence of unseen matter implied in the old Schmidt model, which needed a component of about the same mass in addition to the disk component?

M. Schmidt: I do not think there was a halo, or a round component, in that model - was there? (Laughter.) It's rather long ago - I do remember that in those models, 1956 (Bull. Astr. Inst. Netherl. 13, p. 15) and 1965 (in "Galactic Structure", Stars and Stellar Systems, Vol. V, eds. A. Blaauw and M. Schmidt, Univ. Chicago Press, p. 513), there was the problem of the density gradient in the solar neighbourhood. In my 1956 thesis I had the trouble that the high then prevalent ratio of A/B (19.5/-6.9) made the density gradient so large that clearly disk stars could not supply it. Therefore I probably talked about unknown stars or matter, but not in the present sense.

A. Blaauw: When editing our 1965 book on Galactic Structure, Schmidt and I wondered whether we should add a chapter on the evolution of the Galaxy. We thought of one person who could perhaps write something that might make sense, but we decided that the time just had not come yet.

J.P. Ostriker: In the Proceedings of the 1957 Vatican Symposium ("Stellar Populations", ed. D.J.K. O'Connell S.J., Amsterdam: North Holland/New York: Interscience, 1958; also Specola Astr. Vaticana Ricerche Astron. Vol. 5) there was a section which summarized thoughts at that time about galactic evolution.

Blaauw: In "Galactic Structure" I wrote a chapter on Stellar Populations, which summarized definitions and introduced tables about ages and metallicity etc. - but it did not go beyond that; it did not conclude to evolutionary scenarios.

R.H. Sanders: Another historical point. In the book on "Galactic Structure" the Galactic Centre was mentioned in the final chapter by Woltjer, who discussed the 3-kpc arm discovered in 1956 by Hugo van Woerden (and others: H. van Woerden, G.W. Rougoor and J.H. Oort, Comptes-rendus Acad. Sci. Paris 244, 1691). Woltjer suggested that its radial motion might be due to an explosion, and estimated that the explosion energy would have to be of order 10^{58} erg. Some years later Prendergast and I did a numerical calculation; we found after a lot of computing that it took 10^{58} erg to produce the 3-kpc arm. Now I think most of us believe what Frank Kerr first suggested (in I.A.U. Symposium 31, "Radio Astronomy and the Galactic System", ed. H. van Woerden, London: Academic Press, p. 239): that the 3-kpc arm has a more natural explanation by non-circular motions in the potential of a broad oval distortion or something like that --- (Ostriker: You would not want to say: a bar?) --- a bar, O.K., but it is interesting that, because of the unique point source and the remarkable infra-red and radio-continuum structure at the Galactic Centre, we are now coming back to thinking that there might be something unconventional going on there, though on a much smaller scale.

SUBJECT INDEX

Underlining indicates the first page of papers in which reference to the subject is frequently made.

OBJECT INDEX

Underlining indicates first page of a paper in which the object is referred to frequently.

NAME INDEX

Underlining indicates author or co-author of paper in this volume.
Brackets indicate photograph only.